evolution FOURTH EDITION

Companion Website

The **Evolution** Companion Website provides you with a range of valuable study and review tools to help you master the material presented in the textbook. Available free of charge, the site is designed to help you understand the concepts and learn the terminology introduced in each chapter, analyze real-world research, and work with simulations of evolutionary systems.

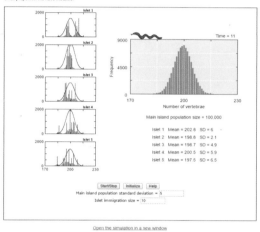

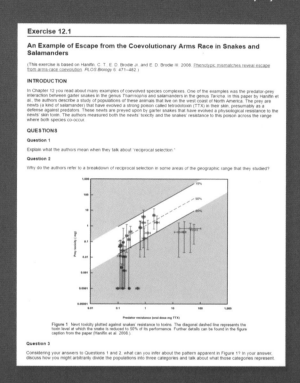

Features of the Companion Website

Data Analysis Exercises: These inquiry-based exercises challenge you to think as a scientist and to analyze and interpret experimental data. Based on real papers and experiments, these exercises involve answering questions by analyzing the data from the experiments.

Simulation Exercises: These exercises include interactive modules that allow you to explore some of the dynamic processes of evolution. Each exercise poses questions that you answer by running a simulation and observing and analyzing the outcomes.

Online Quizzes: For each chapter of the textbook, the site includes a multiple-choice quiz that covers all the main topics presented in the chapter. Your instructor may assign these quizzes, or they may be made available to you as self-study tools. (Instructor registration is required for student access to the quizzes.)

Flashcards: Flashcards help you learn and review the many new terms introduced in the textbook. Each chapter's set of flashcards includes all of the key terms introduced in the chapter.

Chapter Summaries: Concise overviews of the important concepts and topics covered in each chapter.

Chapter Outlines: A convenient outline of each chapter's sections and sub-sections.

Glossary: A complete online version of the glossary, for quick access to definitions of important terms.

evolution4e.sinauer.com

evolution

FOURTH EDITION

evolution

FOURTH EDITION

DOUGLAS J. FUTUYMA
Stony Brook University

MARK KIRKPATRICK
University of Texas at Austin

 Sinauer Associates, Inc. • Publishers
Sunderland, Massachusetts U S A

Key to the Back Cover

Natural selection acting on reproductive success has produced dramatic head ornaments in many species of birds.

1: great crested grebe (*Podiceps cristatus*); **2:** California quail (*Callipepla californica*); **3:** palm cockatoo (*Probosciger aterrimus*); **4:** Indian peafowl (*Pavo cristatus*); **5:** tufted puffin (*Fratercula cirrhata*); **6:** rufous-crested coquette (*Lophornis delattrei*); **7:** Andean cock-of-the-rock (*Rupicola peruvianus*); **8:** ruff (*Philomachus pugnax*); **9:** hooded merganser (*Lophodytes cucullatus*).

evolution, FOURTH EDITION

Copyright © 2017 by Sinauer Associates, Inc. All rights reserved. This book may not be reproduced in whole or in part without permission from the publisher.

Sinauer Associates, Inc., 23 Plumtree Road, Sunderland, MA 01375 U.S.A.
Phone: 413-549-4300
Fax: 413-549-1118
Email: publish@sinauer.com, orders@sinauer.com
Website: www.sinauer.com

Library of Congress Cataloging-in-Publication Data

Names: Futuyma, Douglas J., 1942- author. | Kirkpatrick, Mark, 1956- author.
Title: Evolution / Douglas J. Futuyma, Stony Brook University, Mark
  Kirkpatrick, University of Texas, Austin.
Description: Fourth edition. | Sunderland, Massachusetts : Sinauer
  Associates, Inc., Publishers, 2017.
Identifiers: LCCN 2017000562 | ISBN 9781605356051 (casebound)
Subjects: LCSH: Evolution (Biology)
Classification: LCC QH366.2 .F87 2017 | DDC 576.8--dc23
LC record available at https://lccn.loc.gov/2017000562

Printed in the U.S.A
8 7 6 5 4 3 2 1

To the members of the Department of Ecology and Evolution at Stony Brook, past and present, in gratitude for many years of support and intellectual sustenance.
DJF

To Sharon, who kept the ship afloat, and to Don, who first pointed the prow in this direction.
MK

Brief Contents

Contents

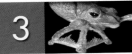

UNIT II   How Evolution Works   77

4

CHAPTER 4
Mutation and Variation   79

5

CHAPTER 5
The Genetical Theory of Natural Selection   103

CHAPTER 6
Phenotypic Evolution   135

CHAPTER 7
Genetic Drift: Evolution at Random   165

CHAPTER 8
Evolution in Space   191

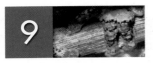

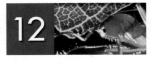

CHAPTER 13
Interactions among Species   321

CHAPTER 14
The Evolution of Genes and Genomes   345

CHAPTER 15
Evolution and Development   369

UNIT IV   Macroevolution and the History of Life   399

CHAPTER 16
Phylogeny: The Unity and Diversity of Life   401

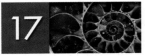

CHAPTER 17
The History of Life   431

CHAPTER 18
The Geography of Evolution   469

Preface

It is thoroughly established that all known organisms descended from a single ancient common ancestor. This means that all characteristics of organisms, in all their glorious diversity, have evolved. Anatomical and cellular traits, biochemical, molecular, neural and developmental processes, life histories and ecological relationships—all can be viewed from the dual perspectives of current mechanism (how they work) and of history (how and why they came to be). The disciplines of organismal biology, including paleobiology, ecology, animal behavior, physiology, and systematics, continue to be central to evolutionary science, but are now being enriched by the genomic revolution, new analytical methods, and new evolutionary theory.

The fourth edition of *Evolution* keeps pace with this explosively developing field. There are now two authors with broadly overlapping but complementary areas of expertise. The organization, content, and style of the book are reworked to such an extent that it is largely a new book. Key changes include:

- Many human examples are used throughout, and there is an all-new chapter on human evolution.
- A new primer in statistics gives a concise and accessible introduction to the field.
- Theoretical concepts are developed in a more informal and inviting style.
- The book has been entirely re-illustrated.

The book is organized into these units:

I. An Idea that Changed the World

Chapter 1 opens with an overview of evolutionary biology and its history. The next two chapters introduce two of the most fundamental ideas in evolution: evolutionary trees (Chapter 2) and the concepts of natural selection and adaptation (Chapter 3).

II. How Evolution Works

The first four chapters of this unit develop genetics and inheritance (Chapter 4), one-locus population genetics (Chapter 5), quantitative genetics (Chapter 6), and genetic drift (Chapter 7). Chapter 8, which is entirely new, discusses spatial patterns and the evolution of dispersal. Chapter 9 then tackles species and speciation in a coherent treatment that has been streamlined relative to the third edition. Every chapter in this unit has been completely rewritten.

III. Products of Evolution: What Natural Selection Has Wrought

This unit treats key aspects of the evolution of phenotypes and genotypes: the all-new Chapter 10 on sexual selection and sexual reproduction, Chapter 11 with a rewritten exposition of the evolution of life histories and ecological niches, Chapter 12 on cooperation and conflict with new topics that include the evolution of virulence in pathogens, Chapter 13 on interactions among species, Chapter 14 on the evolution of genes and genomes, and Chapter 15 on evolution and development. These last two chapters have been rewritten in their entirety.

IV. Macroevolution and the History of Life

Chapter 16 develops the topic of phylogeny in detail. Chapter 17 provides a grand tour through the history of life. We turn to analysis of these historical data in Chapter 18, on biogeography, and Chapter 19, on patterns and causes of changes in biological diversity through time. Concepts drawn from throughout the book culminate in Chapter 20, which treats macroevolution.

V. Evolution and Homo sapiens

Perhaps no topic in biology has captured the imagination of scientists and the public alike than the tremendous recent advances in understanding human evolution. Chapter 21 conveys this excitement with a synthesis of sources that include paleontology, genomics, and cultural anthropology. Our final chapter (22) looks at how evolutionary biology impacts society, including belief systems and our understanding of human behavior.

More than any other science, evolutionary biology has had to prove its validity: in the United States, about half the population does not accept evolution by natural selection, and many of them are college students. *To teach evolution, then, is to teach the nature of science, the habit of reasoning between hypothesis and evidence, and the habit of critical evaluation.* At a time when science and evidence are increasingly misunderstood or even dismissed, we feel it is important to teach students what science is, how it works, and why it is the most reliable way of knowing that has yet been developed. Evolutionary biology is an ideal vehicle for this important function.

ACKNOWLEDGMENTS

Countless colleagues and students have contributed indirectly to this book, through their lectures, publications, conversations, and questions. We are indebted to the reviewers of chapter drafts for abundant, invaluable advice and corrections: Anurag Agrawal, Richard Bambach, Brian Barringer, David Begun, Andrew Brower, Brian Charlesworth, Jerry Coyne, Christopher Dick, Diego Figueroa, John Fleagle, Jacob Gardner, Kenneth Hayes, David Hillis, Gene Hunt, David Innes, David Jablonski, Elizabeth Jockusch, Joel Kingsolver, David Lohman, Greg Mayer, Duane McKenna, Mark McPeek, Monica Medina, Christopher Organ, Sally Otto, David Queller, Kaustuv Roy, Michael Ryan, Ellen Simms, Montgomery Slatkin, David Spiller, Stephen Stearns, Joseph Travis, Mark Welch, Noah Whiteman, Michael Whitlock, David Sloan Wilson, Greg Wray, Stephen Wright, Yaowu Yuan, and Roman Yukilevich. We are very grateful to David Hall, J. Matthew Hoch, and David Houle for writing the Problems and Discussion Topics at the chapter ends.

For advice, references, answers to questions, corrections, and many other favors, Douglas Futuyma is grateful to Richard Bambach, Michael Bell, Jackie Collier, Stefan Cover, Jerry Coyne, Joel Cracraft, Liliana Dávalos-Álvarez, Charles Davis, Christopher Dick, Daniel Dykhuizen, Walter Eanes, John Fleagle, Brenna Henn, Andreas Koenig, Spencer Koury, Harilaos Lessios, Jeffrey Levinton, James Mallet, Ross Nehm, Sally Otto, Joshua Rest, Martin Schoenhals, David Stern, Ian Tattersall, Robert Thacker, and Krishna Veeramah. Douglas Futuyma thanks Rob DeSalle and the American Museum of Natural History for generously providing work space in that marvelous institution, and the Department of Ecology and Evolution at Stony Brook for unending support and intellectual sustenance.

Mark Kirkpatrick is most grateful to the following people for their generous help with countless aspects of this project: Diego Ayala, Claudia Bank, Daniel Bolnick, Andrius Dagilis, Larry Gilbert, Shyamalika Gopalan, Matthew Hahn, David Hall, David Hillis, Robin Hopkins, David Houle, Matheiu Joron, Tom Juenger, Peter Keightley, Marcus Kronfrost, Curtis Lively, James Mallet, Richard Merrill, Michael Miyagi, Nancy Moran, Rasmus Nielsen, Mohamed Noor, Howard Ochman, Kenneth Olson, Sally Otto, Daven Presgraves, Trevor Price, David Queller, Fernando Racimo, Michael Ryan, Sara Sawyer, Dolph Schluter, Michael Shapiro, Stephen Shuster, Montgomery Slatkin, Ammon Thompson, Michael Wade, Stuart West, and Harold Zakon. Mark Kirkpatrick also thanks his colleagues and members of his laboratories at the University of Texas and the University of Montpellier for countless enlightening discussions.

This book was only made possible by the tireless work of many people at Sinauer Associates, consistently done with professionalism and a sense of humor. We are particularly in debt to Joanne Delphia, Laura Green, David McIntyre, Andy Sinauer, and Chris Small.

How to Learn Evolutionary Biology

The great geneticist François Jacob, who won the Nobel Prize in Physiology and Medicine for discovering mechanisms by which gene activity is regulated, wrote that "there are many generalizations in biology, but precious few theories. Among these, the theory of evolution is by far the most important." Why? Because, he said, evolution explains a vast range of biological information and unites all of the biological sciences, from molecular biology to ecology. "In short," he wrote, "it provides a causal explanation of the living world and its heterogeneity."

Jacob did not himself do research on evolution, but like most thoughtful biologists, he recognized its pivotal importance in the biological sciences. Evolution provides an indispensable framework for understanding phenomena that range from the structure and size of genomes to the ecological interactions among different species. And it has many philosophical implications and practical applications, ranging from understanding human diversity and behavior to health and medicine, food production, and environmental science.

Your course on evolution is likely to differ from almost any other course in biology you may have had, and it may present an unfamiliar challenge. Because all organisms, and all their characteristics, are products of a history of evolutionary change, the scope of evolutionary biology is far greater than any other field of biological science. In a course in cell biology, you are expected to learn many factual aspects of cell structure and function, which apply very broadly to various types of cells in almost all organisms. But courses in evolution generally do not emphasize the factual details of the evolution of particular groups of organisms—the amount of information would be impossibly overwhelming. There certainly are some important facts—for example, you should learn about major events in the history of life. But for the most part, your course is likely to emphasize the *general principles* of evolution, especially the *processes of evolutionary change* that apply to most or all organisms, *how we can learn what has happened in the evolutionary past*, and the *most common patterns of change*, those that have characterized many different groups of organisms.

For example, you will learn that natural selection is a consistent, statistical difference between groups of reproducing entities (such as large versus small individuals of a species) in the number of descendants they have. By understanding how a characteristic can affect survival or reproduction, we can arrive at generalizations about how certain characteristics are likely to evolve. For instance, it is easy for us to understand why a feature would be likely to evolve if it made males more attractive to females so that they have more offspring. But evolution by natural selection equally well explains why about half of the human genome consists of repeated DNA sequences that do nothing of value to the human organism! (The reason is that DNA sequences are also reproducing entities, and any sequence that can make more copies of itself will automatically increase more than a sequence that makes fewer copies. This is the essence of natural selection.) So the abstract concept of natural selection has a great range of applications and implications that will make up much of what you will want to learn about evolution.

It is important to learn *how evolutionary hypotheses have been tested*, in other words, what the evidence is for (or against) postulated histories and causes of evolutionary change. Evolutionary biology largely concerns events that happened in the past, so it differs from most other biological disciplines, which analyze the properties and functions of organisms' characteristics without reference to their history. We often must make inferences about past events and about ongoing processes that are difficult to see in action (e.g., differences in the replication rate of different DNA sequences). We make inferences by (1) posing informed hypotheses, then (2) generating predictions (making deductions) from these hypotheses about data that we can actually obtain, and finally (3) judging the validity of each hypothesis by the match between our observations and what we expect to see if the hypothesis were true.

For example, if you imagine that the long tail feathers of males in a species of bird evolved because such males attract more females and therefore have more offspring, you might predict that if you lengthened males' tail feathers, they will mate with more females. (The experiment has been done, with exactly this outcome.) You will find that throughout this book, we develop an idea, or hypothesis, theoretically, and then present one or two examples of empirical (i.e., real-world) studies that biologists have done, which provide evidence supporting the idea. *Understanding the theoretical ideas, and how and why the empirical study provides evidence for them, is the key to learning evolutionary biology.*

It is also the key to understanding how science works. Science isn't merely accumulating facts. In every field, scientists try to develop general principles that explain how natural phenomena work. Often, there are several conceivable explanations. The community of scientists in a field develops fuller understanding by devising alternative hypotheses and thinking of what kind of data would support one while refuting another. There is a competition of ideas (and competition among scientists) that results in a closer approach to reality. We cannot prove that a scientific hypothesis is absolutely true, but we can hope for very high confidence—and no other method of knowing can be shown to come as close. You can have very high confidence that DNA is the basis of inheritance, that human consumption of fossil fuels causes global climate change, and that humans have evolved from the same ancestor as all other animals, and from a much older ancestor of all the living things we know of.

In every field of science, the unknown greatly exceeds the known. Thousands of research papers on evolutionary topics are published each year, and many of them raise new questions even as they attempt to answer old ones. No one, least of all a scientist, should be afraid to say "I don't know" or "I'm not sure." To recognize where our knowledge and understanding are uncertain or lacking is to see where research may be warranted, or where exciting new research trails might be blazed. We hope that some readers will find evolution so rich a subject, so intellectually challenging, so fertile in insights, and so deep in its implications that they will adopt our subject as a career. But all readers, we hope, will find in evolutionary biology the thrill of understanding and the excitement of finding both answers and intriguing new questions about the living world, including ourselves. *Felix, qui potuit rerum cognoscere causas*, wrote Virgil: happy is the person who could learn the nature of things.

Media and Supplements
to accompany *Evolution*, Fourth Edition

FOR STUDENTS

Companion Website (evolution4e.sinauer.com)

Evolution's Companion Website features review and study tools to help students master the material presented in the textbook. Access to the site is free of charge, and requires no passcode. The site includes:

- *Chapter Outlines and Summaries*: Concise overviews of the important topics covered in each chapter.
- *Data Analysis Exercises*: These inquiry-based problems are designed to sharpen the student's ability to reason as a scientist, drawing on data from real experiments and published papers.
- *Simulation Exercises*: Interactive modules that allow students to explore many of the dynamic processes of evolution and answer questions based on the results they observe.
- *Online Quizzes*: Quizzes that cover all the major concepts introduced in each chapter. These quizzes are assignable by the instructor. (Instructor registration is required.)
- *Flashcards*: Easy-to-use flashcard activities that help students learn and review all the key terminology introduced in each chapter.
- The complete *Glossary*

FOR INSTRUCTORS
(Available to qualified adopters)

Instructor's Resource Library

The *Evolution* Instructor's Resource Library includes a variety of resources to help instructors in developing their courses and delivering their lectures. The IRL includes the following:

- *Textbook Figures and Tables*: All of the figures (including photographs) and tables from the textbook are provided as JPEGs, reformatted and relabeled for optimal readability when projected.
- *PowerPoint Presentations*: For each chapter, all of the figures and tables are provided in a ready-to-use PowerPoint presentation that includes titles and full captions.

- *Answers* to the textbook end-of-chapter Problems and Discussion Topics.
- *Quiz Questions*: All of the questions from the Companion Website's online quizzes are provided in Microsoft Word format.
- *Data Analysis and Simulation Exercises*: All of the exercises from the Companion Website are provided as Word documents, with answers, for use in class or as assignments.

Online Quizzing

The Companion Website includes an online quiz for each chapter of the textbook. Via the instructor's area of the companion website, these quizzes can be assigned or opened for use by students as self-quizzes. Custom quizzes can be created using any combination of publisher-provided questions and instructor-created questions. Quiz results are stored in an online gradebook and can be exported. (*Note*: Instructors must register with Sinauer Associates in order for their students to access the quizzes.)

Value Options

eBook

Evolution, Fourth Edition is available as an eBook, in several different formats, including VitalSource, RedShelf, Yuzu, and BryteWave. The eBook can be purchased as either a 180-day rental or a permanent (non-expiring) subscription. All major mobile devices are supported. For details on the eBook platforms offered, please visit www.sinauer.com/ebooks.

Looseleaf Textbook

(ISBN 978-1-60535-696-9)
Evolution is also available in a three-hole punched, looseleaf format. Students can take just the sections they need to class and can easily integrate instructor material with the text.

UNIT I
An Idea that Changed the World

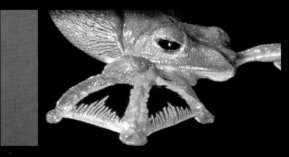

Evolutionary Biology

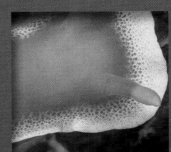

In February 2014, in the West Africa country Sierra Leone, the first cases were reported of the horrifying disease caused by Ebola virus. It rapidly spread to Liberia and Guinea, and within 15 months it had stricken more than 26,000 people and killed more than 11,000.

Among the first questions epidemiologists ask about a new or resurgent infectious disease are where it originated and by what paths it spread. Within 7 months after the start of the Ebola outbreak, a team of health scientists, molecular biologists, and evolutionary biologists had an answer [7]. Based on an evolutionary analysis of the viral genomes from several patients, the researchers concluded that the West Africa virus had almost certainly spread from central Africa about a decade earlier, and that the 2014 outbreak originated from a single person who contracted the virus from another host species, probably a bat. This was an important point, because it indicated that although the virus is readily transmitted from one person to another, it is only rarely contracted by humans from other species.

This was by no means the first time evolutionary methods had been used to trace the origin of an infectious disease. This approach has been routine ever since the origin of the human immunodeficiency virus (HIV), which causes AIDS, was determined in 1989. Two distinct HIVs (HIV-1 and HIV-2) infect humans; the pandemic is caused by HIV-1. Both HIVs are lentiviruses, a group of retroviruses that infect diverse mammals. In monkeys and other primates, the viruses are called simian immunodeficiency viruses, or SIVs (**FIGURE 1.1**). An evolutionary analysis showed that HIV-2 recently evolved from an SIV carried by sooty

This pink nudibranch (*Hypselodoris bullocki*) is a spectacular example of a group of marine mollusks renowned for their unusual shapes and bright coloration. Many nudibranchs contain toxins as a defense against predation and their unusual colors may be an adaptation that warns potential predators not to eat them. The only scientific explanation of such adaptations is the theory of evolution by natural selection.

FIGURE 1.1 (A) Structural model of a human immunodeficiency virus (HIV). (B) The sooty mangabey (*Cercopithecus atys*) and (C) the chimpanzee (*Pan troglodytes*) are the sources of two forms of HIV.

mangabey monkeys, and that HIV-1 evolved from SIV_{cpz}, the virus that infects wild chimpanzees (**FIGURE 1.2**) [9, 25]. The evolutionary analysis showed, moreover, that HIV-1 entered the human population near the beginning of the twentieth century, decades before it spread beyond Africa. It is thought that humans became infected with SIVs by contact with the blood of chimpanzees and mangabeys that they killed for food.

These viruses do not have a fossil record, so how could biologists infer their evolution and spread? They used methods that have been developed to reconstruct evolutionary history, and that are based on understanding the processes of evolutionary change.

Understanding the processes of evolution is highly relevant to human health. For example, the first drug approved to treat HIV-infected people was AZT, in 1987. Within a few years, however, AZT failed to prevent many infected patients from developing AIDS, and it has been necessary to develop other drugs. What happened? Populations of HIV had *adapted* to AZT by *evolving* resistance. Ever since the first antibiotic—penicillin—came into use, bacteria and other pathogenic microbes have rapidly evolved resistance to every antibiotic that has been widely used (**FIGURE 1.3**) [20, 22]. *Staphylococcus aureus*, a bacterium that causes many infections in surgical patients, has evolved resistance to a vast array of antibiotics, starting with penicillin and working its way through many others. Drug-resistant strains of *Neisseria gonorrheae*, the bacterium that causes gonorrhea, have steadily increased in abundance, and many strains of the tuberculosis, pneumonia, and cholera bacteria are highly resistant to antibiotics. Throughout the tropics, the microorganism that causes malaria is now resistant to chloroquine and is becoming resistant to other drugs as well. Worldwide, more than a half million people die yearly from drug-resistant infections. The evolution of antibiotic resistance is a major crisis in public health [3, 22].

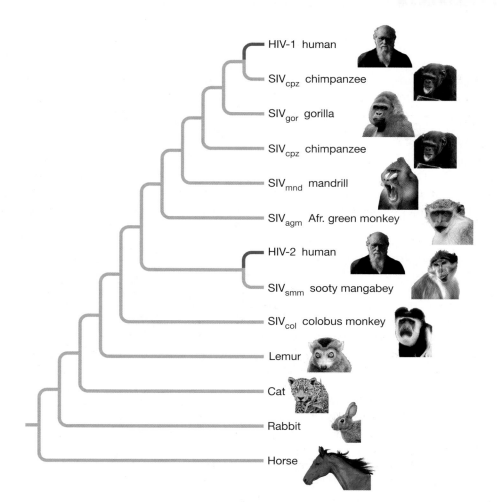

Almost every hospital in the world treats casualties in this battle against changing opponents, but as the use of antibiotics increases, so does the incidence of bacteria that are resistant to those antibiotics; thus any gains made are almost as quickly lost (see Figure 1.3). Why is this happening? Do the drugs cause drug-resistant mutations in the bacteria's genes? Do the mutations occur even without

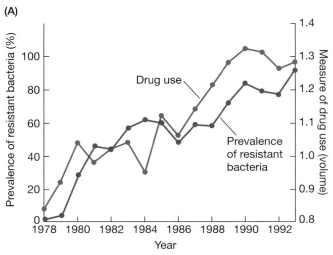

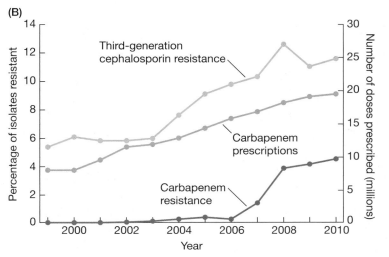

FIGURE 1.3 Evolution of drug resistance. (A) An increase in the use of a penicillin-like antibiotic in a community in Finland between 1978 and 1993 was matched by a dramatic increase in the percentage of antibiotic-resistant isolates of the bacterium *Moraxella catarrhalisis* from middle-ear infections in young children. (B) Resistance of the pneumonia-causing bacterium *Klebsiella pneumoniae* to cephalosporin and carbapenem antibiotics has recently begun to increase in the United States. The use of carbapenems approximately doubled during the period shown. (A after [15]; B after [23].)

exposure to drugs—that is, are they present in unexposed bacterial populations? Do the mutations spread among different species of bacteria? Can the evolution of resistance be prevented by using lower doses of drugs? Higher doses? Combinations of different drugs?

Microbial adaptation to drugs is the same, in principle, as the countless adaptations of every species to its environment, so it is very familiar to evolutionary biologists. The principles and methods of evolutionary biology have provided some answers to these questions about antibiotic resistance, and have shed light on many other problems that affect society. Evolutionary biologists have studied the evolution of insecticide resistance in disease-carrying and crop-destroying insects. They have helped devise methods of nonchemical pest control and have laid the foundations for transferring genetic resistance to diseases and insects from wild plants to crop plants. Evolutionary principles and knowledge are being used in biotechnology to design new drugs and other useful products, and in medical genetics to identify and analyze inherited diseases as well as variation in susceptibility to infectious diseases. In the fields of computer science and artificial intelligence, "evolutionary computation" uses principles taken directly from evolutionary theory to solve mathematically difficult practical problems, such as constructing complex timetables and processing radar data.

The importance of evolutionary biology goes far beyond its practical uses. An evolutionary framework provides answers to many questions about ourselves. How do we account for human variation—the fact that almost everyone is genetically and phenotypically unique? What accounts for behavioral differences between men and women? How did exquisitely complex, useful features such as our hands and our eyes come to exist? What about apparently useless or even potentially harmful characteristics such as our wisdom teeth and appendix? Why do we age, senesce, and eventually die? Evolution raises still larger questions. As soon as Darwin published *On the Origin of Species* in 1859, the evolutionary perspective was perceived to bear on long-standing questions in philosophy. If humans, with all their mental and emotional complexity, originated by natural processes, where do ethics and moral precepts find a foundation and origin? What, if anything, does evolution imply about the meaning and purpose of life? Must one choose between evolution and religious belief?

"Nothing in Biology Makes Sense except in the Light of Evolution"

If you suppose that scientists study evolution by analyzing fossils, you are right—but as the analyses of infectious diseases show, students of evolution also employ many other approaches and address a wide range of questions. Evolutionary biology is concerned with explaining and understanding the diversity of living things and their characteristics: what has been the *history* that produced this diversity, and what have been the *causes* of this history? Some evolutionary scientists try to elucidate the history of viruses, how they became capable of infecting diverse species of animals, and how antibiotic resistance evolves. Others ask similar questions about the origin of humans and human characteristics—or of mammals, plants, beetles, or dinosaurs. And because all features of all organisms have evolved, evolutionary biologists study the evolution of DNA sequences, proteins, biochemical pathways, embryological development, anatomical features, behaviors, life histories, interactions among different species: all of biology. Facing such an overwhelming profusion of subjects, evolutionary scientists aim to develop broad principles and to document common *patterns of evolution*—to arrive at *general principles* that apply to diverse organisms

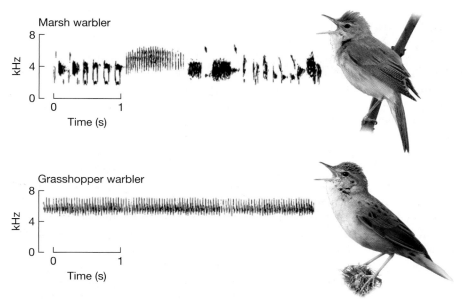

Marsh warbler

Grasshopper warbler

FIGURE 1.4 The song of a male marsh warbler (*Acrocephalus palustris*) is much more complex than the song of a male grasshopper warbler (*Locustella naevia*), which is a simple buzz. The sonograms (diagrams of the song) show frequency in relation to time. The song nucleus in the brain is larger in the marsh warbler than in the grasshopper warbler. Female marsh warblers prefer males with more complex songs. The proximate causes of the song difference include the brain structure; the ultimate causes include natural selection owing to the reproductive success of males whose songs attract more females. (Sonograms from [30].)

and diverse kinds of characteristics. Most of this book attempts to convey these general principles, although we illustrate the principles with studies of particular organisms and characteristics.

Evolutionary biology extends and amplifies the explanation of biological phenomena. It complements studies of the **proximate causes** (immediate, mechanical causes) of biological phenomena—the subject of cell biology, neurobiology, and many other biological disciplines—with analysis of the **ultimate causes** of those phenomena: their historical causes, especially the action of natural selection. If we ask what causes a male bird to sing, the proximate causes include the action of testosterone or other hormones, the structure and action of the singing apparatus (syrinx), and the operation of certain centers in the brain (**FIGURE 1.4**). The ultimate causes lie in the history of events that led to the evolution of singing in the bird's remote ancestors. For example, past individuals whose genes inclined them to sing may have been more successful in attracting females or in driving away competing males, and thus may have transmitted their genes to more descendants than did their less vocal competitors. Proximate and ultimate explanations may interact [14], and together provide more complete understanding than either does alone. As the great evolutionary biologist Theodosius Dobzhansky [5] wrote, "Nothing in biology makes sense except in the light of evolution."

What Is Evolution? Is It Fact or Theory?

The word "evolution" comes from the Latin *evolvere*, "to unfold or unroll"—to reveal or manifest hidden potentialities. Today "evolution" has come to mean, simply, "change." But changes in individual organisms, such as those that transpire in development (ontogeny) are not considered evolution. **Biological** (or **organic**) **evolution** is *inherited change in the properties of groups of organisms over the course of generations.* As Darwin elegantly phrased it, evolution is *descent with modification.*

As the HIV and SIV viruses illustrate, a single group, or **population**, of organisms may be modified over the course of time (e.g., becoming drug-resistant). A population may become subdivided, so that several populations are descended from a *common ancestral population.* If different changes transpire in the several populations, the populations **diverge** —that is, they become different from each other (e.g., as the various HIVs and SIVs have done).

Is evolution a fact, a theory, or a hypothesis? Biologists often speak of the "theory of evolution," but they usually mean by that something quite different from what most nonscientists understand by that phrase. Biologists talk about the "theory of evolution" in the same way that physicists talk about the "theory of gravitation." Scientists are as confident about the reality of evolution as they are of the reality of gravity.

In science, a **hypothesis** is an informed conjecture or statement of what might be true. Most philosophers (and scientists) hold that we do not know anything with absolute certainty. What we call "facts" are in some cases simple, confirmed observations; in other cases, a "fact" is a hypothesis that has acquired so much supporting evidence that we act as if it is true. A hypothesis may be poorly supported at first, but it can gain support to the point that it is effectively a fact. For Copernicus, the revolution of Earth around the Sun was a hypothesis with modest support; for us, this hypothesis has such strong support that we consider it a fact. Occasionally, an accepted "fact" may need to be revised in the face of new evidence; for example, humans have 46 chromosomes, not 48 as once thought.

In everyday use, "theory" refers to an unsupported speculation. Like many words, however, this term has a different meaning in science. Strictly speaking, a **scientific theory** is a comprehensive, coherent body of interconnected statements, based on reasoning and evidence, that explain some aspect of nature—usually many aspects. Thus atomic theory, quantum theory, and the theory of plate tectonics are elaborate schemes of interconnected ideas, strongly supported by evidence, that account for a great variety of phenomena. "Theory" is a term of honor in science; the greatest accomplishment a scientist can aspire to is to develop a valid, successful new theory.

In *The Origin of Species*, Darwin propounded *two major hypotheses*: that organisms have descended, with modification, from common ancestors; and that the chief cause of modification is natural selection acting on hereditary variation. Darwin provided abundant evidence for descent with modification; since then, hundreds of thousands of observations from paleontology, geographic distributions of species, comparative anatomy, embryology, genetics, biochemistry, and molecular biology have confirmed that all known species are related to one another through a history of common ancestry. Thus the hypothesis of descent with modification from common ancestors has long had the status of a scientific fact. (We will describe some of the evidence in Chapters 2 and 22.)

The explanation of how modification occurs and how ancestors give rise to diverse descendants constitutes the scientific theory of evolution. We now know that Darwin's hypothesis that evolution occurs by natural selection acting on hereditary variation was correct. We also know that there are more causes of evolution than Darwin realized and that natural selection and hereditary variation are more complex than he imagined. A body of ideas about the causes of evolution, including mutation, recombination, gene flow, isolation, random genetic drift, the several forms of natural selection, and other factors constitutes our current theory of evolution, or "evolutionary theory." Like all theories in science, it is a work in progress, for we do not entirely know the causes of all of evolution, or of all the biological phenomena that evolutionary biology will have to explain. In evolutionary biology, as in every other scientific discipline, there are "core" principles that have withstood skeptical challenges and are highly unlikely to require revision, and there are "frontier" areas in which research actively continues. Some widely held ideas about frontier subjects may prove to be wrong, but the uncertainty at the frontier does not undermine the core. The main tenets of evolutionary theory—descent with modification from a common ancestor, in part caused by natural selection—are so well supported that almost all biologists confidently accept evolutionary theory as the foundation of the science of life.

The Evolution of Evolutionary Biology

That the past is often the key to the present may be a cliché, but it happens to be true. Just as evolutionary history has shaped today's organisms, and just as social and political history is the key to understanding today's nations and conflicts, so the content of any science or other intellectual discipline cannot be fully understood without reference to its history.

Before Darwin

Darwin's theory of biological evolution is one of the most revolutionary ideas in Western thought, perhaps rivaled only by Newton's and Einstein's theories of physics. It profoundly challenged the prevailing worldview, which had originated largely with Plato and Aristotle, who developed the notion that species have fixed properties. Later, Christians interpreted the biblical account of Genesis literally and concluded that each species had been created individually by God in the same form it has today. (This belief is known as "special creation.") Christian theologians and philosophers argued that since existence is good and God's benevolence is complete, He must have bestowed existence on every creature of which He could conceive. Because order is superior to disorder, God's creation must follow a plan: specifically, a gradation from inanimate objects and barely animate forms of life through plants and invertebrates and up through ever "higher" forms of life. Humankind, being both physical and spiritual in nature, formed the link between animals and angels. This "Great Chain of Being," or *scala naturae* (the scale, or ladder, of nature), must be permanent and unchanging, since change would imply that there had been imperfection in the original creation [16].

As late as the nineteenth century, natural history was justified partly as a way to reveal the plan of creation so that we might appreciate God's wisdom. **Carolus Linnaeus** (1707–1778), who established the framework of modern taxonomy in his *Systema Naturae* (1735), won worldwide fame for his exhaustive classification of plants and animals, undertaken in the hope of discovering the pattern of the creation. Linnaeus classified "related" species into genera, "related" genera into orders, and so on. To him, "relatedness" meant propinquity in the Creator's design.

Belief in the literal truth of the biblical story of creation started to give way in the eighteenth century, when a philosophical movement called the Enlightenment, largely inspired by Newton's explanations of physical phenomena, adopted reason as the major basis of authority and marked the emergence of science. The foundations for evolutionary thought were laid by astronomers, who developed theories of the origin of stars and planets, and by geologists, who amassed evidence that Earth had undergone profound changes, that it had been populated by many creatures now extinct, and that it was very old. The geologists James Hutton and Charles Lyell expounded the principle of **uniformitarianism**, holding that the same processes operated in the past as in the present and that the data of geology should therefore be explained by causes that we can now observe. Darwin was greatly influenced by Lyell's teachings, and he adopted uniformitarianism in his thinking about evolution.

Carolus Linnaeus

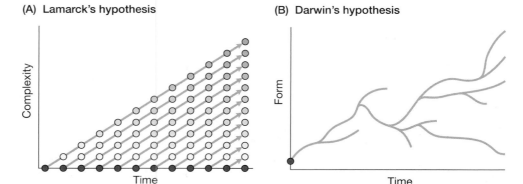

(A) Lamarck's hypothesis

(B) Darwin's hypothesis

FIGURE 1.5 Lamarck's and Darwin's hypotheses of the history of evolution. (A) Under Lamarck's hypothesis, life has originated many times (the red dots). Each lineage that descends from one of these origins becomes more complex. Thus, organisms range from recently originated, simple forms of life to older, more complex forms. (B) Darwin's theory of descent with modification, represented by a phylogenetic tree. From a single ancestor (the red dot), different lineages arise by speciating (splitting) from existing lineages. Some (such as the more central lineages) may undergo less modification from the ancestral condition than others. Darwin supposed that species become different from each other in various features ("form"), not necessarily becoming more complex. (A after [1].)

Jean-Baptiste Pierre Antoine de Monet, Chevalier de Lamarck

In the eighteenth century, several French philosophers and naturalists suggested that species had arisen by natural causes. The most significant pre-Darwinian evolutionary hypothesis was proposed by the **Chevalier de Lamarck** in his *Philosophie Zoologique* (1809). Lamarck hypothesized that different organisms originated separately by spontaneous generation from nonliving matter, starting at the bottom of the chain of being. A "nervous fluid" acts within each species, he said, causing it to progress up the chain. Species originated at different times, so we now see a hierarchy of species because they differ in age (**FIGURE 1.5A**).

Lamarck argued that species differ from one another because they have different needs, and so use certain of their organs and appendages more than others. Just as muscles become strengthened by work, more strongly exercised organs attract and become enlarged by the "nervous fluid." Lamarck, like most people at the time, believed that such alterations, acquired during an individual's lifetime, are inherited—a principle called **inheritance of acquired characteristics**. The theory of evolution based on this principle is called **Lamarckism**. In the most famous example of Lamarck's theory, giraffes must have stretched their necks to reach foliage above them, and so their necks were lengthened. The longer necks were inherited, and over the course of generations, this process was repeated and their necks got longer and longer. This could happen to any and all giraffes, so the entire species could have acquired longer necks because it was composed of individual organisms that changed during their lifetimes (**FIGURE 1.6A**). Lamarck's ideas of how evolution works were wrong, but he deserves credit for being the first to advance a coherent and testable theory of evolution.

Charles Darwin

Charles Robert Darwin (February 12, 1809–April 19, 1882) was the son of an English physician. He briefly studied medicine in Edinburgh, then turned to studying for a career in the clergy at Cambridge University. He believed in the literal truth of the Bible as a young man. He was passionately interested in natural history. In 1831, at the age of 22, his life was forever changed when he was invited to serve as

(A) Lamarck's hypothesis

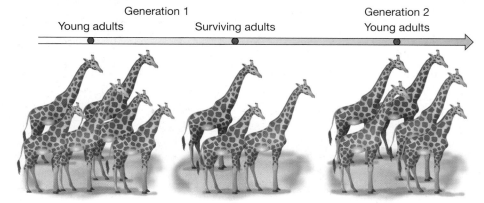

(B) Darwin's hypothesis

FIGURE 1.6 Contrast between Lamarck's and Darwin's hypotheses for how characteristics evolve, shown across two generations. (A) Under Lamarck's hypothesis, traits change within the lifetime of individuals because of their needs, illustrated here by giraffes that need longer necks to reach high leaves. Changes that are acquired during this generation are passed on to the next generation. (B) Under Darwin's hypothesis, there is variation among individuals at the start of each generation. Individuals with certain traits (e.g., a longer neck) have a greater chance of surviving. The variation is inherited, so survivors pass on their traits to the next generation. Darwin was right, but about 50 years would pass before scientists would understand how the inherited variations arise.

a naturalist and captain's companion on the British Navy ship H.M.S. *Beagle*, tasked with charting the coast of South America.

The voyage of the *Beagle* lasted from December 27, 1831, to October 2, 1836. The ship spent several years traveling along the coast of South America, where Darwin observed the natural history of the Brazilian rainforest and the Argentine pampas, then stopped in the Galápagos Islands, which lie on the equator off the coast of Ecuador. In the course of the voyage, Darwin became an accomplished naturalist, collected specimens, made innumerable geological and biological observations, and conceived a new (and correct) theory about the formation of coral atolls.

Soon after Darwin returned, the ornithologist John Gould pointed out that Darwin's specimens of mockingbirds from the Galápagos Islands were so different from one island to another that they represented different species (**FIGURE 1.7**). Darwin then recalled that the giant tortoises, too, differed from one island to the next (**FIGURE 1.8**). These facts, and the similarities between fossil and living mammals that he had found in South America, triggered his conviction that different species had evolved from common ancestors.

Darwin's comfortable finances enabled him to devote the rest of his life exclusively to his scientific work (although he was chronically ill for most of his life after the voyage). He set about amassing evidence of evolution and trying to conceive of its causes. In 1838, at the age of 29, Darwin read an essay by the

Charles Robert Darwin

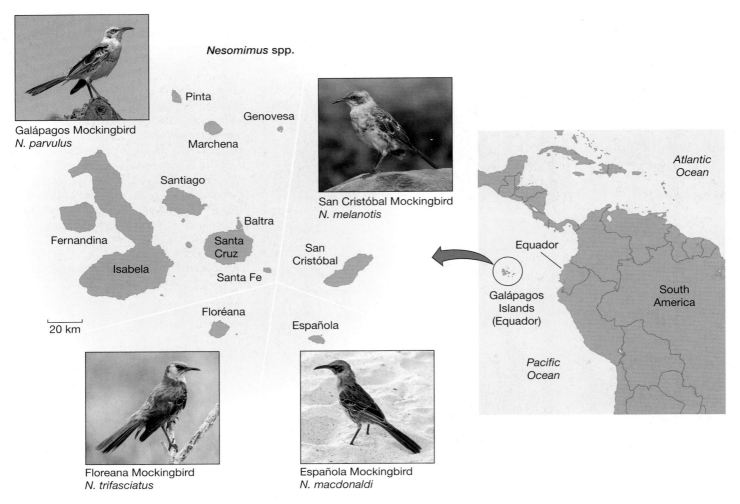

Nesomimus spp.

Galápagos Mockingbird
N. parvulus

Pinta

Genovesa

Marchena

Santiago

Baltra

Fernandina

Santa
Cruz

Isabela

Santa Fe

Floréana

San
Cristóbal Mockingbird
N. melanotis

San
Cristóbal

Española

20 km

Floreana Mockingbird
N. trifasciatus

Española Mockingbird
N. macdonaldi

Atlantic
Ocean

Equador

Galápagos
Islands
(Equador)

South
America

Pacific
Ocean

FIGURE 1.7 Four species of mockingbirds (*Nesomimus*) on different islands in the Galápagos archipelago were among the observations that led Darwin to suspect that different species evolve from a common ancestor.

economist Thomas Malthus. Malthus argued that the rate of human population growth is greater than the rate of increase in the food supply, so that unchecked growth must lead to famine. This essay was the inspiration for Darwin's great idea, one of the most important ideas in the history of thought: natural selection. Darwin wrote in his autobiography that "being well prepared to appreciate the struggle for existence which everywhere goes on from long-continued observation of the habits of animals and plants, it at once struck me that under these circumstances favourable variations would tend to be preserved and unfavourable ones to be destroyed." In other words, of the many individuals that are born, not all survive; and if certain individuals with superior features survived and reproduced more successfully than individuals with inferior features, and if these differences were inherited, the average character of the species would be altered over the course of generations.

Mindful of how controversial the subject would be, Darwin then spent 20 years developing his theory, amassing evidence, and pursuing other researches before publishing his ideas. In 1844 he wrote a private essay outlining his theory, and in 1856 he finally began a book he intended to call *Natural Selection*. He never completed it, for in June 1858 he received a manuscript from a young naturalist,

FIGURE 1.8 Galápagos giant tortoises (*Chelonoidis nigra*) differ in shell shape among islands. Some subspecies, especially those that occupy humid highlands with low vegetation, have a domed shell (A), whereas those in dry lowland habitats tend to have a "saddleback" shell (B) that enables the animal to extend its long neck to reach vegetation higher above the ground.

Alfred Russel Wallace (1823–1913). Wallace, who was collecting specimens in the Malay Archipelago, had independently conceived of natural selection. Darwin's scientific colleagues presented extracts from his 1844 essay, along with Wallace's manuscript, at a meeting of the major scientific society in London. Darwin immediately set about writing an "abstract" of the book he had intended. The 490-page result, titled *On the Origin of Species by Means of Natural Selection, or The Preservation of Favoured Races in the Struggle for Life*, was published on November 24, 1859; it instantly made Darwin, by now 50 years old, both a celebrity and a figure of controversy.

For the rest of his life, Darwin continued to read and correspond on an immense range of subjects, to revise *The Origin of Species* ("on" was deleted from the title of later editions), to perform experiments of all sorts (especially on plants), and to publish many more articles and books, of which *The Descent of Man* is the most renowned. Darwin's books reveal an irrepressibly inquisitive man, fascinated with all aspects of nature, creative in devising hypotheses and in bringing evidence to bear on them, and profoundly aware that every biological fact, no matter how seemingly trivial, must fit into a coherent, unified understanding of the world. Wallace made significant further contributions to biology, especially about biogeography, the geographic distribution of species. He always gave credit to Darwin for the concept of natural selection, referring to it as "Mr. Darwin's theory."

Alfred Russel Wallace

Darwin's evolutionary theory

The Origin of Species contains two major theories. The first is Darwin's idea of **descent with modification**. It holds that all species, living and extinct, have descended, without interruption, from one or a few original forms of life (**FIGURE 1.5B**). Species that diverge from a common ancestor are at first very similar but accumulate differences over great spans of time, so that they may come to differ radically from one another. Darwin's conception of the course of evolution is profoundly different from Lamarck's, in which the concept of common ancestry plays almost no role.

The second theory in *The Origin of Species* is **natural selection**, which Darwin proposed is the chief cause of evolutionary change. He summarized it in the following way: "If variations useful to any organic being ever occur, assuredly individuals thus characterized will have the best chance of being preserved in the struggle for life; and from the strong principle of inheritance, these will tend to produce offspring similarly characterized. This principle of preservation, or the survival of the fittest, I have called natural selection." Unlike Lamarck's *transformational theory*, in which individual organisms change, Darwin's is a *variational theory* of change, in which the **frequency** of a variant form (i.e., the proportion of individuals with that variant feature) increases within a population from generation to generation (**FIGURE 1.6B**). Darwin proposed (as did Wallace) that fitter individuals differ only slightly from the norm of the population, but that a feature such as body size gradually evolves to become more and more different because new, slightly more extreme, advantageous variants continue to arise.

Darwin's theory of evolution includes five distinct components [18]:

1. *Evolution as such* is the simple proposition that the characteristics of organisms change over time. Darwin was not the first to have this idea, but he so convincingly marshaled the evidence for evolution that most scientists soon accepted that it has indeed occurred.

2. *Common descent*: Differing radically from Lamarck, Darwin was the first to argue that species had diverged from common ancestors and that species could be portrayed as one great family tree representing actual ancestry (see Figure 1.5B).

3. *Gradualism* is Darwin's proposition that the differences between even radically different organisms have evolved by small steps through intermediate forms, not by leaps ("saltations").

4. *Populational change* is Darwin's hypothesis that evolution occurs by changes in the proportions (frequencies) of different variant kinds of individuals within a population (see Figure 1.6B). This profoundly important, completely original idea contrasts with the sudden origin of new species by saltation and with Lamarckian transformation of individuals. For Darwin, the average was a statistical abstraction; there exist only varied individuals, and there are no fixed limits to the variation that a species may undergo [10, 18].

5. *Natural selection* was Darwin's brilliant hypothesis, independently conceived by Wallace, that accounts for **adaptations**, features that appear "designed" to fit organisms to their environment. *Because it provided an entirely natural, mechanistic explanation for adaptive design that had previously been attributed to a divine intelligence, the concept of natural selection revolutionized not only biology, but Western thought as a whole.*

Darwin proposed that the various species that descend from a common ancestor evolve different features because those features are adaptive under different "conditions of life"—different habitats or habits. Moreover, the pressure of competition favors the use of different foods or habitats by different species. He believed that no matter how extensively a species has diverged from its ancestor, new hereditary variations continue to arise, so that given enough time, there is no evident limit to the amount of divergence that can occur.

Where, though, do these hereditary variations come from? This was the great gap in Darwin's theory, and he never filled it. The problem was serious because, according to the prevailing belief in **blending inheritance**, variation should decrease, not increase. Because offspring are often intermediate between their parents in features such as color or size, it was widely believed that characteristics are inherited like fluids, such as paints of different colors. (This notion persists today when people speak of having Italian or Indian "blood.") Blending white and black paints produces gray, but mixing two gray paints yields more gray, not black or white. Darwin never knew that Gregor Mendel had solved the problem in a paper that was published in 1866, but not widely noticed until 1900. Mendel's theory of **particulate inheritance** proposed that inheritance is based not on blending fluids, but on particles that pass unaltered from generation to generation—so that variation can persist. The concept of "mutation" in such particles (later called genes) was developed only after 1900 and was not clarified until considerably later.

The Origin of Species is extraordinarily rich in insights and implications. Darwin supported his hypotheses with an astonishingly broad variety of information, from variation in domesticated species to embryology to geographic patterns in the distribution of species. And he showed, or at least glimpsed, how research in every biological subject—taxonomy, paleontology, anatomy, embryology, biogeography, physiology, behavior, ecology—could be advanced and reinterpreted in the light of evolution.

Evolutionary biology after Darwin

Although The Origin of Species raised enormous controversy, by the 1870s most scientists accepted the historical reality of evolution by descent, with modification, from common ancestors. This theory provided a new framework for exploring and interpreting the history and diversification of life, a project that was especially promoted by the German zoologist Ernst Haeckel. Thus the late nineteenth and early twentieth centuries were a "golden age" of paleontology, comparative morphology, and comparative embryology, during which a great deal of information on evolution in the fossil record and on relationships among organisms was amassed [2]. But the consensus did not extend to Darwin's theory of the cause of evolution, natural selection. For about 60 years after the publication of The Origin of Species, all but a few faithful Darwinians rejected natural selection, and numerous theories were proposed in its stead. These theories included neo-Lamarckian, orthogenetic, and mutationist theories [1].

Neo-Lamarckism includes several theories based on the old idea of inheritance of modifications acquired during an organism's lifetime. In a famous experiment, the German biologist August Weismann cut off the tails of mice for many generations and showed that this mutilation had no effect on the tail length of their descendants. Extensive subsequent research has provided no evidence that specific mutations can be induced by environmental conditions under which they would be advantageous.

Theories of **orthogenesis**, or "straight-line evolution," held that the variation that arises is directed toward fixed goals, so that a species evolves in a predetermined direction by some kind of internal drive, without the aid of natural selection. Some paleontologists held that such trends need not be adaptive and could even drive species toward extinction (**FIGURE 1.9**). None of the proponents of orthogenesis ever proposed a mechanism for it.

FIGURE 1.9 The extinct Irish elk (*Megaloceros giganteus*) had such enormous antlers that it was cited as an example of orthogenetic "momentum" that drove the species to evolve a maladaptive feature that caused its extinction. Since the 1940s, evolutionary biologists have rejected this idea. The huge antlers probably resulted from the animal's overall large size and from natural selection caused by competition among males for females.

Ronald A. Fisher

J. B. S. Haldane

Sewall Wright

Mutationist theories were advanced by some geneticists who observed that discretely different new phenotypes can arise by a process of mutation. They supposed that such mutant forms constituted new species and thus believed that natural selection was not necessary to account for the origin of species. The last influential mutationist was Richard Goldschmidt (1940, [8]), an accomplished geneticist who nevertheless erroneously argued that the origin of new species and higher taxa is entirely different in kind from evolutionary change within species. New species or genera, he said, originate by sudden, drastic changes that reorganize the whole genome. Although most such reorganizations would be deleterious, a few "hopeful monsters" would be the progenitors of new forms of life.

The evolutionary synthesis

These anti-Darwinian ideas were refuted in the 1930s and 1940s by the geneticists, systematists, and paleontologists who reconciled Darwin's theory with the facts of genetics [19, 28]. The consensus they forged is known as the **evolutionary synthesis,** or **modern synthesis**, and its chief principle, that adaptive evolution is caused by natural selection acting on particulate (Mendelian) genetic variation, is often referred to as **neo-Darwinism**.[1] Ronald A. Fisher and John B. S. Haldane in England and Sewall Wright in the United States developed a mathematical theory of population genetics, which showed that mutation and natural selection together cause adaptive evolution: mutation is not an alternative to natural selection, but is rather its raw material. The study of genetic variation and change in natural populations was pioneered in Russia by Sergei Chetverikov and continued by Theodosius Dobzhansky, who moved from Russia to the United States. In his influential book *Genetics and the Origin of Species* (1937, [4]), Dobzhansky conveyed the ideas of the population geneticists to other biologists, thus influencing their appreciation of the genetic basis of evolution. Other major contributors to the synthesis included the zoologists Ernst Mayr, in *Systematics and the Origin of Species* (1942, [17]), and Bernhard Rensch, in *Evolution Above the Species Level* (1959, [24]); the botanist G. Ledyard Stebbins, in *Variation and Evolution in Plants* (1950, [29]); and the paleontologist George Gaylord Simpson, in *Tempo and Mode in Evolution* (1944, [26]) and its successor, *The Major Features of Evolution* (1953, [27]). These authors argued persuasively that mutation, gene flow or migration, natural selection, and genetic drift are the major causes of evolution within species (which Dobzhansky called **microevolution**)—and that continued over long periods of time, these same causes account for the origin of new species and for **macroevolution**: the evolution of the major alterations that distinguish **higher taxa** (genera, families, orders, and classes). The principal claims of the evolutionary synthesis are the foundations of modern evolutionary biology.

Although some of these principles have been extended, clarified, or modified since the 1940s, most evolutionary biologists today accept them as substantially valid. They are summarized in **BOX 1A**.

Evolutionary biology since the synthesis

Since the evolutionary synthesis, a great deal of research has tested and elaborated its basic principles. These principles have largely been supported. Progress in evolutionary biology has modified some of these ideas and many extensions of these ideas, and it has spurred additional theory to account for new phenomena as they

---

[1] "Neo-Darwinism" properly refers to Weismann's strict version of Darwin's theory of evolution by natural selection. Darwin had admitted a role for inheritance of acquired characteristics, but Weismann rejected this. Today, "neo-Darwinism" is often used to mean the theory articulated in the evolutionary synthesis.

G. Ledyard Stebbins, George Gaylord Simpson, and Theodosius Dobzhansky

Motoo Kimura

were discovered. Since James Watson and Francis Crick established the structure of DNA in 1953, advances in genetics, molecular biology, and molecular and information technology have revolutionized the study of evolution.

New molecular and computational technology has enabled new fields of evolutionary study to develop, among them *molecular evolution* (analysis of the processes and history of changes in genes). The leaders of the evolutionary synthesis had maintained that almost all features of organisms are adaptive, and evolved by natural selection. But this principle was challenged by the **neutral theory of molecular evolution**, developed by Motoo Kimura (1983, [13]), who argued that most of the evolution of DNA sequences occurs by chance (**genetic drift**) rather than by natural selection. *Evolutionary developmental biology,* growing out of comparative embryology and based partly on molecular genetics, is devoted to understanding how the evolution of developmental processes underlies the evolution of morphological features at all levels, from cells to whole organisms. Because the entire *genome*—the full DNA complement of an organism—can now be sequenced, molecular evolutionary studies have expanded into *evolutionary genomics*, which is concerned with variation and evolution in multiple genes or even entire genomes. Genomic data are enabling biologists to determine phylogenetic relationships with ever-greater confidence; they are revealing the genetic bases of adaptive characteristics of species and how and when they were modified by natural selection, and they are revealing the history of populations and their distributions over the globe. The histories of species are written in their genes.

The advances in these new fields are complemented by vigorous research, new discoveries, and new ideas about long-standing topics in evolutionary biology, such as the evolution of adaptations and of new species. Since the mid-1960s, evolutionary theory has expanded into areas such as ecology, animal behavior, and reproductive biology. Detailed theories that explain the evolution of particular kinds of characteristics such as life span, ecological distribution, and social behavior were pioneered by the evolutionary theoreticians William Hamilton and John Maynard Smith in England and George Williams in the United States. The study of macroevolution has been renewed by provocative interpretations of the fossil record and by new methods for studying phylogenetic relationships. Research in evolutionary biology is progressing more rapidly than ever before.

Since Darwin's time, research on evolution, and in biology more broadly, has transformed evolutionary biology. Were Darwin to reappear today, he would understand very few scientific papers about evolution. *Modern evolutionary biology*

BOX 1A

Fundamental Principles of Biological Evolution

These are fundamental principles of evolution that emerged from the modern synthesis. Much of the rest of this book is devoted to explaining and building on them. Some statements, marked by an asterisk (*), have to be qualified to some degree, in light of later research.

1. *An individual's* **phenotype** *(its observed traits) is distinct from its* **genotype** *(its DNA).* Phenotypic differences among individuals are caused by both genetic differences and environmental effects.

2. *Acquired characteristics are not inherited.**

3. *Hereditary variations are based on particles—the genes.** This is true for traits with continuous variation (e.g., body size) as well as those with discrete variation (e.g., eye color).

4. *Genetic variation arises by random mutation. Mutations do not arise in response to need.* Variation that arises by mutation is amplified by recombination of alleles at different loci.

5. *Evolution is a change of a population, not of an individual.* The elementary process of evolution is a change across generations in the frequencies of alleles or genotypes, which can change the frequencies of phenotypes.

6. *Changes in allele frequencies may be random or nonrandom.* Natural selection results from differences among individuals in survival and reproduction, and causes nonrandom changes. Genetic drift causes random changes.

7. *Natural selection can account for both slight and great differences among species.* Even a low intensity of natural selection can cause substantial evolution

ary change over time. Adaptations are traits that have been shaped by natural selection.

8. *Natural selection can alter populations beyond the original range of variation* when changes in allele frequencies generate new combinations of genes.

9. *Populations usually have considerable genetic variation.* Many populations evolve rapidly, to some degree, when environmental conditions change, and do not have to wait for new favorable mutations.

10. *The differences between species evolve by rather small steps,* and are often based on differences at many genes that accumulated over many generations.*

11. *Species are groups of interbreeding or potentially interbreeding individuals that do not exchange genes with other such groups.** Species are not defined simply by phenotypic differences. Rather, they represent separately evolving "gene pools."

12. **Speciation** (the origin of two species from a single ancestor species) *usually occurs by the genetic differentiation of geographically isolated populations.** Species have genetic differences that prevent interbreeding if they are no longer geographically separated.

13. *Higher taxa arise by the sequential accumulation of small differences,* rather than by the sudden appearance of drastically new types by mutation.

14. *All organisms form a great Tree of Life* (or **phylogeny**) that evolved by the branching of common ancestors into diverse lineages, chiefly by speciation. All forms of life descended from a single common ancestor that lived in the remote past.

does not equal Darwinism, and any antievolutionary critiques of Darwin that do not take into account modern research are irrelevant to our understanding of evolution today.

How Evolution Is Studied

Evolutionary biology is a more historical science than most other biological disciplines, for one of its goals is to determine what the history of life has been and what has caused those historical events.

Occasionally we can document an evolutionary change as it occurs or piece together records to reconstruct a recent change, just as we do when studying human history (see Chapter 3). Usually, however, we must *infer* evolutionary history and its causes by interpreting less direct evidence. Some historical events are inferred from fossils, the province of paleontology (see Chapters 17 and 19). Other evolutionary events are inferred from comparisons among living organisms or by studying their phylogenetic relationships, which provide a framework that enables us to draw conclusions about the historical evolution of their phenotypic characteristics and even their genes (see Chapters 2 and 16).

The causes of evolution, such as genetic drift and natural selection, are often studied by comparing data, such as patterns of variation in genes, with theoretical models (see Chapters 4–8). They are also studied by the methods of *experimental evolution*, in which laboratory populations of rapidly reproducing organisms adapt to an environment (e.g., a stressful temperature) designed by an investigator (see Chapter 6). The adaptive reasons for certain characteristics (e.g., birdsong) may be inferred from experimental and other functional studies, from their "fit" to a theoretical design (e.g., the heart fits a "pump" design), or by comparing many populations or species to see if the characteristic is correlated with a specific environmental factor or way of life (see Chapters 10–13). Certain patterns of variation in DNA sequences can tell us if natural selection has affected evolutionary changes in genes (see Chapters 5, 7, and 14).

When we make inferences about history, or about past causes of change such as natural selection, we do not see the changes occurring, nor do we observe the causes in action. But *throughout science, causes are not seen; rather, they are inferred*. All of chemistry, for example, concerns invisible atoms and orbitals that govern the association of atoms into molecules. These theoretically postulated entities and their behavior have been confirmed because *the theory that employs them makes predictions (hypotheses) that are matched by observed data*. We know that DNA replicates semiconservatively not because anyone has ever seen DNA do that, but because the outcome of a famous experiment (and of later ones) matched the prediction made by the hypothesis.

This hypothetico-deductive method, in which hypotheses are tested (and are rejected, modified, or provisionally accepted), has been a powerful tool throughout the sciences and is the basis of much evolutionary research. For example, would you predict that the DNA in mitochondria carries more mutations that are harmful to males than to females? There is no obvious biochemical reason to expect this, but evolutionary theory makes such a prediction. The mitochondria of both males and females are inherited from the mother; the mitochondria in males are not inherited via sperm and are thus at a "dead end." If a mutation in mitochondrial DNA reduces the survival or reproduction of females, it is unlikely to be transmitted to subsequent generations, but the transmission of a mutation will not be affected if it is similarly harmful only to males, because males do not transmit the DNA anyway. So, male-deleterious mitochondrial mutations are expected to accumulate. This prediction, from the theory of natural selection at the level of the gene, has been verified: mitochondrial variants commonly affect male, but not female, fertility in humans and other animals, and they cause variation in reproductive gene expression in male fruit flies [6,

Ernst Mayr, George C. Williams, John Maynard Smith

William D. Hamilton

12]. This example illustrates how evolutionary hypotheses can be tested, and it also shows how they can predict and reveal aspects of biology we would not otherwise have expected.

Philosophical Issues

Thousands of pages have been written about the philosophical and social implications of evolution. Darwin argued that every characteristic of a species can vary and can be altered radically, given enough time. Thus he rejected the emphasis on distinct "types" that Western philosophy had inherited from Plato and Aristotle and put variation in its place. Darwin also helped replace a static conception of the world—one virtually identical to the Creator's perfect creation—with a world of ceaseless change. It was Darwin who extended to living things, including the human species, the principle that change, not stasis, is the natural order. In contrast to traditional views that elevated the human species to a special position, distinct from other living things, Darwin began the trend to see humans as part of the natural world, a species of animal (though a very remarkable species, to be sure!) subject to the same processes as others, including natural selection.

Darwin has been credited with making biology a science, for he proposed to replace supernatural explanations in biology with purely natural causes. His theory of random, purposeless variation acted on by blind, purposeless natural selection provided a revolutionary new kind of answer to almost all questions that begin with "Why?" Before Darwin, both philosophers and people in general answered questions such as "Why do plants have flowers?" or "Why are there apple trees?"—or diseases, or sexual reproduction—by imagining the possible purpose that God could have had in creating them. This kind of explanation was made completely superfluous by Darwin's theory of natural selection. The adaptations of organisms—long cited as the most conspicuous evidence of intelligent design in the universe—could be explained by purely mechanistic causes. For evolutionary biologists, the pink petals of a magnolia's flower have a *function* (attracting pollinating insects) but not a *purpose*. The flower was not designed in order to propagate the species, much less to delight us with its beauty, but instead came into existence because magnolias with brightly colored flowers reproduced more prolifically than magnolias with duller flowers. The unsettling implication of this purely material explanation is that, except in the case of human behavior, we need not invoke, nor can we find any evidence for, any design, goal, or purpose anywhere in the natural world.

All of modern science employs the way of thought that Darwin applied to biology. Geologists do not seek the purpose of earthquakes or plate tectonics, nor chemists the purpose of hydrogen bonds. The concept of purpose plays no part in scientific explanation.

Ethics, religion, and evolution

In the world of science, the reality of evolution has not been in doubt for more than 100 years, but evolution remains an exceedingly controversial subject in the United States and a few other countries. The **creationist movement** opposes the teaching of evolution in public schools, or at least demands "equal time" for creationist beliefs. Such opposition arises from the fear that evolutionary science denies the existence of God, and consequently, that it denies any basis for rules of moral or ethical conduct.

Science, including evolutionary biology, is silent on the existence of a supernatural being or a human soul, because these hypotheses cannot be tested. Many people, including some priests, ministers, rabbis, and evolutionary biologists, hold both religious beliefs and belief in evolution (see Chapter 22). But to explain phenomena in the natural world, science must assume that only natural causes operate, just as most people do in everyday affairs: we assume that there is a material cause when our car or computer or heart malfunctions. Supernatural explanations for observable phenomena often do conflict with naturalistic, scientific explanation. A literal reading of some passages in the Bible is incompatible with the principles of physics, geology, and other natural sciences. Our knowledge of the history and mechanisms of evolution is certainly incompatible with a literal reading of the creation stories in the Bible's Book of Genesis—just as it is incompatible with hundreds of other creation myths people have devised.

Wherever ethical and moral principles are to be found, it is probably not in science, and surely not in evolutionary biology. Opponents of evolution have charged that evolution by natural selection justifies the principle that "might makes right." But evolutionary theory cannot provide any such precept for behavior. Like any other science, evolutionary biology describes how the world is, not how it should be. The supposition that what is "natural" is "good" is called by philosophers the *naturalistic fallacy*.

Various animals have evolved behaviors that we give names such as cooperation, monogamy, competition, infanticide, and the like. Whether or not these behaviors ought to be—and whether or not they are—moral, is not a scientific question. The natural world is amoral—the concepts of "moral" and "immoral" simply do not apply outside the realm of human behavior. Despite this, the concepts of natural selection and evolutionary progress were taken as a "law of nature" by which Marx justified class struggle, by which the Social Darwinists of the late eighteenth and early nineteenth centuries justified economic competition and imperialism, and by which the biologist Julian Huxley justified humanitarianism [11, 21]. Most philosophers consider all these ideas to be indefensible instances of the naturalistic fallacy. Infanticide by lions and langur monkeys does not justify infanticide in humans; monogamy in penguins does not imply that humans should do the same. Evolution provides no basis for human ethics.

Go to the
————————— **Evolution Companion Website** —————————
EVOLUTION4E.SINAUER.COM
for data analysis and simulation exercises, quizzes, and more.

SUMMARY

- Evolution is the unifying theory of the biological sciences. Evolutionary biology aims to discover the history of life, the causes of the diversity and characteristics of organisms, and the mechanisms that underlie evolutionary change.

- Charles Darwin's major work, *On the Origin of Species*, published in 1859, contains two major hypotheses: first, that all organisms have descended, with modification, from common ancestral forms of life, and second, that the chief agent of modification is natural selection.

- Darwin's hypothesis that all species have descended with modification from common ancestors is supported by so much evidence that it has become as well established a fact as any in biology. His theory of natural selection as the chief cause of evolution was not broadly supported until the evolutionary synthesis that occurred in the 1930s and 1940s.

- Modern evolutionary biology is based on the evolutionary synthesis, which united Darwin's ideas with Mendelian genetics. The major causes of evolution within species are those that change the frequencies of alleles, and hence of the phenotypes they may affect. Different populations of a species may experience different genetic changes, and ultimately become different species. Over long time periods, many slight changes accumulate to yield large genetic and phenotypic differences among species and their ancient ancestors.

- Evolutionary biology makes important contributions to other biological disciplines and to social concerns in areas such as medicine, agriculture, computer science, and our understanding of ourselves.

- The implications of Darwin's theory, which revolutionized Western thought, include the ideas that change, rather than stasis, is the natural order; that biological phenomena, including those seemingly designed, can be explained by purely material causes rather than by divine creation; and that no evidence for purpose or goals can be found in the living world, other than in human actions.

- Like other sciences, evolutionary biology cannot be used to justify beliefs about ethics or morality. Nor can it prove or disprove theological hypotheses such as the existence of a deity. Many people hold that evolution is compatible with religious belief. However, evolution is incompatible with a literal interpretation of some passages in the Bible. Evolutionary biology and other sciences can test and reject claims for supernatural causes of observed phenomena.

TERMS AND CONCEPTS

adaptation
blending inheritance
creationist movement
descent with modification
diverge
evolution (biological evolution; organic evolution)

evolutionary synthesis (modern synthesis)
frequency
genetic drift
genotype
higher taxa
hypothesis
inheritance of acquired characteristics

Lamarckism
macroevolution
microevolution
mutationist theories
natural selection
neo-Darwinism
neo-Lamarckism
neutral theory of molecular evolution
orthogenesis

particulate inheritance
phenotype
phylogeny
population
proximate cause
scientific theory
speciation
ultimate cause
uniformitarianism

SUGGESTIONS FOR FURTHER READING

The readings at the end of each chapter include major works that provide a comprehensive treatment and an entry into the professional literature. The references cited in each chapter also serve this important function.

No one should fail to read at least part of Darwin's *On the Origin of Species by Means of Natural Selection, or The Preservation of Favoured Races in the Struggle for Life*, in either the first edition (1859) or the sixth edition (1872), in which Darwin deleted "On" from the title. After some adjustment to the

Victorian prose, you will be enthralled by the craft, detail, completeness, and insight in Darwin's arguments. It is an astonishing book.

The Cambridge Encyclopedia of Darwin and Evolutionary Thought (ed. M. Ruse, Cambridge University Press, Cambridge, 2013) is a set of outstanding essays on the history of evolutionary thought and on the influence of Darwin and of evolution on science and other areas such as literature, theology, and philosophy.

Among biographies of Darwin, the best include Janet Browne's superb two-volume work, *Charles Darwin: Voyaging* and *Charles Darwin: The Power of Place* (Knopf, New York, 1995 and 2002, respectively); and *Darwin,* by A. Desmond and J. Moore (Warner Books, New York, 1991), which emphasizes the role played by the religious, philosophical, and intellectual climate of nineteenth-century England on the development of Darwin's scientific theories. An enjoyable popular biography is *The Reluctant Mr. Darwin: An Intimate Portrait of Charles Darwin and the Making of His Theory of Evolution,* by David Quammen (W. W. Norton, New York, 2006).

Important works on the history of evolutionary biology include P. J. Bowler, *Evolution: The History of an Idea* (University of California Press, Berkeley, 2003); E. Mayr, *The Growth of Biological Thought: Diversity, Evolution, and Inheritance* (Harvard University Press, Cambridge, MA, 1982), a detailed, comprehensive history of systematics, evolutionary biology, and genetics that bears the personal stamp of one of the major figures in the evolutionary synthesis; and E. Mayr and W. B. Provine (eds.), *The Evolutionary Synthesis: Perspectives on the Unification of Biology* (Harvard University Press, Cambridge, MA, 1980), which contains essays by historians and biologists, including some of the major contributors to the synthesis.

A few works treating the philosophical and practical implications of evolution are D. C. Dennett, *Darwin's Dangerous Idea: Evolution and the Meanings of Life* (Simon & Schuster, New York, 1995); M. Ruse, *The Philosophy of Human Evolution* (Cambridge University Press, Cambridge, 2012); P. Gluckman, A. Beedle, and M. Hanson, *Principles of Evolutionary Medicine* (Oxford University Press, Oxford, 2009), and A. Poiani (ed.), *Pragmatic Evolution* (Cambridge University Press, Cambridge, 2012).

Books that expose the fallacies of creationism and explain the nature of science and of evolutionary biology include B. J. Alters and S. M. Alters, *Defending Evolution: A Guide to the Creation/Evolution Controversy* (Jones and Bartlett, Sudbury, MA, 2001); M. Pigliucci, *Denying Evolution: Creationism, Scientism, and the Nature of Science* (Sinauer Associates, Sunderland, MA, 2002); and E. C. Scott, *Evolution versus Creationism: An Introduction,* second edition (University of California Press, Berkeley, 2009). The evidence for evolution is presented in two outstanding books, *Why Evolution Is True,* by J. A. Coyne (Viking, New York, 2009), and *The Greatest Show on Earth: The Evidence for Evolution,* by Richard Dawkins (Free Press, New York, 2009).

Darwin's birth, in 1809, was marked in 2009 by many bicentennial celebrations and books. Among these is *Evolution Since Darwin: The First 150 Years*, edited by M. A. Bell et al. (Sinauer Associates, Sunderland, MA, 2010), a collection of essays by historians and evolutionary biologists who summarize the state of knowledge and current research directions in all the subfields of evolutionary biology.

WEBSITES

Several excellent websites provide good introductions to evolution;
most of them also include material on teaching evolution and on creationism.

The National Center for Science Education is devoted to defending the teaching of evolution and climate science. Its website (ncse.com) should be the first stop for anyone who wants to learn about creationism, evidence for evolution, relationships between evolution and religion, and any other aspect of the social controversy about evolution.

"Understanding Evolution" (http://evolution.berkeley.edu) is an excellent site developed by the Museum of Paleontology at the University of California, Berkeley.

Outstanding videos on evolution are at www.hhmi.org/biointeractive/evolution-collection.

The National Academy of Sciences of the U.S.A., the members of which are leaders in science, has a website devoted to evolution (www.nationalacademies.org/evolution) and has published an excellent 70-page booklet, *Science, Evolution, and Creationism* (2008), that can be accessed for free through the website or purchased at low cost (order at www.nap.edu).

"Darwin Online" (http://darwin-online.org.uk), compiled by John van Wyhe, provides all of Darwin's writings, including many translations of *The Origin of Species* and his other books into other languages. van Wyhe has also created "Wallace Online" (http://wallace-online.org), a similar website on Alfred Russel Wallace.

PROBLEMS AND DISCUSSION TOPICS

1. Theodosius Dobzhansky wrote "Nothing in biology makes sense except in the light of evolution." What did he mean by this? How does evolution unify the biological sciences? What other principles might do so?

2. Analyze this Ralph Waldo Emerson couplet:

   Striving to be man, the worm
   Mounts through all the spires of form.

   What pre-Darwinian concepts does it express? What fault in it would a Darwinian find?

3. Human immunodeficiency virus entered human populations after evolving from a simian immunodeficiency virus. Nikolaas Tinbergen (1963)[1] proposed explaining shifts in traits from two perspectives: dynamic versus static, and proximate versus ultimate. This framework can be used to understand the evolution of a trait in four ways: (i) causation (proximate/static): the mechanism of the trait as it works in the present; (ii) survival value (ultimate/static): how function of the trait enhances survival or reproduction; (iii) ontogeny (proximate/dynamic): the development of the trait in an individual; and (iv) evolution (ultimate/dynamic): the phylogenetic history of the trait. Use these categories to discuss the causes for the virus shifting to humans from other primates.

4. Joseph Dalton Hooker and Charles Lyell convinced Darwin that the concept of natural selection should be presented to the Linnean Society and read an excerpt from his abstract along with Alfred Russel Wallace's 1858 manuscript. Since Wallace was still in the Malay Archipelago, he did not take part in the decision to make this joint presentation. Critics later pointed out that this was unfair to Wallace (and some even accused Darwin of stealing some of Wallace's ideas). Do some additional background reading and discuss whether the arrangement was fair, how the concept of natural selection would have been received if Darwin hadn't been involved, and how Wallace's 1858 manuscript influenced Darwin's subsequent publication of *On the Origin of Species*.

5. The two revolutionary hypotheses proposed by Darwin in *On the Origin of Species* were descent with modification and natural selection as the main mechanism of evolution. How did Darwin's ideas contrast with the prevailing notions of the origins of species at the time?

6. Some scientists vigorously rejected Darwin's ideas when *On the Origin of Species* was published. Richard Owen (1860), perhaps the most respected biologist in England, wrote (among many other objections): "Are all the recognised organic forms of the present date, so differentiated, so complex, so superior to conceivable primordial simplicity of form and structure, as to testify to the effects of Natural Selection continuously operating through untold time? Unquestionably not. The most numerous living beings ... are precisely those which offer such simplicity of form and structure, as best agrees... with that ideal prototype from which...vegetable and animal life might have diverged." How might Darwin, or you, argue against Owen's logic?

7. During the evolutionary synthesis, biologists conclusively identified natural selection, gene flow, genetic drift, and mutation as the major causes of evolution within species. Using the scientific definition of evolution, explain how these forces cause populations, species, and higher taxa to evolve.

8. Drawing on sources available in a good library, discuss how the "Darwinian revolution" affected one of the following fields: philosophy, literature, psychology, or economics.

---

[1] Tinbergen, N. (1963). On aims and methods of ethology. *Zeitschrift für Tierpsychologie* 20: 410–433.

The Tree of Life

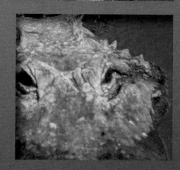

In 1915, Mildred Hoge, studying the genetics of the fruit fly *Drosophila melanogaster*, found a mutation that she designated *eyeless*, because it reduced or eliminated the flies' eyes [15]. The normal form of the *eyeless* gene, then, must be necessary for eye development. Years later, similar mutations were found in mice and humans, and eventually it was discovered that the DNA sequences of the mammals' genes were similar to that of the fly's gene. In 1995, Georg Halder and his colleagues reported a remarkable experiment [12]. By inserting an extra copy of the normal form of a fly's *eyeless* gene into *Drosophila* larvae, they induced the development of almost perfect miniature extra eyes—on the wings, legs, and elsewhere on the body of the adult fly. But more astonishingly, the researchers obtained exactly the same effect when they inserted the mouse version of the gene. The mouse gene (now called *Pax6*) caused the flies to develop eyes. Not mouse eyes, however, but almost perfect fly eyes (**FIGURE 2.1**). The same result was later obtained with the human gene.

Halder and his colleagues noted that more than 2000 different genes are thought to contribute to developing a *Drosophila* eye. The normal *eyeless* gene is near the start of a chain of command: it activates other genes, which activate yet others, and so on, to produce all the details of the eye. This experiment shows that even though insect and vertebrate eyes are radically different in structure, and are produced by somewhat different sets of genes, the system that activates these genes is very similar in insects and vertebrates. The fly genome responds to the signal from the mammalian *Pax6* gene just as it does to

A great egret (*Ardea alba*) stands on an American alligator (*Alligator mississippiensis*). Although it is certainly not obvious, birds and crocodilians are each others' closest living relatives, having descended from a common ancestor that lived more than 200

FIGURE 2.1 Eyes were induced to develop in abnormal places in fruit flies by inserting extra copies of either the fly *eyeless* gene (A) or the corresponding mouse gene, *Pax6* (B). Photo (A) shows extra eyes on the fly's mouthparts (arrows). Photo (B) shows eye induced by expression of the mouse gene in the fly's leg region. (From [12]. Photos courtesy of W. Gehring and G. Halder.)

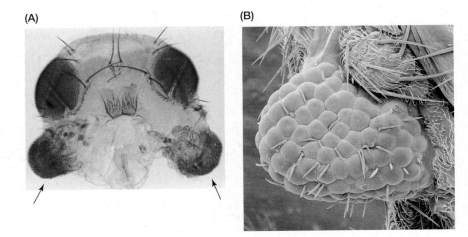

the equivalent fly gene. This is only one of many systems of interacting genes that are now known to be shared by insects and vertebrates—and indeed by all animals.

In a more extreme variation on this theme, more than one-third of the genes in the genome of yeast (a fungus) have recognizable similarity to human genes. In a recent experiment, biologists studied 414 genes that are necessary for yeast to survive, replacing them with the equivalent human genes [18]. Almost half of the human gene replacements enabled the yeast to survive!

Only one natural cause can explain these amazing results: these organisms share fundamental characteristics because they—and their genes—have descended from a common ancestor in the distant past. Throughout biology and health science, common ancestry between humans and all other living things is a fundamental principle that explains countless facts. Indeed, it is the basis for using nonhuman species as models in human biomedical research. Descent from common ancestors is one of the two great themes in Darwin's *On the Origin of Species*. It is one of the two greatest principles of evolution.

The Tree of Life, from Darwin to Today

When Darwin recounted his visit to the Galápagos Islands in *The Voyage of the Beagle* (1845), he wrote about the finches (**FIGURE 2.2**) that, "Seeing this gradation and variety of structure in one small, intimately related group of birds, one might really fancy that from an original paucity of birds in this archipelago, one species has been taken and modified for different ends." Likewise, he came to suspect that the different forms of mockingbirds and of tortoises had descended from a single ancestor. But this thought, logically extended, suggested that those mockingbirds' ancestor itself had descended from an older ancestor that could have given rise to yet other descendants—the South American mockingbirds, for instance. In 1837, a year after he returned from the voyage, Darwin sketched a branching diagram representing the idea of descent from common ancestors (**FIGURE 2.3**). By the time he published *On the Origin of Species* (1859, [5]), he could write, "I doubt not that the theory of descent with modification embraces all the members of the same class." Extending the logic, he went on: "Analogy would lead me one step further, namely to the belief that all animals and plants have descended from some one prototype." And finally, in one of the most daring thoughts anyone has ever had: "I should infer from analogy that probably all the organic beings which have ever lived on this earth have descended from some one primordial form."

Extending his early sketch (see Figure 2.3) into a great metaphor, Darwin proposed that all species, extant and extinct, form a great "Tree of Life," or **phylogenetic**

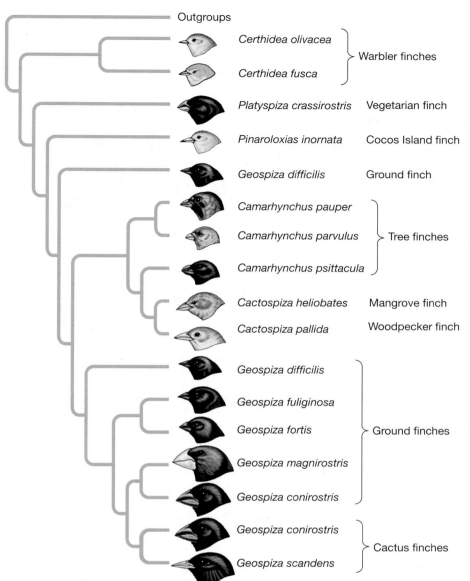

FIGURE 2.2 The finches in the Galápagos Islands and Cocos Island that have become known as Galápagos finches (also sometimes referred to as Darwin's finches). The bills of these species are adapted to their diverse feeding habits. Some hybridization among species may have affected apparent relationships, as well as anomalies such as the occurrence of *G. difficilis* on two branches. The outgroups are genera of finches distributed in South America and the West Indies. (After [10, 20].)

tree. Closely adjacent twigs represent living species derived only recently from their **common ancestors** (shared ancestors). Twigs on more distant branches represent species derived from more ancient common ancestors. Darwin expressed this metaphor in some of his most poetic (and very Victorian) language:

The affinities of all the beings of the same class have sometimes been represented by a great tree. I believe this simile largely speaks the truth. The green and budding twigs may represent existing species; and those produced during former years may represent the long succession of extinct species. At each period of growth all the growing twigs have tried to branch out on all sides, and to overtop and kill the surrounding twigs and branches, in the same manner as species and groups of species have at all times overmastered other species in the great battle for life. The limbs divided into great branches, and these into lesser and lesser branches, were themselves once, when the tree was young, budding twigs; and this connection of the former and present buds by ramifying branches may well represent the classification of all extinct and living species in groups subordinate to groups. Of the many twigs which flourished when the tree was a mere bush, only two or three, now grown into great branches, yet survive and bear the other branches; so with the species

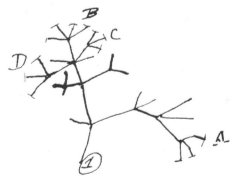

FIGURE 2.3 Darwin's first speculative diagram of a phylogenetic tree, in an 1837 notebook. The numeral 1 represented the ancestor of groups A–D.

FIGURE 2.4 The tree of life. This phylogeny of thousands of species is based on DNA sequences. The root is in the center, and branches are reflected into a circular figure for the sake of compact display. Note the position of the human species ("You are here"). To zoom in on branches of interest, visit www.zo.utexas.edu/faculty/antisense/DownloadfilesToL.html. (From [14], courtesy of D. M. Hillis.)

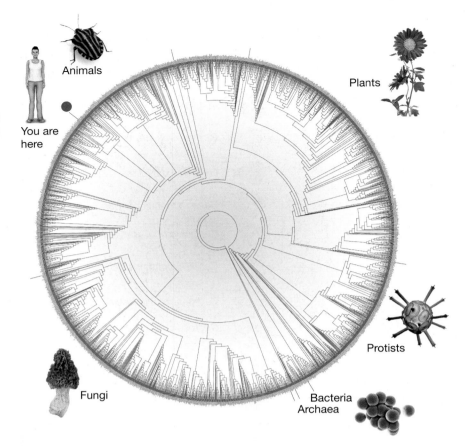

which lived during long-past geological periods, very few have left living and modified descendants. From the very first growth of the tree, many a limb and branch has decayed and dropped off; and these fallen branches of various sizes may represent those whole orders, families, and genera which have now no living representatives, and which are known to us only in a fossil state. As we here and there see a thin, straggling branch springing from a fork low down in a tree, and which by some chance has been favoured and is still alive on its summit, so we occasionally see an animal like the Ornithorhynchus or Lepidosiren,* which in some small degree connects by its affinities two large branches of life, and which has apparently been saved from fatal competition by having inhabited a protected station. As buds give rise by growth to fresh buds, and these, if vigorous, branch out and overtop on all sides many a feebler branch, so by generation I believe it has been with the great Tree of Life, which fills with its dead and broken branches the crust of the earth, and covers the surface with its ever-branching and beautiful ramifications.

Today, biologists agree that Darwin was right: all the organisms we know of have descended from a single ancestral form of life that lived between 4 and 3.7 billion years ago. And thanks to research by hundreds of biologists, we can draw an increasingly complete picture of the history by which the millions of living species, and a great many extinct ones, evolved from common ancestors (**FIGURE 2.4**). Some highlights of this amazing history are shown in **FIGURE 2.5**. The first cellular organisms that we know of were prokaryotes that evolved into two great groups (A in Figure 2.5), the Bacteria and Archaea. Eukaryotes

---

Ornithorhynchus, the duck-billed platypus, is a primitive, egg-laying mammal. *Lepidosiren* is a genus of living lungfishes, a group that is closely related to the ancestor of the tetrapod (four-legged) vertebrates, and which is known from ancient fossils.

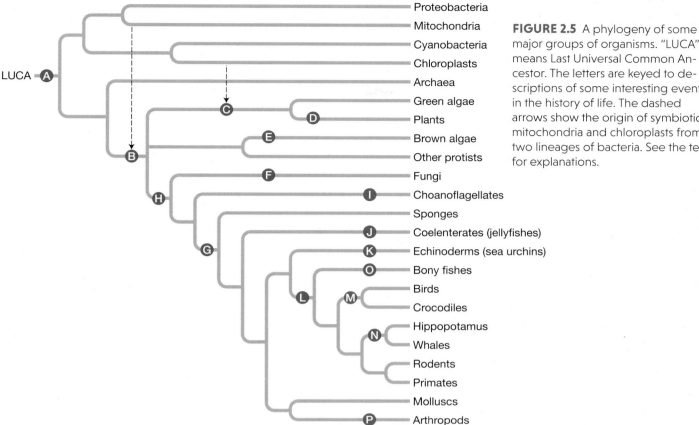

FIGURE 2.5 A phylogeny of some major groups of organisms. "LUCA" means Last Universal Common Ancestor. The letters are keyed to descriptions of some interesting events in the history of life. The dashed arrows show the origin of symbiotic mitochondria and chloroplasts from two lineages of bacteria. See the text for explanations.

evolved from a symbiotic association (B) between an archaean and a bacterium that evolved into the mitochondrion. So the mitochondria in each of our cells are more closely related to *E. coli* bacteria than they are to the nuclei in the same cells! Early eukaryotes evolved into diverse lineages. One of these became the green algae, which acquired symbiotic photosynthetic cyanobacteria that evolved into chloroplasts (C). Green algae gave rise to the true plants (D).

Remarkably, complex multicellular life forms evolved many times. Among these groups are plants, brown algae (E), some fungi (F), and animals (G). Fungi and animals stem from a single ancestor (H), so we are more closely related to mushrooms than to plants. The closest relatives of animals are single-celled choanoflagellates (I), which closely resemble some of the cells in sponges (see Figure 17.6).

Several groups of animals have no heads. Some, such as jellyfishes, never had one in their evolutionary past (J). Others had a head but then lost it. The echinoderms, such as starfishes, evolved radial symmetry from ancestors that were bilaterally symmetric, and in doing so became headless (K).

The vertebrates (L) are most closely related to echinoderms (K) in this tree. Among the tetrapod (four-limbed) vertebrates, research using fossils, comparative anatomy, and DNA has revealed some unexpected relations. Birds are more closely related to crocodiles than to any other living animals (M). Primates—including people—are more closely related to rodents than to most other orders of mammals. Whales are clearly mammals rather than fishes, and it turns out that they are related to hippopotamuses (N).

The diversity of species within different groups is wildly uneven. With about 33,000 species, bony fishes (O) are the most diverse group of vertebrates. All the other vertebrates combined sum to about 30,000 known species. But the diversity

Classification, Taxonomic Practice, and Nomenclature

The scheme of classification that is used today was developed by the Swedish botanist Carolus Linnaeus (1707–1778). Linnaeus introduced *binomial nomenclature*, a system of two-part names consisting of a genus name and a specific epithet (such as *Homo sapiens*). He proposed a system of grouping species in a *hierarchical classification* of groups nested within larger groups (such as genera nested within families). The levels of classification—such as kingdom, phylum, class, order, family, genus, and species—are referred to as *taxonomic categories*, whereas a particular group of organisms assigned to any of these levels is a **taxon** (plural: *taxa*). Higher taxa are those above the species level. Thus the species rhesus monkey (*Macaca mulatta*) is placed in the genus *Macaca*, in the family Cercopithecidae; *Macaca* and Cercopithecidae are higher taxa that exemplify the taxonomic categories genus (plural: genera) and family, respectively. Several intermediate taxonomic categories, such as superfamily and subspecies, are sometimes used in addition to the more familiar and universal ones.

To ensure that names are standardized, taxonomy has developed rules of procedure. For example, the genus name and specific epithet ordinarily agree in gender: *Rattus norvegicus*, not *Rattus norvegica*, for the brown rat. No two species of animals, or of plants, can bear the same name. The valid name of a taxon is the oldest available name that has been applied to it. Thus it sometimes happens that two authors independently describe the same species under different names; in this case, the valid name is the older one. Sometimes two or more species have masqueraded under one name; in this case, the name is applied to the species that the author used in his or her description. To prevent the obvious ambiguity that could arise in this way, it is standard practice for the author to designate a single specimen (the *type specimen*, or *holotype*) as the "name bearer" so that later workers can determine which of several similar species

rightfully bears the name. The holotype, usually accompanied by other specimens (*paratypes*) that exemplify the range of variation, is deposited and carefully preserved in a museum or herbarium.

The rules for naming higher taxa are not all as strict as those for species and genera. In zoology (and increasingly in botany), names of subfamilies, families, and sometimes orders are formed from the name of the type genus (the first genus described). Most family names of plants end in -aceae. In zoology, subfamily names end in -inae and family names in -idae. Thus *Columba* (Latin for "pigeon"), the genus of the familiar pigeon, is the type genus of the family Columbidae and the subfamily Columbinae; *Rosa* (rose) is the type genus of the family Rosaceae. Names of genera and species are always italicized; taxa above the genus level are not italicized, but are always capitalized.

Systematists today rely on phylogeny when classifying organisms. A **monophyletic** taxon is one that includes all the named descendants of a particular common ancestor (for example, the traditional class Aves, which includes all birds, is monophyletic). Most systematists today hold the opinion that classifications should consist of monophyletic taxa only and thus reflect phylogenetic relationships. A **paraphyletic** taxon includes some, but not all, of the descendants from a particular ancestor. (The traditional class Reptilia is paraphyletic because it did not include the birds, which share a common ancestor with dinosaurs and crocodiles. Reptilia is monophyletic if we abolish the class Aves and include the birds in the class Reptilia.) A **polyphyletic** taxon includes species that do not exclusively share a common ancestor. (The falcons, hawks, and eagles were included in the order Falconiformes, but DNA evidence indicates that falcons are more closely related to parrots and songbirds. They are now recognized as a distinct order from hawks and eagles, which are now called Accipitriformes.)

prize goes to the single insect order Coleoptera: there are about 350,000 known species of beetles (P). Among the families of flowering plants (angiosperms), the sunflowers (23,000 species) and orchids (19,500 species) are fantastically diverse. At the other end of the diversity spectrum, the most ancient lineage of all angiosperms is now represented by just a single species, *Amborella trichopoda*, which survives only in remote rainforests on the South Pacific island of New Caledonia. (Notice that here and elsewhere we use terms from taxonomy, the classification of organisms. **BOX 2A** provides a review of some important aspects of classification.)

These are only a few of the fascinating glimpses into the history of life that phylogenetic studies have revealed. We must ask, though, how the relationships among diverse species can be determined, how events in evolutionary history, such as the echinoderms' losing their heads, can be inferred from these relationships, and what else we can learn from phylogenetic studies. This chapter and Chapter 16 delve into these questions.

Phylogenetic Trees

There are two major processes in the evolution of a **higher taxon**, which is a named group of organisms above the level of species. These are **anagenesis**, which is evolutionary change of features within a single lineage (species), and **cladogenesis**, or branching of a lineage into two or more descendant lineages. Following cladogenesis, anagenesis in each of the descendant lineages results in their becoming more different from each other (**divergence**, or **divergent evolution**). A **phylogeny** is the history of the events by which species or other taxa have successively arisen from common ancestors. The branching diagram that portrays this history is called a phylogenetic tree (**FIGURE 2.6**). Other kinds of evolutionary events are sometimes also represented in such diagrams, such as extinction (e.g., taxon F in Figure 2.6) and reticulation, which occurs when two lineages merge or form a hybrid descendant, so that the tree has a netlike structure. We will focus on branching trees here.

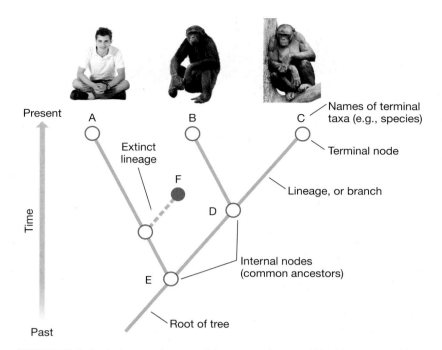

FIGURE 2.6 A phylogenetic tree of three taxa, human (A), chimpanzee (B), and bonobo (C), illustrating major phylogenetic terms. The time scale in most phylogenetic trees is a relative one, but the tree always implies the passage of time from the root of the tree toward the tips of the branches.

The phylogenetic tree in Figure 2.6 shows three living species: human (A), chimpanzee (B), and bonobo (C). (A similar tree could also represent three higher taxa, such as lizards [A], crocodiles [B], and birds [C].) Each segment in the tree is a **lineage**, or *branch*, which may split at an internal *branch point* or *node* (such as D), representing the formation of two descendant lineages (B and C) by speciation from their common ancestor. All the descendants of any one ancestor form a **clade** (also called a *monophyletic group*); thus B and C form a clade that is "nested" within the larger clade A + B + C. Two clades that originate from a common ancestor are called **sister groups**. (If B and C are species, they are sister species.)

The tree in Figure 2.6 represents the *genealogical relationships* among the taxa, meaning the temporal order of branching by which they have originated from the common ancestor (E in this case). The lineage leading to the **most recent common ancestor** (**MRCA**) of all the species in the phylogeny is called the **root** of the tree. Thus a tree has an implicit *time scale* from past (at the root) to more recent time (e.g., the present). This time scale (which is often omitted from published phylogenetic diagrams) is often relative, implying only the order of branching. In some cases, however, an absolute time scale is used, and branch points are drawn to match the dates at which the branching events are thought to have occurred. As we will see, phylogenetic trees can convey information not only about the relationships among species and their time of divergence, but also about evolutionary changes in phenotypic and genetic characteristics and geographic distributions.

The order of branching in a phylogenetic tree defines which species are more closely and which are more distantly related. Two species are more closely related to each other than to a third species if they are derived from a more recent common ancestor. By analogy, two siblings are more closely related to each other than they are to a cousin because they share more recent common ancestors (their parents) with each other than with their cousin (a grandparent).

Closeness of relationship is not the same as similarity. A person might more closely resemble her cousin than her sister with respect to eye color or many other features,

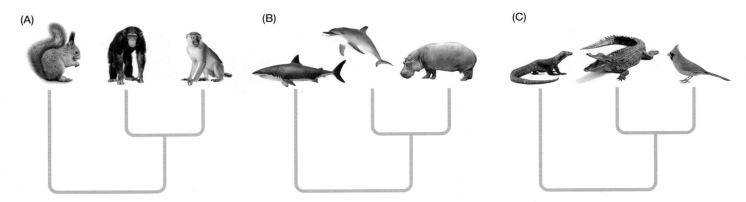

FIGURE 2.7 Similarity versus relationship. (A) Chimpanzees and monkeys are more closely related to each other than to rodents, and they are also more similar. (B) Dolphins are closely related to hippopotamuses and other mammals, even though they superficially look more like sharks. (C) Crocodiles and birds share a more recent common ancestor than either does with lizards, but birds look very different because they have undergone more evolutionary changes than crocodiles.

but she is still more closely related to her sister. Likewise, two closely related species may be less similar to each other than one is to a more distantly related species (**FIGURE 2.7**). For instance, dolphins are more closely related to hippopotamuses and humans than they are to sharks, even though they resemble sharks in some ways. Dolphins have independently evolved fins and a body form adapted for swimming. Crocodiles and lizards are superficially more similar to each other than they are to birds, but crocodiles are more closely related to birds than they are to lizards. The MRCA of lizards, crocodiles, and birds certainly had a lizardlike body form, but birds evolved more differences from that ancestral form than crocodiles did. *Even though certain aspects of similarity may be used as data to determine the relationships among species, a phylogeny portrays relationship (common ancestry), not similarity.*

A phylogenetic tree may be drawn in any of several equivalent ways. The junctions may be angular (**FIGURE 2.8A**) or rectangular (**FIGURE 2.8B**). Figure 2.8B illustrates three equivalent trees that differ in the orientation of the implied time axis. Figure 2.8A shows that the clades arising from a branch point may

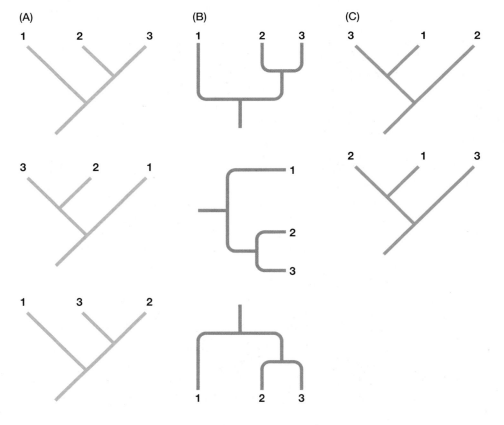

FIGURE 2.8 (A) These three trees are equivalent: they all show that species 2 and 3 are the closest relatives. (B) These three trees are equivalent to each other, but they differ in the direction in which the flow of time is shown (upward, to the right, and down). These trees are also equivalent to those in (A) since again species 2 and 3 are the closest relatives. The trees in (A) and (B) differ in how the shapes of the branching events are shown. (C) These two trees are not equivalent to each other, or to the trees in (A) and (B), because they show different species as each other's closest relatives.

be rotated without any change in the diagram's meaning. *Relationships among the taxa are defined by the order of branching, not by the linear order of the tips of the tree.* However, the trees in **FIGURE 2.8C** represent relationships different from the relationships portrayed in Figures 2.8A and B.

The lengths of the branches in a phylogenetic tree may or may not have any meaning, depending on what information a researcher means to convey. If the tree conveys only branching order, the relative lengths of branches have no significance. If the tree is accompanied by an absolute time scale, however, the positions of branch points indicate when those events occurred. In some phylogenies, the length of a branch indicates the number of evolutionary changes (e.g., DNA nucleotide substitutions) that occurred on that branch.

Inferring phylogenies: An introduction

It can be difficult to determine phylogenetic relationships, and so evolutionary biologists are developing increasingly sophisticated methods. We will touch on some of the difficulties and methods in Chapter 16. At this point we consider one simplified approach, in order to convey the basic ideas.

Our estimate of how taxa are related to one another is based on characteristics that are **homologous** among the taxa, such as the forelimb bones of tetrapod (four-legged) vertebrates (**FIGURE 2.9**). Features are homologous among species if they have been inherited from common ancestors. Homology describes not only morphological and other phenotypic features, but also DNA sequences.

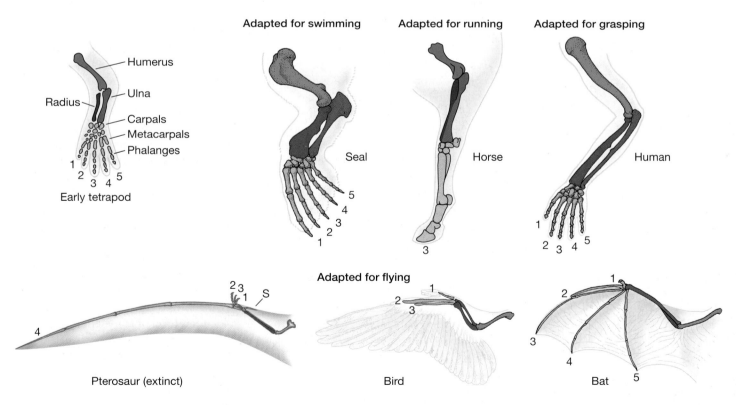

FIGURE 2.9 Forelimb skeletons of some tetrapod vertebrates. Compared with the "ground plan," as seen in the early tetrapod, bones have been lost or fused (e.g., horse, bird) or modified in relative size and shape. Modifications for swimming evolved in the seal, for running in the horse, for grasping in the human, and for flight in the bird, bat, and pterosaur. All the bones shown are homologous among these organisms except for the sesamoid bone (S) in the pterosaur; this bone has a different developmental origin from the rest of the limb skeleton. (After [7].)

In some cases, fossils provide very important information about the evolutionary history of a group, including relationships among its members (see Chapter 16). But many groups of organisms have left no fossil record at all, and even in the best cases, the fossil record is incomplete. We will concern ourselves mostly with how to infer phylogenies from data on living organisms. Each trait of an organism (e.g., the number of toes on a hindlimb) is called a **character**, which may have various **character states** (e.g., five toes in humans, three in rhinoceroses, one in horses). All kinds of phenotypic characters have been used, especially morphological features (which are usually the only features we can use for fossilized extinct taxa). Phylogenetic study has been revolutionized by DNA sequencing, which reveals variation at thousands or even millions of base pair positions in homologous DNA sequences. Each position ("site") on one strand of the double helix represents a character, and the identity of its nucleotide base (A, T, C, or G) represents a character state.

Homologous character states that are shared among species provide evidence of common ancestry if they evolved only once. Using DNA sequence data, we begin our discussion of how phylogenies are estimated with an example. Imagine that we want to find the phylogeny showing the evolutionary relations among three species of squirrels in the genus *Sciurus*. We have sequenced part of the hemoglobin gene from an individual of each of four species: the eastern gray squirrel (*Sciurus carolinensis*), western gray squirrel (*S. griseus*), fox squirrel (*S. niger*), and ground squirrel (*Spermophilus citellus*) (**FIGURE 2.10A**). The homologous sequence fragments from the hemoglobin gene are shown in **FIGURE 2.10B**. The ground squirrel serves as an **outgroup**. This is a taxon that we are quite sure (based on prior evidence) is more distantly related to the three species of interest. Those three species are the **ingroup**. The outgroup/ingroup distinction immediately gives us a basic framework for the phylogenetic tree: the outgroup and the ingroup form two branches from the common ancestor of all the species. Given this framework, there are three possible evolutionary trees for these species (**FIGURE 2.10C**).

We know from many studies of the hemoglobin gene that changes to the sequence are rare over short evolutionary time spans. This means that if we compare possible phylogenies, those that require fewer evolutionary changes are more likely to reflect actual relationships than are those with more changes. That logic makes it simple to find the evolutionary tree that most likely represents the history of the four DNA sequences and hence the four species of squirrels.

Look at tree 1 in Figure 2.10C. At site 3, the eastern and western gray squirrels share an A, and they differ from the other two species, which share a T. Starting with the DNA sequence at the root of the tree, the evolution from T to A (shown by the red bar on the tree) happened in the common ancestor of these two species. At site 9, the evolution from A to T occurred in the ancestor of the fox squirrel (shown by the blue bar). Tree 2 therefore involves two evolutionary changes.

Now consider tree 2. At site 3, there were two changes from T to A (shown by the two red bars) and again one change at site 9, for a total of three changes. The same conclusion applies to tree 3: at least three changes must have occurred to produce the data at the tips of the tree, that is, the DNA sequences from the four species. (We can imagine other scenarios for where and when changes occurred on the tree, but they all require at least three changes.)

To sum up, the phylogeny that requires the fewest evolutionary changes is tree 1. Given our assumption that evolutionary changes to the hemoglobin sequence are rare, this is the most likely phylogeny. This logic for estimating phylogenies is called **parsimony**. A final question you may have is how we could possibly have known the DNA sequence of the ancestor at the root of the tree. For example, that species could have had an A rather than a T at site 1. But if it did, all three phylogenies require at

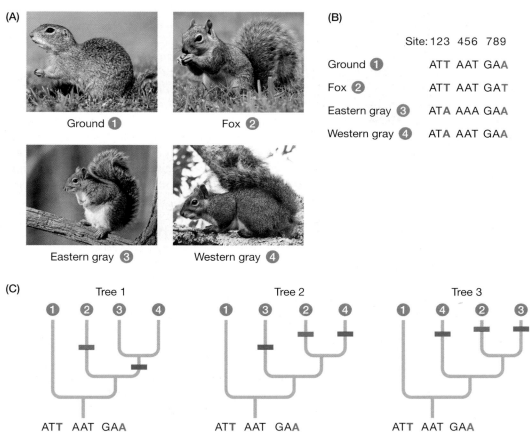

FIGURE 2.10 A simple example illustrates the logic of one method of inferring phylogenies. (A) Four species of squirrels. The aim is to determine relationships among three species of *Sciurus* (species 2, 3, and 4). The ground squirrel (*Spermophilus citellus*; species 1) is an outgroup. (B) Hypothetical sequences of a small part of a hemoglobin gene in the four species. Note the differences among the species at sites 3 and 9. (C) There are three possible relationships (trees 1, 2, and 3) among the three ingroup taxa (fox, eastern gray, and western gray squirrels). In tree 1, the red bar indicates the single evolutionary change at site 3 from T to A in the ancestor of species 3 and 4 (eastern and western gray squirrels). The blue bar in the species 2 lineage marks evolution from A to T at site 9. Trees 2 and 3 would require us to suppose that evolutionary changes happened twice at site 3, shown by two red bars. Based on the assumption that each base pair difference among the species evolved only once, tree 1 is the correct tree.

least three evolutionary changes. (Convince yourself by trying it out!) So the most likely hypothesis is that the ancestor had the sequence shown at the root, and that species then gave rise to four living species by the phylogeny shown in tree 1.

This example touches on two points that we will explore in detail in Chapter 16. First, the logic behind our approach here is to find the most likely tree. In this example, it is the tree in which a change at any given base happens only once. With other cases, however, that is no longer true. (The same mutation is likely to happen more than once when mutation rates are very high, the evolutionary time scale is very long, or there are many species in the phylogeny.) Second, the evolutionary tree of the hemoglobin gene probably reflects the evolutionary tree of the squirrel species, but there are situations in which it will not. For example, if two distantly related squirrel species hybridized in the past, the hemoglobin gene from one species might have spread through the other. That would cause the sequences of their hemoglobin genes to make the species seem more closely related than they really are.

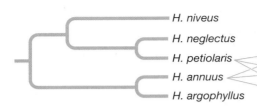

H. niveus
H. neglectus
H. petiolaris —— H. anomalus (sand dune)
H. annuus —— H. deserticola (desert floor)
H. argophyllus —— H. paradoxus (salt marsh)

FIGURE 2.11 Hybrid origin of some diploid species of sunflowers. The phylogeny, based on sequences of chloroplast DNA and nuclear ribosomal DNA, shows that *Helianthus anomalus*, *H. paradoxus*, and *H. deserticola* have arisen from hybrids between *H. annuus* and *H. petiolaris*. (After [11].)

Sometimes the information about DNA sequences or other characteristics simply is insufficient to resolve the relationships among taxa. (Often the tree will then be shown with a *polytomy*, a node from which three or more lineages emerge.) That often is the case if successive speciation events happened so rapidly that there was not enough time for many mutations to become fixed in between successive branching points.

Variations on the Phylogenetic Theme

Branches of a phylogenetic tree sometimes rejoin

The results of phylogenetic studies are often consistent with the assumption that the various lineages that arise from common ancestors remain separate and diverge from each other—that the tree consists only of bifurcations. But branches sometimes rejoin, in whole or in part, so that relationships among organisms may form a network rather than just a branching tree. For example, some species have evolved from hybrid crosses between two different ancestors, a pattern that is especially common in plants (**FIGURE 2.11**). In these cases of **hybrid speciation**, various phenotypic features and DNA markers throughout the genome reveal two ancestral sources.

More commonly, analysis of one or a few genes suggests a radically different phylogeny than most other genes. For example, aphids are obviously insects, based on both their morphology and almost all DNA sequences. A few species of aphids, unlike almost all other animals, can synthesize carotenoid pigments. A phylogenetic analysis of the genes that enable this biosynthesis placed these aphids among the fungi—clear evidence that they acquired these genes from a fungus (**FIGURE 2.12**). In contrast to "vertical" inheritance of genes by offspring from parents, such nonreproductive passage of genes among organisms is **horizontal gene transfer** (**HGT**; see Chapter 4). The genome of most eukaryotes,

FIGURE 2.12 Genes that encode the enzymes that synthesize carotenoid compounds are found in one group of aphids, but not in any other animal that has been studied. This phylogeny of copies of a gene found in aphids and of the homologous gene in fungi shows that the ancestor of these aphids acquired the gene from a fungus. The photo shows pea aphids (*Acyrthosiphon pisum*) that have this gene. (After [27]; photo courtesy of Nancy Moran, University of Texas at Austin.)

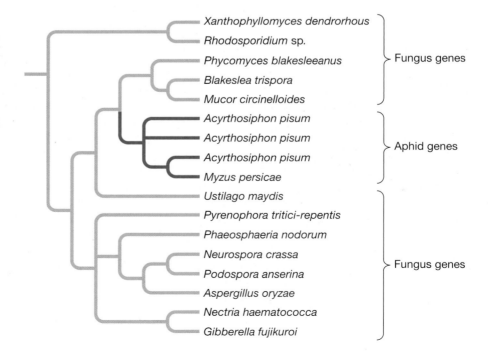

Xanthophyllomyces dendrorhous
Rhodosporidium sp.
Phycomyces blakesleeanus   } Fungus genes
Blakeslea trispora
Mucor circinelloides
Acyrthosiphon pisum
Acyrthosiphon pisum
Acyrthosiphon pisum   } Aphid genes
Myzus persicae
Ustilago maydis
Pyrenophora tritici-repentis
Phaeosphaeria nodorum
Neurospora crassa   } Fungus genes
Podospora anserina
Aspergillus oryzae
Nectria haematococca
Gibberella fujikuroi

including humans, includes at least a few genes that have been horizontally acquired during their ancestry [4].

Among prokaryotes, both during the early evolution of life and among living bacteria, HGT has played a major evolutionary role [1, 6]. Bacteria acquire genes from other species by many mechanisms, including transfer of plasmids and other mobile genetic elements, and natural transformation: uptake of DNA that has been released into the environment by the death of other bacterial cells. HGT among some bacteria can be so frequent that their relationships may look more netlike than treelike. HGT has enabled some bacteria to metabolize new nutritional substrates and to adapt to toxic environments, including antibiotics. Methicillin-resistant *Staphylococcus aureus* (MRSA) is a dangerous pathogen, especially prevalent in hospitals, that has evolved resistance to almost all antibiotics, partly by HGT from other bacteria. Pathogenic *Clostridium difficile* and *Escherichia coli* are among the other species that have acquired antibiotic resistance by HGT [17]. A major concern is that human pathogens can acquire antibiotic resistance from the bacteria that inhabit cattle and other livestock that are routinely treated with antibiotics in order to promote rapid growth [25].

Not only organisms have "phylogenies"

So far we have been concerned with inferring phylogenetic trees of species. But the same methods can shed light on the history of any diverse objects that have arisen by a history of divergence from common ancestors. For example, different copies of a gene, whether within a single species or in more than one species, have a history of descent from common ancestral genes. (We already have encountered this notion in using hypothetical squirrel genes to understand a basic phylogenetic method.) A branching tree that portrays the history of DNA sequences of a gene (**haplotypes**) is often called a **gene tree** or a **gene genealogy** (see Chapter 7). For example, the tree that portrays a gene acquired by aphids from fungi (see Figure 2.12) is a gene tree. A more usual kind of gene tree is illustrated for the mitochondrial cytochrome *b* gene in MacGillivray's warbler (**FIGURE 2.13**). It shows that most of the haplotypes in Mexican populations of this species are more closely related to one another than they are

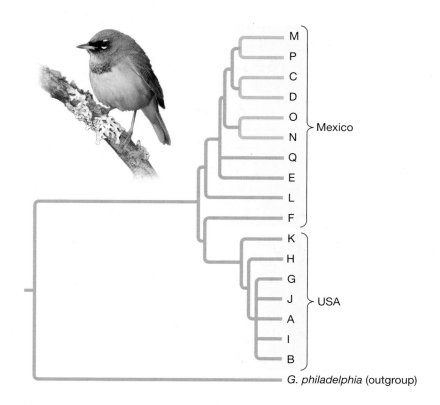

FIGURE 2.13 A gene tree showing the relationships among haplotypes (different sequences) of the mitochondrial cytochrome *b* gene in MacGillivray's warbler (*Geothlypis tolmiei*), using a sequence from the mourning warbler (*G. philadelphia*) as an outgroup. Haplotypes are more closely related within a region than between regions, implying that there is little mixture between warbler populations in Mexico and the United States. (After [26].)

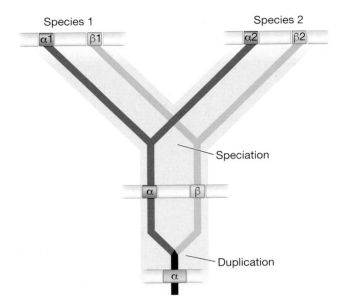

Species 1   Species 2

Speciation

Duplication

FIGURE 2.14 When gene duplication is followed by speciation, two types of relationship exist among the four copies of the locus. The loci in different species that descended from the same locus in their most recent ancestor (α1 and α2, shown in red, or β1 and β2, shown in green) are called orthologous. Loci in the same species or in different species that descended from different duplicate genes in the ancestral species are shown as differing in color, such as α1 and β2. They are called paralogous genes. Homologous loci in different species are more closely related than are paralogous loci within the same species.

to haplotypes in birds from the United States. For several reasons (such as HGT), different genes sometimes have had different phylogenetic histories—different gene trees (see Chapter 16). Thus, a gene tree can differ from the **species tree**, the phylogeny of the species from which the genes are sampled.

Organisms vary greatly in the number of functional genes in their genome; for example, eukaryotes usually have far more genes than prokaryotes (see Chapter 14). One of the most important processes by which genomes have increased in size is **gene duplication** (see Chapter 8). A new copy of a locus (say, β) arises by duplication of a pre-existing gene (α), so that a single gene locus in an ancestor is represented by two loci in the descendant. These two genes will subsequently undergo different evolutionary changes in sequence and can therefore be distinguished. If two species (1 and 2) both inherit the duplicated pair α, β from their common ancestor, the relationships among the genes represent two forms of homology, and so warrant different terms. The genes that originate from an ancestral gene duplication are **paralogous**, whereas the genes that diverge from a common ancestral gene by phylogenetic splitting at the organismal level are **orthologous** (i.e., homologous in the usual sense) (**FIGURE 2.14**). This process may occur repeatedly over evolutionary time, generating a **gene family**. The history of gene duplication and sequence divergence—the relationships among the orthologous and paralogous genes in two or more species—can be determined by standard phylogenetic methods. In the human genome, for example, the 12 members of the globin gene family include genes that encode myoglobin and several α- and β-hemoglobin chains (**FIGURE 2.15**). The origin of myoglobin and hemoglobin by duplication of an ancestral globin gene occurred

FIGURE 2.15 Phylogeny of genes in the globin family in the human genome. Myoglobin consists of a single protein unit, whereas mammalian hemoglobins consist of four subunits, two each from the α and β subfamilies. Each branch point on the tree denotes a gene duplication event; some of these events are marked with estimates of when the duplication occurred. The origin of hemoglobin and myoglobin from a common ancestral gene occurred in the ancestor of all vertebrates, but the α and β hemoglobin subfamilies originated by duplication in an ancestor of the jawed vertebrates. The duplication of the β-hemoglobin into two genes occurred in the ancestor of placental mammals, since the A_γ, G_γ, and ε genes are lacking in monotremes and marsupials. In some instances, one of the pairs of genes formed by duplication became a nonfunctional pseudogene, symbolized by ψ. (After [13, 22].)

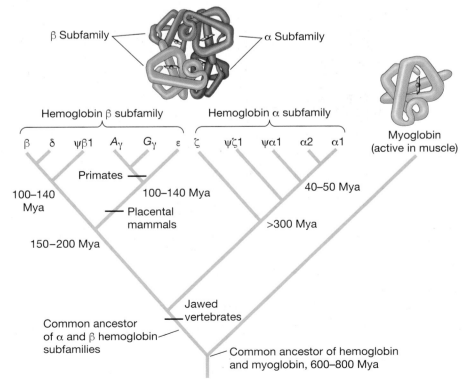

during the ancestry of the vertebrates, all of which have both genes. The α- and β-hemoglobins arose by gene duplication in the ancestor of jawed vertebrates, all of which have a functional hemoglobin composed of both α- and β–chains, whereas the jawless vertebrates (e.g., lampreys) have only a single hemoglobin chain. The origin of the other globin genes can be similarly traced based on their sequences and phylogenetic distribution. A more extended description of the evolution of gene families, and of genome size, is provided in Chapter 14.

Cells give rise to lineages of cells by division, and these lineages can be traced by the somatic mutations that arise and are inherited by descendant cells. Biologists are beginning to use the "phylogeny" of cells to trace the developmental history of the brain and other organs [24], and phylogenies of tumor cells are important for studying the source and spread of metastatic cancers [28]. And the applications of phylogenetic methods extend beyond biology. French, Spanish, and the other Romance languages evolved from Latin, an example of nongenetic *cultural evolution*. Students of cultural evolution are increasingly using phylogenetic methods, borrowed from evolutionary biology, to analyze the history of languages and other cultural traits (see Chapter 16).

Phylogenetic Insights into Evolutionary History

Phylogenetic studies, sometimes in concert with information from the fossil record, enable biologists to piece together the evolutionary history of organisms and their characteristics, ranging from DNA sequences to geographic distributions. They document patterns of evolution—aspects of change that are common to many groups of organisms. Some of these patterns were already known to Darwin and his followers, but have been studied in depth using phylogenetic and other methods.

Inferring the history of character evolution

One of the most important uses of phylogenetic information is to reconstruct the history of evolutionary change in interesting characteristics by "mapping" character states on the phylogeny and *inferring the state in each common ancestor*, right back to the root of the entire tree. In the simplest methods, we assign to ancestors those character states that require us to postulate the fewest evolutionary changes for which we lack independent evidence. This method enables us to infer when (i.e., on which branch or segment of a phylogeny) changes in characters occurred, and thus to trace their history.

Humans, for example, have nonopposable first (great) toes, while the orangutan, gorilla, and chimpanzee have opposable first toes (like our thumbs). In **FIGURE 2.16**

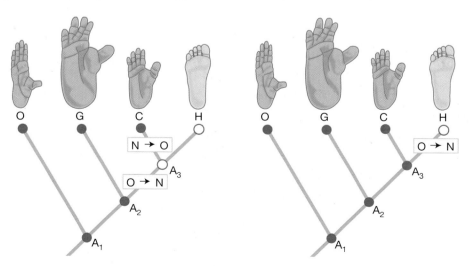

FIGURE 2.16 Inferring ancestral character states. Two possible histories of the evolution of opposable (O) versus nonopposable (N) toes in the Hominoidea (O, orangutan; G, gorilla; C, chimpanzee; H, human) are shown. At left, if nonopposable toes (open circles) are hypothesized for A_3, the common ancestor of chimpanzee and human, two state changes must be postulated. At right, opposable toes are hypothesized for A_3, and only one change need be postulated. Assuming that character state changes are rare, the more likely hypothesis is that humans evolved from an ancestor with opposable toes.

we consider two possible evolutionary histories. The common ancestors are labeled A_1, A_2, and A_3, from older to younger. If we assume that A_1 and A_2 had opposable toes and that A_3, the immediate common ancestor of chimpanzees and humans, had nonopposable toes, we have to postulate two changes, with the chimpanzee reverting to the ancestral state (see left figure). If, however, we assume that A_3, like A_1 and A_2, had opposable toes, we need to infer only one evolutionary change, namely the shift to nonopposable toes in the human lineage that is shown in the right figure. If we assume that changes between these states are very rare, the tree with the fewest changes is the most likely. This leads to the conclusion that the common ancestor of humans and chimpanzees had opposable toes.

Estimating time of divergence

In the 1970s, when DNA sequencing was very difficult, researchers compared the amino acid sequences of proteins in pairs of species that were known to have diverged from their common ancestors at various times in the past. For example, pigs and cows belong to groups that are first recorded as fossils in the Eocene, about 50 million years ago (Mya), so they diverged from their common ancestor at least that long ago. When a few proteins were sequenced from different species, and the corresponding DNA sequence differences were plotted against such estimated divergence times, a close relationship was found (**FIGURE 2.17**). That is, the proportion of base pairs that differ between homologous DNA sequences in two species increases with the amount of time that has elapsed since the species originated from their common ancestor. As long as the increase is linear with time (as shown in Figure 2.17), the difference in sequence can serve as **molecular clock**. Figure 2.17 shows that if you were to sequence the same genes for two species of mammals and find 45 nucleotide differences, you could read horizontally across to the best-fit line, and then down to the time axis. Even if the mammal species belong to lineages that lack a fossil record, you might estimate that they diverged about 74 Mya—as long as you assume that the genes in these lineages have evolved at the same rate as in the mammals with fossil records that were used to determine the original correlation of sequence difference with time. The fossils have been used to *calibrate* the rate, r, at which these genes have evolved in mammals. If r is roughly constant within a clade of organisms, the expected difference D between two species, each evolving at rate r, is $D = 2rt$, where t is time since they split from their common ancestor; hence t is estimated as $t = D/2r$. (The factor 2 appears because the genes have evolved along each of the two lineages that descended from the most recent common ancestor.)

Rates of evolution differ among the different positions in codons and among different genes in the genome (see Chapter 7). Rates of sequence evolution also differ

FIGURE 2.17 This plot of base pair differences against time since divergence was some of the earliest evidence that the rate of sequence evolution might be approximately constant. Each point represents a pair of living mammal species whose most recent common ancestor, based on fossil evidence, occurred at the time indicated on the x-axis. (The fossil would indicate the minimal age of the lineage to which a living species belongs.) The y-axis shows the number of base pair differences between the species, inferred from the amino acid sequences of seven proteins. The three green circles represent pairs of primate species, which have diverged more slowly than other mammal groups. The arrows show how we would estimate that a pair of species with 45 base pair differences shared a common ancestor about 74 Mya. (After [21].)

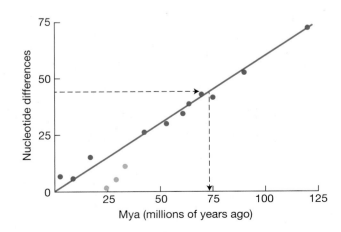

among groups of organisms, especially distantly related taxa: there is not a universal molecular clock. For example, sequence evolution among hominoid primates (apes, including humans) has been slower than among other primates and mammals (see Figure 2.17). Differences in generation time and in mutation rate are among the factors that have been proposed to explain why rates of sequence evolution vary among taxa [8, 9, 23].

Patterns of evolution

Data on morphological and other characteristics of organisms were used to infer phylogenetic inferences before DNA sequences were available, and these characteristics are still used in some studies. A phylogenetic perspective on the diversity of organisms and their characteristics enables biologists to trace patterns of evolution of various characteristics. The inferred patterns provide massive evidence that species have evolved from common organisms; that is, they are very strong evidence for the fact of evolution (**BOX 2B**).

MOST FEATURES OF ORGANISMS HAVE BEEN MODIFIED FROM PRE-EXISTING FEATURES   Phylogenetic analysis is based on homologous features, those derived from common ancestors. It is made possible by one of the most important principles of evolution: the features of organisms almost always evolve from pre-existing features of their ancestors; they do not arise de novo, from nothing. By analyzing homologous characters, biologists have documented many fascinating evolutionary changes in form and function. The middle-ear bones of mammals evolved from jaw bones of reptiles (see Chapter 20). The wings of birds, bats, and pterosaurs are highly modified forelimbs (see Figure 2.9); they do not arise from the shoulders (as in angels and dragons), presumably because the ancestors of these animals had no shoulder structures that could be modified for flight. Homologous morphological characters in different species generally have similar genetic and developmental underpinnings, but these foundations have sometimes undergone greater divergence than have the finished products. Likewise, existing proteins have been modified from ancestral proteins and have new functions (see Chapter 14).

A *character* may be homologous among species (e.g., toes), but a given *character state* may not be (e.g., a certain number of toes). The pentadactyl (five-toed) state is homologous in humans and crocodiles (both have an unbroken history of pentadactyly as far back as their common ancestor), but the three-toed state in guinea pigs and rhinoceroses is not homologous, for this condition has evolved independently in these animals by modification from a five-toed ancestral state.

Determining whether or not characters of two species are homologous can be difficult. The most common criteria for *hypothesizing* homology of anatomical characters are correspondence of *position* relative to other parts of the body and correspondence of *structure* (the parts of which a complex feature is composed). Correspondence of shape or of function is not a useful criterion for homology (consider the forelimbs of a horse and an eagle). Embryological studies are often important for hypothesizing homology. For example, the structural correspondence between the hindlimbs of birds and crocodiles is more evident in the embryo than in the adult because many of the bird's bones become fused as development proceeds. Homology between DNA sequences is determined by finding an alignment that maximizes the match between nucleotides; often there are "gaps," caused by past deletions or duplications (see Chapter 16).

RATES OF CHARACTER EVOLUTION DIFFER   Like DNA sequences, different phenotypic characters evolve at different rates, as is evident from the simple observation that any two species differ in some features but not in others. Some characters, often

BOX 2B

Evidence for Evolution

Both before and since Darwin, systematists have classified organisms by comparing characteristics among them. Darwin drew on much of this information as evidence for his theory of descent from common ancestors. Since Darwin's time, the amount of comparative information has increased greatly, and today it includes data not only from the traditional realms of morphology and embryology, but also from cell biology, biochemistry, and molecular biology.

All of this information is consistent with Darwin's hypothesis that living organisms have descended from common ancestors. Indeed, innumerable biological observations are hard to reconcile with the alternative hypothesis, that species have been individually created by a supernatural being, unless that being is credited with arbitrariness, whimsy, or a devious intent to make organisms *look* as if they have evolved. From the comparative data amassed by systematists, we can identify several patterns that confirm the historical reality of evolution and which make sense only if evolution has occurred.

1. **The hierarchical organization of life.** Linnaeus discovered that organisms fall "naturally" into the hierarchical system of groups-within-groups. A historical process of branching and divergence will yield objects that can be hierarchically ordered, but few other processes will do so. For instance, languages can be classified in a hierarchical manner, but elements and minerals cannot.

2. **Homology.** Similarity of structure despite differences in function follows from the hypothesis that the characteristics of organisms have been modified from the characteristics of their ancestors, but it is hard to reconcile with the hypothesis of intelligent design. Design does not require that the same bony elements form the frame of the hands of primates, the digging forelimbs

of moles, the wings of bats, birds, and pterosaurs, and the flippers of whales and penguins (see Figure 2.9). Modification of pre-existing structures, not design, explains why the stings of wasps and bees are modified ovipositors and why only females possess them. All proteins are composed of "left-handed" (L) amino acids, even though the "right-handed" (D) optical isomers would work just as well if proteins were composed only of those. But once the ancestors of all living things adopted L amino acids, their descendants were committed to them; introducing D amino acids would be as disadvantageous as driving on the right in the United Kingdom or on the left in the United States. Likewise, the nearly universal, arbitrary genetic code makes sense only as a consequence of common ancestry.

3. **Embryological similarities.** Homologous characters include some features that appear during development, but would be unnecessary if the development of an organism were not a modification of its ancestors' ontogeny. For example, tooth primordia appear and then are lost in the jaws of fetal anteaters. Early in development, human embryos briefly display branchial pouches similar to the gill slits of fish embryos, and they have a long tail that mostly undergoes cell death and is lost.

4. **Vestigial characters.** The adaptations of organisms have long been, and still are, cited by creationists as evidence of the Creator's wise beneficence, but no such claim can be made for the features, displayed by almost every species, that served a function in the species' ancestors, but do so no longer. Cave-dwelling fishes and other animals display eyes in every stage of degeneration. Flightless beetles retain rudi-

called **conservative characters**, are retained with little or no change over long periods among the many descendants of an ancestor. For example, humans retain the pentadactyl (five-toed) limb that first evolved in early amphibians (see Figure 2.9). All amphibians and reptiles have paired systemic aortic arches, and all mammals have only the left systemic arch. Body size, in contrast, evolves more rapidly; within orders of mammals, it may vary at least 100-fold.

Evolution of different characters at different rates within a lineage is called **mosaic evolution** (**FIGURE 2.18**). It is one of the most important principles of evolution, for it says that a species evolves not as a whole, but piecemeal: many of its features evolve more or less independently. Every species is a mosaic of *plesiomorphic* (ancestral, or "primitive") and *apomorphic* (derived, or "advanced") characters. For example, the amphibian lineage leading to frogs split from the lineage

BOX 2B (continued)

mentarywings, concealed in some species beneath fused wing covers that would not permit the wings to be spread even if there were reason to do so. In *The Descent of Man*, Darwin listed a dozen vestigial features in the human body, including the appendix, the coccyx (four fused tail vertebrae), and the posterior molars, or wisdom teeth, that fail to erupt, or do so aberrantly, in many people. At the molecular level, every eukaryote's genome contains numerous non-functional DNA sequences, including pseudogenes: sequences that retain some similarity to the functional genes from which they have been derived (see Chapter 13).

5. **Convergence.** There are many examples, such as the eyes of vertebrates and cephalopod molluscs, in which functionally similar features actually differ profoundly in structure (see Figure 2.20). Such differences are expected if structures are modified from very different ancestral features, but are inconsistent with the notion that an omnipotent Creator, who should be able to adhere to an optimal design, provided them. Likewise, evolutionary history is a logical explanation (and creation is not) for cases in which different organisms use very different structures for the same function, such as the various modified structures that enable different vines to climb.

6. **Suboptimal design.** Evolutionary history explains many features that no intelligent engineer would be expected to design. For example, the paths followed by food and air cross in the human pharynx, so that we risk choking on food. The human eye has a "blind spot," which you can find at about 15° to the right or left of your line of sight. It is caused by the functionally nonsensical arrangement of the axons of the retinal cells, which run forward into the eye and then converge into the optic nerve, which interrupts the retina

by extending back through it toward the brain (see Figure 2.20).

7. **Geographic distributions.** The study of systematics includes the geographic distributions of species and higher taxa. This subject, known as biogeography, is treated in Chapter 18. Suffice it to say that the distributions of many taxa make sense only if they have arisen from common ancestors. For example, islands have few species, even though the habitats there are suitable for a great many species that occur only on continents. We know this because many continental species thrive on islands to which humans have inadvertently carried them. They must have originated on the continent, but failed to colonize the islands without human aid.

8. **Intermediate forms.** The hypothesis of evolution by successive small changes predicts the innumerable cases in which characteristics vary by degrees among species and higher taxa. Among living species of birds, we see gradations in beaks; among snakes, some retain a vestige of a pelvic girdle and others have lost it altogether. At the molecular level, the difference among DNA sequences for the same protein ranges from almost none among very closely related species through increasing degrees of difference as we compare more remotely related taxa.

For each of these lines of evidence, hundreds or thousands of examples could be cited from studies of living species. Even if there were no fossil record, the evidence from living species would be more than sufficient to demonstrate the historical reality of evolution: all organisms have descended, with modification, from common ancestors. We can be even more confident than Darwin and assert that all organisms we know of are descended from a single original form of life.

---

leading to mammals before the mammalian orders diversified, so in terms of *order of branching*, frogs are an older branch than cows or humans. In that sense, frogs might be assumed to be more ancestral. But frogs have some ancestral features (e.g., five toes on the hind foot, multiple bones in the lower jaw) and some features (e.g., lack of teeth in the lower jaw) that are more derived than those of many mammals, in that they have changed further from the ancestral state. Moreover, numerous differences among frog species have evolved in the recent past. For example, some frogs give birth to live young. Humans also have both ancestral characters (e.g., five fingers; teeth in the lower jaw) and derived characters compared with those of frogs (e.g., a single lower jawbone, a much more complex brain). *Because of mosaic evolution, it is inaccurate or even wrong to consider one living species more "advanced" than another.*

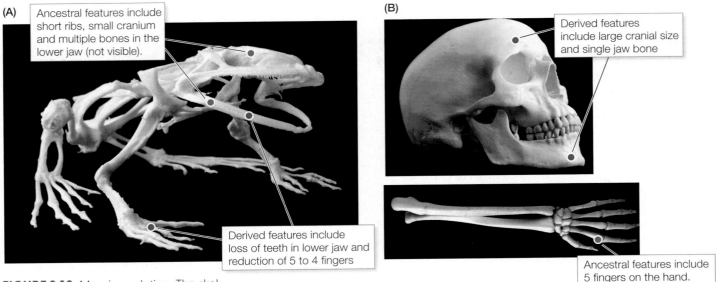

(A) Ancestral features include short ribs, small cranium and multiple bones in the lower jaw (not visible).

Derived features include loss of teeth in lower jaw and reduction of 5 to 4 fingers

(B) Derived features include large cranial size and single jaw bone

Ancestral features include 5 fingers on the hand.

FIGURE 2.18 Mosaic evolution. The skeleton of a frog (A) shows ancestral ("primitive") characters such as short ribs, a small cranium, and multiple bones in the lower jaw (not visible in photo). In mammals such as humans (B), the ribs connect to a breast bone, the lower jaw has a single bone, and the enlarged cranium houses a large brain—derived characters. Characters that are more derived in frogs than in humans include loss of teeth in the lower jaw, reduction from five to four fingers, fused tibia and fibula in the hind leg, and fusion of tail bones into a rod (urostyle).

EVOLUTION IS OFTEN GRADUAL   Darwin argued that evolution proceeds by small successive changes (*gradualism*) rather than by large "leaps" (*saltations*). How often phenotypes evolve by discrete rather than gradual change is debated. Many higher taxa that diverged in the distant past (e.g., the animal phyla; many orders of insects and of mammals) are very different and are not bridged by intermediate forms, either among living species or in the fossil record. However, the fossil record does document intermediates in the evolution of some higher taxa (see Chapters 17 and 20). Gradations among living species are very common, as we would expect if characters evolve gradually. For example, the length and shape of the bill differ greatly among species of sandpipers, but the most extreme forms are bridged by species with intermediate bills (**FIGURE 2.19**).

FIGURE 2.19 Graded differences in bill length in these closely related members of the sandpiper family suggest that evolution has been gradual. (A) Upland sandpiper (*Bartramia longicauda*). (B) Whimbrel (*Numenius phaeopus*). (C) Eurasian curlew (*Numenius arquata*). (D) Long-billed curlew (*Numenius americanus*).

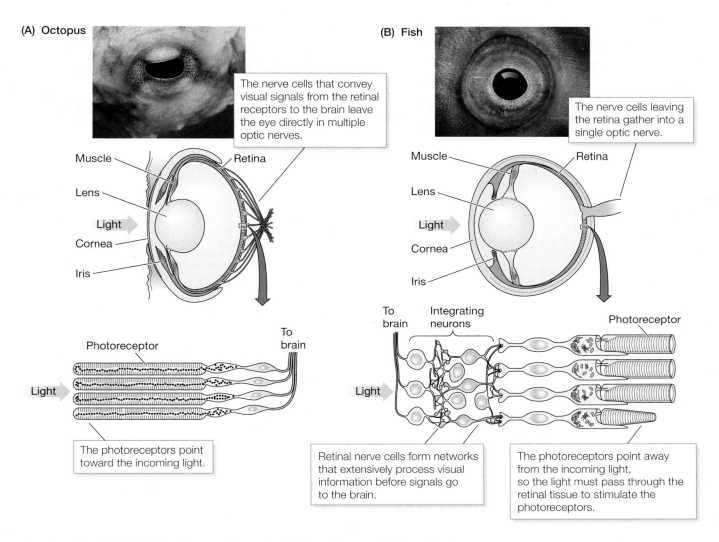

(A) Octopus

The nerve cells that convey visual signals from the retinal receptors to the brain leave the eye directly in multiple optic nerves.

Muscle — Retina
Lens
Light
Cornea
Iris

Photoreceptor
To brain
Light

The photoreceptors point toward the incoming light.

(B) Fish

The nerve cells leaving the retina gather into a single optic nerve.

Muscle — Retina
Lens
Light
Cornea
Iris

To brain
Integrating neurons
Photoreceptor

Light

Retinal nerve cells form networks that extensively process visual information before signals go to the brain.

The photoreceptors point away from the incoming light, so the light must pass through the retinal tissue to stimulate the photoreceptors.

FIGURE 2.20 The eyes of (A) octopus (cephalopod mollusc) and (B) a vertebrate are an extraordinary example of convergent evolution. Despite the many similarities in the two eyes, note the several differences, including interruption of the retina by the optic nerve in the vertebrate, but not in the cephalopod. (A after [34, 35]; B after [33].)

HOMOPLASY IS COMMON Homoplasy—the independent evolution of a character or character state in different taxa—includes **convergent evolution (convergence)**, **parallel evolution (parallelism)**, and **evolutionary reversal**. The eyes of vertebrates and cephalopod molluscs (such as squids and octopuses) are a spectacular example of convergence. Both have a lens and a retina, but their many profound differences indicate that they evolved independently from ancestors without eyes. For example, the axons arise from the back of the retinal cells in cephalopods, but from the front in vertebrates (**FIGURE 2.20**).

Parallel evolution is a term that has been used to describe cases in which independent evolution of a character state is thought to have similar genetic and developmental bases, especially in closely related species. For example, mutational change in a specific gene, *Pitx1*, is the basis of independent loss of the pelvic girdle and fins in many freshwater populations of a small fish, the three-spined stickleback (see Chapter 15). But the distinction between parallel evolution and convergent evolution may not be very meaningful because, as we will see, the same gene often contributes to similar evolutionary changes in distantly related organisms.

Evolutionary reversals constitute a return from a derived character state to a more ancestral state [29]. For example, winged insects evolved from wingless ancestors, but many lineages of insects have lost their wings in the course of subsequent evolution. It was long assumed that complex characters, once lost, are unlikely to be regained, a principle known as *Dollo's law*. However, there are

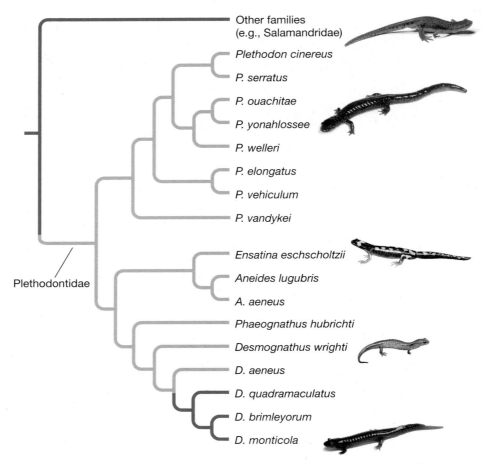

FIGURE 2.21 A violation of Dollo's law is illustrated by the larval stage in salamanders. Red lineages have an aquatic larval stage; green lineages lack this stage and undergo direct development to the adult form. Most families of salamanders (e.g., Salamandridae) have an aquatic larval stage, an ancestral feature of salamanders. It is absent—a derived state—in most members of the family Plethodontidae (green lineages). But one lineage of this family has aquatic larvae, as shown in red: *Desmognathus quadramaculatus*, *D. brimleyorum*, and *D. monticola*. Because this lineage is phylogenetically nested within a large group of taxa that lack the larval stage, we can infer that the aquatic larval stage has re-evolved. (After [2].)

exceptions to Dollo's law [3]. The ancestral life history pattern of salamanders includes an aquatic larval stage with features such as gills and parts of the skeleton that differ from the adult stage. The aquatic larval stage, characteristic of most salamanders, has been lost in the evolution of the terrestrial subfamily Plethodontinae, but phylogenetic analysis showed that it has been regained in one lineage of this subfamily, the dusky salamanders (*Desmognathus*; **FIGURE 2.21**). However, the terrestrial plethodontines develop certain features of the aquatic larval skeleton as they develop in the egg, even though they have adult features when they hatch [19]. This observation suggests that the genetic and developmental potential for producing a "lost" character may persist for a long time, and be capable of again generating the lost phenotype under some conditions.

Convergent features are often adaptations by different lineages to similar environmental conditions. In fact, a correlation between a particular convergent character in different groups and a feature of those organisms' environment or

FIGURE 2.22 Examples of convergent evolution. Many groups of birds have independently evolved long, slender bills for feeding on nectar produced at the base of long tubular flowers. (A) Hummingbirds, family Trochilidae. This violet sabrewing (*Campylopterus hemileucurus*) is from Costa Rica. (B) Sunbirds, family Nectariniidae. The greater double-collared sunbird (*Nectarinia afra*) is native to South Africa. Bird-pollinated plants also have converged, in flower characteristics. A long tubular flower, often red or orange, has evolved independently in many groups of bird-pollinated plants. (C) *Erythrina*, a member of the pea family, Fabaceae. (D) Many species of *Aloe* (Asphodelaceae) are visited by sunbirds in Africa and the Middle East.

niche is often the best initial evidence of the feature's adaptive significance. For example, a long, thin beak has evolved independently in at least six different lineages of nectar-feeding birds. Such a beak enables these birds to reach nectar in the bottom of the long tubular flowers in which they often feed (**FIGURE 2.22A,B**). Likewise, long tubular flowers have evolved independently in many lineages of bird-pollinated plants (**FIGURE 2.22C,D**).

Convergence is also observed at the molecular level. Cardiac glycosides (CGs) are toxic compounds that are synthesized and used for defense by several lineages of plants (e.g., milkweeds, family Apocynaceae) and animals (e.g., toads, family Bufonidae). They inhibit the sodium-potassium pump protein, and so upset cell membrane potentials by disrupting ion transport. Many insects that feed on plants with CGs are resistant to them and actually achieve protection by storing them in their own tissues. Toads are resistant to their own toxin, and resistance has also evolved independently in some rodents, hedgehogs, and four lineages of snakes and lizards that eat toads or CG-containing insects (**FIGURE**

FIGURE 2.23 Rampant convergent evolution in a protein. A diagram of part of the sodium-potassium pump protein, showing the four positions at which amino acids in the extracellular domain (open circles) have been independently substituted in diverse animals. For example, a change from glutamine to arginine occurred independently in four lineages (red arrows). Amino acids are designated by their single-letter abbreviations: D, aspartic acid; E, glutamate; G, glycine; H, histidine; L, leucine; N, asparagine; Q, glutamine; R, arginine. (After [32].)

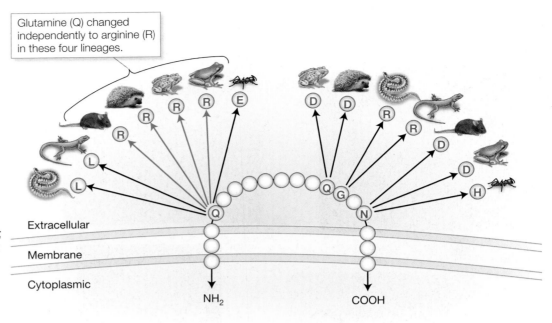

Glutamine (Q) changed independently to arginine (R) in these four lineages.

Extracellular

Membrane

Cytoplasmic

NH₂

COOH

2.23). In all of these cases, only four of the amino acids that compose the protein have been replaced, with exactly the same amino acid substitutions in some cases. The Australasian lineage of *Varanus* lizards, including the giant Komodo dragon, has lost resistance, and the two amino acid positions have reverted precisely to their ancestral states (glutamine and glycine) [32].

PHYLOGENIES DESCRIBE PATTERNS OF DIVERSIFICATION If the time of each branching point in a phylogeny has been estimated by a calibrated molecular clock, the phylogeny may suggest whether new lineages arose steadily over a long period, or episodically, in one or more bursts of diversification. Divergent evolution of numerous related lineages within a relatively short time is called *evolutionary radiation*. In most cases, the lineages become modified for different ways of life, and the evolutionary radiation may be called an **adaptive radiation** [31]. The characteristics of the members of an evolutionary radiation usually do not show a trend in any one direction. Evolutionary radiation, rather than sustained, directional evolutionary trends, is probably the most common pattern of long-term evolution. The most famous example is the adaptive radiation of finches in the Galápagos Islands. These finches, descendants of a single ancestor that colonized the archipelago from South America, differ in the morphology of the bill, which provides adaptation to different diets (see Figure 2.2). The vangas, a family of birds restricted to Madagascar, provide an even more dramatic example, in which species differ greatly in bill morphology, foraging behavior, diet, and habitat (**FIGURE 2.24**) [16, 30]. Another example of adaptive radiation is the Hawaiian silverswords and their close relatives, members of the sunflower family. They occupy habitats ranging from exposed lava rock to wet forest, and their growth forms include shrubs, vines, trees, and creeping mats (**FIGURE 2.25**).

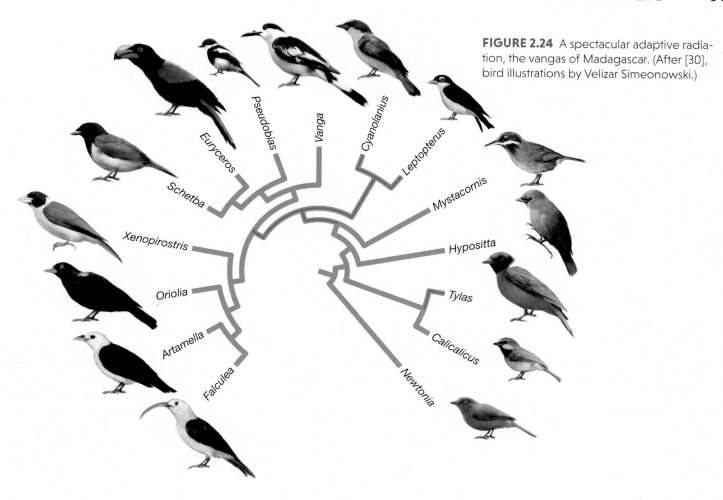

FIGURE 2.24 A spectacular adaptive radiation, the vangas of Madagascar. (After [30], bird illustrations by Velizar Simeonowski.)

Euryceros
Pseudobias
Vanga
Cyanolanius
Leptopterus
Schetba
Mystacornis
Xenopirostris
Hypositta
Oriolia
Tylas
Artamella
Calicalicus
Falculea
Newtonia

(A) *Argyroxiphium sandwicense*

(B) *Wilkesia gymnoxiphium*

(C) *Dubautia menziesii*

FIGURE 2.25 Some members of the Hawaiian silversword alliance: closely related species with different growth forms. (A) *Argyroxiphium sandwicense*, a rosette plant that lacks a stem except when flowering (as it is here). (B) *Wilkesia gymnoxiphium*, a stemmed rosette plant. (C) *Dubautia menziesii*, a small shrub.

SUMMARY

■ Modern biology has affirmed Darwin's hypothesis that all the organisms we know of, present and past, have descended from one ancient common ancestor. Current understanding of the history by which diverse groups have originated reveals fascinating events, such as the symbiotic origin of eukaryotic cells and the multiple origins of multicellular organisms from single-celled ancestors.

■ A phylogeny is the history of the events by which species or other taxa have successively originated from common ancestors. It may be depicted by a phylogenetic tree, in which each branch point (node) represents the division of an ancestral lineage into two or more lineages. Closely related species have more recent common ancestors than distantly related species. The group of species descended from a particular common ancestor is a monophyletic group, or clade; a phylogenetic tree portrays nested sets of monophyletic groups.

■ The phylogeny of a focal group of species can be readily estimated by using characters that change so rarely that those species that share a derived ("advanced") character state can safely be assumed to have inherited it from their common ancestor. A character state that occurs within the group of species can be judged to be derived rather than ancestral if it does not occur among other lineages (outgroups) that are related to the focal group.

■ In some cases, a phylogeny is not strictly dichotomous (branching), but may include reticulation (joining of separate lineages into one). This can occur if some species have originated by hybridization between different ancestral species or if genes have moved "horizontally" between organisms.

■ Phylogenetic methods can be used to describe the history not only of species, but also of DNA sequences, gene families, tumors and other cell lineages, and cultural traits such as languages.

■ Phylogenetic analyses have many uses. An important one is inferring the history of evolution of interesting characters by "mapping" changes in a character onto a phylogeny that has been derived from other data. Such systematic studies have yielded information on common patterns and principles of character evolution.

■ The rate of evolution of DNA sequences can be shown in some cases to be fairly constant (providing an approximate molecular clock), such that sequences in different lineages diverge at a roughly constant rate. The absolute rate of sequence evolution can sometimes be calibrated if the ages of fossils of some lineages are known. The rate of sequence evolution can then be used to estimate the absolute age of some evolutionary events, such as the origin of other taxa.

■ New features almost always evolve from preexisting characters. Homologous characters in different organisms are those that have been inherited from their common ancestors, with or without evolutionary change.

■ Different characters commonly evolve at different rates (mosaic evolution).

■ Homoplasy, including convergent evolution and reversal, is often a result of similar adaptations in different lineages.

■ In an adaptive radiation, numerous related lineages arise in a relatively short time and evolve in many different directions as they adapt to different habitats or ways of life. Radiation, rather than directional trends, is perhaps the most common pattern of long-term evolution.

TERMS AND CONCEPTS

adaptive radiation
anagenesis
character
character state
clade
cladogenesis
common ancestor
conservative
    character
convergence
    (convergent
    evolution)
divergence
    (divergent
    evolution)

evolutionary
    reversal (reversal)
gene duplication
gene family
gene tree (gene
    genealogy)
haplotype
higher taxon
homology
    (homologous)
homoplasy
    (homoplasious)
horizontal (lateral)
    gene transfer
    (HGT)

hybrid speciation
ingroup
lineage
molecular clock
monophyletic
mosaic evolution
most recent
    common ancestor
    (MRCA)
orthology
outgroup
parallel evolution
paralogy
paraphyletic

parsimony
phylogenetic tree
phylogeny
polyphyletic
root
sister group
species tree
taxon (plural: taxa)

SUGGESTIONS FOR FURTHER READING

Tree Thinking: An Introduction to Phylogenetic Biology, by D. A. Baum and S. D. Smith (Roberts and Company, Greenwood Village, CO, 2012), is a comprehensive introduction to the concepts, methods, and uses of phylogenetics in biology, for non-specialists.

These journal articles provide introductions to some of the topics in this chapter:

Omland, K. E., L. G. Cook, and M. D. Crisp. 2008. Tree thinking for all biology: The problem with reading phylogenies as ladders of progress. *BioEssays* 30: 854–867.

Bromham, L., and D. Penny. 2003. The modern molecular clock. *Nat. Rev. Genet.* 4: 216–224.

Pagel, M. 1999. Inferring the historical patterns of biological evolution. *Philos. Trans. R. Soc., B* 352: 519–529.

PROBLEMS AND DISCUSSION TOPICS

1. Suppose species 1, 2, and 3 are endemic to a group of islands (such as the Galápagos) and are all descended from species 4 on the mainland (which will serve as an outgroup; its very large population size means that no new mutations have become fixed in its population in the time since the islands were colonized). We sequence a gene and find ten nucleotide sites that differ among the four species (among many other loci that do not vary). The nucleotide bases at these sites are

   Species 1: GCTGATGAGT

   Species 2: ATCAATGAGT

   Species 3: GTTGCAACGT

   Species 4: GTCAATGACA

   Estimate the phylogeny of these taxa by plotting the changes on each of the three possible unrooted trees and determining which tree requires the fewest evolutionary changes.

2. Suppose the species in the previous question are birds that differ in diet: species 1 and 3 are insectivorous (they eat insects), and species 2 and 4 are frugivorous (they eat fruit). We also happen to know that another frugivorous species, species 5, is a mainland relative of species 4. Given your best estimate of the phylogenetic history, what has been the probable history of the evolution of diet in this clade of birds?

3. Phylogenetic information is the basis for describing patterns of evolution, yet some examples of patterns were presented without phylogenetic trees in the text. Consider the following examples and discuss what phylogenetic evidence or inference was left unstated: (a) The fusion of hindlimb bones during embryonic development of birds is a derived trait, not an ancestral trait, relative to the unfused condition in crocodiles. (b) Pentadactyly (five digits) is homologous in humans and crocodiles. (c) The sting of a wasp is derived from an ovipositor but is modified in both structure and function. (d) Insects evolved wings, but the character was lost for many wingless insect groups. (e) Frogs have some traits that are very similar to those of their deep ancestors (five toes on the hindlimb, multiple bones in the lower jaw) but others that are relatively advanced (lack of teeth in the lower jaw).

4. There is evidence that many of the differences in DNA sequence among species are not adaptive. Other differences among species, both in DNA and in morphology, are adaptive (as you will see in Chapters 3, 5, and 7). Do adaptive and non-adaptive variations differ in their usefulness for phylogenetic inference? Can you think of ways in which knowledge of a character's adaptive function would influence your judgment of whether or not that character provides evidence for relationships among taxa?

5. It is possible for two different genes to imply different phylogenetic relationships among a group of species. What are the possible reasons for this? If there is only one true history of formation of these species, what might we do in order to determine which (if either) gene accurately portrays that history? Is it possible for both phylogenetic trees to be accurate even if there has been only one history of species divergence?

Natural Selection and Adaptation

Some people are thrilled by snakes, while others are repelled and fearful, but no one can deny that they are fascinating animals. Depending on the species, they can crawl, burrow, swim, climb trees, and even glide, all without benefit of legs. Perhaps most amazing, they can swallow prey, whole, that are much larger than their heads (**FIGURE 3.1A**). Snakes can do this because, unlike humans, they have movable skull bones. Their lower jawbones (mandibles) are joined to a long, movable bone so that they can drop away from the skull, and their front ends are not fused, but are joined by a stretchable ligament. The tooth-bearing maxilla bones of the upper jaw can be flexed outward, further increasing the mouth opening. The upper and lower jaw bones on both sides can be independently moved forward and backward to pull the prey into the throat (**FIGURE 3.1B**). Rattlesnakes and other vipers take this apparatus a step further: their maxilla is short and bears only a long, hollow fang—a natural hypodermic needle—to which a duct leads from the massive poison gland (a modified salivary gland). The fang lies against the roof of the mouth when the mouth is closed. When the snake opens its mouth, the short maxilla is rotated 90 degrees, so that the fang is fully erected (**FIGURE 3.1C,D**).

Snakes' skulls, like many anatomical features, are complex mechanisms that look as if they had been designed by engineers to perform a specified function. They are said to be adapted to the animal's way of life: swallowing large prey whole. Every species has features—*adaptations*—that are thought to enhance survival in its environment. For example, cacti that grow in arid environments lack leaves and have thick, sometimes globular stems that reduce the ratio of

Wallace's flying frog (*Rhacophorus nigropalmatus*), which inhabits the rain forest canopy in southeastern Asia, glides between trees with the aid of its extensive toe webbing. This adaptation resulted from natural selection among individuals in ancestral populations that varied genetically in the extent of their webbing. Modification of ancestral features to serve new functions—such as gliding—is a common theme in evolution.

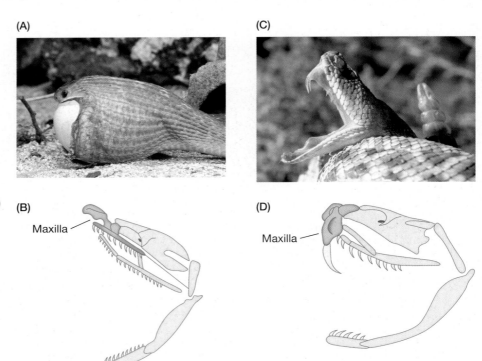

FIGURE 3.1 (A) Most snakes, such as this egg-eating snake (*Dasypeltis*), can eat prey much larger than their heads. (B) This ability is enabled by loose connections among many skull bones. The movable bones of the upper jaw are shown in blue. (C) The head of a red diamondback rattlesnake (*Crotalus ruber*) in strike mode. (D) In vipers such as rattlesnakes, rotation of the shortened maxilla erects its single tooth, a hollow fang. (B and D after [41].)

Maxilla

Maxilla

surface area (over which water is lost by evaporation) to volume (**FIGURE 3.2**). In some species, a coat of hair that reflects sunlight reduces body temperature.

An **adaptation** is a characteristic that enhances the survival or reproduction of organisms that bear it, relative to alternative character states. Adaptations have evolved by natural selection, which is the centerpiece of *On the Origin of Species* and of evolutionary theory, and is perhaps the most important idea in biology. It is also one of the most important ideas in the history of human thought—"Darwin's dangerous idea," as the philosopher Daniel Dennett [10] has called it—for it explains the apparent design of the living world without recourse to a supernatural, omnipotent designer.

For hundreds of years, it seemed that adaptive design could be explained only by an intelligent designer. In fact, this "argument from design" was considered one of the strongest proofs of the existence of God. The Reverend William Paley wrote

FIGURE 3.2 Many plants that live in arid environments have adaptations to reduce water loss, as in these cacti, which lack leaves. The stems, where photosynthesis takes place, are thick, with a low ratio of surface area to volume. This adaptation is taken to an extreme in almost globular barrel cacti (A; *Echinocactus grusonii*). Some cacti, such as this cholla (B; *Opuntia bigelovii*), have hairs that reflect light and so reduce the temperature of the plant body.

in *Natural Theology* [39] that, just as the intricacy of a watch implies an intelligent, purposeful watchmaker, so every aspect of living nature, such as the human eye, displays "every indication of contrivance, every manifestation of design, which exists in the watch," and must, likewise, have had a designer. When Darwin offered a purely natural, materialistic alternative to the argument from design, he not only shook the foundations of theology and philosophy, but also *brought every aspect of the study of life into the realm of science*. His alternative to intelligent design was design by the completely mindless process of natural selection. This process cannot have a goal, any more than erosion has the goal of forming canyons, for *the future cannot cause material events in the present*. Thus the concepts of goals or purposes have no place in biology (or in any of the other natural sciences), except in studies of human behavior.

Adaptive biological processes *appear* to have goals: animals engage in many adaptive behaviors, and a morphological feature, such as a flower, develops toward a suitable shape and stops developing when that shape is attained. We may loosely describe such features by *teleological* statements, which express goals (e.g., "She studied *in order* to pass the exam"; "Wasps sting *to defend themselves* from predators"). But no conscious anticipation of the future resides in the cell divisions that shape a flower or, as far as we can tell, in the behavior of wasps or birds. Rather, the apparent goal-directedness is caused by the operation of a program—coded or prearranged information, residing in DNA sequences—that when activated by external or internal stimuli controls a process [33]. A program likewise resides in a computer chip, but whereas that program has been shaped by an intelligent designer, the information in DNA has been shaped by a historical process of natural selection. Modern biology views the development, physiology, and behavior of organisms as the results of purely mechanical processes, resulting from interactions between programmed instructions and environmental conditions or triggers.

Adaptive Evolution Observed

Darwin could not point to any cases in which evolutionary change of a population or species had actually been observed, and he supposed that evolution was much too slow for us to see it in action. But today we can cite hundreds of examples of adaptive evolution of morphological, physiological, and behavioral traits that have been directly observed. Adaptive evolution can be rapid, especially in species that have been introduced into new regions or subjected to human alterations of their environment [26, 40, 47]. For instance, several species of insects, such as the soapberry bug (*Jadera haematoloma*) [5, 6], have adapted rapidly to new food plants. The bug feeds on seeds of plants in the soapberry family (Sapindaceae) by piercing the enveloping seedpod with its slender beak. In the last 50 years, related species of Asian trees have been so abundantly planted in Texas and Florida that the bug populations feed mostly on these species. Compared with the original native host plants, the Asian tree species in Texas has a larger pod, and the Asian tree species in Florida has a much smaller pod. Corresponding to this difference, soapberry bug populations in Texas have evolved a longer beak, and those in Florida, a shorter beak (**FIGURE 3.3**).

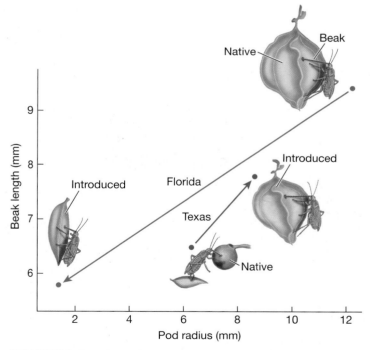

FIGURE 3.3 Soapberry bugs (*Jadera haematoloma*) and the seedpods of their native and introduced host plants in Texas and Florida, drawn to scale. The bug's beak is the needlelike organ projecting from the head at a right angle to the body. The average pod radius of each host species is plotted against the average beak length of associated *Jadera* populations. Beak length has evolved rapidly as an adaptation to the new host plants. (After [5].)

FIGURE 3.4 Cumulative numbers of arthropod pest species known to have evolved resistance to five classes of insecticides. The upper curve shows the total number of insecticide-resistant species. The number has increased since these data were tabulated. (After [34].)

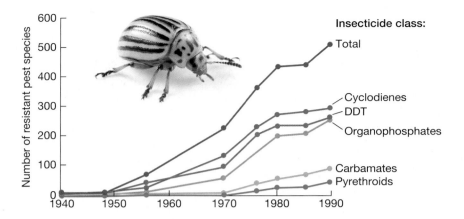

We have already seen (in Chapter 1) that bacteria can evolve resistance to antibiotics very rapidly. Similarly, resistance to chemical pesticides has evolved in hundreds of species of insects (**FIGURE 3.4**), and many species of weeds have evolved resistance to herbicides within 10–20 years of field exposure. Copper, zinc, and other heavy metals are toxic to plants, but in several species of grasses and other plants, metal-tolerant populations have evolved where soils have been contaminated by mine works less than 100 years old. When tolerant and nontolerant genotypes of a species are grown in competition with other plant species in the absence of the metal, the growth of the tolerant genotypes is often much lower than that of the nontolerant genotypes, implying that adaptation has costly side effects [1, 32].

Commercial overexploitation has severely depleted populations of many species of fish and has resulted in evolutionary changes as well [30]. In many species there has been a trend toward earlier sexual maturation at a smaller size, as we predict when larger age classes are more subject to predation (see Chapter 14). In some species, such as Atlantic cod (*Gadus morhua*), these changes clearly have a genetic basis (**FIGURE 3.5A**). Similarly, trophy hunting for bighorn sheep (*Ovis canadensis*)

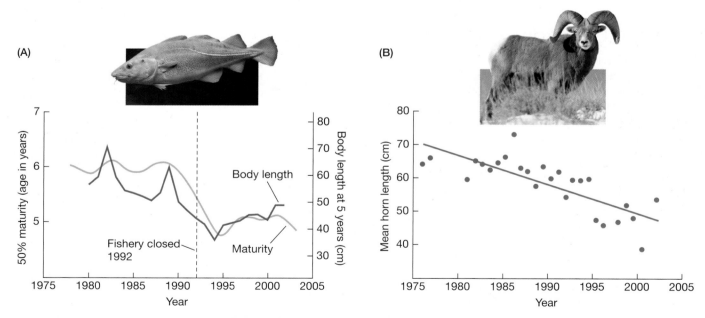

FIGURE 3.5 Evolutionary changes caused by human harvesting. (A) The age at which 50 percent of Atlantic cod reached maturity declined until 1994 when the fishery was closed because of overfishing. Body length at 5 years of age also declined until 1994 (B) Mean horn length of 4-year-old bighorn sheep rams declined because of selection imposed by hunting. (A after [38]; B after [8].)

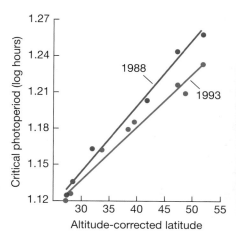

FIGURE 3.6 In North America, the critical photoperiod for entering diapause, in relation to latitude, has decreased in the pitcher-plant mosquito (*Wyeomyia smithii*), as shown by these data from 1988 and 1993. The larvae of this mosquito develop only in the water that collects in the tubular leaves of the pitcher plant *Sarracenia purpurea*. Pitcher plants obtain nutrients from the decaying bodies of other species of insects that fall in and are trapped. (After [3].)

with the largest horns has resulted in the evolution of smaller horns (**FIGURE 3.5B**). In both instances, the very quality that adds value to the resource has been diminished by the response to selection.

Some species show adaptation to the ongoing climate change caused by human production of CO_2 and other greenhouse gases. In many insects, the cue for entering diapause, a state of low metabolic activity that is necessary for surviving the winter, is a critical photoperiod (day length). Northern populations are genetically programmed to enter diapause at a longer day length than southern populations because cold weather arrives at northern latitudes sooner, when days are still long. William Bradshaw and Christina Holzapfel [3] sampled populations of the pitcher-plant mosquito (*Wyeomyia smithii*) from southern Canada to Florida four times between 1972 and 1996 and experimentally measured the day length at which the insects entered diapause. They found that during this time, the critical photoperiod became shorter: the insects became programmed to enter diapause later in autumn (**FIGURE 3.6**). The change was greatest in the most northern populations, as expected because the increase in temperature has been greater at higher latitudes. The speed of evolution was amazing, having taken as little as 5 years.

These evolutionary changes can be so rapid because populations in altered environments, especially those altered by human activities, can experience stringent natural selection, and because they contain genetic variation in many characteristics—a necessary ingredient of evolution.

Natural Selection

The meaning of natural selection

In *The Origin of Species*, Darwin introduced natural selection with these words:

> *Can it …be thought improbable, seeing that variations useful to man have undoubtedly occurred [in domesticated animals and plants], that other variations useful in some way to each being in the great and complex battle of life, should sometimes occur in the course of thousands of generations? If such do occur, can we doubt (remembering that many more individuals are born than can possibly survive), that individuals having any advantage, however slight, over others, would have the best chance of surviving and procreating their kind? On the other hand, we may feel sure that any variation in the least degree injurious would be rigidly destroyed. This preservation of favoured variations and the rejection of injurious variations, I call Natural Selection.*

FIGURE 3.7 Two genotypes of a plant are growing together. Genotype A has a fitness of 3, while genotype B has a fitness of 4. Both genotypes start with 10 individuals. (A) The population size of genotype B grows much more rapidly. (B) Plotting the frequencies of the two genotypes shows that genotype B, which starts at a frequency of 0.5, makes up almost 90% of the population just 7 generations later.

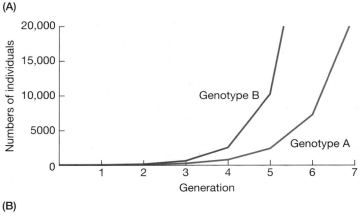

(A)

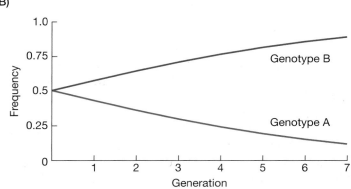

(B)

Among several slightly different definitions of natural selection used by biologists today [12], we use this one: **natural selection** is *any consistent difference in fitness among different classes of biological entities.* A simple way to think of **fitness** is as the number of offspring an individual leaves in the next generation. Suppose, for example, that in a species of annual plant, only 1 of every 1000 seeds survives to reproductive age, and that those that survive produce an average of 3000 seeds. The *average* fitness of that type of individual is 0.001 × 3000 = 3. The components of fitness are survival and reproduction. Fitness is sometimes called **reproductive success**, which includes survival because organisms do not reproduce when they are dead.

If *evolution* by natural selection is to occur, there must be a change in the population across generations, and this requires that the phenotypic differences among the entities be *inherited*. Thus, evolution by natural selection occurs if (1) there is a correlation between an individual's phenotype and its fitness, and (2) variation in the phenotype is correlated between parents and their offspring. Suppose, for example, that in an asexually reproducing annual plant, genotypes A and B differ in a characteristic that affects their fitness (e.g., susceptibility versus resistance to a herbicide), and that their average fitnesses are 3 and 4, respectively. If these values are constant from generation to generation, genotype B increases in number much faster than A, and will make up the great majority of the population within a few generations (**FIGURE 3.7**). We say that the **frequency** (proportion) of genotype B has increased (and conversely, that the frequency of A has declined). In sexually reproducing organisms, fitness is more complicated. Males vary in survival and reproduction. In particular, they vary in mating success, which Darwin called **sexual selection**. (In some species, females also experience sexual selection.) In sexual species, moreover, individuals' genes replicate, but because of recombination their genotypes do not. So it can be useful to think about the fitness of a type of gene (i.e., an allele), and consequently of selection among genes, even though the "entities" that differ in survival and reproduction in most discourse about evolution are individual organisms with different phenotypes (**individual selection**). Evolution by natural selection in sexually reproducing populations entails changes in the frequencies of alleles at the locus (or loci) that underlies variation in the phenotypic characteristic that influences fitness.

We will almost always discuss natural selection among genes and among heritable individual phenotypes because selection has no lasting evolutionary effect without inheritance. Most of our discussion will assume that inheritance of a trait is based on genes. However, many of the principles of evolution by natural selection also apply if inheritance is epigenetic (based on, for example, differences in DNA methylation; see Chapter 4) or is based on cultural transmission, especially from parents to offspring. *Culture* has been defined as "information capable of affecting individuals' behavior that they acquire from other members of their species through teaching, imitation, and other forms of social transmission" [44].

We must be very careful to understand that natural selection is not an agent or active power, and certainly not a purposeful one, even though the language we use often seems to personify it, or suggest that it is an agent. Darwin coined the term "natural selection" to parallel the selection that breeders of crops and domestic animals use to improve desirable characteristics. In later editions of *The Origin of Species*, he wrote that "it has been said that I speak of natural selection as an active power or Deity; but who objects to an author speaking of the attraction of gravity as ruling the movements of the planets? Every one knows what is meant and is implied by such metaphorical expressions; and they are almost necessary for brevity." Likewise, evolutionary biologists often say that selection "favors" a certain characteristic, or they refer to selection as a "force." This is metaphorical language, used for brevity. *Natural selection is a name for statistical differences in reproductive success among genes, organisms, or populations—and nothing more.*

Natural selection and chance

"Natural selection" is not synonymous with "evolution." Natural selection can occur without any evolutionary change, as when natural selection maintains the status quo by eliminating deviants from the optimal phenotype. And processes other than natural selection can cause evolution.

One of those processes is *genetic drift*: random fluctuations in the frequencies of genotypes within a population. (Genetic drift is the subject of Chapter 7.) **Neutral alleles** are those that do not alter fitness: the average reproductive success does not differ between individuals that carry one neutral allele or the other. The frequencies of these neutral alleles may change in a population by genetic drift. If this occurs, the bearers of one allele have had a greater rate of increase than the bearers of the other allele, but natural selection has not occurred, because the genotypes do not differ *consistently* in fitness: the alternative allele could just as well have been the one to increase. There is no *average* difference between the alleles, no *bias* toward the increase of one relative to the other. Fitness differences, in contrast, are *average* differences, *biases*, differences in the *probability* of reproductive success. *Natural selection is the antithesis of chance.* In practice, we can ascribe genetic changes to natural selection rather than random genetic drift only if we measure numerous individuals of each genotype or phenotype, and find an average difference in reproductive success.

The effective environment depends on the organism

The environmental factors that impose natural selection on a species are greatly influenced by the characteristics of the species itself: the evolutionary history of a species affects its relationship to the environment [31]. The branching structure of

(A)

(B)

FIGURE 3.8 The evolutionary histories of some animals have made them less reliant on vision, so selection for visual acuity has been relaxed. (A) Army ants (genus *Eciton*) rely almost entirely on chemical information. In these ants, the compound eyes have been highly reduced, consisting of a single unit (ommatidium) rather than the many that compose most insects' eyes. (B) Similarly, burrowing blind snakes (Typhlopidae) have highly reduced eyes that perceive light but cannot form an image.

trees in a forest is important for many tree-nesting birds such as orioles, but almost irrelevant for ground-nesting species such as partridges; the viscosity of water, which varies with temperature, is much more important for a ciliate than for a fish. To some extent, organisms construct their ecological niches [36], literally (as does a beaver) or more metaphorically. Organisms "screen off" some aspects of their environment, which may then cease to exert natural selection. Many species of ants, rodents, and other animals have become so reliant on chemical signals that they have become blind, because natural selection for sight has become reduced or even negative: well-developed eyes may be disadvantageous if they conflict with other important functions (**FIGURE 3.8**). Likewise, humans have lost the function of many olfactory receptor genes, having become so much more reliant on vision than smell.

Levels of Selection

By "natural selection," both Darwin and present-day evolutionary biologists usually mean consistent differences in fitness among phenotypically and genetically different *individual organisms* within populations. But our definition of natural selection applies to any classes of variable entities that can change in number. Selection can occur among genes, cell types, individual organisms, populations, or species, a hierarchy of **levels of selection**.

Natural selection at the level of the gene (**genic selection**) is illustrated by transposable elements, which replicate and proliferate within the genome, irrespective of whether their proliferation affects the organisms for good or ill. Transposable elements are among the many kinds of **selfish genetic elements**, which are transmitted at a higher rate than the rest of an individual's genome and may be detrimental (or at least not advantageous) to the organism [4, 27]. Some selfish alleles exhibit **segregation distortion**, and are passed to a heterozygous individual's gametes more than 50 percent of the time. Segregation distortion can result from **meiotic drive** (in which meiosis does not follow Mendel's laws; see Chapter 12) and from other processes that happen after the gametes are formed.

An example of a selfish genetic element that exhibits segregation distortion is the *t* locus of the house mouse (*Mus musculus*). In a male heterozygous for a *t* allele and for the normal allele *T,* the *t* allele kills sperm that carry the normal allele. As a result, more than 90 percent of the male's sperm carry *t.* Embryos that are *tt* homozygotes, however, die or are sterile. Despite these disadvantages to the individual, segregation distortion is so great that the disadvantageous *t* allele reaches a high frequency in many populations of mice.

Selfish genetic elements forcefully illustrate the nature of natural selection: it is nothing more than differential reproductive success (of genes in this case), which need not result in adaptation or improvement in any sense. Selection among individuals is at a "higher level" than selection among genes [37]. Selection at the gene level may act in opposition to individual selection: it may be harmful to individual organisms, and might even cause the extinction of populations or species.

Selfish genes and unselfish behaviors

Evolutionary geneticists have long recognized that natural selection will cause an allele to increase in frequency if it consistently leaves more copies of itself to subsequent generations, no matter how it causes its greater success. For example, plants that produce more pollen are likely to fertilize more ovules, so any allele that increases pollen production is likely to spread. J. B. S. Haldane wrote in 1932 [24] that "No sufferer from hay fever will doubt that more pollen is produced than is needed to assure that almost every ovule should be fertilised." In the same book, he wrote that "in a beehive the workers [which do not reproduce] and young queens are samples of the same set of genotypes, so any form of behaviour in the former (however suicidal it may be) which is of advantage to the hive will promote the survival of the latter, and thus tend to spread through the species."

The key issue is that it is often useful think of selection among genes, based on the effects that change their frequencies—whether these effects are on the number of pollen grains, behavior that enhances the survival of relatives that share the same gene, or many other biological features. In a sense, then, any gene that has successfully increased in frequency is a *selfish gene,* as biologist Richard Dawkins has famously written [9]. The evolution of many puzzling features of organisms can be understood by considering the rates at which different variants of a gene that affects the trait would change in frequency over the course of generations.

An important example of this approach is the topic that Haldane addresses in accounting for the behavior of worker bees: what he called "socially valuable but individually disadvantageous characters." Many such **altruistic traits** are best explained by the principle Haldane described, which has come to be called **kin selection**. An allele for altruistic behavior can increase in frequency in a population if the beneficiaries of the behavior are usually related to the individual who performs it. Since the altruist's relatives are more likely to carry copies of the altruistic allele than are members of the population at large, when the altruist enhances the fitness of its relatives, even at some cost to its own fitness, it can increase the frequency of the allele. We may therefore define kin selection as a form of selection in which alleles differ in fitness by influencing the effect of their bearers on the reproductive success of individuals (kin) who carry the same allele by common descent. The simplest example of a trait that has evolved by kin selection is parental care: alleles that enhance a parent's care-giving behavior have increased in frequency because they promote the survival of identical copies of those same alleles that the offspring carry.

FIGURE 3.9 A popular myth about the self-sacrificial behavior of lemmings holds that they rush en masse into the sea to prevent overpopulation. This cartoon illustrates the "cheater" principle and shows why such altruistic behavior would not be expected to evolve. (Cartoon © Mark Godfrey/www.Cartoonstock.com.)

Selection of organisms and groups

Do oysters have a high reproductive rate "to ensure the survival of the species," as we often hear? Do antelopes with sharp horns refrain from physical combat because combat would lead to the species' extinction? Is there any truth to the myth that lemmings (small Arctic rodents) commit suicide by drowning, in order to relieve the pressure of high population density on the food supply (**FIGURE 3.9**)?

If traits evolve by individual selection—by the replacement of less fit by more fit individuals, generation by generation—the possibility of future extinction cannot possibly affect the course of evolution. The process of natural selection lacks forethought (or any thought at all): the future cannot affect the present. It is unlikely that kin selection would result in the evolution of suicide in lemmings, since the entire population, not just the suicides' relatives, would benefit from the food made available. An altruistic trait cannot evolve if it reduces the fitness of an individual that bears it, even if it benefits the population or species as a whole. An altruistic genotype amid selfish genotypes would necessarily decline in frequency, simply because it would leave fewer offspring per capita than the others. Conversely, if a population were to consist of altruistic genotypes, a selfish mutant—a "cheater"—would increase to fixation, even if a population of such selfish organisms had a higher risk of extinction.

So it would seem impossible that a trait could evolve that benefits the population at a cost to the individual. However, there is one conceivable way it might evolve, namely by **group selection**: differential production or survival of groups that differ in genetic composition. For instance, populations made up of selfish genotypes, such as those with high reproductive rates that exhaust their food supply, might have a higher extinction rate than populations made up of altruistic genotypes that have lower reproductive rates. If so, then the species as a whole might evolve altruism through the greater survival of groups of altruistic individuals, even though individual selection within each group would act in the opposite direction (**FIGURE 3.10A**).

This hypothesis of group selection was criticized by George Williams in his influential book *Adaptation and Natural Selection* [51]. Williams argued that supposed adaptations that benefit the population or species, rather than the individual, can be plausibly explained by benefit to the individual or the individual's genes, or may not be adaptations at all. For example, females of many species lay fewer eggs when population densities are high and food is scarce, not to ensure a sufficient food supply for the good of the species, but simply because they cannot form as many eggs. Williams based his opposition to group selection on a simple argument. Individual organisms are much more numerous than the populations into which they are aggregated, and they turn over—are born and die—much more rapidly than populations are formed or become extinct. Thus the rate of replacement of less fit (altruistic) by more fit (selfish) individuals is potentially much greater than the rate of replacement of less fit by more fit populations, so individual selection will generally prevail over group selection (**FIGURE 3.10B**). Among evolutionary biologists, the majority view is that *few characteristics have evolved because they benefit the population or species*, and that cooperation and seeming altruism are most likely to have evolved by other causes, especially kin selection. Some prominent biologists, however, hold that group selection is important in evolution [11], as we will describe in Chapter 12.

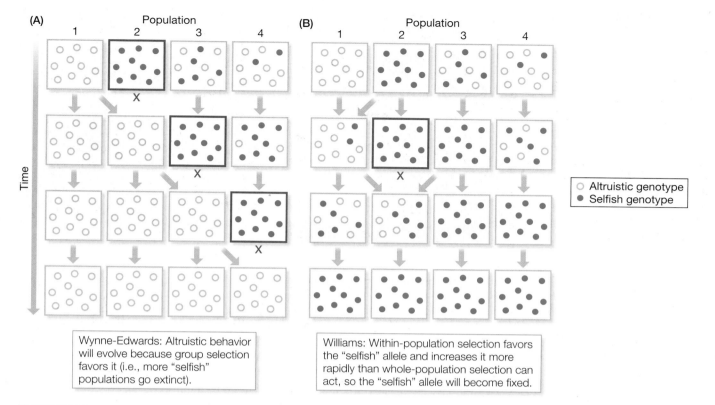

FIGURE 3.10 Conflict between group and individual selection. The rectangles represent four populations of a species (1–4), traced through four time intervals; each circle is an individual organism in a population: open if the individual is an altruistic genotype, filled if it is a selfish genotype. Some new populations are founded by colonists from established populations (shown by diagonal arrows), and some populations become extinct (marked by X). Individuals with the selfish genotype are assumed to have higher fitness than altruistic individuals. (A) An altruistic trait may evolve by group selection if the rate of extinction of populations of the selfish genotype is very high. (B) Williams's argument: Because individual selection operates so much more rapidly than group selection, the selfish genotype increases rapidly within populations and may spread by gene flow into populations of altruists, and replaces them. Thus the selfish genotype becomes fixed, even if it increases the chance of population extinction.

Within figure (A): Wynne-Edwards: Altruistic behavior will evolve because group selection favors it (i.e., more "selfish" populations go extinct).

Within figure (B): Williams: Within-population selection favors the "selfish" allele and increases it more rapidly than whole-population selection can act, so the "selfish" allele will become fixed.

Legend: ○ Altruistic genotype ● Selfish genotype

Species selection

Selection among groups of organisms is called **species selection** when the groups involved are species and there is a correlation between some characteristic and the rate of speciation or extinction [19, 28, 42]. Species selection does not shape adaptations of organisms, but it does affect the *disparity*—the diversity of biological characteristics—of the world's organisms. The consequence of species selection is that the proportion of species that have one character state rather than another changes over time (**FIGURE 3.11**). A likely example of the effects of species selection is the prevalence of sexual species compared with closely related asexual forms. Many groups of plants and animals have given rise to asexually reproducing lineages, but with some interesting exceptions, asexual lineages tend to be young, as indicated by their close genetic similarity to sexual forms. This observation implies that asexual forms have a higher rate of extinction than sexual populations, since few asexual forms that arose long ago have persisted (see Chapter 10) [35].

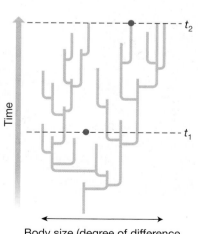

FIGURE 3.11 Species selection caused by a correlation between speciation rate and a morphological character, such as body size (x-axis). Larger-bodied species persist longer before becoming extinct, and so give rise to large-bodied species more often than small-bodied species produce other small-bodied species. The lower extinction rate of lineages with large body sizes is analogous to a lower mortality rate of individual organisms in individual selection. The character value, averaged across species (red dots), is greater at time t_2 (upper dashed line) than at time t_1 (lower dashed line). (After [18].)

Body size (degree of difference from ancestral form)

FIGURE 3.12 The long sharp bill of the kea (*Nestor notabilis*) evolved for functions such as cracking seeds, but it can be used for many other things, such as slicing into sheep skin and ripping windshield wipers, rubber gaskets, and other removable pieces from parked automobiles. At several sites in New Zealand, tourists are warned to protect their cars against keas. Why keas do this is not clear.

The Nature of Adaptations

Adaptation is a central concept in biology. The word has two related meanings. "Adaptation" means the evolutionary *process* by which, over the course of generations, organisms are altered to become improved with respect to features that affect survival or reproduction. "An adaptation" is a *characteristic* of an organism that evolved by natural selection. Both meanings are difficult to define precisely [29, 43]. Most evolutionary biologists think that for a character to be regarded as an adaptation, it must be a derived character that conferred higher fitness than the ancestral character state from which it evolved [25].

A **preadaptation** is a feature that fortuitously serves a new function. For instance, parrots have strong, sharp beaks, used for feeding on fruits and seeds. When domesticated sheep were introduced into New Zealand, some were attacked by an indigenous parrot, the kea (*Nestor notabilis*), which pierced the sheeps' skin and fed on their fat (**FIGURE 3.12**). The kea's beak happened to be useful for this new activity. Such a feature, if co-opted for a new function during evolution, is sometimes called an **exaptation** [20]. For example, the wings of auks are exaptations for swimming: these birds "fly" under water as well as in air (**FIGURE 3.13A**). An exaptation may be further modified by selection so that the modifications are adaptations for the feature's new function: the wings of penguins have been modified into flippers

(A)

(B)

FIGURE 3.13 Exaptation and adaptation. (A) The wing might be called an exaptation for underwater "flight" in members of the auk family, such as this Atlantic puffin (*Fratercula arctica*). (B) The modifications of the wing for efficient underwater locomotion in penguins (these are Humboldt penguins, *Spheniscus humboldti*) may be considered adaptations.

that enhance swimming but cannot support flight in air (**FIGURE 3.13B**). Exaptation is a very common early stage in the evolution of new adaptations.

Selection of and selection for

To say that a feature is adaptive is unsatisfying unless we have some idea of what it is adaptive for: by what mechanism did it increase fitness? What is its function?

In the child's "selection toy" pictured in **FIGURE 3.14**, the holes in each partition are smaller than in the one above. Balls of several sizes, when placed in the top compartment, fall through the holes in the partitions. If the smallest balls in the toy are all red, and the larger ones are all other colors, the toy will select the small, red balls. Thus we must distinguish *selection of objects* from *selection for properties* [48]. Balls are selected *for* the property of small size—that is, *because of* their small size. They are not selected for their color, or because of their color; nonetheless, there is selection *of* red balls. Natural selection may similarly be considered a sieve that selects *for* a certain body size, mating behavior, or other feature. There may be incidental selection *of* other features that are correlated with that feature. We will return several times in the book to the theme that selection on one trait has side effects on others.

The importance of this semantic point is that when we speak of the **function** of a feature, we imply that there has been natural selection *for* the feature itself: that the feature *caused* its bearers to have higher fitness. The feature may have side effects, other consequences that were not its function, and *for* which there was no selection. For instance, a fish species may be selected for coloration that makes it less conspicuous to predators. The *function* of the coloration, then, is predator avoidance. An *effect* of this evolutionary change might well be a lower likelihood that the population will become extinct, but *avoidance of extinction is not a cause of evolution* of the coloration.

Recognizing adaptations

Not all traits are adaptations. There are at least four other possible explanations of organisms' characteristics. First, a trait may be a necessary consequence of physics or chemistry. Hemoglobin gives blood a red color, but the redness is not an adaptation; it is a by-product of the protein's structure. (However, this feature has been co-opted for various functional roles in the evolution of many species of vertebrates, such as the white-winged chough [**FIGURE 3.15**].)

FIGURE 3.14 A child's toy that selects small balls, which drop through smaller and smaller holes from top to bottom. In this case there is selection *of* red balls, which happen to be the smallest, but selection is *for* small size. (After [48].)

(A)

(B)

FIGURE 3.15 (A) The Australian white-winged chough (*Corcorax melanorhamphos*) normally has predominantly yellow eyes. (B) During aggressive displays, the bird shows brilliant red, bulging eyes, using the red color that is a nonadaptive property of hemoglobin for an adaptive function.

Second, the trait may have evolved by other mechanisms (such as random genetic drift) rather than by natural selection (see Chapter 7).

Third, the feature may have evolved not because it conferred an adaptive advantage, but because it was correlated with another feature that did. (As we will see, genetic linkage and pleiotropy—the phenotypic effect of a gene on multiple characters—are important causes of such correlations.)

Fourth, a character state may be a consequence of phylogenetic history. Darwin saw clearly that a feature might be beneficial, yet not have evolved for the function it serves today, or for any function at all: "The sutures in the skulls of young mammals have been advanced as a beautiful adaptation for aiding parturition [birth], and no doubt they facilitate, or may be indispensable for this act; but as sutures occur in the skulls of young birds and reptiles, which have only to escape from a broken egg, we may infer that this structure has arisen from the laws of growth, and has been taken advantage of in the parturition of the higher animals" (*On the Origin of Species*, chapter 6). Whether or not we should postulate that a trait is an adaptation depends on such insights. For example, we know that pigmentation varies in many species of birds, so it makes sense to ask whether there is an adaptive reason for color differences among closely related species. But it is not sensible to ask whether it is adaptive for a hummingbird to have four toes rather than five, because the ancestor of birds lost the fifth toe and it has never been regained in any bird since. Five toes are probably not an option for hummingbirds.

For all these reasons, many authors hold that we should not assume that a feature is an adaptation unless the evidence favors this interpretation [51]. This is not to deny that a great many of an organism's features, probably the majority, are adaptations. Several methods are used to infer that a feature is an adaptation for some particular function. We will note these methods only briefly and incompletely at this point, exemplifying them more extensively in later chapters. The approaches described here apply to phenotypic characters; in Chapter 5 we will describe how selection can be inferred from DNA sequence data.

COMPLEXITY   Even if we cannot immediately guess the function of a feature, *we often suspect it has an adaptive function if it is complex*, for complexity cannot evolve except by natural selection. For example, a peculiar, highly vascularized structure called a pecten projects in front of the retina in the eyes of birds (**FIGURE 3.16**). Only recently has evidence been developed to show that the pecten supplies oxygen to the retina, but it has always been assumed to play some important functional role because of its complexity and because it is ubiquitous among bird species.

DESIGN   The function of a character is often inferred from its *correspondence with the design* an engineer might use to accomplish some task, or with the *predictions of a model* about its function. For instance, many plants that grow in hot environments have leaves that are finely divided into leaflets, or which tear along fracture lines (**FIGURE 3.17**). These features conform to a model in which the thin, hot "boundary layer" of air at the surface of a leaf is more readily dissipated by wind passing over a small than a large surface, so that a divided leaf is more effectively cooled. The fields of functional morphology and ecological physiology are concerned with analyses of this kind.

EXPERIMENTS   Experiments may show that a feature enhances survival or reproduction, or enhances performance (e.g.,

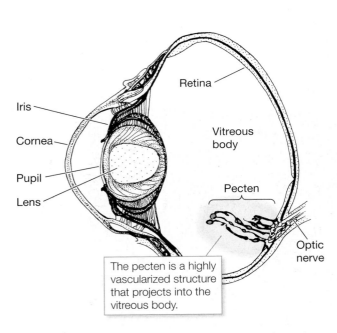

The pecten is a highly vascularized structure that projects into the vitreous body.

FIGURE 3.16 The pecten of a bird's eye, shown in sagittal section. About 30 hypotheses were proposed for the pecten's function. It was finally shown to supply oxygen to the retina. (After [17].)

(A)

(B)

FIGURE 3.17 Functional morphological analyses have shown that small surfaces shed the hot "boundary layer" of air that forms around them more readily than do large surfaces. Many tropical and desert-dwelling plants have large leaves that are broken up into leaflets, as in *Acacia karroo* (A), or split into small sections, as in the banana (B). The form of these leaves is therefore believed to be an adaptation for reducing leaf temperature.

locomotion or defense) in a way that is likely to increase fitness, relative to individuals with other features. For example, several floral characters have evolved convergently in the many plant lineages that have shifted from insect pollination to bird pollination (**FIGURE 3.18A**). Maria Castellanos and colleagues tested the hypothesis that some of these features are advantageous because they facilitate bird pollination, and others because they discourage bees, which are less effective pollinators because they comb much of the pollen into a mass that they feed to their larvae [7]. The researchers surgically altered several features of flowers on a bee-pollinated plant to resemble those of related hummingbird-pollinated species (**FIGURE 3.18B–E**). They then measured pollen transfer from the altered flowers by bumblebees and hummingbirds. The researchers concluded that the lower "lip" typical of bee-pollinated flowers, which bees use as a landing platform (see Figure 3.18B), has been reduced or lost in some bird-pollinated species because its absence discourages bees (see Figure 3.18C). The projecting anthers of bird-pollinated plants also seem to be an "anti-bee" adaptation (see Figure 3.18D), and the narrowly constricted corolla tube (see Figure 3.18E) is both "pro-bird" and "anti-bee": it forces hummingbirds to remove more pollen, but prevents bees from easily obtaining nectar.

THE COMPARATIVE METHOD   A powerful means of inferring the adaptive significance of a feature is the **comparative method**, which consists of *comparing sets of species to pose or test hypotheses on adaptation and other evolutionary phenomena* [13]. This method takes advantage of "natural evolutionary experiments" provided by convergent evolution. If a feature evolves independently in many lineages because of a similar selection pressure, we can often infer the function of that feature by determining the ecological or other selective factor with which it is correlated. For instance, a long, slender beak has evolved in at least six lineages of birds that feed on nectar, and many plants that are pollinated by such birds have independently evolved attractive red or orange coloration and a tubular form that restricts access by bees (see Figure 2.22). Among fishes, open-water, fast-moving

FIGURE 3.18 Experimental test of a hypothesis of adaptation. (A) Bee-pollinated *Penstemon strictus* (top) and hummingbird-pollinated *P. barbatus* (bottom). (B–E) Experimental modifications of flowers of a bee-pollinated species (*Penstemon strictus*) to mimic features of hummingbird-pollinated species of *Penstemon*. (B) The normal flower of *P. strictus*. Modifications included (C) removal of the lower lip "landing platform," (D) re-attaching stamens so that the anthers project from the flower, and (E) constriction to form a narrower corolla tube. (B–E after [7].)

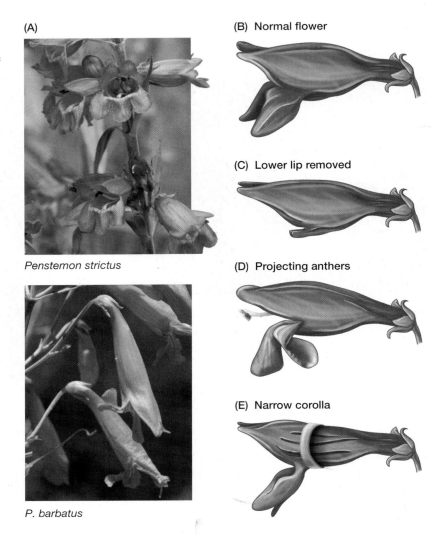

(A)

Penstemon strictus

P. barbatus

(B) Normal flower

(C) Lower lip removed

(D) Projecting anthers

(E) Narrow corolla

predators in many families commonly have a streamlined shape, a forked tail fin, and a slender tail base (caudal peduncle), whereas fishes that live in complicated environments, such as among corals or vegetation, have a deep, compressed body that enables them to change direction rapidly (**FIGURE 3.19**).

Biologists often predict such correlations by postulating, perhaps on the basis of a model, the adaptive features we would expect to evolve repeatedly in response to a given selective factor. For example, in species in which a female mates with multiple males, the several males' sperm compete to fertilize eggs. Males that produce more abundant sperm should therefore have a reproductive advantage. In primates, the quantity of sperm produced is correlated with the size of the testes, so large testes should be expected to provide a greater reproductive advantage in polygamous than in monogamous species. Paul Harvey and Mark Pagel compiled data from prior publications on the mating behavior and testes size of various primates [25]. They confirmed that, as predicted, the weight of the testes, relative to body weight, is significantly higher among polygamous than monogamous taxa (**FIGURE 3.20**).

An important aspect of this example is that although all the data needed to test this hypothesis already existed, the relationship between the two variables was not known until Harvey and Pagel compiled the data, because no one had had any reason to do so until an adaptive hypothesis had been formulated. Hypotheses about adaptation can be fruitful because they suggest investigations that would not otherwise occur to us.

Also, notice that because the consistent relationship between testes size and mating system was not known a priori, the hypothesis generated a *prediction*. The

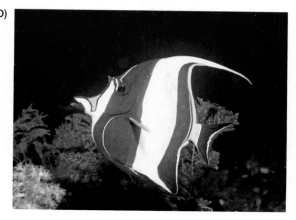

FIGURE 3.19 Body form of fishes is adapted for different modes of swimming. Open-water, fast-moving predators such as (A) jacks (Carangidae) and (B) swordfishes (Xiphiidae) have a slender body, narrow caudal peduncle, and narrow, forked tail fin. Fishes that maneuver in small spaces have a deep body, as in (C) angelfishes (Pomacanthidae) and (D) the Moorish idol (Zanclidae), both inhabitants of coral reefs. (C and D courtesy of Michael D. Bryant.)

predictions made by evolutionary theory, like those in many other scientific disciplines, are usually predictions of what we will find when we collect data. (Prediction in evolutionary theory does *not* usually mean that we predict the future course of evolution of a species.) Predictions of what we will find, deduced from hypotheses, constitute the **hypothetico-deductive method**, of which Darwin was one of the first effective exponents [16, 45].

Imperfections and Constraints

Darwin noted that "natural selection will not produce absolute perfection, nor do we always meet, as far as we can judge, with this high standard in nature" (*On the Origin of Species*, chapter 6). Selection can fix only those genetic variants with a higher fitness than other genetic variants in a particular population at a particular time. It cannot fix the best of all conceivable variants if they do not arise, or have not yet arisen, and the best possible variants often fall short of perfection because of various constraints [14, 49]. Among these constraints are **trade-offs** (following the maxim "There is no such thing as a free lunch"). For example, with

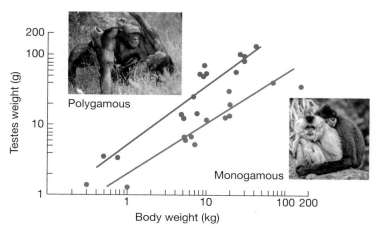

FIGURE 3.20 Relationship between weight of the testes and body weight in polygamous and monogamous primate taxa. The data support the prediction, based on the theory of natural selection, that males in polygamous species have relatively larger testes, which produce more sperm, than do males in monogamous species. The photos show polygamous mating bonobos (*Pan paniscus*) and a monogamous pair of yellow-cheeked gibbons (*Nomaseus gabriellae*). (After [25].)

FIGURE 3.21 Almost all mammals, including the long-necked giraffe and the short-necked aquatic dugong, have seven neck vertebrae, a likely example of a phylogenetic constraint.

a fixed amount of available energy or nutrients, a plant species might evolve higher seed numbers, but only by reducing the size of its seeds or some other part of its structure. If genotypes differ in reproductive output, one would see a negative correlation between seed number and seed size: the greater the number, the lower the average size (see Chapter 11).

In some cases, it appears that adaptations have not evolved because of a shortage of suitable mutations [2]. For example, although some species of grasses have rapidly evolved tolerance to heavy metals in the vicinity of mines, other species have not. In large samples of seeds collected from grass populations on normal soils, far from copper mines, a small percentage of seeds produced copper-tolerant seedlings in every one of eight species that have evolved copper tolerance near mines, but in none of seven species that have failed to do so [2]. These species apparently *lack the genetic variation* in tolerance that would be necessary for adaptive evolution. Lack of suitable genetic variation may explain cases of so-called *phylogenetic constraints,* in which species retain nonadaptive features or are unable to evolve adaptive traits. It makes adaptive sense that birds such as swans have more neck vertebrae than birds with shorter necks. But almost all mammals have seven neck vertebrae, including giraffes, the aquatic dugongs, and whales, despite the extreme difference in the lengths of their necks (**FIGURE 3.21**). Individual mice and humans with an aberrant number of cervical vertebrae show various skeletal abnormalities and a high incidence of embryonic cancer—harmful side effects that probably prevented the evolution, in other mammals, of what might otherwise have been advantageous changes in vertebral number [15, 50].

Natural Selection and the Evolution of Diversity

A mechanic uses a variety of different wrenches because each is suited to a different task. Likewise, any characteristic of an organism is likely to be advantageous under some circumstances but not others. That is, the optimal feature, the character that maximizes fitness, depends on the context in which it functions. A simple, even obvious, example is provided by many instances of cryptic coloration (camouflage) in animals, whereby colors and patterns that match the background lower the likelihood that an animal will be detected by predators. For example, darker populations of many species of animals inhabit areas with darker rocks than do pale populations. In a species of pocket mouse, this difference is based on a single gene, *Mc1r* (see Figure 6.29).

Both on the land and in the sea, the variety of different environments organisms face is immense. There are major differences in physical conditions among geographic regions and over even short distances, in which a species may encounter different sets of prey, predators, parasites, and competitors. Different parts of the human body are different environments for bacteria, and support very different, diverse bacterial communities. Any of these variables may be relevant to a particular species and impose natural selection on many of its features, so natural selection is the ultimate cause of divergence among populations and species: it is the source of the immense diversity of life.

Darwin, in considering why the various species descended from a common ancestor should become different from one another, drew special attention to the role of competition for limiting resources, such as food. He postulated that

(A)

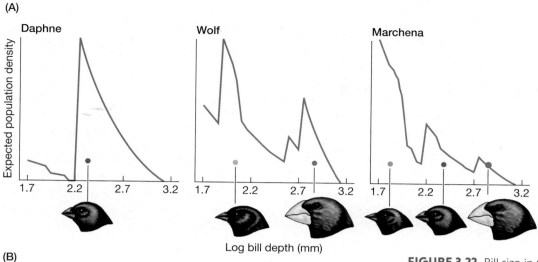

(B)

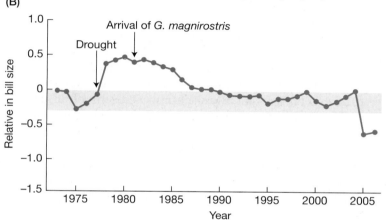

FIGURE 3.22 Bill size in Galápagos ground finches (*Geospiza*) is adapted to feeding on seeds, but competition among species affects what kinds of seeds a species eats. (A) Because of differences in abundance of plant species with different seed sizes, different islands would be expected to differ in the density that various populations of finches would be expected to sustain, as a function of their bill size. For example, Wolf Island has only two abundant kinds of seeds, one small and the other large; the jagged curve shows the theoretical population density of a finch population, depending on its log bill depth (a measure of size). This island has two species of finches, with the predicted small and large bill depths. (B) On the island of Daphne Major, the average bill size of *G. fortis* increased after a 1977 drought that made smaller seeds less abundant than large seeds. Bill size then evolved back to its original level until the population of the large ground finch (*G. magnirostris*) became large enough to deplete the supply of large seeds. (A after [46]; B after [22].)

if different closely related species coexist, those individuals that use different resources from the other species would suffer less competition and have higher fitness. Consequently, the species will diverge from the others. It took nearly a century for biologists to show that Darwin was right. Today we know many cases of what is now called **character displacement**: divergence of species as a consequence of their interaction (see Chapter 13). For example, Peter and Rosemary Grant and their collaborators have studied certain of the ground finches in the Galápagos Islands for more than 35 years (see Figure 2.2) [21, 23]. Among the seed-eating ground finches, those with larger, deeper bills feed more efficiently on larger, harder seeds. Species with different bill depth differ accordingly in diet, and the species that coexist on any island differ, matching the availability of different seeds (**FIGURE 3.22A**). In a population of one species, *Geospiza fortis*, there was high mortality of individuals with smaller bills during a drought, in 1977, that caused a dearth of plants with small seeds. The result was an increase in average bill size (**FIGURE 3.22B**). A few years later, *Geospiza magnirostris*, with the large bill denoted by its name, invaded the island and slowly grew in numbers until, in 2004, it depleted the supply of large seeds. The *G. fortis* population then evolved smaller average bill size, as Darwin would have predicted.

The finch example shows the first stages of the evolution of diversity that is seen in adaptive radiations, such as those described in Chapter 2 (see Figures 2.2, 2.24, 2.25). In each of those cases, the morphological differences are associated with using different resources. The huge diversity seen among higher taxa, such as the immense variety of flowers among plant families and of bills, legs, and wings among the families and orders of birds, may be ascribed partly to the same principle.

FIGURE 3.23 (A) An orchid (*Ophrys scolopax*) that is pollinated by "pseudocopulation." (B) Male bees of certain species are attracted to the flower by its scent (which mimics a female bee's sex pheromone) and color pattern (which imperfectly mimics a bee), and "mate" with it. The male bee shown here carries the yellow pollen mass of a previously visited orchid flower on its forehead.

(A)

(B)

What Not To Expect of Natural Selection

Selection at the level of genes and individual organisms is inherently "selfish": the gene or genotype with the highest rate of increase spreads at the expense of others. The variety of selfish behaviors that organisms inflict on conspecific individuals, ranging from territory defense to parasitism and infanticide, is truly stunning. Indeed, cooperation among organisms requires special explanations, such as kin selection (see Chapter 12). Natural selection—or simply "nature"—has often been invoked to justify codes of human behavior that we might agree are admirable and others that are pernicious. But natural selection is just a name for differences among organisms or genes in reproductive success. Therefore it cannot be described as moral or immoral, just or unjust, kind or cruel, any more than wind, erosion, or entropy can. Hence it cannot be used as a justification or model for human morality or ethics.

Because the principle of kin selection cannot operate across species, "natural selection cannot possibly produce any modification in a species exclusively for the good of another species" (Darwin, *On the Origin of Species*, chapter 6). If a species exhibits behavior that benefits another species, either the behavior is profitable to the individuals performing it, as in bees that obtain food from the flowers they pollinate, or else one species is duped by another, as are insects that copulate with sexually deceptive orchids (**FIGURE 3.23**).

The equilibrium we may observe in ecological communities—the so-called balance of nature—likewise does not reflect any striving for harmony [52]. We observe coexistence of predators and prey not because predators restrain themselves, but because prey species are well enough defended to persist, or because the abundance of predators is limited by some factor other than food supply. Nitrogen and mineral nutrients are rapidly and "efficiently" recycled within tropical rainforests not because ecosystems are selected for or strive for efficiency, but because under competition for sparse nutrients, microorganisms have evolved to decompose litter rapidly, while plants have evolved to capture the nutrients released by decomposition. Selection of individual organisms for their ability to capture nutrients has the *effect*, in aggregate, of producing a dynamic that we measure as ecosystem "efficiency."

Go to the
Evolution Companion Website
EVOLUTION4E.SINAUER.COM
for data analysis and simulation exercises, quizzes, and more.

SUMMARY

- A feature is an adaptation for a particular function if it has evolved by natural selection for that function by enhancing the relative rate of increase—the fitness—of biological entities with that feature.

- Because many characteristics are genetically variable in natural populations, they may evolve rapidly if selection pressures change. Especially because humans drastically alter environments and move species into new environments, many historical examples of rapid adaptive evolution have been documented, often within much less than a century.

- Natural selection is a consistent difference in fitness among phenotypically different biological entities. It is the antithesis of chance. Natural selection may occur at different levels, such as genes, individual organisms, populations, and species.

- Selection at the level of genes or organisms is likely to be the most important because the numbers and turnover rates of these entities are greater than those of populations or species. Therefore most features are unlikely to have evolved by group selection, the one form of selection that could in theory promote the evolution of features that benefit the species even though they are disadvantageous to the individual organism. Both genic and individual selection can be viewed as fitness differences among genes, with "selfish genes" being those that prevail.

- Species selection is a correlation between a trait and the rate of speciation or extinction. It can result in variation among clades in diversity of species.

- Not all features are adaptations. Methods for identifying and elucidating adaptations include studies of function and design, experimental studies of the correspondence between fitness and variation within species, and correlations between the traits of species and environmental or other features (the comparative method). Phylogenetic information may be necessary for proper use of the comparative method.

- Organisms may not have perfect adaptations because of functional compromises or trade-offs, or because mutations enabling perfect adaptation have not been available.

- As a consequence of adaptation of species to different environments and ways of life, natural selection is the basis of adaptive radiations and adaptive diversity. Competition for resources is one of many factors that can select for differences among species.

- Natural selection need not promote harmony or balance in nature, and utterly lacking any moral content, it provides no foundation for morality or ethics in human behavior.

TERMS AND CONCEPTS
---

adaptation
altruistic trait
character
    displacement
comparative
    method
exaptation
fitness

frequency
function (vs. effect)
genic selection
group selection
    (= interdemic
    selection)
hypothetico-
    deductive method
individual selection

kin selection
levels of selection
meiotic drive
natural selection
neutral allele
preadaptation
reproductive
    success

segregation
    distortion
selfish genetic
    element
sexual selection
species selection
trade-off

SUGGESTIONS FOR FURTHER READING

Adaptation and Natural Selection by G. C. Williams (Princeton University Press, Princeton, NJ, 1966) is a classic, and still worth reading for its clear, insightful analysis of individual and group selection.

The Selfish Gene (Oxford University Press, Oxford, 1989) and *The Blind Watchmaker* (W. W. Norton, New York, 1986), both by R. Dawkins, explore the nature of natural selection in depth, as well as treating many other topics in a vivid style for general audiences.

The Evolution Explosion: How Humans Cause Rapid Evolutionary Change by S. R. Palumbi (W. W. Norton, New York, 2001) is an informa-

tion-packed treatment of this important topic, written for a general audience.

Levels of selection and related topics are treated at an advanced level in two books by philosophers of science: *The Nature of Selection: Evolutionary Theory in Philosophical Focus* by E. Sober (MIT Press, Cambridge, MA, 1984), and *Evolution and the Levels of Selection* by S. Okasha (Oxford University Press, Oxford, 2006). Biologists treat this topic in *Levels of Selection in Evolution*, edited by L. Keller (Princeton University Press, Princeton, NJ, 1992).

PROBLEMS AND DISCUSSION TOPICS

1. Discuss criteria or measurements by which you might conclude that a population is better adapted after a certain evolutionary change than before.

2. Consider the first copy of an allele for insecticide resistance that arises by mutation in a population of insects exposed to an insecticide. Is this mutation an adaptation? If, after some generations, we find that most of the population is resistant, is the resistance an adaptation? If we discover genetic variation for insecticide resistance in a population that has had no experience of insecticides, is the variation an adaptation? If an insect population is polymorphic for two alleles, each of which confers resistance against one of two pesticides that are alternately applied, is the variation an adaptation? Or is each of the two resistance traits an adaptation?

3. It is often proposed that a feature that is advantageous to individual organisms is the reason for the great number of species in certain clades. For example, wings have been postulated to be a cause of the great diversity of winged insects compared with the few species of primitively wingless insects. How could an individually advantageous feature cause greater species diversity? How can one test a hypothesis that a certain feature has caused the great diversity of certain groups of organisms?

4. Provide an adaptive and a nonadaptive hypothesis for the evolutionary loss of useless organs, such as eyes in many cave-dwelling animals. How might these hypotheses be tested?

5. Could natural selection, at any level of organization, ever cause the extinction of a population or species?

6. If natural selection has no foresight, how can it explain features that seem to prepare organisms for future events? For example, deciduous trees at high latitudes drop their leaves before winter arrives, male birds establish territories before females arrive in the spring, and animals such as squirrels and jays store food as winter approaches.

7. An exaptation is a pre-existing trait used for a new, seemingly adaptive function. The term was coined by Stephen Jay Gould and Elisabeth Vrba, to improve clarity of language when discussing the co-opting of a trait for a new function—and to distinguish this from "preadaptation," as used by George Gaylord Simpson, referring to a structure that undergoes a change of function followed by tinkering by natural selection. Both terms are used by biologists, with subtly different meanings. Find some examples of pre-existing traits being used by organisms for a new function and discuss whether exaptation or preadaptation would be an appropriate label. Many criticisms exist for both terms. Find some examples of these criticisms and discuss whether they apply to your examples.

UNIT II
How Evolution Works

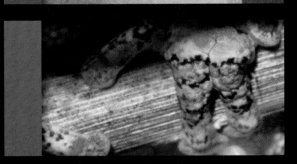

Mutation and Variation

The people around you differ in the color of their hair, the shape of their noses, the length of their fingers, and in countless other ways. If you knew their DNA sequences, you would also see that they differ from you at millions of places in their genomes. Variation among individuals is universal across all species on Earth (**FIGURE 4.1**).

Unlike physics and chemistry, evolution depends on variation. All electrons are identical, but no two living organisms are. Without that variation, evolution—and so life itself—would not be possible.

Understanding how traits and genes vary and how this variation is inherited is fundamental to understanding evolution. Genetics also provides us with a vast trove of information about the history of life on Earth and about the evolutionary factors acting on living species. Before delving into those topics, this chapter starts with a short review of key concepts in genetics that should be familiar to you from earlier courses in biology.

The Machinery of Inheritance

The genetic material of almost all organisms on Earth is **DNA (deoxyribonucleic acid)**. This is a very long molecule made up of pairs of bases. Each base takes one of four forms: adenine (A), guanine (G), cytosine (C), or thymine (T). In each pair of bases, an A is matched with a T, or a C is matched with a G. The average chromosome in humans has more than 100 million **base pairs** (abbreviated **bp**), and the entire human genome consists of 3.2 billion bp. Our genome

The zigzag nerite (*Neritina communis*), a marine snail that lives among mangroves in the Western Pacific, has extremely variable shells.

FIGURE 4.1 The strawberry poison dart frog (*Dendrobates pumilio*) has conspicuous coloration that warns predators it is toxic. Why this species is so variable, however, is not understood. Central questions in evolutionary biology include what maintains variation, and how variation is shaped by selection and other evolutionary factors. (Frog top views from [25b].)

is neither exceptionally big or small. Some bacteria have genomes that are thousands of times smaller, with less than 200,000 bp. At the other extreme are some plants, salamanders, and protozoa that have genomes that are hundreds of times larger than ours. A surprising observation is that there is little correlation between the complexity of an organism and the size of its genome. We'll return to this puzzling fact in Chapter 14.

Some viruses use **RNA** (**ribonucleic acid**) rather than DNA for their genetic material. To replicate themselves, these viruses convert their genome into DNA by a process called reverse transcription. This DNA is then inserted into the genome of the host cell that the virus has infected, and offspring viruses are made using that cell's biochemical machinery. Thus, despite the difference in the genetic material of RNA- and DNA-based life forms, they share much of the apparatus that expresses their genes.

An organism's genetic material is carried by one or more **chromosomes**. Chromosomes in eukaryotes are long strings of DNA bases bound together with proteins. In diploid species such as humans, chromosomes come in pairs, one inherited from each parent. In prokaryotes, chromosomes are unpaired (haploid). **Genes** are segments of chromosomes that perform a function. Many code for proteins that comprise tissues and catalyze reactions. A smaller number have other functions, for example coding for the RNA of ribosomes and the microRNAs that are important to gene regulation. The human genome has roughly 20,000 protein-coding genes, some plants and fish have many more, and some bacteria have hundreds of times fewer.

To make a protein, the cellular machinery reads a gene's DNA in sets of three bases. These sets, called **codons**, represent the amino acids that make up the protein. The **genetic code** is a set of rules that relates the codons to the amino acids they represent (**FIGURE 4.2**). A profound and wonderful fact is that the genetic code is shared by virtually all life on Earth, from viruses to bacteria to pineapples to humans. This is powerful evidence that all life evolved from a single common ancestor.

Since there are four types of DNA bases, there are 4 × 4 × 4 = 64 different codons. But because there are only 20 types of amino acids, most amino acids are represented by more than one codon. For example, the codons CCT, CCC, CCA, and CCG all specify the amino acid proline. Changes to a codon that do not alter an amino acid, for example from GAG to GAA, are called **synonymous** (or "silent"). In contrast, changes to a codon that do alter an amino acid, such as GAG to GTG, are called **nonsynonymous** (or "replacement"). The contrast between these two types of changes is illustrated with the β-hemoglobin gene from humans in **FIGURE 4.3**. We'll see shortly that the nonsynonymous change in β-hemoglobin leads to interesting evolutionary consequences.

Second letter

		T		C		A		G	
T	TTT TTC	Phenyl- alanine	TCT TCC TCA TCG	Serine	TAT TAC	Tyrosine	TGT TGC	Cysteine	
	TTA TTG	Leucine			TAA TAG	Stop codon Stop codon	TGA	Stop codon	
							TGG	Tryptophan	
C	CTT CTC CTA CTG	Leucine	CCT CCC CCA CCG	Proline	CAT CAC	Histidine	CGT CGC CGA CGG	Arginine	
					CAA CAG	Glutamine			
A	ATT ATC ATA	Isoleucine	ACT ACC ACA ACG	Threonine	AAT AAC	Asparagine	AGT AGC	Serine	
	ATG	Methionine; start codon			AAA AAG	Lysine	AGA AGG	Arginine	
G	GTT GTC GTA GTG	Valine	GCT GCC GCA GCG	Alanine	GAT GAC	Aspartic acid	GGT GGC GGA GGG	Glycine	
					GAA GAG	Glutamic acid			

First letter (rows: T, C, A, G)

FIGURE 4.2 The universal genetic code relates the three DNA bases in a codon to the amino acid in the protein made by the gene. All organisms on Earth use this code or a minor variant of it.

Proteins are synthesized in three steps (**FIGURE 4.4**). A cell first *transcribes* the DNA from a gene into RNA. This immature form of the gene's message is called pre-mRNA. That molecule is then *spliced* so that parts of the molecule are removed to form a mature mRNA. Finally, the message in the mRNA is *translated* into the string of amino acids that make up the protein. As a result of the splicing step, a gene has segments of DNA that do not code for any amino acids in the final protein. The segments of the gene that do code for amino acids are called **exons**, while the noncoding segments between the exons are called **introns**. Many eukaryotic genes are spliced in more than one way, removing some of the exons (as well as the introns) from the pre-mRNA. This alternative splicing allows a single gene to code for more than one protein. Introns and exons evolve quite differently, which gives us important cues about how selection and other factors cause genes and genomes to evolve (see Chapters 5 and 14). In humans, a typical gene has 7 introns and 1400 bp in its exons, and on average the introns are 17 times larger than the exons.

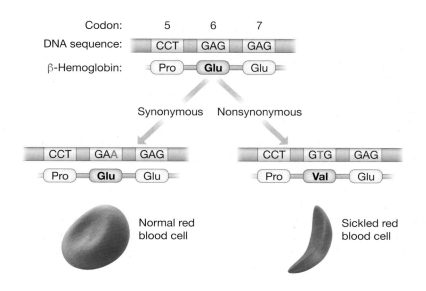

FIGURE 4.3 Synonymous changes to a DNA sequence do not alter the amino acids in a protein, but nonsynonymous changes do. At top is the DNA sequence for three codons of the *A* allele of the β-hemoglobin gene in humans. Beneath each codon is an abbreviation for the amino acid it codes for. The sixth codon, GAG, codes for glutamic acid (Glu). A change in that codon from GAG to GAA does not alter the amino acid, so this is a synonymous change. A change from GAG to GTG, however, replaces the glutamic acid with valine (Val), so this is a nonsynonymous change. The change to the GTG codon produces the *S* allele, which results in the sickle-cell condition.

FIGURE 4.4 In the first step of assembling a protein, a gene's DNA sequence is transcribed into pre-mRNA. This molecule is spliced to remove the introns (and often some of the exons) to produce mature mRNA. Many pre-mRNAs are spliced in more than one way, yielding different mRNAs. The mRNA is then translated into the protein.

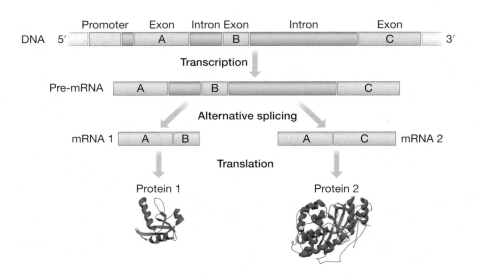

Like the genetic code, the cellular machinery for transcribing DNA and translating mRNA is almost universal across life on Earth. The DNA or mRNA from a sea urchin is translated into a protein if it is injected into a bacterium. This universality is the basis of genetic engineering. The "golden rice" strain was developed by inserting a bacterial gene that allows the plant to synthesize a precursor to vitamin A [26]. A deficiency of that vitamin kills 2 million people each year, many of whom live in countries where rice is a major part of the diet. Golden rice has the potential to relieve much of that suffering. This advance is only possible because rice and bacteria share genetic machinery that they both inherited from a common ancestor that lived more than 1.5 billion years ago (see Chapter 2).

Genes make up only a small part of the genome in eukaryotes. In humans, for example, 98 percent of the DNA does not code for any gene product. A small fraction of this **noncoding DNA** affects how coding genes are expressed. The vast bulk of noncoding DNA, however, does not have an obvious function. The genomes of prokaryotes are very different, and typically only about 20 percent of their genome is noncoding. Chapter 14 will return to the fascinating observation that so much of the eukaryotic genome is noncoding DNA.

The Inheritance of Variation

The variation we see among individuals of a species (such as our own) are differences in **phenotypes**, or observable characteristics. Natural selection acts on phenotypes, but that process only results in evolution if at least some of the variation in phenotypes is transmitted between generations. We resemble our parents more than we do passing strangers on the street. The familiar patterns of inheritance are the result of differences in the **genotypes**, encoded by DNA. To understand how evolution works, it is crucial to understand how this remarkable mechanism of inheritance works.

The basic unit of genetic inheritance is a **locus** (plural: *loci*), which is the more formal term for what we sometimes call a gene. A locus is a section of chromosome, often one that produces a gene product such as a protein. The DNA sequence at a given locus often varies among the chromosomes carried by different individuals, and if it does we say that the locus is **polymorphic**.

The different variants at a locus are called **alleles**. A specific DNA base in the genome that varies among individuals is called a **single nucleotide polymorphism**, or **SNP** (pronounced "snip"). The **allele frequency** tells us how often a variant occurs at a locus or a DNA base in a population (**FIGURE 4.5**). Some alleles are very rare. For example, one of the alleles that causes albinism (a condition in which melanic skin pigments are missing) occurs at a frequency of about 0.0002 in Europeans. Other alleles occur at much higher frequencies. A person's blood type depends on the alleles that affect the surface of the red blood cells. In the United States, the frequency of the A blood type allele is about 0.4 (that is, about 40 percent of chromosomes carry that allele).

How is this variation transmitted from one generation to the next? Most eukaryotic species on Earth reproduce sexually: it takes both a mother and a father to produce an offspring. Sex mixes the genes of the parents to produce genetic combinations in the offspring not found in the parents. In organisms with meiosis (such as ourselves), this mixing involves two basic genetic processes, segregation and recombination. Organisms without meiosis, such as bacteria and viruses, do not have segregation, but most of them still mix their genes by some form of recombination.

Gene mixing by segregation

Segregation is the selection of one of the two copies of a locus when a gamete is made during meiosis. The fusion of an egg and sperm brings together the copy from the mother with that from the father. A result is that the offspring can have a genotype unlike either of its parents. A mother with genotype A_1A_1 (homozygous for the A_1 allele) and a father with genotype A_2A_2 (homozygous for the A_2 allele) will have entirely heterozygous A_1A_2 offspring. Segregation does not occur in organisms that do not have meiosis, such as bacteria and viruses.

The mixing of genes caused by segregation changes the proportions of genotypes in a population. Think of a population in which half the individuals are A_1A_1 homozygotes and half are A_2A_2 homozygotes. When sperm and eggs are produced by meiosis, half will carry the A_1 allele and half will carry the A_2 allele. If sperm and eggs meet at random, the chance that an A_1A_1 offspring is produced equals the chance that an egg carrying an A_1 (= 1/2) is fertilized by a sperm that also carries an A_1 (= 1/2). Thus the chance that an A_1A_1 offspring is produced is $1/2 \times 1/2 = 1/4$. Likewise, the frequency of A_2A_2 homozygotes in the offspring is 1/4. To find the frequency of A_1A_2 heterozygotes in the offspring, we add up the chance that a sperm with an A_1 fertilizes an egg with an A_2 (= $1/2 \times 1/2 = 1/4$) and the chance that a sperm with an A_2 fertilizes an egg with an A_1 (= $1/2 \times 1/2 = 1/4$). The frequency of A_1A_2 offspring is therefore 1/2.

Looking at these numbers, we see that the frequency of genotypes changed from one generation to the next. Heterozygotes are absent in the parents but make up half of the offspring. What has not changed, however, are the frequencies of the A_1 and A_2 alleles, which are equal to 1/2 in both generations. As a result, when the offspring mate to produce a third generation, the frequencies of the A_1A_1, A_1A_2, and A_2A_2 genotypes will again be 1/4, 1/2, and 1/4. Thus the population is at an equilibrium: once the population reaches that state, no further change in genotype frequencies will happen.

This example is a special case of the **Hardy-Weinberg equilibrium**, which tells us the relative proportions of genotypes in a population when segregation is the only factor that changes genotype frequencies. We just looked at the situation in which the frequencies of the two alleles are equal to 1/2. The more general situation is illustrated in **FIGURE 4.6**. We now let the frequency of allele A_2 be any number between 0 and 1, and we represent that frequency by the symbol p. Since there are only two alleles at the locus in this example, their frequencies must sum to 1, and so the frequency of allele A_1 is $(1 - p)$. Following the logic used in the earlier example, we find that the chance of an A_2A_2 offspring being produced is equal to

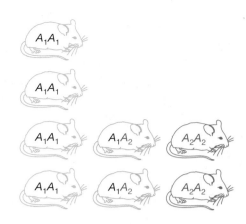

FIGURE 4.5 Allele frequencies and genotype frequencies. In this population, the frequency of the A_1A_1 homozygote genotype is 1/2, the frequency of the A_1A_2 heterozygote genotype is 1/4, and the frequency of the A_2A_2 homozygote genotype is 1/4. Ten of the 16 copies of the gene are the A_1 allele, so its allele frequency is $p = 10/16 = 0.625$.

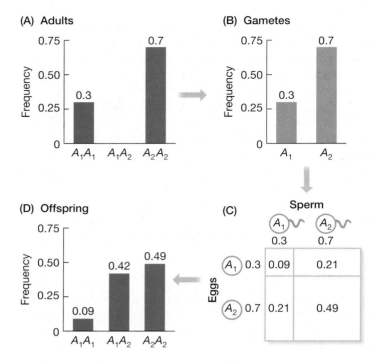

FIGURE 4.6 Hardy-Weinberg equilibrium results from the random union of gametes. (A) In this example, 30% of the adults are A_1A_1 homozygotes and 70% are A_2A_2 homozygotes. The frequency of allele A_2 is $p = 0.7$ and the frequency of allele A_1 is $1 - p = 0.3$. The population is not in Hardy-Weinberg equilibrium: there are too few heterozygotes. (B) The frequencies of the alleles in the sperm and eggs are the same as in the adults. (C) The numbers inside the box are the frequencies of genotypes in the offspring that are formed by random fertilization between sperm and eggs. (D) The genotype frequencies among the offspring are found by adding up the cells inside the box: the frequency of A_1A_1 homozygotes is $(1 - p)^2 = 0.09$, the frequency of A_1A_2 heterozygotes is $2p(1 - p) = 0.42$, and the frequency of A_2A_2 homozygotes is $p^2 = 0.49$. The genotype frequencies are different than they were in the parents (compare panels A and D), and are now in Hardy-Weinberg equilibrium. The allele frequencies, however, have not changed.

the probability that a sperm carries the A_2 allele times the probability that an egg also carries the A_2 allele, or p^2. Likewise, the frequency of A_1A_1 offspring is $(1 - p)^2$, and the frequency of A_1A_2 offspring is $2p(1 - p)$.

Putting those results together gives us the Hardy-Weinberg proportions:

Genotype:	A_1A_1	A_1A_2	A_2A_2
Frequency:	$(1 - p)^2$	$2p(1 - p)$	p^2

An example with $p = 0.7$ is shown in Figure 4.6.

The key conditions for the Hardy-Weinberg equilibrium are:
- An infinite population size
- No natural selection
- No mutation
- No movement between populations
- Random mating

No real population meets all of these conditions. Thus no real population is expected to be exactly in Hardy-Weinberg equilibrium (although the discrepancy may be very small). The Hardy-Weinberg equilibrium nevertheless plays a key role in evolutionary biology. It is the foundation for mathematical models of evolution, and it provides a null model for analyzing data. If a population is not in Hardy-Weinberg equilibrium, then one of the conditions listed above has been violated.

The β-hemoglobin locus in humans has two alleles called *A* and *S*. They differ by a single base in the sixth codon (see Figure 4.3). Here are the frequencies of the genotypes at this locus among 654 adults from Musoma, Tanganyika (Africa), and the frequencies expected if the population was at Hardy-Weinberg equilibrium [1a]:

Genotype:	*AA*	*AS*	*SS*
Number of adults:	400	249	5
Observed frequency:	0.612	0.381	0.008
Hardy-Weinberg expectation:	0.643	0.317	0.039

The observed frequency of heterozygotes is higher and the frequencies of both homozygotes are lower than what the Hardy-Weinberg equilibrium predicts. The discrepancies are small but statistically significant. They result from differences in survival: the *AA* and *SS* genotypes do not survive as well as the *AS* genotype. This is one of the most famous examples of natural selection acting on our own species, and we will discuss it further in the next chapter. It illustrates how the Hardy-Weinberg equilibrium is used to investigate selection and other evolutionary processes.

Gene mixing by recombination

The second type of genetic mixing results from **recombination**. This is the process that combines in a gamete a gene copy at one locus that was inherited from the mother with a gene copy at a second locus that was inherited from the father (**FIGURE 4.7**). In eukaryotes, recombination occurs during meiosis. Loci that are carried on different chromosomes recombine by the independent assortment of those chromosomes. Recombination happens between loci on the same chromosome by crossing over, which joins together a piece of a chromosome inherited from the mother with a piece inherited from the father.

The **recombination rate**, symbolized by *r*, is the probability that recombination occurs between a given pair of loci (**FIGURE 4.8**). If the two loci are on different chromosomes, when an individual makes a gamete there is a chance of 1/2 that one of the chromosomes it carries will be from the mother and the other from the father. Here *r* = 1/2, which is the maximum possible value for the recombination rate. At the other extreme, DNA bases that are adjacent on a chromosome have an extremely low chance of recombining. The smallest possible value for the recombination rate is *r* = 0.

When an allele at one locus is found together in a population more often than expected by chance with an allele at a second locus, we say the loci are in **linkage disequilibrium**[1]. The key effect of recombination is to erode linkage disequilibrium. Recombination moves the population toward a state where there is no statistical association between the alleles at the two loci, a situation called *linkage equilibrium*. This is analogous to the Hardy-Weinberg equilibrium at a single locus, where two alleles at the same locus are uncorrelated with each other. Unlike the

---

[1] The term "linkage disequilibrium" is unfortunate and confusing: genes that are not physically linked (on the same chromosome) can be in linkage disequilibrium, while genes that are physically linked can be in linkage equilibrium.

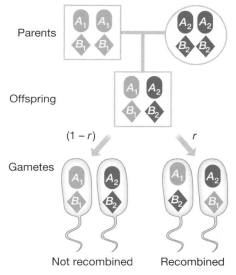

FIGURE 4.7 Recombination randomizes the combinations of alleles at two loci. One locus is shown as an oval and the other as a diamond. These loci may be on the same chromosome or on different chromosomes. Two different alleles are indicated by the different colors. The offspring makes two kinds of gametes (sperm): those that have not recombined the alleles inherited from the parents (left), and those that have (right). The recombination rate, *r*, is the fraction of gametes that have recombined alleles.

Chromosome 1 Chromosome 2

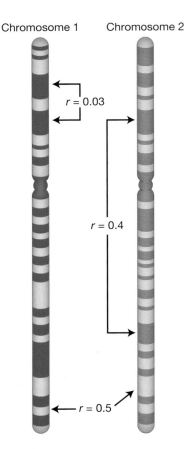

$r = 0.03$

$r = 0.4$

$r = 0.5$

FIGURE 4.8 The recombination rates between three pairs of loci. A pair of loci that are close together on the same chromosome have a low recombination rate (here, $r = 0.03$). A pair that is far apart on the same chromosome has a high recombination rate that approaches 0.5 (here, $r = 0.4$). A pair of loci on different chromosomes has the maximum possible recombination rate, $r = 0.5$.

Hardy-Weinberg equilibrium, linkage equilibrium takes more than one generation to reach. How long it takes depends on the rate of recombination between the loci. Less recombination (smaller r) means the genes at a pair of loci mix more slowly, so linkage equilibrium between them takes longer to reach.

We can be more specific about linkage disequilibrium by introducing a way to measure it. Consider two loci, one with alleles A_1 and A_2, the other with alleles B_1 and B_2. We will use P_{AB} to represent the frequency of gametes carrying both the A_2 and the B_2 alleles, p_A to represent the frequency of gametes with the A_2 allele (no matter which B allele they have), and p_B to represent the frequency of gametes with the B_2 allele (no matter which A allele they have). The measure of linkage disequilibrium is symbolized by D, and is defined as

$$D = P_{AB} - p_A p_B \qquad (4.1)$$

(The term $p_A p_B$ simply means p_A times p_B.) The population is in linkage equilibrium when $D = 0$. If alleles A_2 and B_2 appear together more often than expected by chance, then D is positive. If A_2 and B_2 occur less often than expected, then D is negative.

FIGURE 4.9 compares three populations of gametes. Although the allele frequencies are the same in all three ($p_A = p_B = 1/2$), the populations clearly differ. The differences reflect the effects of linkage disequilibrium. When D is positive, knowing

(A) $D = 1/4$ **(B)** $D = -1/4$ **(C)** $D = 0$

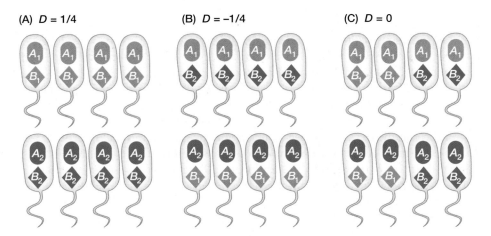

FIGURE 4.9 Three populations of eight gametes that have the same allele frequencies ($p_A = p_B = 1/2$) but have different values of linkage disequilibrium. Linkage disequilibrium is defined the same way regardless of whether the two loci are on the same chromosome or on different chromosomes. (A) When the disequilibrium, D, between alleles A_2 and B_2 is positive, those alleles are found together more often than if they were associated at random. When D is at its maximum possible value ($D = 1/4$), a gamete that carries allele A_2 always carries allele B_2. (B) When disequilibrium is at its smallest possible value ($D = -1/4$), a gamete that carries allele A_2 always carries allele B_1. (C) When a population is at linkage equilibrium ($D = 0$), there is no association between alleles at the two loci. If a sperm carries allele A_2, the chance that it also carries allele B_2 is simply the frequency of B_2 in the population.

that a gamete carries allele A_2 tells us that it is more likely to carry allele B_2 than expected by chance. When D is negative, a gamete that carries allele A_2 is less likely to carry allele B_2. Last, when a population is in linkage equilibrium ($D = 0$), knowing that a gamete carries allele A_2 tells us nothing about which allele it carries at locus B. Linkage disequilibrium is therefore a statistic that measures a property of the population. It has the same meaning whether the two loci are on the same chromosome or on different chromosomes.

The most important role that recombination plays in evolution is through its effects on D. If Mendelian inheritance is the only factor at work, the value of D in the next generation is decreased by a proportion r from its value in the current generation. Thus recombination causes the population to evolve toward linkage equilibrium with $D = 0$. It does so quickly if r is large (near 1/2) and slowly if r is small (near 0). The evolution of D with three different values of r is shown in **FIGURE 4.10**. On average, there is less recombination between pairs of DNA bases on a chromosome when they are closer to each other than when pairs are farther apart. For that reason, D tends to be higher between pairs that are closer (**FIGURE 4.11**). In real populations, linkage equilibrium is never reached exactly because other factors besides recombination are at work (though in many cases D is very close to 0). Nevertheless, linkage equilibrium is a valuable reference just as the Hardy-Weinberg equilibrium is: departures from $D = 0$ can be used to detect those other factors.

Linkage disequilibrium can be produced by natural selection. **Epistasis** is the situation in which the effect of an allele at one locus depends on the allele at a second locus. If some combinations of alleles have high fitness, selection will generate linkage disequilibrium between them. The primrose (*Primula vulgaris*) has an interesting mechanism to avoid fertilizing itself. Populations of this plant have mixtures of individuals with flowers that have the "pin" and the "thrum" phenotypes (**FIGURE 4.12**). In pin plants, the anthers (which produce pollen) are low in the flower, while the stigma (which receives pollen) is high. In thrum plants, this arrangement is reversed. Because the anthers and style are separated in both pin and thrum plants, pollen is rarely transferred between the anthers and stigma of the same flower. The height of the anthers is determined by one locus, and the height of the stigma by another. These loci are in linkage disequilibrium: the allele for low anthers is most often with the allele for a high stigma, as is the allele for high anthers with the allele for a low stigma. Plants with the "wrong" combination of alleles have a stigma and anthers close to each other. They self-fertilize, which produces offspring that survive poorly. Natural selection therefore maintains the linkage disequilibrium.

A second important cause of linkage disequilibrium is the mixing of populations that have different allele frequencies. This situation is conspicuous in countries that have people with ancestries from different geographical regions. For example, the shape of the eyes and the curliness of the hair are determined by different loci. In places that have people of both Asian and African ancestry, seeing the texture of a person's hair tells you what the shape of their eyes is likely to be.

Linkage disequilibrium is important because it affects how genes evolve. In the next chapter we will see that selection on one locus can cause a second

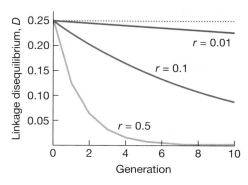

FIGURE 4.10 Recombination causes linkage disequilibrium to decrease. The value of D declines toward 0 rapidly when the recombination rate is large ($r = 0.5$), and slowly when it is small ($r = 0.01$).

FIGURE 4.11 Linkage disequilibrium tends to be higher between pairs of DNA bases that are very close to each other on a chromosome. Shown is a region of 500 thousand base pairs (kb) of a chromosome sampled from 89 humans in eastern Asia. The *x*-axis and *y*-axis are positions along the chromosome. Three sites on the chromosome are labeled. Sites A and B are relatively close (50 kb), and they have high linkage disequilibrium (indicated by red). Sites B and C are farther apart (250 kb), and they have low linkage disequilibrium (indicated by white). Based on [25a].

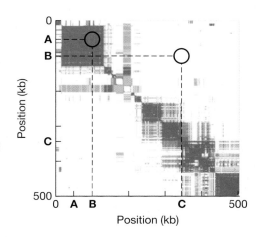

FIGURE 4.12 Variation in flowers of the primrose (*Primula vulgaris*) illustrates how linkage disequilibrium can result from selection. Allele *A*, which produces anthers that are high in the flower, is most often with allele *G*, which produces a stigma that is low. Allele *a*, which produces low anthers, and allele *g*, which produces a high stigma, are together most frequently. Gametes with the *Ag* or the *aG* allele combinations give rise to plants that self-fertilize, producing offspring that survive poorly. This maintains the linkage disequilibrium: the *ag* and *AG* allele combinations are much more common than *Ag* and *aG*. (After [10].)

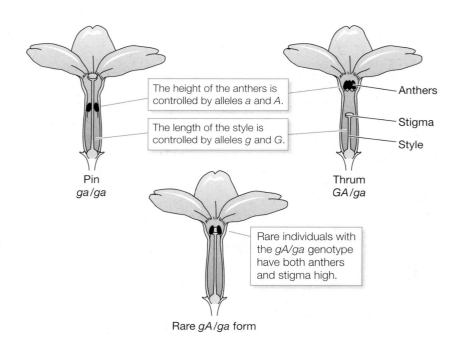

The height of the anthers is controlled by alleles *a* and *A*.

The length of the style is controlled by alleles *g* and *G*.

Anthers
Stigma
Style

Pin
ga/ga

Thrum
GA/ga

Rare individuals with the *gA/ga* genotype have both anthers and stigma high.

Rare *gA/ga* form

locus to evolve if the two loci are in linkage disequilibrium. A second reason that linkage disequilibrium is important is that it allows us to find genes that affect traits of interest. We will see in Chapter 6 how this idea is used to study the genetic basis of traits as diverse as the size of a tomato and adaptation to high elevation in humans.

Gene mixing with asexual inheritance

Almost all eukaryotes reproduce sexually. A small fraction of eukaryotes, all prokaryotes, and all viruses reproduce asexually (without meiosis). But even these organisms have mechanisms that mix their genes. While the mechanisms are very different, their evolutionary consequences are largely the same.

Horizontal gene transfer (**HGT**) is the movement of DNA between different individuals without help from sexual reproduction. It is particularly common in prokaryotes (see Chapter 2), and several mechanisms are involved [5, 19, 20]. HGT can move genes between individuals of the same species and sometimes even between different species. Some bacteria exchange DNA by conjugation, in which two cells exchange DNA while in direct contact. Other bacteria simply take up naked DNA that was left behind by bacteria that died. When a virus replicates inside a bacterium, its offspring viruses sometimes integrate a piece of the host's genome into their own. When those viruses then infect another bacterium (or less commonly, a eukaryote), the virus can insert the gene from the previous host into the DNA of the new host. As we saw in Chapter 2, HGT is important to the evolution of antibiotic resistance in bacteria that are responsible for important human diseases.

Mutation: The Ultimate Source of Variation

The replication of DNA is an exquisitely precise affair, but it is not perfect—errors are made. These **mutations** are the ultimate source of genetic variation in all organisms. Without these errors, there would be no variation, no evolution, and no life. Fundamentally, mutation can be thought of as an inevitable consequence of the Second Law of Thermodynamics, which (among other things) implies that no process can occur without error.

Mutations come in a variety of forms that differ in how much of the genome they affect. We begin with point mutations, which affect only a single DNA base, and end with whole genome duplication, which affects every DNA base in the genome.

Point mutations

The simplest type of mutations are **point mutations**, which occur when a single DNA base is changed from one to another of its four possible states (A, G, C, or T). We have seen that changes to some codons are synonymous (they do not alter an amino acid in a protein) while others are nonsynonymous (they do alter an amino acid). A nonsynonymous mutation in the β-hemoglobin gene results in the *S* allele that appeared earlier in the discussion of the Hardy-Weinberg equilibrium (see Figure 4.3). When homozygous, this single microscopic change to the genome has sweeping effects on development and physiology, causing a medical condition called sickle-cell anemia.

As with the *S* allele in the β-hemoglobin gene, any change to the second position of a codon is nonsynonymous. Most (but not all) changes to the first position are also nonsynonymous. In contrast, most changes to the third base of a codon are synonymous. (You can verify this from Figure 4.2.) We will see in Chapter 14 that as a result, the three positions of codons evolve quite differently from one another.

Much of the eukaryotic genome and some of the prokaryotic genome does not code for any gene product. Point mutations in some noncoding DNA can, however, affect an organism by altering how genes are expressed (see Chapter 15). When a species evolves by natural selection, the genetic changes can involve coding DNA, noncoding DNA, or both.

Structural mutations

Some kinds of mutations affect more than one DNA base. These are **structural mutations**, which can be as small as a few bases or as large as billions of bases. Most happen as errors when chromosomes are replicated. Different kinds of structural mutations are illustrated in **FIGURE 4.13**.

Deletions are a common type of structural mutation that occurs when a segment of a chromosome is left out during replication. A deletion of only three base pairs in a sodium channel gene causes cystic fibrosis, which is one of the most common human genetic disorders in Europe and North America. Other deletions are much larger. The first genome-wide survey of deletions in humans made the surprising discovery that individuals typically have about a dozen deletions that average 465,000 bp in size [21]. While some of those deletions are in noncoding regions, many of them eliminate several genes. Most (but not all) deletions that knock out genes are harmful. **Insertions** are the opposite situation, in which a segment of DNA is added to a chromosome, either from nearby on the same chromosome or elsewhere in the genome. Some insertions cause genetic disease. Huntington's disease is a neurological disorder whose most famous victim was the American folk singer Woody Guthrie. The disease is caused by multiple insertions

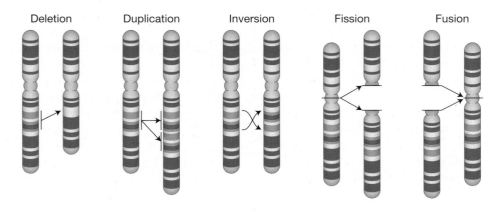

Deletion   Duplication   Inversion   Fission   Fusion

FIGURE 4.13 Five types of structural mutations that alter chromosomes.

(A) *n* = 1

(B) *n* = 23

(C) *n* = 630

FIGURE 4.14 Variation in haploid chromosome numbers. Clockwise from top left: a jack jumper ant (*Myrmecia pilosula*), a human, a fern (*Ophioglossum reticulatum*), and a ciliate (*Oxytricha trifallax*).

(D) *n* = 16,000

of three DNA bases (CAG) into the *huntingtin* gene. Other insertions have played important roles in adaptation.

A **duplication** is a mutation in which a second copy of a gene is inserted into the genome. This process can be repeated, giving rise to a **gene family** with several copies of the original locus. In some cases, the DNA sequences of the duplicate loci diverge and lead to the evolution of new biological functions (see Chapter 14). Less often, duplicates retain the same sequence. This outcome can result from gene conversion, which is an unusual type of mutation in which the DNA sequence of one duplicate in a gene family is replaced by the sequence of another.

Inversions are structural mutations that occur when a chromosome breaks in two places and the middle segment is reinserted in the reverse orientation. Inversions played an important role in the development of evolutionary genetics in the first half of the twentieth century when it was discovered that inversions in fruit flies (*Drosophila*) can be seen under the light microscope. The geneticist Theodosius Dobzhansky, who was one of the leaders of the modern synthesis (see Chapter 1), pioneered the use of inversions to study the evolution of genetic differences within and between species [6]. With the advent of powerful DNA sequencing technologies in the twenty-first century, it has become clear that inversions are a common feature in the evolution of many species. The genomes of humans and chimpanzees differ in about 1,500 chromosome inversions that became fixed in one lineage or the other since our last common ancestor roughly 7 million years ago (Mya) [9]. A **reciprocal translocation** is the exchange of chromosome segments between two nonhomologous chromosomes. Translocation heterozygotes can have reduced fertility, which contributes to the genetic isolation between some closely related species (see Chapter 9).

Fusions are structural changes in which two nonhomologous chromosomes are joined. **Fissions** are the opposite type of mutation, in which one chromosome breaks into two. Fissions and fusions are responsible for changes in the number of chromosomes in the genome. Organisms show a bewildering range of chromosome numbers (**FIGURE 4.14**). A highly venomous ant from Australia called the jack jumper has only a single chromosome [4]. At the opposite extreme, a fern (*Ophioglossum reticulatum*) has 630 pairs of chromosomes [12], while a ciliate (*Oxytricha trifallax*) has about 16,000 pairs of very small chromosomes [24]! This variation results from different histories of fissions and fusions. We still understand little about how and why these differences evolved. Surprisingly, changes in chromosome number and structure often have no obvious phenotypic effects.

The final and most extreme type of mutation is **whole genome duplication**. Occasionally meiosis produces a gamete that carries the entire diploid genome, rather than a haploid with just one of each pair of chromosomes. If two of these unreduced gametes meet and fertilize each other, an offspring is produced that has four copies of each chromosome. This genetic result, which is called **tetraploidy**, happens much more frequently in plants than animals. Later rounds of genome duplication can lead to even more complicated complements of chromosomes. One of the interesting consequences is that the offspring typically cannot interbreed with their parental population. Thus whole genome duplication can produce a new species with a single mutation, as we'll discuss further in Chapter 9.

Rates and Effects of Mutations

Since there are so many kinds of mutations, it is not surprising that their rates and effects vary tremendously [7, 22].

Mutation rates

DNA replication is extremely accurate in eukaryotes and bacteria. Each time an *Escherichia coli* cell divides, there is roughly a chance of only 1 out of 2×10^{10} that a given DNA base in a daughter cell will carry a new point mutation [15]. (Is there any human action that is so precise?) The probability that an offspring carries a new mutation is called the **mutation rate**, which is symbolized by μ. The mutation rate at a single base in *E. coli* is therefore $\mu = 1 / (2 \times 10^{10})$.

Mutation rates vary greatly among species (**FIGURE 4.15**). Each time a person makes an egg or sperm, roughly 1 out of 10^8 of the DNA bases carries a new mutation (and so $\mu = 10^{-8}$ per bp) [11]. The mutation rates in RNA viruses, like those responsible for AIDS, Ebola, and influenza, are thousands of times higher ($\mu = 10^{-3}$ to 10^{-5} per bp). These extremely high mutation rates may result in part from natural selection that favors rapid evolution of viruses to evade host defenses [22]. **BOX 4A** describes how mutation rates are estimated.

The concept of a mutation rate applies not just to a single DNA base but also to an entire gene. Per-locus mutation rates are higher than they are for single bases, simply because the locus carries a mutation if any of its many base pairs mutates. These rates vary greatly, both between loci within a species and between species. Mutation rates for protein-coding loci in eukaryotes such as humans and flies are typically in the range of $\mu = 10^{-5}$ to 10^{-7}. We can also consider the mutation rate across the entire genome. Although the human mutation rate per DNA base is very small ($\mu = 10^{-8}$), our genomes have a large number of bases (3×10^9). As a result, every time we make a gamete, it carries roughly 30 new mutations scattered throughout the genome [11].

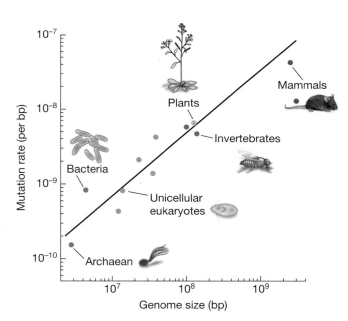

FIGURE 4.15 Mutation rates vary greatly. Organisms with larger genomes tend to have higher mutation rates per base pair per generation. (That pattern does not hold for viruses, however.) The points for Archaea and Bacteria represent averages of many species, and there is considerable variation within those groups. An intriguing hypothesis to explain the correlation between genome size and mutation rate seen here has been proposed by Michael Lynch [16].

Estimating Mutation Rates

Mutation is fundamental to genetics and evolution, and generations of scientists have devoted their careers to its study. Several strategies have been invented to estimate mutation rates [11, 13]. The first approach used was *phenotype screening*. Here the investigator examines ("screens") a large number of individuals in a laboratory population, looking for changes in a character caused by a new mutation (for example, changes in the eye color of the fruit fly *Drosophila melanogaster*). The mutation rate, μ, is simply estimated by counting the number of new mutations found by the total number of individuals. There are several drawbacks of this method; among them, we cannot estimate mutation rates at individual DNA bases, and only some kinds of organisms can be used.

Another strategy became possible with the advent of DNA sequencing. The *phylogenetic method* exploits the fact that if a segment of chromosome has no effect on fitness, it will accumulate mutations at a rate of μ per generation (as we will see in Chapter 7). Two species that last shared a common ancestor t generations ago are therefore expected to differ in that segment by $2\mu t$ mutations. (The factor of 2 appears because mutations have accumulated in each of the two species since their common ancestor.) By sequencing a neutrally evolving piece of chromosome in two species, we can count the number of mutations by which they differ. If we know when their last common ancestor lived (for example, from fossils) and the number of generations per year, we can then calculate μ. Weaknesses of this approach are that it can be applied only to neutrally evolving regions of the genome, and requires that we know the number of generations since the common ancestor.

A related method is called *mutation accumulation*. Here several laboratory populations are established from a single founding population (or individual, in the case of asexual species). Each population is maintained under conditions that largely eliminate natural selection so that mutation is the only evolutionary process at work. After many generations, individuals from each population are sequenced, and mutation rate is estimated by the same calculation used in the phylogenetic method. An advantage of this approach is that we can be much more confident about the number of generations since the common ancestor and the assumption of neutral evolution. Disadvantages are that there are fewer mutations to study (because they accumulate over a much shorter time) and that only some species are suitable.

The conceptually simplest method for estimating mutation rates became possible recently as DNA sequencing became relatively inexpensive. In the *direct method*, we sequence the DNA of parents and offspring and look for differences caused by mutations. This approach is free of the assumptions that are important weaknesses of the other two methods. But it too has limitations. The chance of a mutation in any specific gene is extremely small, and so at present the direct method can only give estimates for mutation rates that are averages over large regions of the genome.

Effects of mutations

Mutations affect virtually all aspects of an organism, ranging from the ability of a bacterium to metabolize a new substrate to the ability of a person to learn language. Despite the huge range of effects that mutations have, they show two general features.

The first is **pleiotropy**, which occurs when a single mutation affects multiple traits. An extreme example is a type of dwarfism called achondroplasia (**FIGURE 4.16**), which results from a mutation in a single gene that interferes with the conversion of cartilage to bone during development. This decreases the size of many bones in the body, particularly in the arms and legs. It also decreases longevity and affects many physiological traits.

Virtually all mutations that have phenotypic effects show pleiotropy. Often the effects are on seemingly unrelated traits. Pleiotropy therefore plays a key role in evolution: genetic changes that alter one aspect of an organism invariably have side effects on other aspects (see Chapters 5 and 6).

FIGURE 4.16 The actor Peter Dinklage (shown here in *Game of Thrones*) has achondroplasia. This condition is caused by a mutation in the gene for receptor protein FGFR3, which interferes with bone formation during development. All physical dimensions of the body are affected, particularly the long bones of the arms and legs. This mutation is a dramatic example of pleiotropy.

A second general feature of mutations involves their effects on an organism's **fitness**, that is, the number of offspring it leaves to the next generation. While many mutations have no detectable effect on survival or reproduction, most of those that do are **deleterious** (that is, harmful to survival or reproduction). On average, each human gamete carries one new deleterious mutation in addition to many that were inherited from the previous generation [11]. (We will see in Chapter 5 that this has important implications for health.) Much less often, mutations are **beneficial**, meaning that they increase fitness (**FIGURE 4.17**). It is easy to understand why this is so by thinking about an enzyme that catalyzes a biochemical reaction. Enzymes are remarkable molecules that are able to bind to a very specific substrate and trigger a precise biochemical reaction. Most changes to an enzyme's amino acid sequence alter its chemical properties, destroying the enzyme's ability to perform its function. In humans, mutations that change a protein typically have deleterious effects that decrease fitness by less than 1 percent, but occasionally their effects are more severe [8]. Most deleterious mutations are partly recessive, meaning that their negative effects are more than twice as harmful when they are homozygous as when heterozygous.

Because most mutations are deleterious, natural selection favors lower mutation rates (at least in organisms with sexual reproduction, like almost all eukaryotes). Several factors apparently prevent mutation rates from evolving to even lower levels than they already are [17]. So it is a strange and wonderful fact that life itself, which is the glorious product of natural selection, only exists because natural

(A)

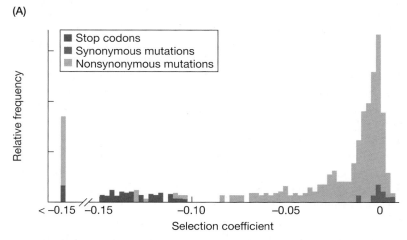

(B)

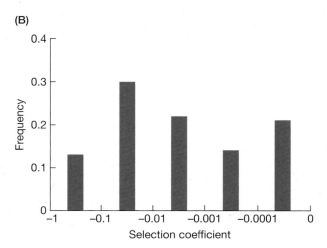

FIGURE 4.17 The effects of new mutations on survival and reproduction. (A) The effects of 560 mutations in yeast (*Saccharomyces cerevisiae*). A "selection coefficient" of −0.1, for example, means that the mutation decreases survival by 10%. These mutations all lie within a small segment of the gene for the heat shock protein Hsp90. The large majority of mutations have negative effects, indicating they are deleterious. Mutations that generate a stop codon destroy the protein's function and are highly delete-rious. Synonymous mutations have almost no effect on fitness. Very few mutations are beneficial (with positive selection coefficients). (B) Selection coefficients for nonsynonymous mutations in humans. Again, mutations that are more deleterious are farther to the left. Roughly 12 percent have large deleterious effects that decrease fitness by 10 percent (that is, a selection coefficient of −0.1) or more. The method used in this study cannot detect beneficial mutations. (A after [1b]; B after [7].)

FIGURE 4.18 The homeotic mutation *Antennapedia* in flies. (A) Head of a normal *Drosophila melanogaster*. (B) Head of a fly with the mutation, which converts the antennae into legs.

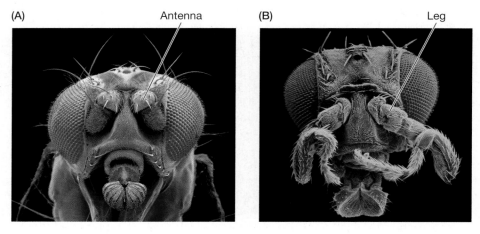

(A) Antenna   (B) Leg

selection is not efficient enough to drive mutation rates to zero. Were that to happen, life on Earth would be deprived of the fuel needed for adaptation, leading inevitably to its extinction.

Fortunately, a small fraction of mutations are beneficial. These are the mutations that spread by natural selection, and allow organisms to adapt. These changes are therefore much more common as differences among species than they are among new mutations that appear within a species. We will discuss many examples in the next chapter.

A particularly dramatic type of mutation are *homeotic mutations* that transform one body part into another (**FIGURE 4.18**). The effects of these mutations result from the disruption of complex developmental pathways that evolved through a large number of much smaller evolutionary changes (see Chapter 15). Mutations do occasionally produce large beneficial effects, but there is no evidence that complex structures (like the leg shown in Figure 4.18) originate by single mutations. Goldschmidt's idea of "hopeful monsters" that we discussed in Chapter 1 was an evolutionary hypothesis that has been tested and proved wrong.

Germ line mutations and somatic mutations

Skin cancer is caused by mutations that result from exposure to sunlight. Those mutations, however, cannot be transmitted to the next generation. Early in development, many groups of animals set aside a small group of cells to form the **germ line**, which then produces the gametes when the individual is sexually mature. The rest of the cells in the early embryo go on to form the **soma**, consisting of all the other tissues in the organism. These somatic cells leave no descendants to the next generation. This explains why somatic mutations, like those responsible for skin cancer, are not transmitted to the next generation.

Somatic mutations still have important evolutionary consequences in animals that have a germ line. Cancer decreases an organism's chance of surviving and reproducing. Natural selection can favor mutations in the germ line that improve DNA repair because they decrease the chance that mutations will occur in somatic cells and produce a cancer.

Plants and some animals (such as sponges and corals) do not have a germ line. Instead, gametes are made from somatic cells that divided many times as an individual was growing. In these species, somatic mutations can be passed to a gamete and then on to the next generation.

Is Mutation Random?

The word "random" often appears with the word "mutation." But are mutations in fact random, and if so in what sense? This is an important question in evolution. But it is also a tricky one, because the word "random" has different meanings.

There is substantial variation between the mutation rates of different regions of the genome, and between different DNA bases. Transition mutations are those between A and G, and between C and T. Transversion mutations are all the other kinds (for example, between A and C). Even though there are twice as many possible transversion mutations, transition mutations are more common. In short, not all mutations are equally likely. In that sense, mutation is not random.

Mutation is random, however, in a different and fundamental sense. Mutations are random with respect to what will improve survival and reproduction. New conditions do not increase the frequency of mutations that are beneficial in those conditions. Geneticists have confirmed this fact experimentally, and it is important for the following reason. We could imagine a world in which organisms respond to increasing temperatures (for example) by increasing the frequency of mutations that make them more heat tolerant. This kind of "directed mutation" would allow organisms to adapt more rapidly to the changing conditions.

An elegant test of the directed mutation hypothesis is the famed *replica plate experiment* done by Joshua and Esther Lederberg [14]. The bacterium *Escherichia coli* is attacked by a virus called T1, but some cells carry a mutation that allows them to resist the attack. The Lederbergs began by inoculating a petri plate with a culture of *E. coli* that they started from a single cell that was not resistant to the virus. After incubation, this "master plate" had a few hundred colonies on its surface, each of which grew from a single cell in the original culture. A disk covered in velvet was gently pressed to the surface of the plate, then gently pressed to the surface of several other "replica plates" that had been sterilized. The velvet picked up a few cells from each colony on the master plate, then transferred them to the replica plates. The brilliance of this experimental design is that the spatial positions of the colonies was the same on the master and the replica plates.

The next step of the experiment was to treat all of the plates with a culture of the T1 virus. As expected, the virus killed almost all of the colonies. A small number of colonies, however, survived. The Lederbergs showed that those colonies carried a mutation for phage resistance: after they were transferred to new plates and attacked again with T1 virus, those colonies still survived.

What can this experiment tell us about the randomness of mutation? Consider the two hypotheses shown in **FIGURE 4.19**. Under hypothesis 1, the addition of the virus induces resistance mutations in the bacteria. In that case, some colonies should survive and some should not, depending on whether they are able to make the adaptive mutation before they are killed. But crucially, there is no reason that the particular colonies that survive on the replica plates should be the same. We predict that the positions of the surviving colonies will vary between the plates.

Under hypothesis 2, resistance mutations occur randomly, whether the virus is present or not. In this case, some of the colonies on the master plate will have the mutation, even though they have never been in contact with the virus. Those colonies will appear in the same locations on the replica plates. When the virus is added to the replica plates, we predict that the colonies that survive will be in the same locations.

The results of the Lederbergs' experiment were clear: the same colonies survived the viral attacks. This shows clearly that the mutation for resistance to the virus occurred in colonies on the master plate, before the mutation was favorable to the bacteria. This experiment, and many more like it, give strong support to the hypothesis that mutation is random with respect to what the environment favors.

A final point is that the environment can (and does) have important effects on mutation rates. Exposure to radiation can increase a person's mutation rate. But the mutations that result will not be particularly good at improving our chance of surviving the effects of radiation. A species becomes better at surviving in its

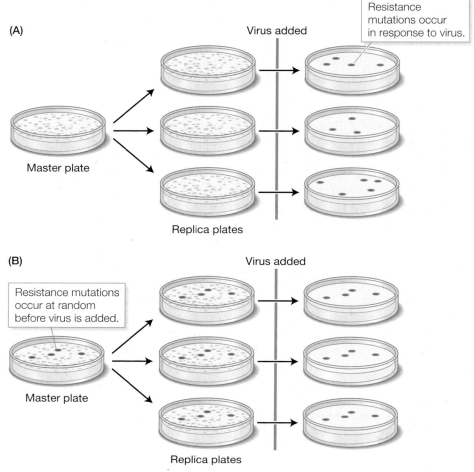

FIGURE 4.19 The replica plate experiment shows that mutations in *E. coli* for resistance to T1 viruses occur before exposure to the virus, rather than being induced by that exposure [14]. The experiment begins with a master agar plate that has numerous colonies of *E. coli* that were all derived from a single cell. Replica agar plates are made by first pressing a velvet-covered disk against the master plate, then against sterilized plates. This transfers some cells from each colony to each replica plate in the same spatial configuration as they are on the master plate. A culture of T1 viruses is then added to the replica plates. (A) Under the hypothesis of directed mutation, exposure to the virus induces the bacteria to generate resistance mutations. This hypothesis predicts that those colonies that are successful in generating these beneficial mutations will appear at different places on the replica plates. (B) The alternative hypothesis is that resistance mutations occur spontaneously before the virus is added. In this case, when the virus is added, the colonies that already have resistance mutations will appear in the same places on the replica plates. The experiment confirms the second hypothesis: mutations appear at random with respect to what the environment favors.

environment because natural selection picks out the favorable mutations and causes them to spread. How that happens is the key topic of the next chapter.

Nongenetic Inheritance

The vast majority of inherited changes involve alterations of the DNA (or RNA) sequence of a genome. In some organisms, however, other mechanisms also contribute to inheritance and so can play a role in evolution [2].

Epigenetic inheritance is caused by inherited changes to chromosomes that do not alter the DNA sequence [3]. Instead, these changes affect the organism by altering how genes are expressed. Several mechanisms are involved. Two of the four DNA bases (cytosine and adenine) can be *methylated*, which is a biochemical

 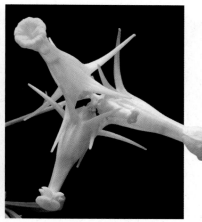

Standard *Linaria vulgaris*   *peloria* mutant

FIGURE 4.20 The *peloria* mutant of toadflax (*Linaria vulgaris*) results from an epigenetic mutation. One of the five petals normally has a long nectar-bearing spur, but in the *peloria* mutant all five petals have this form, transforming the flower's symmetry from bilateral to radial. The mutation can result from either a change in the DNA's sequence or its methylation pattern. (Courtesy of R. Grant-Downton.)

change to their structure (**FIGURE 4.20**). Likewise, the histones that bind to DNA to form eukaryotic chromosomes can be biochemically modified. Both kinds of changes alter how genes are expressed, and both can be transmitted across generations. Most epigenetic changes are not stable and dissipate after a few generations. Epigenetic inheritance can be important in the short term, but it does not make major contributions to long-term evolutionary change.

Genes are not the only way that mothers affect their offspring. **Maternal effects** occur when the genotype or phenotype of the mother directly influences the phenotype of her offspring. The direction that a snail shell coils is determined by the genotype of the snail's mother rather than the individual's own genotype (**FIGURE 4.21**). In mammals, including humans, the amount of milk that an infant receives from its mother has important effects on development. In turn, the amount of milk

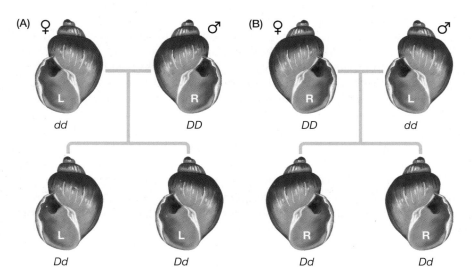

FIGURE 4.21 The direction of coiling in the snail *Lymnaea peregra* is determined by the genotype of an individual's mother, not its own genotype [23]. The shell can coil either to the left (L) or to the right (R). The locus that affects coiling has two alleles, *D* and *d*. (A) All offspring of a *dd* female with an L shell also have L shells. (B) All offspring of a *DD* female with an R shell also have R shells. Although the offspring genotypes in the two panels are the same, their shell phenotypes differ, and are determined by their mother's genotype. This is an example of a maternal effect.

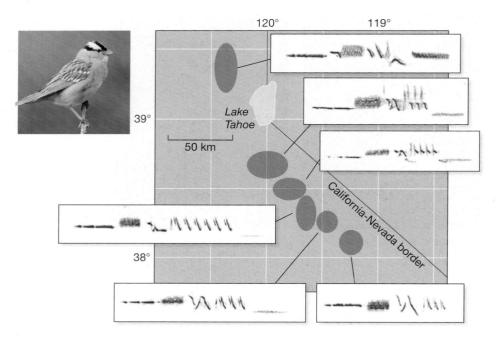

FIGURE 4.22 Song dialects of white-crowned sparrows (*Zonotrichia leucophrys*) illustrate cultural inheritance. Young males learn the local song dialect by hearing the songs of their father and other males. The sonograms (shown inside boxes) plot sound frequency against time for a typical song in each population [18].

provided depends on the mother's physiological condition as well as her genotype. Maternal effects like these are transmitted across only a few generations. They can contribute to the resemblance between mothers and their offspring, but (unlike genetic inheritance) do not contribute to major evolutionary change.

Traits with **cultural inheritance** are transmitted by behavior and learning. In humans, many traits, including language, religion, and dietary preferences, are strongly influenced by cultural inheritance (see Chapter 21). In songbirds, the songs of many species are partly determined by what chicks hear during a critical developmental period (**FIGURE 4.22**). An important difference between cultural inheritance and other forms of nongenetic inheritance is that traits can be transmitted between unrelated individuals (as well as between parents and offspring). That fact is exploited by the advertising industry and politicians in their efforts to persuade us to buy their products and to vote in certain ways.

Nongenetic inheritance has been crucial to the evolution of some traits in some species. Human society would not be possible without cultural inheritance. For the vast majority of life on Earth, however, evolutionary change results only from genetic inheritance. For that reason, we concentrate on genetic change in this book. There are, however, many fascinating aspects of nongenetic inheritance that are studied in fields that include linguistics, psychology, and behavior.

Go to the
Evolution Companion Website
EVOLUTION4E.SINAUER.COM
for data analysis and simulation exercises, quizzes, and more.

SUMMARY

- Genetic variation is produced when the genes of two parents are mixed during sexual reproduction by segregation and recombination. In asexual species, genes are mixed only by recombination, and less often than in sexual species.

- The Hardy-Weinberg equilibrium occurs under idealized conditions in which no evolutionary forces are acting. At this equilibrium, the three genotypes at a locus with two alleles are in the ratios $p^2 :: 2p(1 - p) :: (1 - p)^2$. Deviations from those proportions can be used to detect selection and other evolutionary factors.

- Linkage disequilibrium occurs when alleles at two loci occur together more often than expected by chance. It is eroded by recombination, and increased by selection and other evolutionary forces.

- Horizontal gene transfer (HGT) is the movement of genes between organisms by mechanisms that do not involve meiosis, and is particularly important to prokaryotes. It can move genes between individuals of the same species and of different species. HGT is important to human health because it is the major pathway by which bacteria evolve resistance to antibiotics.

- Mutation, which is an error that occurs when DNA or RNA is replicated, is critical to evolution because it is the ultimate source of genetic variation. Mutations can affect anywhere from a single base to a large piece of chromosome. Mutation rates vary greatly among species.

- Mutations in coding regions are called synonymous if they do not change the protein, and are called nonsynonymous if they do. This distinction has important evolutionary implications.

- Many mutations have no measurable effect on survival or reproduction. Those that do are typically deleterious. A small fraction are advantageous, and their spread leads to adaptive evolution.

- Mutations that affect one trait virtually always have pleiotropic effects, meaning that they also affect other traits.

- In species with separate somas and germ lines, a mutation can be inherited if it alters a gene in a cell in the germ line. Mutations to somatic cells leave no descendants to the next generation.

- Experiments show that mutation is random with respect to what will increase fitness.

- There are several mechanisms of nongenetic inheritance. One is cultural inheritance, an essential part of human civilization.

TERMS AND CONCEPTS

allele
allele frequency
beneficial mutation
bp (base pair)
chromosome
coding region
codon
cultural inheritance
deleterious mutation
deletion
DNA (deoxyribonucleic acid)
duplication

epigenetic inheritance
epistasis
exon
fission
fitness
fusion
gene
gene family
genetic code
germ line
Hardy-Weinberg equilibrium
horizontal gene transfer (HGT)

insertion
intron
inversion
linkage disequilibrium
locus
maternal effect
mutation
mutation rate
noncoding DNA
nonsynonymous
phenotype
pleiotropy
point mutation
polymorphic

recombination
recombination rate
RNA (ribonucleic acid)
segregation
SNP (single nucleotide polymorphism)
soma
structural mutation
synonymous
tetraploidy
whole genome duplication

SUGGESTIONS FOR FURTHER READING

The chapter introduced some of the basic concepts of population genetics that we will build on in the next few chapters. A very good introduction to this field is *Principles of Population Genetics* by D. L. Hartl and A. G. Clark (Sinauer, Sunderland, MA, 2007).

The large literature on mutation is nicely summarized in a review by F. A. Kondrashov and A. S. Kondrashov, "Measurements of spontaneous rates of mutations in the recent past and the near future" (*Philos. Trans. Roy. Soc. B* 365:1169–1176, 2010). P. D. Keightley's "Rates and fitness consequences of new mutations in humans" (*Genetics* 190: 295–304, 2012) focuses on the rates and effects of new mutations in humans. The interesting question of how mutation rates evolve is explored by P. D. Sniegowski and colleagues in "The evolution of mutation rates: separating causes from consequences" (*Bioessays* 22: 1057–1066, 2000) and by M. Lynch in "Evolution of the mutation rate" (*Trends Genet.* 26: 345–352, 2010).

The causes, consequences, and uses of linkage disequilibrium in humans and other species are reviewed by M. Slatkin in "Linkage disequilibrium—understanding the evolutionary past and mapping the medical future" (*Nat. Rev. Genet.* 9: 477–485, 2008).

PROBLEMS AND DISCUSSION TOPICS

1. Below are the DNA sequences that encode the first eight amino acids for five alleles of the Adh protein in *Drosophila pseudoobscura*. Nucleotides that differ from the first sequence are shown by a lowercase letter.

   ATGTCTCTCACCAACAAGAACGTC
   ATGgCTCTCACCAACAAGAACGTC
   ATGTCgCTCACCAACAAGAACGTC
   ATGTCTtTgACCAACAAGAACGTC
   ATGTCTCTCACCAACAAGAACGTg

   a. What are the first eight amino acids for each of these five DNA sequences?

   b. For each of the five polymorphic sites, indicate whether the site represents a synonymous or nonsynonymous polymorphism.

   c. The fourth sequence shown above has two mutational differences from the first sequence. Specifically, the third codon is TTG versus CTC in the first sequence. These two codons are two mutational steps away from each other. Supposing that the CTC sequence gave rise to the TTG sequence, do you think it is more likely that the one-difference intermediate was TTC or CTG?

   d. In general, synonymous polymorphisms tend to be more common than nonsynonymous polymorphisms. Why might that be?

2. The replica plate experiment shows that mutations are random. However, certain environmental stresses (e.g., high temperature, high salt, low pH) can increase the mutation rate.

   a. Does this indicate that mutations are nonrandom, since they increase in response to cell stressors?

   b. Does increasing the mutation rate increase the probability that an individual mutation will be adaptive?

   c. Does increasing the mutation rate increase the probability that a cell will experience an adaptive mutation?

3. A species of daisy has hermaphroditic flowers (i.e., each flower produces both male and female gametes). Researchers genotyped 1000 individuals at a SNP in three populations. The numbers of each genotype in each population were:

Genotype	Population 1	Population 2	Population 3
TT	90	200	50
TC	420	200	500
CC	490	600	450

a. For each population, calculate the allele frequencies and determine whether the population is currently at Hardy-Weinberg equilibrium.

b. For populations not at Hardy-Weinberg equilibrium, indicate whether there is an excess of homozygous or heterozygous genotypes.

c. For populations not at Hardy-Weinberg equilibrium, indicate how many generations of random mating it would take for the population to reach equilibrium.

4. A sample of 100 individuals is genotyped at loci A and B. The following numbers of two-locus genotypes are obtained:

Genotype	Number of individuals
$A_1A_1B_1B_1$	36
$A_1A_2B_1B_2$	48
$A_2A_2B_2B_2$	16

a. What is surprising about the observed number of two-locus genotypes?

b. Determine whether each locus is at Hardy-Weinberg equilibrium. What does your answer tell you about this population?

c. Do you think the two loci are close to one another, or far apart in the genome?

5. Cancers result from mutations in somatic cells, and these mutations therefore are not passed on to gametes. However, some families have much higher rates of cancer than average, showing that there are heritable factors that contribute to the risk of developing cancer. Discuss the roles that somatic mutations and germ line mutations play in producing cancer.

6. The *Dscam* locus in *Drosophila melanogaster* has 24 exons. Four of these exons are able to undergo alternative splicing. Exon 4 has 12 possible splice variants, exon 6 has 48 variants, exon 9 has 33 variants, and exon 17 has 2 variants. If all splicing combinations are possible, how many different Dscam protein sequences could be encoded by a single allele at this locus?

5

The Genetical Theory of Natural Selection

Natural selection explains many of the most fascinating things in nature, from the genetic code to the complexities of the human brain. Natural selection is fundamentally a simple concept. But its explanatory power expands when we see how it acts, and how the evolutionary outcome depends on genetics and development. When we understand these sides of the evolutionary process, we can begin to answer a cascade of fascinating questions: How do cooperative and selfish behaviors evolve? Why do some species reproduce sexually and others asexually? Are different populations of a species likely to evolve the same way if they experience the same environmental challenge? What explains the extraordinary display of the peacock, the immense fecundity of oysters, the brevity of a mayfly's life, the pregnancy of the male sea horse, the abundance of transposable elements in our own genome?

Darwin realized that a complete understanding of evolution requires understanding the mechanism of inheritance. One of the biggest frustrations of his life was that he never was able to learn that mechanism. The field of genetics began after Darwin died, when Mendel's work was rediscovered in 1900. Modern evolutionary genetics started to develop with the synthesis of Mendelian genetics and Darwin's theory of natural selection, and the discovery of other forces that can also cause evolutionary change. The keystone to understanding how evolution works is the "genetical theory of natural selection," to quote the title of a famous 1930 book by the pioneering population geneticist R. A. Fisher.

The brilliant color patterns of tropical butterflies in the genus *Heliconius* warn predatory birds that they are poisonous. Butterflies with color patterns that are common have an advantage because birds quickly learn to avoid them after tasting a few. Rare color patterns (as seen in this hybrid individual) are disadvantageous because birds have not yet learned to avoid them

(A)

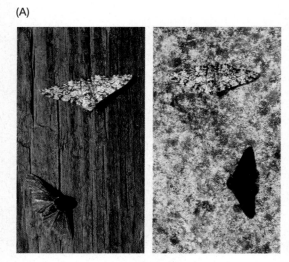

(B)

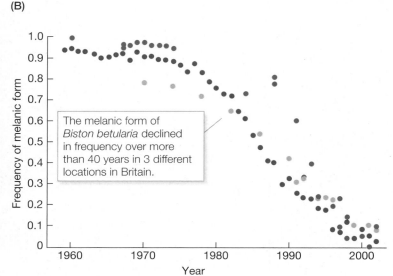

> The melanic form of *Biston betularia* declined in frequency over more than 40 years in 3 different locations in Britain.

FIGURE 5.1 Industrial melanism in the peppered moth is the most famous example of evolution by natural selection directly observed in the wild. (A) The pale gray "typical" form and the melanic form on a tree trunk darkened by air pollution (left) and on a normal, nonblackened trunk (right). (B) The decline in the frequency of the melanic form in three British localities, indicated by dots of different colors, as air pollution decreased during the late twentieth century. (B after [11].)

Natural Selection and Evolution in Real Time

The English have long had an interest in (some might say an obsession with) butterflies and moths. For hundreds of years, amateur collectors have roamed the countryside in search of specimens. They were surprised when a melanic (dark) color form of the peppered moth (*Biston betularia*) was first discovered in 1848 and then rapidly spread in central England (**FIGURE 5.1**). Recent research shows that the dark coloration results from a single mutation that likely occurred around 1819 [20].

Why did the melanic mutation in the peppered moth suddenly spread? During the mid-1800s, England was transformed by the Industrial Revolution, which was powered by burning vast quantities of coal. Before then, peppered moths had been well camouflaged: their light coloration was a beautiful match to the bark and lichen on the trees where the moths rest during the day. But the burning of coal produced soot that killed the lichens and blackened the tree trunks, suddenly making moths with the typical coloration very conspicuous. Moths with the melanic allele, however, were camouflaged in the new environment. That increased their survival, and caused the melanic allele to spread nearly to fixation by the late 1800s [40, 41]. In 1924, J. B. S. Haldane calculated from the speed of its spread that the melanic mutation must have given moths a 50 percent survival advantage [17]. Experiments done by Bernard Kettlewell in the 1950s suggested that predation by birds was responsible.

Starting in the 1960s, coal burning decreased dramatically and trees slowly lost their coatings of soot. The change provided an unintended experiment that verified the hypothesis that pollution gave melanic moths a survival advantage. Between 1960 and 2000, the frequency of melanic moths fell from nearly 100 percent to nearly 0 percent throughout much of central England (see Figure 5.1B) [12, 30].

Several scientists raised questions about whether bird predation was really the cause of the melanic mutation's rise and fall. Michael Majerus took up the challenge of answering this question in the 2000s. Over 7 years, he did a series of experiments involving nearly 5000 moths. Tragically, Majerus died before the work was published, but his colleagues finished the job [12]. The results decisively confirmed Kettlewell's conclusions: bird predation was the overriding reason for the rapid evolution of melanism in the 1800s and its decline 100 years later.

The spread of melanism in the peppered moth was the first time that evolution had been documented in real time, and Haldane's calculations showed that natural selection acting on the melanic allele can explain its rapid spread. The dozens of

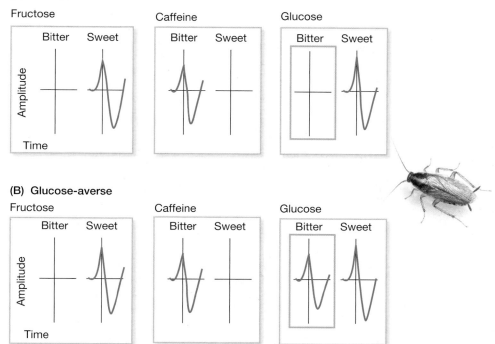

(A) Wild type

(B) Glucose-averse

FIGURE 5.2 German cockroaches have evolved an aversion to the glucose used to trap them. (A) Cockroaches have four types of taste receptors that respond to different tastes. The neural impulses shown here illustrate how two of these receptors respond to two kinds of sugar (fructose and glucose) and to a bitter stimulus (caffeine). In wild-type cockroaches, the bitter receptor is triggered by caffeine but not by either sugar. This causes the cockroaches to avoid caffeine. (B) In cockroaches that have the glucose-aversion mutation, the bitter receptor is also triggered by glucose (but not fructose). This causes the cockroaches to avoid traps that use glucose as an attractant. (After [42].)

studies that have been done on the genetics and ecology of the peppered moth give us one of the most detailed understandings of evolution by natural selection in any species. The peppered moth story also shows that evolutionary biology, like physics and chemistry, is a rigorous science in which scientists propose and test hypotheses.

While the peppered moth is the most famous example of evolution by natural selection that has been directly observed, hundreds of other cases have also been studied. One is close to home. If you've ever had cockroaches in your kitchen, you may be familiar with the traps used to control them. Cockroaches are attracted to the traps by glucose that is laced with poison. Some populations of the German cockroach (*Blattella germanica*), which is found in kitchens around the world, have evolved an aversion to glucose in just a few years. This change is caused by a single mutation that rewires the neural receptors that cockroaches use to taste their food [36, 42]. One type of receptor fires when a cockroach tastes something bitter, causing the cockroach to avoid it. In cockroaches that carry the mutation, the receptor for bitter taste also fires when it is exposed to glucose (**FIGURE 5.2**), effectively repelling the cockroaches from the traps.

Further evidence of the power of selection to cause rapid evolution can be seen at your local supermarket. Almost all the food there results from the remarkable evolutionary changes resulting from **artificial selection**, the selective breeding by humans of animals and plants. When prehistoric farmers harvested their crops, the plants that were most productive contributed the most seeds to the harvest. Some of those seeds were used to plant the next year's crop, resulting in plants that were more productive than the previous generation. Selective breeding of domesticated animals

FIGURE 5.3 Comparison of three strains of broiler chickens at different ages. In the first strain, artificial selection was stopped in 1957, in the second it stopped in 1978, and in the third it stopped in 2005. Birds from the three strains were grown under identical conditions. The average weight of a 56-day-old chicken increased more than four-fold as the result of the 48 years of additional artificial selection between 1957 and 2005. (From [46].)

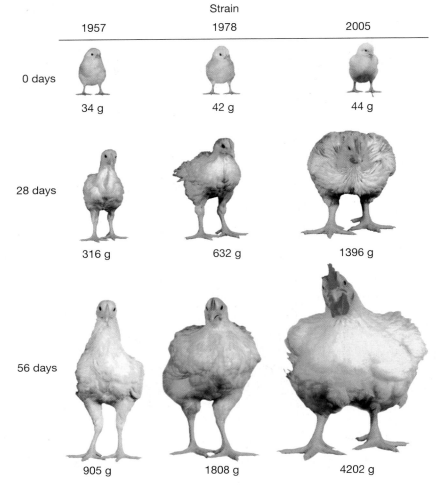

and crops was likely unconscious at first, but it is now practiced with sophisticated statistical methods, and it results in spectacularly rapid evolution (**FIGURE 5.3**).

Evolution by Selection and Inheritance

As we stressed in Chapter 3, selection results when there is a consistent relation between a phenotype and fitness. Evolution by selection also requires a second ingredient: inheritance. In the simplest terms,

If:
- there is a correlation between a phenotypic trait and the number of offspring that individuals leave to the next generation, and
- there is a correlation between the phenotype of a trait in parents and in their offspring,

Then:
- the trait will evolve.

The first condition means that selection on the trait is occurring, favoring individuals with more extreme phenotypes than average. The second condition means that at least some of the phenotypic variation is inherited, causing the individuals with those more extreme phenotypes to pass on their traits to their offspring. The term "correlation" is used here in the statistical sense described in the Appendix.

The logic behind this idea is shown graphically in **FIGURE 5.4**. Of Darwin's many intellectual breakthroughs, this was his most brilliant. This chapter is about evolution that results from selection acting on individuals; Chapter 12 will explore how this concept applies to other entities.

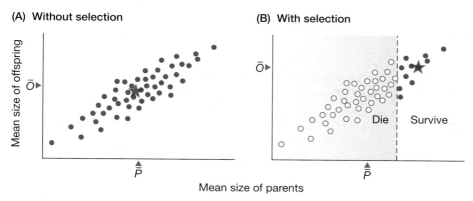

(A) Without selection

Mean size of offspring

$\bar{O}$ ▶

(B) With selection

$\bar{O}$ ▶

Die   Survive

$\bar{P}$

Mean size of parents

FIGURE 5.4 Evolution results when there is a correlation between the phenotypes of parents and offspring and a correlation between the phenotypes of parents and their fitness. Each dot represents a family. The horizontal axis shows the mean trait (in this case, size) of the two parents, and the vertical axis the mean trait of their offspring. In both panels, there is a correlation between the traits of parents and their offspring. (A) With no selection, there is no evolution. The mean value of all offspring in the next generation, $\bar{O}$, is the same as the mean value of all parents in the current generation, $\bar{P}$. No change has happened. (B) When selection acts, evolution results. In this example, only the largest individuals survive and become parents (solid circles to the right of the vertical dashed line) that leave offspring. The mean trait in the offspring of the next generation is larger than the mean of the parents before selection acted in the current generation.

A key point is that natural selection and evolution are not the same thing. If selection on a trait occurs but the trait is not inherited, then evolution will not happen. You will see in later chapters that natural selection is not the only factor that can cause evolution, and so evolution can also occur even if selection does not. But if the twin conditions listed above are met, then the population will change across generations. To make that basic idea more quantitative, let's now consider how to measure selection, account for inheritance, and predict the outcome of evolution.

Fitness: The Currency of Selection

To understand how evolution by selection works, we need a way to measure selection. An individual's **absolute fitness** is the *number of zygotes (offspring) produced over its lifetime*. If an immature butterfly is eaten by a bird, its fitness is 0. If an oak tree survives many years, its fitness could easily be in the millions. (Just think of the number of acorns produced by a big tree.) It is also useful to consider the fitness of an allele, a genotype, or a phenotype. In those cases, fitness is the *average* of the fitnesses of the individuals with that allele, genotype, or phenotype. We use the symbol W to represent absolute fitness.

Over the life cycle of any organism, many events affect fitness. An embryo may develop to sexual maturity or die before that happens. If it survives to maturity, the individual may or may not be able to mate. If he or she does successfully mate, the individual's gametes may or may not be successful at fertilization. These various events are called **fitness components** (**FIGURE 5.5**). They can be divided even more finely, or grouped together, depending on the application. Often it is useful to simplify our thinking by viewing fitness as the product of just two fitness components:

$$W = \text{(probability that the individual survives to maturity)} \atop \times \text{(expected number of offspring if the individual does survive)} \quad (5.1)$$

The strength of selection is determined by fitness differences. It is the relative (or proportional) differences that matter. To see why this is so, consider an individual

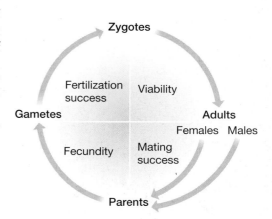

FIGURE 5.5 Selection can affect fitness at different points during the life cycle. This sketch shows the life cycle of a sexually reproducing species with four fitness components that affect the number of descendants that an individual leaves to the next generation. (After [10].)

who leaves two offspring to the next generation (a fitness of 2). That individual enjoys a huge fitness advantage if all others in the population have a fitness of 1, but it is at a selective disadvantage if all others have a fitness of 4.

We therefore often work with **relative fitness**, which is the absolute fitness divided by a fitness reference that we agree on. The choice of the fitness reference is a matter of convenience, and changes depending on the situation under consideration. Say there are two alleles, A_1 and A_2, at a locus. We use the symbol W_{11} to represent the absolute fitness of A_1A_1 homozygotes, W_{12} for that of A_1A_2 heterozygotes, and W_{22} for that of A_2A_2 homozygotes. We write the relative fitnesses of the three genotypes the same way, but using a lowercase w (for example, the relative fitness of A_1A_2 is w_{12}). If we agree to use the A_1A_1 homozygote as our fitness reference, then the relative fitness are:

$$\begin{aligned} w_{11} &= W_{11}/W_{11} = 1 \\ w_{12} &= W_{12}/W_{11} \\ w_{22} &= W_{22}/W_{11} \end{aligned}$$

(5.2)

Relative fitnesses play a critical role in determining the speed and outcome of evolution by natural selection, as you will now see.

Positive Selection: The Spread of Beneficial Mutations

Whenever one allele has higher fitness than another, natural selection will favor its spread through the population. This is called **positive selection**. A hypothetical example is shown in **FIGURE 5.6**. A population of mice has two alleles that affect the coat color. Allele A_2 causes darker fur than allele A_1, and it is initially present at a frequency of 0.5. Hawks prey on the mice and kill half of the light-colored A_1A_1 and one-fourth of the intermediate-colored A_1A_2 individuals. The

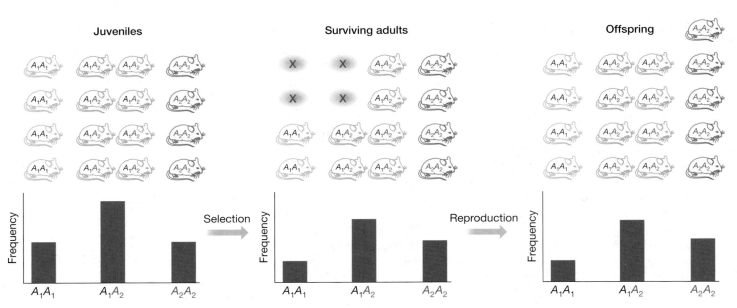

FIGURE 5.6 Positive selection causes the allele with higher fitness to increase in frequency. In this population of mice, allele A_2 is at a frequency of $p = 0.5$ at the start of the generation. Predation kills half of the A_1A_1 homozygotes and one-fourth of the A_1A_2 heterozygotes, but all of the A_2A_2 homozygotes survive. The frequency of A_2 increases to $p = 0.58$ in the surviving adults. The frequency of A_2 is also $p = 0.58$ at the start of the next generation, and random mating has restored the genotype frequencies to Hardy-Weinberg proportions.

BOX 5A

Evolution by Selection on a Single Locus

This box explores the evolutionary change that results from selection on a single locus with two alleles, A_1 and A_2. We will start with the simplest case, in which the heterozygote has intermediate fitness, as is often the case, and we will use the A_1A_1 genotype as our fitness reference. The A_2 allele increases survival, and for each copy of A_2 that an individual carries, its fitness increases by a proportion s. The frequency of A_2 is p, and we assume the population is in Hardy-Weinberg equilibrium. The relative fitnesses and frequencies of the three genotypes are therefore:

Genotype	A_1A_1	A_1A_2	A_2A_2
Relative fitness	1	$1 + s$	$1 + 2s$
Frequency at birth	$(1 - p)^2$	$2p(1 - p)$	p^2

Our goal is find the frequency of A_2 in the next generation. The first step is to calculate the frequencies of the three genotypes among surviving adults after selection has acted. Those are:

Genotype	A_1A_1	A_1A_2	A_2A_2
Frequency in adults	$\dfrac{(1 - p)^2}{\overline{w}}$	$\dfrac{2p(1 - p)(1 + s)}{\overline{w}}$	$\dfrac{p^2(1 + 2s)}{\overline{w}}$

In each numerator, we see the product of the genotype's frequency at birth and its relative fitness. In the denominators is $\overline{w}$, a factor that is needed to ensure that all the frequencies sum to 1 (which is required by the definition of frequencies). That number has an important interpretation: it is the *mean fitness* of the population. It is given by the sum of the three numerators:

$$\overline{w} = (1 - p)^2 + 2p(1 - p)(1 + s) + p^2(1 + 2s) = 1 + 2sp \quad \text{(A1)}$$

We now have what we need to find the allele frequencies at the start of the next generation. Using the logic sketched in Figure 4.6, the frequency of allele A_2 is given

by the frequency among surviving adults of A_2A_2 individuals plus half the frequency of A_1A_2 individuals. After a bit of algebra, we find the answer is:

$$p' = [1 + s(1 + p)]\, p\, /\, \overline{w} \quad \text{(A2)}$$

We can learn more by asking how much allele frequencies change from one generation to the next. Using our results from above, we find that the change in the frequency of allele A_2 caused by one generation of selection is

$$\Delta p = p' - p = sp(1 - p)\, /\, \overline{w} \approx sp(1 - p) \quad \text{(A3)}$$

The first two expressions for Δp are exact. The last one is an approximation that is very accurate when s is less than 0.1. It is useful because it is so simple.

Equation A3 has two key implications. The rate that the allele frequency changes—that is, the rate of evolution—caused by selection is proportional to two quantities. The first is s, the selection coefficient. Logically enough, if $s = 0$, then all genotypes have the same fitness, there is no selection, and therefore no evolutionary change. The second factor that governs the rate of evolution is the quantity $p(1 - p)$, which represents the genetic variation at the locus. Notice that $p(1 - p)$ is 0 if either $p = 0$ or $(1 - p) = 0$. In either case, there is only one allele in the population and therefore no genetic variation. Variation is maximized when both alleles are equally frequent (see Figure 5.8).

We've now seen how to calculate the change in allele frequencies for the particular case of the relative fitnesses shown above. Fitnesses that follow that pattern are common for alleles with small fitness effects. For other situations, including cases in which heterozygotes have a fitness higher or lower than both homozygotes, we calculate the allele frequency change using the same logic used here. The results, however, are a bit more complicated than Equation A3. For more details, see [18].

dark A_2A_2 homozygotes are well camouflaged, and all of them survive. Selection has increased the frequency of A_2 to 0.58. When the survivors reproduce, this new allele frequency is passed to the next generation.

The rate at which an allele's frequency changes—that is, the speed of evolution—is determined by the relative fitness advantage of the favored allele. The mathematical theory behind that fact is explained in **BOX 5A**.

FIGURE 5.7 shows examples of **selective sweeps**, the situation in which a beneficial mutation spreads through a population. The population is initially **fixed** for allele A_1, which means that the allele is at a frequency of 1.0. We will use the A_1A_1 homozygote for the fitness reference. Mutation then produces a new allele A_2 that

FIGURE 5.7 The spread of three beneficial mutations that have different selection coefficients, *s*. Stronger selection leads to faster evolution.

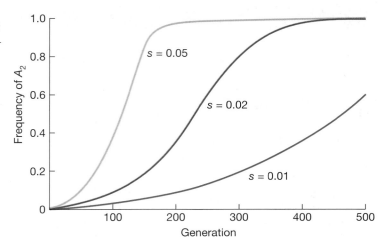

is beneficial. Each copy of this allele that an individual carries increases fitness by a fraction *s*. This means that A_1A_2 heterozygotes have relative fitness $w_{12} = (1 + s)$. The A_2A_2 genotype has the highest fitness of all, and its fitness is increased by a factor of $(1 + 2s)$ relative to the A_1A_1 homozygotes. In the mouse example of Figure 5.6, the selection coefficient favoring the A_2 allele is $s = 0.5$: each copy of that allele increases survival by 50 percent. That is extremely strong selection compared with what is seen in most natural populations, where fitness effects of alleles are typically many times smaller.

The number *s* is called the **selection coefficient**. It is a natural measure of the strength of selection that favors the beneficial allele. For the moment we are assuming selection coefficients are constant in time and that the heterozygote's fitness is intermediate between the two homozygotes, but we will see shortly what happens when those assumptions are not met. Note that selection coefficients depend on which genotype is chosen to be the fitness reference. In the current situation, it is most convenient to use the A_1A_1 homozygotes as the standard, but later in this chapter we will use other genotypes as the reference.

The rate of adaptation

We can predict the course of evolution if we know the current state of the population and the strength of selection. We will use *p* to represent the frequency of the A_2 allele, and so the frequency of the A_1 allele is $1 - p$. The evolutionary change per generation is measured by the change in the frequency of A_2 from the beginning of the current generation to the beginning of the next. Using Δp to represent that change, Box 5A shows that

$$\Delta p \approx s\, p\, (1 - p) \tag{5.3}$$

(Equation 5.3 is an approximation that is accurate when *s* is less than 0.1, which is the case for the large majority of alleles in natural populations. The exact version of the equation is a bit more complicated; see Box 5A.)

Equation 5.3 is beautiful in its simplicity, and it carries important messages. On the left side is Δp, which is the rate at which the allele frequency evolves. The right side shows that the rate is the product of two quantities. The first is the selection coefficient *s*, which measures the strength of selection. When $s = 0$, there is no selection acting, $\Delta p = 0$, and there is no evolution. The second quantity on the right side of Equation 5.3 is $p(1 - p)$, which is a natural measure of genetic variation. If A_2 is absent from the population then $p = 0$, while if A_1 is absent then $(1 - p) = 0$. In either case, there is no genetic variation at this locus. Consistent with that fact,

(A) *p* = 0.05 (B) *p* = 0.5

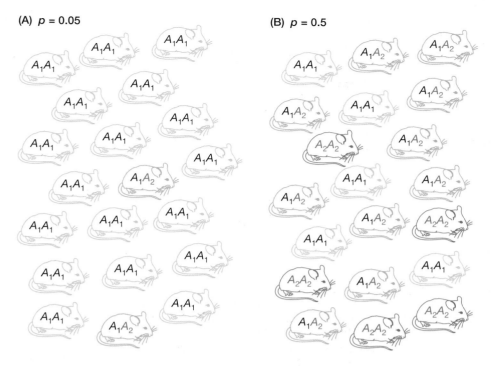

(C) *p* = 0.95

FIGURE 5.8 The amount of genetic variation at a locus depends on the allele frequencies. Genetic variation is greatest when allele frequencies are intermediate. There is little variation when the frequency *p* of allele A_2 is near 0 or near 1. Variation is maximized when *p* = 0.5.

genetic variation as measured by $p(1 - p)$ equals 0. The opposite situation is when variation is maximized, which happens when $p = (1 - p) = 0.5$ (**FIGURE 5.8**).

The key conclusion is this: *the rate of evolution is proportional to the strength of selection and the amount of genetic variation.* In the absence of either of those two ingredients, there is no evolution by selection.

Knowing Δp tells us what the population will look like in the next generation. The frequency of A_2 in the next generation is equal to its current frequency plus the evolutionary change. Using p' to represent the frequency of A_2 at the start of generation 2,

$$p' = p + \Delta p \qquad (5.4)$$

Now let's look further into the future. We can take the frequency of A_2 at the start of generation 2 and substitute that number for p on the right side of Equations 5.3 and 5.4 to find the allele frequency at the start of generation 3. Repeating that process lets us trace the trajectories of the allele frequency through time (see Figure 5.7). The trajectories are S-shaped. They change slowly when p is near 0 or near 1, and most rapidly when p is near 0.5. Although the strength of selection is constant, the genetic variation is not: variation is small when either allele is rare, and large when the alleles are about equally frequent (see Figure 5.8).

When a beneficial allele spreads by selection, the final outcome is that it becomes fixed. That is, it reaches a frequency of 1, and the other allele is eliminated. The conclusion is that positive selection ultimately eliminates genetic variation. That means other evolutionary factors must be responsible for maintaining all the genetic variation that we see in nature. Mutation is one, and we will see shortly that there are also others.

A beneficial allele spreads through a population more quickly if it is more strongly selected. There is a simple rule of thumb that gives the time needed for an allele to spread most of the way through the population when heterozygotes have intermediate fitness. A beneficial allele will increase in frequency from 10 percent to 90 percent in roughly 4/*s* generations (**FIGURE 5.9**). For example, if A_2 increases

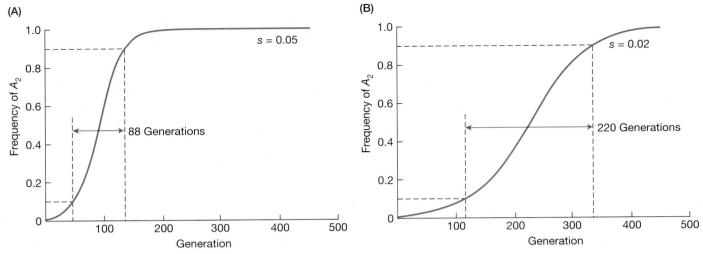

FIGURE 5.9 The time needed for a beneficial allele to spread through a population is inversely proportional to the strength of selection. When the mutation has a selection coefficient s and heterozygotes have intermediate fitness, it takes roughly $4/s$ generations for the mutation to increase in frequency from 10 percent to 90 percent. (A) The trajectory for a mutation that increases fitness by 5 percent. Here $s = 0.05$, and the time needed will be about $4/s = 80$ generations. The exact time taken is a bit longer: 88 generations. (B) The trajectory for a mutation that increases fitness by 2 percent. Now $s = 0.02$, and the time needed will be about $4/s = 200$ generations. The exact time needed is 220 generations.

fitness by 0.1 percent, then $s = 0.001$ and it will take about 4000 generations for A_2 to spread through the population. This shows that a minute fitness advantage will cause large evolutionary change in a geologically short period of time. If selection is 100 times stronger, then the beneficial allele has a 10 percent fitness advantage and $s = 0.1$. The allele frequency changes 100 times faster, and the favorable allele spreads through most of the population in only about 40 generations. To show that there is a fitness difference of 10 percent would require measuring fitness on hundreds of individuals. Yet even a difference of that size can cause evolution fast enough to be directly observed. For the peppered moth, with a generation time of 1 year, a noticeable change in the population took just a few years. For a bacterium, whose generation time might be measured in hours, that change can take just a few days.

The rate at which positive selection causes an allele frequency to evolve depends on dominance. An allele is dominant if it causes the same phenotypic effect when heterozygous as when homozygous. (Dominance does *not* refer to an allele's frequency in a population.) Up to this point we have been discussing alleles that do not have dominance: the heterozygote's fitness has been intermediate between the fitnesses of the two homozygotes. If allele A_2 is dominant and beneficial, it will initially spread even more quickly than what we saw previously (**FIGURE 5.10**). When A_2 is rare (as when it first appears by mutation), almost all copies of it are in heterozygotes. That fact follows from the Hardy-Weinberg ratios. For example, with an allele frequency of 0.01, the frequency of heterozygotes is $2 \times 0.01 \times 0.99 = 0.0198$, while homozygotes are almost 200 times more rare: $(0.01)^2 = 0.0001$. A dominant beneficial mutation spreads more rapidly when rare because heterozygotes share the full fitness benefit that the homozygotes have. The situation reverses, however, when allele A_2 is common. Now the low-fitness A_1A_1 homozygotes are rare, and almost all individuals have high fitness. Because there is so little variation in

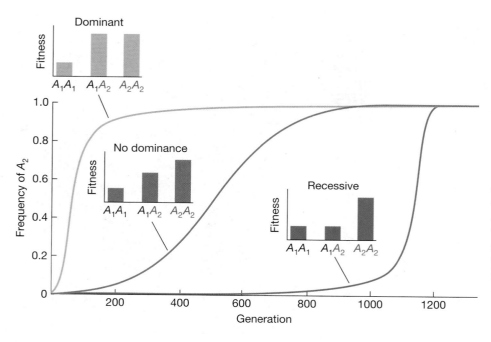

FIGURE 5.10 Dominance affects the evolutionary trajectories of beneficial alleles. When the beneficial allele is dominant, its frequency increases very rapidly while it is rare, then much more slowly when it is common. A recessive beneficial mutation spreads very slowly when it is rare, but rapidly when it is frequent. When there is no dominance (that is, the heterozygote has a fitness intermediate between the homozygotes), the trajectory is intermediate between the dominant and recessive cases. In these trajectories, the fitness of the A_2A_2 homozygote is 1.01 relative to that of the A_1A_1 homozygote, and the beneficial A_2 mutant was introduced at a frequency of 1 percent. The trajectories differ only in the relative fitness of the heterozygote.

fitness, selection has little power to increase the frequency of A_2 further, and the rate of evolution slows dramatically.

The converse situation occurs when the beneficial A_2 allele is recessive. In that case, when A_2 is rare, its frequency increases very slowly: only the exceedingly rare A_2A_2 homozygotes enjoy the fitness advantage. But when the allele is common, selection is now much more effective than it is for a dominant allele that is common (see Figure 5.10).

Thus far, we have focused on positive selection favoring the spread of a beneficial mutation. The same logic and same equations (5.3 and 5.4) also apply to **deleterious mutations** (those that decrease fitness). In this case, the selection coefficient s is negative, and the change in allele frequency Δp is negative (from Equation 5.3). Many genetic diseases in humans are caused by mutations that are nearly or completely recessive. Because they are at low frequency, almost all copies are in heterozygotes who have fitness close or equal to that of individuals who do not carry the mutation. Selection is therefore very ineffective at removing these disease-causing mutations from the population.

We've now seen that if we know the fitnesses of the genotypes, we can predict how many generations it takes for a beneficial mutation to spread. We can turn the tables around by asking: If we know how many generations it took for an allele to spread, how strong was selection? This was the strategy used by J. B. S. Haldane to estimate that the melanic allele of the peppered moth had a selection coefficient of $s = 0.5$.

The peppered moth is not the only example of rapid evolution in response to intense selection. Many other animals have evolved melanic forms in polluted urban environments, so many that the phenomenon has a name: industrial melanism [30]. Other cases of strong natural selection and rapid evolutionary change include bacteria evolving resistance to antibiotics, insects evolving resistance to pesticides, and plants evolving resistance to herbicides. You may have noticed a theme here: in these examples, selection results from a change to the environment caused by humans. With antibiotic resistance in bacteria, the change was intentional (when we administered the drugs), while with industrial melanism it was not. But whether it intends to or not, human activity is rapidly changing the environment of many, in fact probably most, organisms on the planet. This is causing strong selection and

FIGURE 5.11 The region in Europe where lactase persistence is common overlaps with archeological sites where early Neolithic people had cattle. Dark green areas on the map show where the lactase persistence allele is at high frequency. The dashed curve shows the geographic range of the Funnelbeaker culture, which was a pastoral society. Dairy products were likely consumed by these people roughly 4500 years ago. Inset: The diet of northern Europeans is still rich with dairy products. (After [4].)

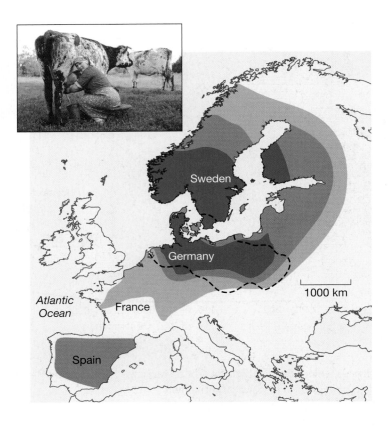

rapid evolution in a wide range of species [34]. While adaptation to changes to the environment caused by humans may help some species avoid extinction (see Chapter 6), most are likely to become extinct (see Chapter 17).

Many genes in humans have also evolved recently in response to strong natural selection. When humans domesticated cattle, cows' milk suddenly offered a rich source of energy that previously had been available only to infants nursing on their mothers. Milk contains a sugar called lactose. Children and other young mammals derive energy from lactose by digesting it with the enzyme lactase. In our ancestors, lactase was produced during infancy to digest the mother's milk, but was not produced in adults because there was no lactose in their diet. Things changed with the arrival of domesticated livestock in northern Europe about 10,000 years ago (**FIGURE 5.11**). Suddenly, natural selection favored lactase in adults so they could digest the milk of cows and other domesticated animals [4]. A mutation that causes *lactase persistence*, in which lactase is produced throughout life, appeared in Europe about 4000 years ago and spread very rapidly by natural selection [32]. Genetic evidence suggests that the selection coefficient was very large, perhaps $s = 0.1$ or even larger [6]. In northern Europe today, about 98 percent of adults show lactase persistence. The remaining 2 percent of the population are lactose intolerant: they do not produce the enzyme as adults and so have trouble digesting milk and cheese. In other populations that independently invented dairy farming, such as the Maasai in East Africa, other mutations with similar effects spread—a striking example of *parallel evolution* [21]. In parts of the world where dairy products are rarely consumed, lactase persistence is uncommon (for example, less than 10 percent in Southeast Asia).

Light skin color is a second example of strong natural selection causing parallel adaptation in humans. Modern humans first spread out of Africa roughly 60,000 years ago (see Chapter 21). Those that moved north lived under reduced sunlight, which generated new selective pressures. The synthesis of vitamin D, which is

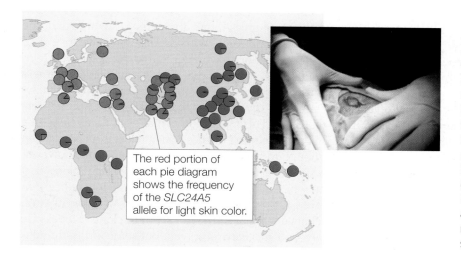

The red portion of each pie diagram shows the frequency of the *SLC24A5* allele for light skin color.

FIGURE 5.12 The "golden" gene (technically, a locus called *SLC24A5*) contributes to light skin color in European populations. Lighter skin results from a nonsynonymous mutation that changes an alanine to a threonine in the gene's protein product. Patterns of DNA polymorphisms around this locus show that strong selection caused this allele to spread when humans spread from Africa into Europe. The cause of the selection was likely that reduced sunlight in the north led to a deficiency in vitamin D. Light skin color in other populations (for example, Asians) results from mutations at other loci. Inset: The arm of this book's second author after removal of a basal cell carcinoma. Increased risk of skin cancer is a negative side effect of light skin. (After [13].)

critical to calcium metabolism, requires ultraviolet (UV) light. If not enough UV light penetrates the skin, children develop a disease called rickets. Mutations that decrease skin pigmentation likely had a selective advantage in the north because they enhanced vitamin D synthesis [39]. Mutations at several loci that lighten the color of the skin appeared and spread to high frequency under strong selection (**FIGURE 5.12**). Different alleles at different loci were selected in different populations: the alleles that cause light skin color in Asia are different than those that cause it in Europe. It is no accident that human populations farthest from the equator have very light skin: it is an adaptation. But like many adaptations, light skin comes with a cost. The increased pigmentation in African skin provides an increased sun protection factor (SPF) of 10 compared with white skin. People with light skin suffer higher rates of skin cancer.

Chance and adaptation: The probability that a beneficial mutation spreads

Our discussion so far might lead you to think that if a beneficial mutation appears in a population, it is certain to spread by natural selection and become fixed. But that is not true: chance also plays a role. Consider an adult who is heterozygous for a mutation that increases the probability of survival. If he or she has two offspring, there is a 50 percent chance that the mutation will not be passed to the first offspring, and a 50 percent chance that it won't be passed to the second offspring. No matter how large a survival benefit the mutation gives, there is a 25 percent chance that it will leave no descendants in the next generation. This example shows how an allele can be lost through the random segregation of alleles during meiosis. Other factors also contribute to the chance that the allele will be lost: even if it increases survival *on average*, any particular individual who carries the allele might not survive.

This example shows just one aspect of an important evolutionary process called *random genetic drift*. As you will see in Chapter 7, drift is particularly important in small populations, but it plays a role in very large ones as well. When an allele first appears in a population by mutation, it is represented by only a single copy. It may be lost by chance then, or in a later generation while it is still rare. Population genetic theory shows that the probability that a single copy of a new beneficial mutation will survive and become fixed in the population is

$$\Pr(\text{fixation}) = 2s \qquad (5.5)$$

where s is the selection coefficient that gives the relative fitness advantage of the mutant heterozygote relative to the homozygote with the original allele.[1]

To clarify these ideas, consider a situation in which A_1A_1 individuals have a relative fitness of $w_{11} = 1$. A new mutation A_2 appears that gives heterozygotes a fitness $w_{12} = 1.01$, and so $s = 0.01$. Beneficial mutations that increase fitness this much are rare—most have even smaller effects. Nevertheless, the mutation has only a 2 percent chance of becoming established. The conclusion is that even an allele with a large fitness advantage has an overwhelming chance of being lost by chance when it first appears. If it is lucky enough to survive the first few generations, however, then it is much more likely to survive and spread throughout the population.

The conclusion is that even when a mutation increases fitness, it is not certain that natural selection will cause it to spread to fixation. For a type of mutation that recurs again and again, chance will have little impact in the long run because eventually one of the mutations will survive and spread. But some kinds of mutations are exceedingly rare, and may occur only once in very long periods of evolutionary time. In those cases, the chance loss of beneficial mutations is a major limitation to adaptation.

Evolutionary Side Effects

Natural selection often has side effects. These result from **genetic correlations**, which occur when two traits tend to be inherited together. One cause of genetic correlations is pleiotropy (see Chapter 4). If you are taller than average, you are also likely to have feet that are larger than average. Height and foot size are genetically correlated because alleles that make individuals large for one trait tend also to make them large for the other.

You can see immediately how this kind of genetic correlation will cause evolutionary side effects. Imagine, for example, that natural selection favors shorter individuals. As alleles for short height spread, height will decrease—and so will foot size, which might or might not be advantageous. Early in the evolution of mammals, selection fixed alleles that enabled females to nurse their young. Among them were alleles for nipples that are used to deliver the milk. Those alleles, however, are also expressed in males. Nipples in males evolved as a pleiotropic side effect of natural selection acting on females.

An allele that increases fitness through its effect on one trait sometimes decreases fitness because of its effect on another trait. We then say there is an **evolutionary trade-off**. When there is a trade-off, natural selection favors the allele that has the highest fitness overall. As that allele spreads, it will increase some fitness components (early reproduction, for example) but have negative effects on others (survivorship). Recall that the alleles for light skin color that are common in northern populations of humans are beneficial (they increase vitamin D production) but also have tradeoffs (they make skin cancer more likely). You will see in Chapter 11 that trade-offs are an important cause of senescence in humans and other species.

Soay sheep are a primitive breed that lives wild on a remote island off the coast of Scotland. Most males have large, curled horns that are important to their mating success. About 13 percent of males, though, develop tiny vestigial horns (**FIGURE 5.13**). This variation results from polymorphism at a single locus. Many of the males that are homozygous for an allele called Ho^P grow vestigial horns, while

---

[1] Equation 5.5 is an approximation that is very accurate when s is smaller than 0.1 and the size of the population is much larger than $1/s$ individuals. Chapter 7 discusses what happens when the population size is smaller than that.

heterozygotes and homozygotes for the Ho^+ allele grow normal horns. These alleles have strong pleiotropic effects on fitness components [23]. Vestigial horns decrease mating success but increase survival. As a result of this trade-off, heterozygous males have the highest overall fitness. We will explore the evolutionary outcome of this situation shortly.

Hitchhiking: When one allele goes for a ride with another

You saw in Chapter 4 that an allele at one locus is sometimes found together with an allele at a second locus more often than expected by chance. When that happens, we say that the two loci are in **linkage disequilibrium**. Recall from Chapter 4 that linkage disequilibrium is a statistical property of the population, like an allele frequency, and can occur whether or not the loci are physically linked.

An important consequence of linkage disequilibrium is **hitchhiking**. This happens when an allele at one locus spreads by natural selection acting on a second locus that is in linkage disequilibrium with the first. **FIGURE 5.14** shows a hypothetical population of the grove snail (*Cepaea nemoralis*), which has separate loci that affect color and banding pattern on its shell. There is linkage disequilibrium in this population. The allele for an unbanded shell is correlated with the allele for a pink shell: all of the yellow shells are banded, but only half of the pink shells are. If all the banded snails die (for example, because birds prey on them), all the survivors will be pink as well as unbanded. Thus the frequencies of the shell color alleles changed because of selection on the banding alleles. While it sounds paradoxical, one implication

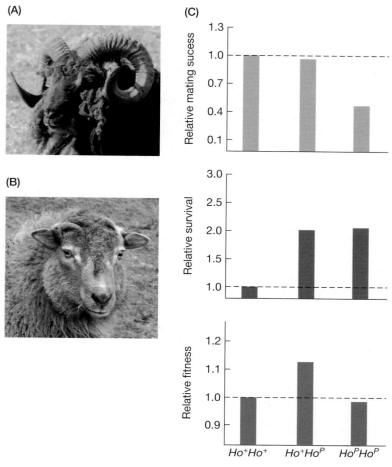

FIGURE 5.13 Genetic trade-offs between survival and mating success maintain a dramatic polymorphism in the size of horns in male Soay sheep. (A) A homozygous Ho^+Ho^+ male with normal horns. (B) A homozygous Ho^PHo^P male with vestigial horns. (C) Ho^PHo^P males have the lowest mating success, while Ho^+Ho^+ males have lowest survival. The net result is that Ho^+Ho^P heterozygotes have the highest overall fitness. (A, B from [23]; C after [23].)

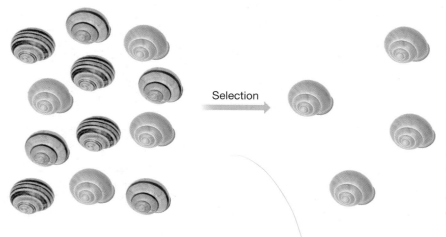

Selection

FIGURE 5.14 An allele can spread by hitchhiking if it is associated with another allele that is a target of selection. In this hypothetical population of the grove snail (*Cepaea nemoralis*), alleles for banding of the shell are in linkage disequilibrium with alleles for shell color. All yellow shells are banded, but only half of the pink shells are. If all snails with banded shells die (for example, as the result of bird predation), the survivors will be unbanded as well as pink. The frequency of shell color has changed even though selection acted only on banding. Note that linkage disequilibrium can occur between two loci even if they are not physically linked to each other.

of hitchhiking is that an allele that does not affect fitness can spread by natural selection.

Population genetic theory quantifies how much change in an allele's frequency will result from hitchhiking. The simplest situation is where two alleles at locus A are under selection. Allele A_2 has a selection coefficient s_A such that the relative fitnesses are $w_{11} = 1$, $w_{12} = (1 + s_A)$, and $w_{22} = (1 + 2s_A)$. The two alleles at locus B have no fitness effects. The change in the frequency of allele B_2 in one generation is then

$$\Delta p_B \approx s_A D \tag{5.6}$$

where D is the linkage disequilibrium between allele A_2 and allele B_2. (Recall from Chapter 4 that $D = P_{AB} - p_A p_B$, where P_{AB} is the frequency of gametes that carry both alleles A_2 and B_2, p_A is the frequency of allele A_2, and p_B is the frequency of allele B_2.)

If Equation 5.6 looks vaguely familiar, it is because it resembles Equation 5.3. The rate of evolution at locus B depends on two quantities. The first is the strength of selection acting on locus A, which is measured by the selection coefficient s_A. The second quantity is the linkage disequilibrium, D, which plays a role analogous to the term $p(1 - p)$ that represents genetic variation in Equation 5.3. Equation 5.6 tells us that hitchhiking will happen at locus B only if it is in disequilibrium with locus A (that is, D does not equal 0). Equation 5.6, like Equation 5.3, is an approximation that is very accurate for selection coefficients less than 0.1.

Hitchhiking is responsible for the evolution of genes that themselves do not impact survival or fecundity, but that do have other effects. In some environments, mutations spread in bacterial populations that drastically increase mutation rates throughout the genome. They do so because they generate mutations at other loci that are beneficial, and as those mutations spread, the mutator allele hitchhikes along with them.

Evolutionary biologists exploit hitchhiking to find genes that have recently evolved by positive selection. When a beneficial mutation first appears, it is in perfect linkage disequilibrium with all the other alleles on its chromosome (**FIGURE 5.15**). As the mutation spreads, recombination breaks down the disequilibrium. The breakdown is most rapid between the selected locus and distant sites on the chromosome (simply because there is more recombination between distant sites than neighboring sites). Sites close to the selected locus do not have a chance to recombine before the mutation becomes fixed. As a result, those sites carry the same alleles that were on the original chromosome where the mutation appeared. All

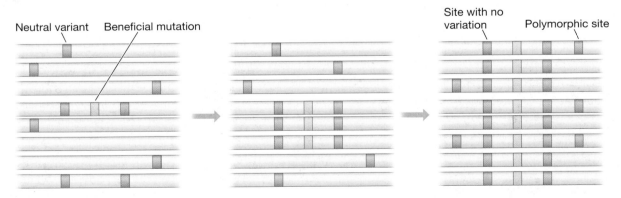

FIGURE 5.15 When a beneficial mutation spreads to fixation, the selective sweep eliminates polymorphism at nearby regions of the chromosome. The beneficial mutation (in yellow) first appears on a chromosome that has selectively neutral variants in its DNA sequence at two sites nearby (in blue). As the mutation spreads to higher frequency, the neutral variants hitchhike with it to higher frequency. When the mutation becomes fixed, genetic variation is eliminated in the region nearby. Regions of the chromosome further from the beneficial mutation retain variation because recombination joins together chromosomes that carry the beneficial mutation with chromosomes that carry different neutral variants.

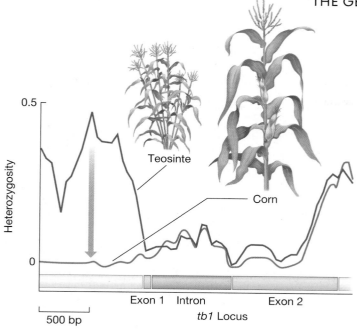

FIGURE 5.16 A selective sweep that occurred during the domestication of corn. The ancestor of corn, which looked very much like the living teosinte plant, was short and bushy. Domestication in Central America roughly 10 thousand years ago selected for corn plants that branched less. This favored a mutation in a region of chromosome that regulates expression of the gene *tb1*, which controls shoot branching. As the mutation spread, heterozygosity nearby on the chromosome was drastically reduced. Other evidence pinpoints the mutation in a noncoding region about 60 kb upstream (to the left) of *tb1* [38]. (After [43].)

variation at those sites is eliminated. Polymorphism remains, however, at regions farther away on the chromosome. This distinctive pattern is one of the telltale signs used by evolutionary geneticists to find evidence of recent adaptation [33].

This pattern is exactly what we see in corn (maize; *Zea mays mays*) (**FIGURE 5.16**). Starting a few thousand years ago, this crop was domesticated in Central America from a wild plant called teosinte. Early farmers bred the teosinte plants that were most productive and easiest to harvest, which generated strong selection on several traits. Among them was the plant's growth form. Teosinte is bushy, while corn grows as a single tall stalk. This major difference is largely the result of a single mutation that ancient farmers unconsciously selected and that is now fixed in corn [43]. The location of the selective sweep is revealed by a region of chromosome that has much lower heterozygosity in domesticated maize than in teosinte.

The pattern of polymorphism along the chromosomes we see in Figures 5.15 and 5.16 results when a mutation is favored by selection when it first appears. In some situations, an allele that is present in the population is initially not favored, but then suddenly becomes beneficial when conditions change. In this case, we say selection is acting on **standing genetic variation**. Before the change, different copies of the mutation will have had time to recombine onto chromosomes with different combinations of alleles at other sites. As a result, when the selected allele reaches fixation, only a very small region of the chromosome around the selected site shows reduced polymorphism (**FIGURE 5.17**). Evolutionary geneticists use the contrast between the patterns seen in Figures 5.15 and 5.17 to determine whether or not selection acted on standing genetic variation in loci that have recently experienced adaptive evolution.

Many other methods are also used to find genes that recently experienced or are under ongoing selection. A key to many of those methods is that they let us distinguish the action of selection from random genetic drift. We will therefore put off exploring those approaches to Chapter 7, where we discuss random factors and how they interact with selection.

When Selection Preserves Variation

Chapter 4 introduced the β-hemoglobin locus. Most humans are homozygous for the *A* allele. Populations in some regions of Africa and Asia also have appreciable frequencies of the *S* allele. Individuals who are *SS* homozygotes suffer from a

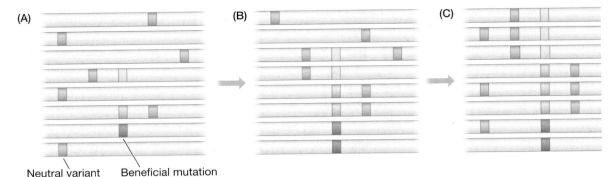

Neutral variant   Beneficial mutation

FIGURE 5.17 Changes in the environment can suddenly give a fitness advantage to a mutation that is already present in a population—a situation called selection on standing genetic variation. (A) Three copies of the mutation (yellow, orange, and red) have already recombined onto chromosomes with different neutral variants at other sites (blue). (B) As the copies of the beneficial mutation spread to higher frequency, the neutral variants linked to each of them hitchhike to higher frequency. (C) When the beneficial mutation becomes fixed, the region of reduced variation nearby on the chromosome is much smaller than if the mutation had a fitness advantage when it first appeared (compare with Figure 5.15). Patterns of variation along the chromosome can therefore be used to determine whether adaptation resulted from selection on standing genetic variation or from a mutation that was beneficial initially.

debilitating condition called sickle-cell anemia, which drastically decreases survival. But the *S* allele has not been eliminated by natural selection. Why?

The β-hemoglobin locus is under **balancing selection**, which is selection that maintains genetic variation with a population. Balancing selection is fundamentally different from selection on beneficial and deleterious alleles, which acts to remove genetic variation. To understand how selection can preserve variation, we will now explore the biology of the β-hemoglobin locus further.

Overdominance

FIGURE 5.18 shows that the *S* allele has high frequency in regions where malaria is common. Malaria is a disease caused by a protozoan parasite (*Plasmodium*) that is transmitted by mosquitoes in tropical regions around the world. About 500,000 people die each year from malaria, making it the most deadly infectious disease on Earth.

Alleles at several loci make people partially resistant to malaria. One of those is the β-hemoglobin locus. Individuals who are *AS* heterozygotes survive the malarial parasite better than *AA* homozygotes [7]. In this situation, it is most convenient

FIGURE 5.18 An allele that protects against malaria in humans is most frequent where the disease is common. The frequency of the hemoglobin *S* allele in Africa (A) is highly correlated with the incidence of malaria (B). Other loci provide protection against malaria in other human populations. (After [35].)

(A) Frequency of *S* allele

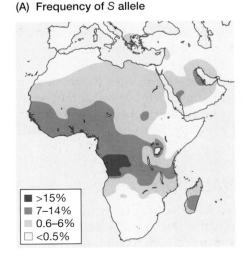

- >15%
- 7–14%
- 0.6–6%
- <0.5%

(B) Incidence of malaria

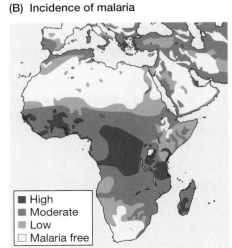

- High
- Moderate
- Low
- Malaria free

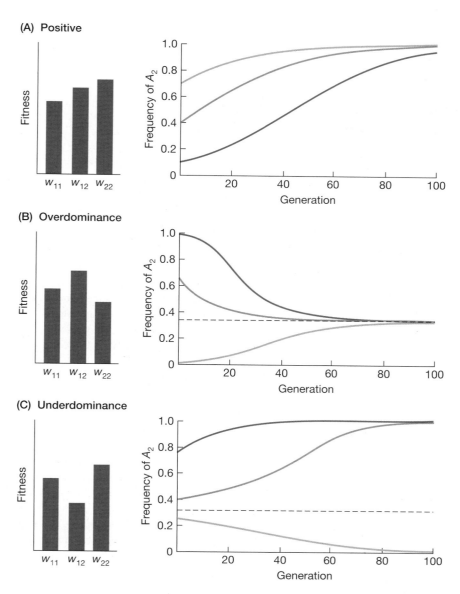

(A) Positive

(B) Overdominance

(C) Underdominance

FIGURE 5.19 Allele frequency trajectories under three kinds of selection. In each case, trajectories are shown for three different starting frequencies of the A_2 allele. (A) Positive selection favors allele A_2, which spreads to fixation from any starting frequency. In this example, A_2 increases fitness by $s = 0.05$, and the relative fitnesses are $w_{11} = 1$, $w_{12} = 1.05$, and $w_{22} = 1.1$. (B) With overdominance, the A_1A_2 heterozygote has highest fitness. Starting at any allele frequency, A_2 evolves to a polymorphic equilibrium. Selection preserves genetic variation. In this example, the relative fitnesses are $w_{11} = 0.9$, $w_{12} = 1$, and $w_{22} = 0.8$, and the final frequency of allele A_2 is 0.33 (dashed line). (C) With underdominance, the A_1A_2 heterozygote has the lowest fitness. Populations evolve to different final allele frequencies depending on where they start. In this example, the relative fitnesses are $w_{11} = 1.1$, $w_{12} = 1$, and $w_{22} = 1.2$. If allele A_2 starts at a frequency below 0.33 (dashed line), it is lost, but if it starts above that frequency it spreads to fixation.

to use the heterozygote as the fitness reference. Estimates for the relative survival of the three genotypes in Nigeria (a country in equatorial West Africa) are:

Genotype	*AA*	*AS*	*SS*
Relative survival	0.88	1	0.14

Although these data are now 40 years old, recent research shows that malaria continues to cause strong selection on the β-hemoglobin locus [14].

This is the most famous example of **overdominance**, which occurs when the heterozygote has higher fitness than both homozygotes. (Another example comes from the Soay sheep discussed earlier; see Figure 5.13.) Overdominance leads to a kind of evolutionary outcome that we have not yet discussed. The population evolves to a stable **polymorphic equilibrium**, which means that both alleles are maintained. Given the relative fitnesses of the *AA* and *SS* homozygotes (and again using the *AS* heterozygote as the fitness reference), a mathematical model like that in Box 5A tells us that any population that has genetic variation will evolve to the same final allele frequency (**FIGURE 5.19**). Ultimately the population reaches an equilibrium at which the frequency of the *S* allele is

$$\hat{p} = \frac{1 - w_{AA}}{2 - w_{AA} - w_{SS}} \tag{5.7}$$

This formula holds for any values of w_{AA} and w_{SS} that lie between 0 and 1.

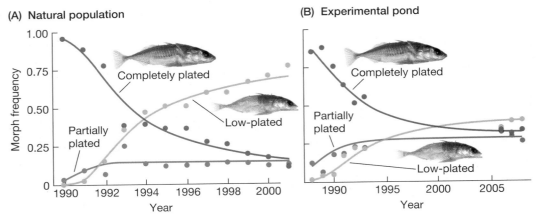

FIGURE 5.20 Strong selection on the bony plates that run along the sides of the three-spined stickleback (*Gasterosteus aculeatus*) causes rapid evolution at the *Eda* locus, which controls the plating. (A) Evolution of the frequencies of the three morphs for bony plates following colonization of a lake in Alaska by a marine population. The curves show the morph frequencies predicted when the relative fitnesses are $w_{11} = 1$, $w_{12} = 0.98$, and $w_{22} = 0.12$, where allele A_1 at the *Eda* locus results in low plates and allele A_2 results in high plates. Allele A_1 has the highest fitness and is predicted to spread to fixation, eliminating the completely and partially plated morphs. (B) The evolution of morph frequencies in an experimental freshwater pond following the introduction of plated sticklebacks. The curves are the predicted trajectories when the fitnesses are $w_{11} = 0.83$, $w_{12} = 1$, and $w_{22} = 0.43$. This situation corresponds to overdominance, which results in a stable polymorphism in which all three morphs are present. (After [26]; photos from [3b].)

We can use that formula and the data on survivorship to predict how the allele frequencies will evolve. The data show that $w_{AA} = 0.88$ and $w_{SS} = 0.14$. Plugging those values into Equation 5.7 predicts that the equilibrium frequency of the S allele will be $\hat{p} = 0.12$. That agrees well with the allele frequency along the coast of West Africa near the equator (see Figure 5.18). The evolutionary genetics of sickle-cell anemia have three general messages about evolution: overdominance maintains genetic variation, population genetic theory makes testable predictions about evolution, and humans (like all other species) are still evolving by natural selection.

The three-spined stickleback (*Gasterosteus aculeatus*) is a small fish that lives in oceans along the coasts of North America, Europe, and Asia. Remarkably, it has independently colonized thousands of freshwater streams and lakes around the world. Michael Bell and colleagues followed one of these invasive freshwater populations for 12 years in a lake in Alaska [5]. The bony plates that are typical of marine sticklebacks became much less frequent (**FIGURE 5.20A**). Later research by Arnaud Le Rouzic and colleagues suggested that most of this change resulted from evolution at a single locus called *Eda* [26]. They estimated the strength of selection on *Eda* by finding the relative fitnesses that produce the observed changes in the morph frequencies. The results suggested very strong positive selection that favored the homozygote for the low-plated allele (adapted to fresh water): its fitness was estimated to be more than eight times that of homozygotes for the high-plated allele (adapted to marine environments). They then introduced plated sticklebacks into a freshwater pond, which they monitored over the next 21 years. Here the fish evolved rapidly to a polymorphic equilibrium indicative of strong overdominance (**FIGURE 5.20B**). This example shows again how overdominance maintains genetic variation. Furthermore, the difference in the results from the natural lake population and the experimental pond underlines the key point that fitnesses depend on the environment.

Other forms of balancing selection

Overdominance is one form of balancing selection. A second type can occur with **frequency-dependent selection**, which occurs when the fitnesses of alleles

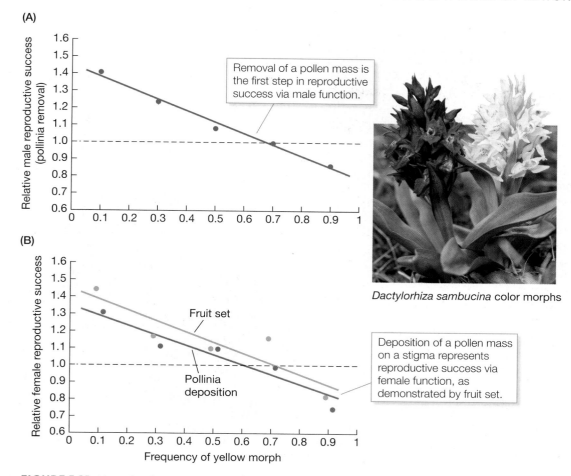

(A)

Removal of a pollen mass is the first step in reproductive success via male function.

(B)

Fruit set

Pollinia deposition

Deposition of a pollen mass on a stigma represents reproductive success via female function, as demonstrated by fruit set.

Dactylorhiza sambucina color morphs

FIGURE 5.21 Negative frequency-dependent selection maintains a polymorphism for flower color in the alpine elderflower orchid (*Dactylorhiza sambucina*), which is polymorphic for yellow and purple flowers (inset). Experimental gardens were planted with different proportions of the two colors. (A) Yellow flowers have declining relative fitness through male function (the amount of pollen removed) as yellow becomes more common in the population. (B) Yellow flowers also have declining relative fitness through female function (the number of fruits produced) as the frequency of yellow increases. Fitness in (A) and (B) is measured relative to purple flowers. The dashed lines indicate where yellow and purple flowers have equal fitness. (After [16].)

change depending on their own frequencies. In some cases, an allele gets a fitness advantage when it is rare, a situation called *negative frequency dependence*. This can maintain polymorphism because when one allele becomes rare, selection causes its frequency to increase. In the alpine elderflower orchid (*Dactylorhiza sambucina*), about half the plants have purple flowers and about half have yellow flowers (**FIGURE 5.21**). Ten experimental arrays of 50 plants, with varying proportions of the two colors, were put out into the field. Results showed that pollinating bumblebees visited flowers of the rare color more frequently, which increased the number of their seeds that were fertilized and the amount of pollen taken from them and carried to other flowers. Why did the bees prefer the rare color? Like many orchids, *Dactylorhiza* does not reward pollinators with nectar or pollen. The bees may have learned more quickly to avoid the orchids with the common flower color.

A third form of balancing selection results when different genotypes specialize on different ecological niches. In effect, each genotype is partly shielded from competition with other genotypes, and so has its own ecological carrying capacity.

FIGURE 5.22 A malaria mosquito is polymorphic for chromosome inversions in Cameroon, a country in Africa. (A) The mosquito *Anopheles funestus* is one of the species that transmits malaria. (B) A photo of part of chromosome 3. Inversion 3Ra is highlighted in green. (C) The pie diagrams show the frequencies of the three genotypes in villages along a highway: white are homozygotes without the chromosome inversion (SS), black are homozygotes with the inversion (II), and blue are heterozygotes (SI). Three groups of populations are visible. In the north, which is hot and dry savannah, the inversion is almost entirely absent. In the south, which is lowland rainforest, the inversion is at almost 100 percent frequency. In the center of the country, which is a mountainous highland, the inversion is at intermediate frequencies. (B courtesy of Igor V. Sharakhov, after [37]; C from [2].)

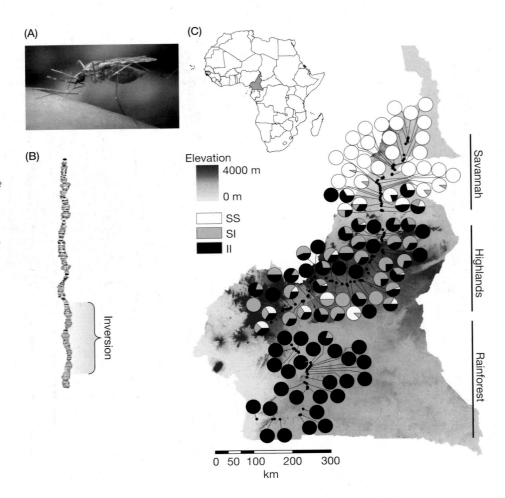

Genetic variation that is maintained this way is called **multiple niche polymorphism**. Many herbivorous insects have "host races" that specialize on different species of host plant. The pea aphid (*Acyrthosiphon pisum*) has host races adapted to different crop plants. Each host selects for different alleles at several loci, which maintains polymorphism in the aphid [22].

A final type of selection that can maintain genetic variation occurs when fitnesses vary in space. This situation is similar to multiple niche polymorphism, but it is usually not considered to be a form of balancing selection because it does not occur within a single population. **FIGURE 5.22** shows a map of the frequency of a chromosome inversion that occurs in a mosquito that lives in Cameroon, Africa. The inversion is near fixation in the south, nearly absent in the north, and present at intermediate frequencies in the center of the country. Analysis of these data suggests that fitnesses vary along a north-south axis: in the south, chromosomes with the inversion have higher fitness than those without it, while the reverse is true in the north [2]. Chapter 8 explores in detail the consequences of selection that varies in space.

Perhaps surprisingly, selection that fluctuates in time does not generally maintain genetic variation. If selection favors one allele in some generations and a different allele in others, in many cases the allele that has the highest fitness on average will spread to fixation. Likewise, the evolutionary trade-offs discussed earlier in the chapter do not typically maintain polymorphism. The horn polymorphism in Soay sheep is unusual. More commonly, despite trade-offs, one allele is on average more fit than the other, and it will spread to fixation.

H. argophyllus

FIGURE 5.23 Comparing the chromosomes of sunflower species reveals chromosome translocations. After *Helianthus argophyllus* speciated from *H. annua*, chromosomes 12 and 16 exchanged pieces (translocations). Arrows show how the pieces were rearranged. The green and blue lines join homologous genes on the chromosomes of *H. annua* (above) and *H. agrophyllus* (below). The left-hand piece of chromosome 12 was joined with the left-hand end of chromosome 16 (segments 1 and 2 on the *H. argophyllus* chromosome). The piece from chromosome 16 also flipped end to end (segment 2). The right-hand piece of chromosome 12 was joined with the right-hand piece of chromosome 16 (segments 3 and 4). Experimental crosses show that heterozygotes for these translocations have low fitness. (After [3].)

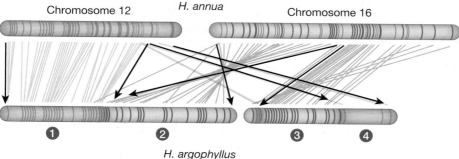

Chromosome 12   *H. annua*   Chromosome 16

H. argophyllus

Selection That Favors the Most Common

Balancing selection preserves genetic variation, and in most cases the population will evolve to the same allele frequency no matter where it begins. A different picture emerges from two kinds of selection that favor whichever allele is most common. Here variation is eliminated, and which allele spreads to fixation depends on the initial allele frequency in the population. These next two kinds of selection therefore lead to *historical contingency*: the outcome of evolution is determined by where the population begins.

Underdominance: When heterozygotes suffer

The familiar sunflower is just one of about 70 species of sunflowers found across North America. They differ in many ways, for example in their growth forms and the habitats where they grow. Their chromosomes also show differences: closely related species differ in the number of chromosomes they have and in the order of genes along the chromosomes (**FIGURE 5.23**). These differences are the result of chromosomal inversions and translocations (discussed in Chapter 4) that became fixed during the evolution of these species.

Did these chromosomal changes become established because they increased fitness and spread as advantageous mutations? Surprisingly, many of them *decreased* fitness when they first appeared. Sunflower species can be hybridized, which allows measurement of the fitness of individuals with different combinations of chromosomes. Heterozygotes for some chromosome rearrangements have lower fertility than either homozygote because their chromosomes fail to pair correctly during meiosis, leading to infertility [25]. When a new rearrangement is still at low frequency, almost all of its copies are in these low fitness heterozygotes. Thus selection acts to eliminate a new chromosome rearrangement when it is still rare.

The situation in which heterozygotes have lowest fitness is called **underdominance**. As its name implies, underdominance is the opposite of overdominance: it eliminates rather than preserves genetic variation. The evolutionary outcome

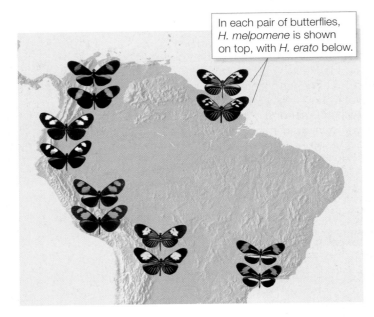

In each pair of butterflies, *H. melpomene* is shown on top, with *H. erato* below.

FIGURE 5.24 Positive frequency-dependent selection favors whatever color pattern is locally most common in populations of poisonous butterflies. The butterflies *Heliconius melpomene* and *H. erato* each show extraordinary geographical variation in coloration, and their colors vary in parallel. Both species gain a fitness advantage by resembling the other because birds are more likely to associate their coloration with distastefulness and so avoid attacking them. The birds learn to avoid a color pattern more quickly when it is common. This results in positive frequency-dependent selection, and eliminates variation in color within populations of both species. (Photos courtesy of Andrew Brower.)

is shown in Figure 5.19C. The key feature is that if an allele's frequency is below a threshold value, then selection tends to drive it out of the population. However, if the allele frequency is above the threshold, selection favors it to increase further. The value of the threshold is determined by the relative fitnesses of the two homozygotes.[2]

If some chromosome rearrangements are underdominant, how did they become fixed in different species? Since selection is clearly not the answer, something else must be responsible. We will return to this mystery when we discuss random genetic drift (see Chapter 7) and meiotic drive (see Chapter 12).

Positive frequency-dependent selection

Earlier we discussed negative frequency-dependent selection, which preserves variation by favoring an allele when it is rare. Frequency-dependent selection can also favor the most common allele, a situation called *positive frequency dependence*. The butterfly *Heliconius erato* is a single species with many geographic races that differ dramatically in their coloration (**FIGURE 5.24**). This species is distasteful, and birds that eat a butterfly quickly learn to avoid it. In field experiments, butterflies were marked and released with two color patterns: the local pattern and a pattern from one of the neighboring races [31]. Butterflies with the local pattern survive more than twice as well as those with the "wrong" color pattern, which generates a selection coefficient of about $s = 0.17$ at each of the three major loci that control the differences in color pattern among the races. Other data show that these survival differences result from bird predation: the birds avoid the color pattern they know, but attack butterflies that have unfamiliar colors. As a result, strong selection favors whatever color is locally common. Within a population there is typically little variation for color: positive frequency-dependent selection eliminates polymorphism. The same form of selection occurs on the closely related *Heliconius melpomene*, which mimics and co-occurs with *H. erato*.

The Evolution of a Population's Mean Fitness

Many of our questions about evolution concern adaptations such as the feathers of birds and the brains of humans. But we can also ask: How does fitness itself evolve?

The **mean fitness** of a population, which is abbreviated with the symbol $\overline{w}$, is simply the average of the fitnesses of the individuals in it. We can easily calculate $\overline{w}$ if we know the fitnesses and the frequencies of the three genotypes at a locus. (Mean fitness can be calculated using either relative or absolute fitnesses, but in either case the symbol $\overline{w}$ is used.) As selection causes the allele frequency p to evolve, the mean fitness $\overline{w}$ evolves. The evolution of $\overline{w}$ follows simple but important rules when fitnesses are constant in time and other evolutionary factors (such as mutation) are weak relative to selection. These principles were discovered by R. A. Fisher and Sewall Wright, two of the founders of population genetics (see Chapter 1).

---

[2] This type of threshold is called an unstable equilibrium. If the allele frequency lies exactly at this equilibrium, it will not change, but the slightest deviation will cause the allele frequency to evolve either to 0 or to 1. In contrast, overdominance produces a stable equilibrium toward which allele frequencies converge.

The fundamental theorem of natural selection and the adaptive landscape

Fisher showed mathematically that evolution by natural selection causes $\overline{w}$ to increase through time. When fitness is normalized so that $\overline{w} = 1$, Fisher found that the increase in mean fitness per generation is simply equal to the genetic variance for fitness itself.[3] Fisher named this principle the **fundamental theorem of natural selection**. Because a variance can never be negative, the important message from this result is that natural selection causes populations to evolve so that they become better adapted to their environment: the average survival and reproduction of individuals increase through time.

During the spread of a beneficial allele, its frequency increases slowly when the allele is rare, then accelerates as the allele frequency nears 0.5, and then slows again as the allele nears fixation (see Figures 5.7 and 5.10). We can understand that rhythm of change in terms of the fundamental theorem. When the beneficial allele A_2 is rare, almost all of the population is homozygous for the A_1 allele (see Figure 5.8). There is little variation among individuals in fitness, so $\overline{w}$ increases slowly. The variance for fitness among individuals is maximized when $p = 0.5$, and so the change in $\overline{w}$ per generation is greatest. The variance in fitness is again small as A_2 nears fixation, so the rate of increase in $\overline{w}$ is again small.

Fisher's fundamental theorem leads to the question of just how much genetic variation for fitness exists in natural populations. The answer depends on the species, the time, and the place, but it seems that the genetic variance in relative fitness may often be a few percentages [8, 9]. All else being equal, the fundamental theorem would lead us to expect that the mean fitness of species should increase by a few percent per generation. But all else is not equal: what selection gives, other evolutionary forces take away. The fitness gains made by selection are continuously offset by environments that change in space and time, deleterious mutations, and other factors.

A complementary perspective on the evolution of fitness was developed by Wright. He plotted the mean fitness, $\overline{w}$, against the allele frequency, p. This plot, which Wright called the **adaptive landscape**, tells us how the population will evolve. His key insight was that *selection causes populations to evolve uphill on the landscape* (**FIGURE 5.25**). Wright proved mathematically that the allele frequency will change at a rate

$$\Delta p = \frac{1}{2}p(1-p)\frac{d}{dp}\ln(\overline{w}) \tag{5.8}$$

On the right side of this equation, you will recognize $p(1 - p)$ as a measure of genetic variation: it equals 0 when there is only one allele in the population (that is, $p = 0$ or $p = 1$), and is maximized when the two alleles are equally common [that is, $p = (1 - p) = 0.5$]. The last term on the right is a derivative that is equal to the slope of the adaptive landscape. This measures the direction and strength of selection. When the slope is positive, selection causes A_2 to spread, and when it is negative, it causes A_2 to be lost. When the slope of the landscape is 0, selection does not favor either allele, and the population is at equilibrium.

The adaptive landscape explains the differences in the allele frequency trajectories that we saw earlier among positive, overdominant, and underdominant selection (see Figure 5.19). With positive selection, the slope of the adaptive landscape is always positive, and selection causes the advantageous allele to spread until it reaches the peak of the landscape at the far right, where $p = 1$. With overdominance, the landscape has a peak at an intermediate allele frequency. No matter what the initial allele frequency is, selection pushes the population to the adaptive peak, where it reaches the polymorphic equilibrium. Underdominant selection produces

---

[3] Technically, it is the "additive" genetic variance for fitness that matters. Additive genetic variance is discussed in Chapter 6.

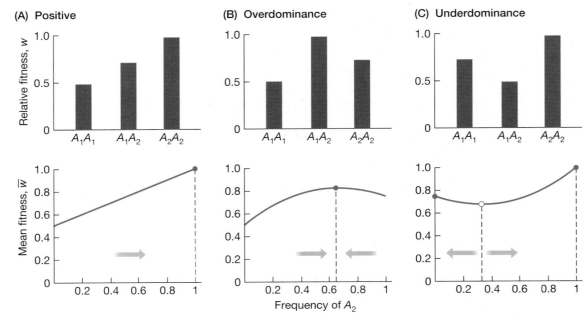

FIGURE 5.25 Wright's adaptive landscape under positive, overdominant, and underdominant selection. For each case, the fitnesses of the three genotypes are shown at the top, and the adaptive landscape that results from them at the bottom. The landscapes are the population's mean fitness, $\overline{w}$, plotted against the frequency of allele A_2. From a given starting point, the allele frequency evolves in the direction that increases $\overline{w}$. The arrows show the direction of allele frequency change. The vertical dashed lines correspond to peaks and valleys on the landscapes. (A) Positive selection in which allele A_2 is favored. The allele spreads until it is fixed, which corresponds to the frequency at which mean fitness is maximized (the solid circle). (B) With overdominance (heterozygote advantage), the population evolves to a stable polymorphic equilibrium, in this case with A_2 at a frequency of 0.66. The mean fitness, $\overline{w}$, is again maximized at this frequency (the solid circle). (C) With underdominance (heterozygote disadvantage), if the frequency of A_2 starts below a threshold (here at 0.33), its frequency declines until it is lost. If it starts above the threshold, the frequency of A_2 increases until it becomes fixed. The threshold represents an unstable equilibrium that minimizes mean fitness, $\overline{w}$ (the open circle). The loss of A_2 and fixation of A_2 correspond to two local peaks on the adaptive landscape (solid circles).

an adaptive valley. If the allele frequency starts to the left of the low point, selection drives the allele frequency toward 0. However, if the allele frequency starts to the right of the valley, it evolves uphill toward 1. This visualization of how selection causes gene frequencies to evolve is one of the most famous images in evolutionary biology.

The fundamental theorem and the adaptive landscape make assumptions that do not apply exactly to any natural populations. In many cases, though, they give very good approximations that are useful to guide our thinking about evolution. In other cases, the assumptions are violated in ways that make evolution behave very differently. A particularly important situation where the fundamental theorem and adaptive landscape do not apply is when selection is frequency dependent. In some cases, this can cause the mean fitness of a population to decline [45].

FIGURE 5.26 shows a cartoon of an example. In a population of a bush, each individual makes many fruits. One day, a mutation appears that makes individuals grow a trunk. Their neighbors are shaded out and die, which gives individuals with the mutation more water and nutrients from the soil. The mutation therefore spreads. But this fitness advantage comes at a cost: energy diverted into growing a trunk causes individuals with the mutation to produce fewer fruits than did their ancestors without the mutation. (Just imagine how much more fecund an oak tree could be if all the energy devoted to growing its enormous truck were converted

(A)

Ancestral population

FIGURE 5.26 Frequency-dependent selection resulting from competition can cause the mean fitness of a population to decline. A mutation in a population of bushes produces a trunk. The mutation gains a fitness advantage by shading neighbors, but it reduces fecundity. As the mutation spreads, the mean fitness (represented by the number of fruit) declines. See the text for further details.

(B)

Mutation introduced that causes plants to grow trunks, shading out original shrub population

(C)

Population has fewer and smaller fruit per individual when compared to ancestral population

into acorns.) As the mutation spreads, the mean fitness of individuals in the population, shown in the cartoon by the number of fruits they produce, declines.

In this example, the mutation spreads because individuals with it have higher fitness than their neighbors. That advantage, however, depends on the frequency of the mutation. When the mutation first appears, it has high fitness because all the neighboring individuals are small and easily shaded. As the mutation becomes more common, tall plants are often shaded by other tall plants, and so the population's mean fitness declines.

Many animals and plants have traits that enhance their abilities to compete with others. Even microbes have these kinds of nasty traits. Some bacteria secrete compounds called bacteriocins that kill neighboring cells, giving them more nutrients [44]. The bacteria that make bacteriocins also have genes that protect them from the toxin, but both the poison and the antidote are expensive to make. A benign population of microbes that does not make bacteriocins does not waste that energy. Nevertheless, individuals that make bacteriocins become more frequent because their individual fitness is higher. As they spread, however, the mean fitness of the population declines.

Deleterious Mutations

Some mutations increase fitness: they make individuals more resistant to a disease, say, or better able to digest a new food resource. But the vast majority of mutations that have fitness effects are deleterious. Studies of the fitness effects of mutations suggest that deleterious mutations are at least ten times more common than beneficial mutations (see Figure 4.17). The injection of deleterious alleles into populations by mutation has important evolutionary consequences.

A mutation-selection balance

Deleterious mutation is an important cause of genetic disease in humans and other organisms. The famous American musician Woody Guthrie died at the age of 55 of Huntington's disease. This results from a dominant mutation that causes degeneration of the central nervous system. In the United Kingdom, the mutation is present in about 12 out of 100,000 people. More than 4000 genes have been identified in humans that when mutated cause diseases such as Down syndrome, cystic fibrosis, color blindness, and hemophilia. Selection that acts to remove deleterious mutations from a population is called **purifying selection**.

Why hasn't purifying selection eradicated deleterious mutations that cause these diseases? The answer is that they are being continually reintroduced. This flow of new mutations into the population is offset by natural selection that acts to eliminate them. This situation leads to a **mutation-selection balance**. Here it is most convenient to use the mutation-free homozygote as the fitness reference, and write the fitness of the mutant heterozygote as $(1 - s)$. When the input of the deleterious allele by mutation balances its removal by selection, the deleterious mutation reaches an equilibrium frequency of

$$\hat{p} \approx \frac{\mu}{s} \tag{5.9}$$

On the right side of the equation, μ is the mutation rate, that is, the probability that a copy of the normal allele mutates to a deleterious allele in a given generation. In the denominator is the selection coefficient s, which is the proportional decrease in relative fitness caused by carrying a copy of the mutation.[4]

Equation 5.9 carries the simple and intuitive message that a deleterious mutation will be more common if it appears at a higher rate (larger μ) and has weaker deleterious effects (smaller s). While both numbers vary tremendously among loci and organisms, to make the ideas clear consider a locus that mutates to a deleterious allele at a rate of $\mu = 10^{-6}$ per generation, and this allele decreases relative fitness by $s = 0.01$. At equilibrium, the deleterious allele will have a frequency of $\hat{p} = 0.0001$.

The mutation load

Deleterious mutations decrease survival and/or fecundity, and so they decrease a population's mean fitness. Remarkably, the impact on mean fitness of a mutation is independent of whether the deleterious effect is strong or weak. Equation 5.9 shows that very harmful mutations will only persist at very low frequencies, while mutations with milder effects will reach equilibrium at higher frequencies. The net effect is that highly deleterious and weakly deleterious mutations decrease the population's mean fitness by the same amount.

The **mutation load**, represented by L, is the proportion by which the mean fitness of individuals in the population is reduced by deleterious mutations compared

---

[4] Equation 5.9 is an approximation that applies when $\mu \ll s$ and the mutation is not completely recessive, which is typically the case. The fitness of the mutant homozygote does not matter because it is so rare in this case. A more complex equation applies when those conditions do not hold (see [18]).

with a hypothetical population without mutations. Mutations that reduce individual fitness to different degrees have the same mutation load when they are at mutation-selection equilibrium, because of the balance between their harmfulness and their frequency. This means that the effect of a deleterious mutation on a population's fitness is unaffected by s, that is, the size of the mutation's effect on fitness. The load is determined only by the rate at which mutations enter the population. The load caused by deleterious mutation at a single locus is simply 2μ. For example, with mutation rate of $\mu = 10^{-6}$, the mean fitness of the population is reduced by a very small amount: $L = 0.000002$.

But while mutation rates at individual loci are typically very small, eukaryotes (including humans) have a great many loci. We use U to represent the total mutation rate across the genome for deleterious alleles. That is, U is the average number of new deleterious mutations that are added to the genome each generation. Recent studies suggest that U in humans is about two new deleterious mutations per genome per generation [24]. A result is that each of us carries hundreds of deleterious mutations at various loci scattered throughout the genome [1, 15]. Their combined impact on mean population fitness (i.e., the total mutational load) depends on how the effects of deleterious mutations at different loci combine to determine overall fitness, which is very difficult to estimate accurately. If we assume that mutations have independent effects on fitness, the mutation load is:

$$L = 1 - e^{-U} \qquad (5.10)$$

where e is a mathematical constant approximately equal to 2.7.

The genome-wide mutation rate estimated for humans is $U = 2$, so this equation says that the load is $L = 0.86$. This implies that 86 percent of the potential mean fitness in humans is lost to the effects of deleterious mutations, either by mortality or reduced fertility.

Do deleterious mutations really have that big an impact on human health? The actual effect is likely much smaller because it also depends on factors such as demography and how individuals compete for resources. Furthermore, the assumption that mutations have independent fitness effects may not be correct [27]. Nevertheless, deleterious mutations do contribute to senescence and genetic disease in humans and other species (see Chapter 11). Less than half of eggs that are fertilized lead to a successful birth [29], and it is possible that this high rate of mortality at the earliest stages of development is partly caused by deleterious mutations.

What is the impact of modern health care on deleterious mutations? By saving individuals from genetic diseases, medical intervention does allow mutations to be passed on that would have been eliminated by natural selection earlier in human history. But population genetic theory tells us that the frequency of these mutations will increase slowly, and many generations will pass before they become much more common. Whether the end result will have a major impact on human health is controversial [19, 24, 28].

Go to the
─────────── **Evolution Companion Website** ───────────
EVOLUTION4E.SINAUER.COM
for data analysis and simulation exercises, quizzes, and more.

SUMMARY

- Natural selection is any consistent difference in fitness among different phenotypes or genotypes. Evolution caused by natural selection has been observed directly many times.

- Fitness is the number of offspring that an individual leaves to the next generation, or the average number that an allele, genotype, or phenotype leaves. Selection causes evolution when there is a correlation between a phenotype and fitness, and a correlation for that phenotype between parents and offspring.

- The rate at which a beneficial allele spreads through a population is determined by how strongly it is favored, measured by its selection coefficient, and by the amount of genetic variation at that locus in the population. Even if it is favored by selection, a new beneficial mutation can be lost by chance when it is still rare.

- An allele that has no effect on fitness can spread if it is associated (in linkage disequilibrium) with an allele at another locus that is favored by selection. One consequence is that a selective sweep reduces genetic variation in the region of chromosome near the selected locus.

- Several kinds of selection can act to maintain genetic variation. One is overdominance, the situation in which heterozygotes have highest fitness. Other kinds are negative frequency dependence, multiple niche polymorphism, and spatial variation in selection.

- Positive frequency-dependent selection occurs when fitness increases with the frequency of genotypes or phenotypes in a population. Unlike balancing selection, this situation eliminates variation. Which allele becomes fixed depends on the initial allele frequency.

- With underdominance, heterozygotes have lower fitness than both homozygotes. Underdominance does not preserve genetic variation, and one of the alleles will either be lost or spread to fixation, depending on its initial frequency. An allele that is underdominant can spread when rare only if some evolutionary factor other than selection is at work.

- The mean fitness of a population evolves as allele frequencies change. Fisher's fundamental theorem of natural selection states that selection causes the mean fitness to increase. In Wright's adaptive landscape, allele frequencies change in the direction that increases mean fitness. These conclusions hold only under certain conditions. Evolution can cause a population's mean fitness to decrease when fitnesses are frequency-dependent.

- Deleterious mutations occur frequently. They are maintained in populations by mutation even though selection acts to remove them. Their combined effects across the entire genome contribute to senescence.

TERMS AND CONCEPTS

absolute fitness
adaptive landscape
artificial selection
balancing selection
deleterious mutation
evolutionary trade-off
fitness
fitness component

fixed, fixation
frequency-dependent selection
fundamental theorem of natural selection
genetic correlation
hitchhiking

linkage disequilibrium
mean fitness
multiple niche polymorphism
mutation load
overdominance
polymorphic equilibrium

positive selection
purifying selection
relative fitness
selection coefficient
selective sweep
standing genetic variation
underdominance

SUGGESTIONS FOR FURTHER READING

Many of the concepts discussed in this chapter were first explored by the three people who pioneered the subject of population genetics in the early twentieth century: R. A. Fisher, Sewell Wright, and J. B. S. Haldane. The title of this chapter is taken from that of a book written by Fisher in 1930 (*The Genetical Theory of* Natural Selection, Oxford University Press, Oxford). It is a brilliant but difficult work that is still an inspiration for researchers. W. B. Provine's *The Origins of Theoretical Population Genetics* (University of Chicago Press, Chicago, 1971) is a wonderful history of the period leading up to the rise of population genetics.

Several texts on population genetics give the current view of the field. John Gillespie's *Population Genetics: A Concise Guide* (Johns Hopkins University Press, Baltimore, MD, 2004) is concise and accessible. D. L. Hartl and A. G. Clark's *Principles of Population Genetics* (Sinauer Associates, Sunderland, MA, 2007) is a more detailed exploration. The more recent text by R. Nielsen and M. Slatkin, *An Introduction to Population Genetics: Theory and Applications* (Macmillan Education, 2013), emphasizes how the principles of population genetics are applied to data, including in the study of human evolution. Brian and Deborah Charlesworth's *Elements of Evolutionary Genetics* (Roberts and Co., Greenwood Village, CO, 2010) is an advanced but definitive summary of the field of evolutionary genetics.

Molecular evolution is among the fastest moving areas of biology. The data, analyses, and data-collection technologies change so rapidly that any review we suggest here will soon be dated. Given that caveat, we recommend R. Nielsen's review "Molecular signatures of natural selection" (*Annu. Rev. Genet.* 39: 197–218, 2005) for those interested in learning more about how we use genetic data to find evidence of adaptation. "Strength in small numbers" by S. Tishkoff (*Science* 349: 1282–1283, 2015) and "Signals of recent positive selection in a worldwide sample of human populations" by J. K. Pickrell and colleagues (*Genome Res.* 19: 826–837, 2009) give nice summaries of adaptation in humans.

For a broader perspective on what genetic data can tell us about adaptation, "Molecular spandrels: tests of adaptation at the genetic level" by R. D. H. Barrett and H. E. Hoekstra (*Nat. Rev. Genet.* 12: 767–780, 2011) and "Commentary: When does understanding phenotypic evolution require identification of the underlying genes?" by M. D. Rausher and L. F. Delph (*Evolution* 69: 1655–1664, 2015) give thoughtful perspectives on a complex topic.

PROBLEMS AND DISCUSSION TOPICS

1. If the egg-to-adult survival rates of genotypes A_1A_1, A_1A_2, and A_2A_2 are 90 percent, 85 percent, and 75 percent, and their fecundities are 50, 55, and 70 eggs per female, respectively, what are the absolute fitnesses (W) of these genotypes? Using A_1A_1 as the fitness reference, what are the relative fitnesses (w)? If the frequency of the A_2 allele is $p = 0.5$, what will be its frequency one generation later? What will be the allele frequency when the population reaches equilibrium (stops evolving)?

2. How rapidly would a large population adapt to an environmental change if an advantageous allele were already present at low frequency (say, 1 percent), compared to the situation where it adapts by a newly arisen mutation at that locus? Would both of these events be accompanied by a selective sweep? Would they both be detectable by studying variation in the DNA sequence near the locus after adaptation occurred?

3. Describe a situation in which evolution does not occur even though natural selection is acting on a genetically variable character. (Assume that genetic drift is not occurring.)

4. Imagine a population in which the survival of A_1A_1 homozygotes is 80 percent as great as that of A_1A_2 heterozygotes, while the survival of A_2A_2 homozygotes is 95 percent that of the heterozygotes. What is p, the frequency of the A_1 allele,

at equilibrium? Now suppose the population has reached this equilibrium, but that the environment then changes so that the relative fitnesses of A_1A_1, A_1A_2, and A_2A_2 become 1.0, 0.95, and 0.90. What will p be in the adults after one generation of selection in the new environment?

5. Suppose a species has two generations per year, that adult survival rates of genotypes A_1A_1, A_1A_2, and A_2A_2 are identical, and that the fecundity values are 50, 55, and 70 in the spring generation and 70, 65, and 55 in the fall generation, respectively. Will polymorphism persist, or will one allele become fixed if fecundity values are unchanged for many years? What if the fecundity values are 55, 65, 75 in the spring and 75, 65, 55 in the fall?

6. What hypotheses could account for the observation that more genes have experienced recent adaptive evolution in the chimpanzee genome than in the human genome?

7. Do you expect that natural selection acting within a species would increase the population size of the species? Do you expect that it would increase the rate at which new species arise, thus increasing the number of species?

Phenotypic Evolution

What was the most important advance in human history? Fire? The wheel? The computer? A strong case can be made for the harnessing of evolution by our ancestors some 11,000 years ago. They bred wheat that provided more grain and cows that produced more milk. Selective breeding produced domesticated plants and animals that gave the first farmers vastly more food than their ancestors could have imagined (**FIGURE 6.1**). For the first time in history, humans had abundant resources. They abandoned a nomadic lifestyle and settled in communities that became the earliest villages, towns, and then cities. Much of civilization—buildings, writing, commerce—only then became possible. In short, the genetic modification of plants and animals by selective breeding is the foundation of human civilization. The discovery of the power of selective breeding is simply the discovery that selection and inheritance together can produce large evolutionary changes. And, as Darwin pointed out, evolutionary changes caused by artificial selection of domesticated animals and plants illustrate what natural selection can do in the wild.

Traits such as crop yield in corn, milk production in cows, and body height in humans are examples of **quantitative traits**. These are traits that vary continuously and that are affected by several, sometimes thousands, of loci (and for that reason they are also called **polygenic traits**). **Quantitative genetics** is the study of how quantitative traits are inherited and how they evolve.

This chapter begins by looking at how genes and the environment affect quantitative traits, how selection acts on them, and how fast they evolve in response. The next topic is artificial selection, in which people selectively breed

The common sunflower (*Helianthus annuus*) has been selectively bred for its showy flowers, and to increase its production of oil and seeds. Hundreds of domesticated animal and plant species have been vastly modified by artificial selection.

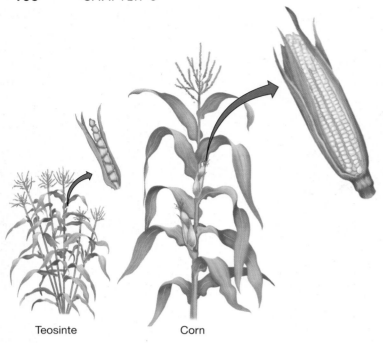

Teosinte   Corn

FIGURE 6.1 Modern corn and its wild ancestor, teosinte, differ in many ways. The striking differences in the pattern of branching result from a small number of genetic changes. Other differences, for example in the size of the cob, involve changes at many loci.

animals and plants to improve them for food production and other purposes. The chapter then describes how correlations between traits alter evolutionary trajectories. The final topic is the genetic basis of quantitative traits. Quantitative genetics is largely based on statistics, and the Appendix provides a quick introduction to the key concepts.

Genotypes and Phenotypes

Variation in DNA is discrete. At any particular site on a chromosome, the DNA can be one of four possible bases (A, C, G, or T). But a quantitative trait like height in humans varies in a continuous way (**FIGURE 6.2**). What is the connection between discrete variation in the DNA and continuous variation in height? Height is affected by thousands of loci [53]. It is also affected by environmental (nongenetic) influences, such as nutrition during early development. Identical twins have slightly different heights for that reason. The distinction between phenotypes and genotypes made in Chapter 5 is particularly important for quantitative traits: the phenotype can be directly seen, but the genotype cannot.

The variation in quantitative traits like height that is visible is measured by the **phenotypic variance**. This is simply the variance in the measurements of the trait in the population. (Variance is a key concept in this chapter, and is explained in the Appendix.) The phenotypic variance results from both genetic and environmental (nongenetic) causes. **FIGURE 6.3** shows how these factors combine to determine

FIGURE 6.2 Height in humans is a classic example of a quantitative trait that varies continuously. These college students are lined up behind signs that show their heights, varying from short on the left to tall on the right. Women are dressed in white and men in blue. This is an example of a normal (or bell-shaped) distribution. Other continuous distributions have different shapes.

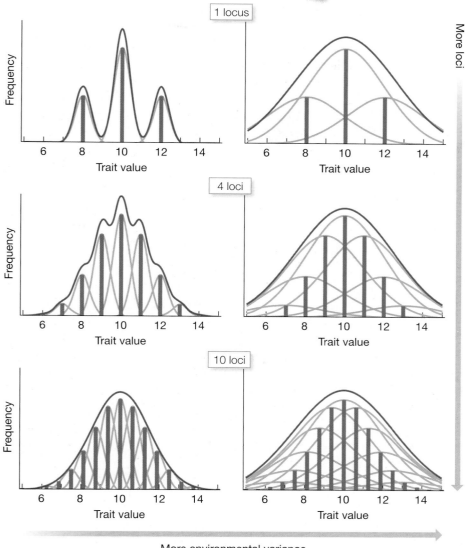

1 locus

4 loci

10 loci

More loci

More environmental variance

FIGURE 6.3 A phenotypic distribution is determined by genetic and environmental factors. The vertical blue bars show the phenotypes that would be produced without any environmental influence. The green curves show the phenotypic distributions for each of the genotypes that result with the addition of environmental variance. The red curve shows the phenotypic distribution for the entire population. Each locus has two alleles with frequencies 1/2. The loci are in linkage equilibrium, and alleles have equal and additive effects on the trait (there is no dominance or epistasis). The phenotypic distributions become smoother with larger numbers of loci. Left panels: Small environmental variance. Right panels: Large environmental variance.

the distribution of phenotypes in a population. The three rows show the situation when different numbers of loci contribute variation to the trait. The left-hand panels show traits that have small amounts of environmental variation, while the right-hand panels show traits for which the genetic and environmental sources of variance are about equally large (a situation typical of many traits). When just a single locus affects the trait, the phenotypic distribution shows distinct categories if environmental variation is small, but the distribution is smooth if there is more environmental variation. When ten loci contribute variation to the trait, the distribution is very smooth even when there is little environmental variation. The combination of a moderate number of loci with some environmental variation explains why so many familiar traits such as body height vary in a smooth, continuous way. Figure 6.3 shows another feature common to quantitative traits: phenotypes often follow a **normal distribution**, which is also known as a Gaussian or bell-shaped distribution.

The mean value of the trait in the population evolves when allele frequencies at the loci change. The change in the mean can be so large that the range of trait values in the population falls entirely outside the range that was present in the

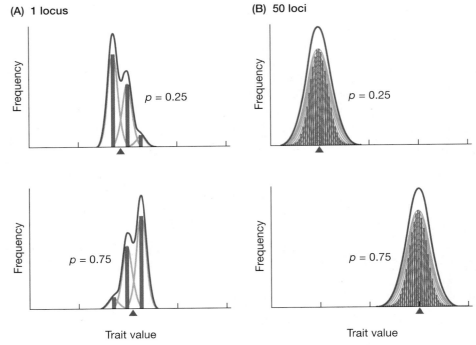

(A) 1 locus

Frequency

$p = 0.25$

Frequency

$p = 0.75$

Trait value

(B) 50 loci

Frequency

$p = 0.25$

Frequency

$p = 0.75$

Trait value

FIGURE 6.4 Large changes in quantitative traits can occur by the evolution of allele frequencies without the addition of new mutations. Colors of bars and lines are as in Figure 6.3. (A) The trait is affected by one locus with two alleles. Each copy of the A_2 allele increases the trait by 3 units, so A_2A_2 individuals are 3 units larger than A_1A_2 individuals, which are 3 units larger than A_1A_1 individuals. In the top graph, the frequency of the A_2 allele is $p = 0.25$, while in the bottom it is $p = 0.75$. The change in the allele frequency causes the mean of the trait to increase by 3 units (red triangles). The two phenotypic distributions overlap substantially. (B) The trait is now affected by 50 loci. One of the alleles at each locus increases the trait value by 0.4 units. For simplicity, we assume the frequency of the allele that increases the trait is the same at all 50 loci. The top and bottom graphs again compare the trait distributions when the allele frequency is $p = 0.25$ and $p = 0.75$. The mean of the trait now increases by 20 units. That change is larger than in (A), even though the effect of each allele is smaller. A key point is that the distributions for $p = 0.25$ and $p = 0.75$ do not overlap.

original population. This is because the new allele frequencies produce significant numbers of some genotypes that previously were rare or absent. **FIGURE 6.4A** shows how the phenotypic distributions change for a trait affected by one locus. There are two alleles, A_1 and A_2. Each copy of the A_2 allele that an individual carries increases the value of the trait by 3 units. With the frequency of allele A_2 at $p = 0.25$, the Hardy-Weinberg ratio tells us that the frequency of the rarest genotype (the $A_2 A_2$ homozygote) is $(0.25)^2$, or 6.25 percent. Thus even the rarest genotype is common enough to be seen. When the allele frequency increases to $p = 0.75$, the frequencies of the genotypes shift and the mean of the trait increases by 3 units, but the distribution of the trait still overlaps substantially with the original distribution.

What if now there are two loci? Imagine that the alleles A_2 and B_2 increase the trait's value by the same amount, so that the genotypes $A_1A_2B_1B_2$, $A_2A_2B_1B_1$, and $A_1A_1B_2B_2$ all have the same phenotype on average. When alleles A_2 and B_2 are both at a frequency of $p = 0.25$, the rarest genotype ($A_2A_2B_2B_2$, which is also the biggest) is present at a frequency of $(0.25)^4$, which is less than 0.4 percent. But when the allele frequency shifts to $p = 0.75$, the frequency of that genotype rises to $(0.75)^4 \approx 32$ percent, which is more than 80 times its initial frequency. When multiple loci affect a single trait, changes in their allele frequencies can drastically change genotype frequencies and so change the distribution of the trait they affect.

The situation is even more extreme with 50 loci (**FIGURE 6.4B**). When the frequency of the allele that increases the trait is $p = 0.25$ at all loci, the frequency of the genotype that is homozygous for the "big" allele at all 50 loci is expected to be $(0.25)^{100} \approx 10^{-60}$. That number is so small that this genotype—the one with the largest size—will never exist, let alone ever be seen. If the allele frequency at all loci increases to $p = 0.75$, the mean increases so much that the new distribution of the trait does not overlap at all with the original distribution. Quantitative traits can therefore evolve to produce entirely new phenotypes, using only alleles that are already in the population, without the introduction of new mutations.

Fitness Functions Describe Selection on Quantitative Traits

The horned lizards of the American Southwest would fit in well at Jurassic Park (**FIGURE 6.5**). Some species have dramatic horns projecting from the back and sides of their heads. The horns help deter predators, such as the fearsome loggerhead shrike. This bird has a remarkable and rather macabre behavior. After catching a horned lizard, the shrike often impales it on a branch (or in a pinch, on barbed wire), where it can eat it later.

Researchers have exploited that behavior to learn how natural selection acts on the size of the lizard's horns [56]. They compared the horn lengths of living lizards with those they found impaled by shrikes. From these data, the researchers were able to estimate how relative survival varies with horn length. A plot of survival against horn length shows that lizards with longer horns survive best (see Figure 6.5C). We don't know exactly how the horns protect the lizard. Perhaps shrikes have difficulty picking up a lizard with large horns, or perhaps the lizard stabs the shrike with its horns if it is caught.

The plot of survival against horn length is an example of a **fitness function**. This quantifies how selection acts on a quantitative trait. The horizontal axis is the value of the trait, and the vertical axis gives the expected fitness for individuals with that phenotype.

(A)

(B)

(C)

(D)

FIGURE 6.5 The fitness function for horn length in the horned lizard (*Phrynosoma mcalli*) has been estimated by comparing horn size in living and dead individuals. (A) A lizard that was caught and impaled on a thorn by a loggerhead shrike (*Lanius ludovicianus*). The arrow indicates one of the rear-most horns on the lizard's head. (B) A lizard that has avoided predation. (C) The fitness function showing how survival varies with the length of the horns. The function was estimated using the frequencies of live and shrike-killed lizards with a given horn length. (D) The loggerhead shrike is a major predator of horned lizards. The lizard shown here apparently did not successfully defend itself. (A courtesy of E. D. Brodie, Jr.; B courtesy of Kevin Young; C after [56].)

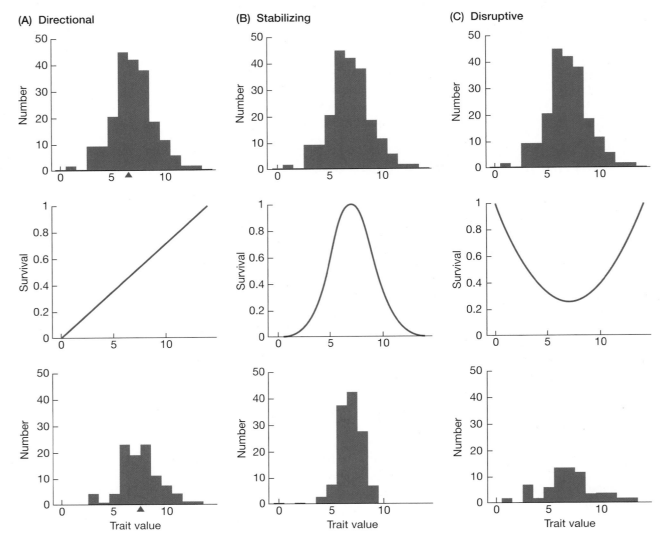

FIGURE 6.6 There are three basic modes of selection on a quantitative trait. (A) Directional selection favors a change in the trait mean, in this case toward a greater value. The top panel shows the distribution of the trait before selection acts, the middle panel shows the survival rate, and the bottom panel shows the distribution of the trait after selection. Triangles show the trait means before and after selection. (B) Stabilizing selection favors individuals near the population mean, which reduces the trait's phenotypic variance. (C) Disruptive selection favors the largest and smallest individuals, which increases the variance. (After [12].)

The horned lizards give an example of **directional selection**, which favors either an increase or a decrease in a trait's mean (**FIGURE 6.6A**). Many of the phenotypic differences we see among species are the result of directional selection. The ancestors of today's horned lizards had smaller horns that were enlarged by directional selection. In other cases, directional selection favors a decrease in a trait mean.

Earlier we emphasized that selection and evolution are two very different things. Selection happens *within* a generation, and may or may not lead to evolution. In 1977 a severe drought hit the Galápagos Islands, and many plants there failed to set seed. One of the seed-eating Galápagos finches (*Geospiza fortis*) was forced to eat new kinds of seeds, and birds with larger bills had higher survival rates (**FIGURE 6.7**). The mean size of survivors' bills was about 0.5 mm larger than the mean size in the population before the drought. The difference, which is highly

FIGURE 6.7 A severe drought in the Galápagos Islands in 1977 produced very strong directonal selection on bills of the Galápagos finch *Geospiza fortis*. Bill depth, which is the distance from the top to the bottom of the bill at its base, has a strong effect on the size of seed the finches are able to eat. (A) The fitness function of survival as a function of bill depth shows there was strong directional selection for deeper bills. The fitness function was estimated from the data in (B) and (C). (B) Distribution in bill depth of 751 *G. fortis* before the drought. (C) Distribution in bill depth of the 90 individuals that survived the drought. The selection differential *S*, which is the change in mean beak size (red triangles) from before to after selection (shown by the arrow), is highly statistically significant. (A after [42]; B and C after [8].)

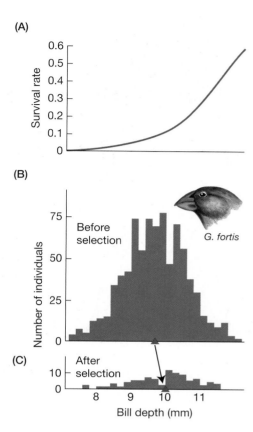

statistically significant, shows there was directional selection. Note that this comparison is between individuals of the same generation, before and after the drought. At this point, no evolution has yet occurred.

A fitness function can also act on the variance of a trait (**FIGURES 6.6B and C**). **Stabilizing selection** favors individuals whose trait values are near the population's mean. After stabilizing selection acts, the phenotypic variance is reduced. This is a common form of selection because the means of many traits are near the values that have the highest fitness (often referred to as the **optimum phenotype**). In that situation, individuals that are much smaller or much larger than the mean have lower fitness. The result is that the tails of the phenotypic distribution are trimmed so the variance is smaller. Birth weight in humans is a classic example. Babies that weight much less or much more than the average at birth have a lower chance of surviving (**FIGURE 6.8A**).

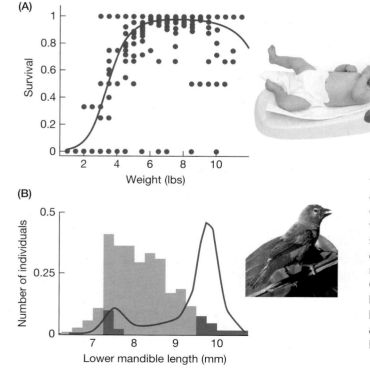

FIGURE 6.8 (A) Birth weight in humans is under stabilizing selection. Infants with birth weights much smaller or much larger than average have a lower probability of surviving to 28 days. Each circle shows the survival rate for a group of infants that had the birth weight shown and the same gestation time. (B) The probability of survival to adulthood in the black-bellied seedcracker (*Pyrenestes ostrinus*) depends on an individual's lower mandible length, a measure of bill size. The fitness function is shown by the curve. Green portions of the histogram show birds that did not survive; blue portions show birds that did. Birds with intermediate-sized bills have the lowest survival, showing that disruptive selection is acting. (A courtesy of Dolph Schluter after data from [42]; B after [47], with photo courtesy of Thomas B. Smith.)

(A)

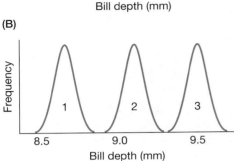

(B)

FIGURE 6.9 The type of selection acting on red crossbills (*Loxia curvirostra*) depends on both the phenotypic distribution in the birds and the cones on which they feed. (A) The fitness function for bill depth was estimated by the rate at which birds can feed. Small bills are efficient at opening the cones of western hemlocks, while large bills are efficient at opening cones of lodgepole pine. Intermediate-sized bills have low fitness. (B) The distributions of bill depth in three hypothetical populations of crossbills. Population 1 experiences directional selection to increase its mean bill size. Population 2 experiences disruptive selection. Population 3 experiences stabilizing selection. (A based on data from [6].)

The opposite situation is called **disruptive selection**. Here the smallest and largest individuals have higher fitness than individuals near the mean. After selection, the phenotypic variance is greater than it was before. Disruptive selection rarely splits a population into two separate groups, but rather makes intermediate individuals less common (**FIGURE 6.8B**).

Selection can alter both the mean and the variance of a trait at the same time. For example, if a trait's distribution after selection has a larger mean and a smaller variance than it did before, then both directional selection and stabilizing selection have acted.

The fitness function and the trait's distribution together determine whether selection is directional, stabilizing, or disruptive. **FIGURE 6.9** shows the fitness function for bill depth in the red crossbill, a bird that specializes in extracting seeds from the cones of pine trees and other conifers. The cones of different conifers vary in size and shape. This generates several peaks in the fitness function for the crossbill, with each peak representing the bill depth that is best for feeding on a particular type of cone. Is selection on bills directional, stabilizing, or disruptive? The answer depends on the distribution of bill depth relative to the fitness function. If most individuals in the population fall in a region where the fitness function is increasing or decreasing, then directional selection acts. If the population lies near a peak in the fitness function, then stabilizing selection acts. Last, if the population lies near a low point, then disruptive selection acts.

Fitness functions are also used to visualize selection acting on more than one trait. In these cases, the fitness function tells us which *combinations* of traits give high or low fitness. An example is shown in **FIGURE 6.10**. The northwestern

FIGURE 6.10 The fitness function for combinations of two traits in the northwestern garter snake (*Thamnophis ordinoides*), based on survival in the field. Snakes vary in their coloration and in their escape behavior. The height of a point on the surface represents the relative survival of individuals with a given combination of values for stripedness and the tendency to reverse course when escaping. Snakes with stripes that escape in a straight line have high fitness, as do snakes without stripes that reverse course. The fitness function shows that correlational selection is acting. (After [10].)

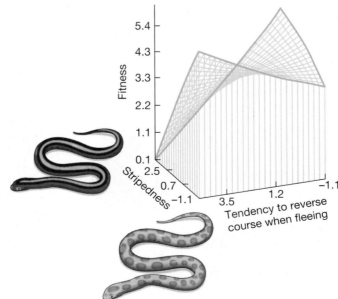

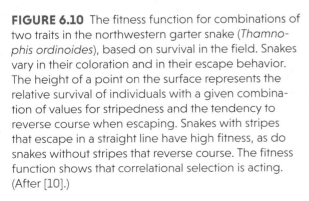

garter snake (*Thamnophis ordinoides*) varies in its color pattern: some individuals are striped, others are not. The snakes also vary in how they react to a predator. Some individuals escape in a straight line, while others often reverse their course. Survival of snakes with different combinations of these traits has been estimated by marking individuals, releasing them, and then recapturing the survivors at a later date [10]. Snakes that are striped and that escape in a straight line have high survival, probably because visual predators (such as birds) have difficulty judging the speed and location of a moving stripe. Snakes that are unstriped and reverse course also survive well, likely because reversals of unstriped snakes confuse the predator. Snakes with the two other combinations of coloration and behavior have lower fitness. Selection that favors particular combinations of traits, as in the garter snake, is called **correlational selection**.

Measuring the Strength of Directional Selection

Many of the questions that evolutionary biologists ask are about how and why the mean values of traits evolve. (Why did whales become so large?) Evolutionary changes in means are often caused by directional selection, and so it is important to be able to quantify its strength.

The **selection gradient** measures the strength of directional selection acting on a quantitative trait. It plays a role analogous to that of the selection coefficient for the alleles at a single locus. The basic recipe for estimating a selection gradient is simple. The data needed are measurements of the trait and of fitness on a set of individuals. Ideally, we would like to use the lifetime fitness. Often that is not possible to measure, so instead we use an important fitness component, such as survival or mating success. Relative fitnesses are calculated by dividing each individual fitness by the mean fitness of all the individuals, and these relative fitness values are plotted against the trait value. Finally, the selection gradient is the slope of the regression line fit through those points. (The Appendix gives a brief introduction to regression.) The selection gradient is symbolized by β, and its units are 1/[units of measurement]. If the trait is measured in millimeters, for example, then β is expressed as per millimeter. If the gradient is positive, then directional selection favors the mean to increase. A negative β implies that selection favors smaller values of the trait. Last, a value of β = 0 means that there is no directional selection acting.

The guppy (*Poecilia reticulata*) is a tropical freshwater fish that is popular among aquarium enthusiasts because males are colorful. Females prefer to mate with males that have more orange on their body (**FIGURE 6.11**). The estimate of the selection gradient from the data shown in the figure is β = 3.8. (In this case β has no units because the trait is measured as a proportion of the body surface.) If they existed, completely orange males would on average have 3.8 times more matings than males with no orange at all.

Evolutionary biologists have estimated the selection gradients acting on many natural populations of animals and plants. **FIGURE 6.12** shows the

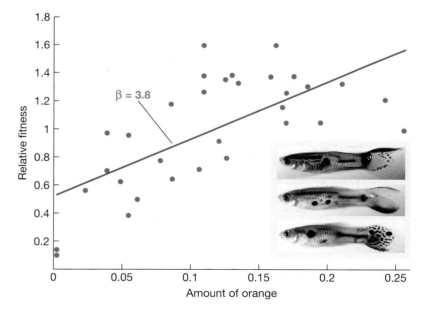

FIGURE 6.11 Selection gradient on orange coloration in male guppies. The horizontal axis shows the proportion of the body that is orange, and the vertical axis shows relative fitness, as measured by attractiveness to females in the lab. The slope of the regression line gives an estimate of the selection gradient: β = 3.8. (After [24]; photos from [26].)

FIGURE 6.12 The distribution of selection gradients acting on size in natural populations of animals and plants show that directional selection is common and at times very strong. The distribution is based on 2819 estimates of β from 143 studies. Negative values of β reflect selection favoring smaller size, and positive values denote selection for larger size. To help visualize what different values of β represent, the insets show how much the phenotypic distribution of a trait is changed by moderately strong (β = 0.5 or −0.5) and very strong (β = 1) directional selection. Green curves show the distributions before selection, and the blue curves show the distributions after. The gradients shown here have been normalized by multiplying each β by the trait's phenotypic standard deviation. This makes β unit-free, which allows comparison of different kinds of traits. (Main panel after [32].)

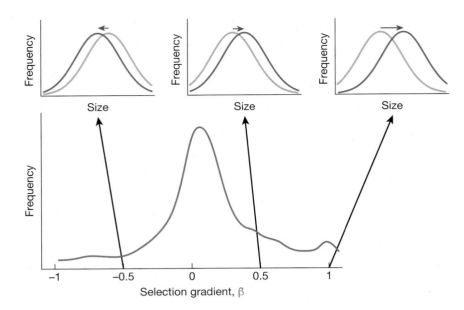

frequency distribution of these gradients. Directional selection is common. In some cases, it is very strong: directional selection can cause the mean of the population to shift by more than one phenotypic standard deviation.

In Chapter 5 we saw that natural selection drives populations uphill on Wright's **adaptive landscape**. This landscape is a plot of the population's mean fitness against the frequency of an allele. The concept of an adaptive landscape also applies to quantitative traits [34]. Here the landscape plots the population's mean fitness against the mean value of the trait, rather than the allele frequency. (It is important to distinguish between the fitness function and the adaptive landscape. The first shows how the phenotype of an *individual* affects its fitness. The second shows how the *mean trait value in a population* affects the population's mean fitness.) When relative fitnesses are constant in time, natural selection causes populations to evolve uphill on this landscape. The mean will stop evolving when either it reaches a peak or the population runs out of genetic variation.

The selection gradient and adaptive landscape are useful tools for visualizing how selection is acting. A second use for them is to test hypotheses about adaptation. If a trait has reached an optimum favored by natural selection, then the population should be at a peak on the adaptive landscape and there should be no directional selection. That idea has been used to study the evolution of clutch size (the number of eggs laid) in birds. Many birds are physiologically able to lay more eggs than they actually do. This seems like an evolutionary paradox: why doesn't natural selection favor them to lay more? In fact, the selection gradient on clutch size is close to zero [46]. Females that lay more eggs than average hatch more chicks, but many of the chicks starve because their parents are unable to feed so many mouths. Females that lay the average number of eggs leave the largest number of surviving offspring to the next generation. We'll look more closely at the evolution of clutch size in Chapter 11.

Evolution by Directional Selection

We saw in Chapter 5 that if selection acts on a trait and if that trait is inherited, then evolution will result. This is a condensed version of the most important point made by Darwin in *The Origin of Species*. We can now go further than what Darwin was able to do: we can predict how *much* evolution will result.

FIGURE 6.13 Schematic of the breeder's equation. Each dot represents a family. The mean size of the two parents is plotted on the x-axis, and the mean size of their offspring on the y-axis. The regression line shows the mean size of offspring that are expected from parents of a given size. The slope of this line is equal to the heritability, h^2. (A) With no selection, there is no evolutionary change, so the mean size of all offspring in the next generation, $\bar{z}'$, is equal to the mean size of all parents in the previous generation, $\bar{z}$. (B) Directional selection occurs. In this example, only parents whose size is larger than a threshold survive. The mean size of the surviving parents is $\bar{z}^*$. Now the mean size of their offspring ($\bar{z}'$) is larger than if selection had not acted: the mean of the population has evolved. (C) The selection differential, S, is the difference in the mean size of individuals before and after selection. The evolutionary change in the mean from one generation to the next is $\Delta\bar{z} = h^2 S$.

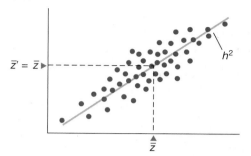

(A) Without selection

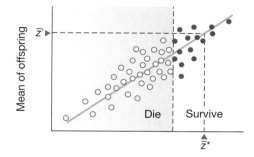

(B) With selection

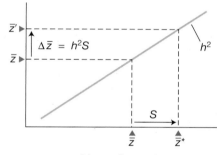

(C) Evolutionary change

Both evolutionary biologists and breeders need to know how much the mean of a trait will evolve if there is directional selection (caused either by nature or by breeders). Happily, the answer is the simplest mathematical relationship that makes sense, and it can be understood with a diagram (**FIGURE 6.13**). The evolutionary change in the mean of a trait from a single generation of selection equals the product of two quantities: the strength of directional selection, and the amount of genetic variation. This is an exact parallel to the discussion in Chapter 5 of selection acting on a single locus (see Equation 5.3).

To make this point more quantitative, let $\bar{z}$ represent the mean of a trait at the start of a generation. Selection acts on the trait, and the survivors breed to produce the next generation. Using $\bar{z}'$ to represent the mean at the start of that new generation, the amount of evolutionary change is just the difference between $\bar{z}'$ and $\bar{z}$, which we symbolize by $\Delta\bar{z}$. As shown in Figure 6.13, that change is predicted to be:

$$\Delta\bar{z} = \bar{z}' - \bar{z} = h^2 S \qquad (6.1)$$

This is the famous **breeder's equation**, which is used to predict how much evolutionary change will result from selective breeding.

On the right side of Equation 6.1 is h^2, which represents the trait's **heritability**. The heritability is equal to the slope of the regression line that relates the value of a trait in two parents to its value in their offspring (**FIGURE 6.14**). The heritability therefore measures the strength of inheritance. If h^2 is 0, then there is no resemblance between offspring and their parents. At the other extreme, if h^2 is 1, then offspring look exactly like the average of their parents. (In this discussion, we assume that resemblance between parents and offspring is caused only by shared genes. Nongenetic factors can also contribute to that resemblance, as when some families live in good environments and others in poor environments. In those situations, a correction is made to remove the environmental effects from the estimate of heritability.)

The second quantity on the right of Equation 6.1 is S, which is the amount of change in the mean of the trait caused by selection within a generation. That is, S equals the difference between the mean of the population after selection, which is written $\bar{z}^*$, and the mean before selection, $\bar{z}$ (see Figure 6.7). If smaller individuals are more likely to survive and reproduce than larger individuals, for example, then S will be negative. A key point is that this difference is the change caused by selection *within* a generation, while $\Delta\bar{z}$ is the evolutionary change *between* one generation and the next. The selection differential is related to the selection gradient by the equation $S = P\beta$, where P is the phenotypic variance.

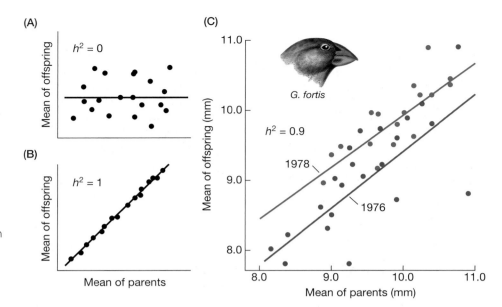

FIGURE 6.14 The plot of the phenotypes in parents and offspring is used to measure heritability. Each point represents the mean of all the offspring in a single family, plotted against the mean of their two parents. At left are two hypothetical cases showing what the plot looks like when (A) the heritability is $h^2 = 0$ and (B) when it is $h^2 = 1$. (C) The plot of parents and offspring for bill depth in the Galápagos finch *Geospiza fortis* in 1976 and 1978. Although offspring were larger in 1978, the slope of the regression between offspring and parents was nearly the same in both years. The heritability, estimated from the slope of the regression, is $h^2 = 0.9$ in both years. (C after [7].)

The essential message from Equation 6.1 is that the rate of evolution depends both on the strength of inheritance, measured by h^2, and the strength of directional selection, measured by S. The trait will not evolve if it is heritable ($h^2 > 0$) but there is no selection acting on it ($S = 0$). Likewise, a trait will not evolve if there is selection ($S \neq 0$) but no heritability ($h^2 = 0$).

A second version of the breeder's equation is mathematically equivalent but often more useful in evolutionary biology:

$$\Delta \bar{z} = G\beta \qquad (6.2)$$

Here G is an important quantity called the **additive genetic variance**. It is the part of the phenotypic variation that is caused by genetic variation and that contributes to the resemblance between parents and offspring. In symbolic form, the additive genetic variance is defined as

$$G = h^2 P \qquad (6.3)$$

where again P is the phenotypic variance of the trait. Equation 6.3 can be rearranged to give

$$h^2 = G/P \qquad (6.4)$$

This shows that the heritability equals the fraction of the phenotypic variance that is due to heritable genetic variation. The rest of the phenotypic variance is contributed by two other sources. The first (and most important) source is nongenetic factors, such as nutrition. These contribute **environmental variance** to the trait, causing individuals with the same genotype to have different phenotypes (see Figure 6.3). The other source is genetic variation that is not additive, caused by dominance and epistasis, which we will discuss shortly.

Genetic analysis of hundreds of species has shown that most quantitative traits are heritable and evolve if selection acts on them [22, 35]. Heritabilities vary among traits and species. The values for morphological traits in vertebrates typically fall in a range between 0.2 and 0.6. That means that much (and sometime most) of the phenotypic variation we see for quantitative traits is genetic in origin and can respond to selection. Traits that are more closely connected to fitness (such as fecundity and longevity) tend to have lower heritabilities than morphological traits because they often have more environmental variance [25].

We now have everything needed to predict the direction and distance that the mean of a trait will evolve in one generation. The heritability h^2 is estimated from the regression of the trait measured in offspring plotted against the trait in their parents (see Figure 6.14). We can also use those measurements to find P, the phenotypic variance of the trait. With those values in hand, Equation 6.3 gives us the additive genetic variance G. The strength of directional selection, measured by the selection gradient β, is estimated by the regression of relative fitness onto the trait value (see Figure 6.11). Finally, the evolutionary change in the mean of the trait, $\Delta \bar{z}$, is simply the product of G and β (see Equation 6.2).

Consider this implication: we can predict the outcome of genetic evolution without knowing anything about the genes that affect the trait! This means that the rate of evolution is not determined by the number of genes that affect the trait (at least in the short term). A second insight is that the rate of evolution is not determined by the population size (again, in the short term). A small population does not evolve more quickly than a large one if the two have the same additive genetic variance G.

When genes interact: Dominance and epistasis

You may be wondering why G is called the "additive" genetic variance. The answer is that there are also other types of genetic variation. It is important to distinguish between them because only the additive genetic variance contributes directly to evolutionary change.

Imagine that the height of a plant is completely determined by variation at a single locus, with no environmental variance. This locus is overdominant (see Chapter 5): both A_1A_1 homozygotes and A_2A_2 homozygotes are 20 cm tall, while A_1A_2 heterozygotes are 25 cm tall. If both the A_1 and A_2 alleles have a frequency of 1/2 and the population is at Hardy-Weinberg equilibrium, then half the plants will be 20 cm tall and half will be 25 cm tall. There is lots of variation in this population, and all of it is caused by genetic differences. Now imagine that all the short plants die, and only the tall plants (the heterozygotes) reproduce. In the next generation, the population looks exactly like it did before selection acted, with equal numbers of short and tall plants.

Why didn't selection cause an evolutionary change? Certainly not because of a lack of genetic variation. Rather, it is because none of it is additive genetic variance. In this example, the genetic variation is of a form called *dominance variance*, which results when the phenotype of heterozygotes is not intermediate between the phenotypes of the homozygotes. Here the two alleles interact: the effect of an allele on an individual's phenotype depends on the other allele that is carried at the same locus. Alleles at different loci can also interact, a situation called epistasis (see Chapter 4), which generates *epistatic variance*. Like dominance variance, epistatic variance does not contribute to evolutionary change.

For the great majority of traits, the additive genetic variance is much larger than the dominance variance and the epistatic variance. (The example of the plants was made extreme to make the concepts clear, and is not typical.) In short, most but not all genetic variation contributes to how fast a population evolves in response to directional selection. The additive genetic variance (as well as the dominance and epistatic variance) can evolve as allele frequencies change. For some traits, the additive variance stays relatively stable, and the trait can evolve at a constant rate for many generations. In other cases, selection fixes alleles at the loci that contribute genetic variation. This causes the additive genetic variance and heritability to decline, slowing and even halting the trait's evolutionary response to selection.

Adaptation from standing genetic variation versus new mutations

As our atmosphere becomes enriched in CO_2, many of Earth's organisms are experiencing directional selection caused by changed temperatures, acidified oceans, and other new conditions. Will they be able to adapt? Many traits will evolve rapidly using genetic variation that already exists, what is called **standing genetic variation** (see Chapter 5). Other traits do not have genetic variation now and cannot evolve until new mutations favored by the new environmental conditions appear. In some cases this happens quickly, but in other cases the critical mutations may not appear for long periods of time. This kind of speed limit to adaptation is particularly common to small populations because fewer new mutations enter a population when there are fewer copies of the genes to mutate. We currently have a poor understanding of how often adaptation is based on standing genetic variation and how often on new mutations [1]. This is a topic of active research.

Can adaptation rescue species from extinction?

Although all species now alive owe their existence to adaptation in the past, the fact that well over 99 percent of species that ever lived are now extinct tells us that evolution does not guarantee survival. When conditions change, what determines whether a species can adapt fast enough to avoid extinction?

We can use mathematical models to explore when an abrupt change in the environment will cause extinction [21]. Imagine that a species is initially at a fitness peak for a quantitative trait. The environment then changes, favoring a new value for the trait and causing the mortality rate to exceed the birth rate. Then the population will decline to extinction unless the trait is heritable and so can evolve towards the new optimum value that maximizes survival. Thus, there is a race between adaptation and extinction. If the species can adapt quickly enough, survival rates will rebound and the species will be rescued. If it cannot, however, the population can fall below a critical threshold size where extinction will occur.

How this race ends depends on several key factors (**FIGURE 6.15A**) [11]. A population is more likely to survive if it has greater standing genetic variation, which will allow it to adapt more quickly. A large initial population size helps survival in several ways: the population size must decline a long way before it is at risk of extinction, and more new beneficial mutations enter the population in each generation. Some species can buffer themselves from the environmental change by adjusting to new conditions physiologically [13].

One approach used to study how these and other factors affect the risk of extinction is experimental evolution [4, 5]. **FIGURE 6.15B** shows results from a laboratory study with yeast. Populations were suddenly subjected to high concentrations of

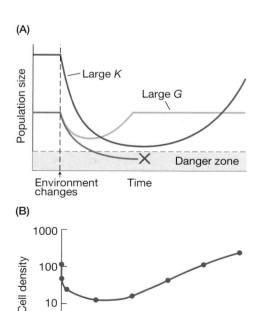

(A)

(B)

FIGURE 6.15 Adaptation can rescue some species but not others from extinction. (A) Simulations of how the size of a population changes in time following an environmental change that suddenly favors a different value of a quantitative trait. The environmental change (vertical dashed line) triggers declines in the population sizes of three species. The population size of the blue species falls below a critical threshold and into the "danger zone" (shaded area), leading to its extinction (marked by the X). The green species has larger genetic variance (G) for the trait, which allows it to adapt more rapidly to the new adaptive peak and avoid extinction. The red species has a larger carrying capacity (equilibrium population size, K). It avoids extinction because it has a longer time to adapt before reaching the danger zone. For simplicity, these simulations assume that the additive genetic variance is the same in all three cases and does not change in time. (B) Evolutionary rescue allows laboratory populations of yeast to avoid extinction following the sudden introduction of salt into their medium at time 0. The trajectory of population size during the decline and recovery is a good match for the "Large K" simulation shown in (A). Note that cell density (on the y-axis) is plotted logarithmically, so the changes in density are very large. (B after [4].)

salt. Initially, population sizes declined as the salt killed off yeast cells more rapidly than they were able to divide. As the populations adapted to the salt, population growth rates became positive. In this case, adaptation happened quickly enough to prevent extinction.

Studies of populations in nature also inform us about which species may or may not survive changing conditions [45]. Genetic variation for several traits was measured in three populations of the partridge-pea (*Chamaecrista fasciculata*) along a north-south transect, and seedlings were then transplanted among the three sites [15]. Plants that were moved farther south experienced warmer and drier environments, which are predicted to occur in their native population by 2050. These transplants showed reduced fitness. Only one of three transplanted populations showed sufficient genetic variation that it will likely be able to adapt to the new conditions. These results suggest that the partridge-pea may not be able to avoid extinction in the face of climate change that is currently happening.

While we expect that adaptation will allow some species to avert extinction as humans change the planet, many (perhaps most) will not be so lucky. The climate changed rapidly many times during the Pleistocene (see inside back cover), with warm periods interspersed by cold glacial periods. Many species survived by colonizing new areas where the altered climate matched the climate they were adapted to. But others did not adapt rapidly enough and became extinct. Currently, global change caused by the burning of fossil fuels is causing the climate to change at rates more than 100 times greater than during the Pleistocene, probably too fast for most species to shift their ranges to favorable regions [41]. It is likely that a large part of life on Earth is together with the partridge-pea on a path to extinction.

Artificial Selection

Humans have been selectively breeding animals and plants for millennia. Long before Equations 6.1 and 6.2 were known, people were genetically improving animals and plants by breeding together the best individuals in their fields and flocks. That process is called **artificial selection**, and it continues to this day as a critical part of modern agriculture.

The next time you walk into a supermarket, think about the animals and plants used to make the food you are about to buy. Virtually all of them have been radically changed by artificial selection, some so much that they only vaguely resemble the wild species that were domesticated by prehistoric farmers. The wheat for a pizza's crust, the tomatoes for its sauce, the cows whose milk makes the Parmesan cheese on top—all of these species have been changed dramatically by human-caused evolution (**FIGURE 6.16**).

Despite centuries of selective breeding, domesticated species continue to evolve in response to artificial selection. Milk production and poultry growth are increasing at 1 to 2 percent per year (see Figure 5.3). This is

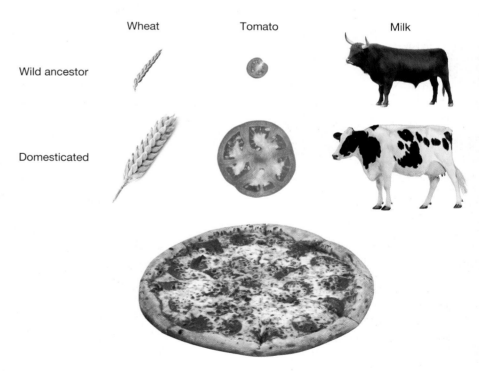

FIGURE 6.16 A familiar food item, showing the sources for three major ingredients as they looked in their wild ancestors before domestication and as they do today.

FIGURE 6.17 The oil content of corn kernels increased over 107 generations of artificial selection. In the initial population, the mean oil content was 4.7 percent, and the highest oil content measured on any corn ear was 6 percent. The average oil content in the most recent generation was 22 percent, more than 4.5 times higher than in the initial population. (Data from [14].)

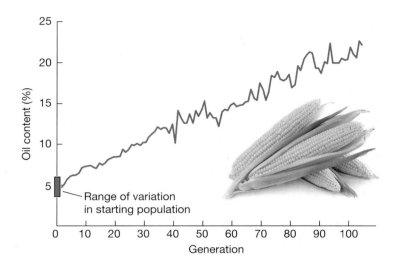

Range of variation in starting population

a critical contribution to society since it allows us to produce more food with the same or even fewer resources. Artificial selection on many species is now done using sophisticated statistical methods that evaluate each individual's genetic potential.

In addition to improving domesticated species, artificial selection is used by biologists to study basic questions about evolution. This research often uses model organisms such as *Drosophila* and *E. coli* because they are convenient and well known genetically. Several general conclusions have emerged are likely to apply to all species:

Almost all traits evolve when selected. Results from hundreds of selection experiments show that most traits in diverse species immediately respond to selection based on standing genetic variation [23]. We will see shortly, however, that there are exceptions to this rule that have significance for our understanding of the limits to adaptation.

Selection can cause a trait to evolve far beyond its original range of variation. We saw early in this chapter that changes in allele frequencies can cause a quantitative trait to evolve far beyond the range of variation that was originally present in the population. An example in real organisms comes from a famous artificial selection experiment on corn that is still continuing after more than 100 years (**FIGURE 6.17**). Early in the experiment, evolutionary change was based on standing genetic variation, but new mutations contributed in later generations.

Large populations evolve faster and farther than small populations. Researchers have used artificial selection to learn what factors affect how populations adapt. One pattern that emerges is that large populations tend to evolve faster and farther (**FIGURE 6.18**). That finding is interesting because there is nothing in Equations 6.1 and 6.2 that suggests population size should have an effect. The explanation is that over the course of several generations, the additive genetic variance can decline, and it tends to do so more rapidly in smaller populations (see Chapter 7). An important conclusion is that species that are already rare are particularly vulnerable to environmental change because they may not adapt as quickly as abundant species.

Strong selection on one trait often has negative side effects on other traits. Over a span of 50 years, artificial selection on dairy cows increased milk production by 1 percent per year, but also caused fertility to decline at about the same rate [23]. This is an example of how selection on one trait often causes evolutionary side effects on other traits, which is our next topic.

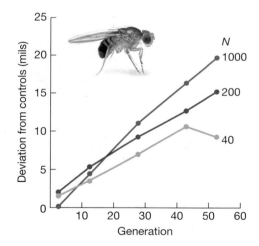

FIGURE 6.18 A selection experiment for increased wing-tip height in *Drosophila melanogaster* shows evolutionary change over 54 generations in populations of different sizes. Shown are the numbers of individuals that are selected to start each generation. (For example, with N = 200 flies, a total of 1000 flies were measured, and the 200 flies with the longest wings were bred to begin the next generation.) The largest population has evolved the fastest and farthest. (After [52].)

Correlated Traits

Traits are correlated: if you have long arms, you probably also have long legs. This kind of correlation is in part heritable and genetic, meaning that individuals with long arms tend to have offspring that have both long arms and long legs. **Genetic correlations** such as these cause evolutionary side effects. When selection acts to increase one trait, it will not only cause the mean of that trait to increase, it will also change the means of other traits that are genetically correlated with it.

These evolutionary side effects are described by an expanded version of Equation 6.2. Let's say that directional selection is acting on two traits. The evolutionary change in trait 1 caused by one generation of selection is

$$\Delta \bar{z}_1 = G_1 \beta_1 + G_{1,2} \beta_2 \tag{6.5}$$

There are two terms on the right side. The first is just as we saw in Equation 6.2: it is the product of G_1, the additive genetic variance for trait 1, and β_1, the selection gradient acting on trait 1.

The second term in Equation 6.5, however, is new. It is the product of $G_{1,2}$, which is the **genetic covariance** between trait 1 and trait 2, and β_2, which is the selection gradient on trait 2. A genetic covariance measures how strongly two traits tend to be inherited together. (A covariance is closely related to a correlation, which is a covariance that has been rescaled so that it ranges from -1 to 1. Covariances and correlations are explained in the Appendix.) A genetic covariance of 0 means that the traits are inherited independently. A positive covariance means that individuals that are larger than average for one trait will tend to have offspring that are larger for both traits. That is typically the case for morphological traits, simply because big individuals tend to be big for all traits. A negative genetic covariance implies the opposite: individuals that are bigger than average for one trait will have offspring that are big for that trait but smaller than average for the second trait. An example is reproductive rate and longevity in fruit flies. Females that lay many eggs survive less well than females that lay few eggs.

The two terms on the right side of Equation 6.5 show that a trait can evolve in two ways. The first way is as a **direct response** to selection, meaning that the trait is evolving as the result of selection acting on it. The second way is as an **indirect response** to selection, meaning that the trait is evolving because of selection on another trait with which it is correlated.

One implication of Equation 6.5 is that a trait can evolve by natural selection even if selection does not act on that trait. If that statement sounds nonsensical at first, consider what happens to trait 1 when the selection gradient on that trait is zero. Its mean will nevertheless evolve if directional selection acts on trait 2 and there is genetic covariance between the traits (that is, neither β_2 nor $G_{1,2}$ are 0). This situation is sketched in **FIGURE 6.19**. Even more remarkable is that selection can cause a trait to evolve in the direction opposite to what selection on that trait favors. For example, if there is weak selection to increase leg length but very strong selection to decrease arm length, both traits can evolve smaller size.

Earlier we discussed selection on bill size in one of the Galápagos finches during an intense drought. Trevor Price and colleagues found that selection favored finches with narrower beaks, likely because they could better crack open new seed types [40]. Nevertheless, the average bill width among birds that survived the drought was larger than it was before the drought. The explanation for this counterintuitive result is that bill width has strong positive correlations with other traits, including body size. Those traits caused indirect selection on beak width that was stronger than the direct selection on beak width.

FIGURE 6.19 Directional selection on one trait can cause another trait to evolve as a correlated response. In these plots, each point represents the values for two traits in a single individual. (A) The two traits are not correlated. Selection acts only on trait 1, and only individuals larger than the threshold shown by the dotted line survive (red points in the middle panel). After selection, the mean of trait 1 has increased, but the mean of trait 2 is unchanged. (B) The two traits have a strong positive correlation. Selection again acts only on trait 1 (middle panel). After selection, the means of both trait 1 and trait 2 have increased. The change in trait 2 is a correlated response to selection.

Side effects like these may explain some evolutionary enigmas. The Mexican tetra (*Astyanax mexicanus*) is a fish that has both surface-dwelling and cave-dwelling populations. Fish from the surface have eyes and can see, but fish from caves have lost their eyes (**FIGURE 6.20**). Why does adaptation to the dark and nutrient-poor environment in caves favor mutations that eliminate sight? Cave fish find prey in the dark using sensory cells on their heads that respond to vibrations in the water. Genetic analysis shows that mutations that increase the responsiveness of this detection system also cause a reduction in the eyes as a correlated side effect [55]. The blind cave fish also illustrate one of the ways that natural selection can cause the loss of a complex structure such as the eye.

Constraints and trade-offs

While the great majority of quantitative traits have standing genetic variation, not all do. Traits that lack variation cannot respond to directional selection, and so we say they have an **evolutionary constraint** that can prevent them from adapting. Species of *Drosophila* that live only in wet tropical habitats have little or no genetic variation that would allow them to adapt to cool and dry habitats. This may explain why their ranges do not expand outward into drier habitats [30, 31].

The cliché that there's no such thing as a free lunch applies to the evolution of many quantitative traits. Say that natural selection favors deer that can run faster. Increased speed puts more stress on the deer's leg bones, which selects for stronger bones. If there is genetic variation for growing thicker bones, that trait can increase. But there's a catch: bones that are thicker are also heavier, which decreases speed.

This is an example of an evolutionary **trade-off**, which occurs when increasing fitness in one way decreases it in another. Trade-offs can be understood at different levels. The trade-off between the strength and weight of a femur results from simple physics: more bone mass increases both strength and weight. A complementary perspective comes from genetic correlations. Bone strength and bone weight are highly correlated, so an evolutionary increase in one necessarily causes an increase in the other. Genetic correlations can therefore cause evolutionary constraints. Even though individual traits show genetic variation, there can be combinations of trait values for which there is little or no variation.

(A)

(B)

FIGURE 6.20 Populations of the Mexican tetra (*Astyanax mexicanus*) that live in streams on the surface have eyes (A), while populations that live in caves have lost their eyes (B).

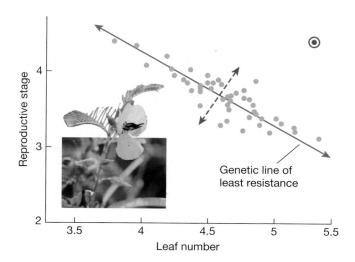

FIGURE 6.21 A negative genetic correlation in the partridge-pea (*Chamaecrista fasciculata*) results in an evolutionary trade-off. Plant size, measured by leaf number, is plotted on the x-axis. Plant growth rate, measured by the reproductive stage, is plotted on the y-axis. Each dot shows the values of those traits for a genotype in a population from the northern United States (Minnesota). The genetic line of least resistance (in blue) is the combination of traits that can evolve rapidly because there is abundant standing genetic variation. There is little variation to evolve in the directions indicated by the red dashed arrows. Climate change is selecting for the combination of traits indicated by the bull's-eye. This population is predicted to become extinct because there is little genetic variation to evolve in the direction favored by selection. (After [16].)

Earlier in the chapter we discussed the partridge-pea, which may not adapt quickly enough to avoid extinction. **FIGURE 6.21** develops that story further. It plots the growth rate against plant size for genotypes sampled from a northern population. The two traits show a strong negative genetic correlation. There is abundant genetic variation that will let the population evolve either more leaves and slower growth, or fewer leaves and faster growth. This is an example of a **genetic line of least resistance**, which is a combination of traits for which a population has abundant genetic variation [43]. In contrast, there is little variation that would allow the population to evolve toward either more leaves and faster growth, or fewer leaves and slower growth. The challenge faced by the partridge-pea is that adaptation to changing climates requires evolving both more leaves and faster growth.

Genetic correlations can evolve as gene frequencies change (just as genetic variances do), so correlations like those in the partridge-pea will only constrain adaptation in the medium to long term if they remain relatively constant. Many morphological traits are highly correlated with overall body size, and genetic correlations between them may be stable over thousands or even millions of years [34]. Other genetic correlations change over shorter time scales.

The hypothesis that genetic correlations can constrain evolutionary change in the short term has been tested in a selection experiment. A butterfly with the curious name of squinting bush brown (*Bicyclus anynana*) has spots on its wings (**FIGURE 6.22**). Artificial selection on the two large spots can change their sizes independently, so they are not constrained. Selection on the colors of two other eyespots was able to make both become more black or both become more golden, but it was unsuccessful in making one eyespot black and the other gold. The color of these two eyespots is constrained to be the same by a genetic correlation [3]. Genetic correlations between traits are a major cause of evolutionary constraints [22, 33, 43]. An important but unanswered question is how often the evolutionary limits seen in short-term experiments like that with the butterfly persist over longer evolutionary time scales.

The causes of genetic correlations

Genetic correlations have two sources. The first is **pleiotropy**, the situation in which a single locus affects more than one trait (see Chapter 4). Many loci affect body size in humans. These genes generate correlations among virtually all morphological traits, since individuals who are large for one body part tend to be large for others. They also generate correlations among other types of traits. For

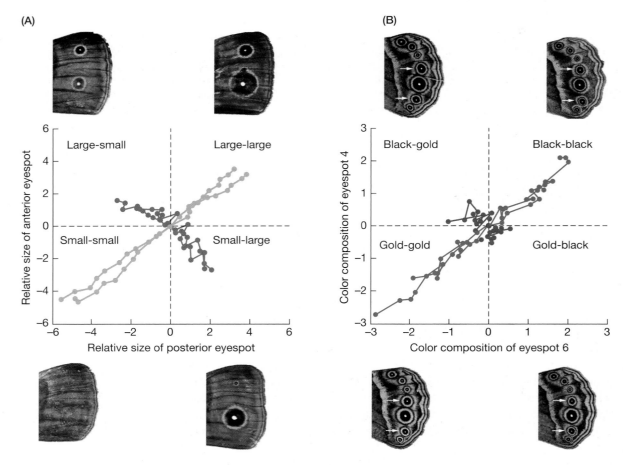

FIGURE 6.22 Artificial selection experiments showing different levels of constraint on the evolution of eyespots in wings of the butterfly *Bicyclus anynana*. (A) Evolutionary trajectories that result from selection on the sizes of the two eyespots on the dorsal side of the wing. The graph shows the results of selection for both eyespots to be large or small (Large-large and Small-small, green trajectories) and for one to be large and the other to be small (Large-small and Small-large, purple trajectories). Each point represents one generation. Rapid evolution occurred for all combinations of traits that were selected, showing there is no evolutionary constraint. (B) Trajectories resulting from selection on the colors of two other eyespots (shown by arrows) on the ventral hind wing. Selection to make both eyespots more black or more gold caused substantial evolutionary change (Black-black and Gold-gold, blue trajectories). In contrast, selection for different colors in the eyespots was largely ineffective (Black-gold and Gold-black, red trajectories). This is an example of an evolutionary constraint caused by a genetic correlation between two traits. (After [1a], courtesy of Cerisse Allen and Paul Brakefield.)

example, smaller individuals tend to have higher metabolic rates per body mass. A second source of genetic correlations is **linkage disequilibrium**, the nonrandom association between alleles at different loci (see Chapter 4).

In some cases, selection favors genetic correlations. We saw earlier that correlational selection on garter snakes favors certain combinations of coloration and predator escape behaviors. These traits are also genetically correlated, and the high fitness combinations of traits are more common than they would otherwise be [9]. While we do not know whether the genetic correlation in garter snakes results from pleiotropy or linkage disequilibrium, in other cases we do. The butterfly *Heliconius numata* has several wing-color morphs that mimic different species of model butterflies. (See Chapter 13 for more on mimicry.) Selection caused by predation favors certain combinations of color elements. Genetic analysis showed that the color morphs are controlled by a small segment of chromosome with 18 genes [29]. The high fitness combinations of alleles that control the colors have been locked

FIGURE 6.23 A wing-color polymorphism in the butterfly *Heliconius numata* is controlled by a small segment of chromosome. Genetic analysis shows that the segment consists of two overlapping inversions that do not recombine. These inversions carry loci with alleles that alter the pattern and coloration of the wings. The different color morphs are favored in different parts of the species' range because they mimic other species of toxic butterflies that are common in those regions. Top: Schematics of the chromosomes showing the changes in gene order produced by the inversions. Bottom: The wing-color patterns produced by the different chromosomes. (After [29].)

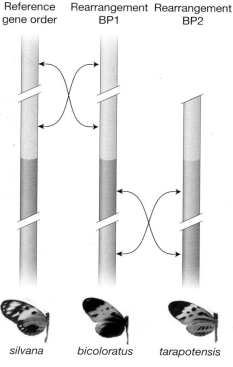

together by a series of chromosome inversions that prevent recombination from producing low fitness color patterns (**FIGURE 6.23**). Selection favored a genetic correlation between the colors controlled by several loci, and the inversions spread because they strengthened that correlation.

Phenotypic Plasticity

Most tadpoles (the larval stage of frogs and toads) live on a diet of algae and detritus. Spadefoot toads (*Spea*), however, have a remarkable trick (**FIGURE 6.24**). When their eggs hatch in ponds where algae are the main food source, they develop into typical omnivorous tadpoles. But when they hatch in ponds with a high density of shrimp and other animal prey, they develop into carnivorous tadpoles with a greatly enlarged head and sharp horny beak [39]. The omnivorous tadpoles have large fat reserves that increase their survival as adults. The carnivorous tadpoles sacrifice these reserves but can develop more rapidly on their diet of animal protein, which allows them to metamorphose at an earlier age. That is adaptive because the conditions that trigger development of carnivores occur in ponds that dry up quickly, and tadpoles die if they have not yet metamorphosed when that happens.

This developmental shift is an example of **phenotypic plasticity**, which occurs when an individual's phenotype changes in response to the environment it experiences. In the case of spadefoot tadpoles, the change is developmental and irreversible. Plasticity can also be physiological and reversible, for example the tanning that light-skinned people show after exposure to ultraviolet (UV) light. Plasticity is seen in a wide range of traits that range from gene expression to morphology and physiology to behavior.

Phenotypic plasticity can be visualized with the **reaction norm**, which is a plot showing how environmental conditions affect how a phenotype is expressed. Reaction norms can differ among genotypes, which means that reaction norms themselves can evolve. Genetic variation in a reaction norm is referred to as **genotype-environment interaction** (or **G×E**, for short). **FIGURE 6.25** shows that reaction norms for increased pigmentation in response to UV light differ between populations of water fleas (*Daphnia*). These differences are adaptive because in some populations survival is increased by plasticity while in other populations it is not.

Not all phenotypic plasticity is adaptive [20]. When people who live at sea level ascend to high elevations, physiological changes are triggered by

FIGURE 6.24 A carnivorous tadpole cannibalizes a typical tadpole of the spadefoot toad *Spea bombifrons*. This is a dramatic example of phenotypic plasticity: tadpoles of this species develop into either typical or carnivorous morphs depending on environmental conditions.

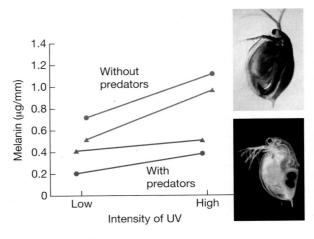

FIGURE 6.25 Reaction norms for pigmentation in the water flea *Daphnia melanica* differ between lakes with and without predators. When water fleas from lakes without predators are exposed to high levels of UV, they develop dark melanic pigmentation that protects their internal organs from the radiation (purple lines). In contrast, water fleas from lakes with predators do not become pigmented under high UV (red lines), which would make them conspicuous and increase the chance that they would be eaten. Phenotypic plasticity in pigmentation is therefore adaptive when predators are absent, while lack of plasticity is adaptive when predators are present. (After [44].)

the decrease in oxygen pressure. Among these changes is an increase in the concentration of red blood cells (RBCs) in the blood. Since RBCs transport oxygen, this change may sound adaptive, but in fact it is not. Increased concentrations of RBCs make the blood more viscous, which slows oxygen delivery and can even trigger medical emergencies. Populations of humans adapted to very high elevation in the Himalayas and Andes have RBC concentrations similar to those of lowland populations. They have adapted to low oxygen pressures by genetic changes to other traits. Increased RBC density is not favored by natural selection at high elevation. In summary, people adapted to living at low elevations show phenotypic plasticity in RBC concentrations when they move to high elevations, but that is a maladaptive response [49].

The Genetic Architecture of Quantitative Traits

We have seen that we can predict the outcome of evolution without knowing anything about the genes that underlie the traits. While that is a tremendous strength of quantitative genetics, there are times when it is important to understand the genetic basis of traits. Questions we would like to answer include: Are the differences among species caused by many or just a few genes? Does the variation in quantitative traits result mainly from genetic variation in the coding or the noncoding regions of the genome? When changing environments generate directional selection, do traits typically respond quickly by evolving with standing genetic variation, or is there a lag until new beneficial mutations occur? When the same phenotypic adaptation evolves independently in different species, are the same or different genes responsible?

Quantitative trait loci

The regions of the genome that affect a quantitative trait are called **quantitative trait loci**, abbreviated as **QTL**. They can range in size from a single nucleotide to a segment of chromosome that contains many genes. Several strategies are used to determine the number, genomic locations, and effects of QTL. Variation in melanism in the peppered moth and sickle cell anemia in humans are caused almost entirely by single loci with alleles that have large effects (see Chapter 4). The inheritance of these traits was discovered by controlled breeding experiments in the moth and by studying inheritance of sickle cell disease in human families. But those research strategies have limitations: without additional data, they do not tell us what the genes are, and they do not work when many genes contribute to the trait.

To make further progress, we use **QTL mapping**. This starts with a genetic map of the species that shows the location of genetically variable markers on chromosomes. Often these markers are single nucleotide polymorphisms, or SNPs. The next step is to genotype a large number of individuals at these markers and measure their values for the trait. Last, the variants that individuals carry at the markers are correlated with the trait phenotype (**FIGURE 6.26**). A significant correlation is evidence that a QTL affecting the trait lies on the chromosome near the genetic marker. (More specifically, a correlation means that the marker and the QTL are in linkage disequilibrium—see Chapter 4.)

The large juicy tomatoes you can see in your local supermarket are very different from their wild ancestors. QTL mapping revealed that alleles at a single QTL change the weight of a tomato by up to 30 percent [18]. To find this QTL, plant geneticists used a *mapping cross* (**FIGURE 6.27**). They hybridized a domesticated

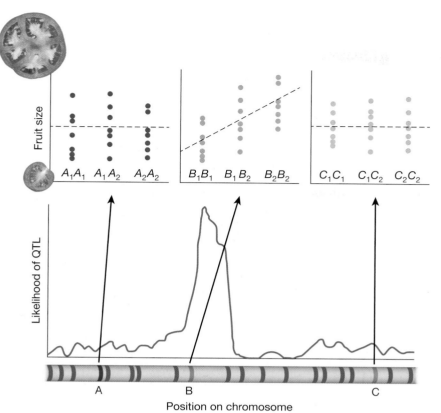

FIGURE 6.26 QTL can be mapped by associating phenotypic variation in a trait with DNA polymorphisms along a chromosome. The positions of genetic markers (for example, SNPs) on a chromosome are shown by the bands on the chromosome at bottom. Individuals are genotyped at each marker, and their phenotypic values are plotted against their genotypes. Examples are shown for three markers. At markers A and C, there is no relation between genotype and phenotype. At marker B, however, individuals with the B_2 allele have a larger phenotype. This suggests that a locus that affects the trait lies near marker B on the chromosome. By combining results across all of the markers, a plot is generated that shows the likelihood of a QTL at each point along the chromosome (blue curve).

tomato with a much smaller wild relative. Several more generations of breeding among the hybrids produced a population whose genomes were a mixture of pieces of chromosomes that came from the two parental species. Correlating marker genotypes with tomato size zeroed in on a region of chromosome 2 with a large effect. Further molecular studies revealed that the locus responsible is a gene called *ORFX* that is expressed early in tomato development. The tomato example shows a basic feature of mapping crosses: they are most powerful when used to find genes that contribute to large differences among populations or species.

There is tremendous interest in finding QTL in humans. Finding genes that affect disease resistance could lead to new therapies, while genes that differ among populations give us insight into how we have adapted to different environments around the planet. Mapping crosses in humans are generally frowned on, so other strategies must be used. One is called a *genome-wide association study,* or *GWAS*. Once again, we look for correlations between the genotypes at genetic markers and phenotypic traits of interest. There are, however, important differences between the GWAS and mapping cross approaches. With GWAS, we are looking for QTL that contribute to genetic variation within a population, while a mapping cross seeks the QTL responsible for differences between populations (or species). Furthermore, because the phenotypic differences are typically smaller

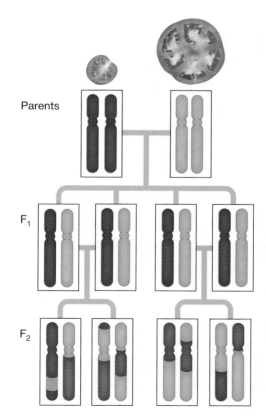

FIGURE 6.27 Mapping crosses are one strategy used to locate QTL. Two parents with very different phenotypes are genotyped at a large number of genetic markers throughout their genomes. The diagram shows a single pair of chromosomes, where red represents the chromosomes that come from the small individual and blue the chromosomes from the large individual. These individuals are crossed to produce an F_1 generation, which is again crossed for one or more additional generations. The offspring from one of these later generations are analyzed by the QTL mapping strategy shown in Figure 6.26.

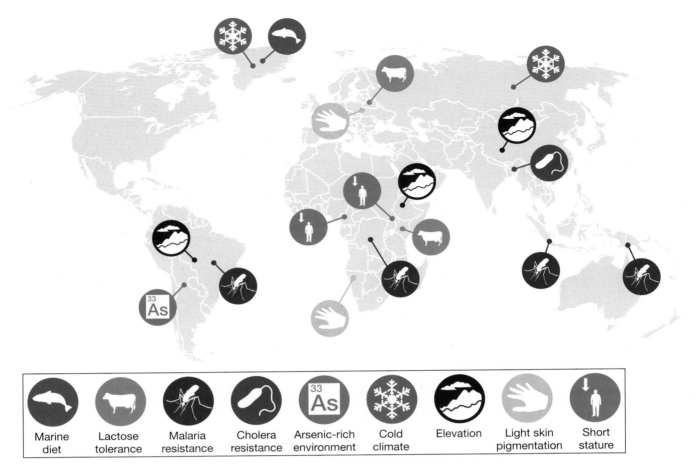

| Marine diet | Lactose tolerance | Malaria resistance | Cholera resistance | Arsenic-rich environment | Cold climate | Elevation | Light skin pigmentation | Short stature |

FIGURE 6.28 QTL involved in adaptive evolution of several traits in humans have been identified by correlating large differences in allele frequencies among populations with distinct phenotypes. (After [51].)

and environmental effects on the phenotype are more difficult to control, much larger sample sizes are needed with GWAS.

Another approach that is used to find human QTL takes advantage of the distinctive traits seen in some populations. The strategy is to scan the genome for loci that have large allele frequency differences among populations with divergent phenotypes. This strategy has discovered QTL that underlie adaptive evolution of traits that include disease resistance, body height, and tolerance to high elevation and cold (**FIGURE 6.28**).

QTL mapping identifies regions of chromosomes that can range in size from a few hundred to many hundreds of thousands of DNA bases. A single QTL often spans several genetic loci. Other research strategies are needed to find which DNA base or bases within a QTL are responsible for the phenotypic variation. The cause is sometimes discovered to be just a single nucleotide. In other cases, several loci or a chromosomal rearrangement are responsible (see Figure 6.23).

The genetics of quantitative traits

One of the most basic questions we can ask about the genetics of quantitative traits is how many loci contribute to their phenotypic variation. The answer is important for several reasons. Population genetics theory tells us that if many loci contribute, then the trait can evolve further before genetic variation is exhausted as alleles become fixed (see Figure 6.4). The dramatic responses to artificial selection in the growth rate in chickens (see Figure 5.3) and in the oil content of corn (see Figure 6.17) show how far and how fast traits can evolve when many loci are involved.

A second reason to ask about the number of loci is that the answer affects strategies for fighting certain diseases. When one or two genes contribute to a disease, it may be possible to exploit knowledge about what those genes are and how they work. Hemophilia is a hereditary disease in which blood clotting is impaired because of a mutation in one of the genes that produce clotting factors. It can now be treated by introducing a working copy of the defective gene [38]. This type of gene therapy may not be feasible for diseases that involve contributions from dozens or even hundreds of loci, such as diabetes [17], heart disease [37], and schizophrenia [19].

We've seen that the vast majority of quantitative traits are heritable, meaning there is standing genetic variation that selection can act on. What maintains that variation? The answer is not entirely clear, but it must involve a combination of the factors that maintain polymorphism at individual loci. Mutation is likely the most important force. Mutation at QTL introduces alleles that are typically deleterious, leading to a mutation-selection balance (see Chapter 5). Although mutation rates at individual loci are usually very small, a considerable amount of additive genetic variance can be generated when there are many QTL. Experiments with *Drosophila* show that mutation typically increases the phenotypic variance of a trait by 0.1 to 1 percent per generation [35]. An equilibrium level of standing genetic variation is reached when selection removes the same amount of variance. In addition to standing variation, new mutations also contribute to the evolution of quantitative traits in the long term. The remote ancestor of the blue whale, the largest animal that has ever lived, was about the size of a cat. That enormous change in body size must have involved many new mutations that appeared as the whale's ancestor evolved to larger and larger sizes.

The QTL responsible for the standing genetic variation within species may be quite different than those responsible for differences among species [27, 50, 54]. A major reason for this discrepancy is that many mutations that contribute to genetic variation for quantitative traits have deleterious pleiotropic effects. When directional selection acts over long periods, for example to produce an animal the size of a blue whale, only those mutations that are largely free of these negative side effects will survive and become fixed. Thus while many alleles may contribute to standing genetic variation within species, a much smaller number may be important to adaptive evolution and contribute to differences among species.

The kinds of loci that contribute to the variation within species may also differ from those responsible for adaptation. A study of the genetics of flower color found that all of the molecular differences among species in color intensity that have been studied result from mutations in transcription factors, which are a type of regulatory locus. By contrast, transcription factors are in the minority of spontaneous mutations that occur within species [50]. In populations of stickleback fishes that have recently adapted to fresh water, some of the genetic changes are in coding regions, but the large majority (perhaps 80 percent) are regulatory [28].

A final question about the genetic basis of quantitative traits is how often convergent evolution of phenotypes, which occurs when two species independently evolve the same trait, involves changes at the same genes [48]. When very

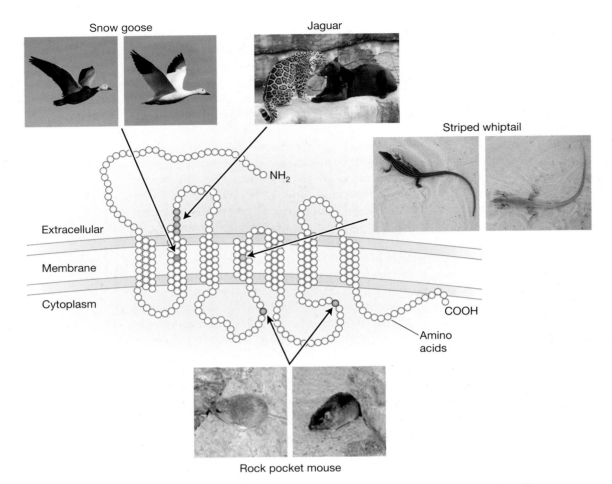

Snow goose

Jaguar

Striped whiptail

NH₂

Extracellular

Membrane

Cytoplasm

COOH

Amino
acids

Rock pocket mouse

FIGURE 6.29 Diagram of the melanocortinin-1 receptor bound to the cell membrane. The arrows point to positions where selection has changed amino acids to produce new coloration in four species of animals. In these cases, the results are polymorphisms within species (shown in the photos). Many other polymorphisms within species and differences between species have also evolved by changes to the *Mc1r* gene. (After [36].)

different animals independently evolve similar coloration, is that the result of changes at the same or at different genes? Surprisingly, the evolution of darker and lighter colors in many species of mammals, birds, and reptiles is often due to changes at just a single locus, the *melanocortin-1 receptor*, or *Mc1r*. The protein produced by this gene regulates the production of pigments in hair, feathers, and scales. Mutations at several different nucleotides in the *Mc1r* gene have produced dark (melanic) coloration (**FIGURE 6.29**). Here parallel phenotypic evolution results from genetic changes that are parallel at the level of the gene but not at the level of the nucleotides.

Go to the
Evolution Companion Website
EVOLUTION4E.SINAUER.COM
for data analysis and simulation exercises, quizzes, and more.

SUMMARY

- An individual's phenotype—the set of its visible traits—is determined by a combination of its genotype and environmental factors.

- Variation in quantitative traits can be caused by just a few or by a very large number of loci. When variation results from many genes, the trait can evolve far past its original range of variation by changes in allele frequencies, without contributions from new mutations.

- A fitness function shows the relation between the value of a trait and the average fitness that individuals with that value have. A fitness function can result in selection that is directional (favoring an increase or decrease of a trait's mean), stabilizing (selection against extreme individuals, which decreases variation in the population), or disruptive (selection against intermediate individuals, which increases variation).

- The force of directional selection on a trait is measured by the selection gradient, which is slope of the regression line that relates relative fitness to the trait value. The selection gradient can be used to predict the rate at which a trait will evolve and to test hypotheses about adaptation.

- The rate at which the mean value of a trait will evolve is given by the breeder's equation, and it depends on the amount of genetic variation (measured either by the additive genetic variance or the heritability) and the strength of directional selection.

- Almost all quantitative traits have standing genetic variation and will evolve when selection acts on them. When selection acts on traits that do not have heritable variation, new mutations must arise before the trait will evolve.

- Artificial selection has been essential to civilization. Selective breeding has caused many species of domesticated animals and plants to evolve dramatically new forms, very different from those of their wild ancestors. The results of artificial selection demonstrate that selection can produce very large changes in relatively short periods of time. Natural selection can do the same in natural populations.

- Genetic covariance (or correlation) between traits causes evolutionary side effects: selection on one trait will cause others to evolve. This can result in trade-offs and constraints, in which adaptation in one trait has negative fitness effects on other traits. Genetic correlations result from pleiotropy and linkage disequilibrium.

- Some traits show phenotypic plasticity, the situation in which the phenotype produced by a genotype is altered by the environment that an individual experiences. Plasticity of some traits has evolved adaptively, but in other cases the response to the environment is not adaptive.

- Genetic variation in quantitative traits can be caused by a small or a large number of quantitative trait loci (QTL). These chromosome regions can be localized by QTL mapping, in which variation at genetic markers is correlated with a trait's phenotypic value.

- The number and types of loci that contribute to additive genetic variation within populations may often be quite different than those involved in adaptive differences among species.

TERMS AND CONCEPTS

adaptive landscape
additive genetic variance
artificial selection
breeder's equation
correlational selection
direct response to selection
directional selection
disruptive selection

environmental variance
evolutionary constraint
fitness function
genetic correlation
genetic covariance
genetic line of least resistance
genotype-environment interaction (G×E)
heritability

indirect response to selection
linkage disequilibrium
normal distribution
optimum phenotype
phenotypic plasticity
phenotypic variance
pleiotropy
polygenic trait

QTL (quantitative trait locus or loci)
QTL mapping
quantitative genetics
quantitative trait
reaction norm
selection gradient
stabilizing selection
standing genetic variation
trade-off

SUGGESTIONS FOR FURTHER READING

There are two classic texts on quantitative genetics. *Introduction to Quantitative Genetics* by D. S. Falconer and T. F. C. Mackay (Longman, Essex, 1996) is written from the perspective of animal breeding but gives a wonderfully clear overview of the basic concepts. *Genetic Analysis of Quantitative Traits* by M. Lynch and J. B. Walsh (Sinauer Associates, Sunderland, MA, 1998) is a comprehensive review of genetic and statistical quantitative genetics. Both books are quite technical and (sadly) now a bit dated, particularly regarding the many advances that have been made in identifying QTL. Quantitative genetic variances and covariances can be estimated in natural (as well as domestic) populations; see "Estimating genetic parameters in natural populations using the 'animal model'" by L. E. B. Kruuk (*Philos. Trans. Roy. Soc. Lond. B* 359: 873–890, 2004) for an overview.

A tremendous amount of effort is being devoted to finding QTL that affect a variety of traits in a variety of species. In "Commentary: When does understanding phenotypic evolution require identification of the underlying genes?", M. D. Rausher and L. F. Delph (*Evolution* 69: 1655–1664, 2015) discuss when this approach can give us valuable insights to evolutionary questions. G. A. Wray's "Genomics and the evolution of phenotypic traits" (*Annu. Rev. Ecol. Evol. Systemat.* 44: 51–72, 2013) gives an excellent overview of how rapid advances in genomics are opening new insights to the genetic basis of quantitative traits and how they evolve.

Much of the interest in quantitative genetics among evolutionary biologists was inspired by research done by R. Lande and colleagues beginning in the 1980s. In "Quantitative genetic analysis of multivariate evolution, applied to brain:body size allometry" (*Evolution* 33: 402–416, 1979), Lande developed the multivariate breeder's equation (Equation 6.1). Lande and S. J. Arnold pioneered methods for estimating selection gradients in "The measurement of selection on correlated characters" (*Evolution* 37: 1210–1226, 1983). J. G. Kingsolver and colleagues provide excellent reviews of selection gradients in natural populations in "The strength of phenotypic selection in natural populations" (*Am. Nat.* 157: 245–261, 2001) and "Phenotypic selection in natural populations: What limits directional selection?" (*Am. Nat.* 177: 346–357, 2011).

Human height is widely used as a model system for methods used to detect QTL that underlie variation in quantitative traits. See P. M. Visscher's "Sizing up human height variation" (*Nat. Genet.* 40: 489–490, 2008) for a review of this topic.

Artificial selection experiments are used by evolutionary biologists to study the evolution of quantitative traits under controlled conditions. W. G. Hill and A. Caballero review this interesting field in "Artificial selection experiments" (*Annu. Rev. Ecol. Systemat.* 23: 287–310, 1992).

PROBLEMS AND DISCUSSION TOPICS

1. In a study of selection on the leg length of migratory locusts, the mean leg length is 18.6 mm, the selection gradient is $\beta = -0.13$/mm, the phenotypic variance is $P = 1.4$ mm2, and the heritability is $h^2 = 0.37$. What is the expected response to selection in the next generation? What do you predict the average leg length will be in the next generation?

2. In the same population of locusts, the mean wing length is 47 mm, the selection gradient on wing length is $\beta = 0.12$/mm, the phenotypic variance for wing length is $P = 3.6$ mm^2, and the heritability of wing length is $h^2 = 0.27$. In addition, we know that the additive genetic covariance between wing length and leg length is 0.6

What is the expected evolutionary change in mean leg length due to selection on both wings and legs? Repeat these calculations to predict what will happen to wing length as a result of the selection on both wings and legs. What do you predict the average wing and leg lengths will be in the next generation?

3. We told you that Figure 6.8A shows stabilizing selection, while Figure 6.8B shows disruptive selection. If the traits shown in this figure have heritabilities greater than 0, do you predict that the mean in the next generation will be equal to the mean of the data shown? Can selection simultaneously be directional and stabilizing? Directional and disruptive? Stabilizing and

4. In some cases, inheritance of a single allele, such as the sex-determining factor, will cause phenotypes to be so different that we can see which individuals carry which allele. List the reasons why this is unusual. Why can't we usually tell at a glance who carries which allele?

5. In Equations 6.1 and 6.2 we wrote the equation for evolutionary change in the mean of a trait $(\Delta \bar{z})$ two different ways. Using the definitions of h^2 and β from the text, investigate the differences between these two equations. Can you write the equation in a form that involves just the quantities P, G, and β? Equations 6.1 and 6.2 seem to suggest that h^2 and G measure inheritance, while β and S measure selection. Given the ways you can rewrite these equations, which of these are the best measures of inheritance and selection?

6. There are many traits for which it seems natural selection should favor an increase every generation, such as survival from birth to reproduction. In most cases, when we look for such increases in natural populations we do not see the predicted change. Make a list of all the reasons we might not see a response to directional selection on such a trait. Include reasons suggested by the material in this chapter, as well as any other reasons you can think of.

7. When the technology for QTL mapping first became available, researchers studying human genetic diseases hoped to discover common alleles that cause increased risk for those diseases. What would be the advantages to studying the causes of diseases that are caused by common, as opposed to rare, alleles? What would be the advantages to treating diseases that are caused by common alleles?

8. The results of QTL mapping studies for human diseases tend to show that disease-causing alleles are either rare or have very small effects on risk. Knowing that this is true, discuss the evolutionary forces that are most likely to be responsible for this state of affairs. Does this observation suggest something about the evolutionary forces that maintain disease risk in human populations?

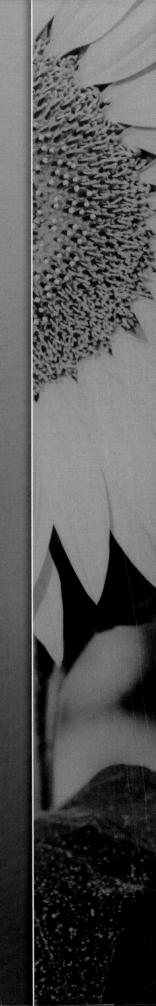

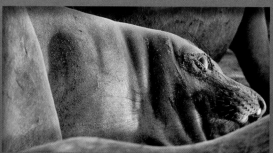

Genetic Drift: Evolution at Random

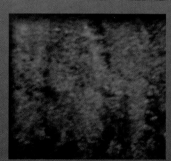

Northern elephant seals are remarkable animals. They spend most of their lives at sea. While hunting fish and squid, the seals dive a mile deep and can hold their breath for more than an hour. During the breeding season, males have dramatic fights, and the most successful mate with dozens of females (**FIGURE 7.1**).

Elephant seals are exceptional in yet another way: they are one of the least genetically variable mammals known [4]. This is a recent state of affairs. When European settlers arrived on the Pacific coast of North America in the 1700s, they found elephant seals living from Alaska south to Mexico. The seals were easily hunted for their blubber and fur, and by the mid-1800s the population had been reduced to fewer than 40 individuals. Fortunately, hunting was banned before the species was driven to extinction. Today there are more than 100,000 northern elephant seals, and the population continues to expand.

Comparison of specimens collected in the 1800s with living individuals shows that elephant seals had much more genetic variation before their population crashed [4, 13]. This loss of variation illustrates one of the effects of random genetic drift, an evolutionary process we have mentioned but not yet discussed in any detail. Drift explains features of the living world that natural selection cannot. It also provides us with tools to estimate population sizes, phylogenies, and other important features of nature.

This chapter opens by looking at drift from two perspectives: forward in time and backward in time. It then turns to the factors that determine the strength of drift, and how drift affects genetic variation. The next topic is how drift interacts

 These females of the northern elephant seal (*Mirounga angustirostrus*) have come ashore to breed. A severe population crash in the nineteenth century caused by hunting dramatically reduced genetic variation in this species, an example of random genetic drift.

FIGURE 7.1 Male elephant seals fighting for control of a harem of females. A small number of males obtain the large majority of matings, contributing to the intensity of random genetic drift.

with natural selection. The chapter then discusses how molecular differences among species accumulate in time. It closes with a look at one of the most exciting subjects of current research in biology: the use of DNA sequences to find evidence for adaptive evolution in humans and other species.

What Is Random Genetic Drift?

Genetic change between generations—that is, evolution—happens even when selection is not at work. We have already seen that mutation changes allele frequencies. Evolution also results from chance events of survival, reproduction, and inheritance. The evolutionary process that results is called **genetic drift**.

The grove snail (*Cepaea nemoralis*) is famous for its highly variable shells (see Figure 5.14). This is a terrestrial species that lives in pastures, which it often shares with cows and sheep. No doubt thousands of snails die each day when they are stepped on by livestock that are oblivious to what is beneath their hooves. In any given field on any given day, some unlucky colors of snails will happen to get crushed more often than other colors. Variation in color is determined by a small set of loci, and so these random deaths cause changes in the allele frequencies in that population. This scenario shows that an individual's genotype and phenotype are not the only factors that determine if he or she leaves genes to the next generation. Chance plays a role too.

Another opportunity for chance to influence the fate of genes comes during meiosis. When an individual is heterozygous at a locus, only one of the two alleles is passed to each gamete that it makes. If you have two children, there is a chance of 1/2 that they will both inherit the same allele from you, and your other allele will leave no descendants in the next generation. So even if all individuals survive and leave the same number of offspring, meiosis itself causes random changes allele frequencies.

We can discover several key features of genetic drift with an experiment. Rather than using living organisms, the experiment simulates evolution on a computer. The advantage of this strategy is that it can perfectly control the conditions. In particular, the simulation is designed so that all individuals have exactly the same chance of surviving and leaving offspring. It simulates a diploid hermaphroditic species, and it follows the changes in the frequency of two alleles at a locus that evolves according these rules:

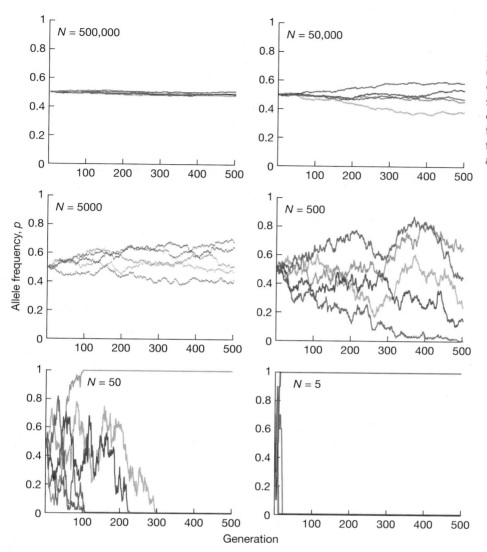

FIGURE 7.2 Computer simulations of drift. In each graph, five identical populations start with two alleles at equal frequencies ($p = 0.5$). Allele frequencies change in later generations by random genetic drift (no selection is acting). Smaller population sizes cause bigger random changes in allele frequencies in each generation, and lead to quicker fixation ($p = 0$ or $p = 1$) of one allele or the other.

1. Each generation starts with N juveniles. This population size is fixed by the limited space that is available.

2. Every individual survives to become an adult, so there is no selection on survival.

3. When adults reproduce, each individual makes 1000 sperm and 1000 eggs. (The actual number does not matter so long as it is reasonably large.) Because each individual makes the same number of gametes, there is no selection on reproductive success. After reproduction, the adults die.

4. Zygotes are formed by random fertilization between sperm and eggs in the pool of $2000N$ gametes. But since space is limited, only N randomly chosen zygotes survive to become juveniles. The life cycle then begins again, and the simulation returns to step 1.

What happens when we run the simulation? **FIGURE 7.2** shows how allele frequencies change over 500 generations in replicate populations that begin with an allele frequency of $p = 0.5$. Frequencies fluctuate randomly. In each generation, allele frequencies change by the random sampling of the genes that start the next generation (step 4 of the simulation). This is the essence of genetic drift.

Figure 7.2 illustrates five fundamental features of drift. The first is that *drift is unbiased*: an allele frequency is as likely to go up as to go down. Natural selection can favor one allele over another, but genetic drift does not.

Second, the figure shows that the *random fluctuations in allele frequency are larger in smaller populations*. That results from a basic law of probability. When you flip a coin, you expect to get heads half the time and tails half the time. If you flip a coin only twice, there is a probability of 1/2 that you will get all heads or all tails, rather than half and half. But if you flip the coin 1000 times, the probability that you will get all heads or all tails is less than 10^{-300}. It is much more likely that about half of the flips will come up heads and half of them tails. Outcomes become more predictable when averaging over a larger number of random events. This is why random genetic drift is stronger in small populations and weaker in large ones.

A third basic feature is that *drift causes genetic variation to be lost*. An allele frequency that fluctuates randomly up and down will eventually reach either $p = 0$ or $p = 1$. (Picture a New Year's Eve partyer staggering along a long train platform with railroad tracks on each side. Sooner or later, he will fall off the platform onto one track or the other.) One of the alleles is then fixed. While the allele that is lost can be reintroduced by mutation, drift by itself causes genetic variation to be lost. The loss is faster in smaller populations because they have larger allele frequency fluctuations.

A fourth basic feature seen in Figure 7.2 is that *drift causes populations that are initially identical to become different*. A useful way to think about this point is to consider the variance in allele frequencies among the populations in our experiment. At generation 0, all populations have an allele frequency of $p = 0.5$, and so the variance among them is zero. After one generation, however, variation among the populations is generated by drift, and the variation grows with time. In quantitative terms, starting with an allele frequency p, the average of the allele frequencies across the replicates in the next generation is also expected to be p, and the variance of allele frequencies across the replicates will be $p(1 - p)/2N$. (This quantity is based on the binomial distribution, which describes how allele frequencies change by genetic drift.) This confirms our earlier conclusion that the variation among populations generated by drift grows more slowly in large populations than in small ones.

Fifth, the figure shows that *an allele can become fixed without the benefit of natural selection*. If we wait long enough, it is certain that one of the two alleles will become fixed and the other lost. A simple rule tells us the probabilities of those two outcomes: if an allele's current frequency is p, then the probability that it becomes fixed is also p. This result implies that a new mutation that has no effect on fitness has a probability of $1/2N$ of ultimately becoming fixed, since it is initially present as single copy among $2N$ copies of the gene in a diploid population.

We can see several key features of drift in an experiment that used the fruit fly *Drosophila melanogaster* [5]. Replicate populations were established, each with eight females and eight males. Each replicate was begun with equal frequencies of two alleles that do not measurably affect survival or reproduction (at least in the lab). The replicate populations were propagated by allowing the parents to reproduce, then choosing eight female and eight male offspring at random to start the next generation. **FIGURE 7.3** shows the results. After one generation, the allele frequencies varied substantially among the replicates. By ten generations, the frequencies were distributed evenly between 0 percent and 100 percent. After 19 generations, more than half of the replicates had lost one allele or the other.

Figure 7.3 shows one more general feature of genetic drift. Across the populations taken as a whole, there are fewer heterozygotes than predicted by the Hardy-Weinberg ratios. This deficit is cause by allele frequency differences among populations. If the *Drosophila* experiment were continued, one or the other allele would become fixed in every population. At that point, roughly half the populations would be

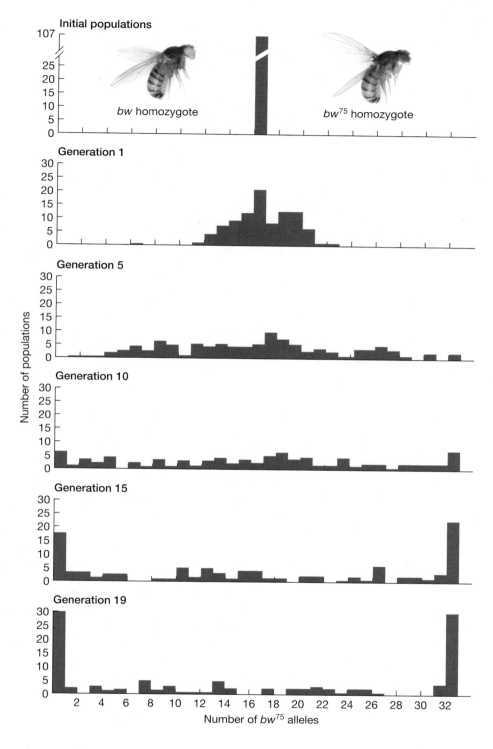

FIGURE 7.3 Random genetic drift in experimental populations of *Drosophila melanogaster*. At a locus that affects eye color, homozygous *bw/bw* flies have white eyes, homozygous *bw*⁷⁵/*bw*⁷⁵ have brown eyes, and heterozygotes are intermediate. Each of 107 populations was founded with 16 heterozygotes. Populations were then propagated with eight males and eight females per generation. Differences among populations accumulated by drift as the experiment went on. By generation 19, the *bw*⁷⁵ allele was lost from 30 populations and was fixed in 28 populations. (Data from [5].)

homozygous for one allele, roughly half would be fixed for the other allele. Within each population, allele frequencies would be in Hardy-Weinberg equilibrium, but the frequency of heterozygotes would be 0 rather than 1/2 as expected if all individuals were in a single population. Thus drift does not cause large departures from Hardy-Weinberg equilibrium within a population, but it does cause a deficit of heterozygotes when a set of diverging populations is considered as a whole.

We can see the effects of drift in natural populations. Researchers collected 2218 individuals of the garden snail (*Cornu aspersum*) living on two adjacent city blocks in

FIGURE 7.4 Allele frequency differences among populations of the garden snail *Cornu aspersum*. The pie diagrams show the frequencies of two alleles at the *GOT-1* locus in groups of snails found on two city blocks. The sizes of the circles are proportional to the numbers of individuals in each group. The "50" allele (shown in blue) is quite common in the left block but almost absent from the right block. That difference almost certainly evolved by drift. (After [35].)

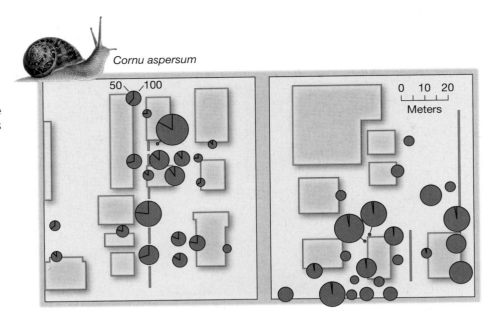

College Station, Texas, and genotyped them at several loci [35]. **FIGURE 7.4** shows the data from one locus. Allele frequencies differ on the two sides of the street. The street acts as a strong barrier to dispersal between the two blocks, and the allele frequency differences between them almost certainly accumulated by drift.

The Genealogy of Genes

For a modest cost, you and your parents can have your genomes analyzed. With those data, you can determine which gene copy at a locus you inherited from your mother and which from your father. With the genomes of your ancestors going back further in time, you could trace the evolutionary history of every gene in your genome. Now imagine we did that for every person now alive. The result would reveal a remarkable fact: for any given locus in the human genome, there was a copy of that gene at some time in the past that was the ancestor of all copies of the gene now carried by all living humans.

To make these ideas more clear, consider a second simulation experiment (**FIGURE 7.5**). Here we follow the genealogy of the 10 gene copies at a locus in a population of 5 individuals. The gene copies may or may not have the same DNA sequence, but we again ensure there is no selection—all individuals have equal chances of surviving and reproducing. In the first generation of the simulation, some gene copies are lost when the individuals carrying them die by random events, while others are lost when by chance they fail to be passed to a gamete during meiosis. Looking at Figure 7.5B, we see that only six of the ten copies of the gene present in generation 0 left descendants to generation 1. By chance alone, four gene copies present in generation 0 will never be represented in future generations. When generation 2 is produced by generation 1, accidents of survival and reproduction eliminate one more of the original 10 lineages (see Figure 7.5C). Once a lineage goes extinct, it can never come back from the dead, so the total number of lineages descending from the original population can only decrease in time.

By the time we reach generation 7, all of the gene copies are the descendants of just one copy that was present in generation 0 (see Figure 7.5D). Looking from generation 7 backward in time, we can trace the ancestry of all the gene copies. That genealogy, shown by the red lines, is called the **gene tree**.

What happens if we run the simulation again? Genetic drift is random, and so different copies of the gene will become the lucky winners that become ancestors of all the future copies. Furthermore, the number of generations that it takes for

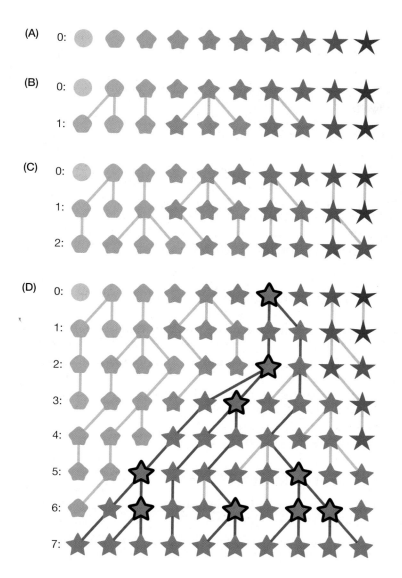

FIGURE 7.5 Evolution of a gene tree. In this example, there is no selection, and differences in the number of descendants left by different copies of the gene are entirely the result of random genetic drift. (A) The population begins at generation 0 with five diploid individuals, so there are a total ten copies of the gene. The shapes and colors are to help distinguish the different copies, which may or may not differ in their DNA sequences. (D) By generation 7, all of the gene copies are descendants from a single copy in generation 0. Their gene tree is shown by the red lines. Coalescence events are shown by gene copies with bold outlines. The most recent common ancestor of all gene copies in generation 7 is in bold at the top of the gene tree in generation 0.

one copy to become the ancestor of all living copies will vary randomly between the simulations.

We are now discussing the genealogy of genes: the paths of their inheritance across generations. Looking backward in time, when the lineages of two gene copies merge we say that they **coalesce**. These events appear in Figure 7.5 as branch points (nodes) on the gene tree that represent the most recent common ancestor of the gene copies. As we have seen, if we go back far enough, we are certain to arrive at the most recent common ancestor of *all* copies of the gene now present in a species[1]. That typically happens in the recent past when the population size is small, and in the remote past when it is very large.

Figure 7.5 illustrates another key point. Although all the gene copies in generation 7 descended from a single copy in generation 0, that does not mean there was

---

[1] The picture is a bit more complicated for loci that recombine. Any given DNA base in a locus has a gene tree of the sort we just described. But recombination between different parts of a locus causes them to have different gene trees. For the purposes of our discussion here, however, we will ignore this complexity because the central principles remain intact. In particular, for any locus there is a most recent common ancestor for all copies of the locus carried by living individuals.

only a single gene copy (or single individual) present then. There were nine other copies in that generation, and it is only by chance that they left no descendants in the long run.

If we could trace the gene tree of all human mitochondrial genomes backward in time, we would eventually arrive at a single genome, which was their most recent common ancestor. Since mitochondria are inherited maternally, the person carrying that ancestral mitochondria would necessarily be a woman. Analysis of variation in mitochondrial DNA among living humans suggests that she lived about 125,000 years ago [33]. She is sometimes called the "Mitochondrial Eve." That name is misleading, however, because many other females were also alive at that time, and they contributed other genes to modern humans. Likewise, tracing the ancestry of all Y chromosomes would lead back to a single male [15].

We can be certain that the ancestor of all mitochondria in living humans was carried by a female and that the ancestor of all Y chromosomes was carried by a male. Thus different genes have different genealogies. They have different common ancestors that lived in different places and at different times. This is a general principle that applies throughout the genome of all sexually reproducing species: each part of the genome has a different genealogy.

This genealogical perspective also extends across species. The mitochondria that each of us carry and the ones carried by all living chimpanzees are the descendants of a mitochondrion that lived perhaps 8 million years ago (Mya) in an ancestor of humans and chimps, before those two lineages split apart. Going back yet further, the common ancestor of the mitochondria in every individual of every eukaryotic species now alive existed some 2 billion years ago. That single mitochondrion left an extraordinary legacy: the DNA that it passed on is now carried by every animal, plant, fungus, and alga on Earth.

How Strong Is Genetic Drift?

Genetic drift is a random process that is always at work. While it is stronger in small populations than in large populations, only an infinitely large population would be immune from drift. The most numerous organism on Earth is a bacterium called *Pelagibacter ubique* that has a population size of 10^{28} individuals [27], a number that is more than one million times larger than the number of stars in the universe. But it is still a finite number, which means that even *Pelagibacter* experiences genetic drift.

The strength of random genetic drift in a population is measured by the **effective population size**, represented with the symbol N_e. This number provides a way to compare the strength of drift in different populations. We've already seen that the size of a population affects drift—it is stronger in smaller populations. Many other factors also affect drift. If most individuals in a population are too young or too old to reproduce, then drift is stronger than it would be if all the individuals were reproductive. Drift is also affected by changes in population size and unequal numbers of reproducing males and females (see Figure 7.1). To account for all these factors, we imagine an idealized hermaphroditic population of constant size in which all individuals have an equal chance of leaving offspring (just as in the two simulations). The effective population size is the number of individuals that would give this idealized population the same strength of random drift as the actual population of interest. A small value of N_e means that drift is strong, while a large value means that drift is weak.

The value of N_e tells us several useful things about a population. We saw earlier that any two copies of a gene share a common ancestor at some point in the past. If the gene is evolving neutrally (that is, with no selection) in a diploid organism, the mathematics of probability tell us that the average time back to this common ancestor is $2N_e$ generations. For *Drosophila melanogaster*, with an effective population size of about 1 million, the ancestor of two gene copies typically existed about 2 million generations ago. For chimpanzees, with an effective population size of about 20,000, the coalescence of two genes typically occurred about 40,000 generations ago [7]. Because of the random nature of drift, however, different pairs of gene copies will show substantial variation around these averages.

Populations that change in size

The evolutionary history of many species includes times when the population size was reduced to small numbers. We began this chapter with the northern elephant seal, whose population crashed in the 1800s and is now rebounding. This is a dramatic example of a **population bottleneck**, the situation in which a population is reduced to a small size for a small number of generations. The term "bottleneck" is a metaphor: we visualize a few individuals squeezing through a period of reduced population size. A population bottleneck causes intense genetic drift for a brief time.

In the northern elephant seal, the bottleneck involved a drastic reduction of the entire species. Reduced population size also occurs during a **founder event**, when a new population is begun from a small number of individuals. Even if that population then grows to a large size, looking backward in time we will see a period in which the ancestors of the present population were few in number. A founder event, like a bottleneck, reduces genetic variation.

The zebra finch (*Taeniopygia guttata*) lives in Australia and the nearby Lesser Sunda Islands, which they colonized from Australia roughly 1 Mya. The birds on the Lesser Sundas show the effects of that founder event: they have much less genetic variation than do Australian birds. This difference can be quantified as the heterozygosity (also called the "nucleotide diversity"), symbolized by π. This is the chance that two chromosomes in a population have different nucleotides at a given site in their DNA sequence. Heterozygosity in zebra finches on the Lesser Sundas is only 20 percent what it is in Australian birds ($\pi = 0.002$ vs. $\pi = 0.010$). The reduced heterozygosity on the Lesser Sundas suggests that population was founded by as few as 9 birds (**FIGURE 7.6**) [2].

Founder events have also shaped genetic variation in human populations. Pennsylvania Amish are a religious community that is closed to intermarrying with people from the outside. They now number about 30,000 individuals, but most are descendants of only a few hundred individuals who arrived in the United States from Europe in the 1700s (that is, about 12 generations ago). One result of a bottleneck or founder event is to cause a random set of rare alleles to become more common. A recessive mutation in the *EVC* locus causes polydactyly, a condition in which individuals have six fingers or toes. The mutation is at a frequency of about 7 percent in Amish populations in the United States, hundreds of times higher than it is elsewhere in Europe or North America [26]. Genealogies show that all copies of the mutation in the Amish descend from a single copy carried either by Mr. or Mrs. Samuel King, who immigrated to America in 1744. By chance, one of those two was carrying the rare polydactyly mutation. Had that person accidentally missed the boat to America, the mutation would be absent from the U.S. Amish population today. Chance is always a factor in evolution.

Modern humans expanded out of Africa starting about 60,000 years ago. Small numbers of individuals colonized new regions and then expanded still further until the entire planet was inhabited. Native

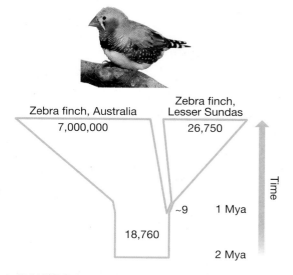

FIGURE 7.6 Estimates of the effective population sizes for the zebra finch (*Taeniopygia guttata*) in Australia and the nearby Lesser Sunda Islands through time. Analysis of genetic variation suggests that the Australian finches are descendants of a population of roughly 19,000 birds that lived some 2 Mya. The much smaller and less genetically diverse Lesser Sunda population was founded by a small number of colonists (perhaps only 9 individuals) that arrived from Australia about 1 Mya. (Based on data from [2].)

(A)

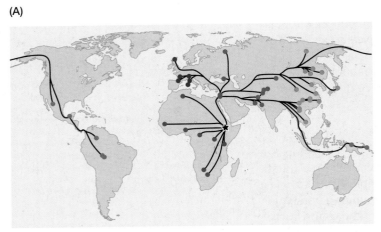

(B)

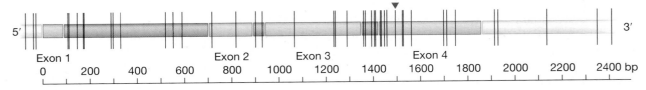

FIGURE 7.7 Genetic variation in humans declines with distance from East Africa, where the ancestors of modern humans lived before expanding out of Africa starting about 60,000 years ago. (A) The routes used by humans to colonize the Earth. (B) Heterozygosity in modern native populations plotted against the distance from Addis Ababa, Ethiopia (shown by the star in [A]) along the colonization routes. Colors indicate the geographical origin of the populations. Heterozygosity is here measured for haplotypes, which are stretches of DNA with multiple SNPs. (A after [21]; B after [20].)

populations now living far from southern Africa therefore experienced many founder events in their past, while those living closer to Africa experienced fewer. Remarkably, the signature of those events can still be seen in our genes. Heterozygosity is highest in Africa and declines as we follow the paths that humans took as they expanded across Earth (**FIGURE 7.7**).

All species experience fluctuations in population size. The effect of these fluctuations is to make the effective population size much closer to the minimum than the maximum population size. During an outbreak of the flu, the viral population grows to an unimaginably large size, but each outbreak starts from a very small number of viral particles. Estimates based on methods discussed in the next section suggest that N_e for the influenza virus is only a few hundred individuals [34].

Drift and Genetic Variation within Species

Genetically speaking, you are very similar to the person next to you. Of the 3.2 billion DNA bases in the human genome, 99.9 percent of them are identical in two randomly chosen individuals. But while only 1 out of 1000 DNA bases differs, the genome is so large that you differ from your neighbor at some 3 million bases.

In humans and other species, polymorphism is spread unevenly across the genome (**FIGURE 7.8**). Polymorphism is typically high in the regions between genes, and within introns. Coding regions (exons) are less variable, particularly the first and second bases of each codon. These patterns are so common across the tree of life that they must result from very general features of evolution.

FIGURE 7.8 Nucleotide variation at the *Adh* locus in *Drosophila melanogaster*. Four exons (green) are separated by introns (blue). The vertical lines show the positions of 43 DNA polymorphisms (SNPs) found in a sample of 11 chromosomes. Of all the SNPs, 17 are in exons, but only 1 of these (shown by the triangle) causes an amino acid change in the alcohol dehydrogenase enzyme. (After [18].)

The explanation is that much of the polymorphism in DNA within species results from random genetic drift acting on selectively neutral mutations. On average, the heterozygosity resulting from neutral mutations evolving by drift in a diploid species is expected to be

$$\pi \approx 4 N_e \mu_n \qquad (7.1)$$

where μ_n is the neutral mutation rate—that is, the chance per generation that the locus mutates to another allele that does not change the organism's fitness. For example, if the total mutation rate at a locus is $\mu = 10^{-6}$ but only 10 percent of mutations are selectively neutral, then the neutral mutation rate is $\mu_n = 0.1 \times 10^{-6} = 10^{-7}$.

Equation 7.1 represents the product of three quantities: the expected number of generations back to the coalescence of two copies of a gene ($2N_e$); the probability that a selectively neutral mutation occurs in a generation (μ_n); and a factor of 2 that accounts for the fact that mutations could occur in either of the two lineages leading back to their most recent common ancestor. (Equation 7.1 is an approximation that is accurate when π is 0.1 or less, which is often the case.) To summarize, polymorphism increases with the effective population size (N_e) and the neutral mutation rate (μ_n).

This simple result explains major patterns in genetic variation seen across the genome. Most mutations that occur in coding regions of the genome are nonsynonymous (they change an amino acid in the protein), and most changes to a protein are deleterious (they decrease survival or reproduction) (see Chapter 4). These mutations are weeded out from the population in a process called **purifying selection**, and they do not contribute to the heterozygosity we observe. Loci that experience purifying selection are said to be under **selective constraint**. The neutral mutation rate for these loci (μ_n) is smaller than the total mutation rate (μ). In contrast, in many noncoding regions, none of the mutations affect fitness, so in those regions the neutral mutation rate is equal to the total mutation rate. Noncoding regions therefore typically have higher heterozygosity, as predicted by Equation 7.1.

The same logic explains patterns of variation at different sites within a coding locus. Many mutations to the third positions of codons are synonymous. Most mutations at the first and second positions, however, are nonsynonymous. They therefore have a lower neutral mutation rate than do third positions, and as a result they are less variable. The *Adh* locus in Figure 7.8 shows this effect: only one of the 17 polymorphisms found in the exons is nonsynonymous.

In sum, at sites of the genome that are free of selection, all mutations are selectively neutral. These mutations are free to drift through the population, and they contribute to heterozygosity as they do. But at sites that experience selection, many or most mutations are deleterious. They are selected out of the population and so contribute very little to heterozygosity. These sites therefore tend to be less genetically variable.

Levels of polymorphism also vary systematically along chromosomes. Regions with high recombination rates tend to be more polymorphic (**FIGURE 7.9**). We saw in Chapter 5 that a selective sweep of a

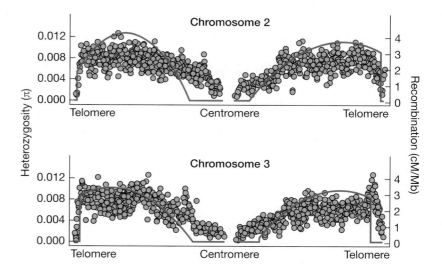

FIGURE 7.9 Heterozygosity (π), shown by the points, along the left and right arms of chromosomes 2 and 3 of *Drosophila melanogaster*. The curves show the recombination rate (in centimorgans [cM] per Mb of DNA) along the chromosome. Heterozygosity is reduced by background selection and selective sweeps in regions of low recombination near the centromeres and telomeres. (After [23].)

FIGURE 7.10 Background selection decreases neutral genetic polymorphism. (A) A population of chromosomes is polymorphic for neutral mutations at five sites (yellow bands). (B) Deleterious mutations appear on two chromosomes (red bands). (C) Selection eliminates chromosomes carrying deleterious mutations. A side effect is that neutral polymorphism is reduced nearby on the chromosome (at sites 2 and 5): the population is now polymorphic at only three sites.

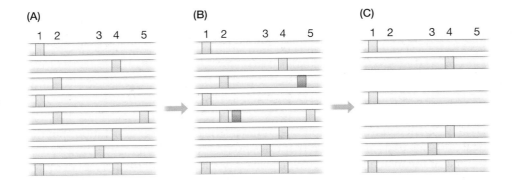

beneficial mutation eliminates variation in the surrounding region of chromosome (see Figure 5.15). In a similar way, when selection eliminates a deleterious mutation, polymorphism is reduced nearby on the chromosome (**FIGURE 7.10**). This effect is called **background selection**. Both selective sweeps and background selection affect larger pieces of a chromosome in regions where recombination rates are low, producing the pattern seen in Figure 7.9. Selective sweeps and background selection reduce the amount of polymorphism at neutral sites of the genome below what Equation 7.1 predicts.

Estimating population size

The relation between heterozygosity and population size suggests a strategy for estimating the effective population size of a species. The idea is to estimate π by sequencing several individuals at sites in the genome that are evolving neutrally. (Introns are often used for this purpose.) We can also estimate the total mutation rate (μ) using the methods discussed in Chapter 4. Because we are focusing on sites that are selectively neutral, the total mutation rate is equal to the neutral mutation rate (μ_n). We find the value of N_e that solves Equation 7.1 for each DNA site, then average those values.

Estimates of the effective population sizes have been made this way for several species (**FIGURE 7.11**). The lowest estimate shown comes from humans: N_e for our species is roughly 10,000. While there are now more than 7 billion people on Earth, our numbers were much smaller just a few thousand generations ago.

This approach can be pushed further to do something even more impressive: we can estimate a species' effective population size from just a single individual [19]. The approach is to sequence its DNA to find the average heterozygosity at a large number of selectively neutral sites, then use the logic just described to estimate N_e. Another remarkable use of DNA polymorphism is to estimate population sizes in the past. The zebra finch discussed earlier now has an effective population

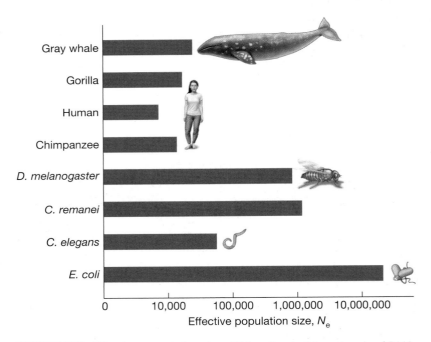

FIGURE 7.11 Effective population sizes (N_e), estimated from levels of DNA polymorphism. The species are ordered by body size; notice that large organisms tend to have smaller effective population sizes. The horizontal scale is logarithmic, so that each vertical line shows a change of N_e by a factor of 10. Humans have a very small N_e: although there are now more than 7 billion people, we descended from only about 10,000 individuals living 100,000 years ago. Two nematode worms are shown. The effective population size of *Caenorhabditis remanei*, which does not self-fertilize, is 20 times larger than that of *C. elegans*, which does self-fertilize. The gut bacterium *Escherichia coli* has an effective population size much larger than that of any of the animals or plants. (Data from [7].)

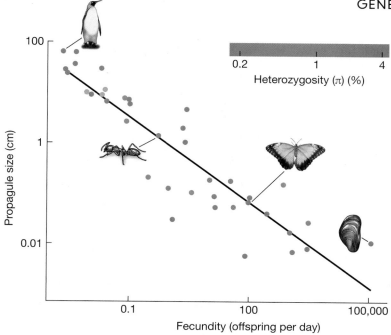

FIGURE 7.12 The heterozygosity averaged across the genome varies among groups of animals. Dots are colored according to their heterozygosity (π) at synonymous sites in coding regions. Ants and vertebrates have low heterozygosity, typically between 0.2 percent and 1 percent. Other groups of animals, including butterflies and bivalves, have higher heterozygosity, typically between 1 percent and 10 percent. These differences result from variation in population sizes, as predicted by Equation 7.1, and from other factors. Fecundity and propagule size (shown on the x- and y-axes) are also strongly correlated with heterozygosity, perhaps because they influence population size, mutation rates, the frequency of selective sweeps, and the strength of background selection. (After [9].)

size of about 7 million individuals in Australia. Analysis of their genetic variation suggests that they are descendants of a much smaller population of roughly 19,000 birds that lived some 2 Mya (see Figure 7.6) [2].

There are large differences in the average values of π among species of animals. Ants and vertebrates tend to have relatively little genetic variation, for example, while butterflies and bivalves (clams and their relatives) are highly heterozygous. Differences in effective population size account for some of this variation, but variation in mutation rates, selective sweeps, background selection, and other factors must contribute as well. Intriguingly, species with high fecundity and small propagules tend to have high heterozygosity (**FIGURE 7.12**). Exactly how those life history factors affect heterozygosity is not yet clear [9].

Genetic Drift and Natural Selection

As an advantageous mutation sweeps through a population, natural selection is not the only process that determines its fate. Genetic drift adds a random component to its trajectory. As a result, the mutation's frequency can increase more rapidly than is expected from selection alone. But sometimes drift causes its frequency to increase less rapidly, or even to decrease.

We can see drift interacting with selection in **FIGURE 7.13**. In these simulations, the favored allele has a selective advantage of $s = 0.01$. In a population of size of $N_e = 500,000$, the allele frequency increases along a smooth trajectory much like those back in Chapter 5 (see Figure 5.7), but with small random fluctuations caused by genetic drift. When the population size is $N_e = 5000$, the fluctuations become more obvious. With a population size of $N_e = 50$, the effects of drift are so strong that the tendency of the advantageous allele to spread is no longer obvious. In fact, that allele is completely lost in two of the five simulations.

These examples illustrate a general point about how genes evolve when both selection and drift are at work. In some cases, the selection is so much stronger than drift that the effects of drift can be largely ignored. In other cases, drift overwhelms selection. A simple rule of thumb tells us if an allele is in one situation or the other. A natural measure for the strength of selection is the selection coefficient s (see Chapter 5). Since large populations have weak drift, a natural measure for the strength of genetic drift is $1/N_e$. We then simply compare these two numbers. The

FIGURE 7.13 Simulations of the spread of beneficial mutations in populations with different effective sizes. The mutations start at a frequency of $p = 0.25$ and have a selective advantage of $s = 0.01$. Each graph shows five replicate populations. With $N_e = 500,000$, allele frequencies follow trajectories like those seen for an infinite population in Figure 5.7. As N_e becomes smaller, the effects of drift become stronger. When the population size is so small that $1/N_e$ is much less than s, there is a high probability that the beneficial allele will be lost by drift, as seen in the graphs with $N_e = 50$ and $N_e = 5$.

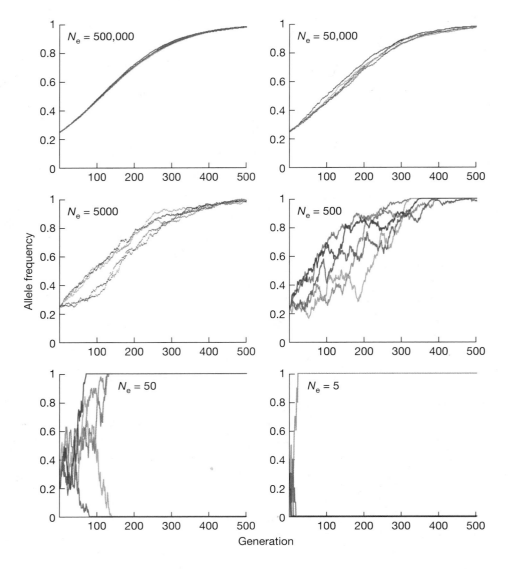

vast majority of alleles fall into one of two categories. Some alleles are strongly selected ($s \gg 1/N_e$), and drift has little effect on their evolution. Other alleles are nearly neutral ($s \ll 1/N_e$), and drift dominates selection in their evolution. These rules explain the differences we see among the graphs of Figure 7.13. A critical point is that it is the relative strengths of drift and selection that matter, not their absolute strengths.

There is an interesting implication of these results: a mutation that is strongly selected in one species can evolve in another as if it were selectively neutral (**FIGURE 7.14**). Some organisms have very large effective population sizes. *Drosophila melanogaster* has an effective population size of about 1 million individuals, and the populations of some bacteria (like the *E. coli* that live in our guts) are bigger still (see Figure 7.11). For these species, selection is much more powerful than drift for alleles that have fitness effect as small as $s = 10^{-5}$. These species have features that reflect exceedingly precise adaptation. One is **codon bias**. Earlier, we said that synonymous mutations are selectively neutral. In fact, different codons that code for the same amino acid can have minute differences in fitness because they affect how accurately and efficiently a gene is transcribed and translated [12]. The genomes of species with very large N_e tend to be biased toward the codons that are most efficient.

Adaptation is less precise in species with smaller population sizes. Gray whales have an effective population size in the tens of thousands (see Figure 7.11). Mutations with fitness effects of $s = 10^{-5}$ evolve in that species as if they were selectively neutral.

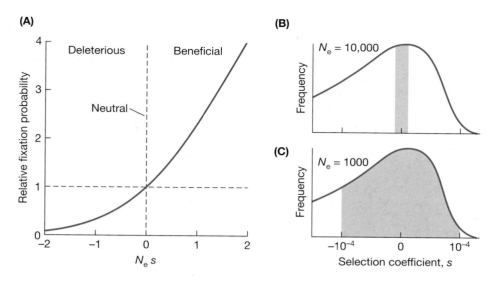

FIGURE 7.14 (A) The probability that a single copy of a mutation becomes fixed depends on its selection coefficient (s) and the effective population size (N_e). Here we assume that the relative fitnesses of A_1A_1, A_1A_2, and A_2A_2 genotypes are 1, $1 + s$, and $1 + 2s$. The x-axis shows the product of N_e and s. The y-axis shows the fixation probability relative to that of a neutral mutation (whose fixation probability is $1/2N_e$). For mutations with $|N_e\,s| \ll 1$, the fixation probability is very close to that of a neutral mutation. (B) Hypothetical distribution of fitness effects for new mutations. When $N_e = 10{,}000$, mutations with fitness effects in the range $-10^{-5} < s < 10^{-5}$ (shown in gray) evolve almost neutrally. (C) With the same distribution of mutation effects as in (B) but with $N_e = 1000$, the range of mutations that evolve almost neutrally expands to $-10^{-4} < s < 10^{-4}$.

The relative contributions that adaptation and drift make to the differences among species vary dramatically among groups of organisms [8, 16]. In flies and bacteria that have very large N_e, about half of the amino acid differences in the proteins of closely related species evolved by **positive selection**, that is, by the fixation of beneficial mutations. The other half of the differences were fixed by drift. The picture is very different for our own species: only 15 percent (and maybe less) of the differences between the proteins of humans and macaque monkeys result from adaptive evolution, and the rest accumulated by drift.

Drift can cause deleterious mutations to spread to fixation. We saw earlier that human effective population size was roughly 10,000 in our recent evolutionary past. As a result, many mutations that reduce fitness by $s = 10^{-5}$ became fixed in our genomes [1]. The fixation of deleterious mutations is a well-known problem in the small populations of animals and plants in zoos [10]. The same problem can also contribute to the extinction of natural populations. The decline of fitness by the fixation of deleterious mutations in a small population is called the **inbreeding load**. (Note that the inbreeding load is distinct from *inbreeding depression*, which is the loss of fitness of offspring from parents that are closely related compared with offspring from unrelated parents; see Chapter 10.) An isolated population of the adder *Vipera berus* had fewer than 40 individuals and was highly homozygous [24]. Females had unusually small litter sizes, and many of their offspring were deformed or stillborn. Twenty adult males were introduced from other populations, left to interbreed with the residents for 4 years, and then removed. Soon after, the population size rebounded dramatically (**FIGURE 7.15**). The snakes from the other populations reintroduced the nondeleterious alleles that had been lost by drift, which led to increased survival of offspring.

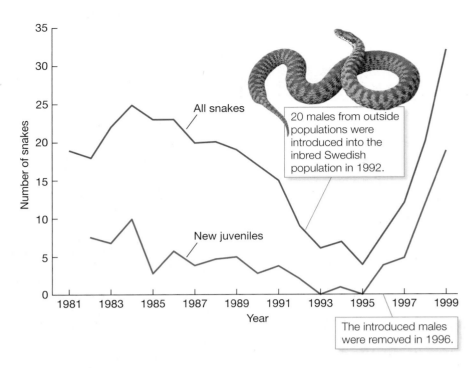

FIGURE 7.15 Population decline and rebound in an inbred population of adders in Sweden. The red line shows the total number of males found in the population each year; the blue line shows the number of juveniles recruited into the population each year. (After [24].)

20 males from outside populations were introduced into the inbred Swedish population in 1992.

The introduced males were removed in 1996.

Crossing an adaptive valley by drift

Adaptation can sometimes result when drift acts against selection. Many chromosome rearrangements are underdominant: heterozygotes have lower fitness than both homozygotes. Recall from Chapter 5 that in this situation the adaptive landscape has a valley. When a new rearrangement appears by mutation, its frequency is low and selection will tend to drive it out of the population. If drift is strong, however, the frequency of the rearrangement can sometimes increase by chance until its frequency is past the low point in the valley (**FIGURE 7.16**). Now selection favors the rearrangement, and it spreads to fixation. The population has crossed the adaptive valley by drift, an event called a *peak shift*. Sewall Wright, one of the founders of modern evolutionary biology (see Chapter 1), believed that peak shifts are important to adaptation. That view, however, is not shared by many evolutionary biologists today.

The chance that drift will cause a population to cross an adaptive valley is highly sensitive to the depth of the valley, which depends on the strength of selection against heterozygotes. A fitness reduction in heterozygotes of even 1 percent is enough to make this outcome virtually impossible except in extremely small populations. Chromosome rearrangements, which are often underdominant, evolve more frequently in weedy plants than in long-lived plants [14]. One hypothesis to explain that pattern is that weeds have frequent population bottlenecks and often self-fertilize, both of which can drastically reduce N_e.

The fate of beneficial mutations in large populations

When a beneficial mutation first appears, it is present as only a single copy. The individual carrying it may leave no offspring, or if it does, the mutation may not be passed through meiosis to its offspring. A mutation that increases fitness has a much greater chance of becoming fixed than does a deleterious mutation, but its fate is still not certain.

A surprising conclusion from population genetics theory is that a single copy of a new beneficial mutation is almost certain to be lost by drift, even if it has a large positive effect on fitness. Say that a mutant heterozygote has a relative fitness advantage of s. Then even in a large population with very weak genetic drift, the probability that the mutation will become fixed is only $2s$. A mutation that improves survival by 1 percent, which is an unusually large advantage, has a 98 percent chance of being lost in the first few generations after it appears. Some

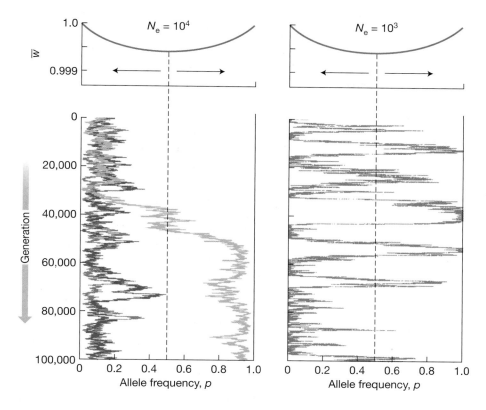

FIGURE 7.16 A peak shift caused by drift is only likely if the population size is extremely small and the adaptive valley is shallow. Top: The adaptive landscape for an underdominant mutation when the fitness of the heterozygote is reduced by $s = 0.001$ relative to the homozygotes. The population's mean fitness, $\bar{w}$, reaches its lowest value of 0.9995 at $p = 0.5$. Arrows show the direction of change in allele frequency, p, that is favored by selection when p is above and below 0.5. Both alleles mutate at a rate of $\mu = 10^{-4}$. Bottom: Simulations of allele frequency trajectories over 100,000 generations. The vertical dashed lines show the bottom of the adaptive valley. Bottom left: $N_e = 10^4$. Here $1/N_e << s$, and selection is stronger than drift. Among the three populations, only one crosses the adaptive valley and evolves to an allele frequency greater than $p = 0.5$. In this case, peak shifts are very rare, even though the fitness valley is very shallow. Bottom right: $N_e = 10^3$. Now $s = 1/N_e$, so selection and drift are about equally strong. The allele frequency in this one simulation tends to stay near the adaptive peaks at $p = 0$ and $p = 1$, but also frequently crosses the adaptive valley.

kinds of beneficial mutations occur repeatedly, and in those cases one of the copies will eventually spread. But other kinds of mutations appear in a population only rarely, or perhaps just once. Their loss by drift can put an evolutionary speed limit on how fast a species can adapt to changing conditions.

The Evolution of Differences among Species

Like other vertebrates, humans, sharks, and carps use hemoglobin to transport oxygen in their blood. Humans extract the oxygen from the air, while sharks and carps extract it from water. You might therefore expect the hemoglobins of sharks and carps to be more similar to each other than they are to that of a human.

Surprisingly, the data tell another story. **FIGURE 7.17** shows the percentage of amino acids that differ among the α-hemoglobins of four species of vertebrates. The hemoglobin in a carp is more similar to the hemoglobin in a human than it is to that

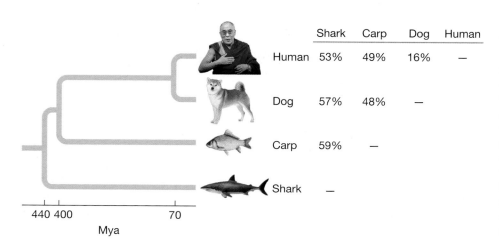

	Shark	Carp	Dog	Human
Human	53%	49%	16%	—
Dog	57%	48%	—	
Carp	59%	—		
Shark	—			

440 400   70

Mya

FIGURE 7.17 Differences in the amino acid sequence of α-hemoglobin are predicted well by phylogenetic relationship but not by ecological similarity. Right: The table shows the percentage of amino acids in α-hemoglobin that differ between pairs of vertebrate species. The shark and the carp are less similar to each other than either is to terrestrial mammals (dog and human). At the left of the table is a phylogeny showing the dates for the most recent common ancestors of the species, which are known from fossils. The percentage of amino acid differences between a pair of species is highly correlated with the age of their most recent common ancestor. (After [17].)

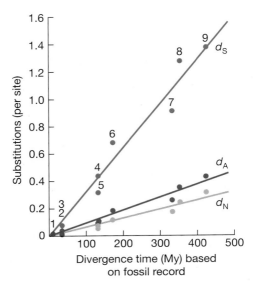

FIGURE 7.18 Molecular clocks based on three kinds of data run at different rates. The x-axis shows time since the most recent common ancestor of humans and nine other vertebrates, as determined from fossils. The y-axis plots the number of amino acid substitutions (d_A), number of synonymous DNA substitutions (d_S), and number of nonsynonymous DNA substitutions (d_N) per site between humans and those species, based on 4198 loci. Humans are compared with: 1 = chimpanzee, 2 = orangutan, 3 = macaque, 4 = mouse, 5 = cow, 6 = opossum, 7 = chicken, 8 = western clawed frog, 9 = zebrafish. (After [29].)

in a shark. But while the protein's similarity does not correlate with the habitat that these animals live in, it does correlate strongly with their phylogenetic relations. Fossils show when the most recent common ancestors of these animals lived. The number of differences between the hemoglobins of two species correlates very well with the age of their most recent ancestors. These data show that changes in hemoglobin molecules accumulate in time at a nearly constant rate. **FIGURE 7.18** shows plots in which differences in DNA and protein sequences between humans and nine other mammals are compared with the time since our most recent common ancestors. Again, there is a very strong correlation between divergence time and sequence differences. The figure also shows that different types of changes accumulate at different rates: synonymous differences build up about five times more rapidly than nonsynonymous differences. But while these two types of changes evolve at different rates, within each type the rate is quite constant.

Genes and proteins that evolve at roughly constant rates provide us with **molecular clocks**. We use these molecules to learn about the evolutionary histories of species that left no fossils. The idea is simply to count up the number of amino acid differences between the hemoglobins of two species, then use a plot like that in Figure 7.18 to estimate when their most recent common ancestor lived (see Figure 2.17). This strategy can be used with many proteins other than hemoglobin, and with DNA sequences as well as amino acid sequences. Extensions of this approach that use more sophisticated statistical analyses are the main tools used to construct the phylogenetic tree of life that links all species on Earth (see Chapters 2 and 17).

The neutral theory of molecular evolution

What could explain the relatively constant evolutionary rates of molecules that evolve in a clocklike way? Motoo Kimura, one of the giants in the history of evolutionary biology, proposed the **neutral theory of molecular evolution**. One of the many predictions that flows from that theory is that random genetic drift will cause genes to evolve at relatively constant rates [17]. Kimura thought that beneficial mutations are so rare that they make only a trivial contribution to the molecular differences among species. Instead, Kimura believed that the vast majority of those differences result from the fixation of neutral mutations (those with $s \ll 1/N_e$).

What does this theory imply about molecular clocks? Consider a locus that is evolving by drift in two species that split apart t generations ago. We sample one copy of the gene from each of the species. In the branch of the gene tree that leads from the first species back to the most recent common ancestor of the two species, on average $\mu_n t$ mutations will have occurred, where once again μ_n is the rate at which neutral mutations occur at the locus per generation. Likewise, on average $\mu_n t$ mutations will have occurred on the branch leading back from the second species. If the amount of time, t, is not too long, there is a negligible chance that a mutation will happen twice at any single DNA site. We therefore expect that the total number of differences between the two copies of the gene, d, will equal the total number of mutations that have occurred along the two branches. That is simply

$$d = 2\mu_n t \tag{7.2}$$

Thus for genes that are accumulating differences by drift, the number of differences between two species is proportional to the time since their most recent common ancestor. This gives us a molecular clock: a linear relation between time and divergence, consistent with what we see in Figure 7.18.

Recall from earlier in this chapter that purifying selection reduces the neutral mutation rate. Equation 7.2 predicts that if most differences among species are selectively neutral, sites in the genome that are free of purifying selection will accumulate differences more rapidly than will sites that experience strong purifying selection. A natural experiment to test this idea comes from **pseudogenes**, which are nonfunctional duplicates of functioning genes. Because pseudogenes do not produce a gene product, we expect that mutations anywhere in their DNA sequences will be selectively neutral, and they should therefore evolve rapidly. Consistent with Kimura's neutral theory, pseudogenes are among the fastest evolving parts of the genome (**FIGURE 7.19**). The neutral theory also predicts that inside coding regions, synonymous differences should accumulate faster than nonsynonymous differences among species. Again, that is what the data show (see Figure 7.18).

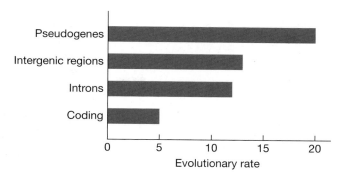

FIGURE 7.19 The relative rates of evolution for different kinds of DNA, estimated from differences between humans and chimpanzees. These rates are measured as the number of differences per nucleotide site, multiplied by 10^3. (After [11].)

The neutral theory does a good job of explaining why parts of the genome that experience little or no purifying selection evolve so fast. We now know, however, that adaptive evolution is much more important to molecular evolution than Kimura believed. We saw earlier that in some organisms, fully half of the amino acid differences between closely related species result from positive selection rather than drift [8, 16].

If adaptive evolution is so common, how can we explain molecular clocks? Fluctuating selection pressures, for example caused by changing environments, can also cause a locus to evolve in a relatively steady, clocklike way. A relatively constant appearance of new beneficial mutations in a population will produce the same result. Clocklike evolution has been seen in experimental populations of *E. coli*, where it was possible to show that the genetic changes resulted from positive selection and not drift [3]. Some genes evolve at very uneven rates and so make poor molecular clocks. But those genes that do evolve at relatively constant rates, whether by selection or drift, give us valuable tools to estimate divergence times among species and to build phylogenies.

In sum, the differences among the DNA sequences of species evolve by both drift and by positive selection. Noncoding regions of the genome that are free of purifying selection accumulate differences quickly, as predicted by the neutral theory of molecular evolution. In coding regions, many sites are under purifying selection. Like neutral sites, they accumulate differences by drift, but more slowly. But other sites in coding regions evolve largely by positive selection, not drift. We will now look at how we can determine which changes occurred by adaptation and which by drift.

Searching the Genes for Signatures of Adaptation

Our genes are books that have recorded the evolutionary histories of our ancestors. Human geneticists have recently made great progress in decoding these books. One important goal is to find genes that show signs of recent adaptive evolution. These will help us understand the genetic basis of adaptation, how we became different from other species, and how differences among human populations evolved.

The biggest challenge in finding signs of recent positive selection in DNA sequences is that other evolutionary processes are also operating. Because genetic drift is always at work, drift provides a null model. If we see that the pattern of polymorphism within a species, or differences among species, differs from what we expect under drift alone, that suggests selection may have played a role. Evolutionary biologists have developed a toolbox of methods to distinguish the signals of selection and drift. We discussed one of the approaches in Chapter 5, where we saw that a selective sweep leaves a region of depressed polymorphism in the region of chromosome near where a selective sweep occurred (see Figures 5.15 and 5.16). Here we will look at three more methods that are widely used to study adaptation at the level of the genes. Other approaches are discussed in the Suggestions for Further Reading [30-32, 36].

Synonymous versus nonsynonymous differences

The *BRCA1* locus has an important place in human health. Women who carry certain mutations in *BRCA1* have greater than a 50 percent chance of developing breast cancer and a 39 percent chance of developing ovarian cancer before age 70 [28]. Some women who find they carry this mutation make the courageous decision to have a double mastectomy rather than risk developing cancer (**FIGURE 7.20**).

This locus also has an interesting evolutionary story: it has evolved rapidly by positive selection in primates, particularly in the last few million years of human history [22]. The signs of positive selection can be seen by comparing the sequences of this gene from two species. We count the number of differences that are nonsynonymous, then divide that number by the number of DNA sites in the gene where nonsynonymous mutations could occur. This gives us the number of nonsynonymous differences per nonsynonymous site, which we refer to as d_N. We then repeat this procedure for the synonymous differences to get d_S, the number of synonymous differences per synonymous site. Finally, we divide the two numbers to find the **d_N/d_S ratio**. We can calculate this ratio using data from just a single individual sampled from each of two species, and for that reason it is widely used.

How does this ratio inform us about selection? For many species, the synonymous sites are expected to evolve neutrally because changes to the DNA sequence at those sites do not alter the protein. Those sites should accumulate differences between species by random genetic drift (as predicted by Equation 7.2). Now consider the nonsynonymous sites. If positive selection is at work, they will evolve more quickly than the neutral synonymous sites, and differences between species will build up more rapidly. Then the d_N/d_S ratio will be greater than 1. Another

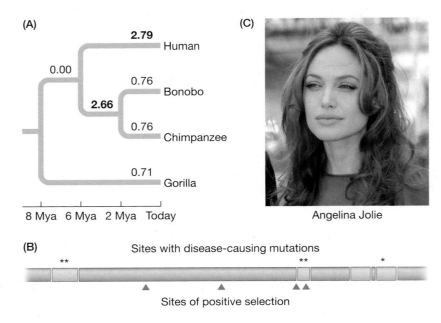

FIGURE 7.20 The *BRCA1* gene, which is responsible for many cases of breast cancer, shows evidence of rapid evolution by positive selection in our recent evolutionary past. (A) A phylogeny of great apes showing the ratio of nonsynonymous to synonymous substitutions, d_N/d_S, along each branch. Ratios in bold are significantly greater than 1, suggesting evolution by positive selection. The d_N/d_S ratios on the branches leading to humans and to chimps have the highest values observed. (B) Schematic of *BRCA1*, showing sites of mutations known to cause disease in humans and sites that have recently evolved under positive selection. The colored bands correspond to domains of the BRCA1 protein. (C) The actress Angelina Jolie made the brave decision to undergo a double mastectomy after learning she had an 80 percent risk of developing breast cancer as a result of her genotype at *BRCA1*. (After [22].)

possibility is that most nonsynonymous mutations are deleterious and are removed by purifying selection. In that case, the d_N/d_S ratio will be smaller than 1. These interpretations of the d_N/d_S ratio are summarized as follows:

$d_N/d_S < 1$ Most nonsynonymous mutations are deleterious and removed by purifying selection

$d_N/d_S \approx 1$ Nonsynonymous and synonymous mutations are evolving largely neutrally

$d_N/d_S > 1$ Many of the nonsynonymous differences between the species were fixed by positive selection

The d_N/d_S ratio for the *BRCA1* gene is greater than 1 along two branches of the primate phylogeny, with the highest ratios occurring on the branches leading to humans and chimps (see Figure 7.20). What caused the positive selection is not certain, but there are tantalizing suggestions. The BRCA1 protein is important to the repair of damaged DNA, which helps explain why mutations in the gene can cause cancer. A plausible hypothesis is that BRCA1 also helps thwart infections by interacting with viral DNA or proteins [22]. Perhaps coevolution with viruses drove the rapid evolution of the *BRCA1* gene, with the unfortunate side effect that it is now more likely to mutate to a form that can trigger a cancer.

The d_N/d_S ratio for *BRCA1* in the human lineage is very unusual: most genes have ratios much smaller than 1. **FIGURE 7.21** shows the distribution of d_N/d_S ratios in a comparison of 15,350 genes in humans and mice. (The d_N/d_S ratio for *BRCA1* in the human lineage is so high that it falls beyond the right end of this distribution.) This distribution suggests that most mutations at nonsynonymous sites are deleterious and are removed by purifying selection, while beneficial mutations are much rarer.

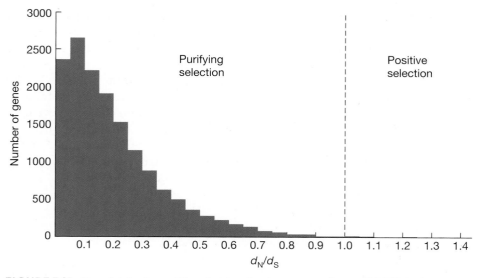

FIGURE 7.21 The distribution of the d_N/d_S ratios in a comparison of 15,350 loci between humans and mice shows the ratio is typically much smaller than 1 (vertical dashed line). This shows that nonsynonymous mutations are typically fixed much more rarely than synonymous mutations. This suggests that most nonsynonymous mutations are deleterious and removed by purifying selection, and that most differences between species are selectively neutral and evolved by drift. Values of d_N/d_S greater than 1 are rare, and suggest positive selection has acted on the locus. (After [29].)

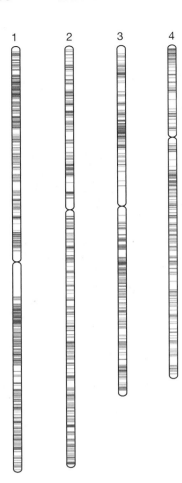

1 2 3 4

FIGURE 7.22 Map of loci on human chromosomes 1–4 where one study found statistically significant signals of recent selection using the MK test. Each vertical bar represents a chromosome. In the entire human genome, 304 genes showed significant signs of recent positive selection (loci marked in blue), while 813 genes showed a statistically significant signal of purifying or balancing selection (loci marked in red). (From [6].)

The d_N/d_S ratio is a conservative test for adaptation: it often has a value less than 1 even when positive selection has occurred. That is because a single nonsynonymous change caused by positive selection will be missed if many other nonsynonymous mutations were deleterious and removed by purifying selection. The next test we discuss is more sensitive to positive selection, but it requires more data than the d_N/d_S ratio.

The MK test

A second strategy to find evidence of positive selection builds on the basic idea of the d_N/d_S ratio by exploiting data on polymorphism within one of the species. The **MK test** is named after John McDonald and Martin Kreitman, who first developed the approach [25].

Looking back at Equations 7.1 and 7.2, we see that the ratio of the number of differences between two species (d) and the heterozygosity within one of the species (π) will be constant for all DNA sites in the genome that are evolving neutrally. If a site is evolving adaptively, however, the d/π ratio will be inflated because additional differences between species will have accumulated by positive selection (making d larger). Conversely, if a site is under purifying selection, the ratio will be depressed because selection will prevent many deleterious mutations from becoming fixed (making d smaller). The MK test assumes that synonymous changes evolve neutrally. It compares the d/π ratio at synonymous DNA sites with the d/π ratio at nonsynonymous sites. If the ratio for nonsynonymous sites is higher, there is evidence of positive selection. If the ratio is lower, there is evidence of purifying (or balancing) selection. An extension of the MK test was used to make the estimates cited earlier for the fraction of amino acid differences among species that result from adaptation and the fraction that result from drift [8, 16].

The MK test has been used to map genes in humans that show evidence of positive and purifying selection (**FIGURE 7.22**). In a comparison of 3377 loci in humans and chimpanzees, one study found that 9 percent of the genes showed evidence of recent positive selection [6]. Many of these loci fall into interesting functional groups. Adaptive evolution has been particularly common in transcription factors and hormone receptor genes. Some of these genetic changes may have been crucial to the emergence of the unique traits we see in modern *Homo sapiens*.

Divergence among populations

In some cases we suspect there has been adaptation to a particular environment, but don't know which gene or genes were involved. In this situation we can use a **genome scan** to look across the genome for signs of positive selection. One type of scan compares allele frequencies in two or more populations. Sites in the genome that have unusually high divergence are suggestive of **local adaptation** resulting from the spread of different alleles in different populations.

The Tibetan Plateau is one of the harshest environments inhabited by humans. It is very cold and very high: many Tibetans live above 4000 m elevation, where oxygen pressure is less than two-thirds what it is at sea level.

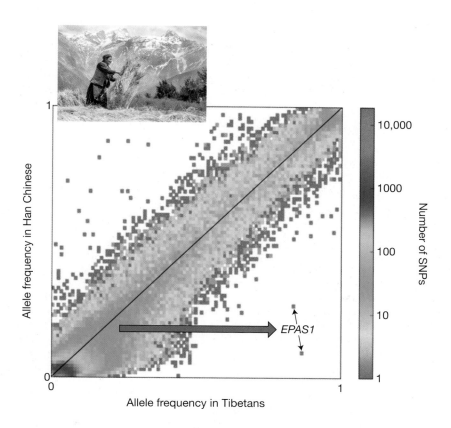

FIGURE 7.23 Tibetans have adapted to low oxygen pressure. These people live at high elevations (over 4000 m), where oxygen pressure is less than two thirds of its value at sea level. The plot compares allele frequencies at SNPs from about 20,000 coding genes in Tibetans and Han Chinese. The numbers of SNPs with the corresponding frequencies are color-coded according to the scale at right. At most SNPs, allele frequencies in the two populations are highly correlated. Alleles at two SNPs in the *EPAS1* gene are outliers: they have increased by positive selection in Tibetans (blue arrow). This rapid evolution is thought to be an adaptation to low oxygen pressure. (After [37].)

FIGURE 7.23 shows a plot comparing frequencies of SNPs (single nucleotide polymorphisms; see Chapter 4) across the genome in Tibetans with those of Han Chinese, who live near sea level [37]. Overall, there is a very strong correlation: alleles that are common in one population are common in the other. Two exceptions stand out at the bottom right of the plot. These are SNPs with alleles that are at about 90 percent frequency in Tibetans but only 10 percent frequency in Han. Those alleles are also at low frequency among Europeans, which shows that the difference between the Tibetans and the Han resulted from evolution in the Tibetan population, not the Han. These data strongly suggest there has been local adaptation in Tibetans at a site on the chromosome near to these SNPs.

The local adaptation hypothesis becomes even stronger when we learn where the two SNPs are in the genome. Both lie in a locus called *EPAS1*, which is expressed in lungs and placenta and appears to be involved in the regulation of red blood cells. Those features suggest that *EPAS1* adapted to the low oxygen pressures at the high elevations of the Tibetan Plateau. Selection must have been very intense. Tibetans colonized the plateau only 2750 years ago, and so the allele at *EPAS1* that is now common there must have swept almost to fixation in less than 150 generations. We'll see in Chapter 21 that the *EPAS1* allele carried by Tibetans has an even more surprising evolutionary story to tell.

SUMMARY

- Random genetic drift is the change in allele frequencies caused by chance events of survival, reproduction, and meiosis. Unlike selection, it does not systematically favor one allele or another.

- Drift is stronger in smaller populations, causing faster evolution.

- Drift tends to erode genetic variation within a population, and causes differences among populations to accumulate.

- Some effects of drift are best understood by looking backward in time at the gene tree that reflects the genealogy of the copies of a gene that are carried by living individuals. Going backward in time along the gene tree, two genes coalesce at their most recent common ancestor. All the copies of a gene in any population, species, or group of species ultimately trace back to a single ancestral gene copy at some point in the past.

- The strength of drift is determined by the effective population size, N_e. A natural measure for the strength of drift is $1/N_e$, which determines the size of the random fluctuations in allele frequencies going forward in time. It also determines the average time in the past of the most recent common ancestor of two gene copies now carried by living individuals.

- N_e is smaller (sometimes much smaller) than the actual population size, as the result of factors that include fluctuating population sizes and unequal sex ratios. Population bottlenecks are short, severe reductions in population size.

- The heterozygosity at a DNA site that is evolving neutrally (with no selection) is expected to be proportional to N_e. This relation can be used to estimate N_e from genetic data. DNA bases at which deleterious mutations occur tend to be less polymorphic. As a result, introns and regions between genes are typically more variable than the coding regions of a genome. For the same reason, the third positions of codons are typically more variable than the first and second positions.

- An allele will evolve largely as if selection is not acting when $s \ll 1/N_e$, while it will evolve largely as if drift is not acting if $s \gg 1/N_e$.

- Because the relative importance of selection and drift depends on the population size, many of the genetic differences among species with large N_e are expected to be adaptive, while in species with small N_e many of the genetic differences result from drift rather than adaptation. Drift has caused many deleterious mutations to be fixed in humans and other species.

- Many (but not all) genes evolve at a relatively constant rate. A constant rate of molecular evolution is called a molecular clock, and it can be used to estimate the time since two species shared a common ancestor. Constant rates are expected when genes evolve neutrally, but relatively constant rates are also seen in some genes that are evolving by positive selection.

- Several methods are used to detect selection acting on DNA sequences. Recent positive selection can be detected using variation within and differences between species. Loci identified this way are of interest because they tell us about the molecular basis of adaptive evolution.

TERMS AND CONCEPTS

background selection
coalesce, coalescent
codon bias
d_N/d_S ratio
effective population size (N_e)

founder event
gene tree
genetic drift
genome scan
inbreeding load
local adaptation

MK test
molecular clock
neutral theory of molecular evolution
population bottleneck

positive selection
pseudogene
purifying selection
selective constraint

SUGGESTIONS FOR FURTHER READING

The human brain is not well adapted to thinking about probability, and so random genetic drift is a difficult concept to grasp. A lucid introduction to the theory is found in J. H. Gillespie's book *Population Genetics: A Concise Guide*

(Johns Hopkins University Press, Baltimore, MD, 2004). The texts suggested at the end of Chapter 5 are also excellent on this topic.

The neutral theory of molecular evolution was wildly controversial when it was first proposed in the 1960s. Since then, a much-tempered version has become woven into the fabric of modern evolutionary biology. Accounts of our understanding of molecular evolution in the age of genomics can be found in "The neutral theory of molecular evolution in the genomic era" by M. Nei and colleagues (*Annu. Rev. Genomics Hum. Genet.* 11: 265–289, 2010) and *Molecular and Genome Evolution* by D. Grauer (Sinauer, Sunderland, MA, 2016).

The development of methods for making inferences about evolution from DNA sequences is an exciting and rapidly changing field of research. Introductions to how different kinds of selection are detected are "Molecular signatures of natural selection" by R. Nielsen (*Annu. Rev. Genet.* 39: 197–218, 2005), *An Introduction to Population Genetics: Theory and Applications* by R. Nielsen and M. Slatkin (Sinauer, Sunderland, MA, 2013), "Population genetic inference from genomic sequence variation" by J. E. Pool and colleagues (*Genome Res.* 20: 291–300, 2010), and "Detecting natural selection in genomic data" by J. J. Vitti and colleagues (*Annu. Rev. Genet.* 47: 97–120, 2013). H. Ellegren and N. Galtier review variation between species in levels of DNA polymorphism in "Determinants of genetic diversity" (*Nat. Rev. Genet.* 17: 422–433, 2016).

PROBLEMS AND DISCUSSION TOPICS

1. Imagine that you have the DNA sequences from the intron of a gene in three species called A, B, and C. Species A and B are most closely related, while C is more distantly related. The sequences of A and B differ by 18 base pairs, A and C differ by 26 base pairs, and B and C differ by 28 base pairs. Fossils show that species A and B diverged about 1.2 Mya, but there is no fossil evidence as to when the most recent common ancestor of all three species lived. Use the genetic data to estimate that date. What assumptions are you making to get this estimate?

2. In a population of 10^5 rabbits that has equal numbers of males and females, how many copies of a gene are there at a locus that is carried on an autosome, at a locus on the X chromosome, and at a locus on the Y chromosome? If all else is equal, which loci do you expect to be the most polymorphic, and which the least? Justify your response. How would your answers change if mutation rates were much higher in males than in females (which is the case in many species)?

3. Consider a species of sparrow that originally lived only in Alaska but recently expanded its range through North America, then Central America, and finally South America. How would you expect heterozygosity for most loci to differ among populations in North America, Central America, and South America? Why? Which of those three regions would you expect to have the most genetically similar populations, and which the most different?

4. The logic of the MK test mentioned in the text is based on the ratio of the number of sites that are different between two species and the number of sites that are polymorphic within one of those species. This ratio is expected to be the same for synonymous and nonsynonymous changes in a locus that is evolving neutrally. A study of a locus in two species of fishes called A and B obtained the following results:

Synonymous changes	Nonsynonymous changes	
Sites different in A and B	13	10
Polymorphic sites in A	11	2
Polymorphic sites in B	7	1

Calculate the ratios of polymorphic sites within a species and the differences between species for synonymous and nonsynonymous changes. What do your results suggest about how this locus is evolving?

5. Two populations of a fly are isolated from all other populations. Population S lives on a very small island and has a population size of 10^3 individuals. Population L lives on a continent and has a population size of 10^8 individuals. There is a beneficial mutation that increases fitness by 1 percent, but initially neither population has a copy of that allele. The mutation rate from the current allele to the beneficial mutation is 5×10^{-7}. Approximately how likely is it that the beneficial mutation will be common in each population after 1000 generations?

6. In Chapter 6 you saw how quantitative traits evolve in response to selection. Quantitative traits can also evolve by random genetic drift. Discuss how you expect the population size might affect the additive genetic variance within a population and the divergence between populations for a quantitative trait.

Evolution in
Space

Clover is said to confer good luck to any person who finds a four-leafed plant, but for herbivores coming across a clover is less fortunate. When white clover (*Trifolium repens*) is eaten, vacuoles in its leaves break open and release compounds that produce cyanide, a chemical that tastes bad and is highly poisonous to herbivores (as well as to secret agents). This protects the clover, since a taste of cyanide quickly persuades most herbivores to feed on a different plant. This is clearly good for clover, but leaves us with a mystery: the genes needed to produce cyanide have been deleted in clover growing in cold climates (**FIGURE 8.1**) [25]. Why should clover lose the genes that help defend it? One hypothesis is that when the leaves freeze, the vacuoles in the leaves break open. Plants with the genes to make cyanide poison themselves, while those missing those genes survive. If this explanation is correct, then cyanide-producing vacuoles are beneficial in warm regions but suicidal in cold ones.

One of the most common features of life on Earth is that species vary geographically. In cases like the white clover, the patterns reflect selection pressures that vary with climate, and therefore in space. Other evolutionary factors also shape the spatial variation in phenotypes. Studying that variation helps us understand the interplay among selection, genetic drift, and the movement of individuals. Patterns of variation in space also give insights into variation in time, that is, evolution itself.

These puffballs (*Calostoma cinnabarina*) are fungi that eject spores which disperse passively on currents of air. Interactions among dispersal, selection, and other evolutionary factors produce patterns of spatial variation in the characteristics of almost all species.

FIGURE 8.1 White clover (*Trifolium repens*) has a smooth gradient (cline) in the frequency of a gene needed to produce cyanide, which protects the plant from herbivores. The white portion of each pie diagram shows the fraction of the population in which the gene *CYP79D15* has been deleted [25]. Frequencies of deletions are much higher in populations growing in cold climates. The solid curve shows the 0°C isotherm for January. (After [8].)

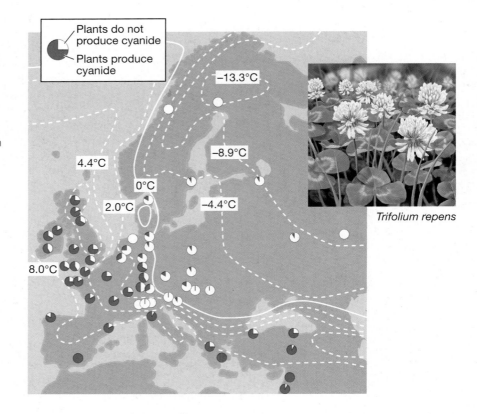

Trifolium repens

Patterns in Space

The geographic variation of cyanide-producing clover is an example of a **cline**, which is a smooth change in space of a trait mean or an allele frequency. Some clines extend over large geographical scales. Moose (*Alces alces*) are much heavier in northern Sweden than they are farther south (**FIGURE 8.2**). Similar clines are seen in many groups of animals: populations in colder climates tend to have greater body mass than those in warmer climates. This pattern is so common that it has a name (Bergmann's rule). In the moose and other homeothermic animals, larger individuals lose heat less quickly in cold climates (because they have lower surface-to-volume ratios), while smaller individuals can dissipate heat more quickly in warm climates. Clines in body size like that in the moose can be hundreds or even thousands of kilometers long.

FIGURE 8.2 Moose (*Alces alces*) are larger in the north of Sweden than in the south. This cline is an example of Bergmann's rule, which says that the body sizes of mammals and birds tend to increase with distance from the equator. (After [31].)

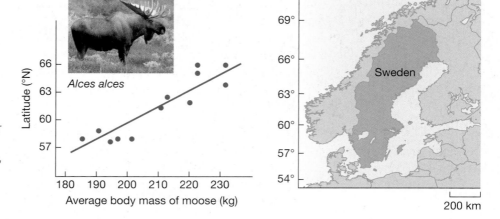

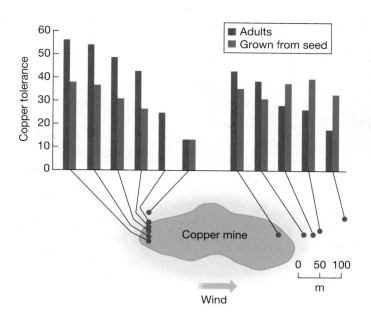

FIGURE 8.3 A grass called common bent (*Agrostis capillaris*) grows on and around an abandoned surface mine in Wales. Soil on the mine has high concentrations of copper, and plants growing there have evolved tolerance of this toxic element. Tolerance declines in clines over very short distances along two transects that run from the mine into the surrounding meadow. On the mine, tolerance is greater among adult plants than plants grown from seed, while off the mine the pattern is reversed. (After [21].)

Other clines are very much shorter. A grass called the common bent (*Agrostis capillaris*) grows on and around an abandoned surface mine in Wales (United Kingdom). An area of about 100 m × 300 m is heavily contaminated with copper, which is lethal to most plants. Remarkably, plants growing on the mine have evolved much higher tolerance to copper than plants growing nearby. Plants and seeds were sampled from two transects, and their copper tolerance was measured by the rates at which their roots grew in solutions with controlled concentrations of copper. The results revealed clines in tolerance that are only dozens of meters long (**FIGURE 8.3**).

This is an example of **local adaptation**, in this case to copper in the soil, working on small spatial scales. Local adaptation of body size in moose and of copper tolerance in the grass produced smooth clines. In other cases, the environment varies in a more patchwork way, and local adaptation results in patterns more like a mosaic than a smooth cline.

Gene Flow

Clines, mosaics, and other patterns can evolve when selection pressures change in space. A second evolutionary force that shapes these patterns is **gene flow**, which is the mixing of alleles from different populations. Gene flow plays two important roles in evolution. First, it equalizes allele frequencies, and so works to erode genetic differences between populations. Natural selection can cause two populations to become either more similar or less similar, but gene flow can only make them more similar. The second effect of gene flow is to introduce new alleles into a population from other populations where they already exist. Here gene flow plays a role similar to that of mutation.

Gene flow results from **dispersal**, which is the movement of individuals and gametes. Some species disperse their genes passively. The pollen of plants such as pines is blown many kilometers by the wind. Many plants have adaptations for dispersing their seeds. Some have fluffy plumes on the seeds that enable dispersal by wind, while others have small hooks that attach the seeds to the fur of mammals or the feathers of birds. Many plant lineages have independently evolved fleshy fruits that are swallowed by animals and later defecated elsewhere. Some animals also disperse passively. Spiders can "balloon." The spider climbs to a high point, such as the top of a plant, and spins out a thread of silk. When it feels the tug of a breeze on the thread, the spider lets go and is carried away. Spiders can reach altitudes of thousands of meters and can be blown downwind hundreds of kilometers. Corals

FIGURE 8.4 The solitary and gregarious forms of the desert locust (*Schistocerca gregaria*, shown at left) differ in color, morphology, and behavior. High population densities cause individuals to develop into the gregarious form, which gathers in huge swarms that disperse long distances (right).

Gregarious

Solitary

and other marine animals that are sedentary as adults have planktonic larvae that are dispersed hundreds or even thousands of kilometers by ocean currents.

In other animals, dispersal is active. Desert locusts (*Schistocerca gregaria*) are grasshoppers that typically live solitary lives. But when they are crowded and resources are scarce, changes in gene expression and hormone profiles radically transform their behavior and morphology [37]. They form vast swarms of hundreds of millions that fly long distances in active search of food (**FIGURE 8.4**). The devastation that locust swarms wreak on crops has been feared by farmers for millennia.

Evolutionary biologists often use the word "migration" as a synonym for "gene flow"—both refer to the mixing of genes between populations. In other contexts, "migration" means something quite different. Many species of salmon are born in streams, then swim to the ocean where they live for several years. When they are ready to reproduce, salmon have the astonishing ability to find their way back to the very stream in which they were born. To a behavioral biologist, the salmon migrate twice in their lives, once to the sea and again a second time when they return to fresh water. But to an evolutionary biologist, there is very little migration here: almost all salmon return to breed in the very population where they were born, and there is hardly any mixing of genes between populations in different streams.

How is gene flow measured?

To understand how gene flow acts, we need a way to measure it. The appropriate measure depends on whether the environment is divided into discrete patches (like islands, lakes, or mountaintops) or is spatially continuous (like a grassland or an ocean). With discrete patches, gene flow is quantified with the **migration rate**. This rate, symbolized by m, is the fraction of individuals in a population that arrives from another population in each generation. If 120 individuals in a population of size 1000 are immigrants, then the migration rate is $m = 0.12$.

The migration rate tells us how quickly gene flow erodes genetic differences between populations. The difference in an allele's frequency before and after migration in a given population is given by the equation

$$\Delta p = m (p_m - p) \tag{8.1}$$

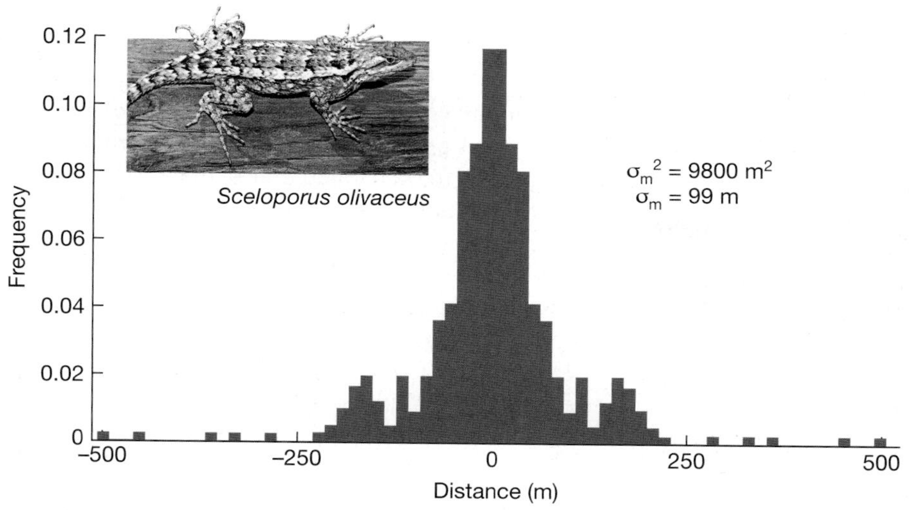
0.12
0.10
0.08
0.06
0.04
0.02
0
Frequency
Sceloporus olivaceus
σm² = 9800 m²
σm = 99 m
–500
–250
0
250
500
Distance (m)

FIGURE 8.5 Distribution of dispersal distances of the Texas spiny lizard (*Sceloporus olivaceus*, inset) in a population in Austin, Texas. The birthplaces of individuals are centered at 0, and distribution shows the places where they established territories and presumably reproduced. (The original data are reported as absolute distances. This figure assumes that half of the individuals dispersed to the left and half to the right, which is why the distribution is exactly symmetrical.) The variance of this distribution equals the dispersal variance, $\sigma_m^2 = 9800$ m^2. This corresponds to an average movement of roughly $\sigma_m = 99$ m per generation. (Based on data from [3].)

where p_m is the allele's frequency in the migrants and p is its frequency in the focal population before migration. The right side of Equation 8.1 shows that the change in allele frequency is proportional to two quantities: the migration rate, m, and the difference in allele frequencies between the local population and the migrants, $(p_m - p)$. For example, say that a population has an allele frequency of $p = 0.25$, and it receives immigrants at a rate $m = 0.1$ from another population whose allele frequency is 0.75. Equation 8.1 tells us that the change in allele frequency is 0.05, so migration will cause the allele frequency in the focal population to increase from $p = 0.25$ to 0.3. The population has become more genetically similar to the population that is the source of the migrants. The higher the migration rate between populations, the quicker genetic differences between them are erased.

In populations that are spatially continuous, there are no distinct populations and so the migration rate cannot be used. Instead, we measure gene flow with the **migration variance**, symbolized by σ_m^2. **FIGURE 8.5** plots the distribution of places where Texas spiny lizards (*Sceloporus olivaceus*) reproduced relative to where they were born. These data were collected by painstakingly marking young lizards where they hatched, then locating them again after they had become sexually mature and established territories where they reproduced [3]. The variance of this distribution gives an estimate of σ_m^2. Roughly speaking, σ_m (the square root of the migration variance) is equal to the average distance between the birthplaces of a parent and its offspring.

Over the course of many generations, migration can cause genes to diffuse across a landscape in a way similar to how a cloud of smoke disperses by diffusion. The migration variance measures the speed of that diffusion. The units of the migration variance depend on how we choose to measure space; for example, σ_m^2 might be in units of square kilometers if we are studying moose but in units of square millimeters if we are studying protozoa. (The Appendix explains variances and how they are calculated.)

The migration rate and migration variance are measured in several ways. Direct methods measure gene flow by following individuals, as in the study of the lizard. This approach is useful if we need a snapshot of dispersal over a short period of time. It does, however, have limitations. The evolutionary effects of gene flow are typically averaged over many generations, and a short-term estimate may not be representative. Another problem is that some individuals do not reproduce successfully after they disperse, so they do not contribute to genetic mixing. And if an individual is not found, we don't know if it dispersed so far that we could not find it, or if it simply died.

These considerations motivate indirect methods that use genetic data to estimate gene flow. The mixing of individuals from populations with different allele frequencies generates a systematic deviation from the Hardy-Weinberg ratios: it

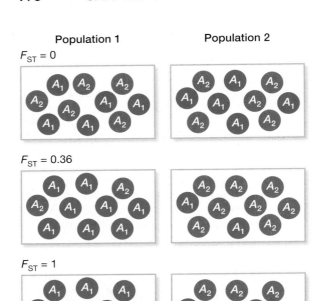

Population 1   **Population 2**

$F_{ST} = 0$

$F_{ST} = 0.36$

$F_{ST} = 1$

FIGURE 8.6 F_{ST} is a statistic used to measure genetic differences between two or more populations. In this schematic, two alleles at a locus are represented by red and blue circles. Top: $F_{ST} = 0$ when allele frequencies are equal in the populations. Middle: $F_{ST} = 0.36$ when allele frequencies are 0.2 and 0.8 in the two populations. Bottom: $F_{ST} = 1$ when the populations are fixed for different alleles.

results in an excess of homozygotes and a deficit of heterozygotes (see Chapter 7). A second effect of migration is to generate linkage disequilibrium (see Chapter 4). In the United States, males who are bald are more likely to have white skin than black skin. That is because the U.S. population includes people with ancestry from northern Europe and from Africa. In northern Europe, alleles for white skin and baldness are common, while in Africa they are rare. The mixing of people from those two populations has caused linkage disequilibrium between alleles that affect baldness and those that affect skin color. Departures from Hardy-Weinberg ratios and excess linkage disequilibrium can be used to estimate gene flow between populations. Other indirect methods for estimating migration rates from genetic data are discussed later in this chapter.

Genetic Divergence between Populations

Genetic differences between populations can be described in several ways. One of the most widely used approaches is based on a statistic called F_{ST}, which measures the fraction of the total genetic variance found across two or more populations that results from genetic differences between them. A value of $F_{ST} = 0$ means that the populations are genetically identical, while a value of $F_{ST} = 1$ means that each population is fixed for a different allele (**FIGURE 8.6**). For a locus with two alleles, it is calculated as

$$F_{ST} = \frac{\text{Var}(p)}{\bar{p}(1-\bar{p})} \qquad (8.2)$$

where $\text{Var}(p)$ is the variance of the allele frequency among populations, and $\bar{p}$ is the mean allele frequency across all the populations.

F_{ST} is often used to measure genetic differences among human populations. The International HapMap Project analyzed differentiation in single nucleotide polymorphisms (SNPs) across the genomes of East Asians, Europeans, and Yoruba from Nigeria [35]. Across all the autosomes, F_{ST} is 0.12. This tells us that only 12 percent of all the genetic variation in these populations is caused by differences among them. A full 88 percent of all the variation in our species can be found within a typical population. The striking phenotypic differences we see among human populations are therefore not representative of the genome as a whole. **FIGURE 8.7** shows how F_{ST} increases as the distance between pairs of populations increases [28]. This pattern is called **isolation-by-distance**. In humans, it reflects the history of how we

FIGURE 8.7 Isolation-by-distance in human populations. The horizontal axis has been corrected for large bodies of water that could not be crossed when humans first spread across Earth. The vertical axis gives F_{ST} estimated from 783 loci. Each dot represents a comparison between a pair of populations from the indicated region(s). (After [28].)

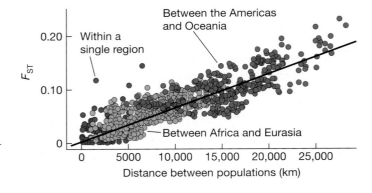

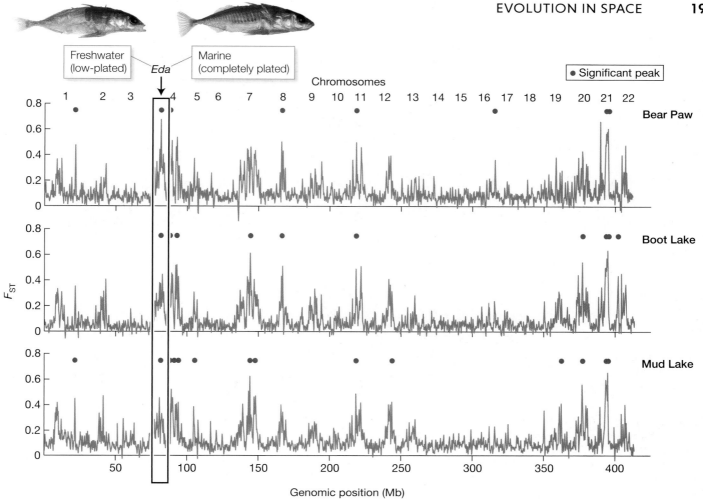

FIGURE 8.8 Divergence measured by F_{ST} between three populations of three-spined sticklebacks (*Gasterosteus aculeatus*) that independently colonized freshwater and the marine populations from which they evolved. The horizontal axis represents locations along the 22 chromosomes (indicated by numbers at top), which have been placed end to end. The vertical axis is F_{ST} at points along the genome. Peaks of significantly greater differentiation are indicated by red dots. One of the strongest regions of differentiation in all three comparisons is in the region on chromosome 4 that carries the *Eda* locus. This gene controls a major polymorphism in bony plates along the side of the fish, which differs between freshwater (low-plated) and marine (completely plated) populations. (After [13]; photos from [1a].)

colonized different parts of Earth. A similar pattern can also result from a balance between gene flow and random genetic drift, as we'll discuss shortly.

In other species, the picture of genetic differentiation is quite different. Populations of the northern dusky salamander (*Desmognathus fuscus*) living near New York City have F_{ST} values more than four times larger than those in human populations spread across the entire planet [23]. The contrast between the relatively low values of F_{ST} in humans and the high values in the salamander underlines the point that simple measures of genetic similarity cannot be used to determine which populations do and do not belong to the same species.

Geographic differentiation varies across the genome. Marine populations of the three-spined stickleback (*Gasterosteus aculeatus*) have invaded thousands of freshwater streams around the Northern Hemisphere. Adapting to a freshwater environment involves many genetic changes. These changes cause high F_{ST} between freshwater and marine populations to develop in regions of the genome with loci that are locally adapted to those very different environments (**FIGURE 8.8**). Repeated divergence in the same genomic regions in different freshwater

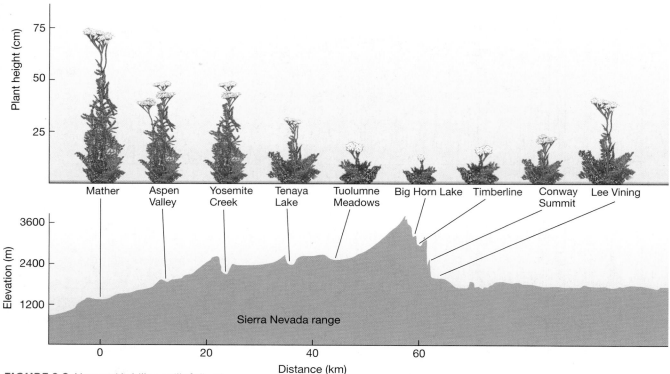

FIGURE 8.9 Yarrow (*Achillea millefolium*) is tall in populations near sea level and short in populations at high elevation. Seeds were sampled along a transect from coastal California into the high mountains inland (bottom), then grown in a common garden near sea level. The drawings (top) show their heights. Plants grown from high-elevation seeds were shorter than those from low-elevation seeds. This shows that some of the difference in height seen in populations growing at different elevations is genetic. (After [7].)

populations is strong evidence for adaptation, and opens the possibility of finding the genes that are involved [13].

Differences between populations can be caused by phenotypic plasticity as well as by genetic divergence (see Chapter 6). A famous study by Clausen, Keck, and Hiesey concerns yarrow (*Achillea millefolium*), a plant that grows from sea level up to tree line in the mountains of California [7]. In their natural habitat, plants at high elevations are much shorter than those at low elevations. This is likely adaptive because tall plants attract more pollinators and have greater fecundity at low elevations, while short plants are able to flower before winter arrives at high elevations. To determine if the differences in height are genetic or result from plasticity, the investigators grew the plants under uniform conditions in a common garden experiment (**FIGURE 8.9**). Plants grown from seeds sampled from high elevations grew to shorter heights than those from lower elevations, but the differences were not as great as what is seen in nature. These results show that both genetic variation and plasticity contribute to the differences seen among populations growing at different elevations. In the yarrow, both genes and plasticity contribute to local adaptation.

Gene Flow and Selection

When selection favors different alleles or phenotypes in different places, a tension develops between local selection, which enhances the genetic differences between populations, and gene flow, which erodes them. Without gene flow, selection would cause whatever alleles have highest fitness at any place to become fixed there. Without selection (or genetic drift), gene flow would make allele frequencies equal everywhere. The clines and other spatial patterns seen in nature are compromises between those extremes.

The tug-of-war between gene flow and selection plays out in the grass growing on the mine in Wales that was discussed earlier. A striking observation comes from comparisons between the tolerance of plants grown from seeds and of adult plants (see Figure 8.3). On the mine, the adults are more tolerant of copper than are plants grown from seed. Far from the mine, the pattern is reversed: plants grown from seed are more tolerant than the adults. These differences result from powerful selection. Among the plants that germinate on the mine, only the few that can tolerate very high levels of copper in the soil survive in each generation. Surviving adults therefore have higher tolerance than the seeds, which have not yet experienced selection in the current generation. Selection works in the opposite direction off the mine. Plants that are tolerant grow more slowly in soil that has no copper, likely because the alleles that make plants tolerant have deleterious pleiotropic effects (see Chapter 4). Gene flow prevents the population that is growing on the mine from becoming fixed for copper-tolerance alleles: each generation, pollen and seeds from the pasture are blown onto the mine. Likewise, gene flow from the mine prevents alleles for tolerance from being completely eliminated in the pasture nearby.

That hypothesis is confirmed by comparing the clines from the two transects on the left and the right in Figure 8.3. The cline on the left is very steep: copper tolerance declines quickly just a few meters from the mine. But the cline on the right is much more gradual. This transect goes downwind from the mine. Pollen from the plants on the mine is blown far by the prevailing winds. That pollen continually introduces alleles for high tolerance into the populations far downwind.

How the compromise between gene flow and local adaptation is struck depends on the relative strengths of selection and migration. The simplest situation is when an island (or other small region) receives migrants from a nearby continent (or other large region). Imagine that the continent is fixed for allele A_1. Different ecological conditions on the island give allele A_2 higher fitness there, so that the relative fitnesses of the A_1A_1, A_1A_2, and A_2A_2 genotypes are 1, $1 + s$, and $1 + 2s$. Migrants from the continent arrive on the island at a rate m per generation.

How do allele frequencies on the island evolve? The answer depends on the relative sizes of the selection coefficient s and the migration rate m. If selection is much stronger than migration ($m \ll s$), then gene flow will be largely overwhelmed and the locally adapted A_2 allele will become nearly fixed. As m grows relative to s, the locally adapted allele will decline in frequency. In the simple case of no dominance, the equilibrium frequency of A_2 on the island is simply

$$P_2 \approx 1 - \frac{m}{s} \tag{8.3}$$

(This is an approximation that is accurate so long as m is much smaller than s.)

We can exploit theoretical results like Equation 8.3 to study selection in natural populations. If we have independent measures of allele frequencies in two populations and the migration rate between them, we can estimate the selection coefficient. The rock pocket mouse (*Chaetodipus intermedius*) lives in the desert southwest of the United States where the landscape is a patchwork of dark fields of lava (much like islands) surrounded by light-colored granite and sand. While most populations of this mouse are light colored, a dark form is common on the lava, where it is camouflaged from the owls that prey on it (**FIGURE 8.10**). The dark coloration is caused by a melanic allele at the *Mc1r* locus (see Figure 6.29). Researchers estimated the migration rate, m, of mice from the granite habitats onto lava with the indirect methods that we will discuss later in this chapter [12]. Using that value and the observed allele frequencies at *Mc1r*, the selection coefficient s favoring the melanic allele on the lava is estimated to be as large as $s = 0.4$. This is extremely strong selection.

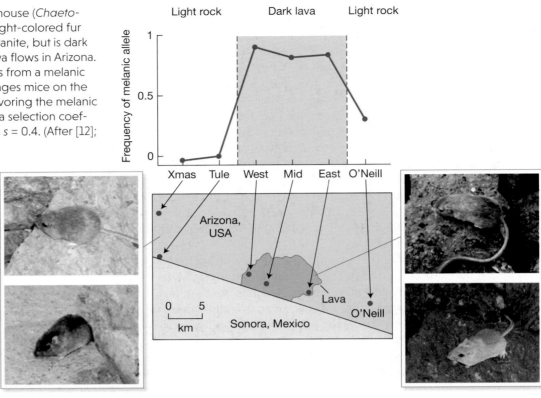

FIGURE 8.10 The rock pocket mouse (*Chaetodipus intermedius*) typically has light-colored fur where it lives on light-colored granite, but is dark colored where it lives on dark lava flows in Arizona. The dark coloration, which results from a melanic allele at the *Mc1r* locus, camouflages mice on the lava from predators. Selection favoring the melanic allele on the lava is intense, with a selection coefficient estimated to be as high as $s = 0.4$. (After [12]; photos from [24].)

A very different outcome occurs when migration onto an island is much stronger than selection ($s \ll m$). The frequency of the locally adapted allele evolves to 0, and the allele is lost entirely. This outcome, in which gene flow overwhelms local adaptation, is called **gene swamping**. Like many insects, populations of the mosquito *Culex pipiens* have evolved resistance to insecticides. On the island of Corsica, however, resistance has not evolved, despite the presence of resistance alleles at low frequency. The best explanation is gene swamping. Patches of habitat on Corsica where insecticide is applied are small. Mosquitoes disperse into those patches from neighboring populations that are not resistant, swamping the resistance alleles that are favored inside the patches [19, 29].

The evolutionary tension between selection and gene flow plays out in continuous habitats as well as on islands. Consider a grass growing in a prairie that has two kinds of soil that meet at a sharp boundary. The soil to the west of the boundary favors allele A_1, while the soil to the east favors allele A_2. To be specific, say that allele A_1 has a relative fitness advantage of s in the west, while A_2 has the same advantage in the east, and there is no dominance. What happens? Without gene flow, A_1 will become fixed everywhere to the west of the boundary, and A_2 will be fixed everywhere to the east. But with gene flow, alleles move across the boundary. This introduces into each habitat the allele that is not favored there, and a cline develops (**FIGURE 8.11**). The cline can be short or long, depending on the relative strengths of gene flow and selection. The width of the cline is

$$w_c = 2.5\sqrt{\sigma_m^2/s} \qquad (8.4)$$

This is the width of region over which allele A_2 increases from a low frequency ($p = 0.1$) to a high frequency ($p = 0.9$).

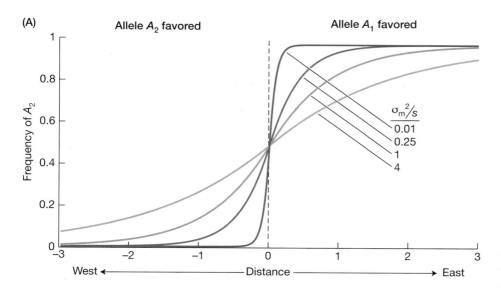

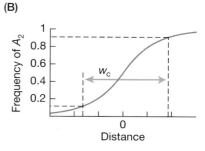

FIGURE 8.11 (A) Clines in allele frequencies predicted by a mathematical model. The horizontal axis is distance along a transect. The vertical axis is the frequency of allele A_2, which has a relative fitness of $1 - s$ to the left (west) of $x = 0$, and fitness $1 + s$ to the right (east). The four curves show the clines for different values of the ratio of the migration variance to the selection coefficient, σ_m^2/s. The clines become flatter as the strength of gene flow increases relative to selection. (B) The cline width, w_c, is the distance over which the allele frequency changes from 0.1 to 0.9.

Equation 8.4 shows that the width of a cline can be used to estimate the strength of selection. Alleles for insecticide resistance in the mosquito *Culex pipiens* form clines in southern France at the boundary between regions that are treated with insecticide and those that are not. Combining estimates for the cline width (w_c) and the migration variance (σ_m^2) with a theoretical prediction similar to that in Equation 8.4 suggests that the insecticide generates very strong selection for the resistance alleles, with values of s up to 0.33 [20].

Clines also develop when the transition between two types of habitats is gradual. In that case, the shape of the cline is typically similar to those in Figure 8.11, even though the clines in that figure result from an abrupt change in selection. The shape of a cline therefore does not tell us much about whether selection varies abruptly or in a smooth gradient.

As we saw with body mass in moose, quantitative traits also have clines. Many quantitative traits experience stabilizing selection with an optimum that varies in space. In that situation, the mean value for a trait evolves to a compromise between what local selection pressures favor and the homogenizing effect of gene flow. The effect of gene flow is again determined by the migration rate (in habitats made of patches) or the migration variance (in a continuous habitat).

Another common situation is when a patch of one type of habitat is embedded in a landscape of another type. As we saw with the mine in Wales and the lava flows in the southwestern United States, these patches can select for alleles that are disadvantageous elsewhere. If the size of the patch is too small, however, gene swamping occurs and the allele favored inside the patch will be driven to extinction. A locally favored allele will be lost in patches that are much smaller than the cline width, w_c, given by Equation 8.4. This sets a limit to the spatial resolution of adaptation. Just as your eyes are unable to pick out details that are too small, selection is unable to cause beneficial alleles to spread if the region in which they are advantageous is too small. The size of the minimal area to which a population can adapt is once again set by the relative strengths of dispersal and selection.

Tension zones

We've just seen that selection can maintain differences between populations that are connected by gene flow when fitness varies in space. While it may sound counterintuitive, in some cases selection can also maintain differences even when it acts the same way everywhere.

FIGURE 8.12 Tension zones are clines that form when there is selection against heterozygotes. The grasshopper *Podisma pedestris* is polymorphic for a chromosome fusion. Where populations with and without the fusion meet in the foothills of the Alps, clines in the frequency of the inversion form because heterozygotes have reduced fertility. The filled part of each pie diagram represents the frequency of the fusion. The cline in the frequency of the fusion is only a few dozen meters wide. (After [2].)

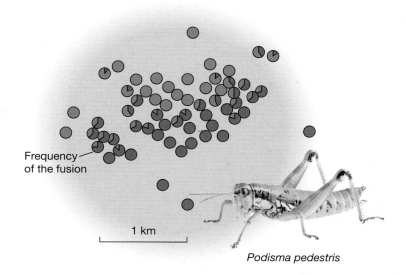

Frequency of the fusion

1 km

Podisma pedestris

Some kinds of chromosomal rearrangements are underdominant (see Chapter 5). Consider what happens if two populations that are geographically isolated become fixed for different forms of the rearrangement. If these populations expand their ranges, come into contact, and begin to interbreed, low-fitness heterozygote offspring are produced. Selection against heterozygotes can prevent the introgression of each rearrangement from the region where it is common into the other region, where it is rare and selected against. As a result, a stable cline can form. In the foothills of the Alps, the grasshopper *Podisma pedestris* has populations that differ by the presence or absence of a chromosome fusion [2]. Steep clines in the frequency of the fusion, only a few dozen meters wide, are found where these populations meet (**FIGURE 8.12**). These clines result from strong selection against fusion heterozygotes, which have low fertility. Clines that result from selection against heterozygotes are called **tension zones**. They often occur in areas where two species hybridize [1b].

Gene Flow and Drift

In Chapter 7 we looked at populations of a terrestrial snail whose allele frequencies have diverged by drift (see Figure 7.4). Groups of snails living on the same city block have similar allele frequencies, but frequencies are quite different between populations that are separated by a barrier (a street). So drift as well as selection can build up differences between populations, and its potential to do that is limited by gene flow.

That intuition is confirmed by population genetics theory. Consider two populations, both with a constant effective size N_e. Each population receives a fraction m of its individuals as migrants from the other population. For a locus that is free of selection and that has been evolving long enough to reach an evolutionary equilibrium, a mathematical model [5] predicts that the genetic divergence between the populations will be

$$F_{ST} = \frac{1}{1 + 16 N_e\, m} \tag{8.5}$$

This equation has an interesting implication. Since N_e is the population size and m is the fraction of each population that migrates, the product $N_e\, m$ is simply the average number of individuals that arrive in each population in each generation.

Equation 8.5 shows that if that number is much smaller than 1, then F_{ST} will be close to 1, meaning that the two populations are expected to have very different allele frequencies. At the other extreme, if the number of migrants per generation ($N_e\, m$) is 1 or larger, then F_{ST} will be close to 0, and the two populations will be genetically very similar. In short, for a selectively neutral locus, a single migrant per generation is sufficient to prevent drift from causing two populations to diverge very much. Remarkably, this conclusion is independent of the population size. (Of course, the picture is very different if selection is at work, as we saw earlier.)

This theoretical fact suggests a way to estimate the amount of migration between populations. The idea is to use neutrally evolving genes to estimate F_{ST}. As explained in Chapter 7, neutral genes can also be used to estimate N_e. (The heterozygosity, π, is proportional to N_e and the mutation rate μ. By estimating π from genetic data, and knowing μ, Equation 7.1 can be used to estimate N_e.) Given values for F_{ST} and N_e, we find the value of m that solves Equation 8.5. That is the basic strategy that was used to estimate the migration rate in the pocket mouse study described earlier [12].

This approach is one of several indirect methods to estimate migration using genetic data. These methods enjoy several advantages over direct methods like those used in the lizard study shown in Figure 8.5. Indirect methods average gene flow over many generations, and are sensitive only to migrants that actually contribute to genetic mixing between the populations. (A limitation of Equation 8.5 is that it is based on assumptions that are often violated in nature, but more sophisticated methods have been developed that relax those assumptions.)

Drift also causes populations that live in continuous habitats to diverge, which can result in a pattern of isolation-by-distance. In many cases, however, distant populations are more genetically similar than would be expected from the amount of gene flow that they currently experience. One explanation is that the populations may not be at an equilibrium. Earlier we saw that humans show a pattern of isolation-by-distance (see Figure 8.7). F_{ST} increases with distance, but populations living on different sides of the globe are still genetically very similar despite the absence of gene flow before the twentieth century. That pattern reflects the history of how humans colonized the planet some 100,000 years ago, rather than an equilibrium between current gene flow and drift. A second factor that can genetically homogenize a species over large spatial scales is a history of frequent extinction and recolonization. In some species, populations that are wiped out by disturbances (such as fires) are replaced by colonists that move into the empty habitat. This generates bouts of high gene flow that decrease divergence, depressing F_{ST} much below what would be expected from typical rates of movement.

Gene flow, local adaptation, and drift

We have seen that two different evolutionary processes—selection and drift—can cause populations to diverge. This can make it challenging to decide whether genetic differences between populations are signs of local adaptation or simply the result of neutral drift.

It is now possible to sample many genes or even entire genomes from different populations of some species. How can those data be used to hunt for genes involved in local adaptation? The most basic approach is simply to scan the genome for regions that show unusually high F_{ST} between two populations. The idea here is that neutrally evolving regions of the genome will show a baseline level of F_{ST} caused by drift, and regions that show much higher divergence may be under local adaptation. Further evidence for local adaptation at these candidate regions can be gleaned using comparisons with additional populations, and by finding genes within the regions that are plausible targets of selection. In the three-spined stickleback that we discussed earlier, comparisons between three independent pairs of stream and marine populations show repeated peaks of high divergence (see Figure 8.8). One of the highest peaks is on chromosome 4 and corresponds to the location of the *Eda* locus that controls the striking differences between the populations in lateral bony plates. This is compelling evidence for local adaptation at both the phenotypic and genomic levels. In humans, the *EPAS1* locus, which contributes to adaptation to high elevations in Tibetans (see Figure 7.23), was also discovered by scanning the genome for regions of high divergence.

Testing the hypothesis of local adaptation in quantitative traits presents new challenges. In body size clines such as that in moose (see Figure 8.2), the evidence for local adaptation is clear: the correlation between body size and latitude in mammals is repeated across many species and has a simple functional interpretation. In other cases, however, the evidence is less clear. One strategy is to compare variation among populations in a quantitative trait of interest with estimates of F_{ST} based on neutral genes. We can then use F_{ST} to predict the amount of variation in the quantitative trait under the null hypothesis of no local adaptation [33, 38]. An excess of variation in the trait between populations suggests local adaptation is at play.

The Evolution of Dispersal

Gene flow is caused by dispersal. But why do individuals disperse? There are often benefits to movement, for example to find food or a mate. But there are costs as well—dispersing can be risky business.

An important evolutionary advantage to dispersal is that it allows individuals to find habitat that is better now or that will be in the future. When the environment changes in time and space, natural selection favors dispersal if individuals that move have higher fitness on average than do those that do not move [22]. If environmental changes wipe out local populations from time to time, any genotype that does not disperse will leave no progeny to the evolutionary future. Weeds such as dandelions live in ephemeral patches of open habitat that are created by disturbances such as fire, grazing, and human activity. These patches are eventually overgrown by larger plants (or paved over by people), driving the population of dandelions living there to extinction. Only dandelions that disperse out of a patch will leave descendants to the distant future. This has selected for seeds that disperse well in the ancestors of dandelions. They evolved seeds with a parachute (technically called a "pappus") that carries seeds to newly opened patches.

The evolution of increased dispersal in response to environmental disturbances has been studied using experimental evolution [10]. Replicated microcosms were established of the nematode *Caenorhabditis elegans* living on agar in petri dishes. Each dish had two patches of food for the worms. The worms could either feed in their current patch or move between patches. This setup allowed the investigator to manipulate environmental disturbance. A single patch in each plate was inoculated with a small number of worms. After the worms had the opportunity to disperse and reproduce, a sample of the offspring was transferred to a new plate. Disturbance was simulated by transferring worms from only one randomly chosen patch, which is analogous to extinction of the other patch. In control treatments, there was no selection favoring dispersal. The experiment was run with a mixture of two worm genotypes that differ in their propensity to disperse. The high-dispersal genotype had lower fecundity than the low-dispersal genotype. The experiment's outcome fits the theoretical expectation (**FIGURE 8.13**). Despite the fecundity disadvantage, in just five generations the high-dispersal genotype had become much more frequent in the treatment that simulated patch extinction. In the control treatments without extinction, it became more rare.

A second factor that can select for dispersal is competition with relatives. To see this, consider a plant that lives in patches of suitable habitat that are so small that each patch is only big enough for a single individual. If a plant drops all its seeds within its own patch, when it dies, many seeds may germinate but only a single offspring will survive and inherit the patch. But if instead the plant disperses its seeds to other patches, it has a chance of leaving more than one surviving offspring.

Dispersing individuals experience less competition than their siblings that stay at home. Genes for dispersal can therefore benefit from kin selection (a topic we will discuss more in Chapter 12).

Another kind of interaction between individuals can also favor dispersal. Mating between closely related individuals often results in offspring that suffer low fitness due to inbreeding depression (see Chapter 10). Many species, including humans, are able to recognize close relatives and avoid mating with them. But other species do not have that ability. They can, however, decrease the risk of inbreeding by dispersing far from where they were born [34].

These benefits to dispersal are offset by costs. Moving is often dangerous. Passive dispersal can land an individual in hostile habitat where there is little or no chance of survival. Active dispersal is also dangerous when patches of good habitat are separated by regions of bad habitat. Several species of salamanders inhabit isolated springs in the hot and dry southwest of the United States (**FIGURE 8.14**). Millions of years ago, the ancestor of these salamanders had a terrestrial phase of the life cycle that could disperse between springs. As the climate became warmer and drier, life on land became increasingly hostile to these salamanders. In response, the salamanders lost the terrestrial phase and are now no longer able to disperse between springs. Ultimately, this may be a form of evolutionary suicide: a species that is endemic to one spring can be driven to extinction by a single catastrophe, either natural or human-caused.

Energetic trade-offs can also select against dispersal. Many species of crickets have two morphs, one with functional wings and the other without (**FIGURE 8.15**). Winged individuals are able to disperse by flying. But they have lower fecundity than the wingless morph, probably because of the large energetic investment required to develop functional flight muscles [39].

Habitats that change in time and space create mosaics of shifting selection pressures, favoring the evolution of increased dispersal at some times and in some places, and decreased dispersal elsewhere. The Glanville fritillary (*Melitaea cinxia*) is a butterfly that lives in meadows that are surrounded by forest. When a new

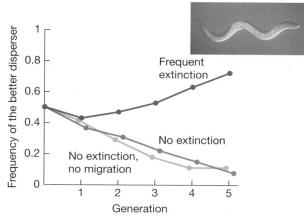

FIGURE 8.13 Dispersal evolved in response to environmental disturbance in a laboratory experiment with the nematode *Caenorhabditis elegans*. Mixtures of two genotypes that differ in their dispersal rates were cultured on agar plates with two patches of the bacteria on which the worms feed. In the experimental treatment, every 3 days nematodes from a randomly chosen patch were used to inoculate a new plate. The old plate was discarded, which simulated extinction of the population in the second patch. There were two control treatments: a single patch with no extinction, and two patches with no extinction. (After [10].)

FIGURE 8.14 The Barton Springs salamander (*Eurycea sosorum*) has lost the terrestrial phase of its life cycle. It cannot survive on land and so is unable to move to another spring. This species is found only in the aquifer that feeds a group of springs in Austin, Texas, that provide water to a large swimming pool (inset). The salamander's entire range is only a few square kilometers.

FIGURE 8.15 Roesel's bush-cricket (*Metrioptera roeselii*) has two morphs, one with normal wings (left) and one with highly reduced wings (right). The morph with normal wings can disperse long distances by flying, but has lower fecundity because of the energy it invests in the growth and maintenance of flight muscles.

meadow appears, for example after a fire, it is colonized by butterflies that disperse from neighboring patches. The colonizers are not a random sample of the population: they are strong fliers that disperse well. Flight strength is strongly heritable, so the population that is established in a new patch has strong fliers for the next several generations. But in later generations, the population evolves toward weaker and weaker flight. Why? Simply because the strongest fliers disperse, leaving the weaker fliers at home [11].

The Glanville fritillary illustrates an interesting consequence of dispersal: it sorts individuals according to how well they disperse. As a result, dispersal can cause evolutionary change even without the help of natural selection. A dramatic example of this effect comes from the cane toad (*Rhinella marina*) [27]. The toad was introduced to Australia from South America in 1935 and has been rapidly expanding its range ever since. Because of the sorting effect, we would predict that toads at the leading edge of the invasion should disperse better than those from older, established populations. That is exactly what is seen: frogs at the leading edge have longer legs, and frogs with longer legs disperse faster than their short-legged conspecifics (**FIGURE 8.16**).

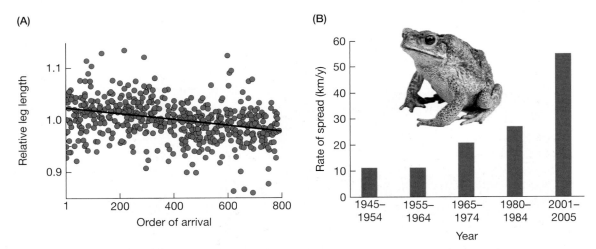

FIGURE 8.16 When a species' range expands, the most rapidly dispersing genotypes automatically become more common at the range edge, causing the rate of spread to increase. (A) During its range expansion in Australia, the first cane toads (*Rhinella marina*) to arrive at a research station had the longest legs, while those that arrived later had shorter legs. Toads with longer legs disperse faster because they jump farther. (B) As a result of the sorting process, the advancing front of the cane toad invasion has accelerated since the toad was introduced in 1935. (After [27].)

The evolution of dispersal rates and distances has a host of downstream effects. Dispersers take their genes with them, and so evolutionary changes to dispersal also affect local adaptation and patterns of geographic variation. Deleterious mutations can become fixed by drift in isolated populations, which can cause immigrants that arrive from populations where they are not fixed to have high-fitness offspring (see Figure 7.15). This effect amplifies the genetic impact of migration [9] and can rescue local populations from extinction [30]. Paleontologists have found that marine snails with planktonic larvae capable of dispersing long distances had larger geographic ranges and survived longer in the geological record than did those without planktonic larvae [14].

The Evolution of Species' Ranges

Humans live over a greater expanse of Earth's surface than any other species. At the other extreme, the world's entire population of the Barton Springs salamander is restricted to just a few square kilometers (see Figure 8.14). The geographic range of a species can evolve, which raises the question of why some species have evolved large ranges and others small ones. In some cases, barriers prevent a species from moving elsewhere. The Barton Springs salamander is unable to live out of water, and so does not have the ability to colonize other springs. On a larger scale, the ranges of many marine species are defined by the edges of continents, and the ranges of many terrestrial species by the edges of oceans.

In many cases, however, there is no obvious barrier that limits a species' range. From an evolutionary perspective, these situations are more difficult to understand. If a tropical plant cannot survive far from the equator because winters are too cold, why don't populations that are at the northern and southern limits of the range evolve greater cold tolerance, and so cause the range to expand outward? If a barnacle cannot live higher in the tidal zone because it reaches its tolerance limits for heat and desiccation there, why doesn't it evolve greater ability to withstand those stresses? If a bird cannot live in the same habitat as another species that is a better competitor, why doesn't it evolve to feed on different foods?

What limits species' ranges is one of the most puzzling questions in evolutionary biology. Several general kinds of explanations have been proposed [32]. Populations may simply lack genetic variation in a trait necessary for adapting to a new environment. For example, populations of two species of rainforest-dwelling *Drosophila* have no detectable genetic variation for desiccation tolerance, which might prevent them from expanding into drier habitats [15, 16]. A second possibility is that gene swamping caused by migration from other parts of a species' range can prevent local adaptation to the extreme conditions at the range edge and prevent the species from expanding outward [17, 19]. Consistent with that idea, cold resistance in the fly *Drosophila birchii* is lower along a steep mountain slope than it is along a shallow slope at the same altitude. Along the steeper slope, populations living at high altitudes are closer to warm-adapted populations at low altitudes, so they may receive more gene flow that prevents them from adapting to cold [4]. Third, biotic interactions can set range boundaries where a species encounters a new competitor, predator, or pathogen.

Global climate change provides a very large (if uncontrolled) experiment that gives insights on how species ranges respond to environmental change. Species might respond in two ways: by changing where they live and by adapting to the new conditions (**FIGURE 8.17**). Reviews of ranges for which there are historical data have found that many (perhaps more than half) shifted in the direction expected from climate change. In the Northern Hemisphere, many northern range limits have expanded farther north, while southern limits are contracting toward the range

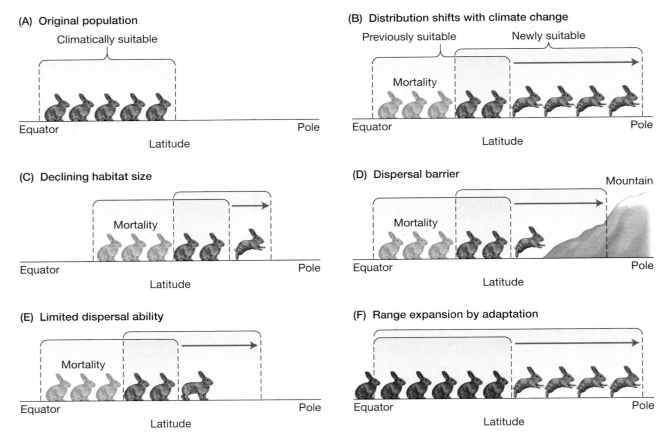

FIGURE 8.17 Species living along an environmental gradient (for example, correlated with latitude) can respond to climate change in several ways. These cartoons depict a species living in the northern hemisphere. (A) Before change occurs, the range is limited to the north and south by physiological factors. (B) With global warming, some species are able to maintain a stable range size by shifting to the north. (C) Other species run out of habitat at their northern limit, causing the range to shrink as the southern limit moves north. (D) Dispersal barriers prevent some species from tracking the moving envelop of suitable habitat. (E) Species with limited dispersal abilities may not be able to keep up with rapidly changing conditions. (F) If species are able to adapt to the new conditions at the south while tracking the envelope of previously suitable conditions, the range will expand. Of the six possibilities described in the figure, only this one includes evolutionary change. This outcome is rarely observed. (After [18].)

centers [6, 26]. Thus, many species are tracking their niche by shifting where they live, as the environment to which they are adapted moves northward and to higher elevations. These are not evolutionary changes, just geographic changes, and in fact, there is no evidence that most species have evolved broader or more extreme climate tolerance. Many other species, however, are not able to keep up with climate change, even by shifting where they live. A variety of environmental challenges prevents them from maintaining a stable range size while moving with the envelope of suitable habitat. As with the weather, predicting how species will respond to the shifting constellation of ecological factors triggered by climate change is very difficult [36]. But the hope that evolutionary adaptation might rescue most species from the challenges of rapid climate change appears to be dim [26].

SUMMARY

- Most species show geographic variation in allele frequencies and the means of phenotypic traits. Clines, which are smooth changes in an allele frequency or trait mean, are a very common pattern.

- Clines and other patterns can result from local adaptation, which results when selection varies in space.

- Gene flow is the mixing of alleles from different populations, eroding differences caused by selection and drift. It results from the dispersal of individuals and their gametes. Gene flow is measured by the migration rate (m) when populations are discrete or patchy, and by the migration variance (σ_m^2) when populations are continuously distributed in space.

- F_{ST} is a statistic commonly used to describe genetic divergence between two or more populations. In many species with broad geographic ranges, F_{ST} increases with the distance between two populations, a pattern called isolation-by-distance. F_{ST} varies across the genome, and genomic regions with high F_{ST} can be used to find loci that are locally adapted.

- When both gene flow and local selection are at work, allele frequencies evolve toward a compromise between them. If gene flow is weak relative to selection, allele frequencies will evolve to what selection favors at each location. If gene flow is relatively strong, allele frequencies will be equalized. Strong gene flow can cause gene swamping, which is when a locally favored allele is lost because migration overwhelms local selection.

- In continuous habitats, the widths of clines are determined by the ratio of the migration variance to the strength of local selection. When there is a patch of habitat that selects for a different allele than that favored outside the patch, the locally adapted allele will be lost by gene swamping if the size of the patch is smaller than a critical size determined by the relative strengths of migration and selection.

- Tension zones are clines in allele frequencies that result from selection against heterozygotes (underdominance) that acts uniformly in space.

- Drift can cause allele frequencies at selectively neutral loci to diverge between populations. Very small rates of migration prevent divergence at neutral loci. The amount of divergence can be used to estimate the amount of gene flow.

- Dispersal rates evolve. Higher dispersal is favored by habitat disturbance that causes extinction of local populations, competition between related individuals, and inbreeding. Lower dispersal is favored because movement is often risky and energetically expensive. In a species that is expanding its range, there is an automatic increase at the range's edge of alleles that enhance dispersal.

- Species ranges evolve. Factors that prevent ranges from expanding outward include dispersal barriers, genetic constraints and gene flow that prevents adaptation to more extreme environments, and competition with other species that have adjacent ranges. Global climate change is causing shifts in the ranges of many species, but there is little evidence that species can generally avoid extinction by adapting to the new conditions.

TERMS AND CONCEPTS

cline
dispersal
gene flow

gene swamping
isolation-by-distance

local adaptation
migration rate
migration variance

tension zone

SUGGESTIONS FOR FURTHER READING

Local adaptation has been studied using a variety of approaches, including geographic surveys, genetic analyses, and mathematical modeling. Reviews by T. Lenormand ("Gene flow and the limits to natural selection," *Trends Ecol. Evol.* 17: 183–189, 2002) and by T. J. Kawecki and D. Ebert ("Conceptual issues in local adaptation," *Ecol. Lett.* 7: 1225–1241,

2004) give excellent perspectives on this rich literature.

The evolution of dispersal involves a fascinating but complex web of evolutionary forces that are discussed in the review "How does it feel to be like a rolling stone? Ten questions about dispersal evolution" by O. Ronce (*Ann. Rev. Ecol. Evol. Syst.* 38: 231, 253, 2007).

B. Charlesworth and colleagues review the effects of gene flow and other evolutionary forces on patterns of neutral variation in DNA in "The effects of genetic and geographic structure on neutral variation" (*Ann. Rev. Ecol. Evol. Syst.* 34: 99–125, 2003). DNA sequences are widely used to study the genetic structure of populations; see "Inference of population structure using multilocus genotype data: Linked loci and correlated allele frequencies" by D. Falush and colleagues (*Genetics* 164: 1567–1587, 2003).

There is a growing literature on how species ranges evolve, and specifically how they are responding to climate change. Two authoritative overviews are "Ecological and evolutionary responses to recent climate change" by C. Parmesan (*Ann. Rev. Ecol. Evol. Syst.* 37: 637–669, 2006) and "Evolution and ecology of species range limits" by J. P. Sexton and colleagues (*Ann. Rev. Ecol. Evol. Syst.* 40: 415–436, 2009).

PROBLEMS AND DISCUSSION TOPICS

1. Suppose that in generation 0, the frequency of allele A_1 in a population of armadillos is 0.4. In each generation, 10 percent of the individuals in that population are migrants from another population that has an allele frequency of 0.6.

   a. Calculate the frequency of A_1 in each of the next two generations (generations 1 and 2).

   b. Is the change in allele frequency in generation 2 greater than, less than, or equal to the change in generation 1? How can you explain that answer?

   c. What will the allele frequency become in this population after many generations?

2. Consider a cricket that has recently colonized a remote oceanic island from a source population on a continent. How do you expect the average size of wings in the island population to compare with the average size on the continent? How do you expect wing size in the island population to evolve over the next several hundred generations?

3. Equation 8.4 gives the equilibrium value of F_{ST} between two populations for a neutrally evolving locus when the populations are of equal size and are exchanging equal numbers of migrants. When there is symmetrical migration among a large number of populations, a different equation holds: $F_{ST} = 1 / (1 + 4 N_e m)$. Suppose you sample individuals from two populations, but you do not know whether these populations exchange migrants only with one another, or whether they are part of a group of many populations that exchange migrants. You genotype the individuals in your samples at several loci

   populations is 0.25. Using the equation given above and Equation 8.4, determine the range of plausible values for the number of migrants that arrive in each population in each generation.

4. Clines in body size have been observed in many species, such as the latitudinal cline in moose shown in Figure 8.2.

   a. Does a cline in body size necessarily result from variation in allele frequencies at loci that affect body size? Why or why not?

   b. How might you determine whether a cline in body size was caused by clines in allele frequencies?

   c. Say there is strong evidence that a latitudinal cline in body size in a squirrel is caused by variation in allele frequencies. Do you think that data showing how rapidly the average body size changes with latitude could by themselves be used to determine how selection varies in space? Why or why not?

5. A species that has a high rate of long-distance dispersal is more likely to colonize new habitat. But that species may also be less likely to adapt to local conditions, because migration will be stronger than local selection pressures for many loci. In light of those considerations, when do you expect that increasing dispersal might result in the evolution of a larger geographic range, and when might it not?

6. It is now common to score many thousands of SNPs in numerous individuals sampled from several populations. (See, for example, the results from stickleback fishes shown in Figure 8.8). Many of these SNPs are neutral and therefore

provide information about rates of migration among populations. SNPs that show unusual patterns of genetic variation may indicate regions of the genome that are subject to selection.

a. Will loci involved in local adaptation, such that different alleles are favored in different populations, show unusually high or unusually low values of F_{ST} between populations? Explain your answer.

b. What could account for SNPs that show high heterozygosity within populations, but unusually low values of F_{ST} between populations?

c. Even when thousands of SNPs are scored, many of the SNPs that are the actual targets of selection are often not genotyped. However, even SNPs that are selectively neutral can be used to detect regions of the genome that are locally adapted. Why?

9

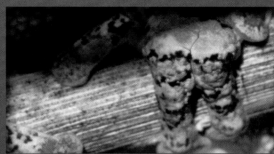

Species and Speciation

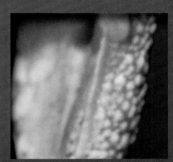

In the Rift Valley of eastern Africa, just south of the equator, lie three great lakes and many smaller ones. Lakes Tanganyika and Malawi are deep and old, having been formed by the separation (rifting) of two continental plates. Lake Victoria, in contrast, is broad and shallow, lying in a basin that was dry only 15,000 years ago. These lakes harbor a few species of catfishes, spiny eels, and other fish families, but more than 90 percent of all the fish species are cichlids, a family that includes species well known to tropical fish hobbyists. Lake Tanganyika has at least 250 species of cichlids, Lake Victoria between 450 and 530 species, and Lake Malawi at least 480 species [89]. (The American Great Lakes, in comparison, have only about 175 species of fishes, of all kinds.) These cichlid fishes are extraordinarily diverse in coloration, form, feeding habits, and habitat use (**FIGURE 9.1**). Different species eat insects, snails, detritus, rock-encrusting algae, aquatic plants, phytoplankton, zooplankton, baby fishes, and larger fishes. Some species are specialized to feed on the scales of other fishes, and one has the gruesome habit of plucking out other fishes' eyes. The teeth of some closely related species differ more than do those of some whole families of fishes. Many of these habits and morphologies have evolved convergently in the different lakes [45]. Phylogenetic analyses show that the 250 cichlids in Lake Tanganyika have evolved from at most 16 original species. The cichlids of Lake Victoria have multiplied faster than any other group of vertebrates on Earth: the 450-plus species evolved from just 5 original ancestral species in perhaps only 15,000 years [95, 105].

A male gray tree frog inflates his vocal sac as he calls to attract females. Female frogs respond almost exclusively to their own species' calls, which are a barrier to interbreeding. Male calls differ between two morphologically indistinguishable species of gray tree frogs in eastern North America. *Hyla chrysoscelis* has 12 pairs of chromosomes, whereas *H. versicolor* is a tetraploid, with 24 pairs.

FIGURE 9.1 Examples of the diversity of cichlid fishes in Lakes Tanganyika (at left) and Malawi (at right). Ecologically and morphologically similar forms have evolved independently in both lakes. (A) Rock-dwelling species with rasping jaws. (B) Open-water fish-eaters. (C) Fleshy-lipped species that suck prey from crevices. (D) Rock-dwellers. (E) Hump-headed species. (F) Slender, striped species. (From [2]).

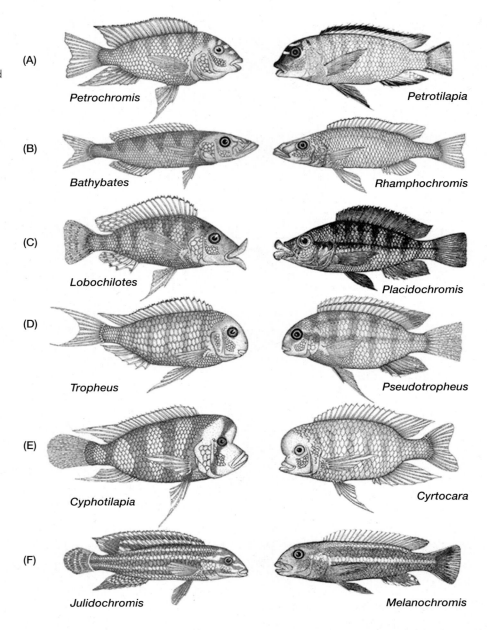

(A) *Petrochromis* — *Petrotilapia*

(B) *Bathybates* — *Rhamphochromis*

(C) *Lobochilotes* — *Placidochromis*

(D) *Tropheus* — *Pseudotropheus*

(E) *Cyphotilapia* — *Cyrtocara*

(F) *Julidochromis* — *Melanochromis*

What caused this explosion of diversity? Do the number and ecological variety of species depend only on current ecological conditions, such as how many different kinds of resources can sustain different species? Or do they reflect the rate at which new species have arisen? Why should the rate of speciation have been so high in this fish family, and only in these lakes? Has speciation been caused by the fishes' mating patterns? By sexual selection on coloration? By adaptation to different ecological niches? By genetic drift? Fundamentally, what we want to know is: How do new species form?

Darwin first came to believe in evolution when he realized that different islands in the Galápagos archipelago harbor different forms of mockingbirds and a variety of similar finches. That these forms were similar, yet subtly different, could most plausibly be explained by supposing that they had descended, with slight modifications, from a common ancestor. Pursuing this reasoning, Darwin concluded that all species of birds—indeed all species of animals, and finally all living things—may have originated by successive branching of lineages throughout the history of life, from a single common ancestor. Modern research has affirmed that this is indeed how the enormous diversity of organisms arose. The forks in the great Tree of Life were caused by **speciation**, the process by which one species gives rise to two.

FIGURE 9.2 Can you distinguish the species? (A, B) Gray and rufous morphs of the eastern screech owl (*Megascops asio*). (C) The western screech owl (*Megascops kennicottii*).

What Are Species?

Several definitions of "species"—which is Latin for "kind"—are used by biologists. It is important to bear in mind that a definition is not true or false, because the definition of a word is a convention. Probably no single definition of "species" suffices for all the contexts in which a species-like concept is used.

For Linnaeus and other early taxonomists, species were simply groups of organisms that could be distinguished. But as knowledge of organisms grew, this criterion became inadequate. For example, two kinds of small owls in eastern North America look very different: one is gray and the other bright reddish brown (**FIGURE 9.2A,B**). Nevertheless, they are clearly the same species: the two forms sound the same, they interbreed, and a brood may include both color forms—which are a simple one-locus polymorphism (with rufous dominant over gray). But the gray form of this species, the eastern screech owl (*Megascops asio*), is almost indistinguishable in appearance from another owl that has a very different voice and that is recognized as a distinct species—the western screech owl (*M. kennicottii*; **FIGURE 9.2C**). The two species can be completely distinguished by mitochondrial DNA [77], indicating that even though they coexist in Texas, there is little or no gene flow between them. They are separate *gene pools*.

Cases such as the screech owls led to the concept of species as groups of individuals that interbreed. Ernst Mayr [55] formalized this idea in what he called the **biological species concept** (**BSC**), defined as follows: "*Species are groups of actually or potentially interbreeding populations, which are reproductively isolated from other such groups.*" **Reproductive isolation** means that any of several biological differences between the groups greatly reduce gene exchange between them, even if they are not geographically separated. The BSC does not require that species be 100 percent reproductively isolated—there can be a little genetic "leakage" between species through hybridization. Although genetic and phenotypic differences do not *define* species according to the BSC, those differences enable us to *recognize* and distinguish them. Note that an inability to form hybrid offspring, or sterility of hybrids, is *not* a necessary criterion of species: it is only one of many ways in which gene exchange may be reduced or prevented.

The biological species concept was developed partly to acknowledge variation, both within a single population (such as the color morphs of the eastern screech owl) and among different geographic populations, which often show evidence of

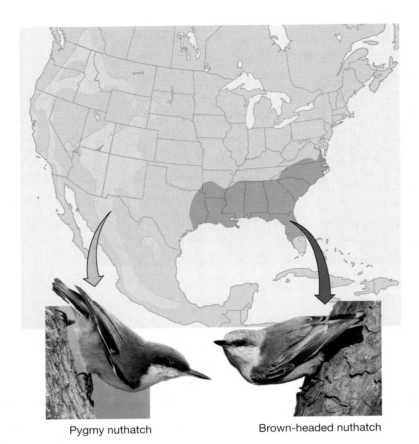

Pygmy nuthatch   Brown-headed nuthatch

FIGURE 9.3 The geographic ranges of the pygmy nuthatch (*Sitta pygmaea*, left), in western North America, and of the brown-headed nuthatch (*Sitta pusilla*, right), in the southeastern United States, are separated by hundreds of miles in which neither bird occurs. They differ in voice and subtly in color pattern. It is difficult to tell if they are different biological species.

interbreeding where they meet. The BSC also recognizes cases of "sibling species" (such as the gray forms of the two screech owls), which are almost identical in appearance and are often discovered by differences in ecology, behavior, chromosomes, or genetic markers. The discovery that the European mosquito *Anopheles maculipennis* is actually a cluster of nine sibling species had great practical importance because some transmit human malaria and others do not [3, 39]. The term "sibling species" differs from **sister species**, which are two species descended from a single ancestral species, and are therefore one another's closest relatives.

The biological species concept is the most widely used definition among biologists, and it can be applied to the majority of sexually reproducing species alive on Earth. It does, though, have limitations. Reproductive isolation evolves gradually, as we will see. So interbreeding versus reproductive isolation is not an either/or, all-or-none distinction. Nevertheless, there are countless examples of closely related forms that occur in the same area, can be distinguished by genetic and phenotypic differences, and interbreed very little or not at all. They are unequivocally distinct, real species.

The greatest practical limitation of the BSC is in determining whether populations that are geographically separated (**allopatric**) belong to the same species (**FIGURE 9.3**). The BSC requires that we make a judgment call as to whether they would interbreed if they came into contact under natural conditions. Climate change in the past and human changes to the environment at present have brought formerly isolated populations together. In some such cases, the populations remained distinct, but in other cases they interbred, showing that they were not fully distinct species. One could test for reproductive isolation experimentally, for example in the lab or garden, but this is impractical or even impossible to do with many species (e.g., giant squids). Moreover, some species that mate under artificial conditions will not do so in nature, and hybrid offspring that are viable and fertile in the lab may not survive in nature. In practice, deciding whether geographically isolated populations are species is at times somewhat arbitrary. Commonly, allopatric populations have been classified as species if their differences in phenotype or in DNA sequence are as great as those usually displayed by species in the same group that are **sympatric** (in the same location) [103]. A similar approach is taken with classifying fossils into species, since paleontologists cannot study the mating behavior or hybrid survival of extinct ammonites or dinosaurs.

Another limitation of the BSC is that it does not apply to organisms that do not reproduce sexually. Bacteria pose particular challenges. Although they do not have meiotic sex, they do exchange genetic material in other ways. Species of bacteria, such as *Escherichia coli* and *Salmonella typhimurium*, were traditionally recognized by differences in their metabolic capabilities. More recently, genetic similarity has been used to group individuals into species. Although bacteria can acquire new genes from even distantly related organisms, most homologous recombination ("sex") occurs within traditionally recognized species [66].

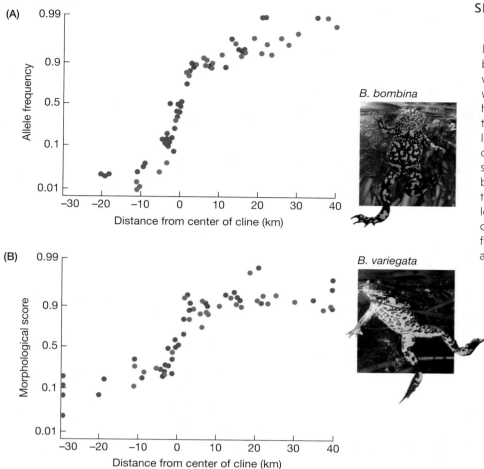

FIGURE 9.4 The eastern European fire-bellied toad (*Bombina bombina*) and the western European yellow-bellied toad (*B. variegata*) meet and interbreed in a narrow hybrid zone. The two species differ in loci that code for enzymes and several morphological features. (A) Average allele frequency at six enzyme loci. (B) A morphological score based on seven characters. Red and blue dots represent two different 60-km transects in Poland. The clines in enzyme loci and morphological features are coincident, suggesting that this hybrid zone was formed by contact between two formerly allopatric populations. (After [100].)

These and other considerations have inspired several alternative species definitions. Some systematists prefer the **phylogenetic species concept** (**PSC**), which emphasizes species as the outcome of evolution—the products of a history of evolutionary divergence. In one widely accepted definition, lineages are different species if they can be distinguished: a phylogenetic species is *an irreducible (basal) cluster of organisms diagnosably different from other such clusters, and within which there is a parental pattern of ancestry and descent* [17].

The phylogenetic and biological species concepts have different uses, and tend to be used by different groups of researchers. The PSC can be useful for classification, because unlike the BSC, it can be applied to allopatric populations, such as those on different islands, in which reproductive isolation is difficult or impossible to assess. Although some systematists use the PSC in classifying organisms, most evolutionary biologists use one or another variant of the BSC, because they view the evolution of reproductive isolation as the key event that enables sexually reproducing lineages to evolve independently and generate biological diversity. Without the evolution of reproductive isolation, there would be only one (or at most a few spatially separated) species of cichlid in each of those African lakes.

No matter which species concept is adopted, some populations of organisms cannot be unambiguously assigned to one species or another, because the features that distinguish species (by any definition) evolve gradually. There exist graded levels of gene exchange among adjacent (**parapatric**) populations and sometimes between more or less distinct populations that are sympatric. Species as recognized by the BSC are ambiguous in **hybrid zones**, which exist where genetically distinct populations meet and interbreed to a limited extent, but in which there exist partial barriers to gene exchange (**FIGURE 9.4**). Hybridization occurs, at least

FIGURE 9.5 Advantageous alleles have spread by introgression between distantly related species of *Heliconius* butterflies in South America. The phylogeny is based on many genes. The DNA sequence of two genes that control color pattern shows that *H. timareta ssp. nov.* acquired the "postman" pattern in the hindwing from *H. melpomene amaryllis*, and that *H. elevatus* acquired the "rayed" hindwing pattern from *H. melpomene aglaope/malleti*. (From [35]; large wing images courtesy of J. Mallet.)

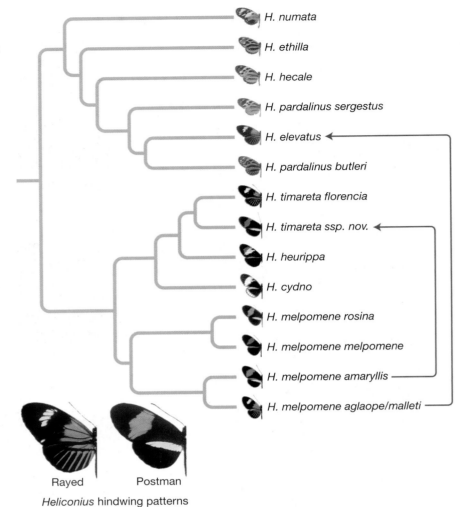

Rayed   Postman

Heliconius hindwing patterns

occasionally, among sympatric species in many groups of plants and animals [51], and genes are sometimes incorporated into the gene pool of one species from another, a process called **introgression** (or introgressive hybridization). Some such genes may enhance adaptation [1]. For instance, *Heliconius* butterflies are distasteful to predators and have warning coloration: predators do not attack butterflies with this pattern after one or two experiences in which they learn to associate the coloration with distastefulness. Alleles that determine part of the color pattern of the wings of certain *Heliconius* species have spread among even distantly related species (**FIGURE 9.5**).

Biological species are seldom distinguished in practice by directly testing their propensity to interbreed or their ability to produce fertile offspring. Indeed, this is usually not necessary. Morphological and other phenotypic characters are the usual evidence used for diagnosing sympatric species (**FIGURE 9.6**), because they can serve as *markers* that indicate reduced gene flow—that is, reproductive isolation—among sympatric populations. If a sample of sympatric organisms falls into two discrete clusters that differ in multiple characters, it is likely to represent two species. In modern studies, genetic markers are often used to reveal the existence of two or more sympatric species. A polymorphic locus that shows few heterozygotes, and so departs strongly from Hardy-Weinberg equilibrium, is a signal that there are likely to be more than one species. (**BOX 9A** provides an example.)

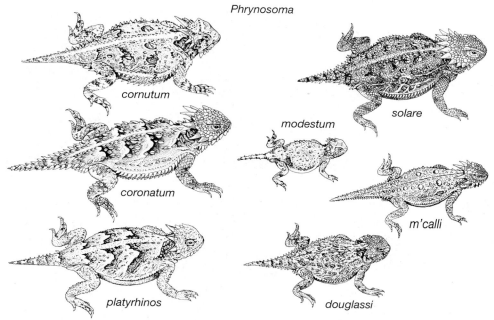

Phrynosoma

cornutum

solare

modestum

coronatum

m'calli

platyrhinos

douglassi

FIGURE 9.6 An example of species distinguished by morphological characters. These seven species of horned lizards (*Phrynosoma*) from western North America can be distinguished by differences in the number, size, and arrangement of horns and scales as well as body size and proportions, color pattern, and habitat. Good scientific drawings can often show detailed features better than photographs can, especially when the critical features are subtle. (From [98].)

BOX 9A

Diagnosis of a New Species

Each species in the leaf beetle genus *Ophraella* feeds on one species or a few related species of plants. *O. notulata*, for example, has been found feeding only on two species of *Iva* along the East Coast of the United States. This species is most readily distinguished from other species of *Ophraella* by the number and pattern of dark stripes on each wing cover.

Some leaf beetles found in Florida closely resembled *O. notulata* but were collected on ragweed, *Ambrosia artemisiifolia*. This host association suggested the possibility that these beetles were a different species. In a broader study of the genus, one of the authors of this book (DJF) collected samples of beetles from both *Ambrosia* and *Iva* throughout Florida and examined them by enzyme electrophoresis [30]. He found consistent differences in allele frequencies between samples from *Iva* and from *Ambrosia* at three loci, even in samples from both plants in the same locality. In the most extreme case, one allele had an overall frequency of 0.968 in *Ambrosia*-derived specimens, but was absent in *Iva*-derived specimens, in which a different allele had a frequency of 0.989. No specimens had heterozygous allele profiles that would suggest hybridization. Later study showed differences in mitochondrial DNA as well. Thus these genetic markers were evidence of two reproductively isolated gene pools.

A careful examination then revealed average differences between *Ambrosia*- and *Iva*-associated beetles in a few

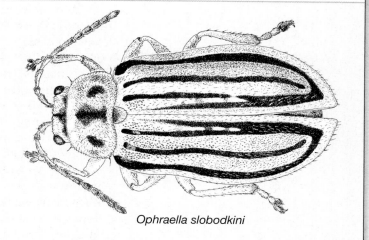

Ophraella slobodkini

morphological characters, such as the shape of one of the mouthparts and the relative length of the legs. Later studies showed that adults and newly hatched larvae strongly prefer their natural host plant (*Ambrosia* or *Iva*) when given a choice, and that the beetles mate preferentially with their own species. In laboratory crosses, viable eggs were obtained by mating female *Ambrosia* beetles with males from *Iva*, but not the reverse. Few of the hybrid larvae survived to adulthood, and none laid viable eggs. Based on all of this evidence, the author concluded that the *Ambrosia*-associated form is a distinct species, and he named it *Ophraella slobodkini* in honor of the ecologist Lawrence Slobodkin.

FIGURE 9.7 Pollinator isolation in monkeyflowers. (A) *Mimulus lewisii* has the broadly splayed petals characteristic of many bee-pollinated flowers. (B) *M. cardinalis* has the red coloration and narrow, tubular form that have evolved independently in many bird-pollinated flowers. (C) Some F₂ hybrids, showing the variation that Schemske and Bradshaw used to analyze the genetic basis of differences between these two species. (From [92].)

(A) *M. lewisii*   (B) *M. cardinalis*

(C)

Reproductive Isolation

Gene flow between biological species is prevented by biological differences called **reproductive isolating barriers** (**RIBs**), also referred to as **isolating mechanisms**. Under the biological species concept, speciation is the evolution of biological barriers to gene flow, and so understanding the evolution of reproductive isolating barriers tells us how new species evolve.

The total degree of reproductive isolation between two species may result from several RIBs that act in sequence—and some potential RIBs may not come into play. For example, the monkeyflower *Mimulus lewisii* is distributed in the Sierra Nevada of California at higher elevations than its close relative *Mimulus cardinalis*, although they both occur at intermediate elevations [92]. *Mimulus lewisii* has pink flowers with a wide corolla and is pollinated by bees, whereas *M. cardinalis* has a narrow, red, tubular corolla and is pollinated by hummingbirds (**FIGURE 9.7**). Although almost no hybrids are found where the species occur together, the species can be readily crossed, and they produce viable, fertile hybrids.

To understand the roles played by different isolating barriers between these species, Douglas Schemske, Toby Bradshaw, and their colleagues performed a massive field experiment. They bred a large number of F₂ hybrids and planted them in an area where the two species coexist. The F₂s have far greater phenotypic variation than the parental species, and they have novel combinations of traits. By amplifying the variation this way, the researchers were able to determine which of 12 floral traits that distinguish the parental species are important to reproductive isolation. They went further by using a quantitative trait loci (QTL) study (see Chapter 6) to reveal the genes underlying those traits [8]. At least four traits affect the type of pollinator that is attracted to a flower, which in turn determines which individuals exchange genes. The difference between the species in some of these traits is based on as few as one to as many as six QTL, so a change to one or a few genes can greatly affect reproductive isolation.

The investigators were able to quantify the contribution that different mechanisms make to reproductive isolation, in sequence (**FIGURE 9.8**). Separation by elevation alone reduces gene exchange by 59 percent. Among plants living at the same elevation, pollinator fidelity alone is 98 percent effective. If a flower receives both species' pollen, the conspecific pollen (that is, pollen from the same species) fertilizes the ovules at least 70 percent of the time. The germination of F_1 hybrid seeds is reduced by 20 percent compared with nonhybrids, but a hybrid seed that does germinate is just as viable. Hybrids produce fewer seeds, however, and they produce much less viable pollen. But because isolation by elevation and pollinator behavior is so great, the later barriers—reduced production, viability, and fertility of hybrids—hardly come into play at all.

The *Mimulus* species illustrate some of the many kinds of RIBs (**TABLE 9.1**). **Prezygotic barriers** reduce the likelihood that hybrids are formed. These include such factors as separation of the species in different habitats, pollination by different animals, mating at different seasons, mating preferentially with conspecifics, and failure of gametes to unite even if mating occurs. **Postzygotic barriers** reduce gene exchange between populations even if hybrid zygotes are produced. They consist of reduced hybrid viability (survival) or reproduction (fertility). Both classes of barriers are often asymmetric: for example, females of species A may be less inclined to mate with males of species B than females of B are to mate with males of A [40], or F_1 hybrids between the species may differ in viability, depending on the direction of the cross [104, 108].

Since prezygotic isolating mechanisms act before postzygotic mechanisms, they have a greater opportunity to restrict gene flow. A second reason why the distinction between prezygotic and postzygotic mechanisms is useful is that different kinds of selection act on them, as we will see shortly.

It is often difficult to tell which isolating barrier was the original cause of speciation. A character difference that contributes to reproductive isolation now may have evolved partly in geographically segregated populations before they became

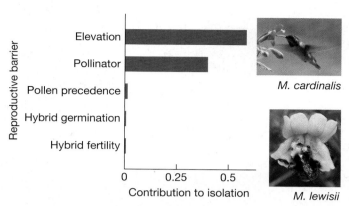

M. cardinalis

M. lewisii

FIGURE 9.8 Relative contributions of successively acting isolating mechanisms between the monkeyflowers *Mimulus lewisii* and *M. cardinalis*. Elevational separation and pollinator isolation account for almost all the reproductive isolation. In places where both species occur, pollinators provided almost complete reproductive isolation. (After [80]; photos courtesy of D. W. Schemske and H. D. Bradshaw, Jr.)

TABLE 9.1   A classification of isolating barriers

I. Premating barriers: features that impede transfer of gametes to members of other species
  A. Ecological isolation: potential mates do not meet
    1. Temporal isolation: species breed at different seasons or times of day
    2. Habitat isolation: species mate and breed in different habitats
    3. Immigrants between divergent populations do not survive long enough to interbreed
  B. Potential mates meet but do not mate
    1. Sexual isolation in animals: individuals prefer mating with members of their own species
    2. Pollinator isolation in plants: pollinators do not transfer pollen between species

II. Postmating prezygotic barriers: mating occurs, but zygotes are not formed
  A. Mechanical isolation: reproductive structures of the sexes do not fit
  B. Copulatory isolation: female is not stimulated by males of the other species
  C. Gametic isolation: failure of fertilization

III. Postzygotic barriers: hybrids are formed but have reduced fitness
  A. Extrinsic: hybrids have low fitness for environmental reasons
    1. Ecological inviability: hybrids are poorly adapted to both of the parental habitats
    2. Behavioral sterility: hybrids are less successful in obtaining mates
  B. Intrinsic: low hybrid fitness is independent of environmental context
    1. Hybrid inviability: reduced survival is due to genetic incompatibility
    2. Hybrid sterility: reduced production of viable gametes

Source: After [15], in part.

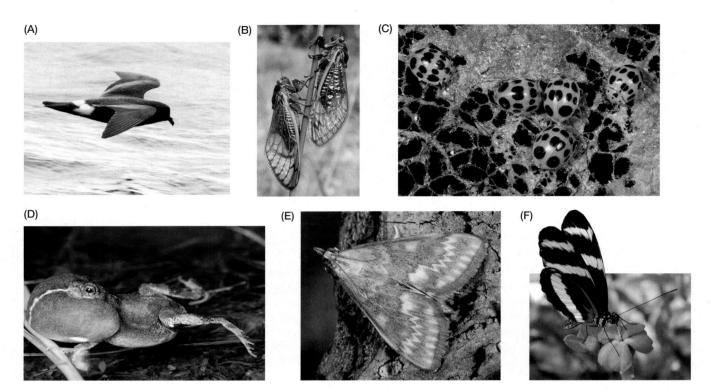

FIGURE 9.9 Prezygotic isolation takes many forms, illustrated by some species that have been extensively studied. (A–C) Three modes of prezygotic isolation. (A) Seasonal isolation: the band-rumped storm-petrel (*Oceanodroma castro*) includes two genetically different populations that mate at different times of year [28]. (B) Temporal isolation: related species of periodical cicadas (*Magicicada*) have either 17- or 13-year life cycles, and rarely emerge in the same year [10]. (C) Ecological isolation: closely related species of ladybird beetles (*Henosepilachna*) feed and mate on different species of plants [41]. (D–F) Examples of sexual isolation based on different sensory modalities. (D) Female *Physalaemus* frogs respond almost exclusively to the calls of conspecific males [88]. A calling male *P. pustulosus* is shown here. (E) In moths and many other animals, sexual isolation is based on different chemical signals. Two forms of the European corn borer (*Ostrinia nubilalis*) are strongly isolated by responses of males to different female sex pheromones [85]. (F) Males of *Heliconius pachinus* recognize conspecific females by their wing color pattern [48].

different species, partly during the process of speciation, and partly after the reproductive barriers evolved. Because genetic differences continue to accumulate long after two species achieve complete reproductive isolation, some of the genes, and even some of the traits, that now confer reproductive isolation may not have been instrumental in forming the species in the first place. Such information can be obtained by studying populations that have achieved reproductive isolation only very recently.

Prezygotic barriers

In many plants and animals, prezygotic barriers are the most important isolating mechanisms. There are many kinds of barriers, depending on the biology of the organism (**FIGURE 9.9**). Species may be temporally isolated by mating at different times of year, or even in different years. **Ecological isolation** results when ecological differences, for example habitat preference, contribute to genetic barriers [64, 93]. For example, two Japanese species of herbivorous ladybird beetles (*Henosepilachna*) feed on different genera of host plants (*Cirsium* and *Caulophyllum*). Each species mates exclusively on its own host plant, and this ecological segregation appears to be the only barrier to gene exchange [41]. **Sexual isolation** is an important barrier to gene flow among sympatric species of animals that frequently encounter each other but simply do not mate. Commonly, females will not respond to inappropriate male vocalizations or other display signals. Many birds, fishes, and jumping spiders are sexually isolated by visual signals. In many groups of animals, sexual isolation is based on differences in sex pheromones.

(A)

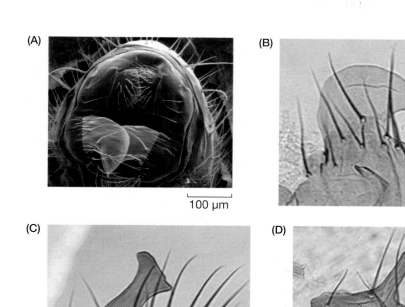

100 μm

(B)

(C)

(D)

FIGURE 9.10 Differences in genitalia can contribute to reproductive isolation between species if copulation between them occurs. (A) The genital arch in male *Drosophila* is involved in transferring sperm to females. Its shape differs among closely related species, as the close-ups show: (B) *D. sechellia*, (C) *D. mauritiana*, (D) *D. simulans*. This morphological feature is almost the only one by which these species differ. (A from [53]; B, C, and D courtesy of J. R. True.)

Gametic isolation occurs when gametes of different species fail to unite. This barrier is important in many externally fertilizing species of marine invertebrates that release eggs and sperm into the water. Because cell surface proteins determine whether or not sperm can adhere to and penetrate an egg, divergence in these proteins can result in gametic isolation [69]. Among species of abalones (large gastropods), the failure of heterospecific eggs and sperm to unite is related to the high rate of divergence in the amino acid sequences of both lysin (the sperm protein that dissolves the egg's vitelline envelope) and the vitelline envelope protein with which it interacts (see Chapter 10) [32]. In cases that fall in between premating and postmating isolation, mating occurs but fertilization does not. In many groups of insects and some other taxa, the genitalia of related species differ in morphology. There is evidence that females terminate mating, and prevent transfer of sperm, if a male's genitalia do not provide suitable tactile stimulation (**FIGURE 9.10**) [22].

Postzygotic barriers

Postzygotic barriers consist of reduced survival or reproductive rates of hybrid zygotes that would otherwise backcross to the parent populations. These barriers can be classified as either extrinsic or intrinsic, depending on whether or not their effect depends on the environment. Intrinsic isolation is based on interactions between genes from two populations, and is often more permanent than extrinsic isolation.

Extrinsic postzygotic isolation is often based on reduced survival because of ecological factors. In some cases, the parent species are adapted to different environments; the hybrid may be poorly adapted to both. A simple example is provided by hybrids between species of *Heliconius* butterflies that are distasteful to birds and have different patterns of warning coloration. Birds learn to associate common color patterns with distastefulness, but are likely to attack butterflies with rare, unfamiliar phenotypes, such as hybrids. Researchers placed artificial butterflies, with wing patterns of two species and their F$_1$ hybrid, in a tropical forest, and scored the number that were damaged by attacking birds [60]. Those with hybrid color patterns were more frequently attacked (**FIGURE 9.11**).

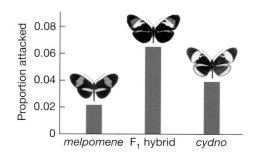

FIGURE 9.11 Model butterflies with the color pattern of the F$_1$ hybrid between *Heliconius melpomene* and *H. cydno* were attacked by birds significantly more frequently than those with the pattern of either parent. The low survival of hybrids is an example of postzygotic isolation caused by an extrinsic factor. (From [60].)

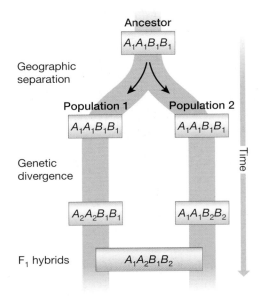

Ancestor
$A_1A_1B_1B_1$

Geographic
separation

Population 1
$A_1A_1B_1B_1$

Population 2
$A_1A_1B_1B_1$

Time

Genetic
divergence

$A_2A_2B_1B_1$

$A_1A_1B_2B_2$

F_1 hybrids
$A_1A_2B_1B_2$

FIGURE 9.12 Dobzhansky-Muller in-compatibilities (DMIs) can evolve when geographically separated populations become fixed for different alleles at two loci. (After [75].)

Postzygotic isolation is *intrinsic* if hybrids suffer high mortality, or are partially or entirely sterile, irrespective of environment. The causes of intrinsic postzygotic isolation and its genetic bases are diverse. Reduced hybrid viability is largely caused by incompatible interactions among genes from the two populations when they occur together in hybrids. Hybrid fertility may be reduced by incompatible genes or by differences in the number or structure of chromosomes. Bear in mind that the genetic differences that cause these effects may have evolved after prezygotic barriers, so we cannot assume that they were the cause of speciation.

Incompatible interactions between genes inherited from the two parents were postulated by Theodosius Dobzhansky in 1937 [21] and by Hermann Muller in 1942 [62], and are often referred to as **Dobzhansky-Muller incompatibilities (DMIs)**. The Dobzhansky-Muller hypothesis is clever because it explains how incompatibilities between populations can originate without ever producing incompatibilities within a population (**FIGURE 9.12**). Imagine that the ancestor of the two species had genotype $A_1A_1B_1B_1$. That species was then divided into two populations by a geographic barrier. In one population, allele A_2 spreads to fixation (perhaps because of adaptation to local conditions). This population is now $A_2A_2B_1B_1$. In the second population, allele B_2 spreads to fixation, so this population becomes $A_1A_1B_2B_2$. During this period, alleles A_2 and B_2 have never been in the same population, so there is no reason they should have been selected to function well together. If they are incompatible, hybrids between the two populations will have low fitness.

A simple example has been described for a cross between strains of the mouse-ear cress *Arabidopsis thaliana* from different regions [6]. Both strains have two paralogous loci (call them α and β), formed by duplication. In one strain, the α locus is nonfunctional, but the β locus is functional. The other strain has a functional α but a nonfunctional β. The F_1 offspring of a cross between the strains are viable, but in the F_2 generation, some recombinant offspring are homozygous for nonfunctional alleles of both α and β genes—a lethal combination.

DMIs between *Drosophila simulans* and *D. mauritiana* cause male F_1 hybrids to be sterile, while females are fertile. The genetics of the hybrid male sterility have been studied with laboratory crosses that produce different combinations of chromosome segments [13]. Two results emerge. The first is that many combinations of chromosomes from the two species reduce male fertility, showing that there are many DMIs throughout their genomes. The second is that male sterility is caused by interactions between the autosomes of *simulans* and the X chromosome of *mauritiana*. This reflects a general phenomenon called **Haldane's rule**: hybrid sterility or hybrid inviability is often limited to the heterogametic sex. (The heterogametic sex is the one with two different sex chromosomes, while the homogametic sex has two sex chromosomes of the same type.) In mammals and most insects, males are XY and thus are the heterogametic sex. In birds and butterflies, the situation is reversed: females have two different kinds of sex chromosomes. Thus male hybrids are frequently sterile in mammals (for example, mules), while female hybrids are frequently sterile in birds.

DMIs have many causes. Gene regulation can be anomalous due to a mismatch between *cis-* and *trans*-regulatory elements from the two species [11]. Intragenomic conflict (see Chapter 12) appears to be a common cause (see below) [18, 75]. DMIs can also be manifestations of cytonuclear incompatibility. For example, hybrids between different geographic populations of a marine copepod have reduced survival and fecundity if their mitochondria and nuclear genome come from different populations (**FIGURE 9.13**) [9].

Many sister species are distinguished by chromosome rearrangements: structural differences between the chromosomes (see Chapter 4). Two common

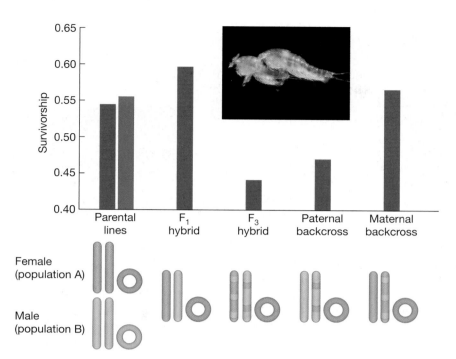

FIGURE 9.13 Crosses show that the low fitness of hybrids between populations of the copepod *Tigriopus californicus* is caused by a genetic mismatch between mitochondrial and nuclear genes. Maternally inherited mitochondria (circles) and nuclear chromosomes inherited from both parents (rods) of populations A and B are colored red and blue, respectively. Crosses produce F_1 hybrids with population A mitochondria. These F_1 offspring have slightly higher survival, showing "hybrid vigor." Crosses then produce F_2 and F_3 hybrids, with recombined nuclear genes. The paternal backcross is produced by mating F_3 females with population B males. These offspring have low fitness, because most of the nuclear genes come from population B and are mismatched to the mitochondrial genes from population A. In contrast, offspring of the maternal backcross, in which most of the nuclear genes come from the same population as the mitochondria, have normal, high survival. (After [9].)

rearrangements are **inversions** and **reciprocal translocations** (see p. 90). Especially in the case of translocations, heterozygotes have reduced fertility compared with homozygotes for either the original or the derived (new) arrangement. For this reason, populations with different chromosome arrangements are nearly or entirely monomorphic, and may form narrow hybrid zones where one "chromosome race" meets and interbreeds with another (**FIGURE 9.14**). The fertility of heterozygotes for chromosome rearrangements may be low either because the rearrangements carry different alleles that create Dobzhansky-Muller incompatibilities, or because of mispairing of chromosomes in meiosis produces gametes that lack certain chromosome regions.

How fast does reproductive isolation evolve?

The time required for reproductive isolation to become strong, after it has started to evolve, varies greatly. The origin of a new species by polyploidy, which is especially common in plants, requires only one or two generations (see p. 232). If

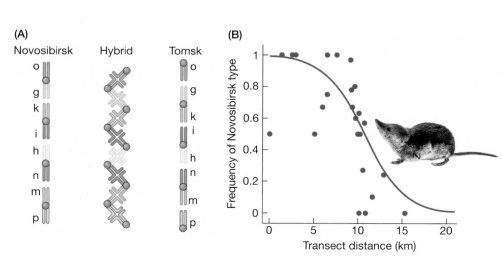

FIGURE 9.14 Two "chromosome races" of the common shrew (*Sorex araneus*) form a very narrow hybrid zone in Siberia. (A) The Novosibirsk and Tomsk races differ by the fusion of some single-armed chromosomes (e.g., o and p in Tomsk) into double-armed chromosomes (e.g., o and g in Novosibirsk). In meiosis in hybrids, the multiple rearrangements cause a chain of nine chromosomes to form, and irregular segregation produces many unbalanced gametes and low fertility. (B) A transect from Novosibirsk to Tomsk shows a cline in the frequency of the Novosibirsk chromosome arrangement less than 9 km wide. The chromosome configuration of either race cannot increase within populations of the other race, probably because meiosis in F_1 hybrids produces gametes that lack some chromosomal regions. (A after [73]; B after [74].)

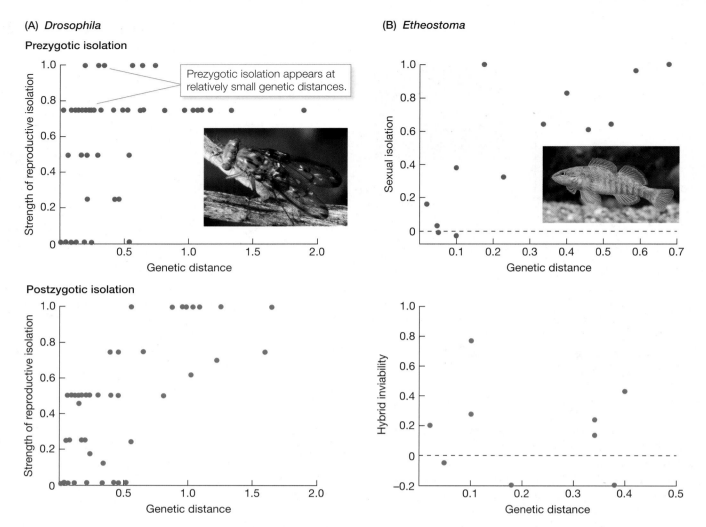

(A) *Drosophila*

Prezygotic isolation

Prezygotic isolation appears at relatively small genetic distances.

y-axis: Strength of reproductive isolation
x-axis: Genetic distance

Postzygotic isolation

y-axis: Strength of reproductive isolation
x-axis: Genetic distance

(B) *Etheostoma*

y-axis: Sexual isolation
x-axis: Genetic distance

y-axis: Hybrid inviability
x-axis: Genetic distance

FIGURE 9.15 Prezygotic isolation evolves faster than postzygotic isolation in flies and fishes. (A) The strength of prezygotic and postzygotic reproductive isolation between pairs of populations and species of *Drosophila* increases with the amount of time since the lineages split. The time is estimated by the genetic distance between each pair. The strength of prezygotic isolation was measured by observing mating between flies in the laboratory. The strength of postzygotic isolation was measured by survival and fertility of hybrid individuals. Comparison of the upper left part of the two graphs reveals that strong prezygotic isolation evolves shortly after isolation (at small genetic distances), while strong postzygotic isolation evolves only later. (B) Similar patterns are seen in a genus of freshwater fishes, the darters (*Etheostoma*). Thirteen pairs of allopatric species were tested for both sexual isolation and the survival of artificially produced hybrids. For both indices, a value of 0 indicates that the pairs are no more isolated than conspecific individuals, and a value of 1 indicates complete reproductive isolation. (A after [14]; B after [59].)

450-plus species of cichlid fishes evolved from 5 ancestral species in Lake Victoria in just 15,000 years (see opening of this chapter), the average time between speciation events was less than 2000 years, which is astonishingly rapid. Molecular clocks (see Chapter 7) can be used to estimate the time back to the most recent common ancestor, giving us the age of their speciation. Based on this approach, sympatric sister species of *Drosophila* are estimated to have taken about 200,000 years to evolve, while it requires 1.1–2.7 million years for allopatric populations to evolve full reproductive isolation [15]. Some populations of birds that have been diverging for 1.5–3 million years form hybrid zones, showing that it take more time for birds than flies to evolve strong reproductive isolation [106].

In many groups of organisms, prezygotic isolation evolves considerably faster than intrinsic postzygotic isolation (**FIGURE 9.15**). Consequently, closely related species are often fully interfertile: many species and genera of birds, even after more than 5 million years of divergence, can form fully viable, fertile hybrids when crossed [76]. In these cases (which may be the rule in many kinds of organisms), postzygotic isolation probably plays a minor role in speciation. It may, however, affect the further evolution of prezygotic isolation (see p. 230), and it may help keep species separate, because prezygotic barriers such as ecological or sexual isolation may not evolve to completion, or can become weaker if habitats change [87]. For example, increasing turbidity in Lake Victoria interfered with female cichlids' ability to see differences in male coloration that are the basis of sexual isolation between some closely related species. The result is that species that were previously well isolated are now hybridizing [96]. In contrast, strong postzygotic isolation, such as complete hybrid sterility, is probably irreversible, and can make species permanent.

The Causes of Speciation

Speciation is the evolution of reproductive isolating barriers. But because these barriers decrease the chance that some individuals mate or that their offspring survive, it might seem paradoxical that they could ever evolve.

The solution to this conundrum is that speciation often starts with a geographic barrier (such as a mountain range) that separates two populations of the same species. Over time, the populations evolve genetic and phenotypic differences, perhaps as they adapt to different ecological conditions. At this stage, there is no reason that genetic differences between the populations, or traits such as mating behavior, should be compatible, because the genes in the two populations are prevented from mixing by the geographic barrier. Sometimes those differences cause prezygotic or postzygotic isolation between the populations if they come back into contact (for example, if the mountain range erodes or if colonists disperse across it). If reproductive isolation is sufficiently complete, two species have evolved from one by the process of **allopatric speciation**. (Remember, we defined reproductive isolation as based on biological differences that reduce gene exchange, not extrinsic barriers such as mountain ranges.)

This scenario illustrates a key point: to initiate speciation, something is needed to restrict free interbreeding between two diverging populations, since interbreeding tends to erase their emerging genetic differences. Most often, that restriction results from geographic separation of the populations, although other mechanisms can have this effect.

We now turn to the question of what causes the evolution of genetic and phenotypic differences between geographically separated populations that result in reproductive isolation. That is, what are the causes of speciation?

ECOLOGICAL SPECIATION   The two monkeyflower species discussed earlier (see Figure 9.7) provide a vivid example of how reproductive isolation can result when natural selection acts differently on two populations [91]. Based on phylogenetic reconstruction of ancestral characteristics in the genus, it is likely that the ancestor of these species resembled *Mimulus lewisii* (see Figure 9.7A): it was bee-pollinated and occupied high elevations. The population that gave rise to *M. cardinalis* colonized lower elevations, where natural selection favored flower traits that attract hummingbirds: red pigments, abundant nectar, and extension of the petals to form a long, tubular corolla that excludes bees but allows hummingbirds to reach the nectar (see Figure 9.7B). Those changes to the elevational distribution and flowers had the effect of strongly decreasing the exchange of pollen (and genes) with the ancestral population, giving rise to the new species.

This scenario is a plausible reconstruction of past events. Biologists have also observed the evolution of reproductive isolation by selection in the laboratory. **BOX 9B** describes an experiment in which laboratory populations of *Drosophila melanogaster* were selected for adaptation to two different environments. In only about 20 generations, the divergently selected subpopulations became substantially reproductively isolated.

BOX 9B

Speciation in the Lab

Can different regimes of natural selection cause populations of a species to become different species? Darwin and many later evolutionary biologists have supposed that this is how speciation usually happens. Indeed, most closely related species have different adaptations to their ecological circumstances (for example, they often are adapted to slightly different habitats or diets), and of course they are reproductively isolated. But that does not provide evidence that the genetic changes underlying their ecological adaptations caused the reproductive isolation.

One way of obtaining relevant evidence is to use experimental evolution. In this approach, we expose a laboratory population to a simplified version of the conditions we suspect might occur in nature. The results determine if real organisms can in principle speciate because of different ecological selection pressures. We can also gain other key insights, for example how long the process might take.

Among many such experiments is one by Diane Dodd [21], who used eight laboratory populations of *Drosophila pseudoobscura*, all of which were founded by flies collected in a single locality in Utah. For 1 year (about 20 generations), four of the populations were reared on each of two larval food media, one based on starch (*st*) and the other on maltose (*ma*). Both media were stressful: Dodd reported that "it initially took several months for the populations to become fully established and healthy." Thus these treatments provided conditions for adaptation to the different diets to occur by natural selection. (It was natural selection, not artificial selection. In artificial selection, the investigator would decide which flies reproduce and which do not. Dodd didn't do that. Instead, she simply put the flies into a stressful environment and let selection take its course.)

After a year, Dodd reared flies from all eight populations on standard *Drosophila* food for one generation (to eliminate any maternal effects of starch or maltose). She then put virgin females and males from a pair of populations together in an observation chamber and recorded how many of each of the possible matings occurred. For instance, in one combination of *st* and *ma* populations, two kinds of "homogamic" matings (female *st* × male *st*, female *ma* × male *ma*) and two "heterogamic" matings (female *st* × male *ma*, female *ma* × male *st*) might occur. Each of the 16 possible pairs of starch-adapted and maltose-adapted populations was tested in this way. In order to be sure than any reproductive isolation could be attributed to the divergent selection, and not just genetic drift in isolated populations, Dodd also counted matings between pairs of populations that had been subjected to the same stressful diet. For every pair of populations, an index of sexual isolation was calculated that ranged from 0, if the proportion of different-population matings equaled the proportion of same-population matings, to 1.0 if no different-population matings

The monkeyflowers and the *Drosophila* experiment illustrate how reproductive isolation can evolve as a side effect of adaptation to different ecological circumstances, a process called **ecological speciation** [64, 93]. A key point is that the RIBs evolve by pleiotropy (see Chapter 4). There was no direct natural selection for isolation between the populations. Rather, selection acted on other traits that happened to cause isolation. (Recall the distinction between "selection for" and "selection of" features, in Chapter 3.) Although speciation, one of the most important elements of evolution, is commonly a consequence of adaptive changes in organisms' characteristics, it is typically not an adaptation itself.

SPECIATION BY GENETIC CONFLICT   Another powerful cause of the evolution of reproductive isolation is **genetic conflict**, which occurs when an allele increases its own transmission to the detriment of other alleles at the same or other loci (see Chapter 12). Many mutations have been found that transmit more copies of themselves to the next generation not by increasing survival or reproduction, but by violating the rules of inheritance. They are transmitted to more than 50 percent of the gametes (a process called **segregation distortion**). These mutations increase in frequency in a population even though they often reduce

Speciation in the Lab (continued)

occurred. (Incidentally, for these tests Dodd clipped a wing tip on flies from one of the two experimental populations, in order to distinguish them. This procedure did not affect the results.)

Here are the numbers of matings for 1 of the 16 pairs of populations adapted to different diets, and 1 of the 16 pairs adapted to the same diet:

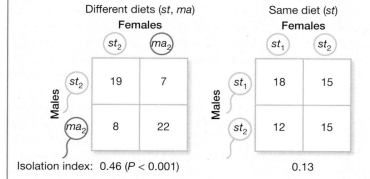

Different diets (*st*, *ma*)

Females

Same diet (*st*)

Females

Isolation index: 0.46 (*P* < 0.001)   0.13

Females of the *st* and *ma* populations, adapted to different diets, were more likely to accept males adapted to the same diet as themselves. In all 16 combinations of different-diet populations, there was a tendency for females to show same-diet preference, and this was statistically significant in 11 combinations. (The notation *P* < 0.001 means that the probability is less than 1 in 1000 that the correlation be-

tween mating and rearing environment could have occurred by chance.) But in none of the pairs of populations adapted to the same diet was there a statistically significant excess of same-population matings.

The sexual isolation index value of 0.46 suggests that in a mere 20 or so generations, these divergently selected laboratory populations had progressed about halfway toward full sexual isolation—in which case speciation would have been completed in the laboratory! This is astonishingly fast, especially in the context of evidence on how long it takes for speciation to occur in nature (see p. 226).

What caused the populations to evolve partial sexual isolation? One possible answer is pleiotropy: some of the same genes that enhance adaptation to starch or maltose might also affect female preference and some feature of males that enables females to distinguish them. Or perhaps the strong selection for alleles that enhance adaptation to the novel diets carried along alleles at closely linked genes that affect male characteristics and female responses to those characteristics.

Dodd did not do further research on these possibilities, and in the 1980s it would not have been possible to identify and obtain the sequences of the relevant genes. That would be a much easier task today. Dodd's experiment is waiting for someone to repeat it and do the genetic detective work.

fertility. Selection therefore favors mutations at other loci that restore full fertility by disabling the segregation distortion caused by the "selfish" mutation.

When this conflict between distorter and a restorer has played out in one population but not another, the populations may be genetically incompatible. This is the basis for strong postzygotic isolation between populations of *Drosophila pseudoobscura* in North America and in Bogotá, Colombia: hybrid males are almost completely sterile [71]. Sterility is the result of a mutation at a locus (*Overdrive*) that reduces male fertility, but that spreads by segregation distortion through the Bogotá population. This population has restorer alleles at other loci that maintain male fertility, but restoration is inadequate in hybrid males. Genetic conflict seems to be an important cause of Dobzhansky-Muller incompatibilities in *Drosophila* and perhaps other groups of organisms. A similar conflict sometimes occurs between nuclear and mitochondrial or chloroplast genes, as in the copepod example described earlier (see Figure 9.13).

Earlier we saw that different species of abalones are reproductively isolated because proteins on the outside of their eggs and sperm have diverged to the point where they do not bind to one another. Divergence may have been caused by sexual conflict: changes in the egg surface that slow down sperm entry are

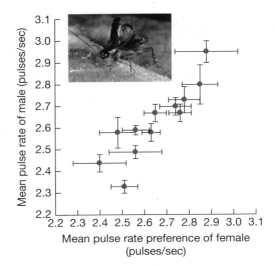

FIGURE 9.16 Evolution of sexual isolation by sexual selection. The pulse rate of the mating call of male crickets (*Laupala cerasina*) and the pulse rate preferred by females both vary among local populations. These differences are genetically based. The confidence intervals around each point show that the preference ranges of females of the most widely different populations would not include the most divergent males. (After [33].)

advantageous because fertilization by more than one sperm kills the egg. Any such changes in the egg will impose selection for sperm that can beat their competitors by penetrating more quickly [99].

SPECIATION BY SEXUAL SELECTION   In many groups of rapidly speciating animals and plants, species differ more in their secondary sexual traits,[1] such as male coloration or vocalization, than in ecologically important traits (see Figure 9.9). In many cases, one sex (let's suppose the female) chooses mates based on variation in these traits. Females impose strong sexual selection, which can drive the rapid evolution of male secondary sexual traits (see Chapter 10). Species that differ in sexually selected male features also commonly differ in female preference, so females recognize and mate preferentially with males of their own species. These patterns suggest that divergent sexual selection can cause rapid evolution of prezygotic isolation between populations [88]. Certain groups of animals, such as cichlid fishes and hummingbirds, have indeed speciated rapidly and show strong sexual selection. Recent phylogenetic analyses of birds suggest that male coloration patterns associated both with sexual selection and species recognition evolve fastest in lineages with high speciation rates [94].

Studies of closely related populations and species provide more direct evidence that sexual selection may cause speciation. For example, male calls and female preferences covary among populations of a Hawaiian cricket (*Laupala cerasina*), to the point that females hardly respond to the calls of the most different population (**FIGURE 9.16**) [33]. Sexual isolation appears to be the sole basis of reproductive isolation between some ecologically indistinguishable species of freshwater fishes called darters (see Figure 9.15B) [52]. Why then does sexual selection vary among populations? In Chapter 10 we will consider some of the factors at work. These include direct benefits to mate preferences, selection acting on pleiotropic effects of preference genes, preferences for mates with "good genes," and ecological factors that make different courtship signals more effective in different environments.

REINFORCEMENT OF REPRODUCTIVE ISOLATION   So far, we have discussed how speciation can result as a side effect of divergent selection. In some cases, natural selection can also directly favor the evolution of prezygotic isolation. Consider two populations that have already evolved some degree of isolation so that hybrids have lower survival or fertility. A female that chooses a male from her own population will leave more descendants than one that makes the mistake of mating with a male from the other population. This creates a selective advantage to an allele for a mating preference that increases the chance of mating within rather than between populations. A "discrimination" allele will be transmitted to more progeny, on average, than a "random-mating" allele.

The evolution of stronger prezygotic isolation because of selection against low-fitness hybrids is called **reinforcement**. Not all types of isolating mechanisms can evolve this way. Alleles that strengthen *prezygotic* isolation gain an advantage because individuals with them have higher fitness than do those that hybridize. But stronger *postzygotic* isolation usually cannot evolve by natural selection. An allele that lowers hybrid fitness cannot increase in frequency, for that would be the antithesis of natural selection. (Exceptions are in organisms such as plants and mammals, in which embryos compete for the mother's nutrients. It can be advantageous for a mother to abort hybrid embryos and allocate resources to nonhybrid

---

[1] Secondary sexual traits are those that differ between the sexes, other than the gonads and reproductive structures.

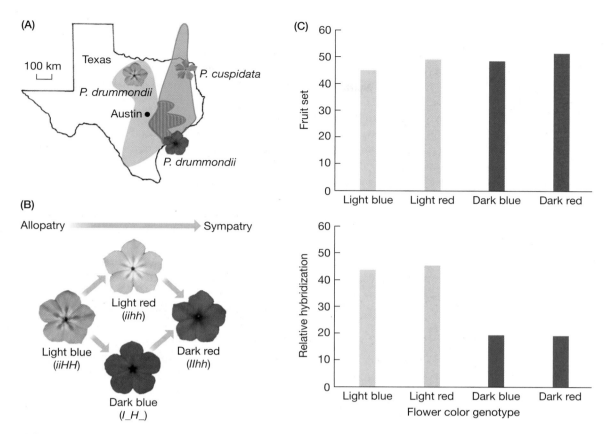

FIGURE 9.17 Reinforcement of reproductive isolation by flower color in *Phlox*. (A) The geographic distributions of *P. cuspidata* and *P. drummondii* overlap in Texas. Allopatric populations of both species are light blue, but populations of *P. drummondii* are dark red where the species is sympatric. (B) The flower color difference in *P. drummondii* is based on two loci. (C) Results of common-garden field experiments, in which all four color types of *P. drummondii* were grown together with *P. cuspidata*. Both parental types (light blue and dark red) and hybrid genotypes with light red and dark blue flowers have equal fruit production (top graph), but differ in the proportion of their offspring that are hybrids with *P. cuspidata* (bottom graph). (A and B from [37]; C after [37].)

offspring that are more likely to pass on her genes [12].) This is one reason why the distinction between prezygotic and postzygotic mechanisms is important: prezygotic mechanisms can evolve by reinforcement, but postzygotic mechanisms generally cannot.

Wildflowers in the genus *Phlox* provide a clear example of reinforcement [37]. Allopatric populations of the two species *Phlox drummondii* and *P. cuspidata* both have light blue flowers (**FIGURE 9.17**). Where their ranges overlap, however, *P. drummondii* has evolved dark red flowers, a difference in color caused by changes at two loci. Because the fertility of hybrids is up to 90 percent lower than that of non-hybrids, the difference in color is strongly favored by selection: pollinators move less pollen between flowers that have different colors, so *P. drummondii* produces fewer low-fitness hybrids when it has dark red flowers rather than light blue flowers in the zone of sympatry.

Selection for reinforcement can occur only when two species continue to inter-breed after some postzygotic isolation (reduced fitness of hybrids) has already evolved between them. If reinforcement is common, we would expect sympatric pairs of species (which could potentially hybridize) to show greater prezygotic isolation than allopatric pairs of species (which have no chance of hybridizing).

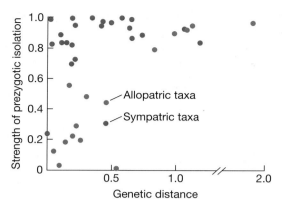

FIGURE 9.18 In *Drosophila*, the strength of prezygotic isolation increases more rapidly between sympatric pairs of species (red dots) than it does between allopatric pairs (blue dots). The genetic distance between members of a pair is an index of the time since divergence began. At small genetic distances, many sympatric pairs show strong isolation, but allopatric pairs do not. (After [14].)

Data from *Drosophila* show exactly the expected pattern. In **FIGURE 9.18**, the strength of prezygotic isolation between a pair of populations or species is plotted against the genetic difference (or genetic "distance," based on molecular differences) between them. Prezygotic isolation here refers to behavioral isolation when females' choice between males of both species was observed in laboratory tests. The genetic distance serves as a molecular clock, indicating time since the species diverged from their common ancestor. Sympatric pairs of species show strong prezygotic isolation at lower genetic distances than do allopatric pairs, meaning they evolved mating discrimination faster. This pattern is what we expect if the sympatric pairs tended to hybridize shortly after speciation, and reinforcement then strengthened their prezygotic isolation. Further analysis reveals additional support for the role of reinforcement. Hybridization between some pairs of sympatric species is asymmetric: the offspring from a cross between a female of species A and a male of species B have lower fitness than those from the reciprocal cross (female B × male A). In these cases, sexual isolation is stronger in the female-male combination that produces less fit hybrids, as we predict from the hypothesis of reinforcement [108].

SPECIATION BY POLYPLOIDY   When a diploid species' entire genome is doubled (see Chapter 4), the result is a tetraploid that has four copies of every chromosome. Tetraploids originate by the union of two "unreduced" gametes—both carrying a full diploid set of chromosomes—that are formed when chromosomes occasionally fail to segregate in meiosis. The polyploid offspring is *autopolyploid* if both unreduced gametes come from the same diploid species, and *allopolyploid* if they come from different diploid species. If similar events happen in tetraploid species, offspring with even more sets of chromosomes (e.g., eight in octoploids) result. The increased number of gene copies in polyploids changes the expression (e.g., amount of gene product) of many genes, and alters many phenotypic traits [50, 67].

Tetraploids typically have complete reproductive isolation from their diploid ancestors, and so are distinct biological species that have arisen in one step [79]. That is because hybrids between a diploid and a tetraploid are triploid: they have one set of chromosomes from the diploid parent and two from the tetraploid parent (in which four homologous chromosomes generally segregate two by two in meiosis). Triploids are largely sterile [38], because their gametes are unbalanced: they have one copy of certain chromosomes and two copies of other chromosomes. Genome doubling is a large mutational event: one of the very rare situations in which mutations of large effect make important contributions to evolution, in this case the origin of new species.

Speciation by polyploidy is rare in animals, but it is quite common in some groups of plants. It accounts for about 15 percent of speciation events in flowering plants [107], and all plants have a polyploid ancestor somewhere in their evolutionary past [67]. Speciation by polyploidy has occurred even very recently. For example, hybridization among three species of goatsbeards (*Tragopogon*) generated new allopolyploid species within the last several centuries, after their accidental introduction to North America from Europe (**FIGURE 9.19**).

How can a new tetraploid build up a population? There is a serious obstacle. The tetraploid species starts out with just one or a few individuals, so often the diploid ancestor is more abundant in its habitat. This can cause the tetraploids to hybridize most often with the diploids, producing triploid offspring with low fitness and pushing the tetraploid toward extinction. A new tetraploid might increase if hybridization were reduced by self-fertilization, vegetative propagation, higher fitness than the diploid, or habitat segregation from the diploid [27, 84]. Many

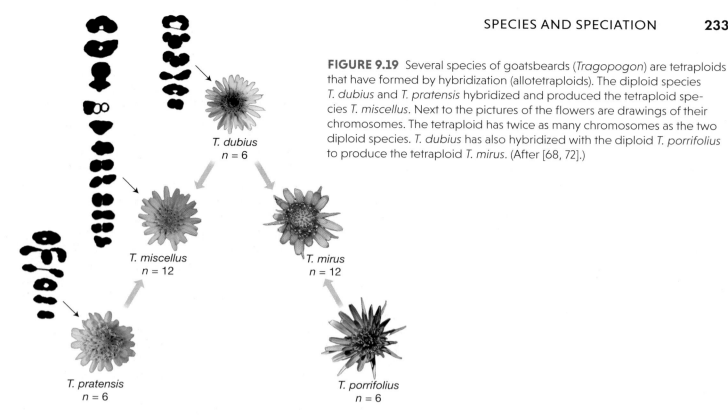

FIGURE 9.19 Several species of goatsbeards (*Tragopogon*) are tetraploids that have formed by hybridization (allotetraploids). The diploid species *T. dubius* and *T. pratensis* hybridized and produced the tetraploid species *T. miscellus*. Next to the pictures of the flowers are drawings of their chromosomes. The tetraploid has twice as many chromosomes as the two diploid species. *T. dubius* has also hybridized with the diploid *T. porrifolius* to produce the tetraploid *T. mirus*. (After [68, 72].)

tetraploid taxa do indeed reproduce by selfing or vegetative propagation, and most differ from their diploid progenitors in habitat and distribution, and so would be segregated from them. The phenotypic differences that are an immediate effect of chromosome doubling may cause such separation. In California, tetraploids and hexaploids of the yarrow *Achillea borealis* grow in wetter and drier habitats, respectively. Justin Ramsey planted seedlings of both forms in dry dunes, as well as "neohexaploids" that had originated de novo from tetraploid parents that he grew [78]. The neohexaploids survived better and flowered earlier than the tetraploids (**FIGURE 9.20**), showing that they would be partly isolated from the tetraploids, by habitat and flowering time, immediately upon their origin.

HYBRID SPECIATION   Interbreeding between populations usually opposes divergence and so makes speciation less likely. Occasionally, however, hybridization generates a new species without help from polyploidy [1]. For example, three species of sunflowers have originated from independent hybridization events

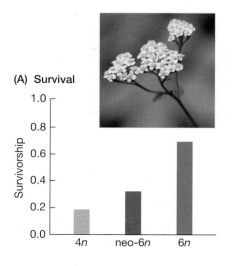

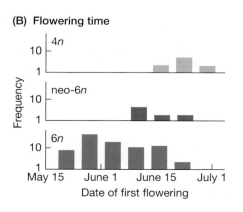

FIGURE 9.20 Differences between a newly formed polyploid and its ancestor may confer ecological differences that could reduce the opportunity for crossing between them. Survival (A) and flowering time (B) of a newly originated hexaploid (neo-6*n*) yarrow (*Achillea borealis*), planted in a dry dune, were intermediate between those of its tetraploid parent (4*n*) and an existing hexaploid (6*n*) species. Increasing the chromosome number immediately changes these characteristics, but the wild hexaploid (6*n*) differs even more from the tetraploid ancestor (4*n*), so the wild hexaploid must have undergone some additional evolutionary change. (After [78].)

between two parental species, *Helianthus annuus* and *H. petiolaris* (see Figure 2.11) [81]. All have the same number of chromosomes. The hybrid species live in drier or saltier habitats than the parents and are genetically incompatible with them. The combination of spatial separation, genetic incompatibilities, and perhaps other mechanisms effectively isolates the hybrid species from their parents.

DNA studies show that all three hybrid species have arisen from the same two parental species, but that they have different combinations of parental genes. The origin of the hybrids has also been confirmed experimentally, by crossing the parental species. Their F_2 hybrids show some of the same distinctive adaptive features as the wild hybrid species [82].

SPECIATION BY RANDOM GENETIC DRIFT   Some closely related species differ by chromosomal rearrangements that contribute to postzygotic isolation because of low fertility of chromosome heterozygotes. In the sunflowers just discussed, about half of the postzygotic barrier between *H. annuus* and *H. petiolaris* is caused by underdominant chromosomal rearrangements [36, 83]. How these isolating mechanisms evolve is puzzling, because underdominant mutations are selected against when they first appear (see Chapter 5). How, then, can these chromosome rearrangements increase and become fixed in one of the two sister species?

One possible answer is random genetic drift. If the population size is so small that genetic drift is stronger than selection, there is a chance that the new rearrangement will increase in frequency even if heterozygotes have decreased fitness. Several factors, including self-fertilization and large fluctuations in population size, make this more probable. Drift is unlikely to establish individual rearrangements that cause very strong postzygotic isolation, because the force of selection against them will be overwhelming.

But even a rearrangement that makes only a small contribution to reproductive isolation at first may later reduce gene exchange with other populations of the other species. For example, heterozygous inversions suppress recombination (see Chapter 4). If a species becomes fixed for a new chromosomal inversion, either by drift or some other mechanism, hybrids with other populations of the species will be heterozygous for the inversion, and genes in that chromosome region will not introgress between the populations. Thus, genetic differences between the populations accumulate more rapidly in the inverted region than in other parts of the genome [63].

One hypothesis for how random genetic drift might trigger the origin of new species is called **founder effect** or **peripatric speciation**. Drift can be particularly strong when a new population is founded by a small number of individuals (see Chapter 7). Under this hypothesis, proposed by Ernst Mayr [56, 58], drift in a new population, founded by a few individuals, fixes rare alleles at certain loci. Alleles at other loci that increase fitness by interacting favorably with these newly fixed alleles increase, resulting in a new combination of genes that may be genetically incompatible with the parent population from which the colony was derived. Mayr suggested that founder effect speciation is an important way that new species arise at the periphery of species ranges, and he offered potential examples from island populations of birds (**FIGURE 9.21**). Founder effect speciation is controversial both for theoretical and empirical reasons [2, 54, 97].

A possible example of this idea is the cytonuclear incompatibility between populations of the copepod described in Figure 9.13. In that case, a deleterious mutant mitochondrial genotype may have been fixed by genetic drift, followed by selection for nuclear alleles that counteract the harmful mitochondrial genotype and restore high fitness. Several investigators have tested the idea that drift can cause the evolution of reproductive isolation by passing laboratory populations of fruit flies through bottlenecks (see Chapter 7), then measuring reproductive isolation

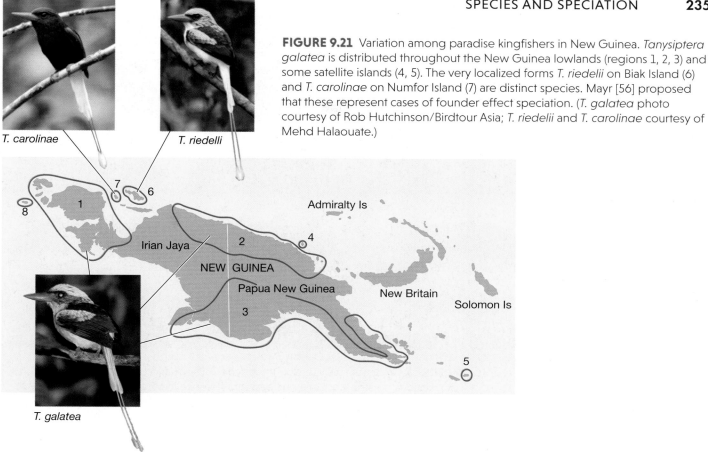

FIGURE 9.21 Variation among paradise kingfishers in New Guinea. *Tanysiptera galatea* is distributed throughout the New Guinea lowlands (regions 1, 2, 3) and some satellite islands (4, 5). The very localized forms *T. riedelii* on Biak Island (6) and *T. carolinae* on Numfor Island (7) are distinct species. Mayr [56] proposed that these represent cases of founder effect speciation. (*T. galatea* photo courtesy of Rob Hutchinson/Birdtour Asia; *T. riedelii* and *T. carolinae* courtesy of Mehd Halaouate.)

several generations later [31, 54]. Typically a few of the replicate populations display sexual isolation from the parent population, although the change is often temporary. There is some disagreement about whether or not to view these results as support for founder effect speciation [15, 101].

The Geography of Speciation

A critical issue in understanding speciation is how the level of gene flow is initially reduced between two populations when they first start to diverge. This is key because populations cannot diverge, and evolve reproductive isolation, if gene flow is high enough to counteract divergence by selection or genetic drift.

The most common way for speciation to begin is with the appearance of a geographic barrier that partly or completely blocks genetic exchange between two populations (**FIGURE 9.22**) [15, 57]. This is allopatric speciation. In other cases, speciation occurs with little or no help from a geographic barrier. The most extreme case is when a single population splits into two reproductively isolated populations while living together, a process called **sympatric speciation**. An intermediate between those two situations is **parapatric speciation**, in which neighboring populations of a single species that exchange genes nevertheless diverge into two species. The critical difference among these situations is that sympatric speciation and parapatric speciation involve **speciation with gene flow**, while allopatric speciation does not. The following sections look into the details of allopatric speciation and speciation with gene flow.

Allopatric speciation

Allopatric speciation is *the evolution of genetic barriers between populations that are geographically separated* by a physical barrier (for example, a mountain range). Allopatry is defined by a severe reduction of movement of individuals or their

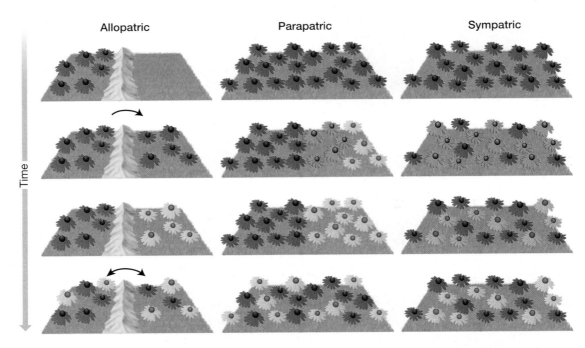

FIGURE 9.22 Schematic showing three types of speciation. In allopatric speciation, populations diverge (shown as increasing difference in color) while separated by a geographic barrier (such as a mountain range). In this drawing, an allopatric population is established by colonization. When the two populations have become so different that reproductive isolation has evolved, the two can coexist without interbreeding even if each form disperses into the range inhabited by the other (shown by the double-headed arrow). In parapatric speciation, neighboring populations diverge while still exchanging genes. In sympatric speciation, two new species emerge from a single ancestor without any geographic isolation.

gametes, not by geographic distance. In species that disperse little or are faithful to one habitat, populations may be "microgeographically" isolated (for example, among patches of favorable habitat along a lakeshore).

Allopatric populations can originate in two ways. One is **vicariance**, which is when a barrier appears and divides a population that was occupying a larger region. When the Isthmus of Panama rose out of the ocean several million years ago, it divided many marine species into Caribbean and Pacific populations. Since then, these populations have evolved into distinct species [44, 49]. Allopatric populations also originate by **dispersal**, when individuals from one population colonize another region. Speciation by dispersal has happened innumerable times when oceanic islands have been colonized from continental populations.

From paleontological and genetic studies (see Chapters 8 and 18), we know that species' geographic ranges change over time, and that populations may become separated and later rejoined. If sufficiently strong isolating barriers have evolved during the period of allopatry, the populations may become sympatric without exchanging genes. Many sister species that today are sympatric have speciated allopatrically and then expanded their ranges. This means that *current sympatry, in itself, is not evidence that speciation occurred sympatrically.*

Many species show partial reproductive isolation among geographic populations. For example, when males and females from different populations of dusky salamanders (*Desmognathus ochrophaeus*) from various localities in the eastern United States are placed together, the level of sexual isolation among them varies. The more geographically distant the populations, the less likely they are to mate (**FIGURE 9.23**) [102]. Similarly, sexual isolation has been shown among allopatric species of darters that occupy different river systems (see Figure 9.15B). Often, allopatric speciation can be related to the geological history of barriers that emerged between populations of a widespread ancestral species, as with the Isthmus of Panama.

Species on islands provide abundant evidence of allopatric speciation. For example, no pairs of sister species of birds occur together on any isolated island smaller than 10,000 km². This observation implies that speciation in birds does not occur on land masses that are too small to provide geographic isolation between populations [16]. A similar pattern is found in many other taxa [43]. As expected, taxa in which gene flow is high (such as bats) have speciated only on very large islands, while taxa in which gene flow is very limited (such as snails) have speciated on small islands (**FIGURE 9.24**).

The role of geographic isolation on islands is obvious, but what kinds of barriers could have produced the great numbers of species that are found on continents? Geographic distributions may be fragmented if populations maintain dependence on specific environmental conditions, such as climate regimes or habitats. For example, a species that is widely distributed at low elevations in a mountain range when the climate is cool may move upward and form separate populations on different mountains when the climate becomes warmer. Exactly this pattern has been found for allopatric sister species of salamanders, which are found in locations with similar climate conditions and are absent from intervening regions with different climate conditions [47]. The number of species of birds, plants, and some other taxa is very high in mountainous regions such as the Andes, where many species have small ranges and are isolated by valleys from their sister species [25].

In allopatric speciation, isolating mechanisms evolve in geographically separated populations. They play a role in restricting gene flow only if the populations come back together, an event called **secondary contact**. This often happens as the range of one or both incipient species expands. The newly formed species can then coexist as distinct populations if they are sufficiently reproductively isolated.

If reproductive isolation is incomplete when secondary contact happens, three outcomes are possible. One is that the populations hybridize so freely that they meld back into a single population, and speciation fails. For example, this happened to incipient species of three-spined sticklebacks (*Gasterosteus aculeatus*) when the habitat changed and ecological selection against hybrids was alleviated

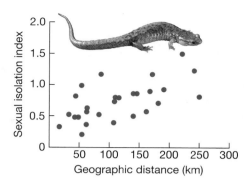

FIGURE 9.23 The strength of sexual isolation among populations of the salamander *Desmognathus ochrophaeus* is correlated with the geographic distance between the populations. The data are based on observations of mating behavior of pairs of salamanders in the laboratory. (After [102].)

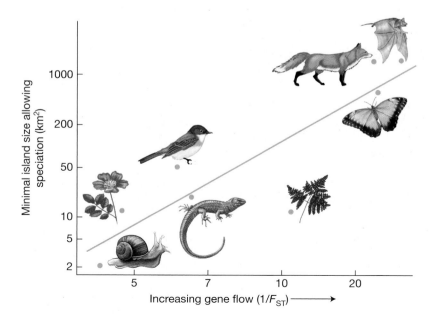

FIGURE 9.24 Speciation is more likely to occur on larger islands and in species with restricted gene flow. The minimal island size allowing speciation is small in taxa with low rates of gene flow, such as snails. Islands must be much larger for speciation to occur in taxa with high rates of gene flow, such as bats. Gene flow is measured here as $1/F_{ST}$ between populations ranging from 10 to 100 km apart. F_{ST} is a measure of genetic differentiation that decreases with greater gene flow (see Chapter 8). (After [43].)

[4]. A second outcome is that a hybrid zone is formed (for example, between the toads in Figure 9.4). Allele frequency clines are produced as alleles mix between the populations. Genetic differentiation may persist—perhaps indefinitely—for some parts of the genome because of selection, but other parts of the genome become homogenized by gene flow. We will return to this phenomenon below. A third outcome can happen when hybrids have low fitness, for example because of genetic incompatibility. Natural selection can then result in reinforcement of prezygotic isolation. Whether or not the newly formed species become sympatric can also depend on their ecological similarity. The two new species often use similar resources and live in similar habitats (see Chapter 13). Competition between them can be intense, and may prevent them from coexisting, or may even result in extinction of one of the species.

Sympatric speciation

The most extreme case of speciation with gene flow is sympatric speciation, which occurs when an ancestral population splits into two species without any geographic isolation (see Figure 9.22). In most scenarios, there is random mating at first. But in some cases gene flow might be reduced by an extrinsic factor (i.e., before any genetic divergence) even without geographic separation. For example, plants growing in different soils might be intermingled, but the soils might induce them to flower at different seasons, creating a temporal barrier to gene flow.

Sympatric speciation is controversial because interbreeding between the populations causes genetic mixing that can prevent the populations from diverging [24, 42]. Imagine a bird species that has disruptive selection on its bill. Birds with long thin bills eat insects, and birds with stout bills eat seeds (**FIGURE 9.25**). Birds with intermediate bills, however, have difficulty finding food and survive poorly. The

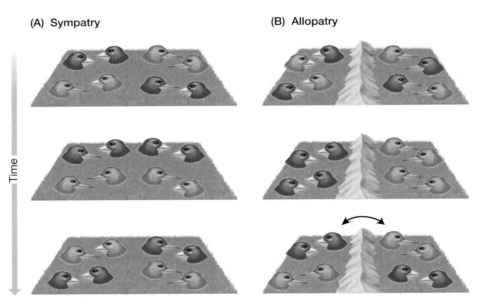

FIGURE 9.25 Sympatric speciation is less likely than allopatric speciation because recombination breaks down genetic combinations that might form new species. In this example, there are two morphs of a species of bird. Individuals with thin bills specialize on insects, while birds with stout bills specialize on seeds, and both forms have high fitness. Birds with intermediate bills, however, do not survive well. The birds choose mates on the basis of color: red mates with red, and blue mates with blue. (A) In sympatry, recombination erodes linkage disequilibrium between the loci that affect color and bill size. This prevents the emergence of two species, for example a red bird with a stout bill and a blue bird with a thin bill, that would be separated by both prezygotic and postzygotic isolation. (B) With allopatry, geographically isolated populations can diverge both in traits that affect prezygotic isolation (such as color) and in traits that affect postzygotic isolation (such as bill size). If the differences are sufficient, the two populations will stay genetically distinct when they come into secondary contact. New species have formed.

birds also vary in color, and they mate *assortatively*: red birds tend to mate with red birds, and blue birds with blue birds. (Assortative mating could arise, for example, if young birds imprint on the color of their parents.) This situation provides the ingredients for speciation. If all the birds with thin bills are blue and all the birds with stout bills are red, the two populations have both prezygotic isolation (caused by assortative mating based on color) and postzygotic isolation (caused by selection against hybrids with intermediate bills). In this case, the alleles for color and bill size are in *linkage disequilibrium* (see Chapter 4), which causes the prezygotic and postzygotic reproductive barriers to reinforce each other.

Why would linkage disequilibrium develop between the color locus and the bill shape locus? Pairs of birds with the same type of bill (either thin or stout) will tend to have offspring with the same bill shape rather than an intermediate, disadvantageous bill. Pairs with the same type of bill are more likely to form if bill shape is correlated with color, the basis for assortative mating. Pairs that have the same color but different bill shapes will often produce progeny with intermediate bills and lower survival. Therefore, selection favors associations between color and bill-shape alleles (for example, red and stout alleles together, and blue and slender alleles together). In this way, selection can favor linkage disequilibrium between loci that contribute to prezygotic barriers and those that contribute to postzygotic barriers.

The difficulty for sympatric speciation is that if there is any continued interbreeding between birds with different combinations of alleles for color and bill shape, the linkage disequilibrium tends to be broken down. If linkage disequilibrium is not very strong to begin with, recombination erodes the buildup of advantageous combinations of alleles that can diverge into distinct populations and ultimately different species. This becomes an even greater problem if mating is based on several loci, because stronger selection is needed for all the color and bill shape loci to stay in linkage disequilibrium [24, 29]. Sympatric speciation can happen under the right conditions (strong disruptive selection and assortative mating), but those conditions are relatively rare. This is the basic reason that sympatric speciation is thought to be much less common than allopatric speciation.

Sympatric speciation is made much easier by a **speciation trait** (sometimes called a "magic trait"), namely a trait that causes *both* ecological divergence and reproductive isolation between the incipient species. This situation may occur frequently in insects that feed on a narrow range of host plants [5, 7]. Many herbivorous insects mate on the plants where they feed, ensuring that matings tend to be between individuals with the same host preference. Natural selection can favor mutations that strengthen the preference for a particular host (perhaps because that host is common), and as these mutations spread they will also strengthen prezygotic isolation from individuals that prefer other kinds of host plants. Reproductive and ecological isolation can build up this way to the point that what was a single species becomes two non-interbreeding populations, that is, new species. Populations that are at early and intermediate stages in this process are called "host races." For example, the ancestor of the fly *Rhagoletis pomonella* laid eggs only in the fruit of hawthorns (**FIGURE 9.26**). In the late nineteenth century, the fly started to infest apple trees in the same areas as the normal host—and it is now known as the apple maggot. The flies mate on the host plant, and now consist of genetically divergent populations that differ in host preference and especially in their mating

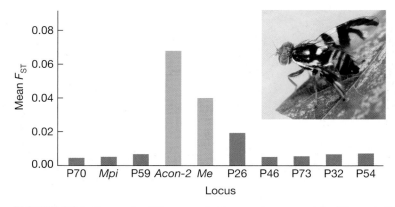

FIGURE 9.26 Genomic differences in the apple maggot fly (*Rhagoletis pomonella*), in which populations associated with different host plants have diverged by natural selection. The difference in allele frequency between the populations is measured as F_{ST} for several loci on one chromosome. Similar patterns were found for loci on the other chromosomes. Loci with significant allele frequency differences between samples from different host populations are shown in green. These loci and several of those shown in purple, are thought to be near genes that contribute to reproductive isolation. (After [61].)

(A)

(B)

(C)

FIGURE 9.27 The palms of Lord Howe Island are among the best examples of sympatric speciation. (A) Lord Howe Island is small and remote, lying between Australia and New Zealand. (B) The kentia palm (*Howea forsteriana*). (C) The curly palm (*H. belmoreana*). (D) The flowering time of the kentia palm overlaps only slightly with that of the curly palm. Each plant has a phase when male flowers open, beginning shortly before the female flowers begin to open. (After [90]; B,C courtesy of W. J. Baker, Royal Botanic Gardens, Kew.)

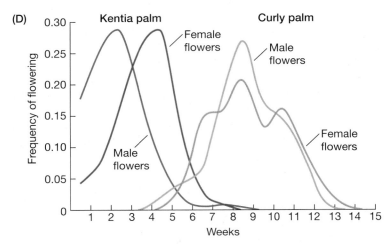

season: their life cycle is timed to match the difference between fruit development of the two plants. Some of the difference in timing probably evolved in distant populations, with those alleles spreading to populations in apple-growing regions, where they enhance adaptation to apple [23].

The best evidence of sympatric speciation is provided by cases in which sister species occupy a small, isolated island or body of water that provides little opportunity for spatial separation of speciating populations. This is the case with several pairs of sister species of plants on Lord Howe Island, a small (15 km²) island in the South Pacific (**FIGURE 9.27**) [70]. More than one-third of the plant species on this island occur nowhere else in the world. Among these endemic species are the curly palm (*Howea belmoreana*) and its sister species, the kentia palm (*H. forsteriana*) [90]. The two species are often found growing in close proximity, and they are wind-pollinated, which makes it unlikely that there was ever a time when they were unable to exchange genes because of a physical barrier. The key difference between these palms is a 6-week difference in peak flowering time: cross-pollination can occur only for a short time, between relatively few plants (see Figure 9.27D). The kentia palm is found more often on calcareous soil than the curly palm. There may have been divergent selection for adaptation to calcareous versus noncalcareous soils. Moreover, the flowering time of the kentia palm is altered if it grows on noncalcareous soil.

How common is sympatric speciation? The answer varies substantially among groups of organisms [7, 15]. There are a fair number of possible cases of sympatric speciation in herbivorous insects, but only one example in birds: a seabird, the band-rumped storm-petrel (*Oceanodroma castro*; see Figure 9.9), which has split into sympatric populations with separate breeding seasons [28].

By contrast, perhaps as many as one-third of the endemic species of plants on Lord Howe Island may have originated there by sympatric speciation [70]. Most evolutionary biologists believe that allopatric speciation is much more common than sympatric speciation. Perhaps, though, many cases of sympatric speciation have gone undetected, because it is usually difficult to rule out a past history of allopatric divergence.

Parapatric speciation

An intermediate between allopatric speciation and sympatric speciation is parapatric speciation, in which neighboring populations diverge while they continue to interbreed (see Figure 9.22). We expect parapatric speciation to be more common than sympatric speciation because it involves less gene flow between the diverging populations.

Many examples have been described in which strongly selected genes and phenotypes differ between populations that interbreed [64]. Among these, a few indicate the evolution of some reproductive isolation that reduces gene exchange. For example, the White Sands region of New Mexico consist of dunes, formed less than 5000 years ago, that differ starkly from the surrounding dark soils. In three species of lizards that are distributed across both soil types, the populations that inhabit the dunes differ in head shape, toe length, and most strikingly in color—all characteristics that are adaptive and are thought to be strongly selected (**FIGURE 9.28**) [86]. In two of the species, a fence lizard (*Sceloporus undulatus*) and an earless lizard (*Holbrookia maculata*), there are strong differences in the frequencies of genetic markers across the boundary between the habitats. These genetic differences imply that gene flow has been reduced by the evolution of partial reproductive isolation. A likely reason is that the pale coloration is associated with differences in the color of ventral blotches that are displayed in sexual and other social encounters. Studies

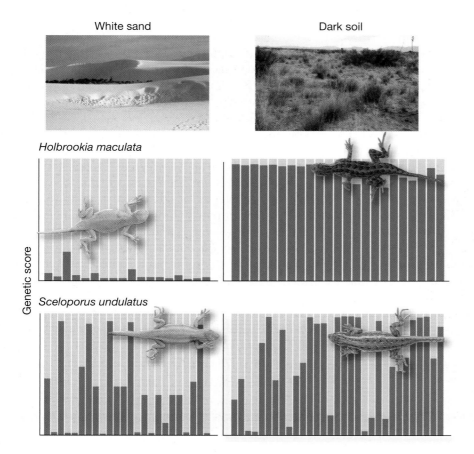

FIGURE 9.28 Incipient speciation with gene flow: divergence of lizards on the white sand and dark soil habitats shown in the photos at top. Multiple nuclear genes show that genotypes differ between habitats strongly in *Holbrookia maculata* and less so in *Sceloporus undulatus*. Each bar represents the genotype of one individual. The proportion of a bar that is green shows the probability, based on a lizard's genotype, that the individual belongs to a distinct white sand population, and the proportion that is blue that it belongs to a distinct dark soil population. Bars with intermediate amounts of green and blue indicate that the individual has a mixed genotype. (After [86].)

of mating behavior showed that partial sexual isolation has evolved in at least one of these species [34]. The level of genetic differentiation between the populations on white sands compared with those on normal soils is lowest in the whiptail lizard (*Aspidoscelis inornata*), the species that is most active and shows the highest level of gene flow among populations.

The Genomics of Speciation

As in all other fields in biology, the genomics revolution has opened up new perspectives on speciation. Genomics can help us determine the number, identities, and genomic locations of **speciation genes**, the loci that contributed to the evolution of reproductive isolation [64, 65].

When two populations or species are partly isolated but continue to hybridize, alleles introgress between them. The introgression, however, is expected to be uneven across the genome. In parts of the genome that are evolving neutrally, gene flow between the populations will tend to make them more similar. In addition, if a mutation appears in one species that is beneficial to both species, it will sweep though both and homogenize that region of the genome between the species. Earlier in this chapter, we saw that genes for warning colors in *Heliconius* butterflies have spread among species this way (see Figure 9.5) [35].

A contrasting picture is expected in genomic regions that carry loci that isolate hybridizing species. Those regions, sometimes called "genomic islands of speciation," are expected to show greater genetic divergence than the rest of the genome. The group of mosquitoes that transmit malaria in Africa (genus *Anopheles*) includes six species that diverged recently and that continue to hybridize [26]. Comparisons of the genomes reveal some regions (particularly inversions) that are similar in some species pairs because they have introgressed following hybridization, like the color pattern genes in *Heliconius* butterflies. But other regions of the *Anopheles* genome show unusually high divergence among species—the signature of speciation genes. In particular, the X chromosome is more genetically different than the autosomes. It has deep gene trees (see Chapter 7), which strongly suggests that genes on the X chromosome were among the first to contribute to reproductive isolation during the speciation process (see Figure 16.11). That pattern is consistent with the observation that sex chromosomes often play a disproportionately large role in the evolution of reproductive isolation [15].

Genomic islands of speciation might provide evidence about the geography of speciation, but this is still uncertain. Genomic islands are predicted to develop during speciation with gene flow. They will also appear when two populations that speciated in allopatry come back into contact and then hybridize [19]. Allopatric populations will show different genomic "islands" at sites where divergent natural selection fixed different alleles that may or may not make a potential contribution to reproductive isolation. If these populations expand and hybridize, divergent selection will maintain the adaptive differences between the hybridizing populations, while neutral regions introgress and become homogenized between the populations. Distinguishing speciation with gene flow from secondary contact presents the same difficulty as determining whether sympatric species originated sympatrically or became sympatric by secondary contact after speciation happened allopatrically. Deciding between those possibilities requires additional evidence.

SUMMARY

- Several definitions of "species" have been proposed. Most evolutionary biologists use the biological species concept, which defines species as groups of actually or potentially interbreeding organisms that are reproductively isolated from other such groups, meaning they do not (or would not) exchange genes even if they encounter each other. Under this definition, speciation is the evolution of reproductive isolation. Some other biologists favor the phylogenetic species concept, according to which species are sets of populations with character states that distinguish them.

- Under any definition of species, the defining qualities (such as reproductive isolation) usually evolve gradually, so some populations cannot be clearly classified as the same or different species.

- The biological differences that constitute reproductive isolation include prezygotic barriers to gene exchange (e.g., ecological or sexual isolation) and postzygotic barriers (hybrid inviability or sterility). Several potential isolating barriers may be discovered between two species. Some of them may have evolved before the others, and been the actual basis of speciation. Some barriers (e.g., postzygotic barriers) may not come into play because an earlier-acting difference already prevents gene exchange.

- Speciation is rapid in some cases, requiring only a few thousand years or even less. Partial reproductive isolation has evolved even in laboratory populations. Occasionally, a new species is generated instantly by whole genome duplication. In other cases, it may take millions of years for populations to evolve reproductive isolation.

- The causes of the evolution of prezygotic reproductive isolation include divergent natural selection arising from ecological factors (ecological speciation) and divergent sexual selection. When hybrids between two divergent populations have low fitness, there is selection for stronger prezygotic isolation, which may result in reinforcement of a prezygotic barrier.

- The causes of evolution of postzygotic isolation are less well understood. Hybrid inviability and sterility are often based on incompatible interactions among two or more genetic loci that diverged between populations by genetic conflict or divergent selection. Hybrid sterility can also be caused by differences in the numbers or arrangements of chromosomes. In some cases, these chromosome differences may have been established by random genetic drift.

- New species sometimes evolve from hybrids between parent species. In many cases, the hybrid species is polyploid.

- Evolutionary biologists agree that allopatric speciation is common. Here a physical barrier separates populations of an ancestral species, and evolutionary changes in one or both populations result in biological barriers to gene flow if the populations come back into contact. One possible mode of allopatric speciation, peripatric or founder effect speciation, is thought to be initiated by genetic drift in a small local population of an ancestral species. This is generally thought to be rare.

- In speciation with gene flow, a species evolves into two species because of strong divergent selection, without a physical barrier between populations. The evolution of reproductive isolation is hindered by ongoing interbreeding (which maintains gene exchange) and recombination (which opposes the buildup of divergent sets of genes and characteristics).

- Sympatric speciation is the evolution of reproductive isolation within an initially randomly mating population. It is the most extreme instance of speciation with gene flow, and requires very strong selection. It is made more likely if traits that are disruptively selected because of their ecological function also automatically reduce gene exchange (e.g., seasonal timing of reproduction).

TERMS AND CONCEPTS

allopatric
allopatric speciation
biological species
    concept (BSC)
dispersal
Dobzhansky-Muller
    incompatibility
    (DMI)

ecological isolation
ecological
    speciation
epistasis
founder effect
genetic conflict
Haldane's rule

hybrid zone
introgression
    (introgressive
    hybridization)
inversion
isolating mechanism
parapatric

parapatric
    speciation
peripatric speciation
phylogenetic
    species concept
    (PSC)
postzygotic barrier
prezygotic barrier

reciprocal translocation	reproductive isolation	sister species	sympatric
reinforcement	secondary contact	speciation	sympatric speciation
reproductive isolating barrier (RIB)	segregation distortion	speciation gene (and speciation trait)	vicariance
	sexual isolation	speciation with gene flow	

SUGGESTIONS FOR FURTHER READING

Speciation, by J. A. Coyne and H. A. Orr (Sinauer Associates, Sunderland, MA, 2004) is the most recent comprehensive book about speciation.

Ecological Speciation, by P. Nosil (Oxford University Press, New York, 2010) is a short but comprehensive treatment of speciation caused by divergent environmental selection.

The Ecology of Adaptive Radiation, by D. Schluter (Oxford University Press, New York, 2000) discusses ecological aspects of speciation and its relationship to adaptive radiation.

Useful overviews and reviews include:

Baack, E., M. C. Melo, L. H. Rieseberg, and D. Ortiz-Barrientos. 2015. The origins of reproductive isolation in plants. *New Phytol.* 207: 968–984.

Bolnick, D. I., and B. M. Fitzpatrick. 2007. Sympatric speciation. *Annu. Rev. Ecol. Evol. Syst.* 38: 459–487.

Harrison, R. G. 2010. Understanding the origin of species: Where have we been? Where are we going? In M. A. Bell, D. J. Futuyma, W. F. Eanes, and J. S. Levinton (eds.), *Evolution Since Darwin: The First 150 Years*, pp. 319–346. Sinauer, Sunderland, MA.

Kirkpatrick, M., and V. Ravigné. 2002. Speciation by natural and sexual selection: models and experiments. *Am. Nat.* 159: S22–S35.

Marie Curie Speciation Network. 2012. What do we need to know about speciation? *Trends Ecol. Evol.* 27: 27–39.

Price, T. D. 2008. *Speciation in Birds*. Roberts and Co., Greenwood Village, CO.

Rieseberg, L. H., and J. H. Willis. 2007. Plant speciation. *Science* 317: 910–914.

Seehausen, O., R. K. Butlin, I Keller, and 5 others. 2014. Genomics and the origin of species. *Nature Rev. Genet.* 15: 176–192.

PROBLEMS AND DISCUSSION TOPICS

1. Some degree of genetic exchange occurs in bacteria, which reproduce mostly asexually. What evolutionary factors should be considered in debating whether or not the biological species concept (BSC) can be applied to bacteria?

2. Suppose the phylogenetic species concept (PSC) were preferred over other species concepts, such as the BSC. What would be the implications for (a) discourse on the evolutionary mechanisms of speciation; (b) studies of species diversity in ecological communities; (c) estimates of species diversity on a worldwide basis; and (d) conservation practices under such legal frameworks as the U.S. Endangered Species Act?

3. How might the fate of two hybridizing populations—that is, whether or not they persist as distinct populations—depend on the kinds of isolating barriers that reduce gene exchange between them?

4. The heritability of an animal's preference for different habitats or host plants might be high or low. How might heritability affect the likelihood of sympatric speciation by divergence in habitat or host preference?

5. Three-spined sticklebacks that have colonized freshwater streams and lakes have repeatedly evolved into similar forms. The pattern can be seen in hundreds of freshwater bodies around the world. Can a single biological species arise more than once (i.e., polyphyletically)? How might this possibility depend on the nature of the reproductive barrier between such a species and its closest relative?

6. If a researcher discovers regions of genome in a set of hybridizing populations that have much higher F_{ST} than the rest of the genome, what alternative hypotheses must be considered before concluding that these regions are "genomic islands of speciation"? How might a

UNIT III
Products of Evolution: What Natural Selection Has Wrought

All About Sex

Many of nature's greatest wonders are about sex. It is impossible not to be astonished by the mating display of a male peacock. Plants are equally remarkable: the flowers of an orchid are to the plant what the peacock's feathers are to the bird.

Animals and plants reproduce in a staggering diversity of ways (**FIGURE 10.1**). Many species are **hermaphroditic**: each individual has both male and female gonads, and mating involves exchanging eggs and sperm (in animals) or mutual pollination (in plants). When a pair of leopard slugs mate, their intimate moment begins as they hang together from a long thread of mucus (see Figure 10.1A). Each of them extrudes its penis, and the two fertilize each other's eggs simultaneously. Some hermaphrodites, like the cactus shown in Figure 10.1C, dispense entirely with the complications of mating by simply fertilizing themselves.

The slipper shell is a snail with the provocative name *Crepidula fornicata* that lives in stacks of several individuals (see Figure 11.14C). A young snail settles on the top of a stack and matures into a male. All the snails beneath him in the stack are also male except the one at the very bottom, which is female. When she dies, the male immediately above her, which is the largest and oldest male in the stack, changes sex to become the new female. Other species that switch sex include the famous clownfishes (see Figure 10.1B). Like the slipper shell, clownfishes live in groups that consist of a single female and several males. When the female dies, the most dominant male changes sex to become the group's female.

Courtship displays provide some of the most spectacular and unexpected sights in nature. This is a male yellow-crowned night heron (*Nyctanassa violacea*).

FIGURE 10.1 Sexual reproduction is remarkably diverse. (A) The leopard slug (*Limax maximus*) is a hermaphrodite. Mating pairs hang from a mucus thread while copulating. (B) Clownfishes (genus *Amphiprion*) are sequential hermaphrodites that first mature as male and later change sex to female. (C) The cactus *Epithelantha micromeris* has flowers that do not open. It is one of the few species that reproduce almost entirely by self-fertilization. (D) Bacteria, such as these *E. coli*, use conjugation and other mechanisms to exchange genetic material.

Even bacteria and viruses have sex of a sort. Although they lack the form of sex based on meiosis that is familiar from animals and plants, they exchange genes using a variety of mechanisms that together are called "parasexuality" [41]. Some bacteria pass DNA between each other by conjugation (see Figure 10.1D). Others can absorb naked DNA left in the environment by bacteria that have died. When several viral particles infect the same host, progeny viruses are produced that carry mixtures of their genotypes. These different processes result in the genetic mixing that is the fundamental evolutionary effect of sex.

Flowering plants cannot go in search of partners. Many enlist help from animals such as bees, bats, and birds that pollinate the plants while visiting their flowers for nectar. Orchids have evolved elaborate features to attract their pollinators, in some cases exploiting those animals' own sexual instincts (see Figure 3.23). Euglossine bees turn the tables on the orchids: males collect the flowers' scents and use them as their own sex pheromones for courting female bees (**FIGURE 10.2**).

Our group of animals, the primates, has its own set of sexual curiosities. Different species have penises that vary dramatically in shape and size. Several have spines. In fact, our recent evolutionary ancestor had a spiny penis. That can be inferred from a chromosome deletion found in humans but not other primates that removes a regulatory element needed for development of the spines [31]. Across the animal kingdom, male genitalia are among the most rapidly evolving kinds of traits [15].

These examples immediately inspire questions. Why do males and females often look so different? Why do organisms have such diverse ways of mating? And why do some species, such as the common dandelion, forego sex altogether?

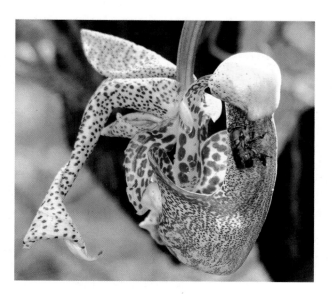

FIGURE 10.2 Many animals include chemical signals called sex pheromones to attract and court mates. In tropical America, male orchid bees (tribe Euglossini) collect scent from certain flowers and use it as their sex pheromone. Here a brilliantly colored male *Euglossa igniventris* collects scent from a *Coryanthes* orchid flower in Panama. (Courtesy of David Roubik.)

What Are Females and Males?

In plants, animals, and some other eukaryotes, the gametes of a species come in two sizes. One, the egg, is large and immobile. The other gamete, the sperm, is small and mobile, either actively or passively. This dimorphism in gamete size, which is called **anisogamy**, likely evolved because there are two ways that a gamete can have high fitness. It can succeed by being large and well provisioned, which inhibits movement. Alternatively, it can succeed by being small and mobile, which enhances its ability to find and fertilize a large immobile gamete. This difference in size leads to a difference in numbers: females typically make far fewer eggs (or ovules) than males make sperm (or pollen). The difference in gamete size is so basic that we use it to define the two sexes. Males are the sex that makes small gametes, and females the sex that makes large ones.

As you saw earlier, some species are hermaphroditic, while others have separate sexes. Ecological conditions can favor one or the other of these mating systems. An important factor favoring hermaphroditism is **reproductive assurance**, which is the increased chance of successful reproduction when potential mates are rare (or even absent). When finding a reproductive partner is difficult, there is an advantage for individuals to be simultaneously male and female. This condition allows them to mate with any other individual they encounter, or (in some species) even with themselves. The need for reproductive assurance is one reason why plants, which cannot actively search for a mate, are hermaphroditic more often than animals.

In species with separate sexes, a variety of mechanisms determine which embryos develop into females and which into males (**FIGURE 10.3**) [4]. In humans and other mammals, sex determination is genetic: females have two X chromosomes, while males have one X and one Y. In birds, the sex chromosomes are reversed: it is the females that have two different sex chromosomes (called the Z and W chromosomes) and the males that are homozygous (with two Z chromosomes). In yet other groups, sex is determined not by chromosomes but by the physical or social environment [4]. Cool temperatures cause the eggs of many reptiles to develop into males, while in warm temperatures they develop into females [23].

Males and females of some animals are so similar that the only way they can be distinguished is by their genitalia. But in other species, including humans, many traits are **sexually dimorphic**, meaning they are expressed differently in males and females. One of the most spectacular examples of sexual dimorphisms is found in

FIGURE 10.3 Animals have a variety of mechanisms that determine an individual's sex. The pie diagrams show the fraction of species in vertebrates and arthropods in which different mechanisms are used. The XY system of sex determination is used in all mammals but is not very common among other vertebrate groups. The most common sexual system in teleost fishes is hermaphroditism. All Hymenoptera (ants, bees, wasps, and their relatives) and a large fraction of Acari (mites) determine sex by haplodiploidy. (After [4].)

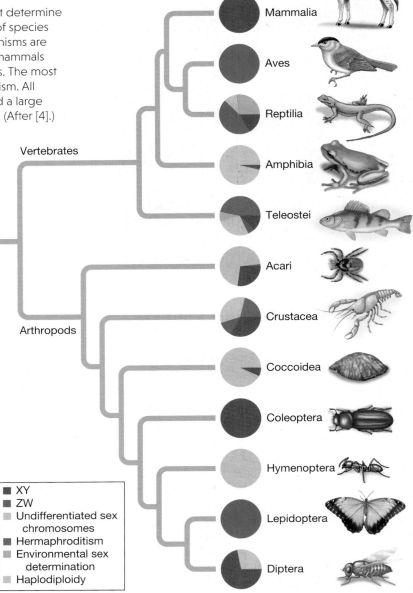

fishes that live deep in the world's oceans [39]. A female anglerfish attracts her prey with a bioluminescent lure that she dangles in front of her enormous mouth. But even more bizarre are the small growths attached to the female's body (**FIGURE 10.4**). These are male anglerfish. When young, a male swims freely until he finds a mature female. He bites onto whatever part of her he can, and then never lets go. Their circulatory systems fuse, and almost all of the male's organs—eyes, gills, digestive system—degenerate. One organ, however, grows much larger: when the metamorphosis is complete, the male's body is almost entirely filled by his testes. To mate, the female releases her eggs and the male his sperm, which fertilize the eggs in the surrounding water.

Sexual dimorphism can evolve when selection favors the expression of a trait to be different in males and females. This situation is called *sexually antagonistic selection*, because an increase in the trait's expression benefits one sex but harms the other. Genetic constraints (see Chapter 6) sometimes prevent a trait from evolving sexual dimorphism, and so the fitnesses of both sexes are permanently compromised.

FIGURE 10.4 Anglerfishes (suborder Ceratioidei) have extreme sexual dimorphism. When a male finds a female, he bites and attaches himself permanently to her. His body fuses with hers and his organs largely degenerate except for his testes, which grow to fill most of his body. This female has two males attached to her (indicated by the arrows).

Sexual Selection

It is easy to understand how some traits have evolved to become sexually dimorphic. An individual without gonads will leave no genes to the next generation. Gonads and genitalia are called **primary sexual traits**. But Darwin pointed out that other kinds of traits, such as the male peacock's spectacular train of feathers, are more difficult to explain. These are **secondary sexual traits**, which differ between the sexes but do not play a direct role in reproduction. A male peacock's spectacular feathers make flight more difficult and attract predators, so we might expect them never to have evolved by natural selection. Interestingly, secondary sexual traits are often among the most rapidly evolving phenotypic characters. In many groups of animals and plants, they are the only traits that can reliably distinguish species.

Darwin reasoned that secondary sexual traits must evolve by something other than selection for survival and the production of gametes. He made several key observations. Traits with extreme sexual dimorphism tend to be exaggerated much more often in males than in females. Often these traits are not expressed in immature males. Most important, these traits are used by males in the mating season when they interact aggressively with other males and court females.

These observations lead Darwin to propose that these traits have evolved by **sexual selection**, which is selection caused by competition for mates among individuals of the same sex. A male that prevents other males from mating will leave more copies of his genes to the next generation than they will. Likewise, a male that attracts many females to mate with him will be more genetically successful than a male that attracts few or none. Sexual selection can cause the evolution of traits that decrease survival if the reproductive advantage they produce compensates for that cost. In short, a trait can evolve by sexual selection if it increases a male's overall fitness, even if it decreases survival.

Sexual selection is one of Darwin's most ingenious ideas. He developed it without the benefit of any data showing whether these traits in fact do decrease survival but increase male reproductive success. Was Darwin right? Many studies since his time have verified that secondary sexual traits have exactly these effects.

A simple and elegant experiment with the long-tailed widowbird (*Euplectes progne*) shows that the male's extremely long tail attracts females [1]. Males establish territories on the African savannah, and females nest on the territory of the male that attracts them. Males were captured on their territories and divided into four groups. The tail feathers of the first group were cut to half of their original

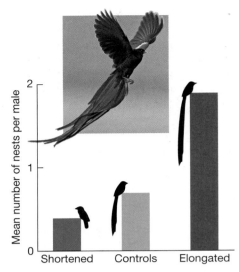

FIGURE 10.5 An experimental test of Darwin's hypothesis that secondary sexual traits increase mating success. Male long-tailed widowbirds (*Euplectes progne*) have extremely long tail feathers. Tails were artificially shortened in one group and lengthened in another. There were two control treatments: no manipulation was done in one group of males, and tails were cut and then reattached in the other group. Results from the two controls were not significantly different, so they are shown together here. Mating success was measured by the number of females nesting on each male's territory. Males with lengthened tails had the greatest success, as predicted by the hypothesis. (After [1].)

length. The clipped feathers were glued onto the tails of males in the second group, increasing their tail length by 50 percent. The third and fourth groups were controls: in one there was no manipulation, and in the other the tails were clipped and then glued back on. All the males were then released, and their mating success was measured by the number of females that nested on their territories. The results were striking. Males with lengthened tails attracted more than twice as many females as males with shortened tails and the control males, showing that long tails are favored by sexual selection on males (**FIGURE 10.5**). This experiment confirms Darwin's logic: exaggerated secondary sexual traits are favored by sexual selection because they increase male mating success.

The second part of Darwin's hypothesis is that exaggerated secondary sexual traits decrease survival. While the hypothesis hasn't been tested with widowbirds, it has been with many other species. One is the túngara frog (*Physalaemus pustulosus*) [36]. Males spend the night crowded together in ponds and puddles, where they attract females with calls that are reminiscent of some 1980s video game. Some males make a simple call known as a whine. Others make more complex calls by adding one or more "chucks" to the end of the whine. Experiments using speakers that play calls with different numbers of chucks show that females are attracted to calls with more chucks. This preference results because the chuck's acoustic frequencies match those to which the female's inner ear is most sensitive.

This presents a puzzle: why don't males always make calls with chucks? The answer is that a male's call attracts more than just females. Many bats are nocturnal predators that can use echolocation (much like sonar) to find their prey. But some bats simply listen for sounds made by their prey. The fringe-lipped bat (*Trachops cirrhosus*) preys on male túngara frogs that it finds using their calls. Experiments show that chucks decrease the survival of males because they make the male easier for the bat to localize (**FIGURE 10.6**). When bats are presented with two speakers, one playing a frog call with chucks and the other with just a simple whine, they more often attack the speaker playing the call with chucks.

Secondary sexual traits experience a tug-of-war between their effects on two fitness components: survival and mating success (**FIGURE 10.7**). These traits evolve to a compromise that maximizes a male's

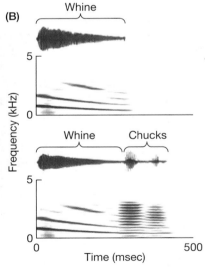

FIGURE 10.6 Calls of the male túngara frog (*Physalaemus pustulosus*) attract females for mating, but they also attract predators. (A) A male is attacked by a fringe-lipped bat (*Trachops cirrhosus*) that has located the frog by his calls. (B) This sonogram of a male's call is a plot of frequency against time. Some males make calls that consist only of a "whine" component, while others add one or more "chuck" components. The chuck is attractive to females, but also more conspicuous to the bats. (B from [36].)

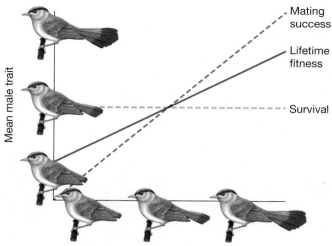

Mean female mating preference

FIGURE 10.7 Selection favors mating displays that maximize a male's lifetime fitness, which is a compromise between what maximizes survival and what maximizes mating success. In this schematic, male survival is maximized by a tail length that optimizes aerodynamics, resulting in stabilizing selection for tails of intermediate length. Male mating success depends on the mean female mating preference (on the *x*-axis). If most females prefer long tails, then males with long tails will have the greatest mating success (at right). Tails will evolve to an equilibrium that is longer than what maximizes survival. If females prefer very short tails (at left), tails will evolve to a length shorter than what maximizes survival. (After [27].)

total lifetime fitness. That optimal value of the trait shifts if the strength of natural or sexual selection changes. Increased predation, for example, can select for decreased male traits. Male crickets (*Teleogryllus oceanicus*) scrape together their wings to make a song that attracts females. The song also attracts a fly (*Ormia ochracea*) that parasitizes the crickets. After the arrival of the fly on the island of Kauai in the Hawaiian Islands, the crickets evolved modified wings that do not produce song after only 20 generations [51].

Attracting females is one way that males increase their mating success. Darwin pointed out there is a second way: males can directly interfere with other males' access to females. We will explore these two modes of sexual selection shortly. But before doing that, we will first consider the basic question of why sexual selection acts more frequently on males than females.

Why are males sexually selected?

Widowbirds and túngara frogs illustrate a very general pattern. In most cases, it is the male's secondary sexual traits that are exaggerated. This shows that sexual selection is more common and intense on males than females. Why should that be?

The solution to this puzzle comes from considering the fundamental differences in the reproductive biology of males and females [3]. Because a male makes a large number of sperm, he is often capable of fertilizing a large number of females. A trait that increases the number of mates that he can acquire is favored by selection and so will spread. This creates the opportunity for selection on traits that increase male mating success. In contrast, a female can often fertilize all her eggs with a single mating. A trait that increases the number of mates she acquires therefore has no fitness advantage.

This logic is called Bateman's principle in honor of the geneticist who demonstrated it experimentally with *Drosophila melanogaster* [7]. Bateman put several males and females together in bottles for several days. These individuals carried genetic markers that allowed Bateman to determine how many of the offspring that later hatched were produced by each parent. He concluded that the number of offspring sired by a male increased in proportion to the number of females he mated. In contrast, the number of offspring produced by a female did not increase with the number of males she mated. Bateman also observed that there was greater variance in reproductive success among males than among females. That suggests there is greater opportunity for sexual selection on males than on females.

FIGURE 10.8 Sex role reversal. (A) Two female red phalaropes (*Phalaropus fulicarius*) fight over the smaller, duller-plumaged male on their breeding ground. In contrast to most birds, female phalaropes court males, which care for the eggs and young in their nests. (B) A male Australian seahorse (*Hippocampus breviceps*) giving birth from his pouch. Males choose which courting females will lay eggs in their pouches.

(A)

(B)

The outcome of sexual selection is strongly influenced by the **operational sex ratio**, which is the relative number of males and females available to mate at any moment [16]. Many females in a population are unavailable because they are developing new eggs, carrying embryos to term, or caring for young. Not so for males, who are often able to remate shortly after their last reproductive bout. Consequently, *females are often a limiting resource for males*. Whenever the number of males and females is unequal, the more common sex must compete for access to the less common sex. Traits that make males more successful in competition for mates are therefore favored by sexual selection, leading to spectacular evolutionary outcomes such as the long tail of the widowbird. There is potential for sexual selection on males even in species that appear to be monogamous. Among passerine birds that are socially monogamous, it is not unusual for more than 10 percent of offspring to be fathered by males that are not the female's social partner (her apparent mate), and some males are particularly sought after as secondary sexual partners [50].

Males are not always the sexually selected sex. The red phalarope (*Phalaropus fulicarius*) shows **sex role reversal**: females are larger and more brightly colored than the males (**FIGURE 10.8A**). After mating, a female lays her eggs in the nest of a male, who is responsible for all of the incubation and rearing of the chicks. While he is preoccupied with those duties, the female can mate with another male and lay more eggs in his nest. Males, by contrast, get no fitness benefit from mating with additional females once their nest is filled with eggs. A similar situation occurs in seahorses. Males have a pouch in which they carry and nurture their developing young, and males are courted by females who seek to lay their eggs in the pouches (**FIGURE 10.8B**). Species with sex role reversal are the exceptions that prove the rule about why sexual selection is more intense on one sex than the other. In these species, Bateman's principle is reversed: mating with more partners increases the fitness of females but not males. Females have therefore evolved bright colors, courtship behaviors, and other traits that increase their mating success.

Sexual selection by male-male competition

The first of two modes of sexual selection identified by Darwin is **male-male competition**, in which males interfere directly with each other. In taxa ranging from beetles to whales, males have evolved horns and other structures to prevent other males from mating. Males of the red deer (*Cervus elaphus*) carry a magnificent set of antlers in the breeding season (**FIGURE 10.9A**). These antlers are no mere ornaments: males use them in fights that are dangerous. About 20 percent of males sustain permanent

(A)

(B)

FIGURE 10.9 Males of many animals have evolved horns and other weapons that they use to fight with each other for reproductive access to females. (A) Male red deer (*Cervus elaphus*) fighting during the breeding season. Successful males guard groups of females with which they mate. (B) Male stag beetles (*Lucanus cervus*) fighting. Males defend the sites where females lay their eggs, and mate with females when they arrive to lay.

injuries, for example blindness in an eye, from fighting. But the payoffs are big for the winners. While nearly half of all males fail to reproduce in a given year, the most successful males sire up to five offspring [12].

Fights like those in the red deer are called **male combat**. Male red deer follow groups of females to prevent other males from mating with them. In other species, such as the stag beetle (*Lucanus cervus*; **FIGURE 10.9B**), males defend critical resources needed by females (for example, sites where they lay eggs or feed), and then mate with females as they arrive. Ecological factors can have an important impact on the potential for sexual selection. When females need resources that are clumped in time or space, there is opportunity for a few males to control the resources and obtain the majority of matings, leading to strong sexual selection [16]. When resources are dispersed, there is less opportunity for males to control them and gain a strong sexual selection advantage.

Just as in sports, there can be more than one winning strategy in male-male competition. This situation opens the door to the evolution of **alternative mating strategies**, which are divergent ways that males of the same species use to acquire matings [33]. Isopods are crustaceans distantly related to crabs. Males of a species called *Paracerceis sculpta* have three genetically determined morphs (**FIGURE 10.10**) [46]. The largest is the alpha morph. These males dominate the other morphs and guard several females in a harem. Males of the beta morph are very similar in size and shape to females. By mimicking females, beta males can slip past alpha males, gain access to the harem, and mate with the real females. The gamma morph is by far the smallest. These males use stealth and speed to sneak past alpha males and gain access to females. Despite their striking differences in size, shape, and behavior, the three morphs have equal reproductive success. This is an example of a polymorphism maintained by negative frequency-dependent selection (see Chapter 5).

A common tactic used by males to interfere with each other's reproduction is **sperm competition** [37]. Even after a male has mated with a

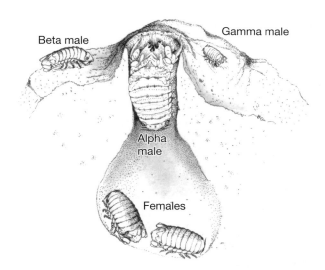

FIGURE 10.10 These isopods (*Paracerceis sculpta*) live on and inside sponges. The large alpha male guards a harem of females by blocking the entrance to the cavity in which they live. A medium-sized beta male (left) and a small gamma male (right) try to gain access to the females by tricking the alpha male. The beta male imitates a female, while the small gamma male seeks to reach the females by slipping past the alpha male. (From [45], illustration by Marco Leon.)

FIGURE 10.11 A male damselfly removes the sperm of a female's previous mates before depositing his own. (A) In this photo of mating damselflies (*Calopteryx splendens*), the blue individual is the male and the gray individual the female. When they mate, claspers on the end of the male's abdomen grasp the female behind her head. She then curls her abdomen so that the opening of her oviduct (at the end of her abdomen) makes contact with his penis (on his thorax, near the base of his wings). They then copulate. (B) A scoop-like structure on the male's penis removes the sperm of males that copulated with the female earlier. A clump of sperm adheres to the structure, at bottom. (B courtesy of J. Waage.)

(A)

(B)
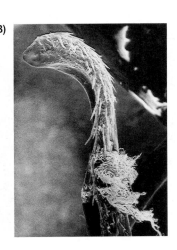

female, there are opportunities for other males to prevent his sperm from fertilizing her. The penis of *Calopteryx* damselflies has a structure that resembles a very small rake or scoop (**FIGURE 10.11**). It plays no role in helping transfer sperm to the female. Its job is to remove the sperm of any males that mated previously with the female. By removing those sperm, the male increases the number of eggs he will fertilize, and decreases the fitness of the other males. This increases the male's relative fitness, and so the structure is favored by sexual selection.

A male can also win at sperm competition by making it more difficult for later males to fertilize his mate's eggs. One strategy is to guard the female against other suitors until she has laid the eggs he fertilized. Another is to deposit in her a sperm plug, which is a physical barrier to later matings. A third strategy to win at sperm competition is simply to make more sperm. Females of some primate species mate with many males, leading to strong sperm competition. Males of those species have evolved large testes to produce copious sperm. As you would predict, species in which females mate with only one or a few males have smaller testes (see Figure 3.20) [21]. The human testes are slightly smaller than average for primates our size, suggesting that our ancestors may have had monogamous tendencies.

A particularly gruesome kind of male-male competition results in **infanticide**. Lions are social animals that live in groups. Occasionally, the dominant male in a group will be ousted by one or more rivals. The new males often kill all the young in the group (**FIGURE 10.12**) [35]. This decreases the number of offspring left to the next generation by previous males, which increases the relative fitness of the new males. Further, it causes the females in the group to become fertile more quickly (since females are not receptive while they are caring for their young). This again enhances the fitness of the new males since they can now mate and have their own offspring.

Male combat, sperm competition, and infanticide are just three of the diverse ways by which males interfere with each other's reproduction. A simple theme unites them all: any mutation or trait that increases the fitness of a male relative to other males is favored. Sexual selection gives an evolutionary advantage to selfish genes, even if they have negative effects on other males, on females, or the species as a whole. Infanticide is a particularly graphic example. We will return to these kinds of sexually antagonistic traits in Chapter 12.

FIGURE 10.12 This male lion has killed a cub after displacing the cub's father and other adult males in the group that he recently joined. Male infanticide has also been described in many other species of mammals.

Sexual selection by female choice

Many extravagant male secondary sexual traits are useless for male-male competition. Instead, their function is to attract females and persuade them to mate. This second mode of sexual selection is called female choice.

Do female animals in fact have the ability to make choices? Many species that show evidence of sexual selection by female choice have very simple nervous systems that are not capable of higher levels of cognition. In the context of sexual selection, the word "choice" is used much more broadly to mean *any phenotype of the female that biases the type of male that she will mate*. A female's mate choice can, for example, be influenced by the colors to which her eye is most sensitive. Experiments like that with the long-tailed widowbirds have confirmed in many species that females do bias their mate choice, so there is no doubt that sexual selection by female choice occurs.

Furthermore, mating preferences can evolve. Females of closely related species have innate preferences for their own species, which shows that those preferences have evolved since those species' most recent common ancestor. A second line of evidence comes from heritability experiments (discussed in Chapter 6) that have established that there is genetic variation for mating preferences in many species [24].

How and why do female mating preferences evolve? When choosing a partner, the biggest mistake a female can make is to mate with another species. Hybrid offspring typically suffer from low viability or fertility. This generates strong selection for the reinforcement of female mating preferences that discriminate against males of other species (see Chapter 9). Other evolutionary factors also cause mating preferences to evolve. Here we look at four of them: direct benefits, pleiotropic effects, good genes, and Fisher's runaway process.

Males of many species provide their mates with **direct benefits**, which are resources that increase the females' survival and reproductive success. Direct benefits come in many forms, including food and care for the offspring. When crickets and katydids mate, the male inseminates the female with a large spermatophore made up of lipids and carbohydrates as well as sperm (**FIGURE 10.13**) [19]. After insemination, the female eats the spermatophore, and those nutrients increase the number of eggs she lays. Since larger males make larger and more nutritious spermatophores, natural selection favors female mating preferences for large males. In some species, the spermatophore has become so large that females can obtain all of their food simply by mating. In some insects and spiders, the nutritional gifts used to seduce females have evolved to the ultimate extreme. After copulation, male redback spiders (*Latrodectus hasselti*) often somersault into their female's mouth and are eaten! In addition to benefiting his mate, a male's suicide enhances his own fitness: males that sacrifice themselves fertilize more eggs than those that do not [2]. Whenever males provide their mates with direct benefits, natural selection favors female preferences for male traits that increase female fitness.

Direct benefits cannot, however, explain the evolution of female mating preferences for some of the most extreme male mating displays in the animal kingdom. In some birds, fishes, insects, and other taxa, males congregate in **leks**, which are arenas in which males do

FIGURE 10.13 Males of some species provide direct benefits to females. This female Mormon cricket (*Anabrus simplex*) will eat the large white spermatophore her mate has placed in her genital opening. (Courtesy of John Alcock.)

FIGURE 10.14 Males of the Andean cock-of-the-rock (*Rupicola peruvianus*) perform mating displays together on leks. Females visit the lek, mate with the male of their choice, and then leave to raise the offspring alone. There is no opportunity for direct benefits to females in lek-breeding species.

nothing but make sexual displays (**FIGURE 10.14**). Their complex courtship display shows off the males' elaborate plumage. Females visit the lek, mate with the male of their choice, and then leave to rear the offspring with no help from their mate. In lekking species, the females receive no direct benefit whatsoever from the males. There must be some other cause for the evolution of female preferences for these elaborate male displays.

One such cause is natural selection acting on **pleiotropic effects** of genes that affect female mate choice [27]. Recall from Chapter 4 that virtually all genes have pleiotropic effects, meaning they affect multiple traits. An allele that changes a female's mating preference will typically also change other things about her. The guppy (*Poecilia reticulata*) is a small tropical freshwater fish with brightly colored males. Females have mating preferences for males with more orange on their body (see Figure 6.11). Experiments show that both females and males are attracted to small orange discs, which they peck at as if to eat [43]. A plausible hypothesis is that guppies evolved an attraction to round orange objects because they feed on orange fruit, and as a side effect females are now attracted to orange males.

Mating preferences that evolved by selection on pleiotropic effects are called **perceptual biases** [44]. These biases are sometimes surprising: experiments with fishes, frogs, and birds show that females of some species are attracted to signals that males of their own species do not even make (**FIGURE 10.15**). In these cases, it appears that the female mating preferences are side effects of features of the sensory system that evolved for reasons unrelated to mating, before the male signals were even present.

Mating preferences can arise as side effects of how the courtship signals interact with the environment. As signals propagate, they are filtered by the environment they pass through. Colors are transmitted differentially through water depending on depth and the amount of sediment in the water, for example, and sounds with different frequencies attenuate at different rates in open habitat and forest. As a result, different mating displays can be favored by sexual selection depending on the habitat where they are performed [44].

A third way that female mating preferences evolve is called the **good genes** mechanism. Some male displays are correlated with traits that increase lifetime

FIGURE 10.15 Some female mating preferences are perceptual biases that evolved before the origin of the male trait on which they act. Males of swordtails (such as this green swordtail, *Xiphophorus helleri*) have dramatic swords that attract females. Males of a closely related species (*Priapella olmecae*) lack the sword, but the females prefer males with a sword that has been surgically added [6]. The phylogeny of these fishes (in green) shows that the preference for the sword evolved before the sword itself did. (*Xiphophorus* photo courtesy of Alexandra Basolo.)

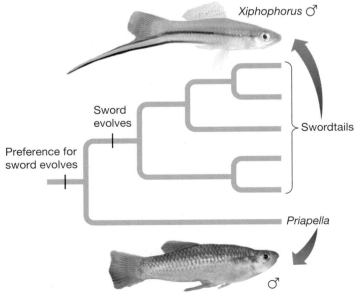
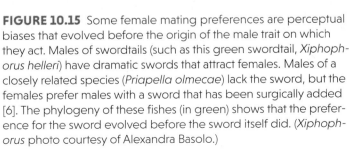

fitness. These kinds of displays are called "indicator traits" because they indicate to females a male's genetic quality. For example, males with alleles that make them good at foraging and resisting parasites may have more energy to grow a long tail. A female that chooses a long-tailed male is therefore mating with a male that also carries good genes for foraging and immunity. The offspring inherit their mothers' preference for long tails and their fathers' good genes. (That is, the preference loci and the good genes loci are in linkage disequilibrium; see Chapter 5.) As natural selection causes the good genes to spread, it also causes preferences for long tails to spread as well.

Evidence for the good genes mechanism comes from experiments with three-spined stickleback fish (*Gasterosteus aculeatus*). Barber and colleagues studied a character preferred by females: intense red coloration of the male's belly [5]. They found that young fish with bright red fathers were more resistant to infection by tapeworms than were their half-siblings that had dull red fathers (**FIGURE 10.16**). The red coloration is based on carotenoid pigments, which are obtained from food and appear to enhance development of an effective immune system. The implication of this result is that female preferences for red males will become correlated with alleles that strengthen the immune system, and stronger preferences will evolve as a side effect of natural selection on immunity.

A final mechanism for the evolution of mating preferences is called **Fisher's runaway**, after the pioneering population geneticist R. A. Fisher (see Chapter 1). Females with preferences for long tails tend to mate with males that have a long tail, and so their offspring tend to have the genes both for a long tail and the preference for a long tail. (That is, the genes for the male trait and the female preference are in linkage disequilibrium.) If the combined forces of natural and sexual selection favor longer tails, then longer tails *and* stronger preferences for longer tails will evolve. The stronger preferences in turn will favor even greater exaggeration of the tails, causing both the long tails and preferences for them to experience an explosive evolutionary runaway. Unlike the other mechanisms we have discussed for the evolution of mating preferences, the runaway process has never been directly observed.

The good genes and runaway mechanisms differ in an important way from direct benefits and pleiotropic effects [27]. With direct benefits and pleiotropic effects, alleles that affect female mating preferences also influence survival and reproduction, so they themselves are targets of natural selection. That is, they evolve by **direct selection** acting on alleles that affect a mating preference. In contrast, with the good genes and Fisher's runaway mechanisms, the preferences evolve by **indirect selection**. Here preference alleles evolve because they are correlated with (in linkage disequilibrium with) alleles at other loci that are the targets of selection. Natural selection acts directly on the "good genes," but not on the preference genes themselves. Theoretical analysis suggests that direct selection on preference genes may often be stronger than indirect selection [26]. While there are many well-documented examples of direct selection acting on preferences, the evidence for the role of indirect selection is less strong. Experiments such as that on three-spined sticklebacks described earlier support the good genes mechanism, but others do not [40].

Female choice and male-male competition are often considered alternative modes of sexual selection, but in fact they can operate together. A male's horns (or other armaments) are frequently tested in combat. Winners of those contests may be more likely to carry good genes, which would favor female preferences for the armaments. If this hypothesis is correct, female preferences and male combat can reinforce each other [9].

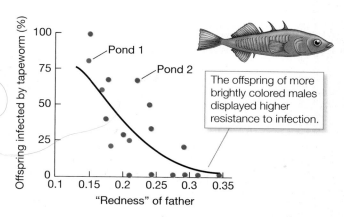

FIGURE 10.16 Evidence for the good genes mechanism for the evolution of female choice. The percentage of young three-spined sticklebacks (*Gasterosteus aculeatus*) that became infected when exposed to tapeworm larvae declined with the intensity of their fathers' red coloration. Red males, which are attractive to females, have alleles that make them resistant to tapeworms and that are passed to their offspring. (After [5].)

Sexual selection in flowering plants

It may sound odd to suggest that sexual selection occurs in plants: males have no weapons, females no nervous system. But individuals of the same sex compete for reproductive access to the other sex, just as in animals.

The vast majority of flowering plants are hermaphrodites. As discussed earlier, the operational sex ratio is a key ingredient that determines which sex experiences sexual selection. The operational sex ratio is often male-biased in plants because even after all of an individual's ovules have been fertilized, it can continue to send out pollen. That is, more individuals are available to act as males (pollen donors) than as females (pollen acceptors). This situation sets the stage for male-male competition. Individuals with more attractive floral displays attract more pollinators. This gives them increased fitness because they export more pollen than do individuals with less attractive displays [8]. You can admire the evolutionary outcome the next time you see a field of wildflowers or smell a rose.

A second opportunity for intense male-male competition in plants happens once a pollen grain arrives on a flower [49]. It is now in a race against the other pollen grains on the same stigma to fertilize an ovule. This situation parallels sperm competition in animals: the evolutionary prize is reproductive success at the expense of other individuals of the same sex.

Sex Ratios

In most species with separate sexes, the **sex ratio**—the relative numbers of males and females—is about equal at birth. In animals such as humans and in plants such as papayas, the sex of an individual is determined by its chromosomes. A human embryo that inherits an X chromosome from the father is female, while an embryo that inherits a Y chromosome is male. Meiosis in males typically transmits the X and the Y chromosome with equal probability, which suggests that the sex ratio might be fixed at 50 percent male. But in fact there are species with sex chromosomes that have unequal sex ratios, and even species that can adjust the sex ratio in their offspring. Females of the Seychelles warbler (*Acrocephalus sechellensis*) produce up to 87 percent daughters when conditions on their territories are good, but only 23 percent daughters when conditions are bad [28]. This shift in the sex ratio is thought to be adaptive because daughters sometimes help their parents raise more offspring when the territory has enough food to support them all. Although the mechanism behind the shift is not known, this warbler does show that sex chromosomes do not always lock in the sex ratio to equal numbers of females and males.

Earlier in the chapter, you saw that in some species sex is determined not by chromosomes but by the physical or social environment [4]. This is called **environmental sex determination**. In turtles and lizards in which sex is determined by the temperature during egg development, the sex ratio can evolve simply through changes in where females lay their eggs.

The Hymenoptera (ants, bees, wasps, and their relatives) use yet another way to determine sex. When a female lays an egg, she can fertilize it with sperm that she has stored from an earlier mating. All eggs that are fertilized develop into females. But the female can also lay an unfertilized egg, which will then develop into a male. Males are haploid, and so this system is called **haplodiploid sex determination**. Females can therefore adjust the sex ratio of their offspring behaviorally, by altering the number of eggs they fertilize. While this system for sex determination may seem bizarre, it is used by 12 percent of animal species (see Figure 10.3). You will see shortly that this enables certain wasps to do something remarkable.

Generation 1   Generation 2

Later generations   Generation 3

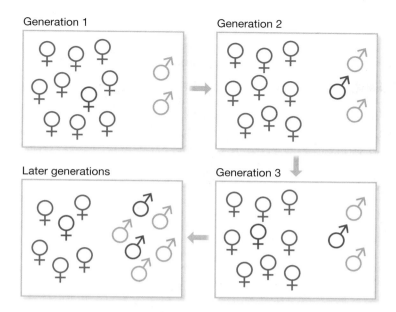

FIGURE 10.17 Selection usually favors a sex ratio with nearly equal numbers of males and females. In this example, females initially produce more daughters than sons, so the sex ratio is biased toward an excess of females. Generation 1: a mutation occurs in one female (red) that causes her to produce only sons. Generation 2: the mutation is now carried by the son of the female that carried the mutation in generation 1. There are only three males, and each of them will be the father to one-third of the individuals in the next generation. Generation 3: the mutation has increased in frequency because each male in the previous generation is father to one-third of the population, and each of his offspring has a probability of 1/2 of inheriting the mutation. Later generations: the mutation will continue to spread until there are equal numbers of males and females. At that equilibrium, the mutation will be more common in males because females with the mutation produce only sons, while males with the mutation produce both sons and daughters that carry the mutation.

If the sex ratio can evolve, why does natural selection seem to favor equal numbers of males and females? At first glance, that situation seems odd: a population can grow faster if there are more females than males, since only females produce offspring. The answer results from the simple fact that every individual has one father and one mother. As a result, whenever there are fewer males than females, a male will on average leave more offspring to the next generation than will a female (**FIGURE 10.17**). Any mutation that causes females to produce more sons therefore has an evolutionary advantage and will spread. The advantage disappears when the sex ratio reaches an equal number of males and females. If a population has an excess of males rather than females, a mutation causing mothers to make more daughters is favored. In short, natural selection pushes the sex ratio to evolve toward equal numbers of males and females.

The equal sex ratio predicted by this theory is seen in many animals and plants, but there are spectacular exceptions. These actually give us a deeper understanding of sex ratios and, more generally, of how natural selection works. Much of our understanding of this topic was developed by William D. Hamilton, who drew attention to the remarkable biology of fig wasps [20]. In these minute insects, the life cycle is tightly bound to the fig trees they pollinate (**FIGURE 10.18**). A fig is a strange inflorescence that resembles a sunflower that has closed in on itself. The interior is hollow, and lined with flowers. When the flowers are mature, a female fig wasp enters the fig through a small opening. There she does two things: she lays her eggs in the flowers, and she pollinates the flowers with pollen that she carried into the fig with her. With those missions accomplished, the female dies.

Some weeks later, a first wave of offspring hatch, and they are all sons. Male fig wasps are the stuff of science fiction. They have no wings or digestive system. Two organs, however, are greatly enlarged: their jaws and their testes. The males literally fight to the death. The next wave of eggs to hatch are all female, and the lone surviving male mates with all of them. The females then exit from the fig, picking up pollen on their way out. They begin another life cycle, leaving the male behind to die.

In these fig wasps, many more females than males are born. What could explain that bias? The answer comes from considering the relationship among the individuals inside a fig: they are typically a single family of brothers and sisters. A

FIGURE 10.18 The fig wasp *Tetrapus costaricensis* (inset) is highly sexually dimorphic. Females lay their eggs inside a fig. When the offspring hatch, the males fight each other, and the winner then mates with the females inside the fig—many or all of whom are his sisters.

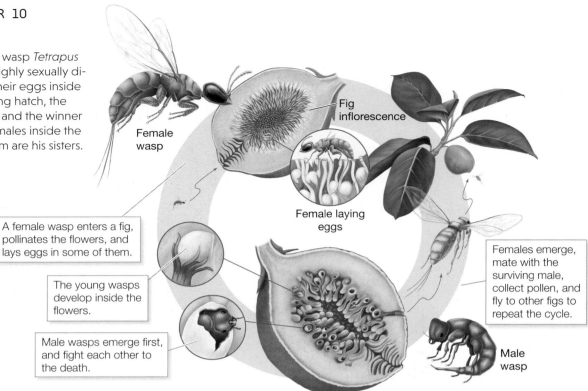

Female wasp

Fig inflorescence

Female laying eggs

A female wasp enters a fig, pollinates the flowers, and lays eggs in some of them.

The young wasps develop inside the flowers.

Male wasps emerge first, and fight each other to the death.

Females emerge, mate with the surviving male, collect pollen, and fly to other figs to repeat the cycle.

Male wasp

mutation that causes females to make more sons will be selected against because only females leave the fig to start the next generation. Here selection favors the families that produce the largest numbers of females, and those are families that have strongly female-biased sex ratios. This prediction differs from the theory of equal sex ratios described earlier because selection is now acting on the family rather than on the individual (see Chapter 12).

The prediction changes, however, if several females lay their eggs in the same fig. As the number of females increases, natural selection favors them to produce sex ratios that are closer and closer to equal numbers of sons and daughters. That is because competition among individuals, rather than families, now occurs within each fig. The theory of equal sex ratios described above then comes into play. A mathematical model can be used to predict how the sex ratio favored by natural selection changes depending on the number of females that lay their eggs in a fig.

This model's prediction has been tested with data from natural populations of fig wasps. The sex ratios produced by females largely agree with the theoretical prediction (**FIGURE 10.19**). Recall that wasps have haplodiploid sex determination, so females can change the sex ratio of their offspring behaviorally. By unknown means, female wasps are somehow able to sense how many other females are in the fig and produce an appropriate number of sons and daughters. The agreement between the theory and experiment is a wonderful example of the predictive power of evolutionary theory.

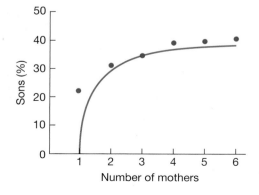

FIGURE 10.19 A mathematical model predicts that natural selection will favor female fig wasps to produce few sons if only one female lays her eggs in a fig, but to produce increasingly equal numbers of sons and daughters as the number of females that lay eggs in the fig increases (shown by the blue curve). In nature, the number of females that enter a fig to lay eggs varies. When only a single female lays her eggs in a fig, fewer than 25 percent of her offspring are male. With increasing numbers of females, the sex ratio tends toward equal numbers of sons and daughters (red dots), in good agreement with the theoretical prediction. (After [22].)

(A)

(B)

(C)

FIGURE 10.20 Asexual reproduction is very rare among animals and plants. Examples include (A) bdelloid rotifers (*Philodina roseola* is shown here), (B) some whiptail lizards (*Aspidoscelis uniparens* is shown here), and (C) the common dandelion (*Taraxacum officinale*).

Why Sex?

The simplest but most profound question about sex is why it exists at all. Sex is not the only option for reproduction. About 1 percent of plant species and 0.1 percent of animal species reproduce by making genetic clones of themselves, a reproductive mode called **parthenogenesis** (**FIGURE 10.20**) [34]. These asexual species are typically found on the tips of the tree of life: their closest relatives reproduce sexually. This implies that most asexual species have a short life expectancy on an evolutionary time scale: if they did survive a long time, we would expect to see groups of related asexual species that are connected to sexual relatives back in the distant past. The most famous exception to this generalization is the bdelloid rotifers. This group of small aquatic animals has lived in sexual abstinence for more than 100 million years [18].

The rarity of parthenogenesis is one of the deep puzzles in evolutionary biology. Parthenogenesis has evolutionary advantages that should make it more common than sexual reproduction. By far the most important of these is the **twofold cost of males**: if all else is equal, the production of males in a sexual population reduces its reproductive potential by a factor of two. Consider the following thought experiment. Females of a sexual species each produce two offspring (**FIGURE 10.21A**). Half of the individuals are males that do not give birth, so the population size is constant. An asexual female then appears, for example by mutation, and she also has two offspring. But since all of her offspring are asexual females, the number of asexual individuals doubles in each generation. In short, a mutation for asexual reproduction enjoys a 100 percent fitness advantage over an allele for sexual reproduction, and it will spread to fixation in only a handful of generations (**FIGURE 10.21B**). The mystery is why that scenario is not constantly playing itself out, very quickly causing all sexual species to evolve parthenogenesis.

Parthenogenesis has other evolutionary advantages as well. Sexual reproduction requires finding a partner, which is sometimes difficult. Weeds such as the

FIGURE 10.21 The biggest evolutionary disadvantage of sexual reproduction is the twofold cost of males. (A) Each female produces two offspring. The sexual females are exactly replacing themselves, and their numbers are stable. Asexual females produce only daughters, and so their numbers double in each generation. (B) If all else is equal, a mutation in a sexual population that causes females to reproduce asexually will spread to fixation in just a few generations. The fact that asexuality is so rare shows that there must be strong advantages to sexual reproduction that compensate for the twofold cost of producing males.

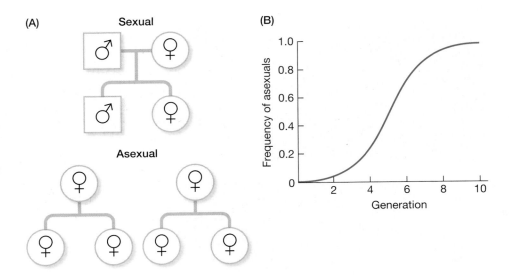

dandelion (see Figure 10.20C), which is very successful at colonizing new patches of habitat, gain an advantage from parthenogenesis through reproductive assurance: in the absence of other individuals, just a single unfertilized seed can start a whole new population. Another advantage is that sexually transmitted diseases, which are suffered by plants as well as animals, are avoided by organisms without sex. Together with the twofold cost of producing males, these further factors would seem to give parthenogenesis an overwhelming evolutionary advantage. Yet the ubiquity of sexual reproduction shows that is not the case.

Since sexual reproduction is so overwhelmingly common in life on Earth, it must have some benefits that offset these drawbacks. The most fundamental feature of sex is that it causes genetic mixing: zygotes carry combinations of alleles inherited from two parents (see Chapter 4). In most eukaryotes, the mixing results from segregation and recombination that happen during meiosis, and then fusion of the sperm and egg. Prokaryotes rely on various kinds of parasexuality we mentioned earlier. The solution to the evolutionary enigma of sex must involve this mixing.

Advantages to sex in changing environments

Clues about the factors that favor sex can be gleaned from looking at species that have both sexual and asexual reproduction. Sex seems to be more common in those species in situations where the environment is changing. Water fleas (genus *Daphnia*) are small crustaceans that live in freshwater lakes. During the summer, the populations are entirely female and reproduce for several generations by parthenogenesis. In the fall, sexual males and females appear. They mate and produce eggs that lie dormant through the winter, then hatch to start the cycle again in the spring. The chemistry and biology of a lake can change substantially from one year to the next, and it seems that sexual reproduction is timed to happen when the environment is most unpredictable.

The timing of sexual reproduction in water fleas suggests that changing environments can favor sex but not why they do so. A popular argument (frequently repeated in introductory biology texts) is that sex is favored because it increases genetic variation. On closer inspection, there are problems with that idea. A useful analogy here is with a card game in which the cards play the role of alleles [34]. Say that you have just won a game with a very good hand. If you had the choice, would you keep those cards for the next game, or would you mix half of the cards you now have with random ones taken from the deck? If the rules of the game don't change, we expect that a set of cards that were successful the last time will also be successful the next time. For the same reason, an individual that has survived selection is likely to carry a good combination of alleles, and mixing them with the alleles of a sexual partner can produce offspring that are genetically less fit. Suddenly the genetic mixing of sexual reproduction does not seem like such a good idea.

One way in which evolution can favor sexual reproduction is the **Red Queen hypothesis**, named in honor of the colorful character in Lewis Carroll's *Through the Looking-Glass* who must run constantly to stay in the same place. According to this hypothesis, all species are in an evolutionary arms race with other species, such as pathogens that are evolving rapidly to defeat their host's defensive systems (see Chapter 13). This creates a shifting adaptive landscape for the host. Recombination can increase the frequency of rare combinations of alleles that are good at defending the host against attacks from its pathogens. Returning to the analogy of the card game, evolutionary changes in the pathogens change the rules about which cards are best. In that case, changing cards can be the winning strategy.

Support for the Red Queen hypothesis comes from the New Zealand mud snail (*Potamopyrgus antipodarum*), which has both sexual and asexual genotypes. Populations that are exposed to higher densities of parasites have higher frequencies of sexually reproducing individuals [29]. Furthermore, because sexual females have lower infection rates, the fitness of sexual females is sometimes more than twice that of asexual females (**FIGURE 10.22**) [48]. This result suggests that the evolutionary benefit of recombination can sometimes compensate for the twofold cost of males.

Parasites and pathogens can also receive an evolutionary benefit from the genetic mixing caused by recombination. People develop immunity to the genotypes of influenza virus they have been exposed to. New outbreaks result when a novel viral genotype is produced that can efficiently infect humans. The genome of the influenza virus is not a single molecule, as in most viruses. Instead, its genome is broken into eight segments of RNA. This arrangement is an adaptation that enhances recombination. You can think of the next outbreak of the flu as a dramatic demonstration of the evolutionary power of recombination.

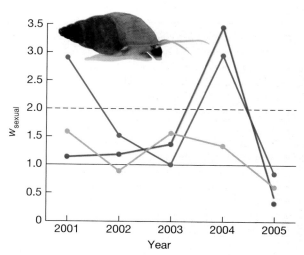

FIGURE 10.22 Attacks by parasites give an evolutionary advantage to sexual reproduction in the New Zealand mud snail (*Potamopyrgus antipodarum*), as predicted by the Red Queen hypothesis. The curves show the relative fitness of sexual females, using asexual females as the fitness reference, in three populations across 5 years. Sexual females usually have higher fitness than asexual females (points above the solid horizontal line). In some years, sexual females are more than twice as fit (points above the dashed horizontal line), showing that they have overcome the twofold cost of producing males. The higher fitness of sexual females results because they have lower rates of infection by parasites. (After [48].)

Selective interference favors sex and recombination

Alleles are not selected independently. The chance that a particular copy of an allele is passed to the next generation depends in part on the rest of the genome in which it is carried. Adaptation can be hampered as a result, particularly if there is not much genetic mixing. This phenomenon is called **selective interference** (also known as the Hill-Robertson effect). Many of the advantages of sex revolve around the fact that sex reduces selective interference because it separates alleles from their genomic backgrounds and allows selection to act more efficiently [17].

One form of selective interference is called **clonal interference**, which happens when two or more beneficial mutations spread through a population at the same time (**FIGURE 10.23**). Consider the outcome in an asexual population if beneficial mutations *A* and *B* appear at two loci in different individuals at about the same time. Selection causes the two clones with those alleles to spread. When they become common, they compete. If clone *A* has higher fitness than clone *B*, it will drive clone *B* to extinction. The genotype *AB*, with the highest fitness, can be established only after a second *B* mutation appears in a genotype that already

(A) Asexual, no recombination

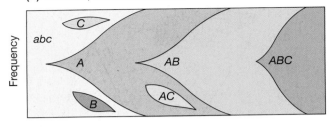

(B) Sexual, with recombination

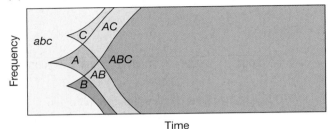

Time

FIGURE 10.23 Clonal interference slows adaptation. (A) In an asexual population that is initially fixed for the genotype *abc*, beneficial mutations A, B, and C occur at those three loci and begin to spread. Mutation A is the most fit, and so it drives the clones that carry mutations B and C to extinction. The population finally gains all three mutations only after B occurs in a genotype that already carries A, and C occurs in a genotype that already carries A and B. (B) In a sexual population, recombination can bring together in a single genotype several mutations that originally appeared in different individuals. This reduces clonal interference and accelerates adaptation. The asexual population is therefore at greater risk of extinction in changing environments. (After [14].)

has the *A* mutation, and that may take a substantial time. In short, spread of the beneficial *A* mutation interferes with the establishment of the beneficial *B* mutation, and adaptation is slowed. The problem of clonal interference is magnified when more than two loci are adapting.

The picture is different in a sexually reproducing population. If mutations *A* and *B* are spreading at the same time, recombination can bring them together in a single individual. The high-fitness *AB* genotype can then spread to fixation without the wait for a second *B* mutation to appear. By avoiding clonal interference, sexual reproduction accelerates adaptation.

A second type of selective interference is called the **ruby-in-the-rubbish effect**, which is the loss of beneficial mutations as the result of their linkage to deleterious mutations [38]. Recall that deleterious mutations are constantly raining down on the genome, and as a result every individual carries many of them (see Chapter 5). When a beneficial mutation appears in an asexual population, one of two things can happen. If the positive fitness benefit that it gives is small relative to the combined negative fitness effects of all the deleterious mutations in the genome where it appeared, the beneficial mutation is doomed (**FIGURE 10.24A**). That is because the fitness of its genotype will be improved by the beneficial mutation, but not enough to compete successfully with other, more fit genotypes in the population.

(A) No recombination

(B) With recombination

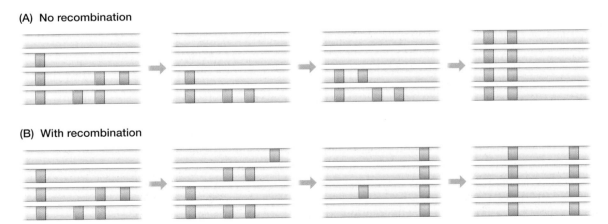

FIGURE 10.24 The ruby-in-the-rubbish effect gives an advantage to sexual populations. (A) In an asexual population with no recombination, the first beneficial mutation (blue) that appears occurs in a genotype that is already carrying two deleterious mutations (orange). The fitness of that genotype is less than the one with no deleterious mutations, and so it is lost. A second beneficial mutation occurs in a genotype with only one deleterious mutation. It spreads to fixation, but by doing so causes fixation of the deleterious mutation. (B) In a sexual population with recombination, the beneficial mutations can recombine away from deleterious mutations on the chromosomes where they first appeared, and into genotypes with no deleterious mutations. That increases the chance they will not be lost and will become fixed without having deleterious mutations hitchhike to fixation with them.

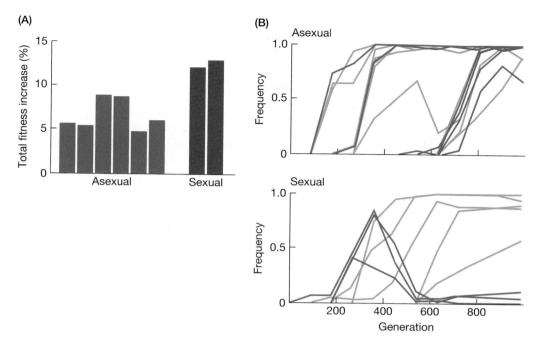

FIGURE 10.25 Experiments with yeast (*Saccharomyces cerevisiae*) show how recombination speeds adaptation. This species can be manipulated in the lab to reproduce either sexually or asexually. (A) After 1000 generations of adaptation to a laboratory environment, the fitness of the sexual populations increased about twice as much as that of the asexual populations. (B) DNA sequencing at different time points followed the frequencies of mutations spreading in an asexual and a sexual population. Separate experiments were used to estimate the fitness effects of those mutations. The trajectories of beneficial mutations are shown in green, and of deleterious mutations in purple. In the asexual populations, eight beneficial mutations had become fixed by the end of the experiment. As they did so, they dragged five deleterious mutations to fixation with them. In the sexual populations, recombination freed the beneficial mutations from the deleterious ones. As a result, all of the deleterious mutations were eliminated. (After [30].)

(Metaphorically, the ruby is thrown out with the rubbish.) The second possibility is that the beneficial mutation has such a positive effect that its genotype is now the most fit in the population. It can then spread to fixation. But as it does so, all the deleterious mutations elsewhere in its genome hitchhike along with it, causing those loci to degenerate. In contrast, a sexual population avoids both of these fates (**FIGURE 10.24B**). Recombination can liberate a beneficial mutation from deleterious mutations at other loci. That increases the chance that the beneficial mutation will not be lost immediately, or drag to fixation bad alleles at other loci.

The ruby-in-the-rubbish effect is seen clearly in experimental populations of yeast that have been manipulated to reproduce either with or without sex [30]. As the yeast adapt, beneficial mutations at many loci begin to sweep through the populations. In asexual populations, the beneficial mutations drag deleterious mutations at other loci along with them to fixation (**FIGURE 10.25**). The picture is different, however, in sexual populations. As beneficial mutations begin to spread, deleterious mutations again start to hitchhike along for the ride. But before they become fixed, recombination uncouples the beneficial mutations from their deleterious passengers. The beneficial mutations then spread to 100 percent frequency, while the deleterious mutations are eliminated. By the end of the experiment shown in Figure 10.25B, five deleterious mutations became fixed in the asexual population but none in the sexual population.

One last form of selective interference is **Muller's ratchet** (named after the Nobel Prize–winning geneticist H. J. Muller), which is the irreversible accumulation of deleterious mutations in an asexual population. In any population, a relatively small number of individuals have the fewest mutations. There is always a chance

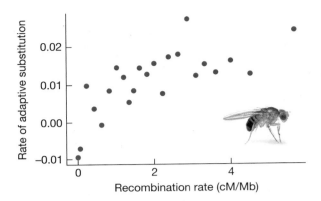

FIGURE 10.26 Selective interference in *Drosophila melanogaster* decreases the rate of adaptive evolution in regions of the genome that have low recombination rates. The *x*-axis shows the recombination rate (in centimorgans [cM] per megabase [Mb] of DNA); regions of the genome with high recombination are to the right. The *y*-axis shows a measure of the rate of adaptive amino acid changes in proteins. Genomic regions with higher recombination rates have higher rates of adaptive evolution. (After [10].)

that none of those individuals will leave any offspring. In a sexual population, recombination can regenerate that class of high-fitness individuals. But that does not happen in an asexual population (except in the unlikely event of back mutations from a deleterious to a beneficial allele). Each time the most fit genotypes fail to reproduce, the population's mean fitness is ratcheted downward. It can never recover, and in principle this process can lead to the extinction of an asexual species.

Asexual species are thought to be so rare because some combination of these forms of selective interference and the Red Queen hypothesis drives them to extinction rapidly, while sexual species are much more likely to survive. The same factors that favor sexual over asexual species can cause selection for changes to recombination rates within the genomes of sexually reproducing species. Recombination rates can evolve by several mechanisms, for example changes in the frequencies and locations of crossovers during meiosis. The details are beyond this text, but the underlying principles are much the same: factors that favor sexual over asexual species tend to favor increased recombination within sexual species.

Selective interference is particularly severe in asexual populations because they have no recombination, but it also occurs in sexual populations. Within a species, some parts of the genome have high recombination rates, while others have low rates. Using data on the polymorphism within species and the differences between species at protein-coding loci in *Drosophila melanogaster*, the rate of adaptive amino acid substitution can be estimated for different parts of the genome. The data show a clear pattern: adaptive evolution is fastest in regions with high recombination rates (**FIGURE 10.26**) [10].

The human sex chromosomes give a graphic example of the evolutionary consequences of giving up recombination. The Y chromosome originated from a recombining X chromosome some 180 million years ago, and then ceased to recombine with the X [13]. From that point onward, the Y chromosome has evolved asexually, like the mitochondria that are passed through the female lineage. Meanwhile, the X chromosome continues to recombine in females. As various forms of selective interference caused the Y chromosome to degenerate, it lost almost all of the 2000 or so genes and more than 60 percent of the DNA carried on the X [32]. As a result, we now see dramatic differences between the X and Y chromosomes (**FIGURE 10.27**).

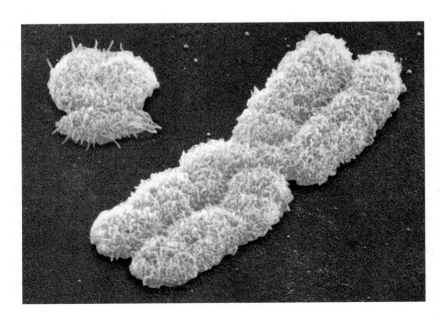

FIGURE 10.27 The human sex chromosomes, as seen in a scanning electron micrograph. The Y chromosome (at left) does not recombine. Selective interference caused it to degenerate from an ancestor that was much like the X chromosome (at right).

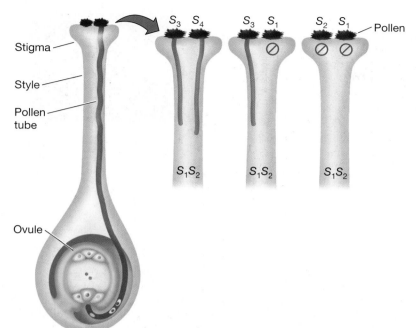

FIGURE 10.28 Several kinds of self-incompatibility in flowering plants prevent ovules from being fertilized by pollen from the same plant. The mechanism shown here is called "gametophytic incompatibility." When a pollen grain lands on a stigma, it begins to grow a tube toward the ovary, where it will fertilize an ovule. A biochemical reaction compares the one allele carried by the pollen with the two alleles carried by the stigma at the *SI* (*self-incompatibility*) locus. If there is a match, the female tissue kills the pollen. This prevents the plant from being fertilized by its own pollen.

Selfing and Outcrossing

Some hermaphroditic plants and animals have evolved a compromise between sexual and asexual reproduction. They produce gametes by meiosis, but then can fertilize themselves. **Self-fertilization**, or "selfing" for short, is particularly common among weedy and colonizing species. Like parthenogenesis, selfing provides reproductive assurance—a single individual can reproduce without a partner.

Most hermaphroditic species have mechanisms that largely or completely prevent self-fertilization. They outcross, that is, mate with other individuals. Animals can avoid selfing behaviorally, but plants cannot. Most flowering plants are hermaphrodites: each individual produces both pollen and ovules. Plants have evolved diverse ways to prevent pollen from fertilizing the ovules of the same individual. In some species, the pollen and ovules on a plant mature at different times. In others, the anthers (floral structures with pollen) are physically separated from the stigmas (structures that receive the pollen), decreasing the chance that an ovule will be fertilized by pollen from the same flower. A remarkable scheme to prevent self-fertilization, called **self-incompatibility**, has evolved several times [11]. When pollen lands on the stigma of a flower, biochemical systems compare the genotypes of the pollen and stigma at the *SI* (*self-incompatibility*) locus. If the stigma and pollen share an allele at this locus, biochemical machinery is triggered that kills the pollen (**FIGURE 10.28**).

Why do these hermaphrodites forego the advantage of reproductive assurance? The major downside of self-fertilization is that offspring can suffer from **inbreeding depression**. This is the loss in fitness shown by offspring whose parents are close relatives compared with offspring whose parents are unrelated. (Inbreeding depression is different than *inbreeding load*, discussed in Chapter 7, which is the decline in a small population's fitness caused by fixation of deleterious mutations.) You saw in Chapter 4 that most mutations are deleterious, and that deleterious mutations tend to be recessive. When an organism self-fertilizes, every deleterious mutation in the genome that is heterozygous has a 50 percent probability of becoming homozygous in the offspring. These homozygous mutations can dramatically decrease the offspring's fitness.

FIGURE 10.29 Inbreeding depression reduces the fitness of offspring produced by matings between relatives. Experimental crosses were made with the white campion (*Silene latifolia*) using plants with different degrees of relatedness. The *x*-axis shows relatedness, ranging from unrelated (at left) to 3/8 related (which is more closely related than first cousins, but less related than full siblings). The *y*-axis measures the probability that the seed germinates, which is a critical fitness component. The fitness of the most inbred individuals is reduced by more than 80 percent compared with offspring produced by unrelated parents. (After [42].)

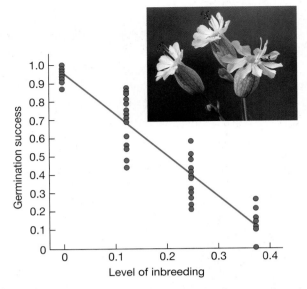

Inbreeding depression also results when close relatives mate. In some species, inbreeding depression is severe. The fitness of offspring produced by matings between close relatives in the white campion (*Silene latifolia*) is reduced by up to 80 percent compared with the offspring of unrelated individuals (**FIGURE 10.29**). Inbreeding depression is also seen in humans. Tay-Sachs disease is a lethal degenerative condition. It is caused by a recessive mutation that is relatively common in Ashkenazi Jewish populations, where more than one-quarter of people who suffer from the disease are children of parents who are first cousins [47]. Inbreeding depression is likely the reason that many societies have social taboos against marriage between close relatives.

Some hermaphroditic species have little inbreeding depression, and they can evolve to reproduce almost entirely by self-fertilization (see Figure 10.1C). But almost all of them occasionally mate with other individuals—obligate selfing is extremely rare. The reason, once again, is selective interference [25]. A population that is entirely selfing suffers from clonal interference for exactly the same reasons that parthenogenetic populations do. Occasional outcrossing allows alleles to escape their genetic backgrounds, which accelerates adaptation and stops the evolution of a complete dependence on reproduction by self-fertilization.

Go to the
———————— **Evolution Companion Website** ————————
EVOLUTION4E.SINAUER.COM
for data analysis and simulation exercises, quizzes, and more.

SUMMARY

- Animals, plants, and microbes have evolved diverse forms of sexual reproduction, which is the mixing of genetic material from different parents. Almost all organisms on Earth engage in some kind of sex.

- Males and females are distinguished by the size of the gametes they make: males make small gametes, and females big ones. Many species are hermaphroditic. In species with separate males and females, sexual dimorphism ranges from minimal to extreme.

- Sexual selection, which is selection caused by competition among individuals of the same sex for mates, leads to the evolution of exaggerated secondary sexual traits that increase mating success but usually decrease survival. Sexual selection acts on males much more often than on females. Males can often increase their fitness by mating with more females, but females typically do not benefit from mating with more males.

- Some unusual species show sex role reversal. Here sexual selection acts on females because the operational sex ratio is female-biased: more females are available to mate than males.

- One of the two major modes of sexual selection is male-male competition, which occurs by male combat, sperm competition, infanticide, and other mechanisms.

- The second major mode of sexual selection is female choice. Female mating preferences evolve as the result of direct benefits that females receive from their mates, pleiotropic effects of preference genes, and the good genes mechanism. Some preferences result from perceptual biases and apparently did not originate by either direct benefits or good genes. A final mechanism for the evolution of female preferences is Fisher's runaway process.

- Sexual selection on plants favors flowers that increase pollinator visitation, production of greater quantities of pollen, and pollen that outcompetes other pollen in fertilizing ovules.

- The sex ratio can evolve in many species. In most situations, selection favors producing equal numbers of males and females. Exceptions occur in organisms such as fig wasps where selection favors those families that produce the largest numbers of daughters.

- The rarity of asexual reproduction is a puzzle because several factors give it an evolutionary advantage over sexual reproduction. The biggest of these is the twofold cost of males suffered by sexual species. Other advantages to asexual reproduction include reproductive assurance and escape from sexually transmitted diseases.

- Recombination gives sexual reproduction several advantages that compensate for its disadvantages and thereby explain why it is so common. The Red Queen hypothesis suggests that sex is favored in changing environments. Recombination is also favored because it reduces selective interference, which is a general term that includes clonal interference, the ruby-in-the-rubbish effect, and Muller's ratchet.

- Inbreeding depression frequently causes the evolution of mechanisms that prevent self-fertilization and mating between close relatives.

TERMS AND CONCEPTS

alternative mating
   strategy
anisogamy
clonal interference
direct benefit
direct selection
environmental sex
   determination
Fisher's runaway
good gene
haplodiploid sex
   determination

hermaphroditic
inbreeding
   depression
indirect selection
infanticide
lek
male combat
male-male
   competition
Muller's ratchet
operational sex ratio

parthenogenesis
perceptual bias
pleiotropic effect
primary sexual trait
Red Queen
   hypothesis
reproductive
   assurance
ruby-in-the-rubbish
   effect
secondary sexual
   trait

selective
   interference
self-fertilization
self-incompatibility
sex ratio
sex role reversal
sexual selection
sexually dimorphic
sperm (pollen)
   competition
two-fold cost of
   males

SUGGESTIONS FOR FURTHER READING

As with so many topics in evolutionary biology, Charles Darwin wrote not only the first but also some of the most insightful thoughts on sexual selection. He laid out the principles of sexual selection in *On the Origin of Species* (John Murray, London, 1859), then elaborated them in his longest book, *The Descent of Man and Selection in Relation to Sex* (John Murray, London, 1871). Both books have many important ideas that still have not been fully explored.

Insects are by far the most diverse group of animals, so it is not surprising that they have a remarkable range of strange and fascinating forms of sexual selection. The book by D. M. Shuker and L. W. Simmons, *The Evolution of Insect Mating Systems* (Oxford University Press, Oxford, 2014) is a recent collection of articles by leading researchers in that field.

Males have evolved a remarkable variety of weapons to fight each other. These weapons, and their implications for humans, are explored (and beautifully illustrated) in a book by D. J. Emlen, *Animal Weapons: The Evolution of Battle* (Henry Holt, New York NY, 2014).

An overview of the diversity and evolution of sex determination mechanisms is given by D. Bachtrog and colleagues in "Sex determination: Why so many ways of doing it?" (*PLOS Biology* 12: e1001899, 2014).

A witty yet scholarly exploration of some of the more interesting and unusual sides of sex in the animal kingdom is Olivia Judson's popular book, *Dr. Tatiana's Sex Advice to All Creation* (Henry Holt, New York NY, 2013).

The definitive work on the evolution of sex ratios and sex allocation is S. West's *Sex Allocation* (Princeton University Press, Princeton NJ, 2009).

The evolution of sexual reproduction and recombination is one of the most fascinating but difficult topics in evolutionary genetics. Lucid introductions have been written by several of the leading researchers in the field: S. P. Otto ("The evolutionary enigma of sex," *Amer. Nat.* 174: S1–S14, 2009), N. H. Barton ("Why sex and recombination?", *Cold Spring Harbor Symp. Quant. Biol.* 74: 187–195, 2009), and C. M. Lively and L. T. Morran ("The ecology of sexual reproduction," *J. Evol. Biol.* 27: 1292–1303).

PROBLEMS AND DISCUSSION TOPICS

1. Populations of some species of fish, insects, and crustaceans consist of both sexually and asexually reproducing individuals. Would you expect such populations to become entirely asexual or sexual? What factors might maintain both reproductive modes? How might studies of these species shed light on the factors that maintain sexual reproduction?

2. Many parthenogenetic species of plants and animals are known to be genetically highly variable. What processes might account for this variation?

3. The text says that asexual, parthenogenetic species are typically found on the tips of the tree of life: their closest relatives reproduce sexually. Explain this pattern.

4. Would you expect sexual selection to increase or decrease adaptation of a population to its environment? Do the pleiotropic effects and good genes mechanism for the evolution of female preferences differ in their implications for adaptation to the environment?

5. In many socially monogamous species of parrots, both sexes are brilliantly colored. Is sexual selection likely to be responsible for the coloration in both sexes? How can there be sexual selection in pair-bonding species with a 1:1 sex ratio, given that every individual presumably

obtains a mate? Which of the types of sexual selection described in this chapter might account for bright coloration in both sexes of these species?

6. In many reptiles, including crocodiles and many turtles, sex is determined by temperature during early development. Many scientists expect that as Earth's climate warms, the sex ratios in these species may become highly biased, further endangering these animals. Do some outside research and conclude whether this concern is warranted and what might happen to these populations as a result.

7. Anisogamy is the term for sexual dimorphism in gamete size. Discuss the evolution of anisogamy from an ancestor in which gametes had equal sizes. What factors would lead to a divergence in gamete size among members of a population? Are females of one lineage homologous with the females of distantly related lineages (for example, female birds and female flowering plants)? That is, how many times might the sex roles have evolved?

8. What aspects of human behavior, physiology, and morphology might be explained by sexual selection? What are the alternative hypotheses, and how might we determine which is (or are)

How to Be Fit

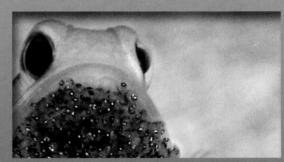

Who has not dreamed of living forever? We know we won't. Someone born in Japan today has a life expectancy of about 83 years. A white American born in 2010 is projected to live about 79 years on average, a black American 75 years. A French woman, Jeanne Calment, is said to have been mentally fully competent when she died in 1997 at the age of 122 years.

But some other species live much longer. Greenland sharks (*Somniosus microcephalus*) were recently found to live for at least 272 years, and the largest individuals are estimated to be almost 400 years old [41]. This is impressive, but hardly compares with a 5065-year-old bristlecone pine in California that is the oldest known unitary ("individual") organism. The quaking aspens that grow on a mountain slope in the western United States appear to be separate trees but are often a clone: a single genetic individual, originating from a single seed, that may be more than 80,000 years old. Some organisms might well be immortal: they show no signs of **senescence**, the intrinsic changes that lower survival and reproduction with age. Some biologists think that clonal plants and fungal mycelia do not senesce, and this may be true of some sponges (estimated at more than 11,000 years old) and corals (more than 4000 years old). In contrast, many plants, insects, and even some mammals live for only 1 year, many small insects for only a few months, and some rotifers for fewer than 20 days (**FIGURE 11.1**). There exists enormous variation among organisms not only in maximum life span, but also in the process of senescence that foreshadows ultimate demise [29, 35, 45, 55].

Parental care in many fishes is a male role, as in this male yellow-headed jawfish (*Opistognathus aurifrons*). Having fertilized his mate's eggs, he protects them until they hatch by holding them in his mouth. This reproductive strategy limits the number of offspring, but enhances their chance of survival.

FIGURE 11.1 Species vary greatly in life span. (A) Bristlecone pines (*Pinus longaeva*), surviving in the punishing environment of desert mountaintops in California, are among the oldest known individual organisms. (B) *Draba verna*, a member of the mustard family, is an annual plant that germinates in early spring, sets seed within a few months, and dies. (C) Asexually propagating corals may not age, and persist for thousands of years. (D) Some rotifers live for only a few weeks.

(A)

(B)

(C)

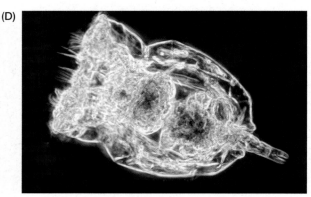

(D)

Clearly, life spans can evolve. It is possible that senescence does not occur in some clonally propagated organisms that retain primordial stem cells, which can differentiate into the organism's diverse cell types [45]. But in most multicellular species, potential life span is one of several components of an organism's life history that are intimately related to fitness, and that have evolved by natural selection. Fecundity (number of offspring) likewise varies. Many bivalves and other marine invertebrates release thousands or millions of tiny eggs in each spawning, but a blue whale (*Balaenoptera musculus*) gives birth to a single offspring that weighs as much as an elephant, and a kiwi (*Apteryx*) lays a single egg that weighs 25 percent as much as its mother (**FIGURE 11.2**). Many species, such as humans, reproduce repeatedly; others, such as century plants (*Agave*) and some species of salmon (*Oncorhynchus*), reproduce only once and then die. Reproductive age may be reached rapidly or slowly. A newly laid egg of *Drosophila melanogaster* may be a reproducing adult 10 days later, and a parthenogenetic aphid may carry an embryo even before she herself is born. In contrast, periodical cicadas feed underground for 13 or 17 years before they emerge, reproduce, and then die within a month.

What accounts for such extraordinary variation in species' survival and reproduction—the very features we would expect to be most intimately related to their fitness? In this chapter we seek to understand how natural selection has shaped **life history** traits (those that affect rates of survival and reproduction at each age) and organisms' ecological niches (the range of conditions they live in and resources they use). These topics are themes in *evolutionary ecology*, the study of how evolution has shaped the interactions between organisms and their environment.

Life History Traits as Components of Fitness

A mutation that increases fecundity or survival will increase individual fitness, as long as it has no other effects, and will therefore become fixed in the population. At first surmise, then, we should expect any species to evolve ever-greater fecundity and an ever-longer life span. The challenge, therefore, is to understand

FIGURE 11.2 Variation in fecundity (number of offspring). (A) Spawning oysters release clouds of minuscule eggs and sperm. (B) A coconut is a single enormous seed, and the coconut palm (*Cocos nucifera*) can produce only a few at a time. (C) Poplars (*Populus*) produce millions of tiny seeds, with fluffy hairs that enable dispersal by wind. (D) This X-ray of a kiwi (*Apteryx*) shows the bird's enormous egg. (D, photo courtesy of Otorohanga Zoological Society.)

how low fecundity or a short life span can evolve by natural selection. In thinking about this, we must be clear about the level at which selection acts on these traits (see Chapter 3). Some biologists used to think that species such as codfishes produce hundreds of thousands of eggs in order to compensate for high mortality and ensure the survival of the species. Likewise, it has sometimes been suggested that animals die of old age to make room for a vigorous new generation. But we have noted that the future persistence or extinction of a species cannot affect, and is irrelevant to, the course of natural selection among individuals (see Chapter 3). So how can individual selection result in low reproductive rates or short life spans?

The life history traits with which we are concerned are the ages at which reproduction begins and ends, fecundity at each age, and the average survival to each possible age [11, 49, 55]. These traits affect the growth rates of populations and are major components of a genotype's fitness. The age to which individuals survive in nature is often shorter than their potential life span, which would be attained only if extrinsic mortality factors, such as predation, disease, and food shortage, were not operating. The maximum life spans cited at the start of this chapter are closer to potential life spans than average realized life spans.

An organism acquires from its environment a certain amount of energy and nutrients, which are allocated among several functions, especially self-maintenance (hence, survival), growth, and reproduction. (The "growth" portion is ultimately allocated to the other two functions.) We expect that there will be **trade-offs** among functions: a fitness benefit of one function that is correlated with a fitness cost of another function. (This is another way of saying "there is no such thing as a free lunch.") The fraction of energy and nutrients allocated to reproduction is sometimes referred to as **reproductive effort**. The trade-off between reproduction and all other functions is often called the **cost of reproduction**.

(A)

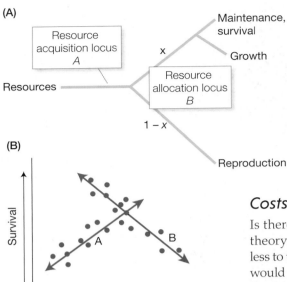

FIGURE 11.3 Factors giving rise to genetic correlations between life history traits such as survival (or growth) and reproduction. (A) Alleles at locus *A* affect the amount of energy or other resources that individuals acquire from the environment. Alleles at locus *B* affect allocation of resources to functions such as growth or survival versus reproduction, in proportions *x* and 1 − *x*. (B) Genotypes that differ in their ability to acquire resources (for example, because of variation at locus *A*) are represented by blue circles. Genotypes that differ in how resources are allocated between survival and reproduction (for example, because of variation at locus *B*) are shown by red circles. The overall genetic correlation between survival and reproduction depends on the relative magnitude of variation in resource acquisition versus resource allocation.

Costs of reproduction

Is there evidence for a cost of reproduction? This concept plays a large role in the theory of life history evolution. Genotypes that allocate more to reproduction and less to themselves may display decreased survival or growth. This allocation trade-off would be manifested as a *negative genetic correlation* (see Chapter 6) between reproduction and survival. If there were also genetic variation in the amount of resources individuals acquired from the environment, however, that variation could give rise to a *positive genetic correlation* between reproduction and survival (**FIGURE 11.3**) [7, 59]. Both kinds of correlation were found in a seed beetle (*Callosobruchus maculatus*) that develops as a larva within a bean [37]. Reproductive adults may continue to feed, but females lay eggs even if they are deprived of food. Variation among families showed a positive genetic correlation if females were given food (hence, variation in acquisition of resource), but a negative genetic correlation between fecundity and survival—evidence of a cost of reproduction—when females were deprived of food.

Costs of reproduction have been detected in many kinds of organisms. Genetic correlations found in a wild population of *Drosophila melanogaster* showed strong trade-offs between the number of eggs females laid when young and both their longevity and their fecundity later in life. However, longer-lived genotypes showed higher fecundity late in life—a rather significant observation, as we will soon see [58]. In a study of brown anoles (*Anolis sagrei*), Robert Cox and Ryan Calsbeek surgically removed the ovaries from wild females and then released them [14]. Even though this species lays only one egg per clutch, these females, prevented from allocating energy and nutrients to reproduction, showed higher growth and survival than did sham-operated females with intact ovaries (**FIGURE 11.4**).

FIGURE 11.4 Evidence of the cost of reproduction. (A) Female *Anolis sagrei* from which ovaries were removed (OVX) grew larger and gained more weight than sham-operated (SHAM) females. SVL is the snout-to-vent length. (B) Over the 2-year study, the proportion of females that survived to the following year was higher for ovariectomized females (blue columns) than for sham-operated females (red columns), which produced eggs. Allocation to reproduction reduced females' growth and survival. (From [14].)

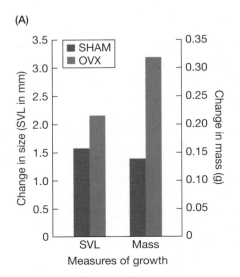

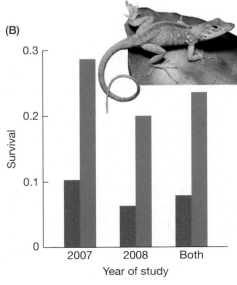

Fitness in age-structured populations

In Chapter 3 we defined fitness for the simple case in which individuals reproduce once and then die (a **semelparous** life history). Fitness was defined as number of offspring produced by an individual (taking into account the probability of surviving to reproductive age). For **iteroparous** species—those in which individuals reproduce more than once—the lifetime reproductive success is found by adding up the reproduction over all the ages at which individuals reproduce.

To make the ideas concrete, take the example of an asexual lizard that starts to reproduce at age 2 and never lives longer than 3 years. We form a **life table** that shows the probability that a newborn will live to age x, symbolized as l_x, and the average fecundity for a female at that age, symbolized as m_x:

x	l_x	m_x	$l_x m_x$
0	1	0	0
1	0.75	0	0
2	0.5	1	0.5
3	0.25	2	0.5
4	0	0	0
			R: 1.0

The last column of the table gives the product of the survival probability and the fecundity for the given age, $l_x m_x$. By adding those values over all ages, we find the expected **lifetime reproductive success**, symbolized as R:

$$R = l_0 m_0 + l_1 m_1 + \ldots = \sum_{x=0} l_x m_x \qquad (11.1)$$

In the life table shown above, $R = 1$. That means each female leaves on average one descendant, and the population is exactly replacing itself. If R is greater than 1, each female leaves more than one offspring, and the population size increases. Lifetime reproductive success is related to the **intrinsic rate of increase**, symbolized as r, which is widely used in ecology. If R is near 1 and time is measured in generations, lifetime reproductive success can be translated into the intrinsic rate of increase by the formula $r \approx \ln(R)$. In a population that is stable in size, $R = 1$ and $r = 0$.

Lifetime reproductive success also is closely related to absolute fitness. For values of R that are near 1, as in this example, lifetime reproductive success is an accurate measure of fitness. But if R is much different than 1, a correction is needed that gives different weights to offspring born at different ages. We will not go into the details of the correction (because the math is complicated), but a simple example illustrates the main point. Consider two asexual lizards that both have a lifetime reproductive success of $R = 2$. The first lizard lives 2 years, produces two offspring, and then dies. The second lizard matures after just 1 year, produces two offspring, and then dies. Its offspring do the same: each matures after 1 year and has two offspring of its own. After 2 years, the first lizard has two descendants, but the second lizard has four. The genes of the second lizard are spreading more quickly in the population. It has higher fitness, even though its lifetime reproductive success is the same as that of the first lizard.

This example illustrates a general principle: in growing populations, natural selection favors earlier reproduction. No species can have a growing population for very long because (as Malthus pointed out) it will exhaust the resources it needs. But some species do spend much of their evolutionary histories in growing populations. For example, weedy plants specialize on colonizing patches of disturbed habitat. A new population increases in size rapidly and disperses seeds to other patches before becoming extinct. This lifestyle favors the evolution of early reproduction, so many weeds have evolved to mature at a young age. Common ragweed (*Ambrosia artemisiifolia*), an American plant that is invading Europe and produces pollen that is a major cause of hay fever, colonizes road edges and other disturbed soils. Its seeds germinate in late spring and it flowers in early fall, just 4–5 months later. In contrast, some oak trees do not mature for several decades.

In a sexually reproducing population, we can also use a life table to find the fitness of an allele. For the survivorship (l_x) and fecundity (m_x) entries, we use the average values in males and females for individuals that carry the allele. Alleles with larger values of R have higher fitness and will spread. (Again, a correction is needed if R is much different than 1.) Suppose a mutation appears in a population that has the life table we just looked at. The mutation increases the average fecundity of both males and females at age 2 from 1 to 1.2 offspring. That increases l_2m_2 from 0.5 to 0.6, and so R increases from 1 to 1.1. The mutation has higher fitness than the other allele at the same locus (which has $R = 1$). The mutation will increase in frequency, causing the life history of the population to evolve and the growth rate R to increase.

Senescence

Our life table illustrates a general point: *natural selection does not act to prolong survival beyond the last age of reproduction.* In the life table above, a mutation that increases the chance of survival to age 4 from 0 to 0.25 has no effect on R, because females do not reproduce at that age. (There are a small number of interesting exceptions. In humans and orca whales, postreproductive parents care for their offspring, so postreproductive survival may be advantageous [15, 19].) But why should reproduction cease? Why do women experience menopause, and older men have lowered sperm production and sex drive? The answer is that, all else being equal, the selective advantage of reproducing declines with age.

Increasing survival and fecundity at earlier ages has a larger effect on fitness than at later ages, simply because predators, disease, and all sorts of accidents make individuals less likely to survive to the later ages. In the discussion of the life table above, we saw that a mutation that increases fecundity at age 2 from 1 to 1.2 increases fitness, R, from 1 to 1.1. Now consider a second mutation that increases fecundity by the same amount at age 3 instead of age 2, increasing l_3m_3 from 0.5 to 0.55. Comparing the life tables for the two mutations, we have:

	First mutation			Second mutation		
x	l_x	m_x	$l_x m_x$	l_x	m_x	$l_x m_x$
0	1	0	0	1	0	0
1	0.75	0	0	0.75	0	0
2	0.5	1.2	0.6	0.5	1	0.5
3	0.25	2	0.5	0.25	2.2	0.55
4	0	0	0	0	0	0.00
			R: 1.1			R: 1.05

We see that the second mutation increases R from 1 to 1.05. This is a smaller increase in fitness than that for the first mutation, because a smaller number of individuals survive to age 3 than to age 2.

The principle that increasing survival and fecundity at earlier ages has a larger effect on fitness than at later ages is the basis of the two major factors responsible for the evolution of senescence and life span. One, first identified by Peter Medawar [36], is **mutation accumulation**: mutations that compromise biological

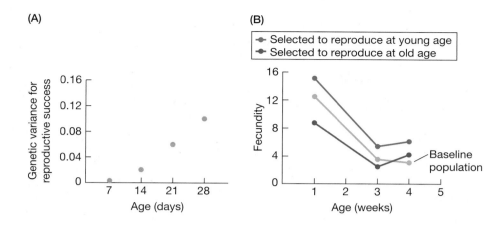

FIGURE 11.5 Evidence for two hypotheses for the evolution of senescence and life span. (A) The genetic variance for reproductive success increases with age in *Drosophila melanogaster*, as expected under the mutation accumulation hypothesis. (B) Several laboratory populations of *Drosophila* were selected for reproduction at a young age (by propagating offspring only from young parents) or for reproduction at an older age (by growing flies only from eggs laid by old females). Flies from the old-selected populations laid fewer eggs when they were young (1 week after reaching adulthood), relative to the nonselected baseline populations. This effect is expected under the antagonistic pleiotropy hypothesis. (A after [27]; B after [44].)

functions reduce fitness less, the later in life they exert these effects. That is, selection against these mutations is weaker, and so they persist at higher frequencies in the population than if they affected younger individuals (see Chapter 5). Mutation accumulation will cause the genetic variance for reproductive success to be greater for older than for younger age classes. Exactly that pattern was found in a study of a laboratory population of *Drosophila melanogaster* (**FIGURE 11.5A**) [27]. Studies in several species of birds and mammals have found that mating between relatives results in greater inbreeding depression expressed at later than at earlier ages [32]. Because inbreeding depression is caused by homozygosity for partially recessive deleterious alleles, this pattern is consistent with mutation accumulation.

Because of allocation trade-offs, alleles that increase reproduction early in life are likely to have a pleiotropic effect on reproduction or survival later in life. The greater fitness impact of early reproduction causes the advantage of reproducing when young to outweigh the pleiotropic disadvantages at greater ages. Therefore, reproduction will be expected to dwindle with age, and perhaps to cease altogether. The selective value of surviving to later ages likewise declines, finally to zero. Based on this principle, George Williams proposed a second factor that can explain senescence and limited life span: **antagonistic pleiotropy** [62]. Williams suggested that a great many genes are likely to affect allocation to reproduction versus self-maintenance—that is, they incur a cost of reproduction. Alleles that increase allocation to reproduction (reproductive effort) early in life will thus reduce function later in life.

Antagonistic pleiotropy can cause a negative relationship between early reproduction and both longevity and later reproduction. This has been found in many selection experiments with *Drosophila*. For example, several investigators have selected *Drosophila* populations for higher reproduction late in life (and therefore were also selecting for long life) (**FIGURE 11.5B**) [43]. The flies evolved higher late-life fecundity, but their egg production at younger ages was lowered—exactly as expected under the pleiotropy hypothesis.

Although both antagonistic pleiotropy and mutation accumulation contribute to senescence, many biologists think antagonistic pleiotropy is often the more important factor. Both factors can affect many genes, so it is unlikely that a single cause of senescence can ever be found.

Evolution of the Population Growth Rate and Density

The values of survival and fecundity in a life table are affected by ecological conditions. We have seen that mutations that increase *R* will spread through a

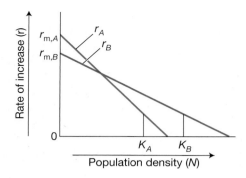

FIGURE 11.6 A model of density-dependent selection of rates of increase. Assume that a population contains two different genotypes, A and B. Their instantaneous per capita rates of increase are r_A and r_B, both of which decline as population density (N) increases. The maximum intrinsic rate of increase for genotype B ($r_{m,B}$)—its growth rate at very low density—is lower than the maximum rate of increase for genotype A ($r_{m,A}$). Genotype B has a selective advantage at high density, however, and it attains a higher equilibrium density (K) than genotype A: K_B is greater than K_A. (After [50].)

population, causing the population to evolve a higher rate of increase. But this potential for population growth is often limited, as resources become depleted or as predation or disease become more common. These factors cause population growth to be **density-dependent**, and constrain population size. In the simplest ecological models, the per capita growth rate of the population (r) declines in proportion to the population's size (**FIGURE 11.6**). If we measure population growth per unit of time, r declines from its maximum possible value, written as r_m, which in most species occurs when density is very low. The reduction of population growth causes population size to reach a stable equilibrium number that is called the **carrying capacity**, symbolized by K.

When a population is near or at carrying capacity, natural selection favors alleles that affect characteristics that increase K [11]. These will often be alleles that increase the ability of individuals to compete with others for limited resources: as the population density approaches equilibrium, a more competitive genotype may sustain positive population growth while inferior competitors decline in density. The more competitive genotype is likely to achieve a higher equilibrium density (K). Experimental *Drosophila* populations, maintained for a long time, evolve higher population densities (**FIGURE 11.7**). Species that are well adapted to crowded conditions near carrying capacity are said to be **K-selected**. Those genotypes, however, may have pleiotropic trade-offs that decrease the population's maximum growth rate when population size is far below the carrying capacity. That is, the population may evolve a lower maximum potential rate of per capita increase (r_m).

As we already noted, however, populations of some species are frequently in a state of rapid, exponential increase (as illustrated in Figure 3.7), so genotypes with higher r have higher fitness. These species are said to be **r-selected**. Life history characteristics that increase r include higher fecundity (m_x), especially at young ages. Genotypes that reproduce at an early age have a shorter generation time, and so a higher rate of increase per unit of time, than do genotypes that defer reproduction to later ages. These characteristics often have trade-off effects that make r-selected species poor competitors. During ecological succession, for example, soil that is newly exposed (e.g., landslides or abandoned crop land) is colonized by rapidly growing, fast-reproducing weeds that are later replaced by more slowly growing, K-selected trees that start to reproduce at a later age but have a long reproductive life span.

Diverse life histories

Some species conform to the "reproduce early, die young" scenario that we have described. These include some semelparous species, such as annual plants and Australian "marsupial mice" (*Antechinus*) that grow fast, reproduce, and die within a

FIGURE 11.7 Experimental evolution of population density in *Drosophila serrata*. The densities of two experimental populations (red and blue dots) increased over 70 generations, implying adaptation to high densities and improved conversion of food (supplied at a constant rate) into flies. The potential rate of increase of this species is so great that without evolutionary change, the population would have reached carrying capacity in fewer than 10 weeks, and would have remained fairly constant in size. (After [5].)

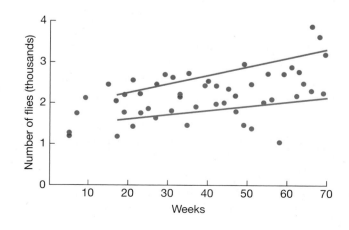

(A)

(B)

FIGURE 11.8 Some semelparous plants reproduce once, after many years, and then die. (A) The cabbage palm *Corypha utan*, of southeastern Asia, may produce up to a million flowers. (B) Bamboos engage in highly synchronous reproduction and then die, resulting in years of great food scarcity for animals that eat the shoots (such as specialized insects and giant pandas) and great food abundance for those that eat the seeds (such as finches). (Photos by D. J. Futuyma.)

year. But many other species, such as bristlecone pines and tortoises, and humans for that matter, do not fit this pattern. Perennial herbs, most trees, most familiar vertebrates, and many other species are iteroparous. Some of them, such as albatrosses and humans, delay reproduction. And some species are semelparous but reproduce at an advanced age (**FIGURE 11.8**). How have these life histories evolved?

Under some circumstances, the cost of early reproduction may exceed its benefit, and fitness may be lower than it would be if reproduction were deferred. Several factors may make deferred reproduction advantageous. For one, fecundity is often correlated with body mass in species that grow throughout life, such as many plants and fishes. In such species, allocating resources to growth, self-maintenance, and self-defense rather than to immediate reproduction is an investment in the much greater fecundity that may be attained later in life (see Figure 11.12A). In this vein, a very important factor is that many species suffer much higher mortality from predation and other ecological factors when they are young and small than when they are older and larger. The vast majority of seedlings of forest trees, for example, die before they reach even modest size; furthermore, the fecundity of trees increases greatly with size and their ability to compete for light and other resources [20]. Mammals do not grow in size very much after they reach reproductive age, but they might provide better parental care, and enhance the survival of their offspring, if they grow in strength or experience with age.

We might expect, then, that in species with high rates of adult survival, especially if young age classes have high mortality, selection favors delayed reproduction and higher reproductive effort later in life. As predicted, species that have low mortality rates as adults also reach reproductive maturity at a later age (**FIGURE 11.9**) [47, 53].

Experiments have also provided strong support for life history theory. David Reznick and colleagues have studied

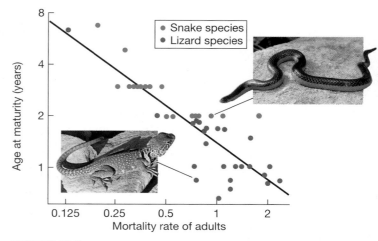

FIGURE 11.9 Among species of snakes (e.g., the worm snake, *Carphophis amoenus*) and lizards (e.g., the collared lizard, *Crotaphytus collaris*), the lower the annual mortality rate of adults, the later they reach reproductive maturity. This pattern conforms to the prediction that delayed onset of reproduction is most likely to evolve in species with high rates of adult survival. (After [53].)

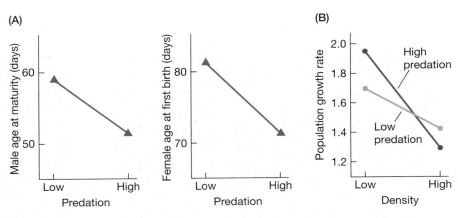

FIGURE 11.10 Life history variation in guppies in Trinidad, where waterfalls separate guppies into two populations: below the waterfall, they are in a high-predation (HP) environment with large predatory fishes; above the waterfall is a low-predation (LP) environment, where guppies reach higher density and there is strong competition for food. (A) Both sexes of guppies from the HP environment start reproducing earlier in life than those from the LP environment. (B) An experiment shows that life history differences between guppies from the HP and LP environments are genetic. When reared in a controlled environment, populations founded with guppies from the LP environment are less sensitive to the effects of density than those from the HP environment: the LP fish are adapted to high density. The population growth rate was estimated from survival and fecundity values measured over a 28-day interval. (A data from [48]; B after [6].)

guppies in Trinidad [6, 48]. In some streams, the cichlid fish *Crenicichla* preys heavily on guppies. In other streams, or above waterfalls, *Crenicichla* is absent and there is much less predation; as a result, the density of the guppy population is higher and there is stronger competition for food. The low-predation (LP) guppy populations are therefore more *K*-selected than are the high-predation (HP) populations that coexist with *Crenicichla*. When guppies from both situations are reared in a common environment, HP populations mature earlier, have more offspring, and devote more resources to larger offspring than do LP populations: features expected to evolve under *r*-selection (**FIGURE 11.10A**). Moreover, experimental populations of guppies moved from HP to previously guppyless LP environments rapidly evolve life history characteristics typical of LP populations. When experimental populations of HP and LP guppies are subjected to different density levels, the growth rate of LP populations is less depressed by high density, showing that these fishes are more adapted to contend with high-density conditions (**FIGURE 11.10B**).

Is it better to reproduce only once or repeatedly? In theory, a semelparous, or "big bang," life history may be favored by selection if the probability of survival increases with body mass and if there is an exponential relationship between body mass and reproductive output [38, 51]. These conditions have been documented in many species of semelparous plants (see Figure 11.8) [38]. Bamboos, agaves, cabbage palms, and other species reproduce only after many years, produce massive numbers of seeds, and then die. Compared with iteroparous species of trees in an Amazonian rainforest, a semelparous tree (*Tachigali vasquezii*) has a very rapid growth rate, which it achieves by producing low-density wood. The rapid growth reduces the risk of dying before maturity, when the tree produces a very high number of seeds. These results together compensate for not reproducing repeatedly [46]. A "big bang" life history is also advantageous if reproduction is so stressful or risky that an individual is unlikely to reproduce more than once. The most famous example is migratory salmon that expend enormous energy and face great hazards in swimming upstream from the ocean in order to spawn in streams. Once they

(A)

(B)

FIGURE 11.11 A "big bang" life history. (A) Coho salmon (*Oncorhynchus kisutch*) swim from the ocean up raging rivers to streams where they spawn. (B) Their physiological and physical condition deteriorates as all their energy goes into reproductive activities, and they die as poor semblances of their former selves.

arrive, they shut down immune defenses and other physiological functions, and expend all their energy in a frenzy of reproduction (**FIGURE 11.11**).

Iteroparity, however, can be advantageous for several reasons. For one, it increases the chance of successful reproduction in fluctuating environments, when reproduction or offspring survival vary from one reproductive season to another. It is also advantageous if adult mortality is low and greater fecundity can be achieved by deferring reproduction to older age classes (**FIGURE 11.12A**). In this case, reproductive effort, in each reproductive episode, is expected to be lower in iteroparous than in semelparous organisms [49]. This pattern has been found in comparisons among species within several taxa. For example, inflorescences make up a lower proportion of plant weight in perennial than in annual species of grasses (**FIGURE 11.12B**) [63].

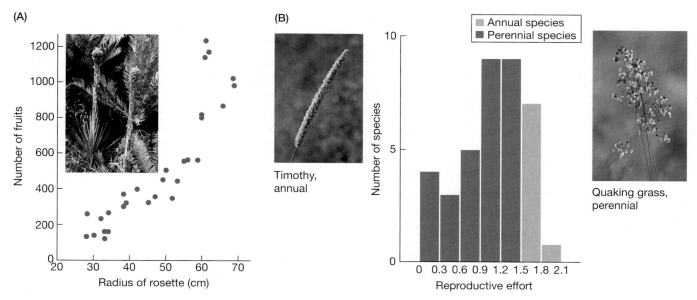

FIGURE 11.12 (A) The number of fruits increases exponentially with plant size (measured by the radius of the rosette of leaves) in the terrestrial bromeliad *Puya dasylirioides*. Larger plants are older. (B) Reproductive effort—measured here by the proportion of biomass allocated to inflorescences—is greater in annual (semelparous) species of grasses, such as *Phleum pratense*, than in perennial (iteroparous) species, such as *Briza media*. (A after [3], photo by D. J. Futuyma; B after [63].)

Number of offspring

All else being equal, a genotype with higher fecundity has higher fitness than one with lower fecundity. Why, then, do some species, such as humans, albatrosses, and kiwis, have so few offspring?

The answer, again, is trade-offs. The British ecologist David Lack proposed that the optimal clutch size for a bird is the number of eggs that yields the greatest number of surviving offspring [33]. The number of surviving offspring from larger broods may be lower than the number from more modest clutches because parents are unable to feed larger broods adequately. This decrease in offspring survival has proved to be one of several costs of large clutch size in birds. Excessively large clutches may also reduce the parents' subsequent clutch size and survival [55]. A modest number of eggs per clutch may therefore result in higher fitness than a greater number.

The great tit (*Parus major*) has been the subject of many ecological studies because it is abundant and will use boxes provided for nesting, which enables researchers to monitor the birds' lives. A long-term study of survival and reproduction was performed in the Netherlands that involved changing brood size by moving some hatchlings among nests [57]. The researchers estimated the effects of these treatments on fitness. Artificially increasing brood size decreased fitness because it lowered survival in the nest, survival from fledging to the next breeding season, and the probability that the parents would lay a second clutch in the same year. Decreasing brood size also reduced fitness, simply because the nests produced fewer fledglings. From these data, the clutch size that would maximize fitness was estimated to be 8.9 eggs, close to the natural mean of 9.2.

At a given level of reproductive effort, there must be a trade-off between the number of offspring and their size, ranging from many but small to large but few. Relative to adult body size, the largest offspring among birds and mammals, or those that require the most care, are in those species that have only a single offspring at a time, such as kiwis, albatrosses, elephants, and humans. Larger offspring are advantageous in species in which, because of their habitat or mode of life, starting life at a large size greatly enhances the chance of survival. Among plants, the wind-dispersed seeds of orchids are microscopic, and can grow only if the embryo becomes associated with a mycorrhizal fungus. At the other extreme, the water-dispersed seed of a coconut palm (*Cocos nucifera*) can weigh up to 20 kg (see Figure 11.2). Seeds are typically larger in plants that germinate in the deep shade of closed forests than in species that germinate in well-lit sites, such as early-successional disturbed environments or light gaps formed by treefalls in forests, because the survival and growth of a seedling under adverse conditions are enhanced by the food stored in a large seed's endosperm (**FIGURE 11.13A**) [18]. Perhaps for this reason, trees and vines generally have larger seeds than forbs (herbs) and grasses, most of which grow in open habitats (**FIGURE 11.13B**) [39].

Life histories and mating strategies

Males as well as females are subject to costs of reproduction, and that fact underlies some interesting variations in life histories. For example, some plants, annelid worms, fishes, and other organisms change sex over the course of the life span (a phenomenon called sequential hermaphroditism). In species that grow in size throughout reproductive life, a sex change can be advantageous if reproductive success increases with size to a greater extent in one sex than in the other (**FIGURE 11.14**). For example, the pollen required to fertilize many ovules requires much less energy to produce than an equivalent number of seeds. Many species of squashes (Cucurbitaceae) and other plants produce male flowers when they are small, and switch to producing female flowers when they become larger and can

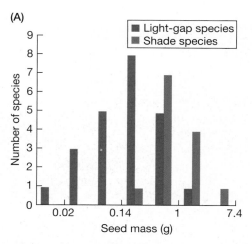

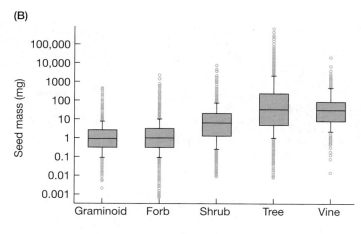

FIGURE 11.13 The size (mass) of individual seeds is correlated with the microhabitat in which the seedlings grow. Larger seeds are thought to be adaptive for species whose seedlings grow slowly because of dim light. (A) Among 40 tree species from tropical moist forests, those that germinate in light gaps have smaller seeds than those that germinate in subcanopy shade.

(B) Seed mass of more than 7200 species of plants. Forest species (trees and vines) have larger seeds, on average, than do the other growth forms, which predominate in open environments. The median of each sample is surrounded by a box representing the 25th to 75th percentiles, and lines indicating the 10th and 90th percentiles. Dots show outliers. (A after [18]; B after [39].)

obtain enough energy to develop more seeds. A female slipper shell (*Crepidula fornicata*) carries a stack of smaller males; when she dies, the biggest male changes sex, having become large enough to produce abundant eggs (see Figure 11.14C). Conversely, many sex-changing fishes are females first. In the bluehead wrasse (*Thalassoma bifasciatum*), some individuals start life as females and later become brightly colored "terminal-phase males" that defend territories and achieve high reproductive success by mating with many females (see Figure 11.14B) [60]. Almost all species of sex-changing fishes, as well as other sex-changing animals, change sex when they have reached about 70 per cent of their maximum size [1], as predicted by mathematical theory [12].

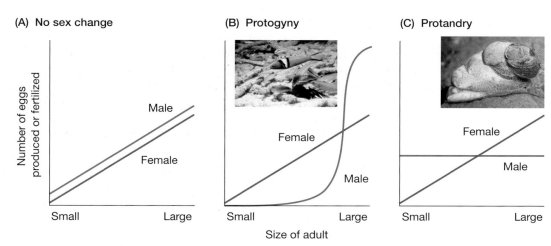

FIGURE 11.14 A model for the evolution of sequential hermaphroditism. (A) When reproductive success increases equally with body size in both sexes, there is no selection for sex change. (B) A switch from female to male (protogyny) is optimal if male reproductive success increases more steeply with size than female reproductive success does. This is the case in the bluehead wrasse

(*Thalassoma bifasciatum*) shown here: females are yellow; males are blue, white, and green. (C) The opposite relationship favors the evolution of protandry, in which males become females when they grow to a large size. For example, a female slipper shell (*Crepidula fornicata*) carries a stack of males; when she dies, the lower-most male becomes female. (After [60].)

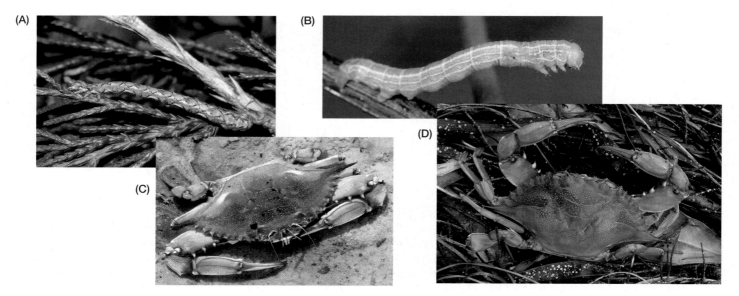

FIGURE 11.15 Specialists and generalists. (A) The larva of the juniper geometer (*Patalene olyzonaria*) feeds only on juniper trees, and closely resembles juniper foliage. (B) The larva of the fall cankerworm (*Alsophila pometaria*) is a generalist that feeds on diverse species of trees. (C) The blue crab (*Callinectes sapidus*) is found in waters with a broader range of salinity than is (D) the less tolerant lady crab (*Ovalipes ocellatus*).

Specialists and Generalists

As noted in the previous section, understanding the evolution of life history variation generally focuses on survival and reproduction components of fitness. But a broader conception of an organism's life history includes many other aspects of its life, such as dispersal (see Chapter 8) and its use of habitats, food, and other resources. For example, if you were to join the growing crowd of moth enthusiasts in eastern North America, you might want to add the juniper geometer (*Patalene olyzonaria*) to your "life list." The best way to do that would be to find caterpillars (**FIGURE 11.15A**) and rear them until they become moths. The only way to find the caterpillars is to search the foliage of juniper trees, because the juniper geometer eats nothing else. In the same habitat, you might find larvae of the fall cankerworm (*Alsophila pometaria*; **FIGURE 11.15B**), a member of the same moth family, on oaks, maples, cherries, elms, hickories, and many other trees. With respect to diet, one species is a specialist and the other a generalist. Such a distinction can be made with reference to many aspects of organisms' ecology. For example, some species of crabs and other marine animals are restricted to full-salinity waters, while related species may be found both in seawater and brackish estuaries (**FIGURE 11.15C,D**). In both of these examples, one species is more specialized than the other; it is said to have a narrower **ecological niche**.

The range of environments that a species can tolerate is often matched by the amount of variation in its habitat. Many tropical species experience less variation in temperature than do species at higher latitudes, where seasonal change is much more pronounced [28]. As expected, temperate-zone species of *Drosophila*, frogs, and other ectothermic animals tend to have broader temperature tolerance than do tropical species (**FIGURE 11.16**) [24, 42, 56]. However, this pattern is based mostly on the greater cold tolerance of high-latitude species; the critical thermal maximum (CT_{max}), above which the animal cannot function, is much the same, regardless of the species' geographic distribution. This has led to the suggestion that CT_{max} is near or at its upper evolutionary limit. If so, tropical species that already live near the upper limit may be especially endangered by global warming [16].

Why have species evolved differences in niche width? We might think that having a broad and versatile niche would usually be advantageous, because environmental factors vary in space and time: it should be advantageous to tolerate changes in salinity or temperature, or to switch to a different host plant if the usual or best host becomes rare. Even if the environment is constant, individuals in every species succumb to unfavorable conditions or to a limited supply of a resource; this was what inspired Darwin to conceive of natural selection. So selection should always favor more versatile genotypes. But no species can occupy all environments or eat all possible food items, and some are exceedingly specialized. How can we account for specialization [17, 22]?

Advantages of specialization

The plasticity that often underlies broad tolerance can be disadvantageous because it has costs [2, 4, 40]. There may be costs both to developing an altered phenotype (e.g., protein synthesis) and to maintaining the ability to do so. Also, acclimation takes time, and is triggered by cues such as temperature or day length. There is a possibility of making a mistake if the cue is somewhat unreliable, or of being in a maladaptive physiological state if the environment quickly reverses.

Conversely, specialization may be advantageous for several reasons. First, interactions with other species may favor specialization. We will see in Chapter 13 that when a species competes with other species for resources such as food or habitat, genotypes that choose or are adapted to a less used resource can have higher fitness. In some cases, it is advantageous to evolve a preference for a safe space, an environment that is relatively free of predators [10]. Herbivorous insects often suffer less predation and parasitism if they reside on some host plant species rather than on others [54].

The most fully supported hypothesis for the advantage of specialization is based on trade-offs, expressed by the aphorism "a jack of all trades is master of none." A specialist is likely to become more effective or efficient than a generalist, in which performance of any one task is likely to be compromised by the characteristics—behavioral, morphological, or physiological—needed to perform other tasks. Elizabeth Bernays proposed that trade-offs in cognitive processing may account for host specialization in some herbivorous insects [8]. She tested this idea with a species of aphid that is a host specialist in the eastern United States but a generalist that uses diverse plants in the West. The eastern specialists were quicker to find a host plant amid a bouquet of nonhost species, and tapped into the phloem for sap faster than the western generalists [9]. Morphological trade-offs have been shown in many organisms. Flowerpiercers are tropical birds, some of which feed on nectar in long, tubular flowers that are adapted for pollination by hummingbirds. These species have an unusual hooked bill with which they hold the flower and punch a hole in its base (**FIGURE 11.17A**). When the hooked tip of the bill is clipped experimentally, the birds are less efficient at obtaining nectar, but they become more proficient at eating berries, which are the main diet of other species of flowerpiercers that have a less developed hook (**FIGURE 11.17B**) [52]. A physiological trade-off has been found between adaptation to salt water and fresh water in a small crustacean, the copepod *Eurytemora affinis*, that has recently invaded the North American Great Lakes from the ocean [34]. Copepods taken from freshwater and saline environments survived better in their "home" salinity (**FIGURE 11.18**), and both samples revealed strong negative genetic correlations in survival at different salinities—evidence of a trade-off.

Specialization without trade-offs

Many studies, however, have not found evidence of trade-offs in performance across different environments. For example, growth and survival of genotypes of herbivorous insects, reared on two or more plants, are seldom negatively

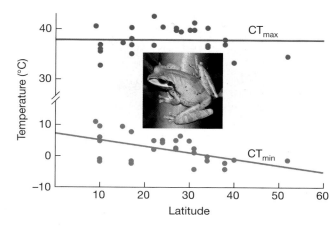

FIGURE 11.16 The critical thermal maximum (CT_{max}) and critical thermal minimum (CT_{min}) of species of frogs at different latitudes. CT is a temperature at which an important function (often measured by locomotion) fails. Temperate-zone (high-latitude) species are better able to function at low temperatures than are tropical species, but there is little difference among species in their thermal maximum. High-latitude species have wider temperature tolerance, being adapted to a more variable thermal environment. (After [24].)

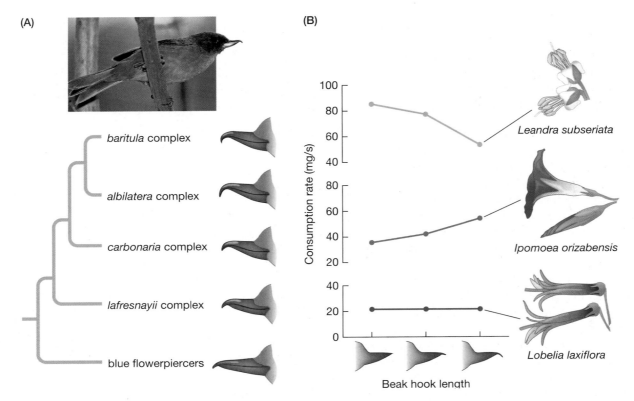

(A)

baritula complex

albilatera complex

carbonaria complex

lafresnayii complex

blue flowerpiercers

(B)

Leandra subseriata

Ipomoea orizabensis

Lobelia laxiflora

Consumption rate (mg/s)

Beak hook length

FIGURE 11.17 Experimental demonstration of a functional trade-off that makes a specialist more efficient than a generalist. (A) Flowerpiercers (*Diglossa*) differ in the length of the maxillary hook. Species with a pronounced hook feed mostly on nectar; those with a less hooked bill feed mostly on fruit. (B) Bills were experimentally modified (as shown by black profiles) in a species with a pronounced hook, and the feeding rate was then assessed when the birds were given fruit (top graph), flowers that required them to use the hook in order to get nectar (middle), and flowers from which they could obtain nectar without using the hook (bottom). Comparing the top and middle graphs, note the trade-off between fruit consumption and flower use. (After [52].)

correlated [17, 23]. But a population could nevertheless evolve to be specialized, for several related reasons.

The strength of selection for an allele that is advantageous in a certain environment depends on how much of the population experiences the environment. Suppose some larvae in a moth population feed on an abundant plant species, some feed on a rarer plant, and after they complete development, the adult moths mix and mate at random. Imagine that a mutation at locus *A* increases fitness of

FIGURE 11.18 Survival in relation to salinity in a saltwater population (St. Lawrence Seaway) and a freshwater population (Lake Michigan) of a copepod (*Eurytemora affinis*). Each is more highly adapted to its normal environment than the other population is, and both suffered higher mortality in the other population's environment. Other data showed that the higher the survival of a genotype in one environment, the lower it was in the other, indicating a trade-off within populations. (After [34].)

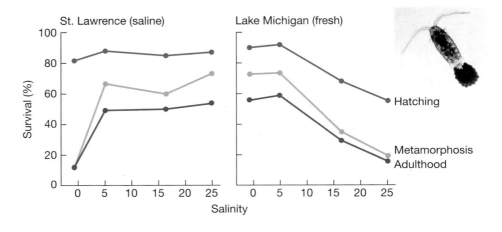

St. Lawrence (saline)

Lake Michigan (fresh)

Survival (%)

Salinity

Hatching

Metamorphosis
Adulthood

individuals that develop on the abundant plant but not of individuals that develop on the rare plant (**FIGURE 11.19A**). A mutation at another locus, *B*, improves fitness of individuals that develop on the rare plant but not on the common plant. Both mutations are advantageous, but the mutation at locus *A* increases in the whole population faster than the mutation at locus *B*, because the abundant plant supports and produces more moths. So although the average fitness in both habitats increases, adaptation to the abundant plant will increase faster than adaptation to the rare plant (**FIGURE 11.19B**). There will therefore be selection for mutations at other loci that increase the insects' preference for the abundant plant, because offspring of females that lay their eggs on that plant will be more likely to survive (**FIGURE 11.19C**). A specialized population will evolve [21, 26].

Moreover, as the population becomes more and more limited to the majority habitat, mutations that disable adaptations to the minority habitat become nearly neutral and may increase by genetic drift, resulting in mutational decay [31]. For this reason, the ability to feed on less common plants may be lost. Due to the accumulation of mutations, fitness-related traits may display greater variation among individuals reared in the minority environment than in the majority environment [61].

Experiments on niche evolution

Many investigators have compared genetic changes in laboratory populations, especially of bacteria, in constant and variable environments [30]. Populations of bacteria maintained in variable environments evolve broader niches, both by the evolution of generalist genotypes and by the maintenance of diverse specialized genotypes. In contrast, specialist genotypes usually become prevalent in constant environments. Under these conditions, both negative correlations in fitness and mutational decay can occur. In populations of *E. coli* that were propagated in Richard Lenski's laboratory for 20,000 generations with glucose as the only energy source, the ability to metabolize other substrates declined, apparently because of antagonistic pleiotropic effects [13]. In another study, an unused character, swimming motility, declined in *Pseudomonas fluorescens* bacteria at a rate that depended on how well fed the bacteria were. When resources were plentiful, mutations that degraded motility were selectively neutral, but when resources were limiting, such mutations were advantageous, probably because less active bacteria saved energy [25]. Thus, unused features may be lost either by selection or by mutation and genetic drift, depending on the population's resources.

Go to the
Evolution Companion Website
EVOLUTION4E.SINAUER.COM
for data analysis and simulation exercises, quizzes, and more.

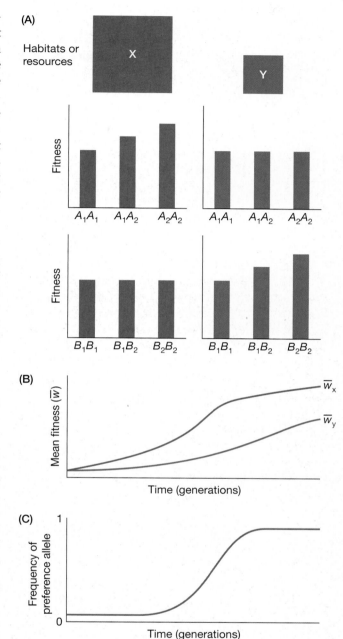

FIGURE 11.19 Adaptation and the evolution of specialization in a population that inhabits a common environment (X) and a rarer environment (Y). (A) Variation at locus *A* and locus *B* affects fitness in environment X and environment Y, respectively. Equal selection coefficients are assumed for both loci. (B) Because the more abundant environment X contributes more individuals to the population in each generation, selection increases the frequency of allele A_2 faster than the frequency of B_2, so mean fitness in environment X ($\overline{w}_X$) increases faster than mean fitness in environment Y ($\overline{w}_Y$). (C) Because the population has become better adapted to environment X than environment Y, selection increases the frequency of an allele that inclines individuals to prefer to use environment X.

SUMMARY

- Adaptations such as reproductive rates and longevity can best be understood from the perspective of individual selection. Life history traits are components of the fitness of individual genotypes, which is the basis for natural selection.

- The major components of fitness are the age-specific values of survival, female fecundity, and male mating success. Natural selection on morphological and other phenotypic characters results chiefly from the effects of those characters on these life history traits.

- An organism allocates energy and resources among several functions, such as reproduction and survival. The trade-off between reproduction and survival, or cost of reproduction, prevents organisms from evolving indefinitely long life spans and infinite fecundity.

- The effect on fitness of changes in survival (l_x) or fecundity (m_x) depends on the age at which such changes are expressed and declines with age. Hence selection for reproduction and survival at advanced ages is weak.

- Consequently, senescence (physiological aging) evolves. Senescence appears to be a result, in part, of the negative pleiotropic effects on later age classes of genes that have advantageous effects on earlier age classes. In addition, more deleterious alleles are expressed at later ages.

- Reproduction at a later age may maximize fitness if juveniles have high mortality, adults have high survival, and large body size greatly increases fecundity. Under these conditions, there can be selection for long life. In populations that are frequently growing in number, selection favors early reproduction and a short generation time. The life history of many species lies on a fast–slow continuum, ranging from rapid maturation, short life, and numerous small offspring to delayed maturation, long life, and fewer but larger offspring.

- The optimal number of offspring is affected by a trade-off between number and the size (mass) of each offspring, and by the optimal reproductive effort at that age—the parent's allocation to reproduction versus continued survival.

- Because lower fecundity and delayed reproduction can evolve, the intrinsic rate of population increase—the maximum rate of increase, which occurs at low density, may evolve to be lower. These features often evolve, especially in stable populations that are limited by resources and are not increasing anyway.

- In addition to survival and reproduction, the life history of a species includes its ecological niche. Species vary in niche width—the range of conditions they tolerate or resources they use. Broad tolerance, often enabled by phenotypic plasticity, has some costs, such as lowered efficiency because of trade-offs between functions. Specialization may evolve because it increases efficiency or because of relaxed selection for fitness in a relatively rare environment or habitat. Mutations that disable features adapted to rare environments may increase by genetic drift.

TERMS AND CONCEPTS

antagonistic pleiotropy

carrying capacity

cost of reproduction

density-dependent population growth

ecological niche

evolutionary ecology

intrinsic rate of increase

iteroparous

K-selection

life table

lifetime reproductive success

mutation accumulation

r-selection

reproductive effort

semelparous

senescence

trade-off

SUGGESTIONS FOR FURTHER READING

The Evolution of Life Histories, by S. C. Stearns (Oxford University Press, Oxford, 1992), and *Life History Evolution*, by D. A. Roff (Sinauer Associates, Sunderland, MA, 2002), are comprehensive treatments of the topics discussed in this chapter. Among theoretical treatments, *Evolution in Age-structured Populations*, by Brian Charlesworth (Cambridge University Press, Cambridge, UK, 1994, is a key reference.

See Robert D. Holt, "Evolution of the ecological niche" (pp. 288–297 in *The Princeton Guide to Evolution*, edited by J. B. Losos, Princeton University Press, Princeton, NJ, 2014) for an outstanding introduction to this topic. The most comprehensive treatment of the evolution of niche width is by D. J. Futuyma and G. Moreno, "Evolution of ecological specialization" (*Annu. Rev. Ecol. Systemat.* 19: 207–233, 1988). A more recent review, focused mostly on herbivorous insects, is "Revisiting the evolution of ecological specialization, with emphasis on insect-plant interactions," by M. L. Forister and colleagues (*Ecology* 93: 981–991, 2012).

PROBLEMS AND DISCUSSION TOPICS

1. Female parasitoid wasps search for insect hosts in which to lay eggs, and they can often discriminate among individual hosts that are more or less suitable for their offspring. Behavioral ecologists have asked whether or not the wasps' willingness to lay eggs in less suitable hosts varies with the female's age. On the basis of life history theory, what pattern of change would you predict? Does life history theory make any other predictions about animal behavior?

2. Suppose that a mutation in a species of annual plant increases allocation to chemical defenses against herbivores, but decreases production of flowers and seeds (i.e., there is an allocation trade-off). What would you have to measure in a field study in order to predict whether or not the frequency of the mutation will increase?

3. In many species of birds and mammals, clutch size is larger in populations at high latitudes than in populations at low latitudes. Species of lizards and snakes at high latitudes often have smaller clutches, however, and are more frequently viviparous (bear live young rather than lay eggs) than are low-latitude species. What selective factors might be responsible for these patterns?

4. Compared with most other mammals, primates and bats have lower fecundity and a later age at first reproduction. Why might that be?

5. Shrimps of the genus *Pandalus* mature as males but later in life change into females. The shrimps are commonly fished with nets that tend to capture large individuals. How do you predict that these shrimps will evolve in response to the selective removal of the largest and oldest individuals in their populations?

6. Suppose you are studying an organism that shows strong signs of declining health and reproduction late in its life. What are two hypotheses that could explain this drop in fitness? How might you distinguish between them experimentally?

7. Why are species of weedy plants more likely to be *r*-selected than *K*-selected? Why are most species with large body size *K*-selected? What other general patterns of lifestyle are associated with either *r*- or *K*-selection?

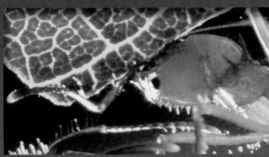

12

Cooperation and Conflict

Look at an ant nest. It is familiar, yet utterly remarkable. Hundreds to millions of individuals—a mother and her many, many nonreproducing daughters—perform a complex ballet of cooperative behaviors to gather food, raise offspring, and defend the nest. Leafcutter ant nests can include tens of millions of individuals, all daughters of a single queen that mated once and then stored sperm in her reproductive tract so that she could fertilize eggs for years afterward. These workers differ in size and form and are specialized for different tasks. Some are soldiers that defend the nest, some cut and bring home pieces of leaves, and some are farmers that chop up the leaves and use them to grow a fungus that provides the colony's food. All these individuals sacrifice their own reproduction to increase the fitness of their queen.

But not all ants are so unselfish. In some species, female workers kill their brothers and nephews, and sometimes even kill their mother [64]. What could possibly lead to the evolution of such extreme forms of altruism and aggression in ants?

As you likely know from personal experience, complex relations within families are not limited to ants. Cooperation and conflict are found at all levels of biological organization [66]. Genes compete against genes, and offspring fight with their parents. Cooperation is also ubiquitous: the functioning of your body depends on harmonious interactions among its cells. The goal of this chapter is to understand when evolution results in cooperation and when it results in conflict.

Leafcutter ants (*Atta*) carry leaf fragments to their subterranean nest, where they are used to grow a fungus that is the ants' only food. Leafcutters are among the thousands of species of social insects with sterile workers that cooperate in

The Costs and Benefits of Interacting

A useful way to think about the interactions among individuals within a species starts with a table that involves just one actor and one recipient [36]. Interactions are classified by how they affect the fitnesses of the two individuals:

Effect on actor

	+	−
+	Mutualistic	Altruistic
−	Selfish	Spiteful

Effect on recipient

If the fitness of both individuals is increased, the action is **mutualistic**. If the actor's fitness increases but the recipient's is harmed, then the action is **selfish**. In the opposite situation, the actor suffers but the recipient benefits, and the action is **altruistic**. Last, if both individuals are harmed, the action is **spiteful**. When the fitness interests of two individuals are different, they are in **conflict**. When one individual's behavior benefits another (as in mutualism and altruism), the behavior is **cooperative**. The evolution of cooperation and conflict depends on when these kinds of behaviors are favored by natural selection. Understanding how they evolve is a major goal in the field of *behavioral ecology*.

Before diving into the details, it is important to understand the vocabulary used in this field. Although conflict and cooperation are studied in organisms ranging from bacteria to plants to vertebrates, most of the research is done on animals. As in much of biology, the language in this field is drawn from everyday speech. When we say that an organism "cooperates" or "cheats," we are describing the fitness effects of its behavior. We do not mean that animals—much less microbes or genes—consciously plan their actions. After all, a "selfish gene" is nothing more than a sequence of DNA base pairs.

Social Interactions and Cooperation

It is far from obvious how natural selection could favor cooperation. An individual can "cheat," meaning that it can benefit from the actions of others without providing benefits to them in return. If a cheater has high fitness in a population of cooperators, say because it conserves resources or reduces the risk of harm, then a mutation that causes individuals to cheat will spread, and cooperation can collapse.

The evolutionary puzzle of cooperation is illustrated by the unicellular slime mold *Dictyostelium discoideum* (**FIGURE 12.1**). This species has an odd life history [71]. When food is scarce, individual cells aggregate to form a "slug." The slug wanders a bit, then transforms into something like a very small mushroom with a spherical cap on top of a stalk. Cells in the cap form spores that disperse. The cells in the stalk die without reproducing, sacrificing themselves for the good of the cells that make the spores. Some cells carry a cheater mutation that makes spores but that avoids contributing to the stalk. In laboratory culture, the frequency of this mutation increases over the course of several life cycles [19].

Why, then, hasn't this mutation spread to fixation in natural populations, ending the cooperative behavior of the stalk-forming cells?

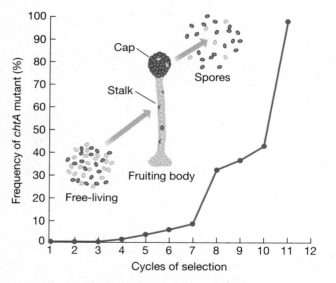

FIGURE 12.1 In the slime mold *Dictyostelium discoideum*, cells with a mutation at the *chtA* locus are cheaters that behave selfishly. In a mixture of wild-type cells (in yellow) and cells with the *chtA* mutation (blue), the mutant cells become concentrated in the cap and so are more likely to form the reproductive spores. Over the course of 11 growth and development cycles, the frequency of the selfish mutant increased in laboratory culture. (After [19].)

By analogy, why don't all ants in a colony reproduce, and why don't all humans cheat on their taxes? Until the 1960s, many biologists supposed that evolution might favor traits that benefit the population or species as a whole, even if they were detrimental to the individual (for example, by forming a stalk or committing suicide to relieve pressure on a scarce food supply). Because selfish cheater genotypes are expected to increase within populations, a trait that benefits the group would have to evolve by selection among groups, rather than by selection among individual organisms within the groups (see Chapter 3). Such **group selection** was thought to involve the increased survival of populations of altruistic individuals, and a high extinction rate of populations of selfish individuals. But simple group selection of this kind is likely to be uncommon, because it requires a high rate of population formation and extinction to counteract the strong fitness advantage of selfish individuals. Evolutionary biologists have discovered a variety of other ways that evolution causes cooperation to evolve at the expense of pure selfishness [47, 68, 81]. A key distinction among them is whether or not interacting individuals are relatives.

Cooperation among Unrelated Individuals

Natural selection favors the evolution of cooperation among unrelated individuals when the fitness costs of cooperation are equaled or surpassed by the direct fitness benefits, that is, an increase in fitness of the individual performing the behavior. An individual that joins a group can lower its risk of predation simply by finding safety in numbers [37]. This explains why birds fly in flocks, fishes swim in schools, and ungulates roam in herds (**FIGURE 12.2**). Predators such as wolves hunt cooperatively, and share prey that a single individual could not capture by itself [16]. These behaviors are beneficial to both the actor and the recipient.

The benefits of cooperation are sometimes delayed. In the lance-tailed manakin (*Chiroxiphia lanceolata*), a subordinate male forms a long-lasting association with an unrelated dominant male. The two males in this team court females by an elaborate display that they perform at the same site year after year (**FIGURE 12.3**). The males leapfrog over each other frenetically, making synchronized vocalizations as they do. Females strongly prefer teams with highly coordinated displays, and they almost always mate with the dominant male [21, 53]. When the dominant male dies, the subordinate inherits the display site, is joined by another male, and becomes the dominant male—although he may have had to wait as long as 13 years to do so. The benefit of joining a dominant male is delayed and uncertain, but a subordinate male has no choice if he is to have any chance of reproductive success.

FIGURE 12.2 European starlings (*Sturnus vulgaris*), threatened by a marsh harrier (*Circus aeruginosus*), form a tight flock. Each individual finds safety in numbers, and the density of starlings is greatest in the center of the flock because the safest place for every individual is in the center, behind as many other birds as possible. The harrier is the larger, isolated bird at right. Courtesy of Dr. Giangiorgio Crisponi.

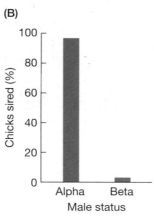

FIGURE 12.3 (A) Two male lance-tailed manakins (*Chiroxiphia lanceolata*) perform a cooperative leapfrogging courtship display to a female. The dominant male obtains all, or almost all, copulations. (B) The reproductive success of dominant (alpha) males, as determined genetically, is much higher than that of subordinate (beta) males. (A after [2]; B after [21].)

Reciprocity

Robert Trivers suggested that cooperation can evolve when one individual provides a fitness benefit to another, as long as the second individual is likely to return the favor later [74]. This kind of cooperation is called **reciprocity**. It can evolve if there are repeated interactions between individuals, if individuals recognize and remember each other, and if the benefit received is great enough to outweigh the cost of providing the benefit to others [47]. Reciprocal cooperation is known in many species of fishes, birds, and mammals [73].

Mathematical models predict the conditions under which reciprocity is favored by natural selection. While cooperative behaviors in most organisms are innate and involve no active thought, those behaviors have close parallels to some kinds of human interactions that have been well studied by economists. The famous evolutionary biologist John Maynard Smith introduced **game theory** from economics to the study of the evolution of social behaviors. Some situations that occur in humans and other animals are described by the "prisoner's dilemma" [4]. Here each of two individuals will do best by acting selfishly, but if both individuals act selfishly, they will do worse than if they both cooperate (**BOX 12A**). Game theory models show that selfish behavior is favored if individuals interact only once, but that repeated interactions can favor cooperative behavior. Thus, reciprocity can be favored when the association between individuals is so long-lasting that the benefits that each partner provides to the other feed back to the individual's own benefit [68].

Vampire bats (*Desmodus rotundus*) risk starvation if they do not get a blood meal every night (**FIGURE 12.4A**). Unrelated individuals form long-term social bonds and regurgitate blood to members of their group [83]. As a result, individuals with friends are less likely to starve. Social bonding also pays off in primates (**FIGURE 12.4B**). Offspring of female yellow baboons (*Papio cynocephalus*) that have strong bonds with other females survive better than offspring of females with weaker bonds [69]. Cooperation was able to evolve in vampire bats and baboons because individuals have long histories of repeated interactions.

Mathematical models also show that, under some conditions, cooperation is enhanced if one of the partners in an interaction punishes selfish individuals:

(A)

(B)

FIGURE 12.4 Mammal species that display reciprocity between individuals. (A) The vampire bat *Desmodus rotundus* forms roosting groups in which members that have fed successfully sometimes feed regurgitated blood to other members of the group. (B) Among primates, such as these yellow baboons (*Papio cynocephalus*), social alliances between individuals are reinforced by activities such as grooming.

Evolutionarily Stable Strategies

An important tool used to understand the evolution of cooperation is the concept of an **evolutionarily stable strategy**, or **ESS**. This is a behavior (or "strategy") with fitness greater than, or at least equal to, that of any other possible behavior if all individuals in the population behave that way. If a mutation causing a different behavior appears in a population that is at an ESS, it will not have a fitness advantage and thus will not spread. The population's behavior is therefore evolutionarily stable.

Theoretical biologists have studied how and when cooperation will evolve by determining the ESS for simplified scenarios that capture the essence of common types of social interactions. A famous example of one of these scenarios is the "prisoner's dilemma." Two gang members are caught and isolated so that they can't communicate with one another during interrogation. The jailers explain to the prisoners their options. If they both defect from their partnership and admit the terrible things they did together, they will each serve 2 years in prison. If they both cooperate with each other by refusing to talk to the authorities, they will each serve only 1 year. The last possibility is that the prisoners do different things. The prisoner who defects will be rewarded by immediate release, while the prisoner who tries to cooperate with his partner by remaining silent will be punished with 3 years in prison. These outcomes are summarized in this table:

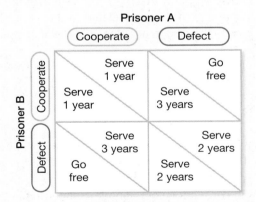

What should they do? Looking at the table, we see that Prisoner A does better if he defects than if he cooperates, no matter what Prisoner B does. The best strategy is therefore to defect, and if both do that they will spend 2 years in prison. The situation is a dilemma, however, because the prisoners could do better if they both cooperated, since then they would each serve only 1 year.

What does the prisoner's dilemma tell us about the evolution of cooperation in animals? In some species, individuals work together to hunt, attract mates, and raise families. We can use a table like this one to show the fitness effects of the different possible behaviors. Given that table, a mathematical analysis can be used to find the ESS, which predicts the behavior we expect to see in that population. If a single interaction between two individuals has the potential outcomes shown in the table, the ESS is for both to defect. We therefore predict that natural selection will not favor cooperation in single encounters between unrelated individuals.

The situation becomes more interesting, however, if the same individuals interact repeatedly. Imagine that a mated male and female are caring for their eggs in a nest. Each day, a predator attempts to take some of the eggs. Should the male and female cooperate by defending the nest, which risks injury, or should they defect and run from the predator? After a few days, the male and female learn whether their partner tends to cooperate or to defect, and each can adjust his or her behavior accordingly. Mathematical analysis shows that this situation is much more favorable to the evolution of cooperation. One behavioral strategy that has high fitness is called "tit for tat" [4]. Here each individual starts by cooperating, and then does whatever the other did in the previous round. Another strategy with high fitness is for each individual to repeat its previous action whenever it has done well in the last few interactions, but to change if not [57]. These theoretical results help explain why humans and other animals are much more likely to cooperate when they interact repeatedly with the same individuals.

punishment alters the ratio of benefit to cost [22, 28]. The punishing partner may impose "sanctions," terminating the relationship by withholding benefits from the other partner. Some of the best evidence is found in eusocial insects, as we will see shortly. One role of the immune system is to kill cancer cells, which are selfish members of the otherwise cooperative society of cells that make up an animal's body.

Shared Genes and the Evolution of Altruism

Many species provide care for their offspring, often at great effort and risk to their own survival. Mammals and birds feed their young and in some species (including our own) teach them how to survive. Grasshoppers invest energy in their eggs and invest time by burying them for safety. Plants endow their embryos with endosperm and surround them with husks, fleshy fruits, and structures that aid dispersal. In short, parents are altruistic. They enhance the fitness of other individuals—their offspring—at a cost to themselves.

Doesn't this altruism violate the selfish principle of natural selection? "The answer is obvious," you reply. "Fitness is measured by successful reproduction, and what would the mother's fitness be if all her offspring died?" That is precisely the solution. A gene can leave more copies of itself to the next generation if it increases the odds that the individual's children will survive.

This logic can be extended to more distant relatives. J. B. S. Haldane, one of the founders of evolutionary genetics, was once asked if he would give his life to save a drowning brother. He replied, "No, but I would to save two brothers or eight cousins." Haldane (who was a genius) had rapidly calculated the conditions under which natural selection will favor saving drowning relatives.

In a groundbreaking paper, William D. Hamilton reasoned that from a gene's point of view, fitness has two components [36]. Consider an allele that causes individuals to act altruistically, for example by saving drowning brothers. An individual carrying the allele can pass copies of it to his or her own children. This is the allele's **direct fitness**. The allele can also pass extra copies of itself to the next generation as the result of the increased fitness of relatives that benefit from the altruistic individual's actions. This is the allele's **indirect fitness**. The allele's **inclusive fitness** is the sum of its direct and indirect fitness. These concepts have been important in understanding not only cooperation, but also parent-offspring conflict, spite, sex ratios, dispersal, cannibalism, genomic imprinting, and other phenomena [47, 81, 1, 7].

The most common way that altruism between related individuals evolves results from **kin selection**, a type of selection based on indirect fitness. An allele that causes an individual to act altruistically decreases the fitness of the actor, but that act increases the fitness of others. If they are related to the actor, then more copies of the allele can be passed to the next generation, and the altruistic behavior can spread through the population.

This logic is formalized in **Hamilton's rule**. It states that an allele that causes an altruistic behavior will spread if the following condition is met:

$$r\,B > C \qquad\qquad (12.1)$$

The left-hand side of this inequality represents the effect of the behavior on the indirect fitness of the allele. The quantity r is the **relatedness**, also known as the coefficient of relationship. (Be aware that r is used elsewhere in this book to represent recombination rates and correlation coefficients.) Relatedness is easiest to calculate when the allele is rare. In that case, r is the probability that if the allele is carried by the actor, then it is also carried by the recipient of the altruistic behavior. B is the *fitness benefit to the recipient*, that is, the average increase in the number of offspring that the recipient will have as a result of the altruistic behavior. From the allele's point of view, the altruistic behavior increases its fitness through the recipient just as if it caused the actor to have $r\,B$ more children of its own. The right-hand side of Inequality 12.1 represents the effect of the behavior on the direct fitness of the allele. C is the *fitness cost to the actor*, that is, the decrease in the number of offspring that individual will have as the result of acting altruistically.

BOX 12B

Calculating Relatedness

The relatedness between the copies of a gene in two individuals, symbolized by r, depends on how those individuals are related and how the gene is inherited. Relatedness is simplest to calculate when an allele for altruism is rare. In that case, r is the probability that if an actor carries the allele, then the recipient also carries it. For the autosomes of a diploid species (Figure 12.B1), the alleles in a mother are related to those in her offspring, with $r = 0.5$. The alleles in a given daughter are also related to those in her brothers and sisters, with $r = 0.5$. Patterns of relatedness are different in hymenopterans, which are haplodiploid (Figure 12.B2). Alleles in a mother (queen) are related to those in her sons and daughters by $r = 0.5$. Males are haploid, have no father, and inherit all of their genes from their mother. If a worker (female) carries a rare allele, the only way her

brother can also carry it is if she inherited the allele from their mother (probability = 1/2) and if the mother passed the allele to her son (probability = 1/2). The alleles in the brother are therefore related to those in the worker by $r = 1/2 \times 1/2 = 0.25$. However, there are two ways that a new queen (the worker's sister) might also carry the worker's allele. The worker might have inherited the allele from their mother (with probability = 1/2), and if so the mother might have passed it to the sister (with probability = 1/2). Alternatively, the worker might have inherited the allele from their father (with probability 1/2). If so, then her sister is certain to carry it also, because males are haploid and always transmit all of their genes to all of their offspring. The alleles in the new queen are therefore related to those in the worker by $r = (1/2 \times 1/2) + 1/2 = 0.75$.

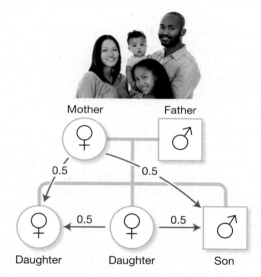

FIGURE 12.B1 Relatedness in diploid species.

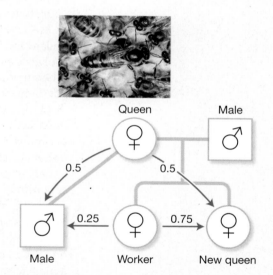

FIGURE 12.B2 Relatedness in haplodiploid species.

The overall effect of an allele on its inclusive fitness is the sum of the allele's indirect and direct effects, which is: $r B - C$.

In short, Hamilton's rule says that an allele will spread if the increase in indirect fitness outweighs the loss of direct fitness caused by the altruistic behavior. In fact, the rule applies to all behaviors, not just altruistic ones. It even works when B or C is negative, as when the actor benefits directly (in which case $C < 0$).

To make these ideas concrete, consider an autosomal locus in a female of a diploid species (**BOX 12B**). If she carries a rare allele for an altruistic behavior, there is a probability of 1/2 that it came from her mother and an equal chance that it came from her father, so a female's relatedness to each of her parents is $r = 1/2$. What about her relatedness to her siblings? No matter which parent the female inherited the allele from, there is a chance of 1/2 that a brother or a sister also inherited the allele, and so she is again related to them by $r = 1/2$. More distant relatives have lower relatedness. For example, the probability that a full cousin also carries the allele is $r = 1/8$. (Do you now understand Haldane's comment about saving drowning brothers and cousins?)

Inequality 12.1 implies that the more distantly related the beneficiaries are to the altruist, the greater the fitness benefit to them must be for the altruistic trait to spread. If an allele causes females to give care to random offspring in the population, it will not increase in frequency. That is because the fitness of all genotypes would be enhanced equally by the altruism, while the allele would still suffer a direct fitness cost. In terms of Inequality 12.1, the relatedness of random offspring to an altruistic female is $r = 0$. If there is any cost to providing care, then $C > 0$ and so the condition for the spread of the allele is not met.

Thus kin selection can favor altruism only if individuals are more likely to help kin than nonkin. Altruism can be directed toward relatives when individuals are able to distinguish related from unrelated individuals. Remarkably, female Mexican free-tailed bats (*Tadarida brasiliensis*) can find their own pups in caves that harbor millions of young bats roosting at a density of 4000 per square meter [52]. The cues used by some species to recognize kin are genetically based, while in others the cues are caused by a shared environmental imprint. Individual colonies of many ants have a distinctive odor. Nestmates cooperate with each other and battle with ants from other nests, and they discriminate between friend and foe using the odors [85].

Even if individuals cannot identify kin, they can preferentially express altruism toward kin if relatives tend to be near each other, and this can enable altruism to evolve. For example, local colonies and troops of many primates, prairie dogs, and other mammals are composed largely of relatives, and these species perform altruistic behaviors, such as giving warning calls if they see a predator [50].

In the wild turkey (*Meleagris gallopavo*), males cooperate in their mating displays (**BOX 12C**). Some males court females solo, but others form teams of brothers that are much more successful [46]. In teams of two males, only the dominant male fathers offspring, and his subordinate brother does not mate. Then why should a male choose to be a subordinate? Consider his options. If he chooses to be a subordinate, he will forgo the matings he could have if he were solo. This is C, the cost of his altruistic behavior. But by displaying with his brother, the subordinate increases his brother's mating success, which represents the fitness benefit to his brother, B. That results in a gain of indirect fitness for the subordinate, $r B$, that more than offsets the cost. Hamilton's rule is satisfied, and so an allele that causes a male to display as a subordinate will spread. On average, each copy of the allele will leave 0.8 extra copies of itself to the next generation as the result of altruism. Box 12C explains the calculation in detail.

The deer mouse *Peromyscus maniculatus* is sexually promiscuous, and sperm of several males compete for fertilizations in a female's reproductive tract. Sperm gain inclusive fitness by teaming up to form aggregates with other sperm from the same male, making it more likely that one of them will fertilize an egg. Kin selection theory predicts there should be no advantage to aggregation in species without sperm competition. That prediction is confirmed: the sperm from the same male do not preferentially aggregate in a closely related monogamous species (*P. polionotus*) [25].

Even bacteria can cooperate. *Pseudomonas aeruginosa* requires iron, which it takes up from its environment by binding iron atoms with proteins called siderophores that the bacteria excrete into their environment. Bacterial cheaters, however, take up iron bound with the siderophores produced by others, and they avoid paying the cost of producing siderophores themselves [34]. The outcome of competition between genotypes that excrete siderophores (cooperators) and genotypes that do

BOX 12C

Altruistic Mating Displays in Turkeys

Wild turkeys (*Meleagris gallopavo*) display in teams of dominant and subordinate brothers. Dominant males have higher fitness (measured by the average number of offspring) than subordinates and males that display solo. Subordinate males increase the reproductive success of their dominant brothers from 0.9 to 7, which gives a benefit to the dominants of $B = 7 - 0.9 = 6.1$ offspring. The cost to a subordinate of teaming up with his brother is the subordinate's loss of direct fitness from not displaying solo: $C = 0.9$ offspring. A subordinate does not mate, so his direct fitness is 0. Genetic analysis shows that, on average, a subordinate and dominant male are related

by $r = 0.42$. (It seems not all pairs are full brothers.) The indirect fitness gained by a subordinate through cooperating with his dominant brother is $r B = 0.42 \times 6.1 = 2.6$ offspring. That value is greater than the cost, C, and so Hamilton's rule is satisfied: an allele that causes a male to join his brother as a subordinate will spread. The altruistic behavior increases the allele's inclusive fitness by $r B - C = 1.7$ offspring. Each of those offspring has a probability of 1/2 of inheriting the allele, so each copy of the allele leaves $1/2 \times 1.7 = 0.8$ extra copies of itself to the next generation as the result of its altruistic behavior.

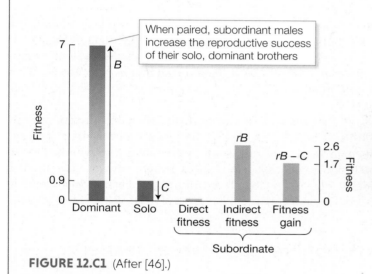

When paired, subordinant males increase the reproductive success of their solo, dominant brothers

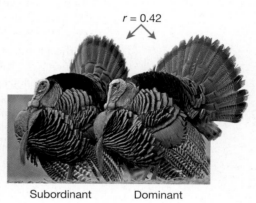

$r = 0.42$

Subordinant   Dominant

FIGURE 12.C1 (After [46].)

not (cheaters) depends on the environment they grow in (**FIGURE 12.5**). The bacteria can be maintained in the lab under conditions to produce either high or low relatedness between them, and either strong or weak competition between relatives. When the bacteria have low relatedness and are in a strongly competitive environment, the cheaters win. But when the bacteria are closely related and are in a weakly competitive environment, the cooperators can drive the cheaters to extinction.

Another pathway to the evolution of altruism is by the "green beard" effect, which occurs when a single gene codes for a phenotypic trait that enables its carrier to recognize and help other individuals with the same trait (for example, a green beard) [20]. This situation is uncommon in nature, but a few cases have been described [31]. One comes from the slime mold that we discussed earlier (see Figure 12.1). The *csA* gene encodes a cell adhesion protein that binds to the same protein in the membrane of other cells. This acts as a green-beard recognition system: cells with *csA* adhere to each other and pull themselves into aggregations. Cells that have the *csA* gene knocked out act as cheaters. If they manage to get into an aggregation with cells that have *csA*, their lower adhesion makes it more likely that

FIGURE 12.5 Evolution of cooperation in an experiment with the bacterium *Pseudomonas aeruginosa*. The cooperator genotype excretes siderophores, which are used by neighboring bacteria to take up iron from the medium. The cheater genotype does not excrete siderophores, but benefits from the siderophores made by others. Bacteria evolved in cultures that were maintained with either low relatedness (low *r*) or high relatedness (high *r*), and with either weak competition (weak comp) or strong competition (strong comp) between relatives. The cooperator genotype increased in frequency when there was high relatedness and weak competition. (After [34].)

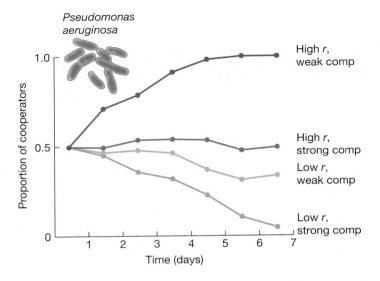

they will end up as spores. Cells with the *csA* gene are more altruistic, and they are able to prevent the cheaters from spreading in the population because they are more effective at recognizing each other and forming aggregations [62].

Spite

A behavior is spiteful if it harms both the actor and the recipient. Spite is the antithesis of altruism, but inclusive fitness theory predicts that spiteful traits can evolve. The conditions needed are that the actor be *less* closely related to the recipient than to an average member of the population, and that *harming* the recipient enhance the fitness of other individuals in the population that are more closely related to the actor [82].

An example of spite comes from bacteriocins, toxins that are secreted by many bacteria and that kill susceptible bacteria [67]. Bacteriocin-producing genotypes are resistant to the toxin because of a resistance gene that is tightly linked to the gene for the toxin. Producing bacteriocin reduces growth. However, genotypes that make bacteriocins increase in laboratory cultures [41]. By killing susceptible cells, they free up resources and enhance the growth of relatives that also carry the producer gene.

Conflict and Cooperation in Close Quarters: The Family

Some interactions within families are the epitome of cooperation, while others are the most extreme forms of conflict imaginable [23, 55]. Evolutionary biology provides unique perspectives on how and why families function as they do.

Conflict between mates

Although males and females must cooperate to produce offspring, conflict between mates is also pervasive [3]. A male can often benefit from mating with a female that is already inseminated since he may father some of her offspring. In contrast, the female often cannot increase her fecundity by mating more than once, but she can become infected, be injured, or lose time if she does. This results in **sexually antagonistic selection**, in which a trait that is favored to increase in one sex is favored to decrease in the other. In many species, males inflict harm on their mates. Groups of male mallard ducks sometimes drown females during forced copulations. Female bedbugs suffer reduced survival from traumatic insemination in which the male mates with his partner by piercing her abdominal wall [72].

Internal fertilization offers an opportunity for mates to manipulate their partners chemically. When *Drosophila melanogaster* mates, the male's ejaculate includes a cocktail of "accessory gland proteins" as well as sperm. These proteins alter the

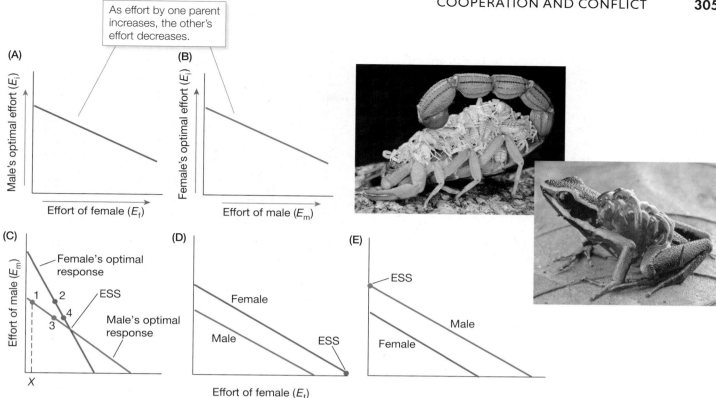

As effort by one parent increases, the other's effort decreases.

(A) Male's optimal effort (E_f) vs Effort of female (E_f)

(B) Female's optimal effort (E_f) vs Effort of male (E_m)

(C) Effort of male (E_m) vs Effort of female (E_f)
Female's optimal response
ESS
Male's optimal response
1 2 4 3
X

(D) Female / Male / ESS

(E) ESS / Male / Female

FIGURE 12.6 Analysis of the evolution of parental care using an evolutionarily stable strategy (ESS) model. (A, B) The optimal parental effort expended by each member of a mated pair declines as the effort expended by its partner increases. (C) Curves for males (blue) and females (red) plotted together. Their intersection marks the ESS. If, for example, the population starts with female effort (E_f) equal to X, male effort (E_m) evolves to point 1. This favors E_f to evolve to point 2 on the female's optimality line, which then favors E_m to evolve to point 3; but then E_f evolves to point 4. Eventually, E_m and E_f evolve to the intersection (the ESS), no matter what the initial conditions are. (D, E) Conditions can occur in which the optimal curves for the sexes do not intersect and the ESS is care by only the female (D), as in this scorpion, or the male (E), as in this poison dart frog (*Epipedobates trivittatus*). (After [17].)

female's reproductive physiology, causing her to lay eggs more rapidly. The fitness advantage to the male is that if the female later remates, his sperm will have already fertilized a larger number of her eggs. But the accessory gland proteins also decrease the female's fitness by shortening her life span [13]. William Rice designed a clever experiment in which male *D. melanogaster* could evolve but females could not [65]. After 30 generations of experimental evolution, the fitness of males had increased compared with controls, but females that mated with these males suffered greater mortality, probably because of enhanced semen toxicity. This result suggests that males and females are continually evolving, but in balance, so that we cannot see the change unless evolution in one sex is prevented.

The evolutionary interests of males and females often conflict. In many bird species that are socially monogamous, a female can copulate with another male, and her partner ends up rearing some offspring that are not his own. Even parental care involves potential conflict. Whether both parents, one parent, or neither care for their young varies greatly among species [15]. Providing care increases offspring survival, which enhances the fitness of both parents and their offspring. But parental care also carries the costs of risk, time, and energy. In species with biparental care, each parent benefits by leaving as much care as possible to the other partner. Decreasing parental care is favored by selection as long as any loss in the fitness of the current offspring is more than offset by the gain in the potential for future offspring. If offspring survival is almost as great with care from just one parent as from two, selection favors individuals that abandon the brood to the care of their partner (**FIGURE 12.6**). Selection favors defection more strongly in the

FIGURE 12.7 Parental care. (A) Great crested grebes (*Podiceps cristatus*) exemplify the many bird species in which both parents care for the young. (B) A male three-spined stickleback (*Gasterosteus aculeatus*) builds and cares for a nest containing egg clutches fathered by him. This activity can attract additional females to mate with him.

parent that pays the greater cost in caring for the offspring (in terms of lost opportunities for further reproduction).

This theory may explain major patterns in parental care [15]. In birds and mammals that must feed their young, parental care is more costly for males than for females, because males could potentially obtain other matings in the time they spend rearing a brood. That may explain why parental care in those animals is generally provided by females or by both mates (**FIGURE 12.7A**). In contrast, many fishes and frogs guard their eggs and young, but most do not feed their offspring. Males can often mate with additional females while they guard the eggs already in their nest (**FIGURE 12.7B**). In contrast, females can increase their reproductive success only by replenishing the massive resources they have spent on producing eggs. To do that, they must abandon the nest to forage. That may explain why in fishes and frogs, males provide parental care more often than females.

Murder in the family

Sometimes an individual's fitness is enhanced by killing the young of its own species [15, 38]. In lions (see Figure 10.12), baboons, and many other social mammals, males kill the offspring of other males, then father their own offspring with their victim's mothers (**FIGURE 12.8**). Selection can favor this behavior for two reasons: it eliminates the genes of competing males, and females become fertile and sexually receptive sooner if they are not nursing young. This behavior occurs mostly in species in which social groups contain more females than males and in which sexual selection is likely to be strong [48]. In this situation, there are few possible fathers of the new offspring from females whose offspring are killed, so the fitness benefit is more likely to go to a murderous male than if there were an even sex ratio.

While the murder of unrelated young might make evolutionary sense, how can we explain the fact that in some species parents kill their own children? Infanticide can be a way of adaptively regulating brood size [55]. A bird's fitness is proportional to the number of its surviving offspring, which equals the number of eggs laid, multiplied by the probability that each egg survives. Survival may decrease as number of eggs increases because of competition among the offspring for food, and because parental care of a large brood can reduce the parent's subsequent reproduction (see Chapter 11). Female mice kill some of the young in their litter

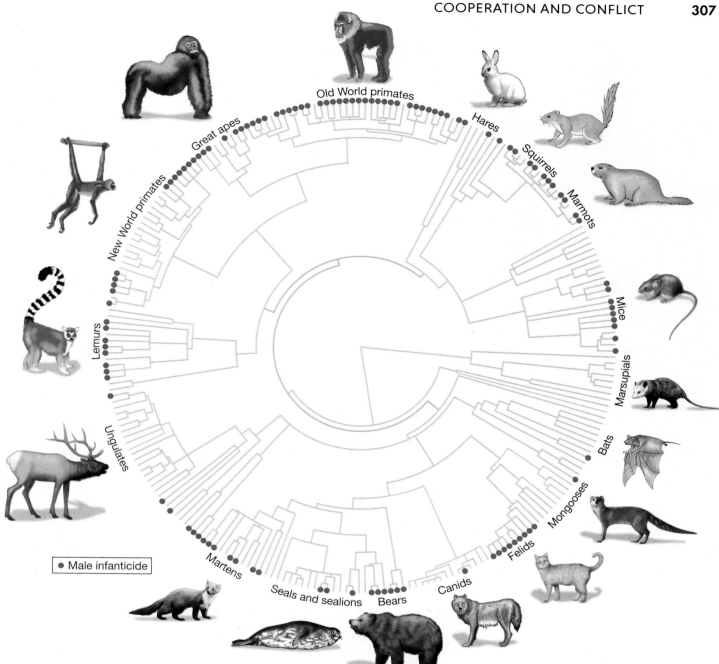

FIGURE 12.8 This circular phylogeny of mammals shows that male infanticide occurs in many species (branches with red circles at the tip). It is most prevalent in social species, less prevalent in solitary species, and least common in monogamous pair-bonding species. (After [48].)

if food is scarce or the litter is too large—behavior that increases the chance that the female will have more offspring in the future. Plants often abort many of their offspring (seeds) and reallocate their limited resources to fewer but larger seeds that have a greater chance of survival.

Aggregation can increase competition among kin for food and other resources [27, 60]. Siblings in a brood may actively fight for resources, and larger individuals may kill smaller siblings. This behavior, called siblicide, is the norm in some species of eagles and boobies. Females lay two eggs, but one of the nestlings always kills the other (**FIGURE 12.9**). The second egg may be the female's evolutionary insurance in case the first is inviable.

The ways in which animals such as these behave toward their family starkly illustrate the point that natural selection lacks any form of morality.

FIGURE 12.9 Siblicide in the brown booby (*Sula leucogaster*). The parent is sheltering a large chick that has forced its sibling out of the nest. The parent ignores its dying chick (foreground). (Photo by John Alcock.)

Parent-offspring conflict

Natural selection typically favors different behaviors in parents and offspring [33, 75]. An offspring can gain indirect fitness benefits by increasing the survival of its siblings, to which it is related by $r = 0.5$. But it has even more to gain by increasing its own survival: an individual is related to itself by $r = 1$. Selection favors an offspring to take more resources from its parents, even if that harms them and decreases the number of other offspring they have, so long as the gain in its direct fitness is larger than its loss of indirect fitness. On the other side of this interaction are the parents. Selection on their inclusive fitness favors them to maximize the number of offspring they have, not to divert extra resources to selfish ones. The result is **parent-offspring conflict**. What is best for a parent conflicts with what is best for an offspring.

In humans, parent-offspring conflict plays out even in the womb [35]. Early in the development of the placenta, cells from the embryo invade the specialized arteries of the mother that supply the embryo with blood. Once there, the cells break down the smooth muscle and nerves in the arterial walls. This prevents the mother from constricting the arteries, and so increases the supply of nutrients to the embryo. In short, the embryo has evolved to extract more resources from its mother than the mother is favored to give.

Further conflicts in the womb involve the father as well as the offspring. A gene called *IGF2* that is expressed in the fetus produces a factor that enhances fetal growth by obtaining more nutrition from the mother. Strangely, only one of the two alleles carried by the fetus is expressed, and it is the one inherited from the father. The product of a second gene, *IGF2R*, degrades the growth factor, and at this locus only the allele inherited from the mother is expressed in the fetus. The expression level of the growth factor in the fetus is therefore determined by the opposing effects of the alleles it inherited from its mother and its father.

The leading hypothesis to explain these observations starts with the idea that the father gains no fitness benefit from the mother's future reproduction if she mates with a different male. Selection therefore favors paternal genes in the fetus that enable it to get more from the mother, even if that decreases the mother's future reproduction. In response, females have evolved to suppress the tactics used by the mates' genes to exploit them. Unfortunately, this war between genes can inflict collateral damage in the form of infant pathologies [29].

Eusocial animals: The ultimate families

The most extreme altruism is found in **eusocial** animals. These are species in which some individuals do not reproduce much or at all themselves, and instead rear the offspring of others, usually their parents. The most familiar examples are found among the ants, bees, and wasps (all in the order Hymenoptera). Eusociality is also found in all species of termites (Isoptera), in several other kinds of arthropods, and in a few species of naked mole-rats (**FIGURE 12.10**) [9, 18, 44, 85].

Eusociality has evolved independently many times in Hymenoptera. In all cases, the ancestors were solitary species in which a single mated female provisioned or reared her offspring by herself [40]. In the eusocial species, reproductive females are called queens, and most of their eggs develop into workers, which are the nonreproductive females that maintain the colony. Some eggs develop into reproductive queens and some into reproductive males. In most species, whether a female becomes a queen or a worker depends on her diet, which is often controlled by the workers, and on how the workers behave toward her.

How did eusociality originate? In some species of bees, some females rear offspring with help from older offspring, while other females are solitary and receive no help. Compared with the reproductive fitness of single females, the inclusive fitness of daughters that help their mother is higher in some cases but lower in

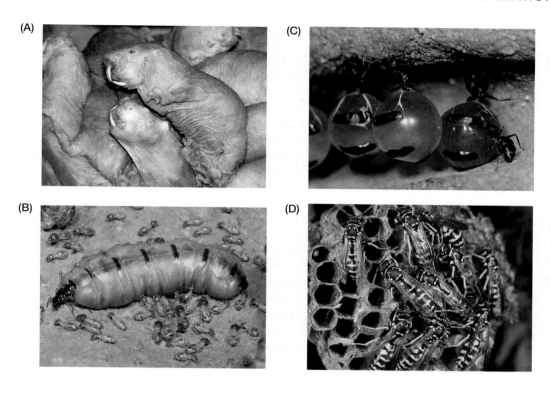

FIGURE 12.10 Some eusocial animals. (A) Several species of naked mole-rats, here *Heterocephalus glaber*, are the only known eusocial mammals. (B) This queen termite is attended by small, sterile workers and large-headed, sterile soldiers. (C) Australian honeypot ants (*Camponotus inflatus*), engorged with nectar, hang from the roof of their nest's larder. These "repletes" regurgitate nectar on demand to their worker nestmates. (D) Paper wasps (*Polistes gallicus*) at their nest.

others (e.g., [43, 86]). So some origins of hymenopteran eusociality may well have been facilitated by kin selection. But ecological and behavioral factors were also important [61, 63]. Eusociality probably evolved frequently in Hymenoptera (compared with other insects) because single females of solitary wasps and bees construct a nest such as a burrow, which requires hard work but provides shelter for the young, and the offspring are helpless larvae that the mother must feed. These hymenopterans were predisposed to sociality because they already had the habit of caring for offspring and because the nest provided a safe place—a fortress—for grown offspring to stay and to interact with their mother and younger siblings [61]. In the case of other eusocial insects, such as termites that live in dead wood, the fortress is also the food source. Another advantage of having helpers is that some can defend the larvae while others forage for food; a single female cannot do both. Moreover, dispersing from the natal nest and constructing a new nest is very risky, and the great majority of young mated females do not succeed. Thus, even a slight increment in inclusive fitness from rearing siblings may have made it advantageous to stay and help mother instead of leaving home.

Kin selection explains many aspects of cooperation and conflict in eusocial hymenopterans [7, 8, 64, 70]. These species are *haplodiploid*: fertilized eggs are diploid and develop into females, while unfertilized eggs are haploid and develop into males. A queen can decide the sex of an offspring by releasing sperm, or not, that she stored when she mated early in life. As a result of this strange genetic system, the coefficients of relationship among relatives differ from those in diploid species. Comparing the family trees in Box 12B, we see that in diploids, $r = 0.5$ between parent and offspring and between full siblings. In haplodiploid species, however, a female is more closely related to her sisters ($r = 0.75$) than she is to her sons and daughters ($r = 0.5$), and she is even less closely related to her brothers ($r = 0.25$). If a colony has only one queen that mated with only one male, workers are rearing brothers and sisters, some of which may become queens.

In a colony of hymenopterans, there is evolutionary conflict over which individuals should reproduce. A worker can gain more fitness by raising her own sons (related to her by $r = 0.5$) than by helping to raise the queen's sons (the worker's brothers, related to her by $r = 0.25$). But the queen gains more inclusive fitness through her own sons (related to her by $r = 0.5$) than through her daughters' sons (related to her by $r = 0.25$). Queens of many species therefore destroy their workers' eggs, and in some species (including honeybees, *Apis mellifera*) workers destroy the eggs of other workers [64, 78]. This is one of the best examples of *policing* of noncooperators in social species, and it illustrates that kin selection can underlie the evolution not only of altruism, but also of selfishness.

There is also conflict between queens and workers about how many males and reproductive females the colony should produce. A queen's fitness is maximized by a 1:1 sex ratio, since she is equally related to her daughters and sons. But workers can control the sex ratio among the larvae destined to be males or queens, by feeding some more than others, or even by killing some of them. In a colony with a single queen, a worker's inclusive fitness is maximized by rearing young queens and males in a 3:1 ratio, because a worker is related by $r = 0.75$ to her queen sisters, but only by $r = 0.25$ to her brothers. In colonies with multiple queens, however, workers should favor a sex ratio closer to 1:1 (because not all females are full sisters and so $r < 0.75$). Data support these predictions [18, 61]. Within some species of ants, wasps, and bees, some colonies have a single queen that mated with a single male, and others have colonies with either multiple queens or a single queen that mated with multiple males. A review of studies of species with this kind of variation found that the prediction about sex ratio was upheld in 18 of the 19 species [61].

Levels of Selection

The basic principle of evolution by natural selection is simple: the entities that make more copies of themselves increase in frequency through time. Usually, the "entities" in question are alleles. But the same principle applies to anything that can replicate—bits of DNA, mitochondria, entire chromosomes, even groups of individuals. In this section we will see how Darwin's concept of natural selection can be applied at these different levels to understand important features of the natural world.

Selfish DNA

Meiosis is a remarkably democratic affair: the two alleles carried by an individual at a locus usually have an equal chance of being passed to an offspring. But consider a mutation that can tilt the odds in its favor so that its chances are greater than 50 percent. Any mutation that can do so will enjoy an evolutionary advantage, a situation called **segregation distortion**. It can spread in the population even if it actually decreases survival or reproduction. Given the strong evolutionary incentive to cheat at inheritance, it is surprising how fair meiosis usually is.

There are, however, mutations that do cheat [47b]. Meiosis is fundamentally different in males and females, and so the ways that mutations are able to break Mendel's laws are quite different in the two sexes (**FIGURE 12.11**). Alleles that cause segregation distortion in males are known from diverse groups, including mammals, insects, and fungi. Some alleles at the *t* locus in the house mouse (*Mus musculus*) are transmitted to about 95 percent of a heterozygous male's sperm. Sperm that carry of these alleles gain an advantage by secreting a toxin that kills other sperm in the testes. These sperm are themselves immune to the toxin because they carry a resistance allele at a second locus. This transmission advantage causes the selfish alleles at the *t* locus to spread, even though they reduce the fertility of males. A killer allele must be inherited together with the resistance allele, or else sperm with the killer allele will commit gametic suicide. That explains why in mice and other species with this kind of segregation distortion, the two loci are always tightly linked, and are often found in regions of the genome with reduced recombination (such as the sex chromosomes).

(A)

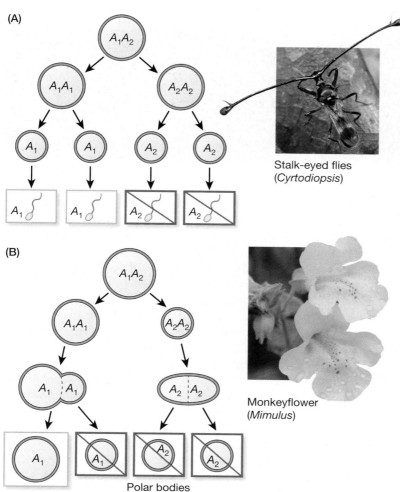

Stalk-eyed flies
(*Cyrtodiopsis*)

(B)

Polar bodies

FIGURE 12.11 In the battle to be passed on to the next generation, some mutations gain an advantage by cheating at the laws of inheritance. The diagrams at left show how a cheating A_1 allele is transmitted during the two meiotic cell divisions and then to the gametes. (A) In males of stalk-eyed flies (genus *Cyrtodiopsis*), segregation distortion results when the sperm that carry an allele (shown as A_1) kill other sperm in the male's testis that do not have that allele [59]. (B) In monkeyflowers (genus *Mimulus*), some chromosomes benefit from segregation distortion [24, 26]. During meiosis, chromosomes with certain DNA sequences in their centromeres (shown as A_1) are more likely to enter ovules, while their rivals end up in polar bodies, which die. (After [47b].)

Monkeyflower
(*Mimulus*)

The loss in male fertility generates a strong evolutionary advantage to mutations at yet other loci to suppress the allele causing segregation distortion. Many distorter systems arise and are then shut down by countermeasures that evolve elsewhere in the genome [10]. This genetic conflict can contribute to speciation (see Chapter 9). Consider a population in which a distorter system arose and then was suppressed by other loci. Now individuals from this population meet and mate with others from a population that does not have the distorter or the suppressor. Some hybrids will inherit the distorter but not the suppressor, reactivating the distorter and once again depressing male fertility. In some taxa, this may be an important source of genetic incompatibilities between populations and species [58].

In females, alleles cheat the laws of inheritance in other ways. Only one of the four products of meiosis in plants and animals becomes a gamete (see Figure 12.11B). Any allele that can increase its odds of ending up in the gamete will enjoy an evolutionary advantage (see Figure 12.11B). Centromeres have the opportunity to do just that [39]. During meiosis, each pair of chromosomes segregates when their centromeres attach to microtubules and then pull themselves toward opposite poles of the cell. The DNA sequence of the centromere and the proteins that bind to it affect how quickly a chromosome moves toward the pole. Depending on the species, chromosomes that segregate more quickly or more slowly have the best chance of ending up in a gamete. This can drive the rapid evolution of the position and genetic sequence of the centromere, and of proteins that interact with it.

Mosquitoes transmit some of the world's most important infectious diseases, including malaria. The secret to controlling these diseases may come from manipulating the rules of inheritance in the mosquitoes. If mutations that make mosquitoes

FIGURE 12.12 In this photograph of the chromosomes that make up the human genome, regions that are rich in the *Alu* transposon fluoresce in green. More than 1 million copies of this element are embedded in the genome. *Alu* is only one of many transposons that together make up half of our DNA. (From [6].)

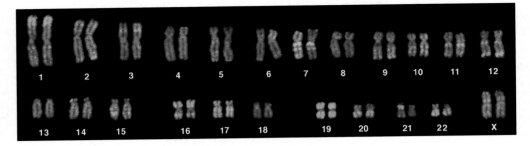

resistant to malaria could somehow be introgressed into natural populations, it might be possible to eradicate the disease. "Gene drive" may now make that possible [30]. A resistance mutation has been genetically engineered that transmits itself from a mosquito to its offspring nearly 100 percent of the time, rather than the usual 50 percent. While this is an exciting prospect for public health, it will be important to evaluate the potential risks of genetically manipulating natural populations.

Transposable elements, or **transposons**, are a type of selfish DNA that is closer to home—they make up almost half our genome (**FIGURE 12.12**). Transposons are short sequences of DNA that are able to insert additional copies of themselves in the genome [49]. They are genetic parasites that do not leave their host, and in fact one hypothesis for their origin is that transposons began as viruses. Transposons have been spectacularly successful, particularly in eukaryotes. Like other parasites, transposons are typically bad for their host. A transposon generates a mutation when it inserts itself into a new place on a chromosome, and many of these mutations are deleterious. Organisms have evolved a variety of mechanisms to combat transposons.

Transposons do not exist to improve the fitness of their host. Instead, they exist simply because selection on the transposons favors those that leave more descendant copies of themselves. Transposons are explored further in Chapter 14, where we will see that they are one of the most important factors in the evolution of genome size in eukaryotes.

Many other kinds of selfish genetic elements have been discovered. A small extra chromosome called PSR (for "paternal sex ratio") in the wasp *Nasonia vitripennis* is transmitted through sperm but not through eggs [79]. When a sperm carrying PSR fertilizes an egg, all the other chromosomes inherited from the father disintegrate, leaving only the maternal set of chromosomes intact. Because diploid eggs develop into females and haploid eggs into males in wasps, the degeneration of male chromosomes converts the female into a male, and PSR is passed to the next generation through his sperm. As PSR spreads in a population, the sex ratio becomes more and more skewed toward males, which in principle could even drive a population to extinction. Natural selection does not always favor traits that make species more likely to survive.

Selfish mitochondria

Thyme is an herb used in cooking that comes from a plant (*Thymus vulgaris*) with an unusual breeding system (**FIGURE 12.13**). Most plants are hermaphrodites—their flowers have both male and female parts. Some individuals, however, are sterile in their male function and reproduce only as females [14]. Sterility is caused by a mutation that is inherited through the cytoplasm, not the nucleus, so it is called *cytoplasmic male sterility* (or *CMS*). Sterility is caused by the CMS^+ allele of a mitochondrial gene. Hermaphrodites carry the sterility allele, and also an allele called R^+ at a nuclear locus that restores male fertility. In some other species, remarkably, all individuals are hermaphrodites and carry both CMS^+ in the mitochondrion and R^+ in the nucleus. Why should one gene exist, only to be counteracted by another?

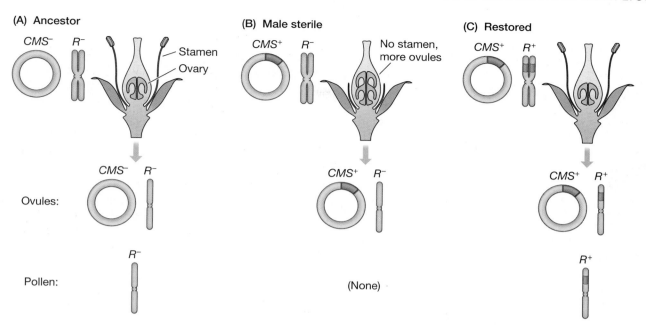

FIGURE 12.13 *Cytoplasmic male sterility (CMS)* in thyme plants (*Thymus vulgaris*) illustrates genetic conflict. Next to each flower is a schematic showing the mitochondrion (the circle) with its genotype at the *CMS* locus, and a pair of nuclear chromosomes that carry the *R* restorer locus. Mitochondria are transmitted only through ovules. (A) Plants that are *CMS⁻* and *R⁻* produce both ovules and pollen. The ovules transmit both the *CMS* and *R* loci, while the pollen transmits only the *R* locus. (B) Plants that are *CMS⁺* and *R⁻* are male sterile. Pollen production is eliminated, which increases the ovule number because the plant reallocates energy and resources from pollen to ovules. This increases the number of copies of the *CMS⁺* allele passed to the next generation, causing male sterility to spread. (C) Plants that carry the *R⁺* allele have their male fertility restored, which causes that allele to spread in populations with the *CMS⁺* allele. A population fixed for alleles *CMS⁺* and *R⁺* may be phenotypically indistinguishable from the ancestral population that was fixed for *CMS⁻* and *R⁻*.

Mitochondria are maternally inherited. Natural selection therefore favors any mutation in mitochondria that increases the number of ovules that females produce. The effect on male reproduction does not matter in the slightest to the mitochondria, since they are not transmitted through pollen. When the *CMS⁺* allele knocks out male reproductive function, resources are diverted from making pollen to making more ovules. This gives the *CMS⁺* allele a fitness advantage, and it spreads in populations of thyme.

But selection on nuclear genes favors a very different outcome. Recall from Chapter 10 that selection on those genes favors a 1:1 ratio of males to females (or male to female gametes). The spread of the *CMS⁺* allele leads to an excess of females in the population. That in turn favors the spread of any mutation in a nuclear gene that cancels the action of the *CMS⁺* allele. The result is an evolutionary arms race between genes on the mitochondria and genes in the nucleus.

Genes that are inherited cytoplasmically often conflict with nuclear genes, as thyme plants illustrate. Mitochondrial mutations that harm males are not selected against [32, 42]. This explains why mitochondrial mutations that cause male-specific diseases are common in humans, fruit flies, and other species (see p. 19).

Group selection

In most situations, competition between individuals for survival and reproduction leads to the evolution of traits that increase each individual's fitness. Under the right conditions, however, selection operating on the phenotypes of *groups* of individuals can lead to the evolution of traits that are not favored by selection acting on differences between individuals within each group [84]. These traits can include altruistic behaviors.

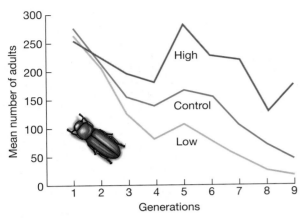

FIGURE 12.14 Evolution of group size in response to group selection among and individual selection within populations of the flour beetle *Tribolium castaneum*. After nine rounds of selection, populations under group selection for high population size (red line) were on average nine times larger than those under group selection for low population size (green line). Individual selection within groups caused the population sizes to decrease in all three treatments. (After [76].)

A classic experiment with the flour beetle *Tribolium castaneum* shows that group selection can cause large evolutionary changes [76]. Three treatments were established, each with 48 small populations (groups) that were maintained in vials of flour. After the beetles reproduced, the population size in each vial was censused. In the first treatment, beetles from the vials with the largest populations were used to establish a new set of 48 vials. In the second treatment, only beetles from the vials with the smallest population sizes were used as founders for the next generation. The third treatment was a control in which all populations contributed equally to the next generation. This procedure was followed for nine generations. A key point is that the experimental selection imposed by the treatments acted only on a property of the group—the population size of the vial. In addition to the group selection imposed by the experiment, individual selection was also at work, favoring individuals that left more offspring within each group.

The results show two striking trends (**FIGURE 12.14**). By the end of the experiment, there were nine times more beetles per group in the treatment that selected for high population size than in the treatment that selected for low size. Clearly group selection had a very strong evolutionary impact. The second pattern is that population size declined in all three treatments. Further research revealed the causes [77]. Larval and adult beetles sometimes eat eggs and pupae. Cannibalism is advantageous to the cannibals, and it increased in frequency as the result of individual selection within each vial. In the treatment that selected for high population size, cannibalism rates were lower, which can be thought of as the evolution of an altruistic behavior. In short, the trend of population size through time in each treatment resulted from an interplay between group selection and individual selection.

To be clear, evolution by group selection results from changes in allele frequencies, just as when selection acts on individuals. The difference between group selection and individual selection is that group selection results from a difference between the rates of survival or reproduction of groups, rather than of individuals. Group selection is closely related to kin selection, and in fact many evolutionary biologists do not distinguish between the two [5, 27, 80]. Although the individuals in each group of beetles were not immediate family members, they were more genetically related to each other than they were to the beetles in other groups. In effect, selection that favors certain groups is favoring certain extended families.

One setting in which group selection has clearly played an important role is the evolution of virulence in pathogens [11]. Each host contains a group of pathogens. Selection on pathogens favors traits that increase the number of hosts that they infect (see Chapter 13). Pathogens face an evolutionary trade-off. If they multiply rapidly within the host, they are more virulent (that is, they kill their host faster), but they infect new hosts more quickly. The influenza virus does this: millions of viral particles can be spewed out in a single sneeze, and infect many unfortunate people nearby. The viral genotypes within a host that replicate fastest are the most likely to infect another host, and so are favored by individual selection. Other pathogens reproduce much more slowly. This prolongs the life of the host, which for these pathogens increases the number of other hosts that they infect over the long term, and so these pathogens are relatively benign. Whether selection favors low or high virulence depends on the biology of the pathogen and the host—for example, on how frequently the pathogen has an opportunity to infect a new host.

A dramatic but entirely accidental "experiment" shows that virulence can evolve rapidly by group selection. In 1950, the myxoma virus was introduced into Australia to control a population explosion of European rabbits (*Oryctolagus cuniculus*) which (ironically) had themselves been introduced earlier [45]. Initially, the virus was lethal and killed more than 99 percent of infected rabbits. But in the years that followed, fewer and fewer rabbits died. The rabbits evolved greater resistance, and the virus evolved lower virulence. The evolution of the virus was studied in laboratory experiments that used a population of rabbits that had never been exposed to the virus. Virulence was measured on a scale from most virulent (= 1) to least virulent (= 5). Between 1952 and 1955, the two most virulent classes made up about 33 percent of the viral isolates. Twenty-five years later, the most virulent classes had declined to about 5 percent of the isolates. This pattern was repeated when the virus was introduced to control rabbits in France in 1952. For the myxoma virus, group selection for decreased virulence was more powerful than the selection among viruses within each rabbit favoring increased virulence. The virus therefore evolved to become more benign.

Cooperation and Major Evolutionary Transitions

The fitness of most parasites and pathogens depends on **horizontal transmission**, that is, infecting other hosts that are not the offspring of their current host (**FIGURE 12.15A**). Parasites and pathogens transmitted this way are sometimes selected to become highly virulent if that increases the probability they will infect a new host.

Endosymbionts are mutualists that live within the cells of their hosts. Some, like mitochondria, are passed by **vertical transmission**, that is, they infect the offspring of their current host (**FIGURE 12.15B**). Here the evolutionary fates of the

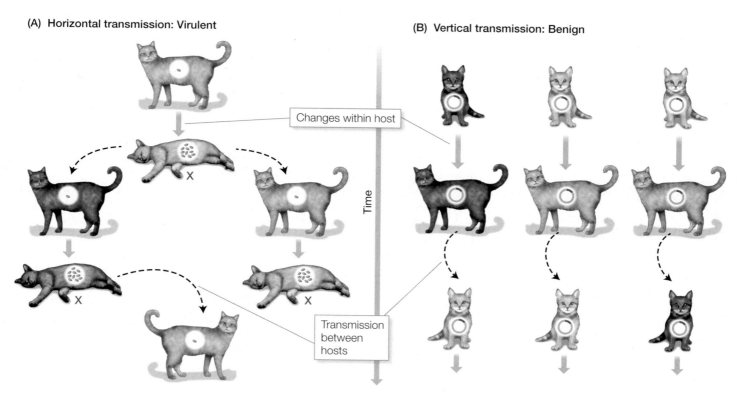

(A) Horizontal transmission: Virulent

(B) Vertical transmission: Benign

Changes within host

Transmission between hosts

Time

FIGURE 12.15 Selection pressures on pathogens and endosymbionts depend on their mode of transmission. (A) Pathogens such as the influenza virus maximize their fitness by multiplying rapidly within their host, which increases their chance of being transmitted horizontally to another host. The rapid replication of the virus can harm and even kill its hosts. (B) Endosymbionts such as mitochondria are transmitted vertically from mothers to their offspring. This kind of endosymbiont maximizes its own fitness by increasing the survival and reproduction of its female hosts.

endosymbiont and host are chained together. The endosymbiont's fitness depends entirely on the fitness of its host. Selection for high proliferation *within* the symbiont population occupying each host is strongly opposed by selection *among* the populations of symbionts that occupy different hosts. On balance, selection at the group level favors mitochondria (and other maternally-transmitted symbionts) that maximize the number of female offspring that their hosts leave to the next generation. In the extreme case, the symbiont may become an essential part of the host, forming a new collective entity [56].

This is what happened in one of the *major transitions* in the history of life on Earth: the evolution of eukaryotes [51]. The key event was the symbiotic union of a bacterial endosymbiont and a host cell, probably an archaean, about 1.5 billion years ago. The bacterium evolved into the mitochondrion [12]. This was the first of many symbioses that formed major new collective entities. A second was the incorporation of blue-green bacteria (cyanobacteria) into a one-celled eukaryote. That enabled the eukaryote to photosynthesize, and it became the ancestor of the green algae and plants. The common interest of the endosymbiont and host genomes resulted in the evolution of a new kind of collective entity, and a higher level of organization.

A third major transition occurred with the origin of multicellular organisms [27, 51, 54]. These organisms are more than just groups of cells. For example, dividing bacteria that remain loosely attached but physiologically independent of each other do not constitute an organism. The cells of a multicellular organism cooperate in ways reminiscent of the ants in a colony: they differentiate into tissues specialized for different tasks that contribute to the fitness of the group (that is, the individual they belong to). Why should unicellular ancestors, in which each cell had a prospect of reproduction, have given rise to multicellular descendants, in which some cells sacrifice their own reproduction?

The fundamental answer is kin selection. If the cell lineages in a multicellular organism arise by mitosis from a unicellular egg or zygote, the genes of cooperative cells that sacrifice reproduction for the good of the cell "colony" are propagated by closely related reproductive cells. However, the coefficient of relationship is reduced if genetically different cells invade the colony, or if mutational differences arise among cells. A mutation that increases the rate of cell division has a selective advantage *within* the colony, but unregulated cell division usually harms the organism, as in cancer. Selection at the level of whole colonies of cells—organisms—therefore opposes selection among cells within colonies.

As a result, mechanisms of policing have evolved that regulate cell division and prevent renegade cell genotypes from disrupting the integrated function of the organism. In animals, selection has resulted in the evolution of a germ line that is segregated from the soma early in development. This organization prevents deleterious mutations in somatic cells from being transmitted by the gametes. Selection for organismal integration may be responsible for the familiar but remarkable fact that almost all multicellular organisms begin life as a single cell, rather than as a group of cells. This feature increases the relatedness among all the cells of the developing organism, reducing genetic variation and competition within the organism and increasing the heritability of fitness. The result, then, has been the emergence of the "individual," and with it, the level of biological organization at which so much of natural selection and evolution take place.

Go to the
Evolution Companion Website
EVOLUTION4E.SINAUER.COM
for data analysis and simulation exercises, quizzes, and more.

SUMMARY

- Many biological phenomena result from conflict or cooperation among organisms or among genes. The evolution of most interactions can be explained best by selection at the level of individual organisms or genes.

- Altruism benefits other individuals and reduces the fitness of the actor, while cooperative behavior need not reduce the actor's fitness. Cooperation can evolve because it is directly beneficial to the actor, although the benefit may be delayed. It can also evolve by reciprocity, based on repeated interactions between individuals in which the fitness interests of the associates are aligned. Cooperative interactions can be maintained in part by "policing," or punishment of cheaters.

- Altruism can evolve by kin selection. An allele's inclusive fitness is the sum of its direct fitness (the average number of copies that a carrier leaves to the next generation) and its indirect fitness (additional copies left by the carrier's relatives as the result of the carrier's behavior). Hamilton's rule describes the conditions for the increase of an allele for altruistic trait in terms of the benefit to the recipient, the cost to the actor, and the coefficient of relationship between them.

- Conflict and kin selection together affect the evolution of many interactions among family members. The genetic benefit of caring for offspring is an increase in the number of current offspring that survive. The genetic cost is the number of additional offspring that the parent is likely to have if she or he abandoned the offspring and reproduced again. Parental care is expected to evolve only if its fitness benefit exceeds its fitness cost. Whether or not one or both parents evolve to provide care can depend on the ratio of fitness costs and benefits for each parent.

- Evolutionary conflicts between parents and offspring are widespread. A parent's fitness may be increased by allocating some resources to its own survival and future offspring rather than to its current offspring. Selection acting on the offspring, however, often favors taking more resources from its parents than is optimal for the parents to give. This principle may be one of several reasons why in some species, parents may reduce their brood size by aborting some embryos or killing some offspring.

- The most extreme examples of cooperation and altruism are in eusocial species, in which some individuals reproduce little or not at all, and instead help relatives rear their offspring. In eusocial insects, nonreproductive workers rear reproductive queens and males, as well as other workers. Many social interactions in these colonies are governed by kin selection and policing by workers.

- Under some conditions, selection acting on groups can cause the evolution of altruism. This form of group selection can be viewed as a type of kin selection. Group selection acting on a pathogen sometimes favors the evolution of decreased virulence when increased host survival increases the number of new hosts that the pathogen infects.

- Conflicts may exist among different genes in a species' genome that are inherited by different pathways. Selection acting on loci that are transmitted through only one sex favors alleles that alter the sex ratio in favor of that sex. The changed sex ratio creates selection at other loci for suppressors that restore the 1:1 sex ratio.

- Kin and group selection explain three of the major transitions in the evolution of life on Earth. Eukaryotes evolved by the union of two organisms, in which the fitness of each depends on the other. The union of such a eukaryote with cyanobacteria produced photosynthetic eukaryotes: algae and plants. Multicellular organisms could evolve only because their cells are nearly genetically identical, and so cooperate due to kin selection.

TERMS AND CONCEPTS

altruism
conflict
cooperation
direct fitness
endosymbiont
eusocial
evolutionarily stable

game theory
group selection
Hamilton's rule
horizontal
   transmission
inclusive fitness
indirect fitness

mutualism
parent-offspring
   conflict
reciprocity
relatedness
segregation
   distortion

sexually antagonistic
   selection
spite
transposon
vertical transmission

SUGGESTIONS FOR FURTHER READING

An Introduction to Behavioural Ecology (Wiley-Blackwell, Oxford, 2012) by N. B. Davies and colleagues is an outstanding introduction to that field. A more general introduction to animal behavior is *Animal Behavior: An Evolutionary Approach* by J. Alcock (Sinauer Associates, Sunderland, MA, 2013).

The evolution of social behavior and its implications for major transitions in evolution are comprehensively treated by A. F. G. Bourke in *Principles of Social Evolution* (Oxford University Press, Oxford, 2011). An excellent set of essays on many aspects of cooperation and conflict is *Levels of Selection*, edited by L. Keller (Princeton University Press, Princeton, NJ, 1999). J. A. R. Marshall's *Social Evolution and Inclusive Fitness Theory* (Princeton University Press, Princeton, NJ, 2015) is a comprehensive synthesis of that topic.

Genetic conflict and selfish genes are reviewed in a book by A. R. Burt and R. Trivers, *Genes in Conflict: The Biology of Selfish Genetic Elements* (Harvard University Press, Cambridge, MA, 2006). Much shorter but excellent are the review articles by J. H. Werren, "Selfish genetic elements, genetic conflict, and evolutionary innovation" (*Proc. Natl. Acad. Sci. USA* 108: 10863–10870, 2011) and W. R. Rice, "Nothing in genetics makes sense except in light of genomic conflict" (*Annu. Rev. Ecol. Evol. Syst.* 44: 217–237, 2013).

The evolutionary "battle of the sexes" is an area of active research. We recommend *Sexual Conflict* by G. Arnqvist and L. Rowe (Princeton University Press, Princeton, NJ, 2005). A concise overview that focuses on genetic aspects is the article by R. Bonduriansky and S. F. Chenoweth, "Intralocus sexual conflict" (*Trends Ecol. Evol.* 24: 280–288, 2009).

The topic of group selection has a rich history. One of the most important contributions to this subject is the famous book by G. C. Williams, *Adaptation and Natural Selection* (Princeton University Press, Princeton, NJ, 1966), which has stimulating thoughts on many other topics as well. More recent discussions include a book by E. Sober and D. S. Wilson, *Unto Others: The Evolution and Psychology of Unselfish Behavior* (Harvard University Press, Cambridge, MA, 1988), which takes a positive view of group selection, and an article by S. A. West and colleagues ("Evolutionary explanations for cooperation," *Current Biology* 17: R661–R672, 2007), who instead emphasize kin selection.

PROBLEMS AND DISCUSSION TOPICS

1. Many species of animals make alarm calls, which warn others in their group that a predator is approaching. Alarm calls also attract the attention of the predator, making it more likely that the individual making the call will be eaten. Why might natural selection favor the evolution of alarm calls in a species? How might you test that hypothesis?

2. Darwin argued that natural selection would never cause a trait to evolve in one species that benefitted another species at a fitness expense to the first species. A study of egrets in Florida found that parents eject their own chicks from their nests, and this behavior feeds alligators living in the swamp below the nests. This steady supply of food improves the health and condition of the alligators. In view of Darwin's logic, what are two hypotheses that might explain why the egrets perform behaviors that benefit alligators?

3. Kin selection explains why organisms provide

the principle of kin selection and the evolution of siblicide?

4. Explain why we expect mitochondria to have more mutations that are harmful to males than to females.

5. What differences do you expect to see in how females behave toward their brothers in haplodiploid species (such as ants) compared with diploid species (such as beetles)?

6. Many clonal marine invertebrates (such as corals and sponges) exhibit fierce competition for space, sometimes leading to death among competitors. At other times, two expanding colonies will merge to form a single, larger colony. What factors might account for decisions either to attack or to fuse with another colony?

7. Some pathogens, such as HIV, can be transmitted both vertically and horizontally. How do you expect their virulence to compare with that of pathogens that are transmitted only horizontally

Interactions among Species

Nearly 20 years before he published *On the Origin of Species*, Charles Darwin started to study orchids, intrigued by the extraordinary features of their flowers. He examined British species, and he grew tropical species, solicited from horticulturists, in his greenhouse. In 1862, in his first book after *On the Origin of Species*, he summarized his studies in *On the Various Contrivances by which British and Foreign Orchids are Fertilised by Insects, and on the Good Effects of Intercrossing*. It is a landmark work, in which Darwin put into practice his principles of natural selection and descent with modification. Parting with the prevailing theological interpretation, that flowers were shaped by God to inspire us with beauty, Darwin showed that the astonishingly diverse and peculiar features of orchid flowers increase the chance that they will attract insects and deposit pollen on them in so precise a way as to ensure cross-pollination. Among these remarkable plants was a species from Madagascar, *Angraecum sesquipedale*, with a nectar-bearing tube up to 30 cm long (**FIGURE 13.1**). Other plants with much shorter nectar spurs are visited by insects with tongues long enough to reach the nectar, so Darwin predicted that there must exist in Madagascar a moth with a similarly long proboscis. One reviewer ridiculed this idea, and indeed the very idea that the features of flowers are useful, but in 1903 a sphinx moth with a proboscis up to 30 cm long was described from Madagascar, and was fittingly named *Xanthopan morganii praedicta*. *Angraecum* and its moth perfectly illustrated Darwin's speculation (in *On the Origin of Species*) that both a flower and a pollinating insect "might slowly become, either simultaneously or one after the other, modified and adapted in the most perfect manner."

Red-billed oxpeckers (*Buphagus erythrorhynchus*) like the one shown here on a Cape buffalo (*Syncerus caffer*) spend most of their time on large African ungulates, where they eat mostly ticks, but also feed at open wounds. Their interaction with the mammals is on the border between mutualism and parasitism.

FIGURE 13.1 A coevolved interaction. The orchid *Angraecum sesquipedale* bears nectar in an exceedingly long spur and is pollinated by the long-tongued sphinx moth *Xanthopan morganii praedicta*. The moth was discovered about 40 years after Darwin predicted its existence. Each of the species in this mutualism is adapted to obtain something from the other.

Coevolution and Interactions among Species

Every species is subjected to natural selection from its biotic environment: the complex of other organisms with which it interacts. Most of these species can be classified as resources (used as nutrition or habitat), competitors (for resources such as food and space), enemies (predators or parasites), or mutualists. In mutualistic interactions, each species obtains a benefit from the other. (**Symbiosis**, meaning "living together," describes intimate associations between species that may be either mutualists or parasite and host. An endosymbiont lives within the other organism's body.) The community of other species with which a species interacts is complex and variable—both the identity and genetic composition of interacting species vary in time and place. Thus, a plant species may be pollinated or attacked by many species of insects, and be inhabited by any of hundreds of species of fungi and bacteria that live on or in its leaves and roots. Similarly, the natural environment of humans includes a variable "human microbiome": the trillions of bacteria, including thousands of species—mostly harmless and some even beneficial—that occupy the gut, skin, nostrils, and other microhabitats [15, 34, 55].

Some of the most familiar examples of natural selection, such as industrial melanism in the peppered moth and the sickle-cell polymorphism in human hemoglobin, entail biological agents (predaceous birds and malarial parasites, respectively) (see Chapter 5). In many such interactions, the evolution of one species has been affected by the other, but not vice versa. **Coevolution**, strictly defined, is reciprocal genetic change in interacting species, owing to natural selection imposed by each on the other. Not all adaptations of one species to other species are necessarily coevolved.

The nature and strength of an interaction between two species may vary depending on genotype, environmental conditions, and other species with which those species interact. For example, populations of the limber pine in areas where squirrels eat the seeds have cones that reduce squirrel depredation, but are also less favorable for the Clark's nutcracker, a bird that the pine depends on for seed dispersal (**FIGURE 13.2**). Thus the selection that species exert on each other may differ among populations, resulting in a geographic mosaic of coevolution that differs from one place to another [73].

The term "coevolution" includes several concepts [28, 72]. In its simplest form, called *specific coevolution,* two species evolve in response to each other (**FIGURE 13.3A**). Darwin's *Angraecum* orchid and its specialized pollinating moth are an example. *Diffuse coevolution* occurs when several species are involved and their effects are not independent (**FIGURE 13.3B**). For example, genetic variation in the resistance of a host to two different species of parasites might be correlated [35]. In *escape-and-radiate coevolution*, a species evolves a defense against enemies and is thereby enabled to radiate into diverse descendant species, to which different enemies may later adapt (**FIGURE 13.3C**).

A few cases have been described in which the phylogeny of a group of organisms matches the phylogeny of a group of its parasites or symbionts. An example is the association between aphids and endosymbiotic bacteria (*Buchnera*) that live in special aphid cells and supply the essential amino acid tryptophan to their hosts. The completely concordant phylogenies of the aphids and bacteria (**FIGURE 13.4**)

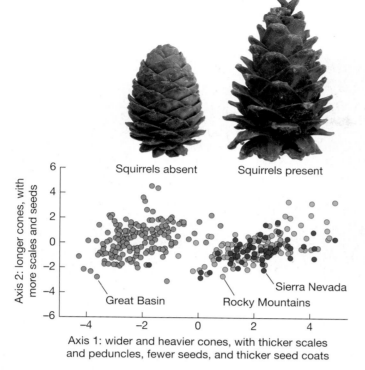

Red squirrel

Clark's nutcracker

FIGURE 13.2 A geographic mosaic of interactions. Typical cones of limber pine (*Pinus flexilis*) populations that (at right) are adapted to resist seed-eating squirrels or (at left) are adapted for seed dispersal by Clark's nutcracker where squirrels are absent. The graph of two variables, each of which combines several measurements of cones and seeds, shows that pines in an area without squirrels (Great Basin, orange dots) differ from those in two areas with squirrels (Sierra Nevada and Rocky Mountains, dark and light green dots, respectively). Each dot represents one tree. (After [67]; pine cone photos from [67].)

show that this association dates from the origin of the aphids, and that the bacteria have diverged in concert with speciation in their hosts. The explanation is simple: the bacteria are transmitted from mother aphids to their offspring just as if they were mitochondria. By themselves, matching phylogenies should not be considered coevolution, because there need not have been any reciprocal adaptation. A match can arise simply because the parasite or endosymbiont has had little or no opportunity to be transmitted between different hosts. The phylogeny

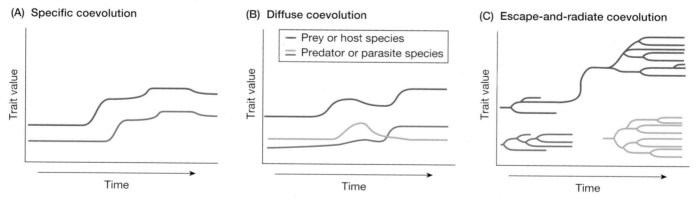

FIGURE 13.3 Three kinds of coevolution. In each graph, the horizontal axis represents evolutionary time, and the vertical axis shows the state of a character in a species of prey or host and one or more species of predators or parasites. (A) Specific coevolution. (B) Diffuse coevolution, in which a prey species interacts with two or more predators, can take many paths. In this case, a prey species becomes better defended against two predators, only one of which (blue curve) becomes better able to capture the prey. (C) Escape-and-radiate coevolution. A prey or host species evolves a major new defense, escapes association with a predator or parasite, and diversifies. Later, a different predator or parasite adapts to the host clade and diversifies.

FIGURE 13.4 (A) *Buchnera aphidicola* bacteria are endosymbionts of aphids. The electron micrograph (at right) shows bacterial cells living inside a specialized aphid cell (bacteriocyte). (B) The phylogeny of endosymbiotic bacteria included under the name *Buchnera aphidicola* is perfectly congruent with that of their aphid hosts. Several related bacteria (names in red) were included as outgroups in this analysis. Names of the aphid hosts of the *Buchnera* lineages are given in green. The estimated ages of the aphid lineages are based on fossils and biogeography. These *Buchnera* lineages are as old as the aphid lineages that carry them. (After [53]; electron micrograph courtesy of N. Moran and J. White.)

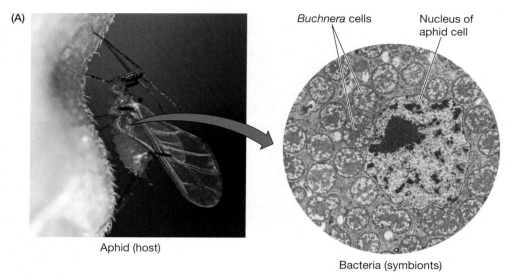

Aphid (host)

Bacteria (symbionts)

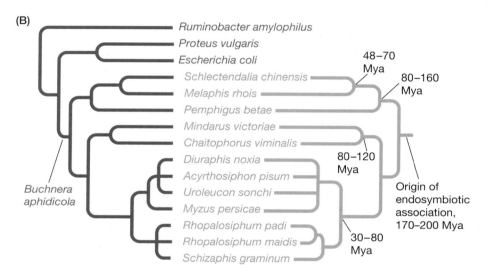

of free-living parasites and mutualists seldom matches the host phylogeny very closely [29, 57, 82].

The Evolution of Enemies and Victims

Interactions between enemies and victims include predators and their prey, parasites and their hosts, and herbivores and their host plants. Such interactions are often unstable, because enemies can extinguish victim populations, or reduce them to the point that the enemy population becomes extinct for lack of food. Many species of Australian marsupials were driven to extinction by introduced foxes and feral cats [19]; a chytrid fungus has extinguished some species of frogs and threatens many other amphibians [12]. Because the future does not affect the action of natural selection (see Chapter 3), the possibility that the prey or host might be killed off does not cause enemies to evolve restraint that might preserve prey populations. Victims and their enemies coexist only if their interactions are stabilized by ecological and evolutionary factors, including adaptations to escape or resist enemies.

(A)

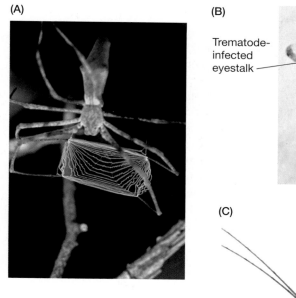

(B)

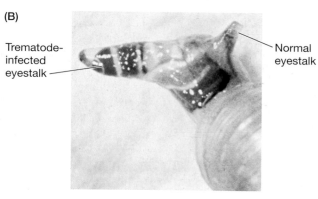

Trematode-infected eyestalk

Normal eyestalk

(C)

FIGURE 13.5 Predators and parasites have evolved many extraordinary adaptations to capture prey or infect hosts, and prey have elaborate counteradaptations. (A) This tropical net-casting spider (*Deinopis subrufa*) holds an expandable web that it uses to quickly envelope slowly flying insects that pass by. (B) The larva of a parasitic trematode (*Leucochloridium*) migrates to the eyestalk of its intermediate host, a land snail, and turns it a bright color to make the snail more visible to the next host in the parasite's life cycle, a snail-eating bird such as a thrush. (C) Katydids (Tettigoniidae) of the genus *Mimetica* have an extraordinary resemblance to leaves, including what looks like leaf venation and damage by herbivores. (B, photo by P. Lewis, courtesy of J. Moore.)

Predators and parasites have evolved some extraordinary adaptations for capturing, subduing, or infecting their victims (**FIGURE 13.5**). Defenses against predation and parasitism can be equally impressive, ranging from cryptic patterning to the most versatile of all defenses: the vertebrate immune system, which can generate antibodies against thousands of foreign compounds. The CRISPR-Cas mechanism in some bacteria is also an elaborate system of recognizing and defending against foreign invaders—that is, viruses. Many such adaptations appear to be directed at a variety of different enemies or prey species, so the coevolution, if any, has probably been diffuse.

R. A. Fisher, one of the founders of evolutionary genetics, suggested that a species' environment, such as the climate, is constantly changing, but "probably more important than the changes in climate will be the evolutionary changes in progress in associated organisms" [25]. This idea is expressed by the **Red Queen hypothesis**, named by paleontologist Leigh Van Valen [79] for the Red Queen whom Alice meets in Lewis Carroll's *Through the Looking-Glass*: each species has to run (i.e., evolve) as fast as possible just to stay in the same place (survive) because interacting species also continue to evolve. The dynamics of Red Queen coevolution may take several forms, including escalation and oscillation. In the long term, the dynamics may lead to indefinite coexistence of enemy and victim species, a switch by the enemy to a different victim species, or extinction of one or both species [1, 56].

An **evolutionary arms race**, also called escalation, may occur if the capture rate of the prey by the predator increases with the difference between the defensive trait of a prey species and a corresponding character in a predator. Then the characteristics of both species that affect their interaction evolve in one direction: for example, greater speed of gazelles and of pursuit predators such as cheetahs (**FIGURE 13.6**). This can lead to extinction or to a stable point when the costs of increasing the trait (e.g., speed, or a plant's defensive chemicals) become too great.

The Japanese camellia (*Camellia japonica*) and the camellia weevil (*Curculio camelliae*) present a dramatic example of escalatory coevolution. The camellia's

FIGURE 13.6 Evolutionary arms race between predator and prey. Selection by predators such as cheetahs has resulted in the evolution of high speed in prey such as the gazelle in this picture. Predators are therefore also under selection for greater speed.

seeds are enclosed by a woody fruit wall (pericarp) that is much thicker in southern than in northern populations (**FIGURE 13.7**). A high proportion of seeds are consumed by larvae of the weevil, which inserts eggs into the seed chamber through a hole that the female bores with her mandibles, located at the end of her long snout, or rostrum. Investigators showed that the weevils' success in boring through to the seed chamber depends on their rostrum length, relative to the thickness of a fruit's pericarp [76]. Although southern weevil populations have a much longer rostrum, the southern plant population is ahead in this conflict, with pericarps thick enough to reduce the weevils' success to less than 50 percent. In the north, weevil populations are ahead—their rostra are long enough to ensure a success rate well over 50 percent. These species may be engaged in an evolutionary arms race.

In some cases, prey species have evolved defenses that can make them as dangerous to predators as predators are to prey. The rough-skinned newt (*Taricha granulosa*) of northwestern North America has one of the most potent known defenses against predation: the neurotoxin tetrodotoxin (TTX). One newt can have enough TTX in its skin to kill 25,000 laboratory mice. The level of TTX varies greatly among geographic populations of the newt. Populations of the garter snake *Thamnophis sirtalis* from outside the range of the newt have almost no resistance to TTX [8, 31]. But snake populations that are sympatric with toxic newts feed on them, and those populations are resistant to TTX. The average level of snake resistance and newt toxicity is not perfectly matched, for some snake populations are resistant to much higher TTX concentrations than any newt possesses (**FIGURE 13.8**). There is no selection for increased resistance in these populations, for the snakes do not vary in survival from eating the highly toxic newts.

Theoretical models of quantitative traits show that in contrast to escalation, oscillations may occur if the capture rate of the prey by the predator depends on a close match between the predator and prey traits [1]. If the prey's trait will evolve in

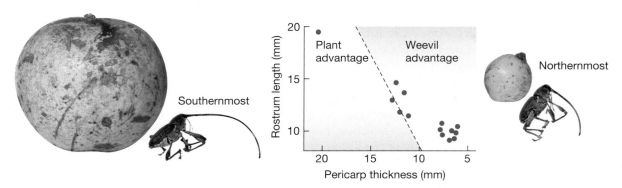

FIGURE 13.7 Imbalance in a coevolutionary conflict. The graph plots the thickness of the pericarp of the Japanese camellia against the rostrum length of the camellia weevil in several populations. To the left of the dashed line, plants are effectively defended against the weevils, while to the right of the line the weevil can effectively feed on the seeds. (After [76]; fruit photos from [76]; weevil photos courtesy of Hiro Toju.)

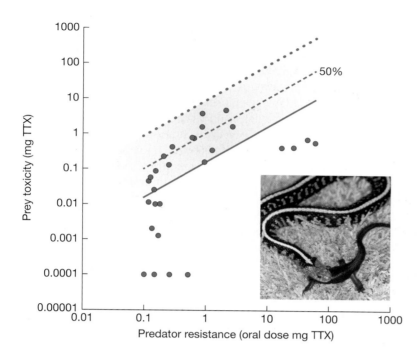

FIGURE 13.8 Toxicity of rough-skinned newts (*Taricha granulosa*) and resistance of garter snakes (*Thamnophis sirtalis*) in several localities. Prey toxicity is the amount of TTX (tetrodotoxin) in the newt; predator resistance is the oral dose of TTX required to reduce the speed of a garter snake by 50 percent. Below the lower boundary (solid blue line), snakes can consume co-occurring newts with no reduction in speed; above the upper boundary (dotted line), toxicity is so high that co-occurring snakes would be completely incapacitated. In general, populations of garter snakes are more resistant where more toxic newts are found, but there is some mismatch: almost half the snake populations fall below the lower boundary, and are therefore much more resistant than they need to be. (After [31]; photo courtesy of Edmund D. Brodie, Jr.)

one direction, the predator's trait will evolve to track it. Eventually, the prey's trait may evolve in the opposite direction as its cost becomes too great, and evolution of the predator's trait will follow. The result may be continuing cycles of change in the characteristics of both species, and these changes may contribute to cycles in population density. Parasite-host interactions can involve oscillations in gene frequencies (see below).

Phenotypic matching is important for some brood-parasitic birds, such as certain species of cuckoos, that lay eggs only in the nests of other bird species. Cuckoo nestlings hatch first and eject their host's eggs from the nest, so the host ends up rearing only the parasite (**FIGURE 13.9A**). Adults of host species treat parasite

(A)

(B)

FIGURE 13.9 Adaptations for and against brood parasitism. (A) A fledgling common cuckoo (*Cuculus canorus*) being fed by its foster parent, a much smaller reed warbler (*Acrocephalus scirpaceus*). (B) Mimetic egg polymorphism in the common cuckoo. The left column shows eggs of six species parasitized by the cuckoo (from top: European robin, pied wagtail, dunnock, reed warbler, meadow pipit, great reed warbler). The right column shows a cuckoo egg laid in the corresponding host's nest. The match is quite close except in the dunnock nest. (B, photo by M. Brooke, courtesy of N. B. Davies.)

nestlings like their own young, but some host species do recognize parasite eggs and either eject them or desert the nest and start a new nest and clutch. Many brood parasites have counteradapted by laying mimetic eggs [65]. Each population of the common cuckoo (*Cuculus canorus*) contains several different genotypes, which prefer different hosts and lay eggs closely resembling those of their preferred hosts (**FIGURE 13.9B**). Nick Davies and Michael Brooke traced the fate of artificial cuckoo eggs placed in the nests of various host species [16]. Bird species that are not parasitized by cuckoos tend not to eject cuckoo eggs. But among the cuckoos' preferred hosts, those species whose eggs are mimicked by cuckoos reject artificial eggs more often than those whose eggs are not mimicked. These species have adapted to brood parasitism by evolving greater discrimination. Moreover, populations of two host species that reject artificial cuckoo eggs in Britain accept them in Iceland, where cuckoos are absent. Thus two evolving traits, host discrimination and cuckoo egg pattern, shape the evolution of this interaction.

Aposematism and mimicry

Diverse animals, such as bees and coral snakes, have evolved warning, or **aposematic**, coloration: bright colors that signal to a potential predator that they are distasteful or dangerous. Predators learn to avoid the color pattern, and so both the predator and the aposematically colored prey benefit. The warning pattern is subject to positive frequency-dependent selection because individuals that deviate from the common pattern, which predators have learned, are likely to be attacked (see Figure 5.24). Thus, a mutation that confers a new aposematic pattern is likely to be disadvantageous. How new aposematic phenotypes evolve is therefore a puzzle. They might be caused by genetic drift in places or at times when selection by predators is relaxed (see p. 130 in Chapter 5) [48].

Mimicry is a form of convergent evolution in which resemblance between different species has evolved because it is advantageous for members of one species to resemble another. The species, then, do not owe their resemblance to common ancestry, but in some cases are so similar that experts have to look very carefully to distinguish them. The most common kind of mimicry is defensive mimicry, which often is based on the aposematic coloration of other species [49]. Two common forms of defensive mimicry are named for the naturalists who first recognized them. In **Batesian mimicry**, a palatable species (a mimic) resembles an unpalatable species (a model). Selection on a mimetic phenotype can depend on both its density, relative to that of a model species, and the degree of unpalatability of the model. A predator that can learn is more likely to avoid eating a butterfly that looks like an unpalatable model if it has had a recent reinforcing experience (e.g., vomiting after eating a butterfly with that pattern). If the predator has recently swallowed a tasty butterfly, however, it will be more, not less, inclined to eat the next butterfly with the same phenotype. Thus the rarer a palatable Batesian mimic is relative to an unpalatable model, the more likely predators are to associate its color pattern with unpalatability, and so the greater the advantage of resembling the model will be. If a rare new phenotype arises that mimics a different model species, it will have higher fitness, and so a mimetic polymorphism can be maintained by negative frequency-dependent selection, as is seen in the African swallowtail *Papilio dardanus* (**FIGURE 13.10**).

The other major form of defensive mimicry is **Müllerian mimicry**, in which two or more unpalatable species are co-mimics (or co-models) and jointly reinforce aversion learning by predators. This hypothesis was proposed by Fritz Müller in 1879 and has been confirmed by experiments in which the survival of distasteful mimetic butterflies was shown to be higher if they closely matched an abundant co-mimic species (**FIGURE 13.11**). This form of mimicry causes positive

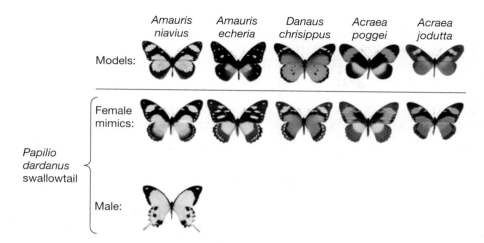

| | *Amauris niavius* | *Amauris echeria* | *Danaus chrisippus* | *Acraea poggei* | *Acraea jodutta* |

Models:

Female mimics:

Papilio dardanus swallowtail

Male:

FIGURE 13.10 Mimetic polymorphism in the African swallowtail butterfly *Papilio dardanus*. Males have only one color form (at bottom), but populations contain several color forms of females (mimics, in the middle row), each of which closely resembles a distantly related distasteful species (models, in the top row). Predators that have attacked a distasteful model learn to avoid butterflies with that color pattern. As the abundance of any specific color morph of *P. dardanus* increases, its fitness tends to decline, because predators are increasingly likely to associate the pattern with a tasty meal rather than a foul taste. (From [75].)

frequency-dependent selection, since common phenotypes will be better recognized and avoided, and deviants from the common pattern will be less likely to survive.

Plants and herbivores

Almost all plants synthesize a variety of secondary compounds (so called because they play little or no role in primary metabolism). Thousands of such compounds have been described, including many that humans have found useful as drugs (e.g., salicylic acid, the active ingredient of aspirin), stimulants (caffeine), condiments (capsaicin, the "hot" element in chili peppers), and in other ways (cannabinol, in marijuana). Families of plants are often characterized by particular groups of similar compounds, such as cardiac glycosides in milkweeds (Apocynaceae) and glucosinolates in mustards (Brassicaceae). Many of these compounds are known to be toxic or repellent to animals, and there is plentiful evidence that insects and other herbivores impose selection for chemical and other defenses.

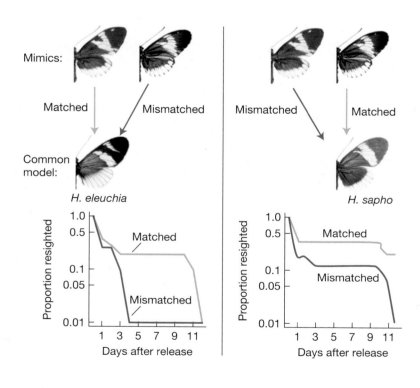

Mimics:

Matched   Mismatched

Common model:

H. eleuchia

Mismatched   Matched

H. sapho

FIGURE 13.11 Müllerian mimicry protects *Heliconius* butterflies. Two color morphs of *Heliconius cydno* mimic two different species of models (*H. sapho* and *H. eleuchia*). The models vary in abundance between localities, and the mimic that matches the common model survives best. Left: In one locality, the model species *H. eleuchia* is most common. Both morphs of the mimic species were marked and released, and their survival was monitored in the following days. The mimic that matched the common model survived best. Right: In another locality, the model species *H. sapho* is more common. The same procedure used in the first experiment showed that again the mimic that matched the locally abundant model survived best. This second experiment confirms that a mimic's survival rate was determined by whether it matched the model that was most frequent, not by an intrinsic advantage of one mimic color pattern over the other. (From [38].)

(A)

Bergapten

Sphondin

(B)

FIGURE 13.12 Secondary chemicals can defend plants against herbivores. (A) The furanocoumarins bergapten and sphondin are among the defensive secondary compounds of wild parsnip (*Pastinaca sativa*), the host plant of a moth larva, the parsnip webworm (*Depressaria pastinacella*). (B) Common milkweed (*Asclepias syriaca*) is genetically variable for latex production. Families of plants with greater latex levels had fewer herbivorous insects and higher fitness, measured by seed production relative to a control without herbivory. (B after [3].)

For example, seed production in wild parsnip (*Pastinaca sativa*) is correlated with genetically variable resistance to the seed-eating parsnip webworm (*Depressaria pastinacella*), based on the concentrations of two furanocoumarin compounds in the seeds (**FIGURE 13.12A**) [5]. Similarly, the fitness of common milkweed (*Asclepias syriaca*) is strongly affected by genetic variation in the production of latex, a gummy white fluid that reduces the abundance and impact of insects on the plant (**FIGURE 13.12B**) [3].

Paul Ehrlich and Peter Raven proposed a scenario of escape-and-radiate coevolution (see Figure 13.3C), in which a plant species that evolves a new and highly effective chemical defense may escape many of its associated herbivores and give rise to a clade of species that share the novel defense [23]. Eventually, though, some insect species from other hosts shift to these plants, adapt to their defense, and give rise to a clade of adapted herbivores.

Subsequent research has provided evidence for this idea [27]. For example, the plant order Brassicales (mustards and relatives) evolved about 92 million years ago (Mya), with the ability to synthesize glucosinolates, the precursors of toxic mustard oils, from certain amino acids. These plants are the almost exclusive larval food of the butterfly subfamily Pierinae, which adapted to Brassicales about 68 Mya by evolving an enzyme that breaks down glucosinolates. Later, new kinds of glucosinolates evolved in one lineage of Brassicales. Soon afterward, two lineages of the butterflies adapted to these novel glucosinolates. In these lineages, different duplications (see Chapter 14) of the gene that encodes the glucosinolate-degrading enzyme led to evolution of enzymes with new functions. These evolutionary innovations were associated with increased proliferation of new species in both the plant and butterfly lineages [22].

Parasite-host interactions and infectious disease

Evolutionary biologists include most pathogenic bacteria and other disease-causing microorganisms among parasites. The two greatest challenges a parasite faces are overcoming the host's defenses and moving from one host to another by vertical transmission from a host parent to its offspring or by horizontal transmission via the environment (see Chapter 12, Figure 12.15). Parasites that reduce the survival or reproduction of their hosts are considered **virulent**. Many parasites are virulent not because it is to their advantage to kill their host, but because their own survival and reproduction require that they consume part of the host, to obtain energy and protein. Some parasites actually prolong the life of their host (and enhance their own reproduction) by interfering with its hormones and effectively castrating it.

Several models of the coevolution of parasites (including pathogens) and their hosts are based on genetic evidence from empirical studies [4, 20]. *Gene-for-gene models* (**FIGURE 13.13A**) are based on interactions between some plants and fungal pathogens [39]. The host has several loci at which an allele encodes a receptor protein that recognizes a cell-surface protein (ligand) of a pathogen and confers resistance. Resistance to pathogens with different ligands depends on the plant's different recognition (receptor) genes. A pathogen can infect (is virulent) if it lacks the ligand or if the plant lacks the corresponding receptor protein. In a population of resistant plants, selection may fix the pathogen genotype that lacks the ligand. In contrast, *matching allele models* (**FIGURE 13.13B**) may assume that a pathogen can infect a host only if it has a protein that matches a cell surface receptor protein of the host, like a key and a lock. In this case, any particular resistance allele will decline in frequency when the pathogen's corresponding infectivity allele has high frequency. As a different resistance allele increases in frequency in the host population, the corresponding infectivity allele increases in the pathogen population. Such frequency-dependent selection can cause cycles or irregular fluctuations in allele frequencies. A matching allele model describes variation in resistance of a freshwater crustacean, the water-flea *Daphnia magna*, to genotypes of the bacterium *Pasteuria ramosa* [45].

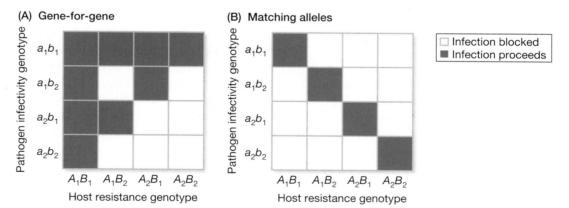

FIGURE 13.13 Two genetic models of coevolution between pathogens and their hosts. The filled cells indicate combinations of host and pathogen genotypes in which the pathogen is able to infect. For simplicity, both a haploid host and a haploid pathogen are assumed. (A) In the gene-for-gene model, the host has two loci at which alleles A_2 and B_2 encode receptor proteins that bind pathogens with surface proteins (ligands) produced by corresponding alleles a_2 and b_2. The plant is resistant, and infection fails, only if either a_2 or b_2 in the pathogen is counteracted by the corresponding allele (A_2 or B_2) in the host. Thus, the pathogen genotype $a_1 b_1$ can infect any host because it lacks ligands to which host proteins can bind. The host genotype $A_1 B_1$ is susceptible to all pathogens because it lacks both binding proteins. In the matching alleles model, both alleles at each locus (A_1 and A_2, B_1 and B_2) are resistance alleles that encode "locks" that can be opened only by the matching "keys" of the pathogen. The pathogen can infect only if it has the matching allele at both the A and B loci. (A after [39]; B after [4].)

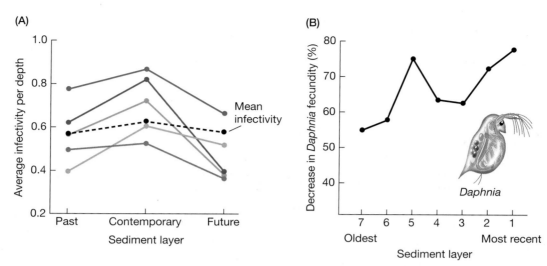

FIGURE 13.14 Parasite-host coevolution. (A) *Daphnia* were hatched from eggs from several layers of pond sediment, dating from different years, and were experimentally exposed to *Pasteuria* from the same ("contemporary"), previous ("past"), or following ("future") sediment layers. The bacteria were generally most successful in infecting contemporary *Daphnia*. Each line presents results for *Daphnia* from a particular sediment layer; the dashed line is the mean infectivity of all trials, some of which, for simplicity, are not shown here. (B) Even though bacteria at any one time were best able to infect contemporary *Daphnia*, their virulence increased over time. The graph shows that bacteria taken from more recent sediment layers were more harmful than bacteria from older layers when they were tested on a standard laboratory strain of *Daphnia*. (After [18].)

Ellen Decaestecker and colleagues described an ingenious "resurrection study" that revealed cycles of genetic change in the freshwater crustacean *Daphnia* and pathogenic *Pasteuria* bacteria [18]. *Daphnia* can produce eggs that remain dormant in pond sediments for many years, and the eggs may harbor *Pasteuria* spores. Decaestecker and colleagues revived eggs and bacteria from different layers of lake sediment, then experimentally cross-infected *Daphnia* from several different years with bacteria from the same ("contemporary") year, a preceding ("past") year, and a subsequent ("future") year. They discovered that the hosts were more frequently infected by contemporary than by past or future bacteria (**FIGURE 13.14A**). These observations indicate that the *Daphnia* population underwent genetic change from year to year and that the bacteria evolved in concert, as in matching allele models of coevolution. The *Daphnia* changed so that they were no longer as easily infected by past bacteria, and the bacteria changed and were able to infect contemporary *Daphnia*. Even though both host and parasite underwent continual cyclic coevolution, the average virulence of the parasite (measured by how much it reduces the host's fecundity) increased over time (**FIGURE 13.14B**).

THE EVOLUTION OF VIRULENCE   In a different study of *Daphnia magna*, Dieter Ebert found that microsporidian parasites (*Pleistophora intestinalis*) that reproduce in the gut produced more spores, and caused greater mortality, when they infected *Daphnia* from their own or nearby populations than when they infected hosts from distant populations (**FIGURE 13.15**) [21]. Thus populations of this parasite are best adapted to their local host population, and have a more virulent effect on sympatric than on allopatric host populations. Like the increasing virulence of the *Pasteuria* bacteria that Decaestecker's group studied, this pattern contradicts the widely held, naïve hypothesis that parasites always evolve to be more benign (also see Chapter 12).

In Chapter 12 we described evolutionary changes in the myxoma virus that was used to control the European rabbit in Australia. In that case, the virus evolved

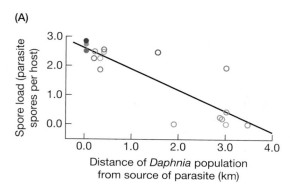

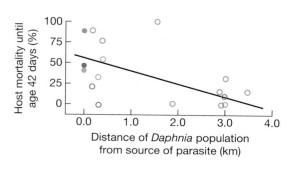

FIGURE 13.15 Fitnesses of three strains of a microsporidian parasite and their effects on various populations of their host species, the water flea *Daphnia magna*. Each strain, represented by a different color, was tested in hosts from its locality of origin (solid symbols) and from localities at various distances away (open symbols). (A) The number of parasite spores produced per host (spore load) was greatest when the parasite infected individuals from its own locality, showing that parasites are best adapted to local host populations. (B) Host mortality was greatest in the parasite's own or nearby host populations, showing that the parasite is most virulent in the host population with which it has coevolved. (After [21].)

a lower level of virulence. This happened by group selection: rabbits harboring a group of highly virulent viruses tended to die before the virus could be transmitted to new hosts. But this is by no means the only possible outcome. Other factors can also determine whether a parasite evolves to be more benign or more virulent [9, 24, 26]. If multiple, unrelated genotypes of parasites occur together within hosts, selection favors the genotype with the highest reproductive rate, which may be highly virulent. This may outweigh group selection for lower virulence. Selection is likely to favor more virulent genotypes in horizontally transmitted parasites than in those that are vertically transmitted from parent hosts to offspring. This hypothesis was supported by an experiment with bacteriophage, in which a phage genotype that reduces its host's growth declined in frequency, and a more "benevolent" genotype increased, when horizontal transmission was prevented [11]. Another factor is that if the host can sustain parasite reproduction for only a short time, selection favors rapid parasite reproduction, which may entail greater virulence. For example, an effective immune system (or medical treatment that rapidly kills the parasite) may sometimes induce the evolution of higher virulence [61]. When chickens were immunized against a virus by a vaccine that did not prevent transmission from infected birds, the virus evolved a higher transmission rate and higher virulence [62].

EVOLUTION AND EPIDEMICS The genetics and evolution of parasite-host interactions are highly relevant to human health, as well as that of other species of concern. Genetic diversity in host populations is important for maintaining resistance to pathogens. Conversely, populations that are inbred or have low genetic diversity may be at risk of infection. For example, in 1970, 85 percent of the hybrid seed corn planted in the United States carried a cytoplasmic genetic factor for male sterility that was considered useful for preventing unintended cross-pollination. Unfortunately, this genetic factor also caused susceptibility to the southern corn leaf blight (*Helminthosporium maydis*), and about 30 percent of the country's corn crop—and up to 100 percent in some places—was lost to this fungus [78]. Widely planting a genetically uniform crop is a prescription for disaster.

Among the greatest threats to human health are emerging pathogens, many of which enter the human population from other species. Phylogenetic analyses are routinely used to trace the origins of new pathogens, such as Ebola virus and the human immunodeficiency viruses (HIV-1 and HIV-2) (see Chapter 16). In some cases, evolutionary change in the pathogen plays a role in its transition to humans [83]. When the origins of a new pathogen can be discovered, it may be possible to determine the genetic basis of the pathogen's adaptation to its new host. For instance, canine parvovirus arose and became pandemic in dogs throughout the world in 1978. Phylogenetic analysis showed that it arose from a virus that infects cats and several other carnivores. Six amino acid changes in the capsid protein of the virus enable it to infect dog cells by specifically binding the canine transferrin receptor. After the virus first entered the dog population, several additional evolutionary changes made it more effective at binding the dog receptor and unable to bind that of its original feline host [36].

Mutualisms

Mutualisms are interactions between species that benefit individuals of both species. However, they exemplify not altruism, but reciprocal exploitation, in which each species obtains something from the other. In *On the Origin of Species*, Darwin challenged his readers to find an instance of a species having been modified solely for the benefit of another species, "for such could not have been produced through natural selection." No one has met Darwin's challenge.

Some mutualisms have arisen from parasitic or other exploitative relationships. Yuccas (*Yucca*), for example, are pollinated only by female yucca moths (*Tegeticula* and *Parategeticula*), which carefully pollinate a yucca flower and then lay eggs in it (**FIGURE 13.16A**). The larvae consume some of the many seeds that develop. Some of the closest relatives of *Tegeticula* simply feed on developing seeds, and one of these species incidentally pollinates the flowers in which it lays its eggs, illustrating what may have been a transitional step from seed predation to mutualism (**FIGURE 13.16B**).

As with intraspecific cooperation (see Chapter 12), there is always the potential for conflict within mutualisms because a genotype that "cheats" by exploiting its partner without paying the cost of providing a benefit in exchange is likely to have a selective advantage. Several possible factors can reduce the fitness of cheater genotypes, and thus maintain a mutualistic relationship. One is simply punishment of cheaters ("sanctions"), to prevent overexploitation [10]. Another possibility is that one or both partner species may be able to choose to reward the most cooperative or beneficial individuals of the other species, or exclude cheaters. Yet another possibility is that selection will favor honest genotypes if the individual's genetic self-interest depends on the fitness of its host or partner [33]. This will be the case if there is a long-term or permanent association between individuals, restricted opportunities to switch to other partners or to use other resources, or vertical transmission of endosymbionts from parents to offspring. For example, the *Buchnera* bacteria that live in the cells of aphids and are vertically transmitted are beneficial mutualists.

The factors that discourage the evolution of cheating have been most studied in legumes and their associated rhizobial bacteria, which convert (fix) atmospheric nitrogen (N_2) to ammonium (NH_4^+) that the plant can use. (Legumes and their rhizobia are extremely important for soil fertility in some regions.) Legumes reward rhizobia by housing them in root nodules and providing them with photosynthate (sugars). In one experiment [40], researchers mimicked cheating rhizobia by replacing normal N_2-containing air with atmosphere that lacked N_2, so that the rhizobia provided less ammonium to soybean plants. The rhizobia on these plants increased far less than in plants that had normal, N_2-fixing rhizobia, because the N_2-deprived plants "punished" their rhizobia, depriving them of oxygen. Other investigators found that plants supplied greater benefits to more beneficial strains

(A)

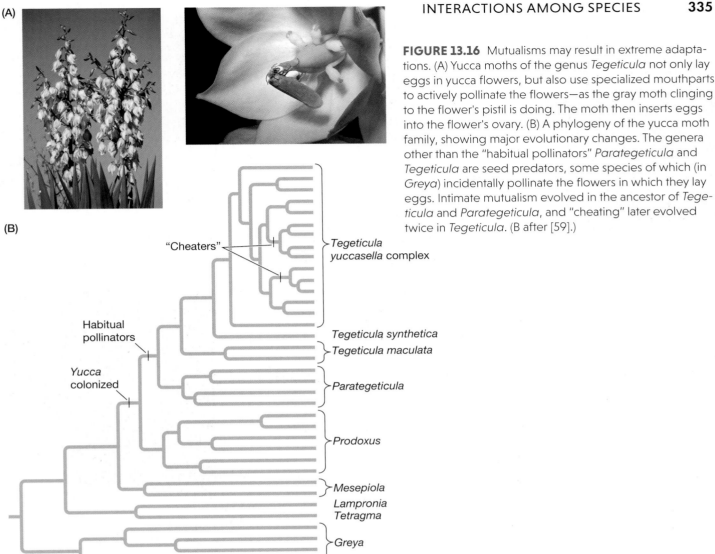

(B)

FIGURE 13.16 Mutualisms may result in extreme adaptations. (A) Yucca moths of the genus *Tegeticula* not only lay eggs in yucca flowers, but also use specialized mouthparts to actively pollinate the flowers—as the gray moth clinging to the flower's pistil is doing. The moth then inserts eggs into the flower's ovary. (B) A phylogeny of the yucca moth family, showing major evolutionary changes. The genera other than the "habitual pollinators" *Parategeticula* and *Tegeticula* are seed predators, some species of which (in *Greya*) incidentally pollinate the flowers in which they lay eggs. Intimate mutualism evolved in the ancestor of *Tegeticula* and *Parategeticula*, and "cheating" later evolved twice in *Tegeticula*. (B after [59].)

of rhizobia, illustrating adaptive "partner choice" [32]. Thus, legume-rhizobia mutualisms may be stable for more than one reason. These mechanisms suggest that if the plants were supplied with excess ammonium and became less dependent on their rhizobial partners, the system might break down. To test this hypothesis, a research team took advantage of an ecological experiment in which some plots were fertilized with nitrogen for 22 years. As they predicted, test plants inoculated with rhizobia from the fertilized plots produced much less biomass than those inoculated from unfertilized plots [81]. The evolution of less beneficial rhizobia may have occurred because fertilized plants relieved sanctions against cheaters or no longer rewarded beneficial strains.

Mutualisms are not always stable over evolutionary time: many species cheat. For instance, many orchids secrete no nectar for their pollinators; some of them, in fact, deceive male insects that accomplish pollination while "copulating" with the flower (see Figure 3.23 in Chapter 3). Two lineages of yucca moths that have evolved from mutualistic ancestors do not pollinate, and they lay so many eggs that the larvae consume most or all of the yucca seeds (see Figure 13.16B) [59].

Let's return to Darwin's extraordinary orchid, and its predicted sphinx moth, with which we began this chapter. Why did the orchid's nectar tube and the moth's proboscis become so long? The answer is that mutualism often is permeated with conflict. Darwin argued that natural selection would cause the insect species to

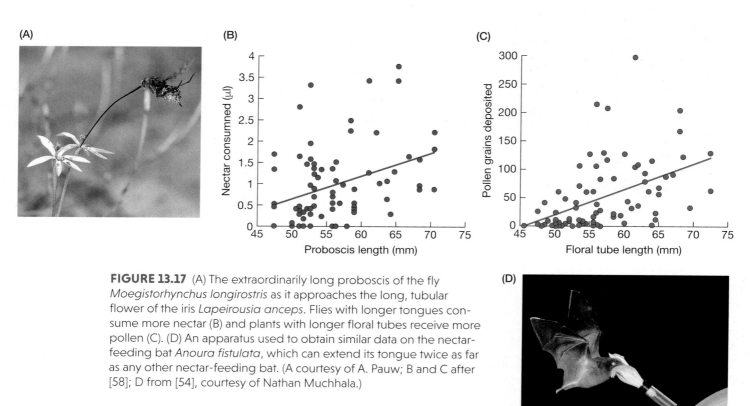

FIGURE 13.17 (A) The extraordinarily long proboscis of the fly *Moegistorhynchus longirostris* as it approaches the long, tubular flower of the iris *Lapeirousia anceps*. Flies with longer tongues consume more nectar (B) and plants with longer floral tubes receive more pollen (C). (D) An apparatus used to obtain similar data on the nectar-feeding bat *Anoura fistulata*, which can extend its tongue twice as far as any other nectar-feeding bat. (A courtesy of A. Pauw; B and C after [58]; D from [54], courtesy of Nathan Muchhala.)

evolve a proboscis long enough to reach the nectar. But why would a very long nectar tube be advantageous to the plant? Because, Darwin suggested, it would force the insect to press its head deeply into the flower and necessarily pick up and deposit pollen. If the insect's proboscis were longer than the tube, its head would not contact the pollen and the plant would not achieve reproduction. So, Darwin suggested, there may be an ongoing "race," in which the plant matches any elongation of the proboscis with an equal or greater elongation of the nectar tube.

One hundred forty-seven years after Darwin presented this hypothesis, two research teams tested and confirmed it,[1] using other plant-pollinator associations that are similarly extreme. In South Africa, Anton Pauw and collaborators, studying an iris with a long corolla tube and a fly with an equally long proboscis, found that flies with longer proboscises consume more nectar and that longer-tubed plants receive more pollen (**FIGURE 13.17A–C**) [58]. In Ecuador, Nathan Muchhala and James Thomson offered long-tongued bats experimentally altered flowers of another long-tubed plant, as well as tubes with sugar water [54]. They found that long-tongued bats delivered and received more pollen when they fed in longer tubes (**FIGURE 13.17D**). Both groups documented an advantage to the plant, and one showed an advantage to the pollinator as well.

Partly because of genomic studies, mutualism is increasingly recognized as an important basis for adaptation and the evolution of biochemical complexity [52]. The best-known examples are the evolution of mitochondria from purple bacteria and of chloroplasts from cyanobacteria (see Figure 2.5). When a new, "compound"

---

[1] Almost all of Darwin's many hypotheses have been fully or partly confirmed by later scientists. The major exception is his theory of heredity.

organism is formed from an intimate symbiosis, the subsequent evolution of both genomes is affected. For example, chloroplasts have fewer than 10 percent as many genes as free-living cyanobacteria, but many of the original cyanobacterial genes have been transferred to the plant nuclear genome. These genes may account for as many as 18 percent of the protein-coding genes of *Arabidopsis* [50].

Some mutualistic symbioses provide one or both partners with new capabilities [52]. For example, many features of bacteria are encoded by phage-borne genes. Bacteria and other microbes have formed intimate mutualisms with diverse multi-cellular organisms, especially animals, which lack the ability to synthesize essential amino acids and vitamins but can obtain some of these nutrients from their microbial partners. Some extreme associations are in sap-sucking homopteran insects (aphids, leafhoppers, cicadas, and relatives), which derive different amino acids from as many as eight different types of coexisting symbionts. Almost all plants and animals, including humans, harbor many kinds of symbionts, whose effects are largely unknown but are the subject of increasing research.

The Evolution of Competitive Interactions

Competition between species plays a huge role in evolutionary theory. In *On the Origin of Species*, Darwin spoke of "divergence of character," explaining that although species arising from a common ancestor will at first be very similar, natural selection will make them more different, because "the more diversified the descendants from any one species become in structure, constitution, and habits, by so much will they be better enabled to seize on many and widely diversified places in the polity of nature, and so be enabled to increase in numbers." (By analogy, if a city has more than enough surgeons but few pharmacists, a student might do better to go to pharmacy school.)

Darwin based his hypothesis on his perception that the population densities of many species are limited, at least at times, by resources such as food, space, or nesting sites. Consequently, competition for resources occurs within many species (intraspecific competition) and between different species if they use the same resources (interspecific competition). Interspecific competition has two major effects. First, two (or more) competing species that use exactly the same resources cannot coexist indefinitely: one will be driven to extinction. Second, competition can impose selection on one or both species. One of the possible results is divergence in resource use. If this happens repeatedly as new species arise in a clade, the result may be adaptive radiation [66].

If two species feed on a variety of food types, those individuals that are most prone to eat the same food as members of the other species may suffer lower fitness because they are competing for a limited supply (**FIGURE 13.18**). Individuals that use other food types, for which they do not compete with the other species, are likely to have higher fitness. Consequently, one or both species may evolve to use somewhat different food types from the other species, and come to overlap less in diet [68, 70]. Such divergence in response to competition between species is called **ecological character displacement**.

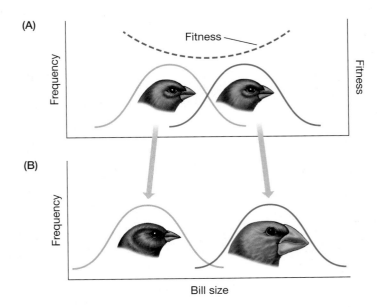

FIGURE 13.18 A model of ecological character displacement in response to competition between two species. The x-axis represents a quantitative phenotypic character, in this case bill size, that is closely correlated with some quality of a resource, such as the average size of the seeds eaten by that phenotype. (A) The frequency distributions of bill size of two species (blue and green) overlap. The dashed curve is the fitness function; fitness is lowest for those in the overlap area, representing individuals of both species that compete for the same resources. (B) Because of selection within each species against phenotypes that compete with the other species, the species undergo divergent evolution. The overlap in their resource use is reduced, and the character they use for processing the resources (bill size) diverges.

FIGURE 13.19 Character displacement in bill size in seed-eating ground finches of the Galápagos Islands. Bill depth is correlated with the size and hardness of the seeds most used by each population; arrowheads show average bill depths. (A) Only *Geospiza fuliginosa* occurs on Los Hermanos, and only *G. fortis* occurs on Daphne Major. (B) The two species coexist on Santa Cruz, where they differ more in bill depth. (After [30]; photos courtesy of Peter R. Grant.)

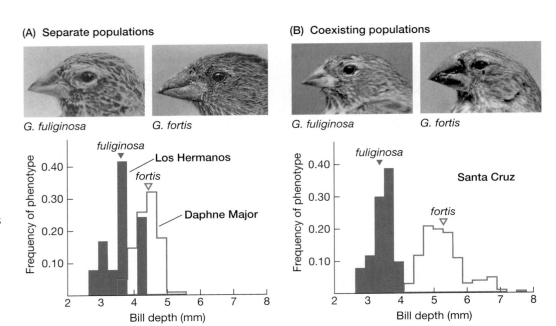

(A) Separate populations

G. fuliginosa   *G. fortis*

(B) Coexisting populations

G. fuliginosa   *G. fortis*

A famous example of this pattern is the Galápagos finches *Geospiza fortis* and *G. fuliginosa*, which differ more in bill size where they coexist than where they occur singly (**FIGURE 13.19**) [30]. The process of character displacement was experimentally demonstrated by Jonathan Losos and his colleagues in the green anole (*Anolis carolinensis*), which is native to the southeastern United States, and the brown anole (*A. sagrei*), a West Indian species that has invaded many islands in southern Florida. Although the two species overlap in perch height, brown anoles typically perch closer to the ground than the more arboreal green anoles. Losos and colleagues experimentally introduced brown anoles to several small islands, but not to others, and discovered that green anoles soon shifted to higher perches [69]. They also found that within 15 years after brown anoles invaded islands, green anoles evolved a higher number of specialized toe scales that enhance their ability to climb trees (**FIGURE 13.20**). Extensive previous research has shown that anoles compete for food and that competition is reduced between species that occupy different structural sites in forested habitats.

Competition for resources can sustain diversity both of species and of genotypes within species [7, 60]. This is the basis of a pattern called **ecological release**, wherein a species or population exhibits greater variation in resource use, and in associated phenotypic characters, if it occurs alone than if it coexists with competing species. For example, pumpkinseed sunfish (*Lepomis gibbosus*) are found in both shallow water and open water in lakes where they are the only species of sunfish, but they are limited to shallow water if another species of sunfish occupies the open water. Pumpkinseeds collected in open versus shallow water have heritable, functionally adaptive differences in body shape and a feeding structure, and differ in diet [64].

Some species compete not only by depleting resources, but also by **interference competition**, whereby individuals suppress competitors by behavioral dominance [2] or by other means, such as poisoning them (as do some plants, fungi, and bacteria). Chinese populations of an unintentionally introduced American goldenrod (*Solidago canadensis*) have evolved enhanced production of a chemical that inhibits the growth of a Chinese plant species more than native American goldenrod populations do [84]. A possible reason for this evolutionary change is that introduced plants may be able to allocate more energy to competitive ability because they have become freed from many of their natural enemies, and can reduce their defenses against herbivores [6]. Some evidence for this hypothesis was found in a natural American population of this goldenrod, in which some plots were kept herbivore-free by insecticide. After 12 years, plants from these plots were more susceptible to

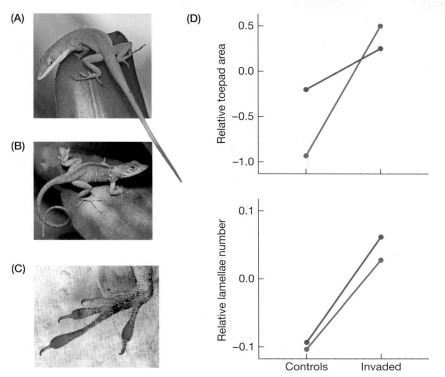

FIGURE 13.20 Results of an experiment in which indigenous green anoles (A) were confined with invasive brown anoles (B) on a small, a medium, and a relatively large island in Florida, and on control islands without the competing species. Evolutionary changes in the size of the toepads and scales (lamellae) specialized for climbing, seen in (C), were monitored over 15 years. (D) Both traits increased significantly on the invaded islands, compared to the controls (circles). The difference persisted when lizards from invaded and control islands were reared under controlled conditions (diamonds), showing that the change is genetic. (D after [69], C courtesy of Yoel Stuart.)

herbivorous beetles than were those from unsprayed plots, and they produced more polyacetylenes, which inhibit growth of competing plants [77]. Competition can result in the exclusion or extinction of some species, a special concern when species are introduced to new areas and may become "invasive" [46].

Evolution and Community Structure

The assemblage of species in a local habitat, such as a lake or forest, is often called a community. The field of community ecology is concerned with questions such as what determines how many species occur together in a local community, how food webs and other interactions among species are structured, and why the number of species differs among habitats and geographic regions. We describe here a few of the many ways in which an evolutionary perspective contributes to answering these questions [13, 51].

Phylogenetic information can cast light on the role of evolutionary history in the species richness and interaction web of a community [13, 63, 74]. The species in a large region (e.g., southeastern North America) belong to clades that, perhaps in the remote past, originated in the region or dispersed into it (see Chapter 18). Phylogenetically conservative traits may affect which clades could, and which could not, persist in a new ecological setting. For example, all species of *Heliconius* butterflies (see Figure 13.11) require passionflowers (Passifloraceae) as food plants for their larvae, and do not exist in places that lack these plants. Because of their previously evolved characteristics, members of some clades, but not others, can succeed in any particular habitat or local area (**FIGURE 13.21**). In a forest in Borneo, the trees that occur together in small plots are more closely related than a random sample of the trees in the entire forest would be, which suggests that closely related species share features that suit them to the particular environmental factors that differ among plots [80]. However, the opposite effect may be seen if specialized enemies destroy seeds and seedlings that are close to adult trees of the same or closely related species, leaving the space available for other species to grow in [42, 71]. Among the 31 species of *Inga* trees studied in a forest in Peru, trees found

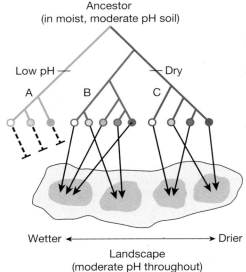

Ancestor
(in moist, moderate pH soil)

Low pH — | — Dry

A   B   C

Wetter ←——————→ Drier

Landscape
(moderate pH throughout)

FIGURE 13.21 Factors that may affect phylogenetic relationships among the members of an ecological community. The phylogeny depicts a hypothetical plant clade, the ancestor of which was adapted to moist soils with intermediate pH. Among its current descendants, one clade (A) has become adapted to, and is now found in, acidic soils, and another (C) is associated with dry soils. In a landscape that includes wetter and drier sites but lacks acidic soils, clade A is not found. Drier sites will generally have species in clade C, and wetter sites species in clade B; the environment acts as a filter, resulting in phylogenetic clustering by habitat. But variation among species in each clade leads to stronger competition and exclusion between the closest relatives, so the closest neighbors are phylogenetically overdispersed: they are not as closely related as might be found in random samples of the species.

growing close to each other were more different in defense characteristics than if they had been sampled at random [41]. The evolution of different defenses that cause each species to be attacked by different enemies has contributed to the great diversity of tree species in tropical forests [14, 37].

The diversity of species in a community also results from evolutionary divergence that reduces competition among multiple species [66]. For example, among sympatric members of the weasel family in both North America and Israel, the spacing of the size of the canine teeth is more regular among species and sexes than expected by chance. The canine teeth are used to kill prey, and the differences in their size are thought to reflect differences in the average size of prey taken by these carnivores (**FIGURE 13.22**) [17].

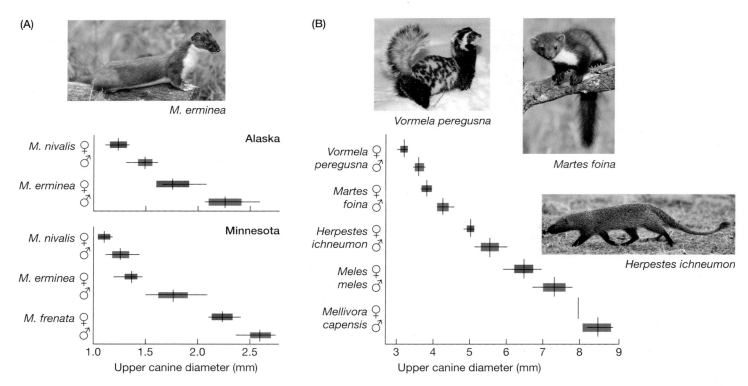

FIGURE 13.22 Size of the canine teeth differs among sympatric species of the weasel family (Mustelidae), and between the sexes in each species, in (A) Alaska and Minnesota and (B) Israel. In all sites, the spacing is more regular than would be expected at random. These differences are thought to be adaptations to feeding on prey that differ in average size. *Herpestes ichneumon*, the Egyptian mongoose, is a member of another family of Carnivora but is ecologically similar to the mustelids. Vertical lines are means, horizontal bars represent ± 1 standard deviation, and horizontal lines are range. (After [17].)

(A) Lower trunks and ground   (B) Twigs

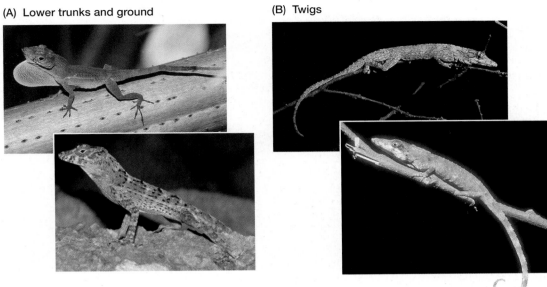

FIGURE 13.23 Convergent morphologies, or "ecomorphs," of *Anolis* lizards in the Greater Antilles, West Indies. (A) *Anolis lineatopus*, from Jamaica (top), and *A. strahmi*, from Hispaniola (bottom), have independently evolved the stout head and body, long hind legs, and short tail associated with living on lower tree trunks and on the ground. (B) *Anolis valencienni*, from Jamaica (top), and *A. insolitus*, from Hispaniola (bottom), are both twig-living anoles that have convergently evolved a more slender head and body, shorter legs, and a longer tail. (Photos by K. de Queiroz and R. Glor, courtesy of J. Losos.)

If two regions have similar environments, we might expect to see convergent evolution not only of individual species but of assemblages of species. For example, communities of plants that have evolved independently in similar environments have similar characteristics. Tropical rainforests throughout the world have tall canopy trees, festooned with ferns and other epiphytes; warm deserts throughout the world have small-leaved or leafless shrubs and cactus-like succulents. Each of these growth forms includes species in diverse phylogenetic lineages.

Convergence is also seen in parallel adaptive radiations, such as the cichlid fishes of the several African Rift Valley lakes (see Figure 9.1). The most extreme example of community-level convergence has been described in the anoles (*Anolis*) of the West Indies [43, 47]. As we noted in the previous section, these insectivorous, mostly arboreal lizards are known to compete for food. Each of the large islands of the Greater Antilles (Cuba, Hispaniola, Jamaica, Puerto Rico) has a monophyletic group of species that have evolved within the island. Many of the species on each island are ecologically and morphologically similar to those on the other islands. They occupy certain microhabitats, such as tree crown, twig, and trunk, and have consistent adaptive morphologies that have evolved independently on each island (**FIGURE 13.23**). However, evolution is not always so repeatable or predictable [44]; the largest of the Greater Antilles (Cuba and Hispaniola) have some ecologically and morphologically unique species of *Anolis*. Ecological niches that are occupied in one region often seem unoccupied in other, climatically similar regions. Blood-feeding (vampire) bats occur in the New World tropics, but not in Africa despite abundant ungulate prey; sea snakes occur in the Indian and Pacific oceans, but are absent from the Atlantic.

Go to the
——————— **Evolution Companion Website** ———————
EVOLUTION4E.SINAUER.COM
for data analysis and simulation exercises, quizzes, and more.

SUMMARY

- Coevolution is reciprocal evolutionary change in two or more species resulting from the interaction among them. Species also display many adaptations to interspecific interactions that appear one-sided rather than reciprocal.

- The phylogenies of certain symbionts and parasites are congruent with the phylogenies of their hosts. This may imply that they diverged in parallel because the symbionts did not disperse between different host lineages. It does not necessarily imply that they coevolved, in the sense of reciprocal adaptation to each other.

- The Red Queen hypothesis states that species may continue to evolve indefinitely because of changes in interacting species. For example, coevolution in predator-prey and parasite-host interactions can theoretically result in a stable genetic equilibrium under some conditions, but often involves an ongoing evolutionary arms race, indefinite fluctuations in genetic composition, or even extinction. Among the many interesting adaptations in predator-prey interactions are aposematism (warning coloration) and mimicry.

- Parasites (including pathogenic microorganisms) may evolve to be more or less virulent depending on the correlation between virulence and the parasite's reproductive rate, the parasite's mode of transmission between hosts (vertical versus horizontal), infection of hosts by single versus multiple parasite genotypes, and group selection. Parasites do not necessarily evolve to be benign. New pathogens sometimes emerge by evolutionary change that enables them to infect new hosts (e.g., humans).

- In mutualism, each species obtains some benefit from the other. This does not entail altruism, and it often involves some conflict. Selection favors genotypes that provide benefits to another species if this action yields benefits to the individual in return. Thus the conditions that favor low virulence in parasites, such as vertical transmission, can also favor the evolution of mutualisms. Mutualisms may be unstable if "cheating" is advantageous, or stable if it is individually advantageous for each partner to provide a benefit to the other.

- Evolutionary responses to competition among species may lead to divergence in resource use and sometimes in morphology (character displacement). Thus, competition is a cause of ecological diversification. However, selection for greater ability to compete can also result in greater aggression, and competitive exclusion of less competitive species.

- Both ongoing evolution and phylogenetic legacies can influence which species coexist in local ecological communities. Phylogenetically conservative characters may be subject to environmental filtering, so that the species in a habitat are phylogenetically clustered; conversely, very closely related species tend to be spatially separated. Because evolutionary history determines the features of species that affect their interactions, it helps explain the networks of interactions among species in a community.

TERMS AND CONCEPTS

aposematic
Batesian mimicry
coevolution
ecological character
 displacement

ecological release
evolutionary arms
 race
interference
 competition

matching allele
 model
mimicry
Müllerian mimicry
mutualism

Red Queen
 hypothesis
symbiosis
virulent

SUGGESTIONS FOR FURTHER READING

J. N. Thompson reviews many aspects of coevolution and provides numerous examples in *The Geographic Mosaic of Coevolution* (University of Chicago Press, Chicago, 2004), and in *Relentless Evolution* (University of Chicago Press, Chicago, 2013), which includes discussion of how interaction networks evolve. Plant-animal interactions are the focus of essays by prominent researchers in *Plant-Animal Interactions: An Evolutionary Approach*, edited by C. M. Herrera and O. Pellmyr (Blackwell Science, Oxford, 2002).

M. E. J. Woolhouse and colleagues provide an outstanding overview of parasite-host coevolution in "Biological and biomedical implications of the co-evolution of pathogens and their hosts" (*Nat. Genet.* 32: 569–577, 2002). "Models of parasite virulence" by S. A. Frank (*Q. Rev. Biol.* 71: 37–78, 1996) is an excellent entry into this subject.

The Ecology of Adaptive Radiation by D. Schluter (Oxford University Press, Oxford, 2000) includes extensive treatment of the evolution of ecological interactions and their role in diversification. "The merging of community ecology and phylogenetic biology" by J. Cavender-Bares and colleagues (*Ecol. Lett.* 12: 693–715, 2009), is an excellent overview of the subject.

PROBLEMS AND DISCUSSION TOPICS

1. How might coevolution between a specialized parasite and a host be affected by the occurrence of other species of parasites?

2. How might phylogenetic analyses of predators and prey, or of parasites and hosts, help determine whether or not there has been an evolutionary arms race?

3. The generation time of a tree species is likely to be 50 to 100 times longer than that of many species of herbivorous insects and parasitic fungi, so a tree's potential rate of evolution should be slower. Why have trees, or other organisms with long generation times, not become extinct as a result of the potentially more rapid evolution of their natural enemies?

4. Design a hypothetical experiment to determine whether greater virulence is advantageous in a horizontally transmitted parasite or in a vertically transmitted parasite.

5. Do you expect that an infectious pathogen such as the bacterium *Staphylococcus aureus* or the HIV virus that causes AIDS will evolve to become more or less virulent? What do you need to know in order to make your best projection? You may want to read about the biology of one such pathogen in order to arrive at an answer.

6. Some authors have suggested that selection by predators may have favored host specialization in herbivorous insects (e.g., Bernays and Graham, 1988, *Ecology* 69(4): 886–892). How might this occur?

7. It seems surprising that certain orchids successfully deceive insects into "copulating" with their flowers. Have these species of insects, evidently failing to perceive the difference between a flower and a female of their own species, failed to adapt? If so, what might account for this failure?

8. In simple ecological models, two resource-limited species cannot coexist stably if they use the same resources. Hence, coexisting species are expected to differ in resource use because of the extinction or exclusion, by competition, of species that are too similar. Therefore, coexisting species could differ either because of this purely ecological process of "sorting" or because of evolutionary divergence in response to competition. How might one distinguish which process has caused an observed pattern? (See Losos, 1992, *Systematic Biology* 41: 403–420, for example.)

9. Suppose that, among related host species that carry related symbionts, the relationship is mutualistic in some pairs and parasitic in others. How would you determine (a) which relationship is mutualistic and which is parasitic, (b) what the direction of evolutionary change has been, and (c) whether the change from one to the other kind of interaction has been a result of evolutionary change in the symbiont, in the host, or both?

The Evolution of Genes and Genomes

Comparing the genomes of people, plants, and protists reveals a startling disconnect between our impressions of their complexity and the DNA that underlies them. The human genome comprises some 3.2 billion base pairs (bp). Humans are obviously more complex organisms than plants, let alone protists. So it comes as a surprise that pine trees typically have about six times more DNA than we do [65], while a single-celled amoeba with the wonderful name *Chaos chaos* has 400 times more [18]. Another way we can quantify the size of a genome is by the number of protein-coding genes it has. That measure also reveals the unexpected. *Homo sapiens* has about 20,000 protein-coding genes. The nematode worm *Caenorhabditis elegans* has about the same number, even though its body has only 1000 cells [28]. By contrast, bread wheat (*Triticum aestivum*) has about five times as many protein-coding genes as we do [10], even though it is unable to write music or program a computer.

These examples highlight a strange fact about life on Earth. There is no simple relation between our sense of the complexity of an organism and the size of its genome, measured either by DNA content or number of genes (**FIGURES 14.1 and 14.2**). One might guess that the genome would be where adaptation rises to the pinnacle of refinement. You will see shortly that the reality is quite the opposite: genomes are the messy outcomes of conflict and cooperation, of selection and random genetic drift, and of all the other evolutionary ingredients that make life the beautiful tangle that we see in nature.

The douc langur (*Pygathrix nemaeus*) is arguably the world's most beautiful primate. It has an unusual adaptation to its diet of leaves: a duplicated gene that allows it to efficiently digest the bacteria that ferment leaves in the monkey's gut.

FIGURE 14.1 Genome size variation. The bars indicate the range of genome size (shown in megabases, Mb; 1 Mb = 10^6 bases) across species within each group. Some protists (which are unicellular eukaryotes) have genomes that are hundreds of times larger than that of any mammal. (After [22, 53].)

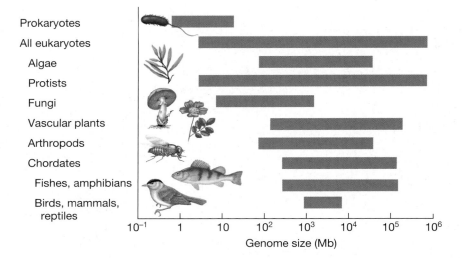

This chapter explores the evolution of the genome by first discussing how new genes originate. It then discusses the different fates of new genes and looks at how genes die. The next topic is the evolution of protein-coding genes by changes to their sequences and their expression. We then consider how and why the number and structure of chromosomes evolve. Finally, we tackle the evolution of genome size and genome content, including the puzzle of why so much DNA has no obvious function.

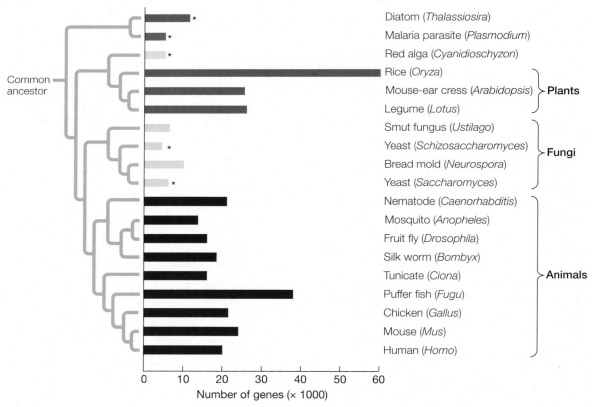

FIGURE 14.2 Numbers of protein-coding genes for several eukaryotes, arranged on a phylogeny. Some patterns make intuitive sense: multicellular eukaryotes with tissue organization (plants and animals) have more protein-coding genes than most unicellular eukaryotes and multicellular eukaryotes without distinct tissues (indicated by asterisks). Other patterns are puzzling: rice has more than twice as many protein-coding genes as the other flowering plants shown, and puffer fishes have almost twice as many as humans. (After [60].)

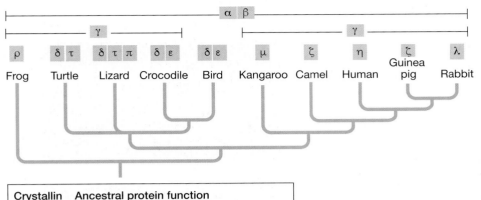

FIGURE 14.3 Crystallin proteins in the lens of vertebrate eyes allow them to focus light. Different groups of vertebrates have different crystallins derived from a wide variety of proteins with other functions. (After [64].)

Crystallin	Ancestral protein function
α	Small heat-shock proteins
β/γ	Related to bacterial stress protein
ρ	NADPH-dependent reductase
δ	Arginosuccinate lyase
τ	α-Enolase
π	Glyceraldehyde phosphate dehydogenase
ε	Lactate dehydrogenase
μ	Similar to bacterial ornithine deaminase
η	Aldehyde dehydrogenase
ζ	Alcohol dehydrogenase
λ	Hydroxyacyl-CoA dehydrogenase

The Birth of a Gene

One of the most fundamental questions we can ask about evolution is: How does biological novelty originate? The answer must lie partly with the origin of genes that have new functions. Like much of evolutionary genetics, this research topic has exploded recently with the arrival of large numbers of genome sequences from diverse branches of the tree of life [12].

The human eye is a remarkable organ. It forms a high-resolution image by focusing light through a transparent lens made of living tissue. The lens is made up largely of proteins called crystallins. Where did they originate? It turns out that many proteins happen to be largely transparent. Crystallins in the vertebrate lens are derived from proteins with a variety of other functions, for example a small heat-shock protein that protects many tissues from different types of stress [14]. Crystallins in other animals are derived from different proteins with yet other functions (**FIGURE 14.3**). In some instances, crystallins continue to function in their original roles while also serving in the lens [56].

The crystallins show how novel biochemical functions can originate by a process central to Darwin's theory: descent with modification. Some of them originated by gene duplication (see Chapter 4). Duplication is the most common way that new genes arise in eukaryotes. At the molecular level, duplications are caused by several mechanisms. One is unequal crossing over, in which recombination happens between different positions on chromosomes that are misaligned during meiosis. Another is replication slippage, in which the DNA polymerase loses its place and copies a segment of chromosome twice. In either event, the resulting chromosomes can carry a gene that is duplicated in tandem with the parental copy. A third way that gene duplicates arise is by **retrotransposition**. Here the messenger RNA from a gene is reverse-transcribed to DNA, which is then integrated into the genome. The result is a duplicate gene that is far from the parental copy, often on a different chromosome. Duplicates that originate this way can be distinguished from their parental genes because they lack introns, which were spliced out of the mRNA before it was reverse-transcribed.

(A)

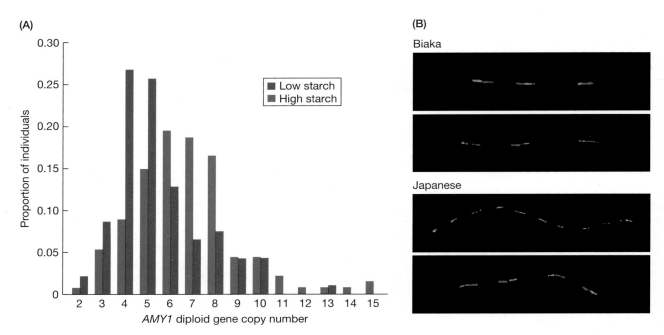

(B)

Biaka

Japanese

FIGURE 14.4 Copy number variation at the *AMY1* locus in humans. This gene codes for amylase, an enzyme that digests starch. Individuals with more copies of the gene have a greater concentration of amylase in their saliva. (A) Distributions of copy number in populations that have low-starch diets (Biaka and Mbuti hunter-gatherers) and high-starch diets (Japanese, European Americans, and Hadza). On average, the populations with a high-starch diet have 6.7 copies of *AMY1*, while those with a low-starch diet have 5.4 copies. (B) Photos of regions of the two chromosome regions that carry duplicates of the *AMY1* locus. Each copy of the gene is labeled with a red and green dye. The Biaka individual shown here has 6 copies, while the Japanese individual has 14 copies of *AMY1*. (A after [55]; B from [55].)

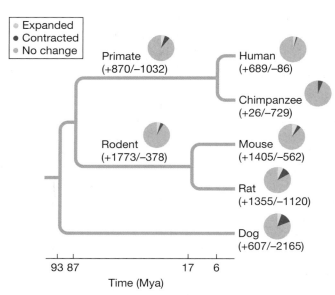

FIGURE 14.5 Gains and losses of genes in several lineages of mammals. The numbers beneath each branch show the numbers of genes gained and lost. The pie charts show the fraction of gene families that have expanded, contracted, or not changed. (After [15a].)

As a chromosome with a new gene duplicate spreads through a population, there is polymorphism in the number of copies of the gene that individuals carry. This situation, called **copy number variation**, is found at the *AMY1* locus in humans [55]. This gene codes for amylase, an enzyme in saliva that breaks down starch. Individuals with more copies of *AMY1* have more amylase in their saliva and digest starch more efficiently. The central African Biaka, who are hunter-gathers with a diet that rich in protein but low in starch, typically have four or five copies of the gene. Japanese, whose rice-heavy diet has abundant starch, typically have six to eight copies of the gene (**FIGURE 14.4**).

Gene duplication plays a key role in genome evolution. In the recent evolutionary past of humans, 1 percent of our genes have been duplicated every million years [15a]. Some 1400 duplicates have been fixed in humans or chimpanzees since we shared a common ancestor roughly 7 million years ago (Mya). More of the base pair differences between humans and chimpanzees have resulted from gene duplication than from changes at single nucleotides. Gene duplication has played major roles in the evolution of other species as well (**FIGURE 14.5**).

Important ecological adaptations can result from gene duplication. The douc langur (*Pygathrix nemaeus*) that we met at the opening of this chapter lives on a diet of leaves. The leaves are fermented in the gut by symbiotic bacteria, much as in cows, and the monkeys gain nutrition by digesting the bacteria. One

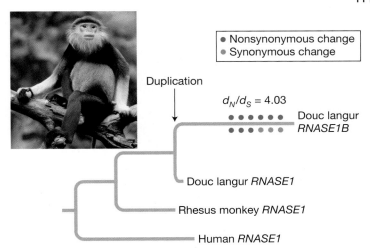

Nonsynonymous change
Synonymous change

Duplication

$d_N/d_S = 4.03$

Douc langur *RNASE1B*

Douc langur *RNASE1*

Rhesus monkey *RNASE1*

Human *RNASE1*

FIGURE 14.6 Duplication and adaptation of the *RNASE1* gene in the douc langur. The gene tree relates the *RNASE1B* duplicate locus in the douc langur to the *RNASE1* locus in the langur, rhesus monkey, and human. Branch lengths are proportional to the number of substitutions in the coding region. Since the duplication event, there have been 12 changes to the coding region of the duplicate, and none to the original copy of the gene. Of the changes in the duplicate, nine are nonsynonymous and three are synonymous. The d_N/d_S ratio (see Chapter 7) is much greater than 1, which is strong evidence that the duplicate has evolved under positive selection. (After [42], based on data from [67].)

of the enzymes that digests the bacteria is encoded by the *RNASE1B* locus, which originated by gene duplication about 4 Mya. This new enzyme rapidly evolved nine amino acid changes (**FIGURE 14.6**) [67]. Those changes allow the enzyme to work in the low-pH environment of the monkey's gut that is needed to ferment the leaves.

Some proteins have repeated **domains** that confer part of their function. The sodium channel that is critical to the firing of vertebrate nerves has four domains with very similar structure (**FIGURE 14.7**). They arose by two rounds of duplication of a gene in a remote ancestor that coded for only a single domain [34].

Mixtures of exons duplicated from genes with different functions can generate new genes with new functions, a process that is called **exon shuffling**. Exon duplication and exon shuffling have played roles in the evolution of many eukaryotic genes [42]. An example is the *jingwei* locus, which is found only in the fruit flies *Drosophila teissieri* and *D. yakuba* (**FIGURE 14.8**). The first three exons of this gene are duplicates of exons in the *yellow-emperor* (*Ymp*) gene found in many *Drosophila* species. The fourth exon, however, is a duplicate of the entire alcohol dehydrogenase (*Adh*) gene that was retrotransposed into an intron of *Ymp*. The new *jingwei* gene shows evidence of rapid evolution by positive selection after it originated about 2 Mya.

(A)

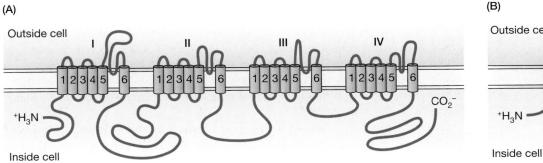

(B)

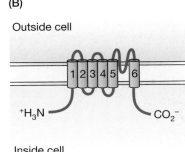

FIGURE 14.7 Vertebrate sodium channels, which are critical to firing neurons, evolved by duplication. (A) The α-subunit of the sodium channel protein has four domains (labeled I–IV). Each domain has six segments that traverse the cell membrane (numbered in each domain). The corresponding segments in each of the four domains have very similar protein sequences. The sodium channel protein evolved by two rounds of duplication of a locus that coded for a single domain. (B) The ancestor of the vertebrate sodium channel had a structure similar to this bacterial sodium channel. (After [27].)

FIGURE 14.8 A new *Drosophila* gene called *jingwei* originated by the retrotransposition of one gene into the intron of another. (A) The *Adh* gene (with exons shown in purple) was retrotransposed into the third intron of the *Ymp* gene (with exons shown in green) about 2 Mya. (B) After the retrotransposition event, the exons downstream of the novel exon degenerated because of the addition of the new stop codon at the end of the *Adh* sequence (red bar). The name *jingwei* comes from a Chinese myth in which a princess metamorphoses into a new form. (After [42].)

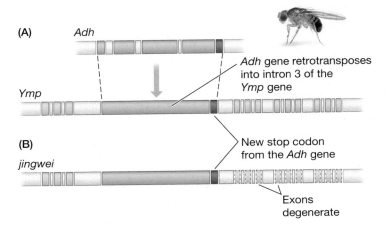

(A) *Adh*

Adh gene retrotransposes into intron 3 of the *Ymp* gene

Ymp

(B)

jingwei

New stop codon from the *Adh* gene

Exons degenerate

Gene duplication happens on a massive scale when mutation produces a tetraploid (see Chapter 4). A tetraploid descendant of a diploid has four copies of each gene rather than two. Tetraploids can arise in two ways (see Chapter 9). The first is by **whole genome duplication**. This can occur when the genome of a single species is doubled (resulting in an *autopolyploid*). Whole genome duplication can also occur when two species hybridize, and the gametes from both of them are mutants with unreduced diploid genotypes (resulting in an *allopolyploid*). Further rounds of hybridization can give rise to species with six, eight, and even more copies of each chromosome. Whole genome duplication is much more common in plants than animals, but it did occur twice in our own remote ancestors, between 650 and 550 Mya [33]. Recent events of polyploidy occurred during the domestication of several important crop plants (including wheat, coffee, and cotton) and were key to improving some of their economically valuable traits [51].

While recombination usually involves the mixing of genes of the same species, occasionally genes from other species are mixed into the gene pool. In eukaryotes, this usually occurs through hybridization between closely related species. Genetic exchange also happens between distantly related organisms by **horizontal gene transfer**, or **HGT** (see Chapter 4). HGT is particularly important to prokaryotes, and is the most common way by which they acquire new genes, including those that confer antibiotic resistance. HGT can vastly speed up adaptation since a new functional gene is acquired in one fell swoop, rather than evolving through many mutations.

HGT has been important in the evolution of the nematodes that are parasites on plants. The worms invade the roots with the help of cellulases, pectate lysases, and other enzymes that break down the cell walls of the plant. These enzymes were acquired by the nematodes from bacteria and fungi [8]. HGT thus opened up an entirely new ecological niche for nematodes, and now enables them to be major pests of crops around the world.

Given the intimacy between the genomes of the mitochondria and the nucleus within a cell, it may not be surprising that extensive HGT has occurred between them [1]. The vast majority of the transfers have been from the mitochondria to the nucleus. The movement of genes from the mitochondria to the nucleus has led to large reductions in the size of the mitochondrial genome. In the most extreme cases, *all* mitochondrial genes have moved to the nucleus. Mitochondria have lost their entire genomes in several unicellular eukaryotes (such as the microsporidia, a group of intracellular parasites).

It might seem impossible that a new gene could originate from DNA that previously had no function. After all, the number of possible combinations of codons is beyond comprehension, and the chance that a random combination might make a

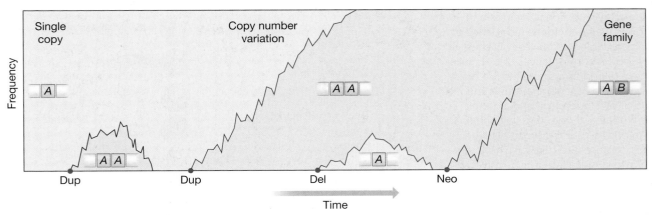

FIGURE 14.9 Fates of gene duplicates. Gene duplication mutations ("Dup") produce chromosomes carrying an extra copy of locus *A*. Copy number variation occurs when a duplicate is polymorphic: chromosomes with different numbers of copies of locus *A* are present in the population. Some duplicates that become fixed retain their original function and serve to upregulate the total amount of gene product that is produced. A functioning duplicate that has become fixed can later be lost by a mutation that deletes it or that renders it a pseudogene ("Del"). A duplicate can also acquire a mutation that leads to neofunctionalization or subfunctionalization ("Neo"). If this mutation becomes fixed, a gene family is produced. (After [31].)

useful protein seems vanishingly small. But as unlikely as that is, recent research has uncovered new genes in organisms ranging from yeast to humans that did indeed originate from noncoding DNA [24, 61]. These are called **de novo genes**, to distinguish them from genes that are born by the much more common routes of gene duplication and HGT. The best-studied examples are in *Drosophila*. Many of their de novo genes are expressed in testis, which is consistent with other evidence that sexual selection on males is a powerful driver of evolutionary change in many animals. The origin of de novo genes may be as simple as a mutation at a single base that by chance turns on transcription of a downstream stretch of DNA that fortuitously codes for a protein that enhances fitness.

No matter how it is born, a new gene starts its life as a single copy in the population. As with any new mutation, by far the most likely outcome is that it will be lost by random drift or selection over the next few generations (see Chapter 7). A very small fraction of new genes manage to spread and become fixed in the species (**FIGURE 14.9**). Even those that do usually do not survive very long—they are later destroyed by mutations, as you will see below. A very small fraction of new genes remain functional and become permanent features of the genome. Often these genes survive because they evolve a new adaptive function, like the *RNASE1* duplicate in the douc langur.

Gene families

The loci that encode hemoglobins are members of a **gene family**, which is a set of loci that arose by duplication and that code for proteins that typically continue to have similar biochemical function. Two or more genes that originated by duplication are said to be **paralogs** (see Figure 2.14). Some gene families, such as the hemoglobins, have many paralogs created by several duplication events that were widely separated in time (see Figure 2.15). In other cases, the rapid growth of a gene family can contribute to a novel adaptation. Male stickleback fishes construct nests in which females lay their eggs. The nest is held together by a glue that the males secrete. This glue is produced by a family of genes called *spiggin* that are recent duplicates of a single gene that made mucus in the ancestor of sticklebacks [35].

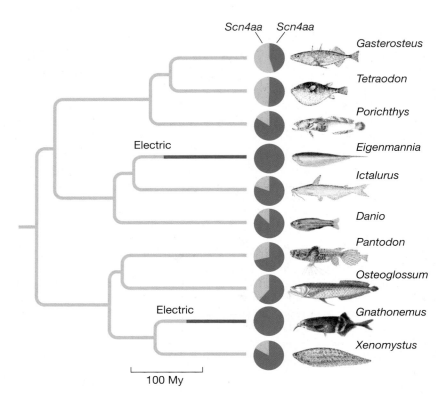

FIGURE 14.10 The neofunctionalization of a sodium channel gene was a key to the independent origin of electric organs in two families of fishes. The phylogeny shows representatives of 10 families of distantly related teleost fishes, and the pie diagrams show the relative expression levels in muscle tissue of two duplicated sodium channel genes, *Scn4aa* and *Scn4ab*, that trigger muscle contraction. In South America an electric organ evolved in the knifefishes (represented here by *Eigenmannia virescens*), while in Africa the organ evolved in elephant-nosed fishes (represented by *Gnathonemus petersii*). In those two families, expression of *Scn4aa* has been lost completely from the muscle. Instead, the gene is expressed in the electric organ, which gives these fishes a remarkable sensory modality that allows them to hunt and communicate in darkness. (After [63]; images from [63].)

A functioning duplicate that becomes fixed in a population can meet one of several ultimate fates [37, 44] (see Figure 14.9). If the duplicate is redundant and does not provide a fitness benefit, it will be lost by deletion or the accumulation of loss-of-function mutations. A second fate is that a duplicate can simply retain its original function. A locus that has been amplified this way now has two (or more) copies in the genome. They can all be favored to continue functioning when there is selection for increased expression of the gene's product. Many insects have evolved resistance to pesticides this way [40]. The insects have enzymes that degrade a broad spectrum of substrates, including insecticides. In the presence of insecticide, duplicates of those genes are favored because they increase gene expression and so enable the insect to detoxify the pesticide more rapidly.

A strange form of inheritance occurs in some families of amplified genes. When duplicates occur in tandem along the chromosome, the DNA replication machinery sometimes gets confused and aligns one copy of a gene with a paralogous copy nearby. *Gene conversion* can then happen between the two genes, causing a mutation in one locus to be copied to the other (see Chapter 4). A mutation at one locus can spread this way through all the paralogs in the gene family. This process is called *concerted evolution*. The ribosomal RNA genes exist in many copies in eukaryote genomes. Gene conversion between the copies keeps their sequences from diverging within species at the same time that they diverge between species [20].

A third fate that can befall a duplicated gene is **neofunctionalization**. Here the duplicate evolves a novel biological function. A crystallin in the lens of the vertebrate eye originated by duplication of a heat-shock gene. Another example is found in electric fishes. All teleost fishes have two duplicates of a sodium channel gene that is expressed in muscles. In most teleost fishes, both genes are involved in triggering muscle contractions. But in two families of fishes (the knifefishes and elephant-nosed fishes), one of the paralogs has independently evolved an entirely different function (**FIGURE 14.10**). Here it plays a key role in firing the electric organ, which is a unique structure that enables these fishes to sense prey and communicate with each other in darkness [2].

Last, a duplicate gene and its parental copy can **subfunctionalize** so that each carries out only some of the roles that the ancestral gene performed. In most vertebrates, the same hemoglobin molecule transports oxygen in the bloodstream at all stages of life. In mammals, however, a duplicate of the β-hemoglobin locus is expressed in the fetus [62]. The fetal hemoglobin has evolved differences that give it a higher affinity for oxygen than the adult hemoglobin. This enables the fetus to strip oxygen from its mother's blood in the placenta. A few months after birth, expression of the fetal hemoglobin is shut off, and the adult hemoglobin takes over the task of oxygen transport. The evolution of the specialized fetal hemoglobin by subfunctionalization was one key to the origin of live birth in mammals.

The Death of a Gene

There are many ways that a gene can die. When a gene is duplicated, the new copy is often dead on arrival. The duplicate many not include the entire gene. Even if it does, if often lacks the regulatory elements needed to express it at the right time and in the right places. Duplicates produced by retrotransposition face an additional challenge. Since they lack introns, there is no opportunity for posttranscriptional regulation that involves splicing.

When a nonfunctioning duplicate is fixed in the population, or a functioning gene becomes nonfunctional, the result is a genetic skeleton called a **pseudogene**. The genomes of many species are littered with pseudogenes. Our own genome has almost as many pseudogenes as functional genes [44]. Although they serve no function, pseudogenes are useful to evolutionary biologists as a sort of natural controlled experiment. The parental gene from which a pseudogene originated continues to evolve under the forces of selection. But the pseudogene does not produce a functioning product, so it is freed from selection. Comparing the sequences of pseudogenes with those of their parental genes reveals that pseudogenes typically evolve much more quickly (see Figure 7.19). This shows that most mutations to functioning genes reduce fitness and are removed from the population by **purifying selection** (that is, selection against deleterious mutations). In contrast, mutations in a pseudogene are selectively neutral and so they are free to drift to fixation.

Deletions are important in shaping the genome (see Chapter 4). Deletions are indiscriminate: they will sometimes eliminate part of a gene, all of a gene, or even a large piece of chromosome that carries many genes. The loss of genes by deletion and pseudogenization contributes to the divergence in the genetic content of the genomes of related species (see Figure 14.5).

It may seem surprising that a deletion can sometimes be beneficial. Attached to the surface of human white blood cells is a protein called CCR5 that plays a role in inflammatory response. Unfortunately, this protein is also a key to human immunodeficiency virus (HIV) infection. The virus binds to a loop of the CCR5 protein and then enters the cell. Some people have an allele called Δ32 at the *CCR5* locus in which 32 bp are deleted. The missing part of the gene codes for the loop in the protein to which the virus attaches. People who are homozygous for the deletion are highly resistant to HIV, while heterozygotes are partly resistant. The deletion has spread in some populations by positive selection that occurred long before HIV first infected humans, probably because of the resistance it confers to other diseases [19].

Deletion plays a key role in **gene trafficking**, which is the movement of genes to new sites in the genome. When a functioning gene duplicate is established at a new location, its parental copy is sometimes lost by deletion. The result is that the gene has moved to a new address in the genome. Although the probability that a gene will move by gene trafficking is very low, over millions of years gene trafficking can cause certain genes to accumulate on certain chromosomes. In fruit

FIGURE 14.11 Gene trafficking in *Drosophila melanogaster*. The height of each bar shows how frequently genes move by retrotransposition from one genome location (shown on the *x*-axis) to another (shown on the *y*-axis). The blue lines separate the autosomes from the X chromosome. Many genes have changed locations. Furthermore, the movement is not random: a large number of genes have moved from the X chromosome to an autosome (shown by the many bars in the back left of the plot). (From [42].)

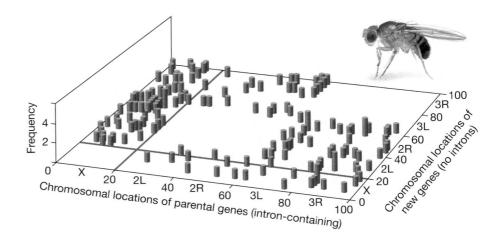

flies (*Drosophila*) there has been a flight of genes from the X chromosome to the autosomes (**FIGURE 14.11**). Several hypotheses might explain this pattern, which is the subject of ongoing research [42].

Evolution of Protein-Coding Genes

Although coding sequences comprise less than 2 percent of the human genome, that slender slice of our genetic material is by far the best understood. Codons provide genetic landmarks that greatly help interpret how the genome evolves.

Evolution of coding regions by genetic drift

A strong evolutionary pattern is seen across protein-coding genes. The DNA bases that appear in the first and second positions of codons tend to evolve slowly. DNA bases in the third positions of codons and in introns tend to evolve much more rapidly (see Figure 16.5). Changes to first and second positions are likely to be nonsynonymous, that is, to change the protein, while changes to third positions or codons are likely to be synonymous, that is, to have no effect on the protein (see Chapters 4 and 7). In short, the DNA bases most likely to affect a gene's product evolve most slowly. The pattern seems puzzling. If most genetic change were caused by natural selection, then we would expect exactly the opposite pattern. Changes that alter a protein have a chance of improving fitness, and so nonsynonymous changes should evolve the most frequently.

But the pattern makes sense if most changes to DNA sequences evolve by random genetic drift rather than by adaptation. Between 70 percent and 97 percent of nonsynonymous mutations are strongly deleterious, depending on the organism [17]. Most of those mutations are eliminated by purifying selection. In contrast, synonymous mutations do not change a gene's product and so are expected to have much smaller effects on fitness. Those mutations have a chance of spreading through the species by random genetic drift, contributing to the differences we see among species. As a result, the DNA bases in a gene that are least likely to have a fitness effect tend to be those that evolve the most rapidly and that show the largest numbers of differences among species.

This pattern lies at the heart of the famous **neutral theory of molecular evolution** developed by Motoo Kimura starting in the 1960s (see Chapter 7). Kimura argued that the vast majority of differences among species and variation within species in DNA sequences are due to mutations that are nearly selectively neutral and have evolved mainly by drift. Purifying selection is constantly at work eliminating deleterious mutations, Kimura thought, but positive selection leading to adaptive differences among species is so rare that it makes only a negligible contribution to molecular evolution. We now understand that the neutral theory is a powerful explanation for some patterns, such as why pseudogenes evolve so quickly. But Kimura's claim that positive selection makes only a trivial contribution to the differences between species turns out to have been somewhat exaggerated, as you will now see.

Evolution of coding regions by positive selection

Random genetic drift is responsible for much of the evolution of DNA sequences—much, but certainly not all. Natural selection occasionally favors changes to a protein, for example when a species encounters a new environment. Then the typical pattern can be reversed, and DNA mutations that change a protein spread more often than those that do not. In the 4 million years following the origin of the *RNASE1B* gene in the douc langur, nine nonsynonymous mutations became fixed, while only three synonymous mutations became fixed (see Figure 14.6). Several lines of evidence show that most or all of the changes to the protein produced by *RNASE1B* were driven by positive selection that improves its new role in digestion [67].

We can get a rough idea of the relative importance of purifying selection, drift, and positive selection using a simple statistic based on a comparison of the DNA sequences of the same gene from two species. We determine d_N, the fraction of sites that differ at nonsynonymous sites (those that do change an amino acid), and d_S, the fraction of sites that differ at synonymous sites (those that do not change an amino acid). Dividing the first fraction by the second gives us the **d_N/d_S ratio**. Imagine that synonymous and nonsynonymous mutations both have very little effect on fitness. Then both types of mutations will have the same chance of drifting through the population to fixation, and so we expect d_N/d_S to be 1. However, if most nonsynonymous mutations are deleterious and are removed by purifying selection, then d_N/d_S will be much less than 1. This is what we see at the great majority of genes (see Figure 7.21).

Occasionally loci show a d_N/d_S ratio greater than 1. This means that more nonsynonymous mutations, which are likely to affect fitness, have been fixed than synonymous mutations, which are likely to be selectively neutral. That suggests the nonsynonymous mutations that fixed had a boost from positive selection. The *RNASE1B* gene in the douc langur discussed earlier shows exactly this pattern.

So some of the genetic differences we see among species were fixed by adaptation, and others by random genetic drift. This raises a fundamental question: How much do these two processes contribute to evolutionary change? The answer depends strongly on the group of organisms [11, 25]. About half of the differences in protein sequences among species of fruit flies (*Drosophila*), mice (*Mus*), and enteric bacteria (*E. coli* and *Salmonella enterica*) evolved by positive selection and half by drift. In our own species, the picture is very different: less than 15 percent of protein evolution in our recent past has been adaptive.[1] Many genes have been found in humans that show evidence of positive selection, but they represent only a small fraction of all the changes that have evolved in the last few million years.

This striking difference between the evolutionary patterns of flies and humans is a consequence of population size. *Drosophila melanogaster* has an effective population size in the millions, while humans have had an effective population size of only about 10,000 over much of our evolutionary history. Consequently, genetic drift has been weaker in *D. melanogaster* and stronger in humans (see Chapter 7). A deleterious mutation in the fly that decreases fitness by $s = 10^{-5}$ will be weeded out by purifying selection. This prevents it from becoming fixed and contributing to differences between species. In humans, however, a mutation with that same selection

---

[1] The relative contributions of selection and drift to the evolution of differences among species are estimated using an extension of the MK test that we discussed in Chapter 7.

coefficient evolves almost as if it is selectively neutral, and so can be fixed by drift. As a result, more evolutionary change is caused by drift and less by positive selection in humans than in *D. melanogaster*. Some microbes have population sizes that are vastly larger than those of fruit flies, and evolution in those organisms is even more strongly dominated by selection. In each generation, moreover, more beneficial mutations enter a big population than a small population, which further tilts the balance of molecular evolution away from drift and toward adaptation.

Earlier we discussed how the evolution of synonymous changes by drift often makes a big contribution to the divergence between the DNA sequences of species. In fact, selection also plays a role in the evolution of synonymous mutations. The different codons that correspond to the same amino acid appear at different frequencies in the genome, a phenomenon called **codon bias**. One cause of this bias is mutation. Mutations from the DNA base G to A and from C to T are more than twice as common as other types of mutations [26, 45]. This tends to favor the accumulation of codons with A and T bases. A second cause of codon bias is natural selection [57]. Because they do not change the protein, synonymous mutations were long thought to be selectively neutral. In fact, they can have minute effects on fitness. The translation of highly expressed genes is most efficient when their codons correspond to transfer RNAs that are common in the cytosol, so selection favors those codons. Selection can also favor codons that produce messages that are less prone to translation errors. The relative strengths of the forces, and so the direction and degree of codon bias, differ among taxa. Codon bias driven by selection tends to be stronger in genes that are highly expressed and in species with very large population sizes, such as free-living microbes. In species with smaller population sizes (such as vertebrates), drift overwhelms whatever selection acts on synonymous mutations, and codon bias is very weak or absent.

Evolution of Gene Expression

The mosquito *Aedes aegypti* is the vector of Zika virus, dengue fever, yellow fever, and chikungunya virus—diseases that together kill some 50,000 people each year. In East Africa, there are two genetically distinct types of mosquitoes (**FIGURE 14.12**). The domestic type specializes in biting humans (whom the mosquitoes infect), while the forest type feeds on other animals. A key adaptation that enables the domestic

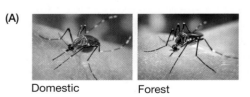

Domestic   Forest

FIGURE 14.12 The mosquito *Aedes aegypti* is the vector that transmits the Zika virus, dengue fever, yellow fever, and chikungunya virus. (A) In East Africa the mosquito has a domestic and a forest form. (B) Among its adaptations to feeding on humans, the domestic form is attracted to human body odor. This adaptation involves changes both to the structure and the expression level of Or4, a receptor on the mosquito's antenna that is sensitive to an odor distinctive to humans. The graph shows results from an experiment in which mosquitoes could respond to the odor of either a human or a guinea pig. Each vertical bar represents the relative preference of a different laboratory colony of mosquitoes that was established from a small number of mosquitoes sampled from Rabai, Kenya. (Colonies are arranged from those that most prefer humans to those that most prefer guinea pigs.) All colonies of the domestic form (red bars) preferred the human odor, while all but one of the colonies of the forest form (blue bars) preferred the guinea pig odor. (After [47].)

type to find humans is its ability to detect human odors [47]. An odorant receptor protein called Or4 that is expressed on the mosquito's antennae has recently evolved in two ways. Domestic mosquitoes carry alleles for the receptor with coding differences that enable it to detect sulcatone, a chemical that is characteristic of human body odor. A second change is that domestic mosquitoes express roughly twice as much of the receptor on their antennae. The combined effect of these changes is to make the domestic form of *Aedes aegypti* very efficient at feeding on humans—and transmitting disease.

In the previous section you saw how a gene adapts by changes to its coding sequence that alter the biochemistry of the protein made by the gene. The evolution of the *Or4* gene illustrates a second pathway to adaptation: changes in expression. Selection can alter how often, when, and where a gene is transcribed, how the transcript is spliced and processed, if and how the transcript is translated into a protein, and how the protein is deployed (see Chapter 15). Several mechanisms are involved. Many evolutionary changes to gene expression come from changes in sites that bind transcription factors. Gene expression can also evolve through changes to alternative splicing patterns, and by epigenetic changes to the DNA and the histones that are bound to it.

When a population adapts to a new environment, regions of the genome that are under positive selection will diverge most rapidly from other populations and closely related species. This fact has been used to study how marine populations of three-spined stickleback fish (*Gasterosteus aculeatus*) have adapted in parallel in several independent colonizations of rivers and streams (see also Figure 8.8) [32]. Of the many places in the genome that show evidence of adaptation, between 40 percent and 80 percent are regulatory, while only 17 percent are in coding regions (**FIGURE 14.13**). In recent human evolution, it appears that adaptive evolution of the nervous system has resulted largely from changes to gene regulation. For example, after modern humans diverged from Neanderthals some 600,000 years ago, a mutation was fixed in an intron of the *FOXP2* gene that changes its expression. That discovery is intriguing because *FOXP2* may have been involved in the evolution of speech [52].

Comparisons among the genomes of distantly related species reveal small regions of noncoding DNA that are much more similar than the rest of the genome. These ultraconserved elements are thought to be under strong purifying selection that constrains them from drifting apart [59]. Ultraconserved elements have been used to identify thousands of noncoding regions in mammalian genomes that may be regulatory elements [41].

Operons are clusters of genes that are transcribed together into a single mRNA. This message can be translated as a single unit or can be spliced into several messages that are translated separately. This setup provides an economical way to regulate expression because all of the genes in an operon are turned on and off together. The favorite food of the gut bacterium *Escherichia coli* is glucose. But if there is no glucose nearby, a set of genes called the lac operon turns on, producing the enzymes needed to feed on lactose. Expression patterns can evolve by adding and removing individual genes from operons [58]. Operons are particularly

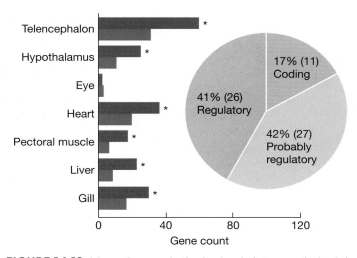

FIGURE 14.13 Many changes in the brain, skeleton, and physiology evolve in three-spined sticklebacks (*Gasterosteus aculeatus*) when they invade fresh water from marine populations. Certain regions of the genome consistently show excess divergence between freshwater and marine populations, indicating they are involved in adaptation to fresh water. The pie chart shows that about one-sixth (17 percent) of these regions involve nonsynonymous changes to coding sequences, while the remainder are in genome regions that are probably or definitely regulatory. The genome regions that show excess sequence divergence are enriched in genes that are expressed differently in freshwater and marine populations. Each pair of bars in the bar graph shows the results from a different tissue. The blue bars show the numbers of genes whose expression levels are expected to be different by chance, and the red bars show the numbers of genes observed. Six of the seven tissues show statistically significant expression differences (indicated by asterisks), showing that gene expression evolves as the fish adapt. The freshwater and marine environments have many differences in addition to salinity, for example in the prey that the fish eat and the parasites that attack them. (After [32].)

important to bacteria and archaea, in which they are widespread in the genome, but they are also found in viruses and a small number of eukaryotes.

Gene Structure

A shocking discovery was made in 1977, in the early days of DNA sequencing. The genes that code for many proteins are broken into pieces. These pieces—the exons—are separated by stretches of noncoding DNA—the introns (see Chapter 4). Further research revealed that most genes in prokaryotes lack introns, while most genes in eukaryotes have them. In humans, about one-fourth of the entire genome is made up of introns (**FIGURE 14.14**).

When a gene with introns is transcribed into messenger RNA, the message is processed by splicing to produce the final message that is then translated into a protein. The splicing removes all the segments of the message corresponding to introns. Splicing can also remove one or more segments that correspond to exons. **Alternative splicing** brings together different combinations of exons from the same locus. As a result, a single gene can produce more than one protein. The current record holder is the gene *Dscam* in *Drosophila*. Alternative splicing of its 95 exons could potentially produce more than 38,000 kinds of proteins from this single gene [69]. Alternative splicing is a major mechanism used by eukaryotes to increase organismal complexity without increasing the size of their genomes, and it may be as important as amino acid changes in the functional diversification of proteins. Changes in alternative splicing can evolve quite quickly. About one-third of alternative splicing events are different between the genes of humans and mice [49], while differences in their coding sequences have evolved only half as fast. Alternative splicing also contributes to phenotypic plasticity (see Chapters 6 and 15). The plant *Arabidopsis* splices some of its genes in different ways depending on the environment in which it grows [68]. In sum, introns are essential to the proper function of the eukaryotic genome. You will see shortly, however, that introns may have originated for nonadaptive reasons.

Another surprising feature of how genes are put together is that they sometimes overlap, so that one stretch of chromosome encodes two different proteins [30]. (Imagine a string of letters that form two different sentences depending on where

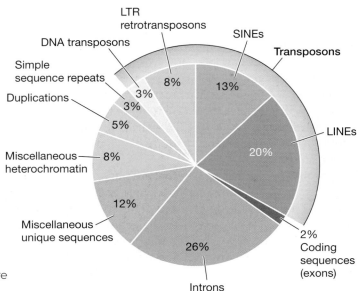

FIGURE 14.14 The ingredients that make up the human genome. Less than 2 percent is devoted to protein-coding sequences. Different kinds of transposable elements (SINEs, LINEs, LTR retrotransposons, and DNA transposons) together make up almost half the genome, while introns make up more than one-fourth. (After [23].)

you start reading.) **Overlapping genes** are found in viruses and bacteria, and much more rarely in eukaryotes, including humans. The selective advantage of overlapping genes is debated. One hypothesis is that the overlapping arrangement allows the genes to share their timing and level of expression. A second idea is that they are favored because they allow the genome to be streamlined.

Chromosome Evolution

Before the rise of molecular genetics in the late twentieth century, much of what we knew about evolutionary genetics came from studying chromosomes using the light microscope. Species differ in their number of chromosomes and in how the genes are arranged on them. The **karyotype** consists of the number and structure of the chromosomes. How do changes in the karyotype evolve? Although many questions are not yet answered, several patterns have emerged.

Fissions, fusions, and the evolution of chromosome number

The most common way for the number of chromosomes to change is when two chromosomes fuse, reducing the haploid chromosome number by one, or when they fission, increasing the number by one. Humans have 23 pairs of chromosomes, but all of the other great apes have 24 pairs. At some point since our lineage split from that of chimpanzees, two chromosomes fused. That mutant chromosome spread throughout our species, resulting in the second-largest chromosome in the human karyotype. We do not yet know what caused it to spread: it may have had a boost from positive selection, or it may have simply drifted to fixation.

We do, however, have a good idea about how fissions and fusions evolve in the house mouse (*Mus musculus*). Here karyotype evolution is on the fast track: changes in chromosome number are evolving at rates hundreds of times faster than in most other mammals [50]. Populations can have anywhere between 11 and 20 chromosome pairs. What accounts for this chromosomal chaos? The answer seems to be meiotic drive and selection that has favored selfish genes rather than the fitness of individuals [54]. During meiosis in females, one haploid set of chromosomes is transmitted to the egg while the other set enters the polar body, where it dies (**FIGURE 14.15**). When a female mouse is heterozygous for a fusion, the fused chromosome competes with the two unfused

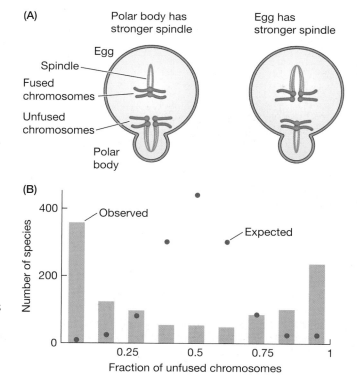

FIGURE 14.15 Female meiotic drive can favor chromosomes that are either fused or unfused. During meiosis, one haploid set of chromosomes enters the egg while the other enters a polar body, which dies. (A) It is hypothesized that in some populations and species, the polar body has a stronger meiotic spindle than the egg (left). In individuals that are heterozygous for a fused chromosome and its two unfused homologues, the unfused chromosomes are more likely to attach to a polar body spindle, giving an advantage to the fused chromosome. In other populations and species, the egg has a stronger spindle, which favors the unfused pair of chromosomes (right). (B) Consistent with the meiotic drive hypothesis, mammals tend to have either most of their chromosomes fused or most unfused. The bars show the frequencies of unfused chromosomes in the genomes of 1170 species of mammals. The dots show the distribution expected if pairs of chromosomes fused independently of each other, which is the null hypothesis in the absence of meiotic drive. (A after [66]; B after [54].)

chromosomes for transmission to the egg. Whether the fused or unfused chromosomes win this battle depends on which of them attaches most strongly to the meiotic spindle of the egg and which to the spindle of a polar body. In some mouse populations, unfused chromosomes attach more often to the egg's spindle, causing the mouse karyotype to evolve rapidly toward 20 pairs of unfused chromosomes. But in other populations, changes to the machinery of cell division turn the tables, and then meiosis favors 11 pairs of fused chromosomes. Remarkably, when this happens it doesn't matter which chromosomes fuse—almost every possible pair of unfused autosomes have been fused in different mouse populations [50]. It appears that the same shifts between karyotypes made up mainly of fused or of unfused chromosomes are playing out at a slower evolutionary tempo across all mammals. As a result, species tend to have either most of their autosomes fused or most unfused (see Figure 14.15).

While the details of how meiotic drive causes fusions and fissions to evolve are a bit complex, the bigger message of this story is simple. Some of the most basic features of the genome, including the number of chromosomes in the karyotype, are not refined adaptations that enhance survival and reproduction. Instead, they are the messy outcomes of competing evolutionary processes that act at different levels of selection.

Inversions and the evolution of chromosome structure

(A)

When geneticists began to study the chromosomes of fruit flies in the 1930s, they saw banding patterns that vary within and among species (**FIGURE 14.16**). Closer study revealed that many of these differences result from chromosome **inversions** (see Chapter 4). A chromosome with an inversion has the same genes as one without it, but they are in a different order.

If chromosomes with and without an inversion have the same genetic content, what could cause an inversion to spread in a species? Several mechanisms

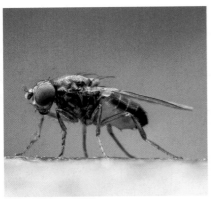

(B)

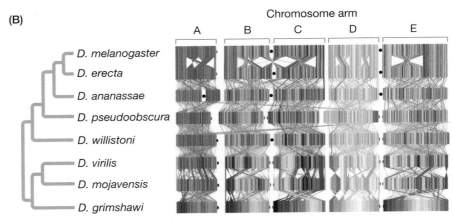

FIGURE 14.16 Chromosome rearrangements in fruit flies. (A) The third chromosome of *Drosophila pseudoobscura* as seen under a light microscope. The banding patterns, which result from staining the DNA, are altered by chromosome rearrangements. This species is famous for being highly polymorphic for chromosome inversions, which were studied intensively by the famous evolutionary geneticist Theodosius Dobzhansky during the mid-twentieth century. Inversion heterozygotes produce loops in their chromosomes, as seen in this picture. (B) Differences in the gene order among eight *Drosophila* species that have resulted from rearrangements. The phylogeny for the species is shown at left. For each species, individual genes are indicated by colored vertical lines, and chromosomal arms are shown in different colors. The lines connecting the genomes of adjacent species link homologous genes. Crossing lines show that two species differ in a chromosome inversion. A translocation can be seen in *D. pseudoobscura* that moved a segment of arm A to arm D. (A, chromosome from [15b]; B from [7].)

FIGURE 14.17 Chromosome inversions in the ruff (*Philomachus pugnax*) are responsible for dramatically different male morphs. (A) The three morphs, which differ in plumage, behavior, and body size, represent alternative mating strategies. "Independent" males are territorial, dominant, and display to attract females; "satellite" males are nonterritorial and submissive; and "faeder" males (which look much like the female shown here) obtain sneaky copulations by mimicking females to avoid aggression from independent males. (B) Satellite and faeder males are heterozygous for a 4.5-Mb chromosomal inversion that independent males do not carry. The inversion spans many loci (indicated by the triangles), some of which affect plumage and sex hormones. Recombination is suppressed between inverted and uninverted chromosomes, which binds together alleles that determine each of the three morph phenotypes. The inversion carried by satellite males has different alleles than the inversion carried by faeder males, which accounts for the differences between those two morphs. (After [39].)

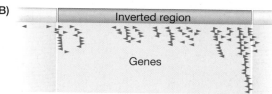

(A) Female   Independent   Satellite

(B)   Inverted region

Genes

are responsible [36]. First, an inversion is produced when a chromosome breaks at two places and the middle segment is then reinserted backward. The break points can disrupt a gene or alter its expression. Occasionally this generates a beneficial mutation that causes the inversion to spread by positive selection. Second, some inversions benefit from meiotic drive. When such an inversion is heterozygous (that is, paired with a chromosome that lacks the inversion), it is transmitted to the gamete more than 50 percent of the time. This favors the inversion to spread just as if it increased survival.

Inversions can also increase in frequency because of their effects on recombination [36]. When an inversion is heterozygous, recombination is blocked in the inverted region of the chromosome. This can bind together a favorable combination of alleles at several loci, causing the inversion to spread. The ruff (*Philomachus pugnax*) is a sandpiper with three male morphs that use different strategies to obtain mates (**FIGURE 14.17**). The morphs differ dramatically in plumage, behavior, and body size. These phenotypes are determined by a chromosome inversion that carries alleles that code for combinations of feather colors and reproductive hormones that determine each morph's mating strategy [39]. (One of the loci is *Mc1r*, which is involved in adaptive changes to coloration in many groups of animals—see Figure 6.29.) Inversions can also be established when they capture alleles that are beneficial in particular environments.

Last, an inversion can spread by random genetic drift. You saw in Chapter 5 that some inversions are underdominant: they decrease fitness when heterozygous. When an underdominant inversion is rare, as when it first appears, selection acts to eliminate it from the population. But if drift causes it to reach a frequency greater than 50 percent, then selection favors it to spread to fixation. Inversions established this way in one population will cause interbreeding with other populations to produce heterozygote offspring with low fitness. This can generate genetic isolation between the populations and contribute to speciation (see Chapter 9).

Evolution of Genome Size and Content

More than 98 percent of our genome does not code for a protein or other gene product (see Figure 14.14). In other eukaryotes, the fraction is even larger. Many scientists think that the noncoding part of a genome is largely "junk DNA" that has no function that is useful to the organism. Where does all this noncoding DNA come from? And is it really junk?

FIGURE 14.18 The relation across the tree of life between a genome's total DNA content and the amount of its sequence that codes for proteins. The three dashed lines indicate the percentage of the genome devoted to coding sequences. Viruses and prokaryotes have little or no noncoding DNA, so nearly 100 percent of their genome is coding sequence. In contrast, the genomes of most animals and plants are made up largely of noncoding DNA, which varies greatly in quantity. In some species, less than 1 percent of the genome is coding sequence. Unicellular eukaryotes show an intermediate pattern. (After [43].)

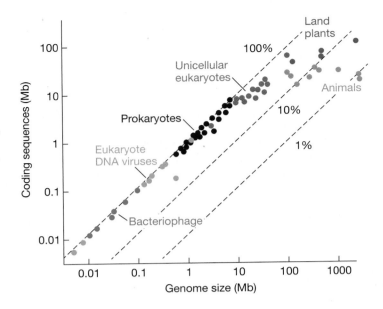

Genomes large and small

The staggering diversity seen across the tree of life extends all the way down to the level of the genome. The record for the smallest known genome is held by a DNA virus (porcine circovirus type 1), with only 1759 bp that code for just three proteins. The smallest genome for a bacterium is an insect symbiont called *Nasuia deltocephalinicola*, with just 137 coding genes and about 112,000 bp [6]. At the other end of the spectrum, the prize for the largest genome is held by the amoeba *Chaos chaos* that you met in this chapter's introduction—it has 400 times more DNA than we do [18]. There is substantial variation in a genome's DNA content even within some species: it differs by up to 30 percent in *Drosophila melanogaster* [9]. The largest number of genes in any organism so far known is bread wheat, with about 100,000 genes, which is five times more than in *Homo sapiens*. Our species, with 3.2 billion DNA bases and about 20,000 protein-coding genes, has neither an unusually large nor small genome for a multicellular organism.

These numbers bring into bright relief an apparent disparity between genome size and any subjective measure of organismal complexity (for example, number of types of tissues). Patterns emerge, however, if we divide organisms into major groups of life and then plot the amount of DNA devoted to coding for proteins against the genome's total size (**FIGURE 14.18**). Across viruses and prokaryotes, there is nearly a one-to-one relation. In animals and plants, however, the correlation is much weaker: species with similar amounts of coding sequence can differ dramatically in the sizes of their genomes.

These patterns begin to make sense with a single key insight. The genomes of most bacteria and viruses consist almost or entirely of coding sequences. In animals and plants, by contrast, most DNA does not code for any protein. The amount of noncoding DNA varies tremendously among species and explains much of the scatter in the genome size of metazoans. The different types of DNA in the human genome are seen in Figure 14.15. Less than 2 percent codes for proteins. About 25 times more—fully half of the genome—is taken up by the DNA parasites that we will now discuss.

Genetic parasites and transposable elements

Transposable elements (**TEs**), also called transposons, are short sequences of DNA that occur in many copies in the genome. The most numerous TE in the human

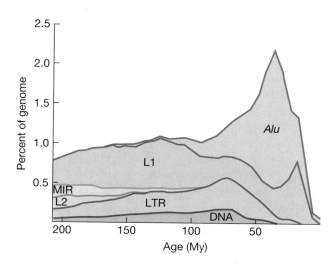

FIGURE 14.19 Age distributions of transposable elements in the human genome. The graph shows the fraction of the current genome that consists of TEs that inserted at a given time in the past. When a new copy of a TE inserts, its sequence begins to diverge from that of other copies. Copies that are more greatly diverged are therefore older, and their ages can be estimated using a molecular clock (see Chapter 7). Age distributions are shown for six different TEs: *Alu* elements, L1 and L2 long interspersed nuclear elements (LINEs), mammalian interspersed repeats (MIRs), long terminal repeats (LTRs), and DNA transposons (DNA). *Alu* first infected the remote ancestors of humans about 175 Mya. They had a burst of activity about 45 Mya, and about 2 percent of our genome is now made up of *Alu* elements that inserted at that time. Lower activity after then inserted additional copies. Taken together, the 1 million copies of *Alu* now make up about 10 percent of our genome. By contrast, L1 elements had a protracted phase of proliferation much earlier, with a second spike roughly 25 Mya. The total insertion rate of all TEs is indicated by the overall height of the plot: it reached a peak about 45 Mya and is now much lower. (After [13].)

genome, which is called *Alu*, is about 300 bp long (see Figure 12.12). Each of us has more than a million copies of *Alu* that together make up more than 10 percent of our genome [4]. *Alu* proliferates by making copies of itself: the DNA sequence is transcribed into RNA, which is then reverse-transcribed into DNA and inserted elsewhere in the genome.

Thus *Alu* and other TEs are parasites that work at the molecular level. They reproduce not to improve the fitness of their host, but simply because they can. By the nature of natural selection, any sequences that copy themselves more prolifically than others will come to make up more of the genome. For this reason, TEs are sometimes referred to as "selfish DNA" (see Chapter 12). The evolutionary origin of TEs is uncertain. One plausible hypothesis is that they are modified viruses that evolved the ability to reproduce without leaving the cell. There are two classes of TEs that differ in how they replicate (perhaps pointing to more than one evolutionary origin). The *Alu* element is an example of a retrotransposon. Its DNA is transcribed into RNA, but that molecule is then retrotranscribed into DNA and reintegrated into the host's genome. Like RNA viruses such as HIV, some retrotransposons code for the reverse transcriptase that they need to retrotranscribe the RNA intermediate into DNA. But TEs such as *Alu* take parasitism to the next level: they don't make reverse transcriptase themselves, but instead use the transcriptase made by other TEs and viruses. The second class of TE is DNA transposons, which do not use an RNA intermediate in their life cycle. They are much rarer in humans, in which they make up about 3 percent of the genome (see Figure 14.14).

TEs can spread through a genome in an evolutionary epidemic, multiplying to vast numbers over short periods of evolutionary time. Among species of *Drosophila*, the fraction of the genome composed of TEs ranges by almost a factor of 10, from 3 percent to 25 percent [16]. The origin of a TE can be dated in two ways. The first approach uses the divergence among different copies as a molecular clock (**FIGURE 14.19**). The second approach uses the age of the most recent common ancestor of the species in which the TE is found. This phylogenetic approach suggests that the *Alu* element infected primates about 65 Mya because it occurs only in primate species that last shared a common ancestor at that time [4]. The molecular clock, in contrast, suggests a much earlier origin (see Figure 14.19).

Like viruses, transposable elements are usually bad news for their host. The insertion of a TE into the host's genome causes a mutation that can disrupt a coding sequence or its proper expression. A second problem for the host is that two copies of a TE at different places in the genome can recombine, causing a chromosomal mutation. (TEs tend to accumulate in regions of the genome that have low recombination rates because there they are less likely to cause this problem and so be selected out of the population.) Host genomes have evolved several ways to fight against the spread of TEs. When they succeed, the host genome is left with the aftermath of the battle: dead TEs that can no longer spread but that still fill up much of the genome. That is the current state of our own genome. Our ancestors evolved ways to shut down most of the movement of TEs. Figure 14.19 shows that currently the activity of human transposons is much lower than in the past. But occasionally a human TE is able to reproduce. When it does, genetic diseases can result [5].

Although transposable elements are largely harmful to their hosts, they occasionally produce a beneficial mutation. Chapter 5 recounts the evolution of melanic coloration in the peppered moth (*Biston betularia*), which is perhaps the most famous example of evolution observed in action. Recent research has revealed that the mutation responsible for the melanism is a transposon whose insertion altered the expression of a gene called *cortex* [29]. On a much grander scale, TEs may be responsible for a fundamental feature of the eukaryotic genome. The molecular machinery used to splice introns out of messenger RNA is also used by TEs. In fact, some introns contribute to their own splicing. These facts suggest that TEs may have been responsible for the evolutionary origin of introns, which are now so essential to gene regulation [44]. If that hypothesis is correct, introns will be the most spectacular example imaginable of a "bug" that was turned into a feature.

Routes to the evolution of the smallest and largest genomes

This chapter opened by looking at the dramatic variation in the sizes of different genomes. While we understand some of the factors responsible, there is still much uncertainty (and debate) among evolutionary biologists about how genome size evolves.

The very smallest genomes are found in viruses (see Figure 14.18). These genomes have very little or no noncoding DNA. They have been highly streamlined by the deletion of genes that are essential to life in free-living organisms, and they reproduce by hijacking gene products produced by their hosts. Many viruses are in a race to replicate, which favors reducing the genome to a minimal size.

The bacterium *Buchnera aphidicola* has evolved a small genome for very different reasons. It has one of the smallest genomes of any bacterium, with only about 500 genes and 600 kb of DNA. *Buchnera* developed a symbiotic relation with aphids more than 160 Mya (see Figure 13.4). Since then, it has lived entirely within its host, and is transmitted vertically from mother aphids to their offspring. Like viruses, *Buchnera* uses metabolic products provided by its hosts, which allows its genome to do without many essential genes. Unlike viruses, replication of these symbionts is typically limited by that of their host, so there is little or no fitness benefit to speed the genome's replication. The population sizes of *Buchnera* are much smaller than those of most free-living prokaryotes, and they experience very strong drift. If deletion of a gene decreases fitness but is not lethal, it can become fixed by chance. Deletions are more frequent than duplications in bacteria, so genome reduction is more likely than expansion. This creates a ratchetlike process in which the genome's size evolves downward until it teeters on the brink of its own destruction: the minimal size necessary for survival [48].

Supercompact genomes have evolved in the world's most abundant organisms by yet another route [3]. Every milliliter of seawater near the surface of temperate and tropical oceans around the world has some 10^6 cells of a planktonic bacterium called *Pelagibacter ubique*. With only 1 Mb of DNA and fewer than 1500 genes, it has the smallest genome of any free-living bacterium known [21]. Nothing about it is superfluous: it has no introns, no transposons, no pseudogenes. *Pelagibacter*'s population size worldwide is some 10^{28} cells (more than 1 million times the number of stars in the universe!), so drift in this species is weak. In contrast to *Buchnera*, the streamlined genome of *Pelagibacter* appears to be the product of adaptation, likely to the very low nutrient levels in its environment. By paring down its genome, the bacterium has reduced its need for the nitrogen and phosphorus needed to replicate the DNA. The tiny genome also allows the cell size to be miniaturized, which may be a further advantage to *Pelagibacter*.

Animals and plants live at the other end of the genomic spectrum, with large numbers of genes and vast amounts of DNA. Much of the variation in gene number in eukaryotes results from whole genome duplication. You saw earlier in this chapter that bread wheat has nearly 100,000 genes, or about five times more than we do. The ancestor of wheat had some 25,000 genes. During domestication, its genome then doubled in size, not once but twice, by allopolyploidy [10]. If the wheat follows the evolutionary path taken by many other plants, we can anticipate that most of the duplicate genes will eventually be lost by deletion or become pseudogenes, and ultimately the wheat's gene number will decline to a level more typical of flowering plants. The total size of the genome can also shrink by deletion. This outcome has occurred many times independently in the crucifers (Brassicaceae) [46].

What explains the large scatter in the DNA content of eukaryotes? As in the prokaryotes, there is a trend for species with more genes to have larger genomes, but the correlation is much weaker (see Figure 14.18). Whole genome duplication is one factor at work. A second is the evolution of noncoding DNA driven by the proliferation of transposable elements. A third major player in the evolution of genome size may be random genetic drift. Animals and plants are much bigger than free-living microbes, and as a result have much smaller population sizes. Their reduced population size makes natural selection less efficient (see Chapter 7). If the deletion of a segment of noncoding DNA produces a fitness benefit but the effect is small relative to the strength of drift, selection will be largely powerless to slim down the genomes of animals and plants [44].

Does that mean the 98 percent of our own genome that is noncoding is actually junk? We are still far from having a clear answer to this fundamental question. Although much of our DNA originated as transposable elements, some of the resulting "junk" now plays key roles in regulating gene expression, and the cell's metabolism has coevolved with the total quantity of DNA in the nucleus. Those factors may diminish the selective benefit of deletions that eliminate noncoding DNA. Like an addict and his drug, eukaryotes may not be able to break their dependence on a bloated genome. But there is also good news in this story. When ancient eukaryotes acquired large amounts of noncoding DNA, it opened up new options for the evolution of gene regulation. That, in turn, may have enabled the origin of complex life forms, including ourselves.

Go to the
Evolution Companion Website
EVOLUTION4E.SINAUER.COM
for data analysis and simulation exercises, quizzes, and more.

SUMMARY

- New genes are an important source of evolutionary novelty and adaptation. The most common origin of new genes in eukaryotes is by gene duplication that happens when there is an error in DNA replication.

- Whole genome duplication is much rarer but has been a key event in the evolution of the genome in many groups of organisms, especially plants.

- New genes can be acquired from unrelated species by horizontal gene transfer. HGT is particularly common in prokaryotes, in which it has enabled the rapid evolution of traits, including antibiotic resistance.

- Most gene duplicates degenerate into nonfunctional pseudogenes. Some duplicates survive, however, and evolve to specialize in one of the functions of the original gene, or to take on a new function. The result is a gene family: a set of loci that originated by gene duplication and that typically have related biochemical roles.

- Chromosomal deletions can eliminate functioning genes. Natural selection can cause a deletion to increase in frequency, for example when the deleted gene codes for a protein that increases the risk of infection.

- Most mutations that change the amino acid of a protein (nonsynonymous mutations) are deleterious and are removed from the population by purifying selection. Mutations in coding sequences that do not change the protein's amino acid sequence (synonymous mutations) have only very weak effects on fitness and evolve largely by random drift. Nonsynonymous mutations that are beneficial are the rarest of all, but their fixation by positive selection is the basis of much adaptive evolution.

- The d_N/d_S ratio provides a rough measure of the relative contributions of drift and selection to the evolution of a gene. The ratio is given by the frequency of differences in the DNA sequence for the gene in two species that are nonsynonymous divided by the frequency of differences that are synonymous. Typically this ratio is less than 1, which is expected when most nonsynonymous changes are eliminated by purifying selection but some synonymous changes have become fixed by drift. Occasionally genes have a d_N/d_S ratio greater than 1, which strongly suggests that nonsynonymous changes have become fixed by positive selection.

- The fraction of protein differences among species caused by adaptive evolution versus genetic drift varies greatly among groups of organisms. About half the differences are adaptive in species with very large population sizes (e.g., *Drosophila* and free-living bacteria). In species with a small effective population size, including humans, drift is much stronger and so the fraction of adaptive differences is much smaller.

- Another important route to adaptation comes from changes in how genes are expressed. Recent research shows that changes in gene expression have been key to adaptation to new environments in many species, including humans.

- In eukaryotes, almost all genes have introns. These allow the mRNA to be spliced in different ways to make a variety of proteins. Chromosome mutations that bring together exons from different genes have yielded new genes with novel functions.

- Chromosome numbers change by fusion and fission. In some cases, fissions and fusions have become fixed not because they increase fitness but because they benefit from meiotic drive.

- Chromosome inversions can spread by several mechanisms. One is when they bind together beneficial combinations of alleles at two or more loci.

- Genome size varies dramatically among species. In viruses and prokaryotes, almost all of the genome is coding sequence. In animals and plants, most of the genome is noncoding, and the quantity of noncoding DNA differs greatly among species.

- Transposable elements, which are genetic parasites, are a major component of the noncoding DNA and account for much of the variation in genome size among species of eukaryotes.

- Whole genome duplication, which is particularly common in plants, is responsible for the large differences in gene number seen among some closely related groups of organisms.

TERMS AND CONCEPTS

alternative splicing
codon bias
copy number variation

de novo gene
d_N/d_S ratio
domain
exon shuffling

gene family
gene trafficking
horizontal gene transfer (HGT)

inversion
karyotype
neofunctionalization

neutral theory
   of molecular
   evolution
operon

overlapping gene
paralog
pseudogene

purifying selection
retrotransposition
subfunctionalization

transposable
   element (TE)
whole genome
   duplication

SUGGESTIONS FOR FURTHER READING

Dan Graur's *Molecular and Genome Evolution* (Sinauer, Sunderland, MA, 2016) gives an authoritative perspective on the topics covered by this chapter (and much more). Michael Lynch's *The Origins of Genome Architecture* (Sinauer, Sunderland, MA, 2007) is another valuable reference. Lynch argues that several

key aspects of the genome are consequences of random genetic drift.

T. Ryan Gregory's website (http://www.gregorylab.org/research/) and his Animal Genome Size Database (http://www.genomesize.com) are fonts of information about genome size.

PROBLEMS AND DISCUSSION TOPICS

1. The origin of genes that have new functions often involves the divergence of gene duplicates. Duplicates can arise via several mechanisms.

   a. One mechanism of gene duplication is retrotransposition, the insertion into the genome of DNA produced by reverse transcription of a messenger RNA. These gene duplicates are often dead on arrival: they are pseudogenes as soon as they are formed. Why are such duplicates so often dead on arrival?

   b. A second mechanism of gene duplication occurs via unequal crossing over during meiosis. Gene duplicates formed this way are functional more often than when they arise by reverse transcription. Why is that?

   c. If a gene duplicate is initially functional, what are its possible ultimate fates? Which is most likely, and why?

2. The ratio of nonsynonymous differences per nonsynonymous site, d_N, to synonymous differences per synonymous site, d_S, can be used to test for positive selection (see Chapter 7). Imagine that in a duplicate pair of loci, one paralog is evolving neutrally while the other is evolving under strong positive selection. What specific data are needed to detect that situation using the d_N/d_S ratio, and what pattern do you expect to see?

3. The human genome contains more than a million copies of the *Alu* transposable element. Comparative genomics reveals that the *Alu* element is found only in the clade of mammals that includes primates, tree shrews, rodents, and rabbits.

   a. What does the observation that the *Alu* transposon is limited to this clade reveal about its origin and method of spread among species?

   b. At many sites in the genome, an *Alu* element is present in humans but absent in chimpanzees, while at many other sites an *Alu* element is present in chimpanzees but absent in humans. What are two hypotheses that could explain this situation? For any particular site, how could the hypotheses be distinguished?

4. Gene trafficking is the movement of loci from one region of the genome to another. You saw that gene trafficking in *Drosophila* has caused many loci that were formerly on the X chromosome to move to the autosomes. Suggest a hypothesis for that observation.

5. Imagine that you have a device that can accurately measure the DNA content of individual cells of living yeast. This device enables you to do artificial selection on the genome size of the yeast. Propose an experiment using artificial selection that could test the hypothesis that the yeast genome size is optimal for yeast fitness versus the alternative hypothesis that transposons have caused the genome to evolve to a size that is larger than what maximizes yeast fitness.

15

Evolution and Development

"In my possession are two little embryos in spirit [alcohol], whose names I have omitted to attach, and at present I am quite unable to say to what class they belong. They may be lizards or small birds, or very young mammalia, so complete is the similarity in the mode of formation... in these animals."

In this passage in *On the Origin of Species*, Darwin quoted the world's leading authority on the embryonic development of animals. Karl Ernst von Baer had found that the early embryonic stages of the most diverse vertebrates had features in common that they would lose later in development, and that the special characteristics that distinguish mammals from reptiles, or primates from rodents, developed only at later stages. Early in development, a human is virtually indistinguishable from an alligator, much less any other mammal (**FIGURE 15.1**). This astonishing fact is one of countless examples of features displayed early in development that added to Darwin's mountain of evidence for the evolution of diverse species from common ancestors. The embryos of birds, anteaters, and baleen whales develop incipient teeth that are resorbed and are lacking in the adults. Human embryos have a tail that usually stops growing and persists only as three fused vertebrae at the end of our spine; but infants are occasionally born with an anatomically complete tail. As in other primates, human fetuses develop a dense coat of hair, although most humans shed it about a month before birth.

Clearly, the developmental processes by which a fertilized egg becomes a differentiated organism, with features such as toes and a tail, are shared among species; and these processes result in yet other puzzling similarities. Some plants

The flowers of most angiosperms bear distinct sepals, petals, stamens, and carpels that differentiate because of the action of combinations of specific regulatory genes. Water lilies, such as *Nymphaea lotus*, are one of several phylogenetically basal angiosperm lineages in which sepals and petals are not well differentiated, and in which the stamens can have a petaloid form. These plants suggest that the gene regulatory networks that determine the identity of organs have evolved gradually.

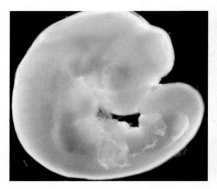
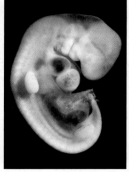
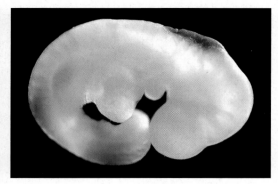

FIGURE 15.1 Many species are very similar as embryos, with the distinctive features of their clade developing only later. Here are embryos of human, alligator, and mouse. Can you tell which is which?[1]

have separate female and male flowers, but in many such species, a flower has the sexual parts of the other sex, in rudimentary form. We see much the same in humans: men have nipples. The bones in a mammal's hindlimb have exactly matching counterparts in the forelimb (femur/humerus, tarsals/carpals), whether in humans or bats. Evidently a developmental process is shared between the sexes, or between limbs. But just as clearly, the developmental processes ultimately diverge—between sexes, between limbs, and among species.

How do we account for the similarities and differences? At one level, we know the answer: homologous similarities are often based on shared genes, and differences on differences in the genes. But that does not tell us how a change in DNA sequence becomes realized as a change in an organism's form and function. Someone trained in architecture, shown a blueprint of St. Paul's Cathedral or the Empire State Building, might well visualize what the completed building looks like, but in itself, the blueprint does not specify that the building must be built from the foundation up, or provide any other information about how to construct it.

In biology, some transitions from genes to phenotypes are simple and clearly understood, such as transcription from DNA to RNA and translation to a phenotype, the protein. The far more complex transition from genes to physical structures such as cells, tissues, and organs is the province of developmental biology. Understanding how these transitions evolve is the task of **evolutionary developmental biology** (often shortened to EDB or "evo-devo"). Especially since the 1980s, when knowledge of molecular genetic mechanisms of development made major advances, EDB has been an active research area that has been filling major gaps in biologists' understanding of organismal diversity and evolution. Research in EDB is concerned with several large questions [85]. First, concerning the evolution of development, what have been the changes in developmental mechanisms that give rise to different phenotypes? A second question, closely related to the first, is how do genetic differences among species map onto phenotypic differences? Third, what is the role of development in either constraining or enhancing evolutionary change in characters? That is, how does development affect "evolvability"? Fourth, how does developmental information help us identify homologous characters, or even define homology? Finally, can understanding development help us understand the origin of novel characteristics? Much of this chapter bears on the first three questions; we will save homology and evolutionary novelty for Chapter 20, on macroevolution.

---

[1] From left to right, they are alligator, human, and mouse.

Comparative Development and Evolution

In the nineteenth and early twentieth centuries, biologists described and compared the embryonic development of diverse animals (and plants, to a much lesser extent) and described ways in which morphogenetic processes in different organisms result in different adult forms. One of the tasks of EDB is to understand these processes, such as growth rates and differentiation of body parts, in modern terms of genetic and molecular processes. As we will see, one of the major explanations for these evolutionary changes is alteration of the time, place, and level of expression (especially transcription) of particular genes or sets of genes. Similar processes enable a single genome to produce different morphologies, depending on environmental signals such as day length, or genetic signals such as sex-determining genes. Developmental causes of phenotypes are *proximate causes*, mechanisms that operate within an individual organism. These causes complement the processes that caused these phenotypes, and these mechanisms, to evolve and to differ among species. These *ultimate causes*, such as natural selection, act at the level of populations across generations; they do not conflict with the mechanistic genetic and developmental processes. For example, embryonic mammals and birds have webbing between the developing digits. This remains in the wings of adult bats and the feet of ducks, but humans and chickens have separated digits because the webbing cells are eliminated by programmed cell death. This is a simple example of how developmental biology helps us understand evolved differences between species. It does not answer why ducks evolved webbed feet and the suppression of the cell death that occurs in most other birds. A likely answer is that effective swimming enhanced fitness in the ancestors of ducks, so that selection favored mutations in the genes that determine the process of cell death and resulted in variant birds that retained some webbing.

Among the first things that scientists learned about development is that species are often more similar as embryos than as adults. Karl Ernst von Baer noted in 1828 that the features common to a higher taxon (such as the Vertebrata) often appear earlier in development than the specific characters of lower-level taxa (such as orders or families) [80]. This generalization is now known as von Baer's law. For example, all tetrapod vertebrate embryos display pharyngeal clefts (gill slits), a notochord, segmentation, and paddlelike limb buds before the features typical of their class or order become apparent (see Figure 15.1). One of Darwin's most enthusiastic supporters, the German biologist Ernst Haeckel, reinterpreted such patterns to mean that "ontogeny recapitulates phylogeny"—that is, that the development of the individual organism (**ontogeny**) repeats the evolutionary history of the adult forms of its ancestors, and could indicate its phylogenetic relationships. By the end of the nineteenth century, however, it was already clear that Haeckel's dictum seldom holds [22]. For example, the pharyngeal clefts and associated branchial arches of embryonic mammals and reptiles never acquire the form typical of adult fishes. Moreover, various features develop at different rates, relative to one another, in descendants than in their ancestors, and embryos and juvenile stages have stage-specific adaptations of their own. Thus ontogeny is not a very useful guide to phylogenetic history.

By the early twentieth century, biologists had identified several common patterns of developmental differences among species—patterns that are now part of the language of evolutionary developmental biology.

Allometric growth, or **allometry**, refers to the *differential rate of growth of different parts or dimensions of an organism* during its ontogeny. For example, during human postnatal growth, the head grows at a slower rate than the body as a whole, and the legs grow at a faster rate. Allometry thus refers to changes in the *shape* of the organism or of certain of its parts, such as the dimensions of a skull or a leaf.

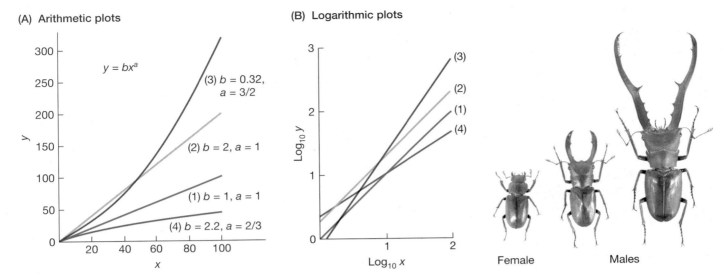

(A) Arithmetic plots

$y = bx^a$

(3) $b = 0.32$, $a = 3/2$

(2) $b = 2$, $a = 1$

(1) $b = 1$, $a = 1$

(4) $b = 2.2$, $a = 2/3$

(B) Logarithmic plots

(3)
(2)
(1)
(4)

Female   Males

FIGURE 15.2 Hypothetical curves showing various allometric growth relationships between two body measurements, y and x, according to the equation $y = bx^a$. (A) Arithmetic plots. The curves 1 and 2 show isometric growth ($a = 1$), in which y is a constant multiple (b) of x. Curves 3 and 4 show positive ($a > 1$) and negative ($a < 1$) allometry, respectively. (B) Logarithmic plots of the same curves have a linear form. The slope differences depend on a. Curves 1 and 2 have slopes equal to 1. When $a > 1$, y increases faster than x. The male stag beetles (*Cyclommatus metallifer*) at right show positive allometry of mandibles, relative to body length. (Photos from [21].)

Such changes account for a great deal of morphological evolution. For example, an increased rate of elongation of the digits accounts for the shape of a bat's wing compared with the forelimbs of other mammals (see Figure 2.9); an elephant's tusks are incisor teeth that have grown faster and for a longer time than the other teeth.

Allometric growth is often described by the equation $y = bx^a$, where y and x are two measurements, such as the height and width of a tooth or the size of the head and the body. In many studies, x is a measure of body size, such as weight, because many structures change disproportionately with overall size. An allometric relationship is often represented logarithmically as $\log y = \log b + a \log x$. The coefficient a describes the relative growth rates of features y and x (**FIGURE 15.2**). If y increases faster than x, as for human leg length relative to body size, $a > 1$ (positive allometry); if y increases more slowly than x, as for human head size, $a < 1$ (negative allometry). Allometric variation is seen both within species (as the beetles in Figure 15.2 illustrate) and among species. For example, species of deer show positive allometry between antler size and body mass; the largest deer, the extinct Irish elk (*Megaloceros giganteus*), had spectacular antlers (see Figure 1.9).

Heterochrony [22, 45] is broadly defined as *an evolutionary change in the timing or rate of developmental events*. One of the best known instances is an evolutionary change called **paedomorphosis**, in which some characteristics of the adult of a species may have a more juvenile form than in the species' ancestor. One way paedomorphosis can happen, called **neoteny**, is seen in the axolotl, a salamander that grows to full size but does not undergo metamorphosis, as most salamanders do. Instead, it reproduces while retaining most of its larval (juvenile) characteristics (**FIGURE 15.3**).

Heterotopy is an evolutionary change in the spatial position of a feature within an organism. Often, it is expressed at an additional, novel position. As we will see, this can result from the expression of specific genes in novel parts of the developing

(A)

(B)

body. Heterotopic differences among species are very common in plants (**FIGURE 15.4A,B**). In vertebrates, many phylogenetically new bones have arisen in tendons or other connective tissues subject to stress. Many dinosaurs had bony tendons in the tail, and the giant panda (*Ailuropoda melanoleuca*) has a novel "thumb" that is not a true jointed digit (see Figure 20.9). Bony elements are conspicuous in the skin of armadillos and crocodiles (**FIGURE 15.4C,D**).

The bodies of many organisms consist of **modules**—distinct units that have distinct genetic specifications, developmental patterns, locations, and interactions with other modules [57]. Some such modules are repeated at various sites on the body and are termed **serially homologous**. In some cases, serially homologous features lack distinct individual identities (e.g., leaves of many plants, teeth of most amphibians and reptiles) and may be considered representatives of a single character. An important evolutionary phenomenon is the acquisition of distinct identities

FIGURE 15.3 Neoteny in salamanders. (A) The tiger salamander (*Ambystoma tigrinum*), like most salamanders, undergoes metamorphosis from an aquatic larva (left; note the gills) to a terrestrial adult (right). (B) The adult axolotl (*Ambystoma mexicanum*), with gills and tail fin, resembles the larva of its terrestrial relative. The axolotl remains aquatic throughout its life.

(A)

Stem

Roots

(B)

(D)

(C)

FIGURE 15.4 Examples of heterotopy. (A) The vine *Monstera deliciosa* has evolved exposed roots that grow from an aerial stem. (B) Entire plants develop on the leaf margins of *Bryophyllum*; they eventually drop off and take root. (C, D) Platelets of bone (osteoderms) develop in the skin of diverse vertebrates, such as armadillos (C) and crocodiles (D).

(A) *Haptodus*

(B) Elephant shrew

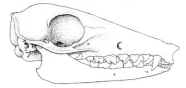

(C) *Prozeuglodon*

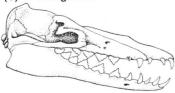

(D) Dolphin

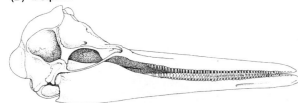

FIGURE 15.5 The teeth of mammals provide an example of the acquisition and loss of individualization. (A) The teeth of mammal ancestors (synapsids), such as the Permian *Haptodus*, are uniform. (B) Teeth became individualized during the evolution of mammals, as illustrated by an elephant shrew. (C, D) Distinct tooth identity was reduced in an Eocene whale (*Prozeuglodon*) and lost altogether in modern toothed whales, such as dolphins. (A after [62]; B–D after [79].)

by such modules, called **individualization** [50, 82]. For instance, during the evolution of mammals, teeth became differentiated into incisors, canines, premolars, and molars, with different functions—suggesting that some different genes are active in the developing primordia of different teeth. Distinct tooth identity was later lost during the evolution of the toothed whales (**FIGURE 15.5**).

Before modern molecular biology, biologists made discoveries about development that remain important today. First, all cells in an organism have the same set of genes, based on replication during mitosis. Second, the differences among cells, tissues, and organs must result from differences in the activity of certain genes. Third, different cells have properties that affect morphogenesis (the development of form). These include growth of individual cells, change in cell shape, adhesion to certain other cells, mitosis in certain dimensions (e.g., forming sheets or masses), cell movement (in animals but not plants), and programmed cell death (apoptosis). Fourth, many aspects of growth and differentiation are affected by chemical signals, especially hormones. For example, metamorphosis in amphibians is triggered by thyroxin, produced in the thyroid gland. The axolotl (see Figure 15.3) does not synthesize thyroxin, and it differentiates into a typical adult form if it is injected with that hormone. (Some other salamanders are irreversibly neotenic and do not respond to thyroxin.) Likewise, cell division and differentiation in plants are controlled by auxins and other hormones.

By performing experiments on embryos, biologists learned, moreover, that certain events in an animal's development depend on preceding events, and that the differentiation of one tissue or organ is often influenced by others. For example, cells in the posterior region of a vertebrate's limb bud induce the formation of limb structures such as digits and muscles by producing signaling molecules (formerly called morphogens). A few scientists developed mathematical models to describe how development could emerge from such chemical interactions. One of the most important models was developed by the mathematician Alan Turing, who invented the prototype of modern computers and famously helped the Allies defeat the Nazis in World War II by breaking the German code. In Turing's models, two chemicals diffuse and interact to produce a morphogen in a spatial pattern that induces a repeated feature such as a structure or pigment. Changing parameters such as diffusion rates produces a variety of patterns that match those seen in real organisms (**FIGURE 15.6**) [32].

A key advance toward modern developmental genetics was François Jacob and Jacques Monod's discovery in 1960 of the bacterial operon, a combination of a regulatory sequence and co-regulated protein-coding genes [29]. Based on this concept, the developmental biologists Roy Britten and Eric Davidson laid the foundations of the modern view, in which eukaryotes' genes have gene regulatory elements, or binding sites, to which proteins bind that initiate or stop transcription [8].[1] Changes in these interactions could result in evolution of altered

---

[1] This chapter is concerned only with eukaryotes, but the evolution of operons in prokaryotes is also a subject of current research.

(A)

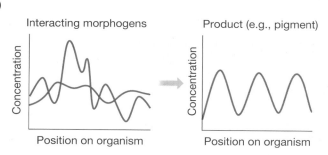

Interacting morphogens — Concentration / Position on organism

Product (e.g., pigment) — Concentration / Position on organism

(B)

Cone snail

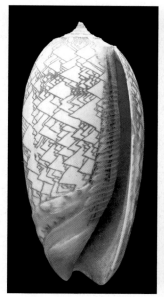
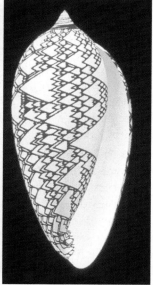

Puffer fish

FIGURE 15.6 (A) Alan Turing's model of diffusion of interacting chemical morphogens (left graph) shows that regular patterns of a product may be produced across a part of a developing embryo (right graph), which may induce the development of regular patterns in the distribution of pigments, hairs, or other features. (B) Patterns of coloration that may result from the reaction-diffusion process described in (A). At left is an olive shell (*Oliva porphyria*) and at right a puffer fish (*Arothron mappa*). A photo of each organism is shown to the left of patterns produced by computer simulations of a reaction-diffusion model. (A after [32]; B shell simulation from [47], fish eye simulation from [65].)

phenotypes. These ideas were very similar to what was later learned about the developmental transition from genes to phenotypes.

Gene Regulation

In eukaryotes, transcription of a protein-coding gene is initiated when RNA polymerase II binds to an upstream region, the promoter. This occurs when certain regulatory proteins—**transcription factors** (**TFs**)—that determine transcription of a gene bind to a short upstream region called an **enhancer** (**FIGURE 15.7**). Several transcription factors form a complex, so different parts (domains) of a TF interact with the enhancer and with other TFs (or other proteins). A single gene often has multiple enhancers, each with a different sequence that binds different transcription factors. Consequently, the gene may be transcribed in different cell types, at certain times, if the cells include different transcription factors that are specific to one enhancer or another. Enhancers are often called **cis-regulatory elements**. *Cis* means that the element regulates a gene on the same stretch of DNA. Transcription factors are **trans-regulatory elements**, meaning that they are encoded by DNA distant from the genes that they regulate.

Much of the research on the evolution of development concerns enhancers and transcription factors, but several other

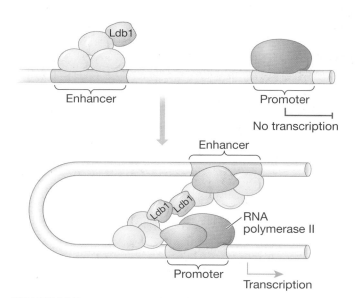

FIGURE 15.7 RNA polymerase II binds to a gene's promoter and initiates transcription of the gene into mRNA, but only after transcription factor proteins bind to both the promoter and an enhancer sequence. The promoter and enhancer may be linked to each other by other transcription factors (labeled Ldb1 here). (From [19], after [13].)

BOX 15A

Some Methods in Developmental Genetics

We briefly describe here some methods commonly used to study genes that underlie development, and their expression as mRNA transcripts. More detailed descriptions can be found at devbio.com; search for topic numbers 2.3 and 2.4. This is the website of a leading textbook, *Developmental Biology* by Scott F. Gilbert and Michael J. F. Barresi [19].

A gene from one organism can be inserted into the genome of another (often a different species), creating a **transgenic** organism. This can be accomplished by micro-injection, usually into a fertilized egg; by electroporation (using a high voltage pulse to "push" the DNA into the egg); or by attaching the gene to a transposable element or to a retrovirus vector, which inserts the gene into the host's genome. Gene function is also studied by gene knockout (or knockdown), which uses RNA interference (RNAi) to prevent transcription. It is also possible to replace the normal gene with a nonfunctional mutated sequence, using CRISPR-Cas targeted mutagenesis (**FIGURE 15.A1A**).

Several methods are used to visualize and measure transcription of specific genes in various cell populations at different times in development. In *in situ hybridization*, a chemical process stabilizes mRNA molecules in the cells in which they are produced. Then a species-specific, single-stranded RNA or DNA probe corresponding to the gene of interest is applied to the specimen, where the probe hybridizes by base pairing with the mRNA of interest. The probe is chemically modified so that it can be detected by a staining procedure or is labeled with a radioisotope so that it can be detected by autoradiography.

Another approach is to use RNA-seq and related methods to measure the entire transcriptome—the expression of all the genes at once. Levels of protein expression can be analyzed using mass spectrometry or antibodies. (The mRNA and protein expression patterns of a given gene may not be identical due to translational regulation.) Antibodies are produced by injecting a mammal (e.g., a rat) with the protein of interest (the antigen). The animal produces antibodies (immunoglobulin molecules) that bind specifically to that protein. Tissue specimens are prepared in a similar way as for in situ hybridization and are incubated with the primary antibody. A secondary antibody, an immunoglobulin that specifically binds to the primary antibody, is then applied to the specimen. The secondary antibody is modified so that it can be detected either by an enzymatic reaction that produces a colored product or by fluorescence (**FIGURE 15.A1B**).

Finally, the transcription patterns of *cis*-regulated genes can be studied using reporter constructs inserted into cultured cells or transgenic individuals. Reporter constructs consist of the regulatory DNA of interest, spliced upstream of a reporter gene that encodes a protein whose expression can be easily visualized under the microscope. One such protein is β-galactosidase, a bacterial enzyme that processes a particular sugar into a blue product. Another is a protein from jellyfishes (green fluorescent protein, GFP) that fluoresces bright green. Because reporter construct analysis requires the use of gene transfer technology, it can be undertaken only in certain well-studied model species, such as *Drosophila*, *Caenorhabditis elegans*, *Arabidopsis*, and mice. **FIGURE 15.A1C** shows the nematode *C. briggsae* expressing a GFP reporter construct containing *cis*-regulatory DNA from the *myo-2* gene, which directs the reporter gene's expression in the pharynx.

In genetic model species, the integration of genomic with genetic, developmental, and functional data provides a

mechanisms also affect the expression of genes and their products. One is **DNA methylation**. In some organisms, such as vertebrates and plants, transcription of a gene is repressed by methylation of certain cytosines that are followed by guanine residues (CpG dinucleotides). The methylated state may be maintained in newly synthesized DNA during cell division by a specific enzyme. Gene expression can also be affected at the level of translation of mRNA to protein. For example, microRNAs, together with proteins, bind to the 3' untranslated region (UTR) of RNA messages and prevent translation. And various posttranslational processes can affect the activity of proteins, such as binding a signaling molecule. Recall also that the various exons of a eukaryotic gene are often differentially spliced into many isoforms that have different functions and may be expressed in different cells

BOX 15A

Some Methods in Developmental Genetics (*continued*)

(A)

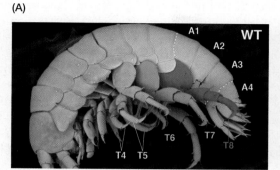

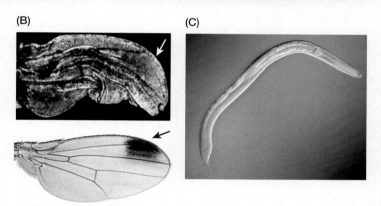

powerful base of knowledge for studies in evolutionary genetics and evolutionary developmental biology. These data are available in public online databases such as FlyBase (http://flybase.org) for *Drosophila melanogaster* and several other databases, such as ENCODE (www.encodeproject.org), for humans. As genomic and transcriptomic data from more species are accumulated and compared with the comprehensive information on model species, more and more investigations of the evolution of species differences become possible.

FIGURE 15.A1 Some methods of studying gene expression in developing animals. (A) CRISPR-Cas has been used to knock out function of various Hox genes in the crustacean *Parhyale*, resulting in homeotic transformation of certain appendages into other appendages. Normal and transformed appendages are shown in the scanning electron micrographs. Thoracic and abdominal segments are labeled T and A, respectively. Appendages on segments T6, T7, and T8 are colored in a wild-type (WT) larva (left photo), and the transformation of these to a T4 or T5 morphology (colored as in the wild type) is shown in a larva in which the *abdA* Hox gene was knocked out (right photo). (B) The top photo shows fluorescent antibody staining of the proteins Yellow (in green) and Ebony (in purple) in the pupal wing of a male *Drosophila biarmipes*. The bottom photo shows the *ebony* gene expressed (as protein) where the pigmented spot is located in the fully developed wing. (C) Green fluorescent protein (GFP) expression (bright green) by a transgenic reporter construct containing *cis*-regulatory DNA from a gene that is expressed in the pharynx of the nematode *Caenorhabditis briggsae*. (A from [42]; B, photos by John True; C, photo by Eric Haag, courtesy of Takao Inoue and Eric Haag.)

(**alternative splicing**). As we discussed in Chapter 14, a large fraction of genes may have alternatively spliced transcripts (isoforms), and the number of isoforms can be very high. These processes are becoming subjects of active evolutionary research.

In interactions among cells, such as those that underlie tissue induction, molecules released by one cell type diffuse and are bound by the extracellular domain of receptor proteins that span the membrane of other cells. This changes the configuration of the protein's intracellular domain, which sets off a signal transduction cascade of protein interactions in the receiving cell, and that may end by activating dormant transcription factors and changing which genes are transcribed in the cell. **BOX 15A** describes some of the methods used to study gene expression during development.

(A)

Haltere

(B)

FIGURE 15.8 A homeotic mutation. (A) A wild-type *Drosophila melanogaster* has a single pair of wings, borne on the second thoracic segment, and a pair of small winglike structures called halteres, borne on the third thoracic segment. (B) In a fly carrying mutations in the *Ultrabithorax* (*Ubx*) gene, the third thoracic segment has been transformed into another second thoracic segment, bearing wings instead of halteres. (Photos courtesy of E. B. Lewis.)

Hox genes and the genetic toolkit

Long before DNA was identified as the basis of heredity, geneticists had described what they called homeotic mutations in *Drosophila* and other species. These are mutations that transform a structure into a different structure. For example, the *Antennapedia* (*Antp*) mutation in *Drosophila* transforms antennae into legs (see Figure 4.18), and the *Ultrabithorax* (*Ubx*) mutation turns halteres (balancers) into wings (**FIGURE 15.8**).

To appreciate the significance of these mutations, we need to be acquainted with a few details about arthropod segments. In arthropods such as crustaceans, the body consists of multiple segments, almost all of which bear a pair of serially homologous appendages with similar basic structure. Appendages on the most anterior segments are modified as mouthparts, and those on the trunk serve for locomotion. In insects, which are descended from crustaceans, three distinct segments make up the thorax, each with a pair of legs, and the posterior segments are legless and compose the abdomen. In most insects, the second and third thoracic segments (T2 and T3) each have a pair of wings. In the true flies (order Diptera), such as *Drosophila*, the T3 wings are modified into small structures, the halteres, that are used for balance rather than flight. The Diptera are one of the insect orders in which the juvenile stage, the larva, has a radically different form than the adult. The adult structures, such as legs, wings, and halteres, develop from special masses of cells, the imaginal discs, within the body of the larva.

Starting in the 1970s, geneticists realized that some of the homeotic mutations in *Drosophila* change the identity of a segment. For example, mutations in the *Ubx* locus change the T3 segment into a second T2 segment, and therefore the halteres into wings (see Figure 15.8). Deleting the gene has the same effect. Mutations of other genes that affect characteristics of the wing affect the duplicated wing just as they affect the normal wing on T2. Thus, it was suggested that the normal *Ubx* gene regulates the transcription of the diverse genes that together produce a T3 segment; if mutation or deletion of the *Ubx* gene causes failure of regulation, the segment develops into a "default" state, T2. *Antp* and *Ubx* are two of eight genes, in two clusters, that control the anterior-posterior identity of body segments, in the same order as the genes' positions along the chromosome (**FIGURE 15.9**). The expression of each of these genes along the anterior-posterior axis of the developing fly corresponds to the segments whose identity the gene affects. These genes encode transcription factors. The part of the genes' sequence that encodes the DNA-binding domain of the protein is now called the homeobox, and the genes are called **homeotic selector genes**, or **Hox genes**.

The Hox genes are part of a **gene regulatory network**—a set of interacting regulatory genes and the genes they regulate—that controls the developmental pathway that specifies the anterior-posterior body pattern (**FIGURE 15.10**). To greatly oversimplify, mRNAs of two of the mother's genes (*bicoid, nanos*) are deposited in the egg. The mRNAs and the proteins they encode form in anterior-posterior concentration gradients. Depending on their concentration, several *gap genes* are transcribed in different broad domains. Gap gene proteins activate *pair-rule genes*, such as *fushi tarazu* (*ftz*), in seven transverse bands. Their protein products bind to and activate a group of *segment polarity genes* that determine the boundaries of 14 segments in the developing larva. The transcription factors produced by gap, pair-rule, and segment polarity genes all act to initiate or repress transcription of the Hox genes (which also repress each other). And as we have seen, the Hox genes that are expressed in each segment directly or indirectly regulate transcription of many genes that determine the segment's form and features. The end points of the pathway are the synthesis of the proteins that define the features of each cell in a particular tissue in a particular part of a segment. Complex regulatory pathways of this kind are characteristic of development of most of the morphological features of multicellular organisms.

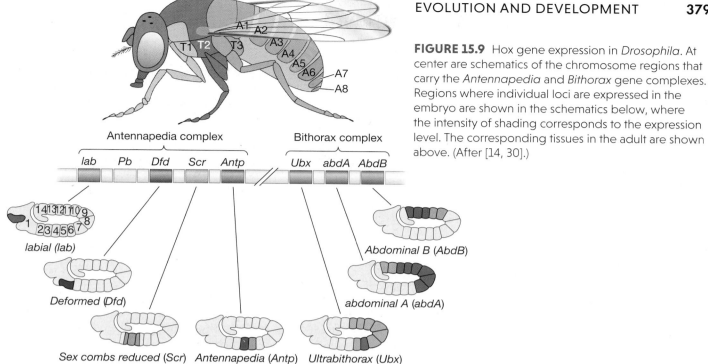

FIGURE 15.9 Hox gene expression in *Drosophila*. At center are schematics of the chromosome regions that carry the *Antennapedia* and *Bithorax* gene complexes. Regions where individual loci are expressed in the embryo are shown in the schematics below, where the intensity of shading corresponds to the expression level. The corresponding tissues in the adult are shown above. (After [14, 30].)

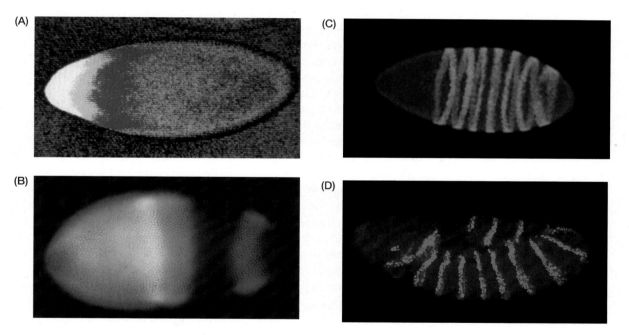

FIGURE 15.10 Example of a developmental network: a simplified model of anterior-posterior pattern formation in the *Drosophila* embryo. Anterior is to the left, and we view the left side of the embryo. Colors show concentrations of protein products. (A) Polarity is first established by maternal effect genes (e.g., *bicoid*, yellow to red) that leave protein concentration gradients in the egg. These proteins are transcription factors that activate expression of (B) gap genes such as *hunchback* (orange) and *Krüppel* (green), which define broad territories in the embryo. The gap gene transcription proteins determine the expression of (C) pair-rule genes such as *ftz*, each of which specifies a region about two segments long. Later in development, the pair-rule genes activate (D) segment polarity genes such as *engrailed* that divide each of these regions into two segment-sized units, each with anterior-posterior polarity. These several groups of genes determine the region where each homeotic selector (Hox) gene is transcribed. These define the identity of each segment, and initiate developmental pathways that result in the segment's various features. (A courtesy of C. Nüsslein-Volhard; B courtesy of C. Rushlow and M. Levine; C courtesy of D. W. Knowles and the Berkeley Drosophila Transcription Network Project, http://bdtnp.lbl.gov/Fly-Net; D courtesy of S. Carroll and S. Paddock.)

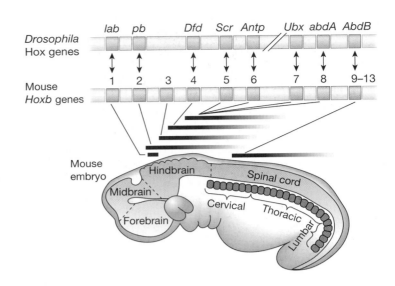

FIGURE 15.11 Segment-specific patterning functions of Hox genes in the vertebrate hindbrain and spinal cord. In this schematic diagram of a mouse embryo, the black horizontal bars indicate segmental patterns of *Hoxb* gene expression, with darker color corresponding to areas of relatively high gene expression. The double-headed arrows connect the genes in the *Hoxb* cluster to the homologous Hox genes in *Drosophila*. (After [44].)

A profoundly important and most surprising discovery was that the Hox genes described in *Drosophila* are also present in mammals, and some of them occur in all animal phyla except sponges. This family of genes evolved in the ancestor of almost all animals, more (probably much more) than 540 million years ago. In some groups, the entire array of Hox genes has been duplicated: mammals and other tetrapods, for example, have four sets, although each set lacks one or more of the genes (**FIGURE 15.11**). To a large degree, these genes play similar developmental roles in the various phyla. For example, they specify features along the anterior-posterior axis of a mammal embryo just as they do in insects. This functional role of the Hox genes has been phylogenetically conserved.

However, it was soon discovered that in addition to their highly conserved function, Hox genes and other genes that encode transcription factors can play many other roles. For example, *Ubx* is expressed in the developing hind leg of some species of *Drosophila* but not others, and is associated with the presence or absence of unicellular hairs on part of the leg. High expression suppresses the development of hairs—a radically different effect of a gene that affects the form of entire body segments! In water striders, insects that skate on the surface of water, *Ubx* controls the length of the middle legs, which are used as propelling oars [58]. The *Ubx* transcription factor can play diverse roles because it can bind the enhancers of diverse genes. For this reason, *Ubx* and some other genes that encode transcription factors are like a hammer or wrench that can be used for a wide range of different tasks. Sean Carroll referred to such genes as a **genetic toolkit** that is shared widely among animals, and can contribute to evolutionary changes in the regulation of diverse genes with diverse developmental roles [9]. The effect of *Ubx* on water strider legs and *Drosophila* hairs represents the evolution of a novel use of a preexisting gene for a new function: it is an example of exaptation (see Chapter 3), or **co-option** [78]. In some instances, genetic pathways, not just single genes, have been co-opted in evolution. For example, a subset of the Hox genes that determine the anterior-posterior differentiation of the vertebrate body (see Figure 15.11) is expressed in the limbs (**FIGURE 15.12A**). These genes are expressed from the base to the tip of the developing limb in the same sequence as their anterior-posterior expression along the body axis (**FIGURE 15.12B**), and they determine the proximal-distal differentiation of the limb (e.g., humerus to radius to digits). The same principles apply to plants. Most species in the potato family (Solanaceae) express the *MADS16* gene, which encodes a transcription factor only in vegetative tissues, where it affects cell shape and division rate. In the genus *Physalis*, known as ground cherries or Chinese lantern plants, the gene is heterotopically expressed in the sepals after pollination, and causes these flower parts to grow into a "balloon" that envelops the fruit (**FIGURE 15.13**) [27].

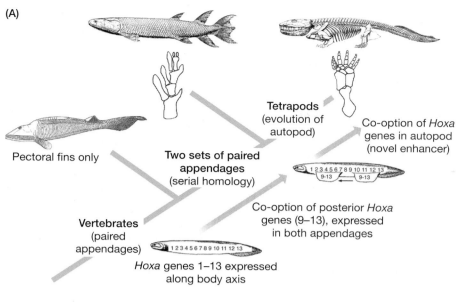

(A)

Pectoral fins only

Vertebrates
(paired
appendages)

Hoxa genes 1–13 expressed
along body axis

Two sets of paired
appendages
(serial homology)

Tetrapods
(evolution of
autopod)

Co-option of *Hoxa*
genes in autopod
(novel enhancer)

Co-option of posterior *Hoxa*
genes (9–13), expressed
in both appendages

FIGURE 15.12 Co-option of developmental pathways in the evolution of novelties. (A) Co-option of the vertebrate *Hoxa* genes during evolution of the tetrapod limb. Ancestrally, Hox genes were expressed only along the anterior-posterior axis of the developing body. The evolution of paired forelimbs and hindlimbs involved novel gene expression, presumably using novel enhancer sequences of *Hoxa* genes 9–13. Autopod refers to the wrist, ankle, and digital bones. (B) The pattern of expression of the *Hoxa* genes in the three regions of the limb (developmental modules) is the same in fore- and hindlimbs, which perform different functions and are referred to as functional modules. Bone names are H, humerus; F, femur; R, radius; T, tibia; MC, metacarpals; and MT, metatarsals. (A from [78]; B from [87].)

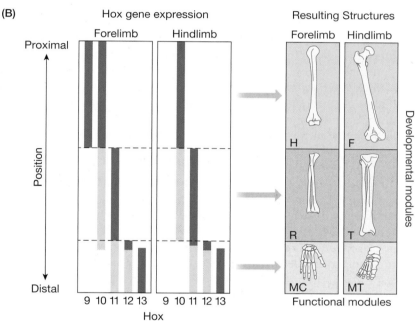

(B)

Hox gene expression

Forelimb   Hindlimb

Proximal

Position

Distal

9 10 11 12 13   9 10 11 12 13

Hox

Resulting Structures

Forelimb   Hindlimb

H   F

R   T

MC   MT

Developmental modules

Functional modules

FIGURE 15.13 (A) The fruit of a tomato plant shows the sepals as small structures at the base of the fruit. This is the condition in most members of the plant family Solanaceae. (B) In the ground cherry genus, *Physalis*, a gene that is ordinarily expressed only in vegetative tissues is transcribed in the sepals after the fruit starts to develop. The sepals form the "Chinese lantern" that envelops and protects the fruit.

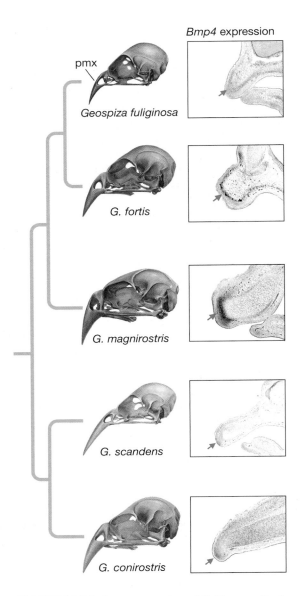

Bmp4 expression

pmx

Geospiza fuliginosa

G. fortis

G. magnirostris

G. scandens

G. conirostris

FIGURE 15.14 Among species of Galápagos finches (*Geospiza*), differences in the depth and length of the premaxilla (pmx) are determined largely by differences in the expression the gene *Bmp4* at a critical stage in development. Darker staining in the region indicated by red arrows shows higher gene expression. The gene shows lower expression in species with more slender, pointed bills (*G. fuliginosa*, *G. scandens*, and *G. conirostris*) at the same stage of development. (From [1]; skull images from [7], reproduced with permission of University of California Press.)

Developmental-Genetic Bases of Phenotypic Evolution

Because early data showed that the proteins of humans and chimpanzees are very similar, Mary-Claire King and Allan Wilson suggested that the many morphological differences between the species are likely to be caused mostly by regulatory differences rather than protein sequence differences [31]. We now know that they were right: differences in gene regulation underlie much morphological evolution. For example, differences in beak size and shape among the Galápagos finches in the genus *Geospiza* are based on growth differences in the prenasal cartilage and later in the premaxillary bone. In a series of studies, Arhat Abzhanov and collaborators have shown that the expression of certain growth-regulating genes during beak development accounts for these differences [1, 41]. The *Bmp4* gene is expressed in the prenasal cartilage earlier and at higher levels in the embryos of species that have deeper, wider beaks (**FIGURE 15.14**). The *calmodulin* gene is expressed more highly in species with elongated beaks. Later in development, differences in expression of three other genes in the premaxilla correlate with beak differences. The expression level of most of these genes has been experimentally altered in chicken embryos and produces the same effects that are seen in the finches. These genes encode transcription factors with well-known functions in craniofacial development. As we will see, evolution of both transcription factors and *cis*-regulatory sequences has been important.

Evolution by cis-regulatory mutations

Many of the mutations that affect morphological variation reside in regulatory sequences [36, 61, 73]. Changes in *cis*-regulatory elements can happen in several ways [64]. Deletion of an enhancer can prevent a characteristic from developing. For example, the pelvic girdle and fins have been reduced independently in many freshwater populations of the three-spined stickleback (*Gasterosteus aculeatus*) that have independently descended from marine ancestors. This change is caused by a recurrent deletion of an enhancer of the *Pitx1* gene (**FIGURE 15.15**) [10]. An example closer to home is the evolutionary deletion of an enhancer of the human androgen receptor gene that determines the gene's expression in the fetal penis. This enhancer is the developmental basis of minute spines on the penis that are present in chimpanzees and many other mammals but are lacking in humans [46].

Many changes in *cis*-regulation evolve by mutational changes in the enhancer's sequence. For example, the dark pigmented spots in the wings of *Drosophila guttifera* correspond to the locations where the *wingless* (*wg*) gene is expressed in the developing wing (**FIGURE 15.16**). The protein product of this gene affects the expression of many other genes. The wings of *D. melanogaster*, which normally lacks dark wing spots, expressed the *D. guttifera* pattern when a specific fragment from the *D. guttifera wg* gene was transferred into the *D. melanogaster* genome. Because this sequence differs only slightly between the species, the researchers concluded that a small number of changes in a single enhancer of the *wg* gene caused the expression of pigmentation genes in the wings of *D. guttifera* [33].

Evolutionary changes in enhancers can increase or decrease their affinity for certain transcription factors, and can sometimes cause them to bind different

(A) Marine

Pitx1 expression
(ventral view)

Dorsal spines

Pelvic spine

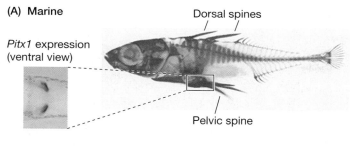

(B) Freshwater

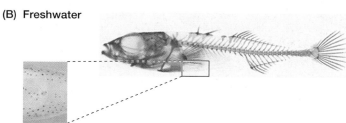

FIGURE 15.15 Loss of expression of *Pitx1* in the pelvis is associated with reduction of the pelvic girdle and fins in freshwater populations of the three-spined stickleback (*Gasterosteus aculeatus*). Adult specimens from (A) marine and (B) freshwater populations. The pelvis and spines associated with fins characterize the marine population but not the freshwater population. In the magnified ventral views of embryos at left, in situ hybridization reveals much greater *Pitx1* expression (purple) in the pelvic area in the marine population than in the freshwater population. (After [71]; photos courtesy of D. M. Kingsley.)

transcription factors. *Drosophila biarmipes* has evolved a large melanized spot on the wing, resulting from a change in one of the enhancers of the *yellow* gene so that it now binds both a repressor protein and a transcription factor produced by the gene *Distal-less* (**FIGURE 15.17**) [4, 20]. This also represents a new expression pattern for *Distal-less*, which plays various other regulatory roles. So the wing spot in *D. biarmipes* is an example of a novel characteristic that is based on evolutionary changes in both *cis-* and *trans*-regulation of a gene, that is, the *yellow* gene.

Where do new *cis*-regulatory elements come from? Some arise de novo. (See Chapter 14, p. 347, on novel genes.) Others originate by duplication and sequence divergence of ancestral sequences. But one source is intriguing: in mammals, some of the binding sites for various transcription factors are embedded within transposable elements [75]. A dramatic example is in the endometrial stromal cells of the placenta in humans and other placental mammals. A gene that encodes the hormone prolactin is expressed in these cells, in response to cAMP [40]. The cAMP-responsive enhancer of this gene is derived from a DNA transposon, MER20. MER20-related sequences are close to the coding sequences of hundreds of genes that are expressed in these cells, and these sequences have been experimentally shown to bind several transcription factors that are important in pregnancy.

Evolution by trans-regulatory mutations

Alterations of genes that encode transcription factors, by changing their binding to enhancers, have also proven to play major roles in phenotypic evolution [11, 39, 84]. For example, insects, which evolved from a crustacean lineage, have legs only on the three thoracic segments. In *Drosophila* embryos, *Ubx* is expressed in the abdomen and inhibits leg development. When biologists caused the gene to be ectopically expressed in the thorax as well, it inhibited the development of embryonic limbs. However, *Ubx* is expressed throughout the segments in the brine shrimp (*Artemia*), which, like other crustaceans, has appendages on most of the body segments. When brine shrimp *Ubx* was transferred into *Drosophila* embryos, it did not suppress embryonic limb development [63]. Researchers found the same effect when they used *Ubx*

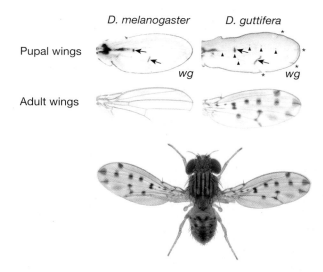

FIGURE 15.16 The spotted wing pattern of *Drosophila guttifera* is based on novel expression of the *wg* gene. The wings of adult *D. melanogaster* and *D. guttifera* are shown below the pupal wings, when pigmentation develops as a result of *wg* expression, visualized by blue color. The gene is expressed in *D. guttifera* not only in the same sites as in *D. melanogaster* (arrows), but also in small sensory structures (arrowheads) and at the tips of the wing veins (asterisks). These sites match the spot pattern in the adult. (From [33]; *D. guttifera* photo courtesy of Nicolas Gompel and Sean Carroll.)

FIGURE 15.17 Evolution of a transcription factor underlies a novel character. The images at top show the wing pigmentation patterns and the expression patterns of *Distal-less* and *yellow* genes in *D. melanogaster* (left), certain other *Drosophila* species (center), and *D. biarmipes* (right). The diagrams at bottom provide a model of the evolution of regulatory changes. Left: In *D. melanogaster*, the enhancer upstream of *yellow* (*y*) does not bind the regulatory protein produced by *Distal-less* (*Dll*). *Dll* has no effect on expression of *y*. Center: In some *Drosophila* species, the enhancer evolved the ability to bind Dll, as highlighted in blue. This caused enhanced expression of *y* in regions of the wing where *Dll* is expressed. Right: The next evolutionary step, found in *D. biarmipes*, involved increased expression of *Dll* in one region of the wing (darker red patch in middle row; blue in the diagram). This caused increased expression of *y* and darker pigmentation in that patch. (From [4].)

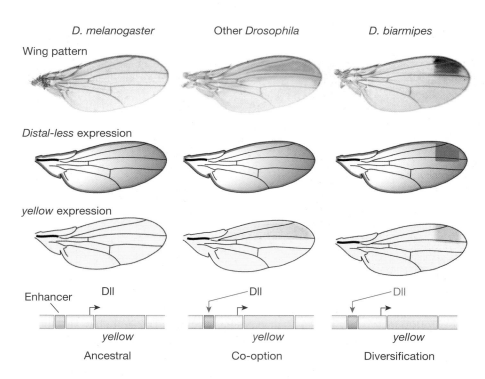

from a velvet worm in the phylum Onychophora, the sister group of the arthropods. The difference between these *Ubx* effects was traced to a new poly-alanine domain that evolved in insects and inhibits transcription of a gene (*Distal-less*) that is necessary for leg development (**FIGURE 15.18**) [17]. Thus, when insects evolved, *Ubx* became a leg repressor.

A more radical change in the function of a transcription factor has been described for *fushi tarazu* (*ftz*). In *Drosophila*, this gene is important in the segmentation of the embryo [37]. In the grasshopper *Schistocerca* and the flour beetle

FIGURE 15.18 In insects, the *Ubx* gene suppresses development of legs on the abdomen, but it lacks this activity in crustaceans and other non-insect arthropods, as well as in the related phylum Onychophora (velvet worms). Part of the Ubx protein sequence is shown for these species; the letters in the sequences represent amino acids (e.g., A = alanine) encoded by the corresponding DNA sequences. The region in brown has high sequence homology among all the taxa. The long sector in blue is a novel poly-alanine sequence that is responsible for repression of legs on the abdomen of the insects. (After [17].)

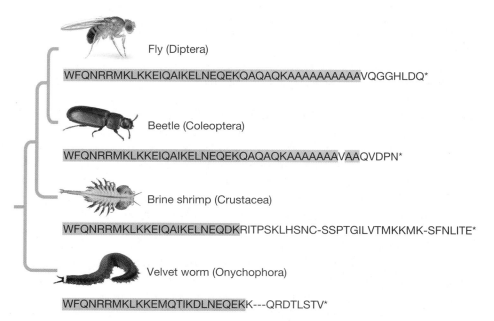

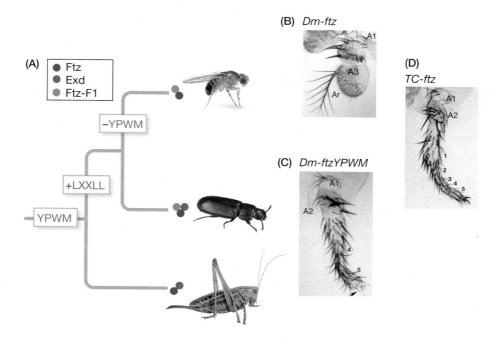

FIGURE 15.19 Evolutionary changes in function of the Ftz (Fushi tarazu) transcription factor (represented by red circle). (A) In ancestral insects, the amino acid sequence YPWM enabled Ftz to interact with another protein (Exd, blue circle) and to help determine segment identity (a homeotic function). This condition is retained in the grasshopper. In the ancestor of beetles and flies, Ftz evolved an additional amino acid motif (LXXLL) that interacts with the cofactor Ftz-F1 (green circle). This condition is seen in beetles, where Ftz and its cofactors affect both segment identity and the formation of segments. In the ancestor of *Drosophila*, the YPWM motif was deleted so that Ftz interacts only with Ftz-F1. This changed the protein's function: it determines segment formation but not identity. The results of these evolutionary changes can be demonstrated by experimentally changing the *ftz* gene sequence in *Drosophila* and seeing the effects of these changes on the fly's antennae. (B) The normal *Drosophila* antenna (labeled *Dm-ftz*) has three short segments (A1, A2, and A3) and a featherlike structure (arista: Ar). (C) When the YPWM motif that was lost from *Drosophila* is added back to the protein (*Dm-ftzYPWM*), the antenna's terminal segment and the arista are partly transformed into a leglike structure. (D) When the fly gene is replaced with a beetle gene (*TC-ftz*), the homeotic transformation to a leg with a five-segmented tarsus (as in many beetles) is complete. The numerals designate specific segments of the tarsus. (Diagram after [39]; antenna images from [37].)

Tribolium, which is more closely related to flies, this gene was found to have a homeotic function instead: it determines organ identity rather than segmentation. Inserting *ftz* from these insects into *Drosophila* caused a transformation of antennae into legs, a homeotic effect that was traced to an amino acid motif (YPWM)[2] in the Ftz protein of the grasshopper and beetle, which is lacking in *Drosophila* (**FIGURE 15.19**). However, the Ftz protein of the beetle and fly has an amino acid motif (LXXLL, where "X" stands for any amino acid) that is required for segmentation function because it interacts with a cofactor protein when it binds to *cis*-regulatory elements. The *ftz* gene of *Drosophila* has lost homeotic function by loss of one short motif and gained segmentation function by acquiring a different motif.

Thus, a transcription factor may retain the amino acid motif that binds to a *cis*-regulatory element, and be highly conservative in evolution, but evolve other functions by changing amino acid motifs that interact with other proteins and cofactors. The result is that gene regulatory networks, sets of interacting genes that determine phenotypic traits, evolve by forming new interacting combinations of transcription factor modules and *cis*-regulatory modules, at various levels in regulatory cascades.

---

[2] Each of the 20 amino acids is represented by a letter.

Overview: The genetics and development of phenotypic evolution

These examples show how evolutionary changes in the presence versus absence of phenotypic traits and in their size, form, and location on the organism can be understood as consequences of evolutionary alterations in gene regulation. The differences in gene expression among species, and consequently in the proteins that determine the form and function of groups of cells and tissues, are the *proximate cause* of the differences in morphology: they are the mechanisms by which the genetic differences among species are expressed as different phenotypes. Moreover, the genotype-phenotype relationship, or "map," may show surprising differences among species; we saw that changes in different genetic pathways account for pigmented wing spots in different lineages of *Drosophila*. (In other cases, similar changes in the same developmental mechanism underlie convergent evolution; for example, mutations in the *melanocortin-1 receptor* gene account for differences in pigmentation in diverse vertebrates [see Figure 6.29].) As we noted earlier, this perspective (sometimes called a structuralist viewpoint) does not conflict in any way with the theory of allele frequency changes under natural selection. Instead, it complements the perspective of natural selection as the major cause—the *ultimate cause*—of phenotypic evolution (see Chapter 1, p. 7). Genetic mutations that produce a phenotypic change that enhances fitness increase and become fixed in populations. What evolutionary developmental biology tries to do is tell us how those mutations act—how they produce the phenotypic alterations that may have increased fitness. These mechanisms will help us understand phenomena that we have so far described in rather abstract genetic terms. For example, genetic correlations between traits can influence their evolution (see Chapter 6). At this time, we cannot predict very well which traits are likely to be genetically correlated within species. A sufficient understanding of the genetic regulatory networks that underlie such correlations will make prediction more feasible, and so can enhance our understanding of how characters respond to natural selection. Thus, developmental biology and population genetics can meet to enhance understanding. We now describe some of the steps evolutionary developmental biologists are taking toward that goal.

Evolvability and Developmental Pathways

Understanding the genetic networks that control development may shed light on **evolvability**, the ability of a characteristic to evolve, especially under directional selection [25, 83]. For example, characters that differ in additive genetic variance differ correspondingly in evolvability, all else being equal (see Chapter 6). But all else may not be equal, because a character may be genetically correlated with other characters. Genetic correlations may constrain a character's ability to respond to directional selection, or they may enhance selection response if the correlation points toward a new and better combination of character states. In particular, functionally interacting features may be more evolvable if they are genetically correlated. For example, the bill of a bird species that picks small insects off leaves may more easily evolve to be shorter or longer if variation of the upper (maxilla) and lower (mandible) parts is correlated (**FIGURE 15.20**). In many animal-pollinated flowers, successful pollination depends on a close match among various dimensions of the flower, such as the length of stamens and styles. These features' evolvability—the response to natural selection on their length—would be enhanced if the structures co-varied.

Are functionally interacting structures especially strongly correlated within populations? The earliest evidence for this idea was presented by the paleontologists Everett Olson and Robert Miller, who called such a pattern "morphological

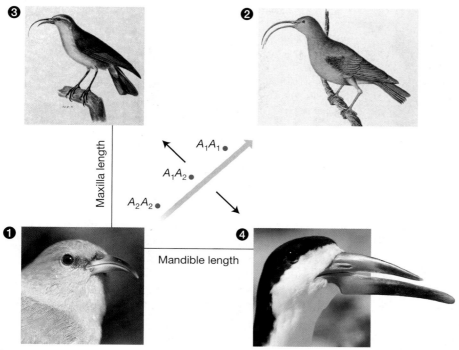

FIGURE 15.20 How pleiotropy might facilitate or constrain evolution. Suppose there is a positive genetic correlation between two features, such as the length of the upper and lower parts (maxilla and mandible, respectively) of a bird's bill. Usually, several or many pleiotropic genes will contribute to the correlation; here only the effects at locus A are shown. Suppose the ancestral state is short for both bill parts (1). Often, selection for a longer bill (arrow pointing up and to the right) will favor both parts maintaining equal length (perhaps because this aids in picking up food). Then the positive genetic correlation enhances the response to selection. But the positive pleiotropic correlation may prevent or slow down the response to selection for a long maxilla and short mandible (arrow pointing up and left), or for a short maxilla and long mandible (arrow pointing down and right). The birds shown are the Hawaiian honeycreepers *Hemignathus virens* (1), *H. obscurus* (2), and *H. lucidus* (3); the latter two species are extinct. Bird 4, the black skimmer (*Rynchops niger*), is very distantly related to the honeycreepers. Skimmers are the only birds with a longer mandible, which is used for snatching fish as the birds fly above the water surface.

integration" [51]. (The term **phenotypic integration** is more frequently used today.) At about the same time, the Russian geneticist Raissa Berg found that correlations among flower structures and among vegetative structures were higher than correlations between floral and vegetative elements [5]. Such observations led others, such as the Austrian zoologist Rupert Riedl [60], to propose that natural selection shapes genetic correlations, so that functionally interdependent features become more strongly integrated. Population genetic models by Günter Wagner and others show that pleiotropic correlations among characters can be shaped by natural selection, so that evolvability itself can evolve [15, 26, 53, 83].

We have seen that a transcription factor often binds to enhancers (*cis*-regulatory sequences) of diverse genes, and that the many enhancers that affect transcription of a gene can bind diverse transcription factors. Gene regulatory networks therefore have great potential for extensive pleiotropy: one TF might affect expression of many genes, or one gene might be responsive to different TFs in different cells. But there is also great opportunity for specificity. In mammals, for example, expression of two *Hox* genes, *Hoxd13* and *Hoxa13*, is required for development of both digits and external genitals. The expression of these genes depends on their enhancers. Some enhancers enable expression in both digital and genitalic primordia, while other enhancers govern expression in one or the other developing structure (**FIGURE 15.21**).

Pleiotropy can evolve if variation in one gene alters the effect of another locus on two or more different characters. For example, a change of a transcription factor could

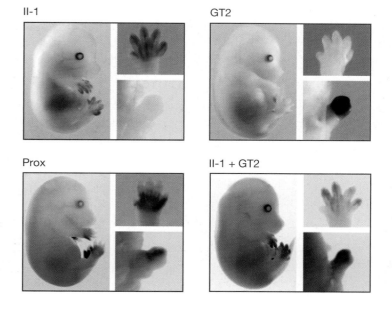

FIGURE 15.21 Different enhancers of the *Hoxd13* gene in mouse embryos vary in whether or not they are pleiotropic. Blue-green staining shows sites of *Hoxd13* expression in transgenic embryos with the II-1, GT2, or Prox enhancer, or with a combination of II-1 and GT2. In each panel, the entire embryo is shown on the left and close-ups of the digits (above) and the genital tubercle (below) are on the right. The Prox enhancer causes the gene to act pleiotropically: it is expressed in both digits and the genital tubercle. Without Prox, enhancers II-1 and GT2 cause gene expression only in the digits or the genital tubercle, respectively. Neither of the latter enhancers is pleiotropic; without Prox, both II-1 and GT2 are needed to express *Hoxd13* in both sites (II-1 + GT2). (From [38].)

FIGURE 15.22 Modularity of characters can evolve by means of changes in pleiotropic effects. Locus *S* affects the size of wing spots of a hypothetical butterfly. (A) In the presence of allele R_1 at a "relationship gene," the alleles S_1 and S_2 have pleiotropic effects on the forewing and hindwing; there appears to be a single character, "spot size." (B) The allele R_2 suppresses the effect of *S* alleles on the forewing, so they affect only the hindwing. Thus, the spot sizes are decoupled and appear to be distinct characters. Solid blue lines indicate pathways between loci and traits they affect; broken lines are pathways that are blocked or inactive. This diagram is based on a genetic study of wing spots in the butterfly *Bicyclus anynana*.

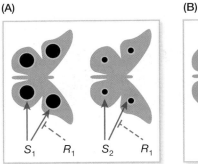

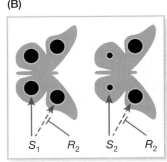

alter another gene's expression in one developing characteristic but not another. Such genes have been referred to as relationship quantitative trait loci (rQTL) [54]. Many genes that are known to interact with other genes may have such effects. For example, a mutation at the *S* locus in the butterfly *Bicyclus anynana* (see Figure 6.22) reduces colored spots ("eyespots") only on the hindwing if the insect has a certain allele at the *R* locus, but has this effect on both the hindwing and the forewing in the presence of a different *R* allele [49].

Suppose, then, that a butterfly population, with alleles S_1 and R_1, originally has large spots on both wings, but that selection by a new predator favors small spots on the hindwing only. A mutation S_2 is pleiotropic: it reduces spots on both wings (**FIGURE 15.22**). This mutation increases fitness, but fitness would be still higher if the forewing were to retain large spots. A mutation R_2 that alters expression of S_2 in the forewing so that the spots are large will be advantageous. The R_2 allele thus differentiates the two wing spots, reducing the pleiotropic effect of the *S* gene, and compensating for its deleterious effect. This scenario illustrates a model in which body parts that shared pleiotropic genes become differentiated so that they become less genetically correlated: they become distinct, individualized characters, or modules [52]. Conversely, selection for phenotypic integration of functionally related traits could increase pleiotropic correlations between characters.

Many studies of phenotypic variation support the ideas of modularity and phenotypic integration [3]. We would expect that when new functional relationships among characteristics evolve, selection would shape new correlation patterns. For example, serially homologous organs, such as the fore- and hindlimbs of vertebrates, are based on similar developmental genetic pathways (e.g., the expression of Hox genes; see Figure 15.12B). Nathan Young and colleagues found that in quadrupedal monkeys, the fore- and hindlimbs are rather similar in length, and the correlations between the lengths of corresponding parts (humerus and femur; radius and tibia; metacarpals and metatarsals) are high (**FIGURE 15.23**) [87]. But apes use their fore- and hindlimbs quite differently; gibbons, for example, have very long arms, used for swinging between branches of trees. The correlations between corresponding bones in the fore- and hindlimbs are lower in apes than in monkeys. Young and colleagues suggest that the reduced pleiotropic integration of fore- and hindlimbs enabled them to evolve more independently (i.e., to become more evolvable). Moreover, this independence may have facilitated the evolution of the unique limbs of humans, the only species of primate in which legs are much longer than arms.

The expression of some developmental genes corresponds to morphological modules that are recognized by patterns of correlation. For example, digit 1 (thumb) in the hand of primates shows more independent variation, both within and among species, than digits 2–5, which are consistently more similar to each other [59]. Corresponding to these two apparent modules, studies of mice show

(A)

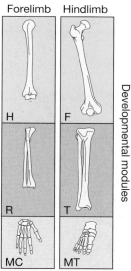

Forelimb Hindlimb

H F

R T

MC MT

Developmental modules

Functional modules

FIGURE 15.23 Developmental and functional modules in the limbs of humans and their relatives. (A) Features that are strongly correlated are considered parts of a module. Both forelimbs (arms) and hindlimbs (legs) have three major sections, marking three developmental modules. Bones in the corresponding sections are labeled H, R, and MC for humerus, radius, and metacarpals in the forelimb, and F, T, and MT for femur, tibia, and metatarsals in the hindlimb. Both limbs have similar patterns of expression of Hox genes (see Figure 15.12B). (B) For each species in the phylogeny, the diagram at the top shows correlations (numbers) between lettered boxes that represent bones in the forelimb (left boxes) and the hindlimb (right boxes). Only statistically significant correlations are shown. For each species, the correlations were combined into an index of overall modularity (shown by the colored bars on the graph), which is greater if the set of correlations among the limb elements is greater. The overall modularity is lower in apes (the four species at left) than in monkeys (the four species at right), chiefly because of lower correlations between elements of the forelimb and hindlimb. (From [87].)

(B)

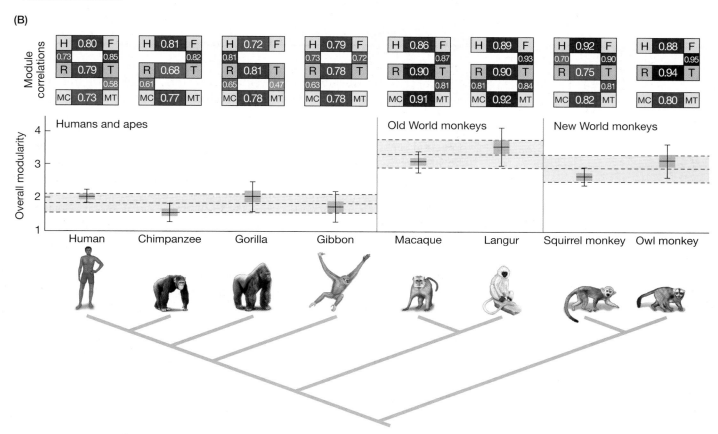

that only two Hox genes are expressed in the developing thumb, whereas these and two other Hox genes are expressed in digits 2–5.

Constraints on Adaptive Evolution

It is easy to imagine organisms that do not exist (**FIGURE 15.24**). Science-fiction writers do so all the time, but biologists can cite many more realistic examples. In no tetrapod vertebrate is the central (third) digit the shortest. No animals can make their own food by photosynthesis (although corals incorporate photosynthetic algae into their bodies). There appear to be **constraints**—restrictions—on what can evolve.

FIGURE 15.24 In a work both amusing and well informed by evolutionary principles, Harald Stümpke imagined an adaptive radiation of "Rhinogradentia," or "snouters," mammals with noses elaborated for diverse functions. *Otopteryx* flies backward, using its ears as wings and its nose as rudder. *Orchidiopsis* feeds on insects attracted to its petal-like nose and ears. Stümpke's fantasy illustrates some of the many conceivable phenotypes that have never evolved. (From [74].)

(A)
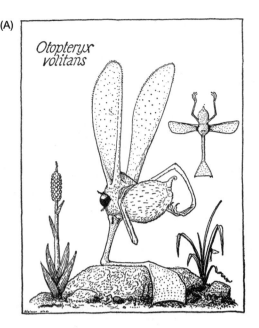

(B)

All organisms are subject to physical constraints. For example, the maximal possible size of insects is thought to be limited by the rate at which oxygen and carbon dioxide can diffuse through the narrow tubes, or tracheae, that conduct gases throughout the body. John Maynard Smith and colleagues defined developmental constraint as "a bias on the production of various phenotypes caused by the structure, character, composition, or dynamics of the developmental system" [43]. The two most common suggested causes of developmental constraint are (1) absence or paucity of phenotypic variation, including the absence of morphogenetic capacity (i.e., lack of required genes, proteins, or developmental pathways), and (2) strong correlations among characters, caused by pleiotropy. Thus, developmental constraints can be considered a form of genetic constraint (see Chapter 6).

Developmental constraints can be revealed by embryological manipulations in the laboratory. In a classic experiment, Pere Alberch and Emily Gale used the mitosis-inhibiting chemical colchicine to inhibit digit development in the limb buds of frogs (*Xenopus*) and salamanders (*Ambystoma*; **FIGURE 15.25A,B**) [2]. The treatment consistently caused preaxial (front) digits to be missing in the frogs, and postaxial (rear) digits to be lost in the salamanders. These results correspond to the different order of digit differentiation in the two taxa: the last digits to form tended to be

FIGURE 15.25 Evidence for developmental constraints. (A) X-ray of the right hind foot of an axolotl salamander (*Ambystoma mexicanum*), showing the normal five-toed condition. (B) The left hind foot of the same individual, which was treated with an inhibitor of mitosis during the limb bud stage. The foot lacks the postaxial toe and some toe segments, and is smaller than the control foot. (C) A normal left hind foot of the four-toed salamander (*Hemidactylium scutatum*) has the same features as the experimentally treated foot of the axolotl. (From [2]; photos courtesy of P. Alberch.)

(A)
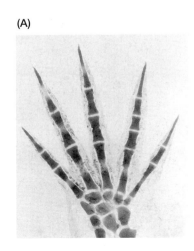

(B)
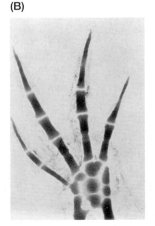

(C)

the most sensitive to the colchicine treatment. Furthermore, the results strongly reflected evolutionary trends: salamanders have often lost postaxial digits during evolution (**FIGURE 15.25C**), and frogs have repeatedly experienced preaxial digit reduction. We do not know if postaxial reduction would ever have been advantageous to any frogs, but any such evolution would have had to overcome a developmental barrier.

Developmental studies are beginning to shed light on some constraints. Among birds, swans have longer necks with more vertebrae than do ducks and most other birds. But almost all mammals—from whales to giraffes—have seven cervical (neck) vertebrae (see Figure 3.21). One might well suppose that a giraffe would profit from more. However, a study of abnormal vertebrae in deceased human fetuses and infants showed that the vertebrae have been homeotically transformed: a cervical vertebra, for example, often has the shape and ribs of a thoracic vertebra. These transformations are typically associated with malformations of the skull and face and of the heart, lungs, and other organs, evidently due to pleiotropy [77]. These harmful pleiotropic effects may have limited cervical vertebra evolution in mammals.

Phenotypic Plasticity and Canalization

The correspondence between genotypic differences and phenotypic differences depends not only on the effects of genotype, but also on environmental conditions that may affect the developmental expression of the genotype.

The **reaction norm** of a genotype is the set of phenotypes that genotype is capable of expressing under different environmental conditions, and can be visualized by plotting the genotype's phenotypic value in two or more environments (**FIGURE 15.26A–C**; see also Figure 6.25). In some cases, a single genotype may produce different phenotypes in response to environmental stimuli, a phenomenon called

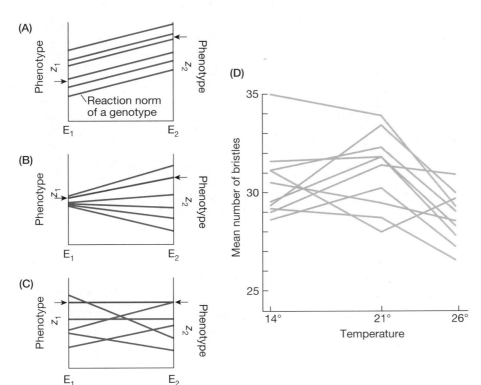

FIGURE 15.26 (A–C) Genotype × environment interaction and the evolution of reaction norms. Each line represents the reaction norm of a genotype—its expression of a phenotypic character (the states expressed are labeled z_1 and z_2) in environments E_1 and E_2. The arrows indicate the optimal phenotype in each environment. (A) The genotypes do not differ in the effect of environment on phenotype; there is no G×E interaction. The optimal norm of reaction cannot evolve in this case because no genotype matches the arrows. (B) The effect of environment on phenotype differs among genotypes; G×E interaction exists. The genotype with the norm of reaction closest to the optima in E_1 and E_2 (red line) will be fixed. New mutations that bring the phenotype closer to the optimum for each environment may be fixed thereafter. (C) Selection may favor a constant phenotype, irrespective of environment, resulting in canalization. (D) The number of bristles on the abdomen of male *Drosophila pseudoobscura* of ten genotypes, each reared at three different temperatures. The variation indicates a G×E interaction. (D after [24].)

phenotypic plasticity (see Chapter 6). In other cases, the phenotype may be quite constant, irrespective of environment (as in Figure 15.26C): it has a flat reaction norm, and is sometimes said to be *canalized*.

When the effect of environmental differences on the phenotype differs from one genotype to another in a population, the reaction norms of the genotypes are not parallel, and the phenotypic variance includes a variance component (referred to as $V_{G \times E}$) that is due to **genotype × environment (G×E) interaction** (see Figure 15.26B). If all the genotypes have parallel reaction norms (see Figure 15.26A), there is no G×E interaction ($V_{G \times E} = 0$). If there is G×E interaction (that is, if reaction norms differ in slope among genotypes in a population), selection could change the average degree of phenotypic plasticity. For example, the effect of temperature on the number of bristles differed among ten genetic strains of *Drosophila pseudoobscura* (**FIGURE 15.26D**). If different bristle numbers were optimal for flies in colder versus warmer environments, it is likely that the ability to develop the right number, depending on temperature, could evolve.

Some environmentally determined phenotypic effects are not adaptive and may be unavoidable: for example, most organisms grow more slowly at lower temperatures, and most of us will weigh more if we eat too much. In many species, however, natural selection has resulted in norms of reaction that most nearly yield the optimal phenotype for the various environments the organism commonly encounters [67, 86]. Phenotypic plasticity includes rapidly reversible changes in morphology, physiology, and behavior as well as "developmental switches" that cannot be reversed during the organism's lifetime (**FIGURE 15.27A**). In some semiaquatic

(A)

Catkins   Caterpillars

(B)

Submerged   Air-water interface   Aerial

FIGURE 15.27 Examples of phenotypic plasticity. (A) Larvae of the geometrid moth *Nemoria arizonaria* that hatch in the spring (left) resemble the oak flowers (catkins) on which they feed. Larvae that hatch in the summer (right) feed on oak leaves and resemble twigs. (B) The form of a leaf of the water-crowfoot *Ranunculus aquatilis* depends on whether it is submerged, aerial, or situated at the air-water interface during development. (A, photos courtesy of Erick Greene; B from [12].)

plants, for instance, the form of a leaf depends on whether it develops below, above, or on the surface of water (**FIGURE 15.27B**). Plastic changes in phenotype result from changes in gene regulation. Environmental stimuli alter levels of hormones or other signaling molecules that affect levels of transcription factors, which in turn alter the expression of downstream genes [23].

For many characteristics, the most adaptive norm of reaction is a constant phenotype, buffered against alteration by the environment (see Figure 15.26C). It may be advantageous, for example, for an animal to attain a fixed body size at maturity or metamorphosis, despite variations in nutrition or temperature that affect the rate of growth. The developmental system underlying the character may then evolve so that it resists environmental influences on the phenotype [66]. This idea was developed independently by the Russian biologist Ivan Schmalhausen [69] and by British developmental biologist Conrad Waddington, who referred to it as **canalization** [16]. Waddington used the concept of canalization to interpret some curious experimental results [81]. A crossvein in the wing of *Drosophila* sometimes fails to develop if the fly is subjected to heat shock in the pupal stage. By selecting and propagating flies that developed a crossveinless condition in response to heat shock, Waddington bred a population in which most individuals were crossveinless when treated with heat. But after further selection, a considerable portion of the population lacked the crossvein even without heat shock, and this condition was heritable. A character state that initially developed in response to the environment had become genetically determined, a phenomenon that Waddington called **genetic assimilation**.

Although this result is reminiscent of the discredited theory of inheritance of acquired characteristics, it has a simple genetic interpretation. Genotypes of flies differ in their susceptibility to the influence of the environment (in this case, temperature)—that is, they differ in their degree of canalization, so that some are more easily deflected into an aberrant developmental pattern. Selection for this developmental pattern favors alleles that channel development into the newly favored pathway. As such alleles accumulate, less environmental stimulus is required to produce the new phenotype. The finding that genetic assimilation does not occur in inbred populations that lack genetic variation supports this interpretation [66]. We encountered an example of genetic assimilation in natural populations in Chapter 6 (see Figure 6.25). In high-elevation lakes, exposed to high levels of ultraviolet radiation (UV), the water flea *Daphnia melanica* can develop a range of low to high pigmentation, as a protection against UV. But dark color makes the *Daphnia* more visible to fish, and two populations in lakes with recently introduced trout show constitutive (unchanging) expression of the low-melanin (light) phenotype. The expression of two genes that affect melanin production is less affected by UV in these populations than in ancestral populations. An environmentally inducible expression has become genetically determined [70].

It is often difficult to tell whether or not constancy of a feature has evolved by natural selection for canalization, because it can also be an automatic effect of complex genetic and developmental pathways [72]. Researchers distinguish environmental canalization, which reduces the effect of environmental variation on the phenotype, from genetic canalization, which reduces the effect of genetic mutations. Stabilizing selection on a characteristic is likely to cause environmental canalization. For example, the floral structures of some animal-pollinated plants, which are thought to be strongly selected for successful pollination, are less variable than leaves are [55]. Genetic canalization is much less likely to evolve, because selection against deleterious mutations is so effective that few individuals deviate from the optimum, so there is little selection for genetic modifiers that prevent the mutations from being phenotypically expressed [28].

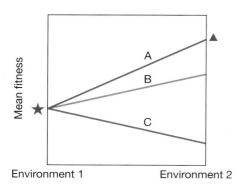

FIGURE 15.28 Phenotypic plasticity may enhance genetic adaptation to a novel environment, or it may not. Lines A, B, and C are three possible reaction norms of a character, expressing the pattern of phenotypic plasticity of three genotypes. The population has occupied Environment 1 for a long time, and has the optimal mean character for that environment (shown by star). If the environment changes to state 2, and the population is mostly composed of genotype A, its phenotypic plasticity will produce the phenotype that is optimal for the new environment (shown by triangle). But if genotype B is the prevalent genotype, its plasticity would only slightly enhance its fitness in Environment 2. If genotype C is prevalent, its phenotypic plasticity would be maladaptive in the new environment, making its fitness lower than if the genotype's phenotype were fixed.

Does phenotypic plasticity contribute to evolution?

In the modern theory of adaptive evolution by natural selection among inherited phenotypes, reaction norms evolve by selection of mutations (in, perhaps, regulatory elements or hormone levels) that differentially affect the expression of the trait in different environments. Mary Jane West-Eberhard [86] and some other biologists have proposed a somewhat different process. They suggest that a new environment can often induce the prevalent genotype to express a novel beneficial phenotype, by phenotypic plasticity. The expression of this phenotype, they suggest, can subsequently become genetically fixed (that is, genetically assimilated) by natural selection [34, 48, 68]. The feature may later be further modified by mutation and natural selection, which West-Eberhard has called "genetic accommodation." According to this hypothesis, phenotypic plasticity paves the way for standard evolution by selection of advantageous mutations that affect the feature. West-Eberhard points out many examples of closely related species pairs in which the reaction norm of one species includes a phenotype that is invariant in the other species [86]. For example, larvae of the sphinx moth *Manduca quinquemaculata* develop black pigmentation at low temperatures and green pigmentation at higher temperatures, whereas larvae in the related species *M. sexta* develop green pigmentation at all temperatures [76]. The still unanswered question is whether or not the green species evolved by genetic assimilation of part of a broader range of colors—a more plastic ancestral reaction norm.

A possible example of evolution by genetic accommodation has been described in tadpoles of spadefoot toads in the genus *Spea*. Like most anuran larvae, these tadpoles generally feed on detritus, but they often switch to feeding on shrimp and other animal prey (see Figure 6.24) [56]. This shift involves phenotypic plasticity: the tadpoles develop a shorter gut and larger jaw muscles. Tadpoles of *Scaphiopus*, the sister genus of *Spea*, normally feed only on the ancestral diet, detritus. But *Scaphiopus* tadpoles raised on shrimp were found to develop more slowly and have shorter guts than those that were fed detritus [35]. This experiment suggests that the phenotype that is adaptive in one genus (*Spea*) could have been produced by a plastic response that was already present, even if not used, in the common ancestor of *Spea* and *Scaphiopus*.

The hypothesis that plasticity is often the first step toward new adaptations is controversial. When a population experiences a new environment, an environmentally induced alteration of the phenotype may or may not be in the right direction (**FIGURE 15.28**) [18]. In the spadefoot toad experiment, shrimp-fed *Scaphiopus* tadpoles developed smaller jaw muscles, not the enlarged muscles seen in *Spea* tadpoles that are adapted to eating animal prey. A stressful environment may change development in ways that are not adaptive. For example, low temperatures reduce

(A)

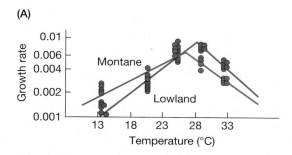

(B)

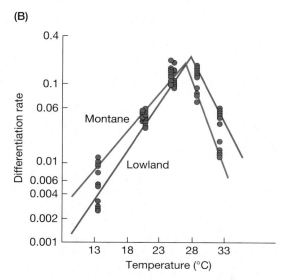

FIGURE 15.29 Populations of green frogs (*Rana clamitans*) at different elevations have evolved different norms of reaction that compensate for differences in environmental temperature. (A) The tadpoles' rate of growth in size, in relation to the temperature at which they were reared in the laboratory. Blue circles are individuals from a montane population, red circles are from a lowland population. Both populations have maximal growth at about the same temperature. At lower temperatures, the montane population has higher growth rate than the lowland population, compensating for the direct effect of temperature on developmental rate. The difference between the populations is reversed if tadpoles are reared at higher temperatures. (B) In the same experiment, similar results were obtained for another measure of development rate. Differentiation rate describes the rapidity with which the tadpoles reach metamorphosis to the frog stage. (After [6].)

the rate of growth and development in plants and ectothermic animals. The tadpoles of populations of green frogs (*Rana clamitans*) at cold, high elevations develop more rapidly than do those from warmer lowlands, when both are reared at low temperatures (**FIGURE 15.29**) [6]. The high-elevation populations have evolved a genetic change (faster growth) that compensates for the plastic environmental effect (slower growth). How often phenotypic plasticity leads to the evolution of new adaptations is not yet known.

SUMMARY

- Evolutionary developmental biology (EDB) seeks to integrate data from comparative embryology and developmental genetics with morphological evolution and population genetics, and to determine how changes in genes are expressed as changes in phenotypes. Understanding the molecular mechanisms by which genes, together with environment, produce phenotypes helps us understand how phenotypes evolve. Mutational changes in the genes that produce a developmental pathway may cause advantageous alterations of the phenotype, so both the phenotype and its underlying genetic network evolve. The mechanistic proximal causes of phenotypes complement the ultimate causes of allele frequency change, such as natural selection, in understanding the evolution of form.

- Many differences among species are due to heterochronic and allometric changes in the relative developmental rates of different body parts or in the rates or durations of different life history stages. Some characteristics have evolved by heterotopy, expression at a novel location on the body. The modularity of morphogenesis in different body parts and in different developmental stages facilitates such changes.

- The vast diversity of multicellular eukaryotes is largely due to diverse uses of a toolkit of genes and developmental pathways that are conserved across wide phyletic ranges.

- Developmental pathways include signaling proteins, transcription factors, cis-regulatory elements and structural genes. Evolutionary change in the regulatory connections among signaling pathways and transcription factors, and between transcription factors and their targets, is believed to underlie much of the phenotypic diversity seen in nature. Morphological variation within and among species may be caused by changes in either regulatory or protein-coding sequences, although regulatory changes may play a larger role.

- A gene may have many cis-regulatory elements (enhancers) that bind different transcription factor proteins and can be expressed in diverse tissues or at different times in development. Some cis-regulatory elements have originated from transposable elements, but most of them have evolved by mutation in their sequence. Changes in their interactions with transcription factor genes can alter the time and place of their activity. Evolution of the coding sequence of a transcription factor can change its developmental function.

- During evolution, genes and developmental pathways have often been co-opted, or recruited, for new functions, a process that is probably responsible for the evolution of many novel morphological traits. This process results from evolutionary changes in functional connections between transcription factors and cis-regulatory elements.

- Modularity among body parts is achieved by patterning mechanisms whose regulation is often specific to certain structures, segments, and life history stages. Modularity helps different parts of the body develop divergent morphologies (e.g., differences among segments). Pleiotropic effects of genes that affect functionally interacting characteristics may evolve, resulting in the evolution of functional modules (phenotypic integration).

- Genetic and developmental constraints can make some imaginable evolutionary changes unlikely to occur.

- Based on changes in the expression of certain genes and developmental pathways in response to environmental signals, a single genotype may be expressed as an array of different phenotypes, the genotype's norm of reaction. Reaction norms are genetically variable, and so can evolve by natural selection. Especially if the environment varies, phenotypic plasticity may evolve. Conversely, selection for a constant phenotype can result in canalization. Genetic assimilation is the genetic fixation of one of the states of a phenotypically plastic character. It is not known how important genetic assimilation is in evolution; nor is it known if adaptation may occur first by a nongenetic phenotypic change that later becomes genetically fixed by natural selection.

TERMS AND CONCEPTS

allometric growth (allometry)	cis-regulatory element	enhancer	gene regulatory network
alternative splicing	co-option	evolutionary developmental biology (EDB)	genetic assimilation
canalization,	constraint		genetic toolkit

genotype ×
  environment (G×E)
  interaction
heterochrony
heterotopy
homeotic selector
  gene

Hox gene
individualization
  interaction
module, modular
neoteny
ontogeny

paedomorphosis
phenotypic
  integration
phenotypic plasticity
reaction norm
serially homologous

trans-regulatory
  element
transcription factor
  (TF)
transgenic

SUGGESTIONS FOR FURTHER READING

An excellent, very readable introduction to current evolutionary developmental biology, emphasizing regulation of gene expression, is *From DNA to Diversity: Molecular Genetics and the Evolution of Animal Design* by Sean B. Carroll, Jennifer K. Grenier, and Scott D. Weatherbee (second edition, Blackwell Science, Malden, MA, 2005). For the more general science reader, S. B. Carroll describes the development of EDB in *Endless Forms Most Beautiful: The New Science of Evo-Devo* (W. W. Norton, New York, 2006).

David L. Stern provides a clear exposition of developmental pathways and their evolution in *Evolution, Development, & The Predictable Genome* (Roberts and Company, Greenwood Village, CO, 2011). Also see M. Rebeiz, N. H. Patel, and V. F. Hinman, "Unravelling the tangled skein: the evolution of transcriptional regulatory networks in development" (*Annu. Rev. Genomics Hum. Genet.* 16: 103–131, 2015).

One of the leading textbooks on developmental biology is *Developmental Biology* (11th edition) by Scott F. Gilbert and Michael J. F. Barresi (Sinauer Associates, Sunderland, MA, 2016). In *Ecological Developmental Biology* (Sinauer Associates, Sunderland, MA, 2009), Gilbert and David Epel treat the evolution and ecology of phenotypic plasticity and related topics.

Rudolph A. Raff, one of the founders of modern EDB, portrayed the field at an early stage in his influential book *The Shape of Life: Genes, Development, and the Evolution of Animal Form* (University of Chicago Press, 1996). Eric H. Davidson, a developmental biologist whose insights helped shape the field, did much the same in *The Regulatory Genome: Gene Regulatory Networks in Development and Evolution* (Academic Press, London, 2006).

An introduction to the evolution of modularity is G. P. Wagner, M. Pavlicev, and J. M. Cheverud, "The road to modularity" (*Nat. Rev. Genet.* 8: 921–931, 2007).

PROBLEMS AND DISCUSSION TOPICS

1. Haeckel's statement that "ontogeny recapitulates phylogeny" differs from what we now call von Baer's law—the generalization that features common to a higher taxon often appear earlier in development than the specific characters of lower-level taxa. Compare and contrast these two ways of thinking about changes in development over the course of animal evolution. Even though we now know that Haeckel's dictum seldom holds, what can we learn about phylogeny from development?

2. If two allometrically related traits show a strong correlation both within and among species, what kinds of experiments would you use to test whether these correlations are due to natural selection or to developmental genetic constraints? (Assume the organisms of interest are easily amenable to laboratory study.) What can we infer about the underlying genetic architecture of traits whose allometric relationships do not vary? What about those that do vary?

3. How might differential expression of and regulation by Hox genes contribute to mosaic evolution in which different segments of an animal body plan evolve different morphologies?

4. If mutations such as those of the *Ubx* gene can drastically change morphology in a single step, why do most evolutionary biologists maintain that modification of existing traits and the evolution of novel characters have generally proceeded by successive small steps?

5. Can convergent (or parallel) evolution of similar morphology in two different lineages involve

DNA sequence evolution in different parts of the same developmental gene? Explain how.

6. In many organisms, such as insects and flowering plants, morphological traits have been lost in some species due to the evolution of developmental arrest: growth and development cease, eliminating the final trait. The remaining tissue can be remodeled or resorbed for other use. What does this observation suggest about the nature of mutations and adaptive evolution in these cases?

7. Development of a morphological structure involves many different types of genes and their products, including transcription factors, signaling proteins, and effectors such as enzymes. When a morphological change occurs in a single mutational step, which types of genes or gene products might be more or less likely to be involved? Within a gene, would such single-step events be more likely to involve mutation to protein-coding sequences or changes in regulation? Why?

8. Almost all extant animals have Hox genes that are involved in body patterning. What can we infer about the common ancestors of living animals based on the presence of these genes? Why do organisms that are incredibly distinct morphologically, with completely different body plans, use so many of the same patterning genes?

UNIT IV
Macroevolution and the History of Life

Phylogeny: The Unity and Diversity of Life

If you compare a chicken with a crocodile, you might find it hard to imagine that they are closely related. And it might strike you as implausible that birds are even more closely related to dinosaurs. In fact, they *are* dinosaurs—in the same major group as *Tyrannosaurus* (**FIGURE 16.1**).

How can biologists know that? How confident can we be in these outrageous claims? Can we be sure that multicellular organisms evolved independently many times, that headless echinoderms evolved from ancestors with heads, that α- and β-hemoglobins evolved by gene duplication in ancient vertebrates? Any biology or anthropology textbook will state that humans and chimpanzees are each other's closest relatives, and will describe how humans evolved in Africa and spread from there throughout the world. How do we learn about the history of life on Earth?

Two sources of information inform us about the past: fossils and living organisms. Paleontology reveals organisms and events we could not have imagined without fossils, as we will describe in the next two chapters. But the record of even the best-fossilized groups of organisms is incomplete, and many kinds of organisms—especially those that lack skeletons or other hard structures—have a very poor fossil record or none at all. Many features we are interested in—behavior, physiology, life history, genome structure, and more—cannot be ascertained from fossils. In these situations, we piece together information from living species, chiefly through analyzing their phylogenetic trees and the history by which differences among species have arisen (see Chapter 2). Phylogeny, moreover,

Skeletal evidence that birds have evolved from theropod dinosaurs has been reinforced by the discovery of feathers in many dinosaurs. These tail feathers of a young mid-Cretaceous coelurosaurian theropod, exquisitely preserved in amber, were found recently in China [37]. Their structure supports a hypothesis about the steps by which feathers evolved. (Courtesy of Royal Saskatchewan Museum [RSM/R. C. McKellar].)

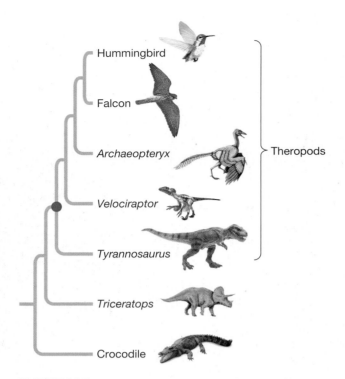

Hummingbird

Falcon

Archaeopteryx

Velociraptor

Tyrannosaurus

Theropods

Triceratops

Crocodile

FIGURE 16.1 Birds are living dinosaurs, falling within the theropod dinosaurs, a group that includes *Tyrannosaurus*. (Their most recent common ancestor is shown by the red circle.) The theropods are distantly related to ornithischian dinosaurs such as *Triceratops* (see Figure 17.26). Extinct dinosaurs, birds, and crocodiles descended from a more ancient ancestor and are included in a group called the archosaurs. (After [3].)

is critically important for interpreting fossils: it provides a basis for telling us that a mammalian skull is relevant to hominid evolution, and that there is an evolutionary connection between dinosaurs and birds.

A phylogenetic perspective on evolution is so important that we introduced the topic in Chapter 2. There, we provided an example of the reasoning by which one might infer relationships among species from very simple data (see Figure 2.10). That example was greatly oversimplified, because real phylogenetic analyses use far more data and take into account many complicating factors (which are interesting in themselves). In this chapter we describe how phylogenetic studies are actually done, and some of what we can learn from a phylogenetic history. Phylogenies help us describe the history of life, they are the basis of a meaningful classification of organisms, and they are necessary for testing many hypotheses about how evolution works.

Inferring Phylogenies

Phylogenetic relationships among species are estimated from similarities and differences. For example, biologists since Darwin have agreed that all animals with vertebrae form a single phylogenetic branch of species that share a common ancestor. This conclusion is based on the supposition that vertebrae were not present in the ancestor of all animals, and instead represent a **derived** character state, in contrast to the **ancestral** state, absence of vertebrae. Furthermore, we suppose that the evolution of a structure as complicated as a vertebra would be so rare that it is unlikely to have happened more than once within the animal kingdom. So a derived character that is shared by a group of species (a **synapomorphy**) is evidence that the species evolved from a common ancestor (**FIGURE 16.2A**). It is strong evidence if we can feel confident that the character evolved only once in the evolutionary history of the organisms we are studying. The set of species that have descended from a common ancestor is called a **monophyletic group** or **clade**. Mammals are a clade of species with synapomorphic characteristics that distinguish them from reptiles and amphibians, such as hair, milk, and a lower jaw consisting of only a single bone on each side. If we are correct in supposing that these and other unique features evolved only once, we must conclude that mammals shared a **most recent common ancestor** (**MRCA**) that was not the ancestor of other vertebrates. Similarly, among the four mammals shown in Figure 16.2A, the rat, human, and beaver share complex derived features not found in the kangaroo, of which the most striking is the placenta. Among our three placental mammals, the rat and beaver share a highly unusual feature: incisors that grow throughout life. These synapomorphies imply the phylogeny shown in the figure. In this figure, the reptile (crocodile) and the amphibian (salamander) are **outgroups**, a term that refers to species more distantly related to the members of a certain clade (mammals, in this case) than species within the clade are to each other. Outgroups help us determine the direction of evolutionary change. For example, these outgroups lack milk and hair, which implies that these were "new" features that evolved in the ancestor of the mammals.

If we now add a bird to our study, we notice that it resembles the human in one interesting respect: both are bipedal, walking on only their hind legs (**FIGURE 16.2B**). But we would not conclude that the human and bird are each other's closest relatives, because the bird has none of the unique characters that the human shares with other mammals. Hair and the other mammalian traits outweigh the single character, bipedalism, that must have evolved independently in the two

(A)

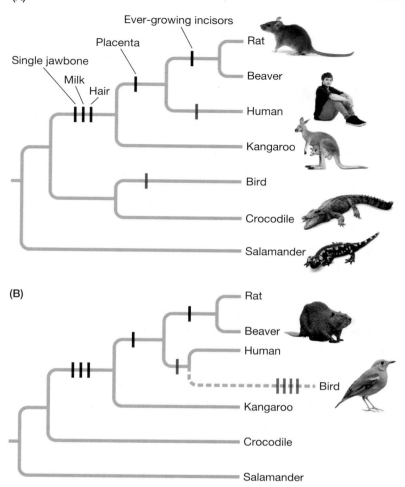

FIGURE 16.2 A simple example of how phylogenetic relationships are based on shared derived characters (shown by black crossbars). (A) Features such as hair, mammary glands, and a single jawbone unite the mammals. Their absence in the outgroups (represented by crocodile, bird, and salamander) shows that they are derived characters. Similarly, placenta and ever-growing incisors are derived characters of groups of mammal species. The red crossbars show the origin of bipedalism, which has evolved independently in the ancestors of human and of birds. (B) The dashed branch leading to the bird shows a convergent character with the human, bipedalism (red crossbar). If we supposed that the bird and human are close relatives because both are bipedal, the four derived characters that the human shares with some or all of the other mammals would have to have been lost (reversed) in the evolution of the bird. These losses are shown by the four blue crossbars on the bird branch. The resulting tree, with its four extra changes, is less parsimonious than the tree in (A).

lineages—an instance of convergent evolution. The character of bipedalism contradicts our supposition that each trait arises only once. But if we place the bird on our tree as the closest relative of the human (as in Figure 16.2B), we must postulate that the four features that the human shares with other mammals all underwent evolutionary reversal in the bird lineage, in which these features are the same as in the crocodile and other reptiles. These four "extra" evolutionary changes contradict our supposition that each trait arises only once and does not change. We must choose between a tree in which one character (bipedalism) violates our supposition and a tree in which at least four characters do so. A method called **parsimony** that is used for estimating phylogenies follows the simple rule of choosing the tree that requires the fewest evolutionary changes, namely the tree in which the fewest traits arise more than once or undergo reversal. By that rule, we accept the tree in which the human and bird are not closely related. (The bird is actually related to the crocodile.)

Notice that the method of parsimony uses *derived* traits as evidence for common ancestry; it does not use shared traits that are ancestral. Animals without a backbone are sometimes called "invertebrates." But invertebrates are not a clade. If they were, all invertebrates would be more closely related (share a MRCA) to each other than to vertebrates. Abundant evidence from fossils, morphology, and DNA sequences shows this is not the case. Insects and sponges are both invertebrates, but insects are more closely related to vertebrates than they are to sponges. Hence, certain similarities among species can give the wrong phylogeny if they are taken at face value. Among the species in Figure 16.2, all have an external tail except the human—but the tail is an ancestral character (shared with fishes!). It does not tell us that all the other species are more closely related to each other than to the human.

Our discussion of these few species and a few characters may convey the basic idea of how we can infer relationships. But in practice, the number of possible phylogenetic trees grows rapidly with the number of species. With 10 species, more than 34 million trees are possible, and with 52 species, the number of possible trees is larger than the number of protons in the universe! So even with the world's fastest computers, it is impossible to consider all possible trees. Sophisticated statistical methods have been developed to enable phylogenetic inference, even with hundreds of species and huge numbers of characters (variable sites in DNA sequences). But this is a mere technical challenge. The process of evolution itself poses difficulties that can make it hard to infer phylogenetic relationships among species.

Why estimating phylogenies can be hard

Most characteristics are not as complex as vertebrae, and are more likely to evolve more than once: we know that many characteristics do evolve repeatedly. (Think of black coloration in various snakes, birds, bears, and black widow spiders—see Figure 6.29.) That is certainly true of individual mutations as well. In practice, no biologist would base a phylogeny only on body color, or only on a single base pair difference among species. Repeated independent mutations and several other evolutionary phenomena can make it difficult to determine relationships and phylogenetic history. We review these difficulties before describing some phylogenetic methods that take them into account.

HOMOPLASY **Homoplasy** refers to the independent evolution of similar traits. It results from convergent evolution, parallel evolution, and **evolutionary reversal** (return to an earlier, ancestral character state) Homoplasy creates problems for estimating phylogenies because the similarity is not caused by shared ancestry, or *homology* (see Chapter 2).

Insect wings illustrate how homoplasy complicates building the phylogeny for a group of species (**FIGURE 16.3**). The bristletails (Microcoryphia) and silverfishes (Zygentoma) are wingless orders of insects that branched off from the lineage that later (about 300 million years ago [Mya]) evolved wings. All the other orders of insects are descended from a winged ancestor. In more than half of these orders, some lineages lost their wings secondarily, and some orders, such as fleas (Siphonaptera) and lice (Phthiraptera), have no winged species at all. These groups have independently undergone reversal to the wingless state. Although lack of wings is a derived state in both fleas and lice, those two groups are not a clade. We know this because we have information about many other morphological characteristics (as well as DNA sequences) in many other kinds of insects. This information shows that lice are related to true bugs (order Hemiptera) and that fleas are related to flies (order Diptera). Likewise, wingless crickets and grasshoppers (order Orthoptera) are closely related to their winged counterparts. By including many characters of many species in an analysis, we avoid being misled by homoplasy.

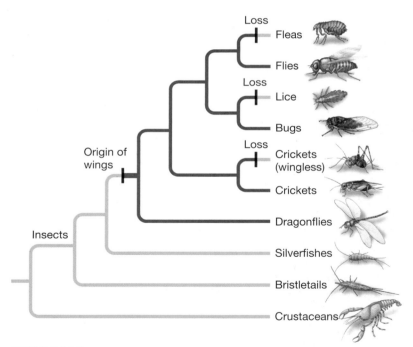

FIGURE 16.3 Winglessness in many insects has resulted from evolutionary reversals. The basal orders of insects, bristletails (Microcoryphia) and silverfishes (Zygentoma), are descended from the wingless ancestor of all insects, and have never had wings. The other orders of insects are descended from an ancestral insect that had wings. In many orders, however, some species have reverted to the wingless condition, such as some species of crickets and grasshoppers (order Orthoptera). Lice (order Phthiraptera) are entirely wingless but are related to winged sucking insects such as true bugs and cicadas (order Hemiptera). The entirely wingless fleas (order Siphonaptera) are related to true flies (order Diptera).

FIGURE 16.4 Convergent evolution: in many plant lineages, bilaterally symmetrical flowers have evolved from radially symmetrical ancestors. Bilateral symmetry is an adaptation to pollination by bees, which are more attracted to such flowers. Examples of three plant families with bilaterally symmetrical flowers (above) are paired with related families (below) that retain radial symmetry, the ancestral condition: (A) Orchidaceae (orchids) and (B) Liliaceae (lilies); (C) Fabaceae (pea family) and (D) Rosaceae (roses, cherries); (E) Violaceae (violet family) and (F) Passifloraceae (passionflower family).

Because many phenotypic traits are genetically variable within species (see Chapter 6), they can readily evolve by natural selection and genetic drift. It is not surprising that size, shape, coloration, and many other traits undergo **convergent evolution** (similar evolutionary changes) in diverse lineages (**FIGURE 16.4**). The same holds for mutations. Because there are only four possible states of a particular site in a DNA sequence, exactly the same mutation will occur repeatedly over sufficiently long periods of evolutionary time, and some mutations will be reversals to the ancestral state. For this reason, a single base pair difference among species provides little reliable evidence about their phylogeny. We require, instead, many differences. The more derived mutations there are that are shared between two species, the less likely it is that they all arose and were fixed twice. Furthermore, adding taxa to the analysis can often help us detect evolutionary reversals. Suppose that the only insects we included in the phylogeny in Figure 16.3 were fleas, dragonflies, silverfishes, and bristletails. We could err by supposing that fleas are more closely related to silverfishes or bristletails, because they all lack wings, than they are to dragonflies. But fleas share many features with some groups of winged insects, as shown in Figure 16.3; for example, fleas and flies both have complete metamorphosis (larval and pupal stages). Including flies and other insects shows us that fleas are not closely related to silverfishes, and also tells us that fleas reverted to the wingless condition from a winged ancestor.

If a site can undergo a substitution twice, it can do so again and again, over a sufficiently long time: it can change from A to C, then from C to T, and even back to A. Thus the number of differences between species may be less than the number

FIGURE 16.5 Different sites in the genome evolve at different rates. Shown here are the proportions of base pairs that differ in the DNA sequences of the mitochondrial gene *COI* between pairs of vertebrate species, plotted against the time since their MRCA (estimated from the fossil record). Sequence differences evolve most rapidly at third positions and most slowly at second positions within codons. Divergence at third positions increases rapidly at first, then levels off as a result of multiple substitutions at the same sites. Thus these positions provide no phylogenetic information for taxa that diverged more than about 75 Mya. More slowly evolving sites in the sequence (such as second positions) are useful for analyzing older relationships. (After [25].)

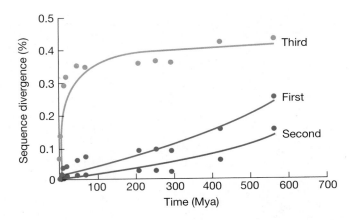

of substitutions that were fixed. As the time since divergence becomes greater, the number of differences begins to saturate, and the number of differences between groups of species that diverged further back in the past will be no greater than those that diverged more recently. One solution to this problem is to use different parts of the genome to estimate different parts of a phylogeny. Rates of molecular evolution vary among DNA sites, among genes, and among lineages of organisms [13]. In protein-coding sequences, for example, the third positions in codons evolve most rapidly, and the second positions most slowly (**FIGURE 16.5**). The reason is that mutations at second positions inevitably cause amino acid substitutions, many of which are eliminated by purifying selection, whereas a large fraction of mutations in third positions are synonymous and selectively neutral (see Chapter 7). Likewise, some proteins evolve much faster than others, largely because of differences in purifying selection. We therefore use more rapidly evolving parts of the genome to estimate phylogenies in the recent past, and more slowly evolving parts for deeper evolutionary time.

RAPID DIVERSIFICATION   If several species arise from a common ancestor over a short time, it can be difficult to determine the phylogenetic relationships among them. A group of such species is called a *radiation* (see Chapter 2). Phylogenetic relationships among species in a radiation are difficult to determine for two related reasons. The simple and obvious reason is that during the short time between two successive speciation events, few new mutations are fixed.

The second reason is that **incomplete lineage sorting**, or **ILS**, may occur during rapid diversification. It is critical to distinguish between a *phylogeny* (the evolutionary relations among species) and a **gene tree** (the genealogical history of a group of gene copies at the same locus—see Chapter 7). Many loci across the genome are polymorphic when a speciation event happens. The gene trees for these loci are sometimes consistent with the species tree. But sometimes they are not: the copies of a gene at a locus sampled from the two most closely related species can have a MRCA further back in time than the copies sampled from more distantly related species. As a result, copies from the most closely related species are less similar—they differ by more mutations—than are copies sampled from more distantly related species (**FIGURE 16.6**). When we have data from multiple loci, the picture can be confusing: different genes suggest different phylogenies.

ILS can be surprisingly common among closely related species. **FIGURE 16.7** shows the gene trees for six different genes that were sequenced in a study of four closely related species of grasshoppers. Each gene displays ILS, with two or more gene lineages persisting from one speciation event through another. Although a single gene can disagree with other genes and with the species phylogeny, combining the information from all the genes yields a good estimate of the species

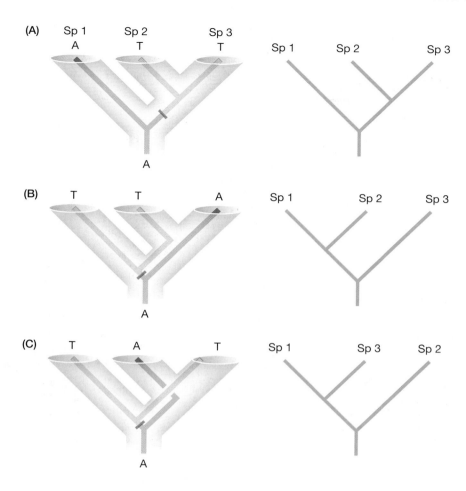

FIGURE 16.6 Incomplete lineage sorting can cause gene trees to have a different topology than the phylogeny of the species from which the genes are sampled. At left, the species trees (phylogenies) for three species are represented by the hollow branches, and the gene trees by the solid colored lines inside them. At right, the gene trees have been straightened out to show the relations between the species that they imply. (A) The ancestor of all three species was fixed for A at a site in the genome. After the first speciation event, a mutation from A to T occurs in one gene copy in the population (red bar). The copies inherited by species 2 and species 3 carry T. The gene tree agrees with the species tree. (B) At a second site in the genome, another mutation changes an A to a T. Copies with the T are inherited by species 1 and species 2. This produces a gene tree that does not agree with the species tree. (C) At a third site, yet another mutation changes an A to a T. This mutation is inherited by species 1 and species 3. The gene tree that results is again inconsistent with the species tree.

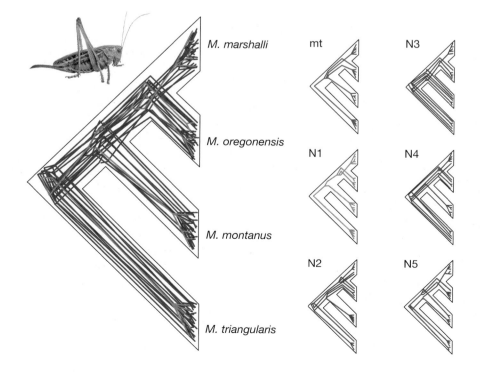

FIGURE 16.7 A phylogeny of four species of grasshoppers (*Melanoplus*) inferred from samples of five gene copies at each of six loci within each species. Gene trees for the mitochondria and five nuclear loci (at right) differ in many ways from the best estimate of the species phylogeny (black outer lines), indicating that each of these four species inherited several gene lineages from the common ancestor of all four species. The diagram at left shows all six gene trees nested within the species phylogeny. (From [4], courtesy of L. L. Knowles.)

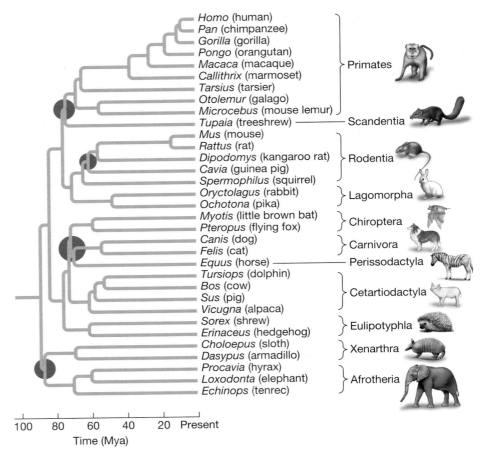

FIGURE 16.8 A phylogeny of many of the higher taxa of mammals, based on 20 million base pairs of genome sequence. Some lineages diverged before the mass extinction 66 Mya, at the end of the Cretaceous period. The red circles show nodes in the phylogeny where relationships among the lineages are uncertain, despite the immense amount of data. (After [32], based on [10].)

tree. Factors that make ILS more frequent, and thus more problematic for estimating phylogenies, are short intervals between speciation events and large population sizes.

Incomplete lineage sorting can be a problem for resolving relationships not only in recent radiations but also among lineages that evolved tens or even hundreds of millions of years ago, if the radiation occurred over a short time. **FIGURE 16.8** shows a phylogeny of many mammals based on DNA sequences from 14,632 genes, amounting to 20 million nucleotides [10]. Most of the relationships confirm earlier studies, and are almost certainly correct. But even genomic data of this scale leave several relationships uncertain (shown by the red circles). For example, the Chiroptera (bats), Carnivora (cats, bears, etc.), and Perissodactyla (horses, rhinoceroses, etc.) are certainly related to one another, but it is still not certain which two of them are closest relatives.

But even though some such cases are difficult, phylogenetic relationships can almost always be resolved with confidence if enough genomic data are used. The cichlid fishes of Lake Victoria in Africa are one of the most famous adaptive radiations. They have the highest speciation rate of any vertebrate group, with more than 450 species having originated in just the last 15,000 years (see

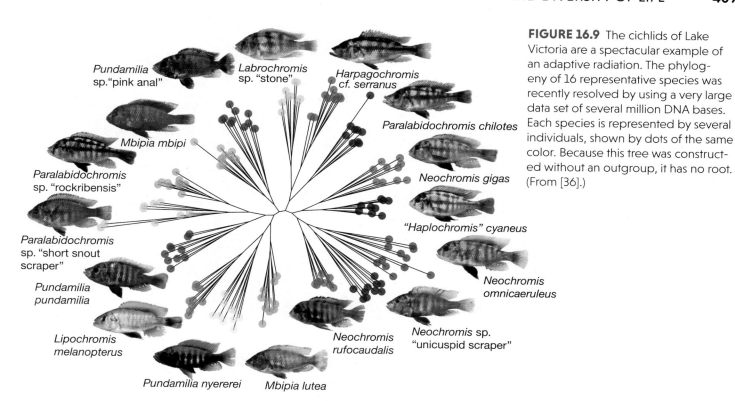

FIGURE 16.9 The cichlids of Lake Victoria are a spectacular example of an adaptive radiation. The phylogeny of 16 representative species was recently resolved by using a very large data set of several million DNA bases. Each species is represented by several individuals, shown by dots of the same color. Because this tree was constructed without an outgroup, it has no root. (From [36].)

Pundamilia sp. "pink anal"

Labrochromis sp. "stone"

Harpagochromis cf. *serranus*

Paralabidochromis chilotes

Mbipia mbipi

Neochromis gigas

Paralabidochromis sp. "rockribensis"

"*Haplochromis*" *cyaneus*

Paralabidochromis sp. "short snout scraper"

Neochromis omnicaeruleus

Pundamilia pundamilia

Neochromis sp. "unicuspid scraper"

Lipochromis melanopterus

Neochromis rufocaudalis

Pundamilia nyererei

Mbipia lutea

Chapter 9). ILS is rampant among them, and until recently it was difficult to determine their phylogeny with any confidence. Using several million base pairs of DNA sequences, however, the evolutionary relations among the species can now be resolved (**FIGURE 16.9**) [36].

INTROGRESSION A final culprit that makes estimating phylogenies difficult is the introgression of genes between different species. In eukaryotes, introgression happens most often by hybridization. Introgression also results from horizontal gene transfer (HGT) between even very distantly related prokaryotes, in which genes are exchanged by a variety of mechanisms (see Chapters 4 and 14). HGT may have been so extensive during the early evolution of prokaryotes that their evolutionary history more closely resembles braided hair than a simple branching tree (**FIGURE 16.10**).

Whatever the mechanism, introgression causes some regions of the genome to have an evolutionary history different from that of the species, and genes sampled from those regions will give a misleading picture of the species tree. An extreme case of introgression is seen in *Anopheles* mosquitoes, some of which transmit malaria [12]. Hybridization has caused extensive genetic mixing, and many parts of the genome have conflicting gene trees (**FIGURE 16.11**). Gene trees on the X chromosome are deeper than those in other parts of the genome, suggesting that they reflect the phylogeny of the species, while gene trees from other parts of the genome reflect more recent hybridization. In Chapter 21 you will see there has been hybridization in our own recent evolutionary past.

Methods for estimating phylogenies

Today, most phylogenies are based on DNA sequences. The data come from a number of genes or even the entire genome. How are these data analyzed?

FIGURE 16.10 An imaginative portrayal of horizontal gene transfer (HGT) during the early history of life. HGT makes the evolutionary relations among species resemble a network rather than a simple branching tree. The early evolution of the Bacteria and Archaea may have involved massive HGT, which continues at a lower rate today. HGT was also responsible for the origin of the eukaryotes and the origin of plants.

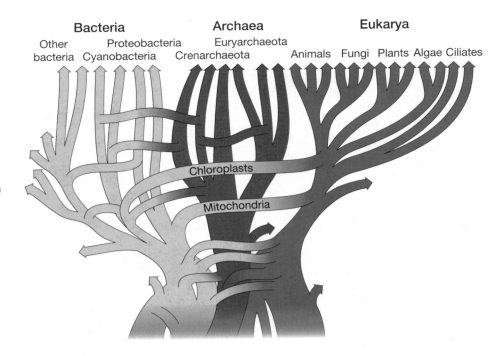

Many methods are used to estimate phylogenies from DNA sequences. The simplest is parsimony, which we described above. The example in **FIGURE 16.12** shows how parsimony works when the data are complicated by homoplasy. We wish to determine the relationships among species 1, 2, and 3. Their true evolutionary history is shown in tree A, but in reality, we do not know that. To estimate the phylogeny, we note that the sequences of the outgroup species suggest that the ancestral sequence was very likely AAA. Under the hypothesis that tree A is correct, at least 4 changes are needed to account for the sequences in the living species. One of the

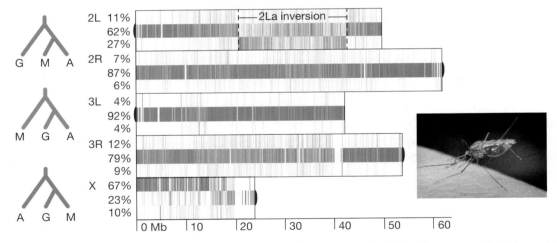

FIGURE 16.11 Gene trees from different parts of the genome give conflicting evidence about the phylogenetic relations of three mosquito species that are important malaria vectors. The different gene trees result from a long history of hybridization, as well as incomplete lineage sorting. Left: The three possible gene trees showing the relations among the species *Anopheles gambiae* (G), *A. melas* (M), and *A. arabiensis* (A). Right: The horizontal panels represent the major chromosome elements of the mosquito's genome: the left and right arms of chromosome 2, the left and right arms of chromosome 3, and the X chromosome.

Each chromosome is divided into many 10-kilobase (kb) windows, and the color of each vertical line shows the gene tree for a given window, with colors corresponding to the gene trees at left. On the left arm of chromosome 3, for example, 92 percent of the gene trees suggest that *A. gambiae* is most closely related to *A. arabiensis*. Gene trees on the X chromosome are deeper (older) than those on the autosomes, which strongly suggests that they reflect the history of speciation. The conclusion is that *A. melas* and *A. arabiensis* are sister species, as shown in the phylogeny at the top left. (From [12].)

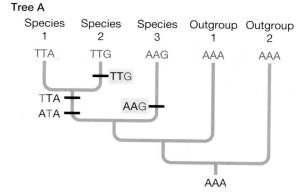

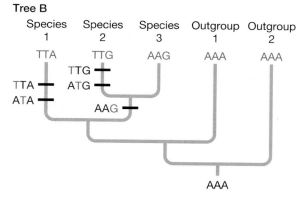

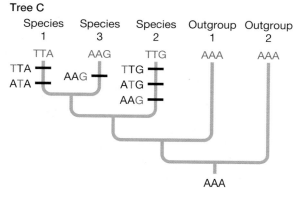

FIGURE 16.12 Parsimony is one of the methods used to estimate phylogenies. The example shown here uses data from three sites in the genomes of five species: the three species (1, 2, and 3) whose relationships are to be determined, and two outgroup species. The outgroups suggest that the ancestral DNA bases were AAA. Changes in the DNA bases that occurred during the evolution of these species are shown by the horizontal bars, with the bases that changed highlighted in blue. Tree A requires at least four changes to account for the data, tree B requires at least five, and tree C requires at least six. Parsimony identifies tree A as the best estimate of the phylogeny because it requires the fewest evolutionary changes.

scenarios that requires 4 changes is the true history shown in tree A. The first and second bases each changed once, but the third base changed 2 times; it independently evolved from A to G in both species 2 and 3. Under the hypothesis that tree B is correct, at least 5 changes must have happened, while tree C would require at least 6. Parsimony therefore identifies tree A as the best estimate of the phylogeny.

While parsimony often gives the correct result, it does a poor job of estimating the phylogeny in some situations, in particular when there is a lot of homoplasy on the phylogeny and when evolutionary rates vary among branches of the tree. Those problems motivate methods that use statistical approaches based on **likelihood**, which is described in the Appendix. These methods start with a set of assumptions about how the characters evolved. With DNA sequences, for example, we might assume that there was a constant substitution rate (a molecular clock; see Chapter 7). Probability theory can then be used to calculate the chance that the species would have the observed DNA sequences for a given order of branching events, set of branch lengths, and substitution rate. Finally, we search for the phylogeny and substitution rate that maximize the probability of the observed data. This is called the maximum likelihood estimate for the phylogeny. More details are given in **BOX 16A**. **Bayesian inference** is a related approach that allows other information (for example, dates from the fossil record) to be incorporated into the estimate (see the Appendix). It also can also estimate phylogenies that are difficult for likelihood because of their size.

BOX 16A

Estimating Trees with Likelihood

One widely used method for estimating phylogenies is based on likelihood. This is a general statistical approach that is described in the Appendix. Here we illustrate how likelihood is used to estimate a phylogeny with a very simple example. This is advanced material that can be considered optional.

The data are from the example used earlier to illustrate parsimony, with the DNA bases at three sites in the genomes of three species (see Figure 16.12). The first step is to calculate the probability that these data would be observed, given the phylogenetic tree. The second step is to find which of all possible phylogenies maximizes that probability, which gives us the maximum likelihood estimate for the phylogeny.

We begin by focusing on just the first of the three bases (**FIGURE 16.A1**). To find the likelihood, we need to make assumptions about how this base evolves. Here we make the simple assumptions that the probability that a substitution occurs (that is, one base replaces another) is constant in time and equal for all possible changes (for example, from C to G, or from A to T). For the moment, we will also assume that we know from using data from outgroups that the base in the MRCA of these three species was an A.

The likelihood of the data depends on three things: the topology (or branching order) of the tree, the lengths of the branches (measured in millions of years), and the substitution rate (that is, the probability per million years that one base will be replaced by another). The three possible topologies are shown in Figure 16.A1. For each of them, we find the lengths of the branches (t_1 and t_2) that make the data most likely. We then choose the topology that has the highest likelihood. For tree A, the likelihood turns out to be given by this rather intimidating equation:

$$L(t_1, t_2) = \frac{1}{64} \exp\left\{-\frac{4}{3}(2t_1, 3t_2)\lambda\right\} \tag{16.A1}$$
$$\times \left[3 + \exp\left\{\frac{4}{3}(t_1 + t_2)\lambda\right\}\right]$$
$$\times \left[3 + \exp\left\{\frac{4}{3}t_1\lambda\right\} - 2\exp\left\{\frac{4}{3}t_2\lambda\right\} + \exp\left\{\frac{4}{3}(t_1 + 2t_2)\lambda\right\} - 2\right]$$

Here λ is the substitution rate. In this example, we'll assume that we know $\lambda = 0.3$, for example from a molecular clock that has been calibrated for this gene in related spe-

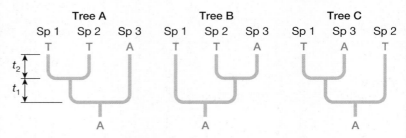

FIGURE 16.A1 The three possible topologies (shapes) for the phylogeny of three species and the base that they each have at a site in the genome. We assume from using data from outgroup species we know that the ancestor of the three species had an A, while species 1 and species 2 have a T. The time from the tree's root to the speciation event is t_1, and the time from that event to the present is t_2.

cies. Other equations (which look quite similar) give the likelihoods for trees B and C. We will not explain here how this equation was derived, but the interested reader can find a clear explanation on p. 194 of [15].

FIGURE 16.A2 shows a plot of the likelihood as the branch lengths of tree A are varied. The maximum value of the likelihood is reached when t_1 (the time from the root to the speciation event between species 1 and species 2) is 2.7 million years, and when t_2 (the time from the speciation event to the present) is 0. The estimate that $t_2 = 0$ makes sense: it implies that the MRCA of species 1 and 2 also had a T, and that there has been no time for either lineage to have a substitution since then.

By evaluating Equation 16.A1 with $t_1 = 2.7$ and $t_2 = 0$, we find that maximum value of the likelihood for tree A is 0.083. Doing similar calculations for trees B and C show that their maximum likelihoods are 0.016, which is about 5 times smaller than the value for tree A. The data therefore suggest that tree A is the actual phylogeny. We are not very confident in this conclusion, though. The difference in the two likelihoods is not statistically significant, which is not surprising since the data come from just a single DNA base.

With more data, however, we become more certain about the phylogeny. Data from two additional DNA bases are shown in Figure 16.12. To make use of them, we assume that the bases have evolved independently and with the same substitution rate. In that case, we can simply multiply the likelihoods for each base calculated from Equation 16.A1 to find the overall likelihood of a given phylogeny. The second base shows exactly the same evo-

BOX 16A

Estimating Trees with Likelihood (*continued*)

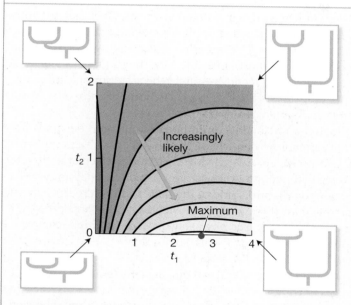

FIGURE 16.A2 Contour plot of the likelihood for different branch lengths (t_1 and t_2) of tree A, based on Equation 16.A1 and using a substitution rate of $\lambda = 0.3$/million years. Combinations of t_1 and t_2 values in the darker regions of the plot yield the lowest likelihoods; those in the lightest region yield the highest likelihoods. The trees in the four corners show the tree shapes for the corresponding values of t_1 and t_2. The maximum value of the likelihood occurs when $t_1 = 2.7$ million years and $t_2 = 0$ (indicated by the red circle).

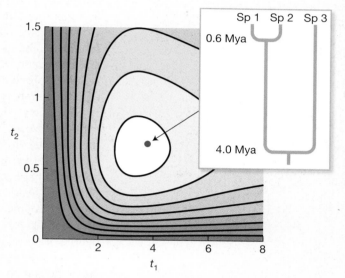

FIGURE 16.A3 Using all the data in Figure 16.12, maximum likelihood estimates that species 1 and species 2 are most closely related, that their MRCA lived 0.6 Mya, and that the MRCA of all three species lived 4 Mya.

lutionary pattern as the first, and so it also lends statistical support to tree A. But the third base is different: because of parallel evolution from A to G in the recent ancestors of species 2 and species 3, this base suggests that tree B is the most likely.

After combining the data from all three sites, we find that the likelihood is greatest for tree A, with $t_1 = 3.4$ million years and $t_2 = 0.6$ million years (FIGURE 16.A3). Thus the estimate for the age of the MRCA of species 1 and species 2 is $t_2 = 0.6$ Mya, and the MRCA of all three species (that is, the root of the tree) is $t_1 + t_2 = 4.0$ Mya.

This example is, of course, tremendously simplified. In practice, we would typically have data from many DNA bases. We would also make more realistic (but complicated) assumptions about evolution (for example, that rates of substitution differ among the four DNA bases, and among different sites in the genome), and we would also use the data to estimate those rates (rather than assuming their

values). Last, rather than assuming what the bases were in the ancestor, we would account for uncertainty in their states, for example by averaging over the probabilities that the ancestor had any one of the four bases at each of the three sites.

But even in those more complex settings, the basic approach is the same: we make assumptions about how evolution works, derive a function that gives the probability of our data given any specific phylogeny, and finally determine which phylogeny is most likely. Comparing these results with those we obtained from parsimony in the main text, we see some of the advantages and disadvantages of the two approaches. Among them, likelihood is able to estimate the ages of nodes in the phylogeny, but it requires us to make explicit assumptions about the evolutionary process and to carry out some moderately complicated calculations.

Likelihood and Bayesian inference have several advantages over parsimony. They are more robust to homoplasy; unlike parsimony, they estimate the lengths of the branches (in terms of time or the number of changes that occurred); they tell us the relative statistical support for different trees (rather than just identifying the most likely tree); and they can be used to simultaneously estimate other quantities (such as substitution rates). For those reasons, likelihood and Bayesian inference are the most widely used approaches for estimating phylogenetic trees from DNA sequences. But they also have limitations. Like any other method, both likelihood and Bayesian inference can give erroneous results if the assumptions are wrong. A second problem is that they are difficult to use with morphological data. Fortunately, all three methods give similar results in many cases.

Once we have estimated a phylogeny, an important question becomes how confident we are in that estimate. When we use likelihood or Bayesian inference, the relative confidence in two alternative phylogenies can be calculated directly. An alternative method called bootstrapping is often used with parsimony. The approach here is to randomly discard some of the data and then reestimate the phylogeny. After doing that many times, if we consistently get the same phylogeny, then we become confident that is the true phylogeny, because multiple, somewhat different data sets yield the same answer. Bootstrapping is often used to assess the degree of confidence in the individual branches of a tree.

There are several ways to test the validity of phylogenetic methods. One is to apply them to phylogenies that are known with certainty: evolutionary histories that have been simulated on a computer, allowing the lineages to branch and their characters to change according to various models of the evolutionary process. The investigators then see whether or not a phylogenetic method using the final characters of the simulated lineages gives an accurate history of their branching. Another test is to apply the method to data on experimental populations of real organisms that have been split into separate lineages by investigators (creating artificial branching events) and allowed to evolve. For example, David Hillis and coworkers successively subdivided lineages of T7 bacteriophage that accumulated DNA sequence differences rapidly over the course of about 300 generations [8, 18]. The investigators then scored the eight resulting lineages for sequence differences and performed a phylogenetic analysis of the data. For this many populations, there are 135,135 possible dichotomous trees (in which each lineage branches into two others), but the phylogenetic analysis correctly found the one true tree. Finally, throughout science, *the chief way of confirming a hypothesis is to see if it agrees with multiple, independent sources of data*. For phylogenetic hypotheses, these sources might be different, unlinked gene sequences, or morphological features and DNA sequences, which evolve largely independently of each other and thus provide independent phylogenetic information. These two kinds of data usually yield similar estimates of phylogeny. For instance, the phylogenetic relationships among higher taxa of vertebrates inferred from DNA sequences are almost all the same as those inferred from morphological features (**FIGURE 16.13**).

PHYLOGENIES FROM PHENOTYPES   Throughout most of the history of systematics, relationships were inferred using morphological data. The vertebrate phylogeny in Figure 16.13 is one of many examples that have largely stood the test of time and the arrival of new DNA sequence data. Even though DNA sequences are now the main source of phylogenetic data, there are situations when DNA sequences are not available and morphology is the only source of information. Nowhere is that more true than in understanding the relationships of fossilized species to one another and to living species. Careful study is needed to discriminate distinct characters and to distinguish derived from ancestral character states.

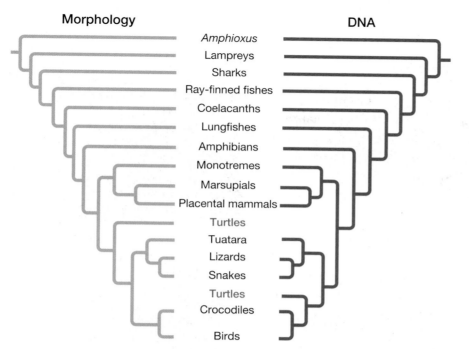

Morphology

Amphioxus
Lampreys
Sharks
Ray-finned fishes
Coelacanths
Lungfishes
Amphibians
Monotremes
Marsupials
Placental mammals
Turtles
Tuatara
Lizards
Snakes
Turtles
Crocodiles
Birds

DNA

FIGURE 16.13 Relationships among major groups of living vertebrates, as estimated from morphological characters (left) and from DNA sequences (right). On the whole, these two sources of information provide similar estimates of the phylogeny. The relationships of turtles were uncertain from morphological data but have been determined from genomic data. Among these taxa, the trees based on the two sources of data differ only with respect to the relationships of the turtles, which are shown in different positions (in blue) in the two trees. (After [7, 24, 34].)

For example, we noted at the start of this chapter that birds are dinosaurs: they are included in a dinosaur clade called the theropods (see Figure 16.1). Many derived characters unify the birds and other theropods, and they differ from those of any other tetrapod vertebrates [27]. The foot of birds and other theropods has the same structure: the fifth toe is absent and the first toe is rotated backward. The fibula is reduced to a thin splint. The pubis has a wide end and is directed backward. There are air sacs in some of the vertebrae, and the long bones are hollow. The clavicles are fused, forming the furcula ("wishbone"). The hand has only three digits. Among the theropods is the famous feathered dinosaur (or early bird), the Jurassic *Archaeopteryx*. Like other extinct theropods, such as *Deinonychus*, *Archaeopteryx* had teeth and a long tail. The hands of *Deinonychus* and *Archaeopteryx* were almost identical (**FIGURE 16.14**).

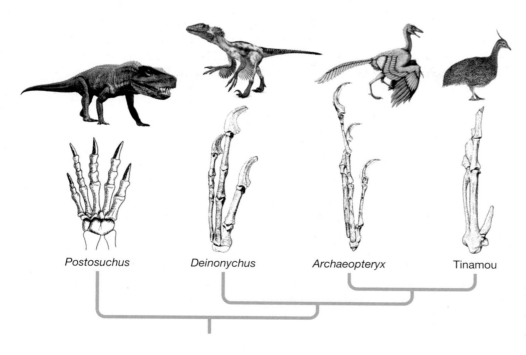

Postosuchus   Deinonychus   Archaeopteryx   Tinamou

FIGURE 16.14 Hand features shared by theropods, including the extinct dinosaurs and living birds. Shown are the left hand of the dinosaur *Deinonychus*, the feathered dinosaur *Archaeopteryx*, and a living bird (a tinamou), along with that of *Postosuchus kirkpatricki*, an extinct archosaur, as an outgroup. The three extinct species share several distinctive derived features: for example, they have the same number of digits and phalanges, the second digit is longest, and the third (leftmost) digit is twisted toward the second digit. The hand of the modern bird is highly modified for flight but retains these features. (*Postosuchus* hand from [5]; other hands from [35].)

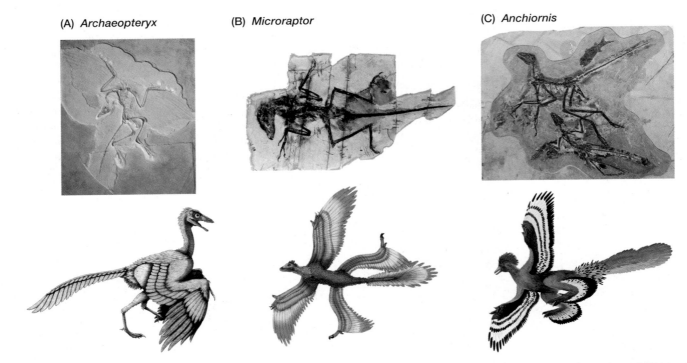

(A) *Archaeopteryx* **(B)** *Microraptor* **(C)** *Anchiornis*

FIGURE 16.15 Close relatives of birds among the feathered dinosaurs. (A) The famous specimen of *Archaeopteryx litho-graphica*, showing the wing feathers and the long bony tail with feathers on both sides (150 Mya). (B) *Microraptor gui* was a four-winged dromaeosaur, closely related to *Velociraptor*, that had long feathers at the back of both the forelimbs and hindlimbs (120–110 Mya). (C) *Anchiornis huxleyi*, from about 155 Mya. The colors in the painting of *Archaeopteryx* are drawn from the art-ist's imagination, but the colors shown in the reconstructions of *Microraptor* and *Anchiornis* are based on analysis of pigment-containing organelles in the fossilized feathers. (B, fossil photo © Mick Ellison, American Museum of Natural History.)

Finally, there are feathers. Since 1996, researchers have described an astonish-ing variety of feathered dinosaurs from China (**FIGURE 16.15**). In some species, the feathers are filaments that coat the body, and likely provided insulation and were used for display rather than flight. At least two extraordinary four-winged dino-saurs, *Microraptor gui* and *Anchiornis huxleyi*, had long feathers on all four limbs. Feathers may have characterized the entire theropod clade (perhaps even *Tyranno-saurus*). All these and many other features have enabled paleontologists to propose phylogenetic relationships among dinosaurs, including the birds.

How Do We Use Phylogenies?

The most basic reason for wanting to know how species are related is simply that a phylogeny shows the family tree of life on Earth. But there are many other moti-vations as well. In the following sections you will see how phylogenies are used to date events in the evolutionary past, to study how genes (and even human cul-tures) have evolved, to study adaptation, and to classify groups of species.

Dating evolutionary events

An important use of phylogenies is that they can tell us when some evolution-ary changes happened—if DNA sequence differences among organisms or genes more or less conform to a molecular clock. We introduced this idea in Chapter 2, and saw in Chapter 7 why genetic differences between two species can accumu-late at a roughly constant rate.

Two major tests for constancy have been used. One is to plot sequence differences between pairs of species (e.g., human, mouse) against the time since the lineages (primate, rodent) diverged (see Figure 2.17). The earliest fossil member of either of the two lineages gives the minimal divergence time. (The lineages are almost certainly older than their earliest fossil.) It may also be possible to determine whether sequence evolution is fairly constant even without information on divergence time.

Walter Fitch suggested a second method for determining constancy, called the **relative rate test** [11]. We know that the time that has elapsed from any common ancestor (i.e., any branch point on a phylogenetic tree) to each of the living species derived from that ancestor is exactly the same. Therefore, if lineages have diverged at a constant rate, the number of changes along all paths of the phylogenetic tree from one descendant species to another through their common ancestor should be about the same (**FIGURE 16.16**).

Fossil-based tests and relative rate tests, when applied to DNA sequence data from various organisms, have shown that rates of sequence evolution of a given gene are often quite similar among taxa, especially if they are fairly closely related, but sometimes they do differ considerably [13, 21]. For example, sequence evolution has been slower in hominoid primates (apes, including humans) than in other primates, in primates than in rodents, and in trees and shrubs than in herbaceous plants. Why do their rates differ? One hypothesis is that sequence evolution is faster in species with short generation times, and therefore more generations per unit of time. This hypothesis assumes that inherited mutations occur in cells destined to give rise to gametes (i.e., the germ line) only during DNA replication. This idea applies to animals, which have a distinct germ line, but might not apply to plants, which may produce flowers from various somatic tissues in which mutations may accumulate with cell division [13].

Despite these complications, divergence times in phylogenies can be reasonably well estimated, given enough sequence data. One study of divergence times was based on a DNA sequence of 59,764 base pairs in 13 species of primates and 6 other (outgroup) mammals, and the rate of sequence evolution was calibrated by four fossil-based divergence times [33]. This analysis suggested that the human-chimpanzee divergence was about 6.6 Mya (range 7.0–6.0), and that this branch diverged from gorilla about 8.6 Mya (range 9.2–7.7) (**FIGURE 16.17**). A more recent study was based on vastly more data—whole-genome sequences—and it used data on human mutation rates to estimate divergence time. (Recall from Chapter 7 that neutral mutations, occurring at a rate μ_n per site, are expected to generate a difference in DNA sequence equal to $2\mu_n t$ for two lineages that separated from their common ancestor t generations ago.) These authors estimated divergence at about 6 Mya for human and chimpanzee and at 10 Mya for divergence of these from gorilla [30]. The two estimates of divergence time use independent methods, and given all the room for error, they are quite close.

Discovering the history of genes and cultures

Almost anything that has the properties of inheritance and variation can be studied with phylogenetic methods. Gene trees can be used to address a wide range of evolutionary questions. For example, an important question is how often adaptation is based on a supply of new mutations (and may therefore be limited by that supply), versus being based on standing genetic variation [1]. If a mutation is beneficial when it first appears and then sweeps rapidly through a population, it generates shallow gene trees in a large region of the chromosome that hitchhikes along with the mutation. Because fewer mutations have had time to accumulate on shallow gene trees, shallow gene trees are visible because they have reduced polymorphism. An example is shown in Figure 5.16. Other mutations are not

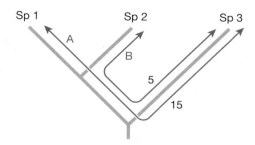

FIGURE 16.16 The relative rate test for the constancy of molecular evolution. Suppose we are confident in the phylogeny of three species, and we compare differences among their DNA sequences. Say that species 1 and species 3 differ by 15 mutations, but species 2 and species 3 differ by only 5 mutations. Then species 1 must have accumulated 10 more mutations along branch A than species 2 did along branch B, because the path between these two species and species 3 is identical except for their separate paths since their common ancestor. Therefore, the rate of molecular evolution was faster in the species 1 lineage than in the species 2 lineage. We can reject the hypothesis of a molecular clock for this gene in these species.

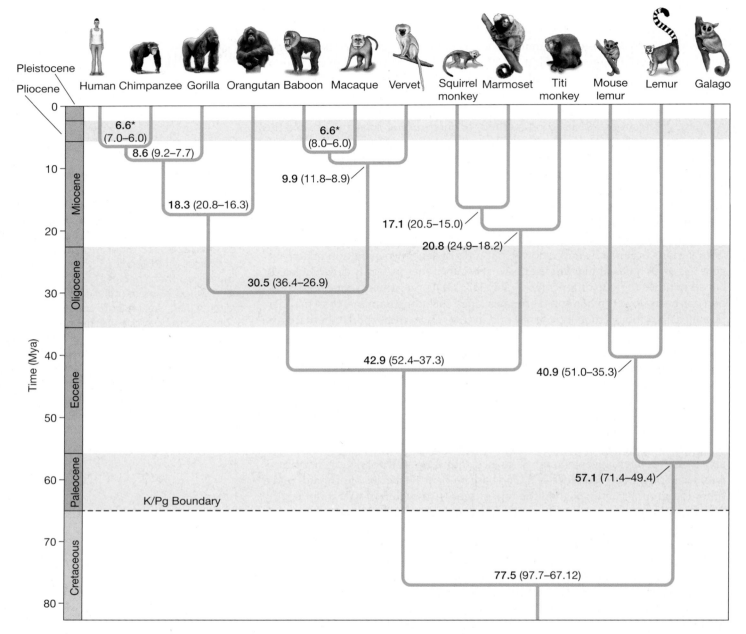

FIGURE 16.17 Estimated divergence times for some lineages of primates, based on maximum likelihood estimates of branch lengths and calibration of the rate of sequence evolution using several fossils. Asterisks denote calibrated nodes (branching points). The best estimates of divergence time are in boldface. The values in parentheses show the error, or range of likely divergence times. (After [33].)

immediately favorable, but become advantageous and sweep through the population later, for example when selection pressures change. These mutations affect the gene trees on a smaller region of the chromosome (compare Figures 5.15 and 5.17) [29]. So clues about whether adaptation resulted from new mutations versus standing genetic variation can be gleaned from the gene trees along a chromosome.

The three-spined stickleback (*Gasterosteus aculeatus*) has independently invaded thousands of rivers and streams from marine populations in the northern Pacific and Atlantic oceans. In most of the freshwater populations, bony armor plates on the side of the body have been greatly reduced, a change that has been traced to the *Ectodysplasin* (*Eda*) locus (**FIGURE 16.18**). In samples from diverse populations, most genes have gene trees that cluster by geographic region, as we would expect. But the *Eda* sequences in almost all fish with reduced (low) plates, whether they are from Atlantic or Pacific coastal regions, form a single branch on the gene tree. This adaptation, then, is based on an allele that has been present in both Atlantic and Pacific marine populations, and has increased in frequency in many different

FIGURE 16.18 Gene trees provide evidence that adaptation in the three-spined stickleback (*Gasterosteus aculeatus*) is based on standing genetic variation. (A) Stained specimens show the ancestral, high-plated morph, found in marine and some freshwater populations, and the low-plated morph found in many freshwater populations in northern Eurasia and North America. The low-plated phenotype is caused by an allele of the *Ectodysplasin* (*Eda*) gene, which encodes a signaling protein that is required for differentiation of ectodermal features. The gene tree of *Eda* sequences at left shows that gene copies from all high-plated fish form one clade (red) and that those from all low-plated fish (blue) form another. This shows that all the low-plated copies have a single origin, so copies of a low-plated allele must have been present at low frequency in ancestral marine populations throughout the northern oceans. The abbreviations designate collection localities. (B) A phylogeny of the populations based on single-nucleotide polymorphisms (SNPs) at 25 other loci shows that several low-plated populations (blue) have evolved independently. (After [6]; photos from [2].)

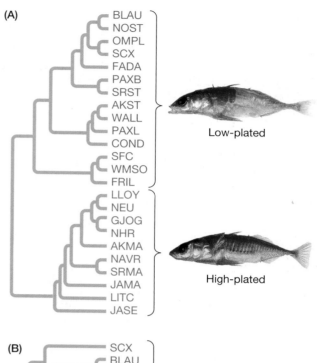

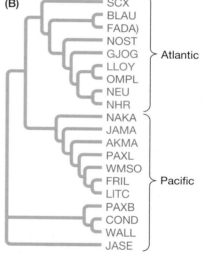

freshwater populations [6]. *Eda* is a classic example of adaptation based on standing genetic variation: the gene tree of *Eda* is old, even in populations that recently adapted to life in fresh water. Much of the genetic divergence between marine and freshwater populations of this species shows a similar pattern [20].

Darwin recognized that languages have diverged from common ancestors, and linguists have recently borrowed phylogenetic methods from biology to trace language histories. Language trees, in turn, have been used to trace the history of other aspects of culture in these populations, in the same way that biologists use DNA-based phylogenies to infer the evolutionary history of organisms' characteristics [23]. For example, diverse societies in the Austronesian language family originated in Taiwan more than 5000 years ago, and spread via the islands of Southeast Asia throughout the Indian and Pacific oceans, from Madagascar to New Zealand and Hawaii. Using a phylogeny based on languages in 84 societies, Thomas Currie and colleagues tested several models of changes in these cultures' political organization (**FIGURE 16.19**) [9]. They concluded that political organization increases in complexity by small incremental changes, but can decrease by either small or large steps. Because cultural inheritance follows different rules than genetic inheritance, using phylogenetic methods to study the evolution of language and culture must be done carefully—but it is exciting.

Reconstructing ancestors

One of the most important uses of phylogenies is that they can enable us to trace the evolution of organisms' characteristics (see Chapter 2). For example, all terrestrial mammals are quadrupedal (walk on all four legs) except humans, the one bipedal twig on the many branches of the mammalian tree. It is far more parsimonious to suppose that the common ancestor of mammals was quadrupedal than to suppose that it walked on two legs and gave rise to a great radiation of quadrupedal descendants, among which one reverted to the bipedal condition. In fact, we can confidently say that all the common ancestors of the many clades of mammals were quadrupedal, except the common ancestor of humans and extinct hominins such as *Australopithecus*. That is, we can mentally reconstruct the state of a character in extinct ancestors. You have encountered many examples in past chapters; for example, we noted in Chapter 15 that primate ancestors of humans had tails and

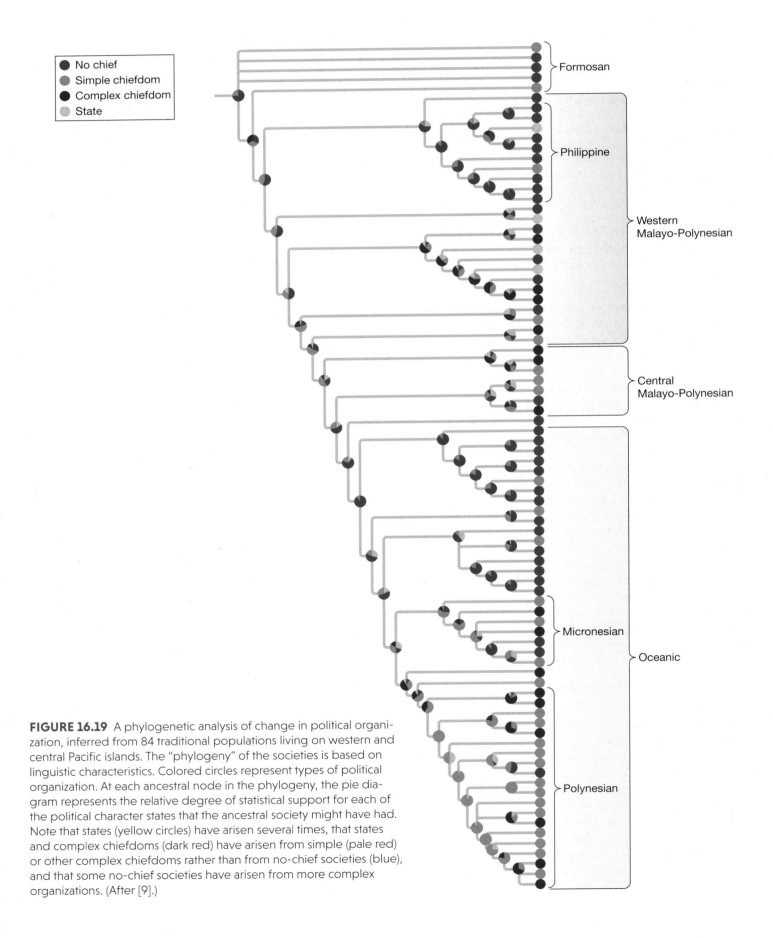

FIGURE 16.19 A phylogenetic analysis of change in political organization, inferred from 84 traditional populations living on western and central Pacific islands. The "phylogeny" of the societies is based on linguistic characteristics. Colored circles represent types of political organization. At each ancestral node in the phylogeny, the pie diagram represents the relative degree of statistical support for each of the political character states that the ancestral society might have had. Note that states (yellow circles) have arisen several times, that states and complex chiefdoms (dark red) have arisen from simple (pale red) or other complex chiefdoms rather than from no-chief societies (blue), and that some no-chief societies have arisen from more complex organizations. (After [9].)

that the ancestors of insects had appendages not only on the thorax but also on the abdomen.

Ancestral state reconstruction is being used to estimate and then synthesize ancestral DNA sequences. The function of the encoded proteins is analyzed, as well as that of proteins that differ by one or more amino acid changes, in order to infer how the function evolved [16]. One example comes from research on opsins, which are proteins involved in vision [38–40]. Vertebrates have several opsins that differ in the wavelength to which they are most sensitive, λ_{max}. Many species, such as the conger eel (*Conger myriaster*), have adapted to dim light, such as in deep water, by substitutions in the opsin genes that shift λ_{max} toward absorption of blue (shorter wavelengths). Some of the same amino acid substitutions, such as A292S (denoting a change from alanine to serine at position 292 in the protein) and D83N (a change from aspartic acid to asparagine at position 83), have repeatedly contributed to this shift (**FIGURE 16.20**). A292S is one of the substitutions in an opsin in the conger eel, with λ_{max} = 486 nanometers (nm). Shozo Yokoyama and colleagues inferred and synthesized the sequence of the ancestral gene, expressed the opsin protein it encoded, and found that the opsin's maximal absorption was at a longer wavelength (λ_{max} = 501 nm). By introducing the mutations in this ancestral sequence to match the conger eel opsin gene, the researchers found that they could recreate the same function (λ_{max} = 486 nm), but only by combining A292S with two other substitutions that also occurred in the conger eel; by themselves, these changes did not change λ_{max}. Apparently, the conger eel's adaptation to dim light is based on epistasis, or synergism, among the three amino acid substitutions. This important conclusion would be difficult to discover except by reconstructing the history of evolution.

The geographic distribution of a population or species can be considered a characteristic, and ancestral state reconstruction can thus trace evolutionary changes in geographic distribution. We will plumb this topic more deeply in Chapter 18. Here we recall an important example with which we began this book: tracing the spread of the deadly Ebola virus. The virus was first described in 1975, in the Democratic Republic of Congo. Twenty-four localized outbreaks occurred during the next 37 years. Then in 2014, a devastating outbreak ravaged West Africa: among more than 26,000 cases, there were more than 11,000 deaths. Viral gene sequences were obtained from the new infections and from samples that had been preserved from earlier outbreaks. By September 2014, evolutionary biologists and epidemiologists had estimated a gene tree of the virus (**FIGURE 16.21**). Most of the West Africa samples were traced to a single common ancestral sequence that was introduced into Sierra Leone from Guinea. This, in turn, was derived from Congo and nearby Gabon.

The key insight was that all the epidemics resulted from a single human infection. This shows that the virus is not easily acquired from the environment, unlike other viruses (such as influenza) that frequently jump between species. That finding, based on a phylogenetic analysis, has dramatic implications for how we might prevent future epidemics.

Studying adaptations: The comparative method

Evidence of convergent evolution has long been seen as a clue to how natural selection has shaped an organism's characteristics. For example, Bergmann's rule is the tendency for populations of many species of mammals and birds in colder climates to have a larger body size than populations of the same species

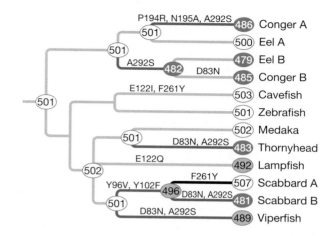

FIGURE 16.20 Evolutionary changes in one of the vertebrate rhodopsins (visual pigments), mapped onto a gene tree. These pigments are adapted for surface, intermediate, or deep-sea light environments, shown by the white, light gray, and dark gray ovals. The number in each oval is the wavelength of maximal absorption, which has been reconstructed for common ancestors, shown at the internal nodes. The wavelength of maximal absorption has become lower in species shown by blue branches. The amino acid substitutions that have occurred along each branch are indicated above the branch. Adaptation of these pigments for vision in dim light has evolved repeatedly, and certain amino acid substitutions have repeatedly played a role. For example, the substitution D83N has occurred on several of the blue branches, and A292S has occurred on all of the branches leading to the dark gray ovals. The A and B pigments of conger and eel are products of paralogous genes. (After [37].)

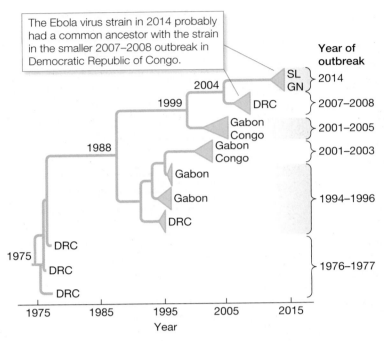

The Ebola virus strain in 2014 probably had a common ancestor with the strain in the smaller 2007–2008 outbreak in Democratic Republic of Congo.

FIGURE 16.21 A gene tree for Ebola virus genotypes in dated outbreaks in West Africa, with dates of common ancestors derived by phylogenetic estimation procedures. The virus strain in the devastating outbreak of 2014 probably had a common ancestor in 2004 with the strain in the smaller 2007–2008 outbreak in the Democratic Republic of Congo (DRC). The other countries that suffered outbreaks were Sierra Leone (SL), Guinea (GN), Gabon, and Congo. (After [14].)

in warmer places (see Figure 8.2). This is considered an adaptation for conserving heat, because larger individuals have a lower ratio of surface area to volume. In Chapter 2 we described several groups of birds, such as hummingbirds and sunbirds, that have independently evolved long slender bills for obtaining nectar from long tubular flowers (see Figure 2.22). Convergent evolution is the foundation of the *comparative method*, which consists of comparing sets of species to test hypotheses about adaptation.

As we saw in Chapter 3, we can often infer the selection pressures responsible for a feature that has evolved independently in many lineages by determining what ecological or other factors are correlated with that trait. We think long slender bills are adaptations for feeding on nectar because that is what these several groups of birds do. The comparative method is also used to test a priori hypotheses (those that are developed before the data are analyzed). For example, in species in which a female mates with multiple males, the several males' sperm compete to fertilize eggs. Our understanding of sexual selection leads to a prediction: males that produce more sperm should have a fitness advantage. In Chapter 3, we saw that in primates, the quantity of sperm produced is correlated with the size of the testes, and that as our hypothesis predicts, males in polygamous species have larger testes, relative to body mass, than males in monogamous species (see Figure 3.20).

One important use of the comparative method is to determine if two features tend to evolve together, which might suggest that having one feature favors the evolution of the second, or that both features are adaptations to the same environmental variable. Do warm-blooded animals tend to have larger brains, perhaps because higher metabolic rates allow growth of more nervous tissue? To answer questions such as that, we need to be wary of a potential complication: the phylogenetic relations of the species. To see why this is so, consider the correlation between the ability to fly and the mode of reproduction in birds and mammals. Most birds fly and all lay eggs, while most mammals do not fly and most give live birth. Does that suggest that those two traits are evolutionarily linked because of some adaptive reason? Almost certainly not. The correlation here results because the ancestor of all birds laid eggs and flew, while the ancestor of all marsupials and placental mammals gave live birth and did not fly. The many species of living birds and mammals have those same characters simply because they inherited them from their ancestors. In statistical terms, we would say that the various species are not independent data points because of their shared evolutionary history.

We therefore need to test for adaptive coupling of characters during evolution, while controlling for phylogenetic relations. One way to do that uses ancestral state reconstruction. The first step is to estimate the states for two characters at each node (or branch point) in the phylogeny. We then ask as we move along each branch of the tree, is a change in the state of one trait correlated with a change in the second? More sophisticated versions of this approach take into account uncertainty in the phylogeny and in the ancestral states. When Paul Harvey and collaborators analyzed the primate data this way, they concluded that there had been many shifts

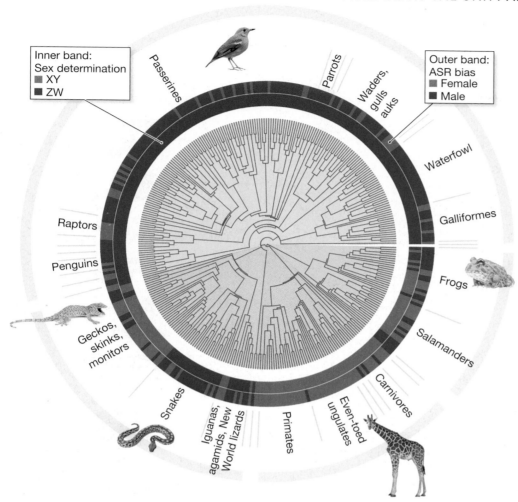

FIGURE 16.22 Comparative analysis of the evolutionary relation between sex determination and adult sex ratio in tetrapods (mammals, birds, lizards, snakes, and amphibians). The inner circle shows the phylogeny that relates the 344 species around the edge. The inner colored band shows whether sex in each species is determined by XY or ZW chromosomes. The outer colored band shows whether the adult sex ratio is female-biased or male-biased. Visually, we see that independent origins of XY sex determination (in red) tend to correlate with a female-biased sex ratio (also in red). A statistical analysis that accounts for the phylogeny confirms that the trend is significant. (From [26].)

between polygamous and monogamous mating systems, and that testes size usually also changed (Harvey and Harcourt 1984). The analysis supported their hypothesis.

Sometimes, evolutionary associations discovered by this method prompt further research. The example in **FIGURE 16.22** reveals a surprising evolutionary correlation between sex determination and sex ratios. In all mammals, males have two different sex chromosomes (the X and Y), while females have two of the same (the X). Furthermore, in adults of most mammals, including humans, the sex ratio is female-biased (with more females than males). In birds, the situation is reversed: it is the females that have two different sex chromosomes (called Z and W), while males have two of the same (the Z). In adults of most birds, the sex ratio is male-biased. By themselves, these data are not convincing evidence for an evolutionary connection between sex chromosomes and sex ratio, for the same reason that we rejected a connection between flight and reproductive mode. But the situation changes if we include lizards and amphibians. In those groups, phylogenetic analysis shows that there have been many independent transitions in both the chromosomal mechanism of sex determination and the adult sex ratio. Statistical analyses that account for the phylogeny show there is a significant trend for XY species to have female-biased adult sex ratios, and ZW species to have male-biased sex ratios [26]. Why this correlation exists is uncertain. One of several hypotheses is that genes on the Y and W chromosome tend to degenerate. This might increase mortality of males in groups with XY sex determination (such as mammals), and of females in groups with ZW sex determination (such as birds). If correct, this hypothesis could help explain the shorter average life spans of human males.

FIGURE 16.23 Classification of some snakes and lizards, based on phylogeny. Squamata includes many lineages of lizards and the snakes (Serpentes). Lizards (sometimes called the "Lacertilia") do not form a monophyletic group, but snakes do. Among the snakes, the families Boidae, Viperidae, Elapidae, and Colubridae are monophyletic groups, each with many genera (only a few of which are shown here). Viperidae, for example, includes the European adder (*Vipera*) and the American rattlesnakes. The timber and western diamondback rattlesnakes are in a group of closely related species that form the genus *Crotalus*. (After [31].)

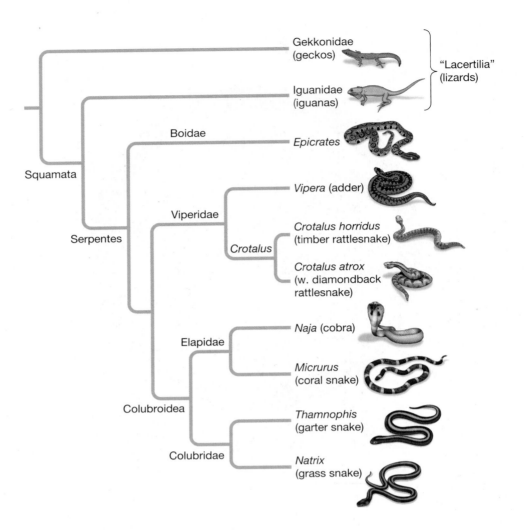

Classification

It is helpful, and often necessary, to name and classify any great variety of objects—be they vehicles, books, or rocks—if we are to think and talk about them. A simple classification of plants, for example, might include trees, shrubs, vines, and herbs; but close examination will show that each of these categories includes wildly different organisms that have in common only their overall form. Perhaps classification by different criteria would summarize more information about them. Carolus Linnaeus, who devised the scheme of classification that is still used today (i.e., a hierarchical classification of groups within groups), used features that he imagined represented propinquity in God's creative scheme. For example, he defined the order Primates by the features "four parallel upper front [incisor] teeth; two pectoral nipples," and on this basis included bats among the Primates. But without an evolutionary framework, naturalists had no objective basis for classifying mammals by their teeth rather than by, say, their color or size. Saying that some species were more closely "related" than others had a metaphorical, not a genealogical, meaning.

Darwin gave classification an entirely different significance. In *On the Origin of Species*, he wrote that when his views on the origin of species are adopted, the term "relationship" among species "will cease to be merely metaphorical" and "our classifications will come to be, as far as they can be so made, genealogies; and will then truly give what may be called the plan of creation."

Many systematists today are fulfilling Darwin's prophecy, by using new data to classify species phylogenetically. This usually means arranging them into

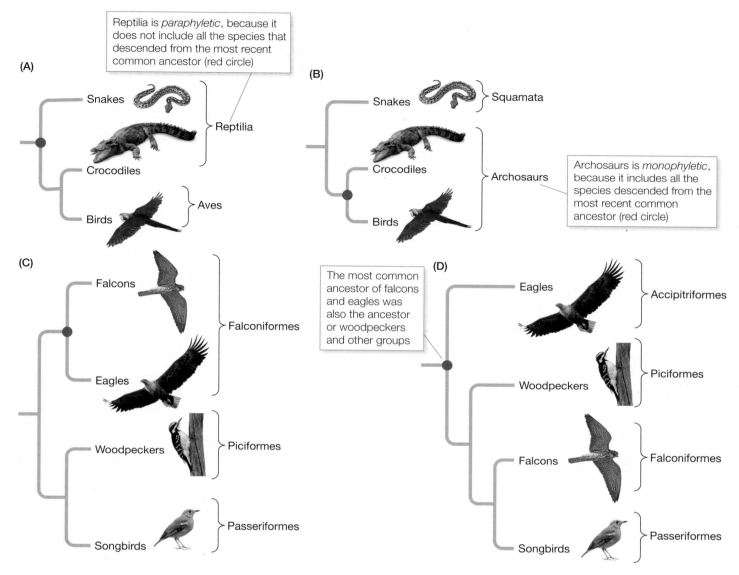

Reptilia is *paraphyletic*, because it does not include all the species that descended from the most recent common ancestor (red circle)

Archosaurs is *monophyletic*, because it includes all the species descended from the most recent common ancestor (red circle)

The most common ancestor of falcons and eagles was also the ancestor or woodpeckers and other groups

FIGURE 16.24 Modern evolutionary biologists prefer to recognize only monophyletic clades as higher taxa. (A) Traditionally, the class Reptilia included snakes and crocodiles but not birds, which were put in the class Aves. This made Reptilia *paraphyletic*, because it did not include all the species that descended from the MRCA of snakes and crocodiles (indicated by the red circle). (B) The names used now are for monophyletic groups: Squamata includes snakes and lizards, and archosaurs include crocodiles and birds. (C) Falcons and eagles were previously thought to be each other's closest relatives and were grouped together in the order Falconiformes. (D) DNA-based studies now show that falcons are more closely related to songbirds than to eagles. The MRCA of falcons and eagles (red circle) was also the ancestor of woodpeckers and other groups. The earlier classification that grouped falcons and eagles together made Falconiformes a *polyphyletic* group because it included species that descended from two distinct ancestors. In current classifications, the orders Falconiformes (falcons) and Accipitriformes (eagles and relatives) are monophyletic.

monophyletic taxa, which include an ancestor and all of its descendants (**FIGURE 16.23**). The Serpentes is a monophyletic group that includes all snakes; the Viperidae is a monophyletic group that includes all those venomous snakes with movable fangs that can be erected, such as vipers and rattlesnakes; the genus *Crotalus* includes most species of rattlesnakes. Some taxa in old classifications have proven not to be monophyletic, but instead were **paraphyletic** or **polyphyletic** (**FIGURE 16.24**). A paraphyletic taxon includes an ancestor and some, but not all, of its descendants. For example, the class Reptilia would be paraphyletic if it excluded birds. Lizards, formerly classified as order Lacertilia, are not a monophyletic group because "Lacertilia" excludes one branch of the lizard tree, the snakes (see Figure 16.23). A polyphyletic taxon includes species from two or more different ancestors, but excludes other descendants that are placed in different taxa. The falcons, hawks, and eagles have similar adaptations for predation, such as a short, hooked bill and strong grasping feet with sharp, curved claws. Because of those shared characters, they were long classified together as the order Falconiformes. However, it has recently become clear from DNA evidence that these similarities result from convergent evolution. Falcons are more closely related to parrots and songbirds, and the hawks and eagles are a distantly related clade, which has been named the order Accipitriformes. Falconiformes, as previously used to include both falcons and eagles, was a polyphyletic order [19, 28]. It is now used to refer only to falcons, a monophyletic group.

Although most evolutionary biologists prefer to give names to taxa that are monophyletic, informal names are sometimes used to refer to paraphyletic assemblages of species. Eukaryotes are a monophyletic group, but "prokaryotes" are paraphyletic because eukaryotes are nested among them. Despite this phylogenetic situation, which was only recently discovered, many biologists continue to use the familiar term "prokaryote" to contrast these organisms with eukaryotes.

No system of classification is without difficulties. The phylogenetic boundaries of a taxonomic rank, such as family, are often arbitrary. For example, the great apes include the Asian orangutans (*Pongo*) and the African gorillas, chimpanzee, bonobo, and human. A "splitter" might recognize two families, Pongidae for the orangutans and Hominidae for the others. A "lumper" might demote those to subfamilies, and combine them all into one family, Hominidae. (This scheme is widely used; see http://tolweb.org/Catarrhini/16293) Another issue is that a taxon composed of extinct species (e.g., Dinosauria) will be paraphyletic if it does not include living descendants of that group (e.g., birds), so the classification may not admit a formal name only for the extinct group. Often, the living members of a group (such as birds), together with their last common ancestor, are called a **crown group**, and the larger clade of related extinct lineages (e.g., the various theropod dinosaurs) is called a **stem group**. Despite these various complications, a genealogical classification is both meaningful and useful, for it usually conveys a lot of information. A zoologist, reading that a species is in the Coleoptera (beetles), immediately knows a great deal about the organism. A geneticist, learning that horseradish (*Armoracia*) is in the Brassicaceae along with *Arabidopsis*, will expect its genome to be rather familiar.

Go to the
Evolution Companion Website
EVOLUTION4E.SINAUER.COM
for data analysis and simulation exercises, quizzes, and more.

SUMMARY

- Phylogenetic relationships can be difficult to determine, and for this reason often require data on many characteristics, such as extensive DNA differences among species. One of the main reasons a phylogeny could be wrong is the repeated, independent evolution of a base pair or other character state by convergent, parallel, or reversed evolution.

- Phylogenies are especially difficult to determine if successive branching events were closely spaced in time, because few evolutionary changes are fixed during short intervals. A related problem is incomplete lineage sorting, resulting in gene trees that differ from the species tree that one may be trying to estimate. Yet another difficulty is introgression caused by hybridization (or horizontal gene transfer).

- A variety of methods are used to estimate phylogenies. The simplest is parsimony, a rule that chooses whichever phylogenetic tree requires the fewest evolutionary changes. Other methods choose among the different possible phylogenies based on their likelihoods or probabilities. Methods differ in their strengths and weaknesses, and in the kinds of data they can analyze. In many cases, different methods return very similar results.

- Branching events in phylogenies can often be dated approximately, using DNA sequence differences that approximately conform to a geologically calibrated molecular clock. The rate of sequence evolution varies among parts of the genome and among clades, and can sometimes vary within clades. Tests are always necessary to confirm rate constancy. The causes of rate differences among groups of organisms are uncertain.

- Phylogenies are useful for inferring histories of genes and other historical changes, such as in human cultures and languages.

- An important use of phylogenies is tracing the history of evolution of characteristics, through ancestral state reconstruction. This approach has been used to synthesize ancestral DNA sequences and proteins, to better understand how their functions have evolved.

- The comparative method uses convergent evolution to test hypotheses about adaptation. Statistical tools use the phylogeny to control for the effects of shared ancestry.

- In modern systematics, classification of organisms is based on their phylogeny. In an ideal classification, each named taxon is monophyletic, including all the species thought to be descendants of a single common ancestor. The classification consists of nested, named monophyletic groups. Such a classification reflects evolutionary history and usually conveys a great deal of information about the species.

TERMS AND CONCEPTS

ancestral
Bayesian inference
clade
convergent evolution
crown group

derived
evolutionary reversal
gene tree
homoplasy
incomplete lineage sorting (ILS)

likelihood
monophyletic group
most recent common ancestor (MRCA)
outgroup

paraphyletic
parsimony
polyphyletic
relative rate test
stem group
synapomorphy

SUGGESTIONS FOR FURTHER READING

Tree Thinking: An Introduction to Phylogenetic Biology, by D. A. Baum and S. D. Smith (Roberts and Company, Greenwood Village, CO, 2012), is a comprehensive introduction to the concepts, methods, and uses of phylogenetics in biology, for nonspecialists.

Molecular Systematics, edited by D. M. Hillis, C. Moritz, and B. K. Mable (Sinauer Associates, Sunderland, MA, 1996), is outdated in some

For deep coverage of phylogenetic analysis, see *Inferring Phylogenies*, by J. Felsenstein (Sinauer Associates, Sunderland, MA, 2004).

The most recent synthesis of phylogenetic studies into a single tree of life is "Synthesis of phylogeny and taxonomy into a comprehensive tree of life" by C. E. Hinchliff, S. A. Smith, J. F. Allman, and 19 others (*Proc. Natl. Acad. Sci. USA* 112: 12764–12769, 2015).

The following are among many review articles on certain aspects of phylogenetic methods and uses in biology:

Model selection in phylogenetics, by J. Sullivan and P. Joyce (*Annu. Rev. Ecol. Evol. Syst.* 36: 445–460, 2005).

Phylogenetic inference using whole genomes, by B. Rannala and Z.-H. Yang, (*Annu. Rev. Genomics Hum. Genet.* 9: 217–231, 2008).

Evolutionary inferences from phylogenies: a review of methods, by B. C O'Meara (*Annu. Rev. Ecol. Evol. Syst.* 43: 267–285, 2012).

High-throughput genomic data in systematics and phylogenetics, by E. M. Lemmon and A. R. Lemmon (*Annu. Rev. Ecol. Evol. Syst.* 44: 99–121, 2013).

Animal phylogeny and its evolutionary implications, by C. W. Dunn, G. Giribet, G. D. Edgecombe, and A. Hejnol (*Annu. Rev. Ecol. Evol. Syst.* 45: 371–395, 2014).

PROBLEMS AND DISCUSSION TOPICS

1. What is the evidence that incomplete lineage sorting (ILS) has affected DNA variation in humans, chimpanzees, and gorillas? How could the authors of the study described in Scally et al. 2012 (*Nature* 483 [7388]: 169–175) tell that ILS had occurred?

2. With improving technology, acquiring DNA sequences from different organisms becomes easier each year. With that in mind, some authors (e.g., Scotland et al. 2003, *Systematic Biology* 52 [4]: 539–548) have suggested that the use of morphological data is less important than DNA sequence data and have called for less emphasis on the use of comparative morphology in building phylogenies. Other authors maintain that despite the explosion of molecular data available, morphology still has an important role in phylogenetics (for examples, see Wiens 2004, *Systematic Biology* 53 [4]: 653–661 and Will and Rubinoff 2004, Cladistics 20 [1]: 47–55). What are the reasons for and against using morphological data in phylogenetic reconstruction? When might morphological data be especially important?

3. A heated debate arose in the mid-twentieth century: some systematists insisted on the identification of monophyletic groups to reconstruct phylogeny and use those for classification or taxa based on their relatedness. Others placed organisms into taxonomic groups based simply on overall morphological similarity (using algorithms in an attempt to remove subjectivity from classification). The major difference in the approaches is that the first uses only apomorphic (derived) characters in its analyses, whereas the second does not distinguish between apomorphic and plesiomorphic (ancestral) characters, because measuring total similarity is the goal. Discuss how this difference in use of characters might result in discrepancy in classification. Which approach is more common today?

4. If a branch on a phylogeny shows few changes in sequence, we can assume that changes are rare, so mutations are unlikely to affect the same nucleotide position more than once. If a branch is "long," with many changes to sequence, multiple mutations at the same nucleotide locus are more likely. How might this mislead researchers working on phylogenetic reconstruction? (Hint: one common complication of this type is called "long branch attraction.") How is this related to the saturation of the curve showing sequence divergence of the third positions in codons in Figure 16.5?

5. Parsimony, maximum likelihood, and Bayesian inference are different analytical techniques for developing phylogenies from DNA sequence data. Why would a researcher choose one method over another? What are the advantages and disadvantages of the three methods?

6. What should a biologist do if she finds that different methods of analyzing the same data (say, parsimony and maximum likelihood) provide different estimates of the relationships among certain taxa? What should she do if the different analytical methods give the same estimate, but the estimate differs depending on which of two different genes has been sequenced? (Hint: your answers do not depend on knowing how maximum likelihood works.)

7. Do the quandaries described in the previous question ever occur? Choose a group of organisms that interests you, find recent phylogenetic studies of this group, and see whether such problems have been encountered. (You can do this using key words, such as "phylogeny" and "[taxon name]," e.g., "[deer]," in any of several literature-search engines that your instructor can suggest.)

8. Phylogenetic reconstruction can be obscured by homoplasy, rapid diversification, and introgression. How can researchers identify those potential complications and ensure that their phylogenetic trees are robust to them?

9. Suppose species 1, 2, and 3 are endemic to a group of islands (such as the Galápagos) and are all descended from species 4 on the mainland (which will serve as an outgroup; its large population size means that no new mutations have become fixed in its population in the time since the islands were colonized). You sequence a gene and find ten nucleotide sites that differ among the four species (among many other loci that do not vary). The nucleotide bases at these sites are:

> Species 1: GCTGATGAGT
>
> Species 2: ATCAATGAGT
>
> Species 3: GTTGCAACGT
>
> Species 4: GTCAATGACA

Estimate the phylogeny of these taxa by plotting the changes on each of the three possible unrooted trees and determining which tree requires the fewest evolutionary changes.

The History of Life

If we could look at Earth 3,500,000,000 years ago, soon after life's beginning, we would find only bacterial cells. Among these cells would be our most remote ancestors, utterly different from ourselves. And if we then time-traveled through life's history toward the present, we would see played out before us a drama grander and more splendid than we can imagine: a planetary stage of many scenes, on which emerge and play—and then, most likely, die—millions and millions of species with features and roles more astonishing than any writer could conceive. In this chapter, we take that journey through time.

The history of evolution can be inferred in two ways that complement each other. As we have seen, phylogenetic inferences from living organisms can tell us a lot about when some lineages evolved and about patterns of change in their characters. But fossils provide direct evidence of some events that living organisms cannot reveal. Fossils tell us of the existence of innumerable creatures that have left no living descendants, of great episodes of extinction and diversification, and of the movements of continents and organisms that explain their present distributions. Without this record, we could not calibrate the speed of DNA divergence, for it is only from the fossil record that we can obtain an absolute time scale for evolutionary events, or evidence of the environmental conditions in which they transpired. The fossil record also provides many details of which we would otherwise have no knowledge. For instance, although we can infer some evolutionary changes in the human lineage from comparisons with other primates, fossils provide other information—such as the sequence of particular anatomical changes—that comparisons with other living species cannot provide.

A sagittal section through the chambered shell of *Cleoniceras*, an early Cretaceous ammonite. Larger chambers were formed as the animal's body grew in size. Ammonites—cephalopods related to squids—were extremely diverse in the Mesozoic, but became entirely extinct at the era's end.

As you read this chapter, you will see examples of some important patterns. Living things have profoundly affected the atmosphere and other physical aspects of Earth. Climates and the distribution of oceans and land masses have changed over time, affecting the geographic distributions of organisms (as we will discuss in Chapter 18). The taxonomic composition of the living world has changed continually as new forms have originated and others have become extinct. As new forms of life originated, the variety of ways of living increased, and a wider variety of habitats became occupied. In fact, the evolution of new kinds of organisms created new habitats. At several times, extinction rates have been particularly high. Especially after these **mass extinctions**, the diversification of higher taxa has sometimes been rapid. Extinct taxa have sometimes been replaced by unrelated but ecologically similar taxa. Of the variety of forms in a higher taxon that were present in the remote past, usually only a few have persisted in the long term, and over time, Earth's biota more and more resembles the biota of today. This magnificent history has transpired over a depth of time that we find hard to comprehend.

Some Geological Fundamentals

The rocks we find at Earth's surface originated as molten material (magma) that is extruded from deep within Earth. Some of this extrusion occurs via volcanoes, but much rock originates as new crust forms at mid-oceanic ridges (**FIGURE 17.1**). Rock formed in this way is called igneous rock. Sedimentary rock is formed by the deposition and solidification of sediments, which are usually formed either by the breakdown of older rocks or by precipitation of minerals from water. Under high temperatures and pressures, both igneous and sedimentary rocks are altered, forming metamorphic rocks. Most fossils are found in sedimentary rocks. A few fossils are found in other situations; for example, insects are found in amber (fossilized plant resin), and some mammoths and other species have been found frozen in permafrost.

The lithosphere, the solid outer layer of Earth bearing both the continents and the crust below the oceans, consists of eight major and several minor plates that

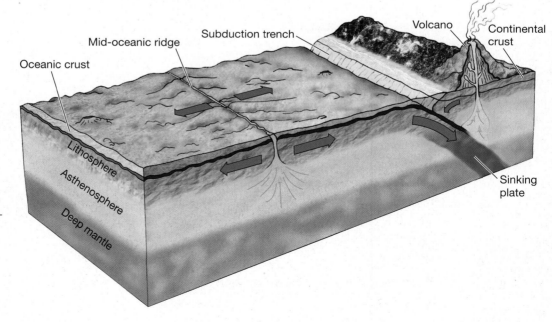

FIGURE 17.1 Plate tectonic processes. At a mid-oceanic ridge, rising magma creates new lithosphere and pushes the existing plates to either side. When moving lithospheric plates meet, one plunges under the other, frequently causing earthquakes and mountain building. Heat generated by this process of subduction melts the lithosphere, causing volcanic activity.

move over the denser, more plastic asthenosphere below. Because the heat of Earth's core sets up convection currents within the asthenosphere, magma from the asthenosphere rises to the surface, cools, and spreads out to form new crust, pushing the existing plates to either side. The plates move at velocities of 5–10 cm per year. Where two plates come together, the leading edge of one may be forced to plunge under the other, rejoining the asthenosphere (subduction). The pressure of these collisions is a major cause of mountain building. When a plate moves over a "hot spot" where magma is rising from the asthenosphere, volcanoes may be born, or a continent may be rifted apart. The Great Lakes of eastern Africa lie in such a rift valley; the Hawaiian Islands are a chain of volcanoes that have been formed by the movement of the Pacific plate over a hot spot (see Chapter 18).

The absolute ages of geological events can often be determined by **radiometric dating**, which measures the decay of certain radioactive elements in minerals that form in igneous rock. (Carbon-14 is used for dating biological materials, such as wood or bone, that are no older than about 75,000 years.) The probability that a radioactive parent atom (e.g., uranium-235) will decay into a stable daughter atom (lead-207) is constant over time. As a result, each element has a specific half-life. The half-life of U-235, for example, is about 0.7 billion years, meaning that in each 0.7-billion-year period, half the U-235 atoms present at the beginning of the period will decay into Pb-207. The ratio of parent to daughter atoms in a rock sample thus provides an estimate of the rock's age. Only igneous rocks can be dated radiometrically, so the age of a fossil-bearing sedimentary rock must be estimated by dating igneous formations above or below it.

Long before radioactivity was discovered—indeed, before Darwin's time—geologists had established the relative ages (i.e., earlier vs. later) of sedimentary rock formations by applying the principle that younger sediments are deposited on top of older ones. Layers of sediment deposited at different times are called **strata**. Different strata have different characteristics, and they often contain distinctive fossils of species that persisted for a short time and are thus the signatures of the age in which they lived. Using such evidence, geologists can match contemporaneous strata in different localities. In many locations, sediment deposition has not been continuous, and sedimentary rocks have eroded; thus any one area usually has a very intermittent geological record, and some time intervals are well represented at only a few localities on Earth. In general, the older the geological age, the less well it is represented in the fossil record because erosion and metamorphism have had more opportunity to take their toll.

Most of the eras and periods of the *geological time scale* (**TABLE 17.1**) were named and ordered before Darwin's time. These geological eras and periods were distinguished, and are still most readily recognized in practice, by distinctive fossil taxa. The absolute times of these boundaries are subject to slight revision as more information accumulates.

Phanerozoic time (whose beginning is marked by the first appearance of diverse animals) is divided into three *eras*, each of which is divided into *periods*. We will frequently refer to these divisions, and to the *epochs* into which the Cenozoic periods are divided. It is useful to learn the sequence of the eras and periods, as well as a few key dates, such as the beginning of the Paleozoic era (and the Cambrian period, 541 million years ago, or 541 Mya), the Mesozoic era (and Triassic period, 252 Mya), the Cenozoic era (and Paleogene or Tertiary period, 66 Mya), and the Pleistocene epoch (2.58 Mya).[1]

---

[1] Commonly used abbreviations for geological time include Gy (billion years) and Gya (billion years ago), My (million years) and Mya (million years ago), Ky (thousand years) and Kya (thousand years ago).

TABLE 17.1 The geological time scale
The Cenozoic era embraces seven epochs, Paleocene through Holocene. The older literature refers to the first five epochs (66–2.58 Mya) as the Tertiary period, and to the Pleistocene and Holocene (or Recent) (2.58 Mya–present) as the Quaternary period. Geologists now recognize, instead, the Paleogene (Paleocene through Oligocene, 66–23 Mya), Neogene (Miocene through Pliocene, 23–2.58 Mya), and Quaternary periods.

Era	Period (abbreviation)	Epoch	Millions of years from start to present	Major events
CENOZOIC	Quaternary (Q)	Holocene	0.012	Continents in modern positions; repeated glaciations and changes of sea level; shifts of geographic distributions; extinctions of large mammals and birds; evolution of *Homo sapiens*, spread out of Africa; rise of agriculture and civilizations
		Pleistocene	2.58	
	Neogene (Ng)	Pliocene	5.33	Continents nearing modern positions; increasingly cool, dry climate; grasslands spread; modern families of mammals and birds; first apes
		Miocene	23.03	
	Paleogene (Pg)	Oligocene	33.9	Radiation of mammals, birds, snakes, angiosperms, pollinating insects, bony fishes
		Eocene	56.0	
		Paleocene	66.0	
MESOZOIC	Cretaceous (K)		145	Most continents separated; continued radiation of dinosaurs; increasing diversity of angiosperms, mammals, birds; mass extinction at end of period, including last ammonoids and nonavian dinosaurs
	Jurassic (J)		201	Continents separating; diverse dinosaurs and other reptiles; first birds; diverse mammals; gymnosperms dominant; evolution of angiosperms; ammonoid radiation; Mesozoic marine revolution
	Triassic (Tr)		252	Continents begin to separate; marine diversity increases; gymnosperms become dominant; diversification of reptiles, including first dinosaurs; transitional mammal-like forms; modern corals, teleost fishes
PALEOZOIC	Permian (P)		299	Continents aggregated into Pangaea; glaciations; low sea level; increasingly "advanced" fishes; diverse orders of insects; amphibians decline; reptiles, including early mammal-like forms, diversify; major mass extinctions, especially of marine life, at end of period
	Carboniferous (C)		359	Gondwana and small northern continents form; extensive forests of early vascular plants, especially lycopsids, sphenopsids, ferns; early orders of winged insects; diverse amphibians; first reptiles
	Devonian (D)		419	Diversification of bony fishes; trilobites diverse; origin of ammonoids, tetrapods, insects, ferns, seed plants; mass extinction late in period
	Silurian (S)		443	Diversification of agnathans; origin of jawed fishes (acanthodians, placoderms, Osteichthyes); earliest terrestrial vascular plants, arthropods
	Ordovician (O)		485	Diversification of echinoderms, other invertebrate phyla, agnathan vertebrates; mass extinction at end of period
	Cambrian (€)		541	Marine animals diversify; first appearance of most animal phyla and many classes within relatively short interval; earliest agnathan vertebrates; diverse algae
PROTEROZOIC	Ediacaran		635	Animal fossils (Ediacaran fauna); inferred lineages of sponges, cnidarians, bilaterians
	Cryogenian		720	Inferred (from DNA) animal lineages
	(others)		2500	Earliest eukaryotes (ca. 1900–1700 Mya)
ARCHEAN			4000	Origin of life in remote past (first fossil evidence at ca. 3500 Mya); diversification of prokaryotes (bacteria and archaea); photosynthesis generates oxygen, replacing oxygen-poor atmosphere; evolution of aerobic respiration

Source: Geological names and dates are from the International Commission on Stratigraphy, http://www.stratigraphy.org

The fossil record

Some short parts of the fossil record in certain localities provide detailed evolutionary histories, and some groups of organisms, such as abundant planktonic protists with hard shells, have left an exceptionally good record. In some respects, such as the temporal distribution of many higher taxa (e.g., phyla and classes), the fossil record is adequate to provide a reasonably good portrait [9]. In some other respects, the fossil record is very incomplete [38]. Consequently, the origins of many taxa have not been well documented. We know that the fossil record is incomplete because continuing exploration constantly yields new discoveries; for instance, most of the discoveries that have documented the origin of birds from dinosaurs have been made in Chinese deposits in the last 20 years.

The incompleteness of the fossil record has several causes. First, many kinds of organisms rarely become fossilized because they are delicate, or lack hard parts, or occupy environments—such as humid forests—where decay is rapid. Second, because sediments generally form in any given locality very episodically, they typically contain only a small fraction of the species that inhabited the region over time. Third, if fossils are to be found, the fossil-bearing sediments must become solidified into rock; the rock must persist for millions of years without being eroded, metamorphosed, or subducted; and the rock must then be exposed and accessible to paleontologists. Finally, the evolutionary changes of interest may not have occurred at the few localities that have strata from the right time; a species that evolved new characteristics elsewhere may appear in a local record fully transformed, after having migrated into the area. Paleontologists agree that the approximately 250,000 described fossil species represent far fewer than 1 percent of the species that lived in the past.

Before Life Began[2]

The current universe came into existence about 14 billion years ago (14 Gya, that is 14,000,000,000 years ago) through an explosion (the "big bang") from an infinitely dense point. Elementary particles formed hydrogen shortly after the big bang, and hydrogen ultimately gave rise to the other chemical elements through nuclear fusion in stars. The collapse of a cloud of dust and gas formed our galaxy fewer than 10 Gya. Material expelled into interstellar space, especially during stellar explosions (supernovas), condensed into second- and third-generation stars, of which the Sun is one. Our solar system was formed about 4.6 Gya, according to radiometric dating of meteorites and moon rocks. Earth is the same age as those bodies, but because of geological processes such as subduction (see Figure 17.1), the oldest known rocks on Earth are younger, dating from about 4 Gya.

Earth was probably formed by the collision and aggregation of many smaller bodies, the impact of which produced enormous heat. Early Earth formed a solid crust as it cooled, releasing gases that *included water vapor but very little oxygen*. As Earth cooled, oceans of liquid water formed, probably by 4.5 Gya, and quickly achieved the salinity of modern oceans. By 4 Gya there were probably many small protocontinents, which gradually aggregated, by plate tectonics, to form large land masses over the next billion years.

---

[2] This chapter differs from most of the others in this book by focusing on factual information rather than general principles of evolution. It contains more information than you may wish to memorize. You may consider it largely as a source of information, or you might simply enjoy reading a sketch of one of the greatest stories of all time. Major events and important points that a well-trained biologist should know are highlighted in italics.

The Emergence of Life

The simplest things that might be described as "living" must have developed as complex aggregations of molecules. These aggregations, of course, would have left no fossil record, so it is only through mathematical theory, laboratory experimentation, and extrapolation from the simplest known living forms that we can hope to develop models of the emergence of life. This is definitely a work in progress.

"Life" is difficult to define. It is generally agreed that an assemblage of molecules is "alive" if it can capture energy from the environment, use that energy to replicate itself, and thus be capable of evolving. (One may argue whether or not viruses are alive, as their energy is supplied by a host organism.) In the living things we know, these functions are performed by nucleic acids, which carry information, and by proteins, which replicate nucleic acids, transduce energy, and generate (and in part constitute) the phenotype. These components are held together in compartments—cells—formed by lipid membranes.

Although living or semi-living things might have originated more than once, *we can be quite sure that all organisms we know of stem from a single common ancestor* because they all share certain features that are arbitrary as far as we can tell [22, 90]. For example, organisms synthesize and use only L optical isomers of amino acids as building blocks of proteins; L and D isomers are equally likely to be formed in abiotic synthesis, but a functional protein can be made only of one type or the other. D isomers could have worked just as well. The genetic code, the machinery of replication and protein synthesis, and basic metabolic reactions are among the other features that are universal among organisms and thus imply that they all stem from the *last universal common ancestor,* or LUCA.

The most difficult problem in accounting for the origin of life is that in known living systems, only nucleic acids replicate, but their replication requires the action of proteins that are encoded by the nucleic acids. Despite this and other obstacles, progress has been made in understanding some of the likely steps in the origin of life [32, 52, 55, 100].

First, *simple organic molecules*, the building blocks of complex organic molecules, *can be produced by abiotic chemical reactions*. Such molecules have been found in space, carbonaceous meteorites, and comets. In a famous experiment, Stanley Miller found that electrical discharges in an atmosphere of methane (CH_4), ammonia (NH_3), hydrogen gas (H_2), and water (H_2O) yield amino acids and compounds such as hydrogen cyanide (HCN) and formaldehyde (H_2CO), which undergo further reactions to yield sugars, amino acids, purines, and pyrimidines.

Next, some such simple molecules must have formed polymers that could replicate. *Once replication originated, evolution by natural selection could occur,* because variants that replicated more prolifically would increase relative to others. The most likely early replicators were short RNA (or RNA-like) molecules. *RNA has catalytic properties, including self-replication.* Some RNA sequences (ribozymes) can cut, splice, and elongate oligonucleotides, and short RNA template sequences can self-catalyze the formation of complementary sequences from free nucleotides.

The first steps in the origin of life probably took place in an "RNA world," in which *catalytic, replicating RNAs underwent evolution by natural selection*. When Sol Spiegelman placed RNAs, RNA polymerase (a catalytic RNA isolated from a virus, phage Qβ), and nucleotide bases in a cell-free medium, different RNA sequences were replicated by the polymerase at different rates, so that their proportions changed [87]. In another experiment, a catalytic RNA (RNA ligase) evolved greater efficiency in ligating an oligonucleotide to itself when it was "grown" in an automated system with RNA polymerase enzymes and reagents (**FIGURE 17.2**) [69].

FIGURE 17.2 Sequence and structure of the catalytic RNA, a ligase, that evolved in a simple laboratory system. The oligonucleotide substrate, shown in blue at left, includes residues that bind to the RNA ligase as shown, as well as a nonbinding loop. Mutations that occurred during the experiment are indicated in red. The mutations that were critical for enhanced function are enclosed in boxes. (After [69].)

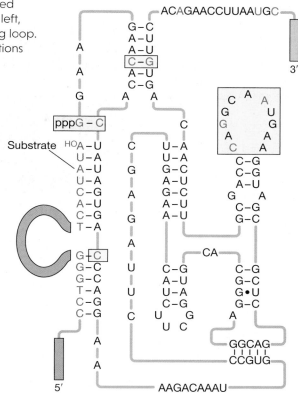

Recent experiments have shown that clay particles with RNA adsorbed onto their surfaces can catalyze the formation of a lipid envelope that can divide into "offspring" envelopes. Protocells might have consisted of such lipid envelopes containing replicating RNA.

Long RNA sequences would not replicate effectively because the mutation rate would be too high for them to maintain any identity. A larger genome might evolve, however, if two or more coupled macromolecules each catalyzed the replication of the other. Replication probably was slow and inexact originally, and only much later acquired the fidelity that modern organisms display.

How protein enzymes evolved is perhaps the greatest unsolved problem. This process may have begun when cofactors, consisting of an amino acid joined to a short oligonucleotide sequence, aided RNA ribozymes in self-replication [62]. Many current coenzymes have nucleotide components. RNA ribozymes can also catalyze the formation of peptide bonds, so the next step may have been the stringing together of several such amino acid–nucleotide cofactors. Ultimately, the ribozyme probably evolved into the ribosome, the oligonucleotide component of the cofactor into transfer RNA, and the strings of amino acids into catalytic proteins. Such ensembles of macromolecules, packaged within lipid membranes, may have been precursors of the first cells—although many other features evolved between that stage and the only cells we know. The origin of cells is often considered the first of the **major evolutionary transitions** in the history of life, evolutionary changes of major magnitude and consequence that often lead to an additional level of organization (**TABLE 17.2**).

TABLE 17.2   Six major transitions in the history of evolution leading to higher-level formations, or groups

Major transition	Group formed	Group transformation
Separate replicators (genes) and formation of cell membranes → genome within cell	Compartmentalized genomes	Evolution of large, complex genomes
Separate unicells → symbiotic unicell	Eukaryotic cells	Evolution of symbiotic organelle and nuclear genomes; transfer of genes between them; formation of "hybrid genomes"
Asexual unicells → sexual unicells	Zygote (sexually reproducing organism)	Evolution of meiosis and (often obligate) sexual reproduction
Unicells → multicellular organism	Multicellular organisms	Evolution of cell and tissue differentiation and of somatic vs. germ cells
Multicellular organisms → eusocial societies	Origin of societies (in only a few lineages)	Evolution of reproductive and nonreproductive castes (e.g., social insects)
Separate species → interspecific mutualistic associations	Origin of interspecific mutualisms	Evolution of physically conjoined partners (e.g., endosymbioses)

Source: After [11], modified from [62].

FIGURE 17.3 (A) Stromatolites formed by living cyanobacteria in Shark Bay, Australia. (B) A 3-billion-year-old stromatolite from Western Australia has the same structure as modern stromatolites. (A by D. J. Futuyma.)

Precambrian Life

Our knowledge of the grand history of life, especially the origin, diversification, and extinction of major groups of organisms, is derived from geological and paleontological evidence and from phylogenetic studies of living organisms that have helped us trace life's history. We will frequently refer to the geological time scale (see Table 17.1).

The Archean, prior to 2.5 Gya (2500 Mya), and the Proterozoic, from 2.5 Gya to 541 Mya, are together referred to as **Precambrian** time. The oldest known rocks formed in the presence of ocean water (3.8 Gy old) and contain carbon deposits that may indicate the existence of life. There is strong evidence of life by 3.4 Gya [12], and debated evidence as far back as 3.5 Gya, in the form of bacteria-like microfossils and layered mounds (stromatolites; **FIGURE 17.3**) with the same structure as those formed today along the edges of warm seas by cyanobacteria (blue-green bacteria).

The early atmosphere had little oxygen, so the earliest organisms were anaerobic. *When photosynthesis evolved* in cyanobacteria and other bacteria, *it introduced oxygen into the atmosphere.* Photosynthesis may have evolved as far back as 3.8 Gya, but the first great increase in atmospheric oxygen was about 2.4 Gya, probably as a result of geological processes that buried large quantities of organic matter and prevented it from being oxidized [48, 57]. As oxygen built up in the atmosphere, many organisms evolved the capacity for aerobic respiration, as well as mechanisms to protect the cell against oxidation.

For about 2 Gy—more than half the history of life—the only life on Earth consisted of two groups of prokaryotes, the Archaea and Bacteria, which are classified as "empires," or "domains." The prokaryotes that descended from the LUCA diversified greatly in their metabolic capacities [19]. Photosynthetic, chemoautotrophic, sulfate-reducing, methanogenic, and other forms soon evolved, and *these forms continue today to be the prime movers of the biogeochemical cycles on which ecosystems depend.* Today many archaea are anaerobic and inhabit extreme environments such as hot springs. (One such species is the source of the DNA polymerase enzyme [Taq polymerase] used for the polymerase chain reaction [PCR] that is the basis of much of modern molecular biology and biotechnology.) The bacteria are extremely diverse in their metabolic capacities, and many are photosynthetic. There was extensive lateral transfer of genes among lineages during the early history of life (see Figure 16.10) [30, 99]. The early phylogenetic history of prokaryotes was more like a network than a simple branching tree, and it still is, to some extent.

A major event in the history of life was the origin of eukaryotes, which are distinguished by such features as a cytoskeleton and a nucleus with multiple linear

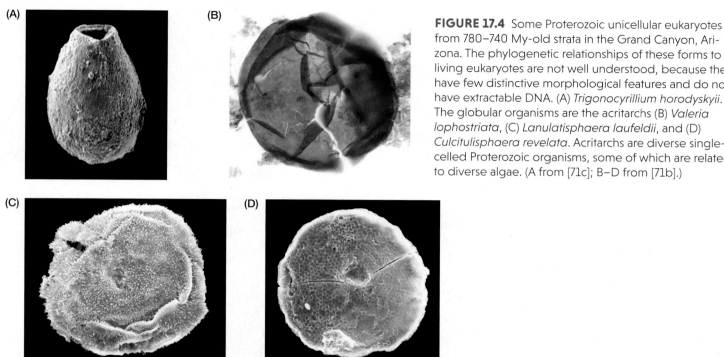

FIGURE 17.4 Some Proterozoic unicellular eukaryotes from 780–740 My-old strata in the Grand Canyon, Arizona. The phylogenetic relationships of these forms to living eukaryotes are not well understood, because they have few distinctive morphological features and do not have extractable DNA. (A) *Trigonocyrillium horodyskyii.* The globular organisms are the acritarchs (B) *Valeria lophostriata*, (C) *Lanulatisphaera laufeldii*, and (D) *Culcitulisphaera revelata*. Acritarchs are diverse single-celled Proterozoic organisms, some of which are related to diverse algae. (A from [71c]; B–D from [71b].)

chromosomes and a mitotic spindle. Most eukaryotes undergo meiosis, the highly organized segregation and recombination of genes that is the basis of sexual reproduction. Almost all eukaryotes have mitochondria (vestigial and nonfunctional in a few), and many have chloroplasts. *Mitochondria and chloroplasts are descended from bacteria* that were ingested, and later became intracellular symbionts (**endosymbionts**) in protoeukaryotes: another major transition in evolution (see Figure 2.5) [59, 62]. These events are the most important cases of endosymbiosis, which has evolved many times in the history of life [67]. Molecular phylogenetic studies show that the Eukarya are nested within a clade of Archaea that has some of the key eukaryote genes, such as those that encode actin, tubulin, and other components of the eukaryotic cell's cytoskeleton [26, 86, 97]. The most recent common ancestor of Archaea and Eukaryota probably was capable of phagocytosis, the likely basis of the capture of the bacteria that became the mitochondrion.

The earliest eukaryote fossils are about 1.8 Gy old, which is consistent with estimates of the date of the common ancestor of eukaryotes derived from DNA sequence comparisons [70]. If there were eukaryotes before then (as some chemical evidence suggests), they left no living descendants. For nearly a billion years after their origin, almost all eukaryotes seem to have been unicellular, and most lineages remain so (**FIGURE 17.4**). Based on cellular characteristics shared among diverse living eukaryotes, the reconstructed last common ancestor of eukaryotes was a highly complex cell, combining components that were derived from its archaeal and bacterial ancestors and components that evolved during the more than 1 Gy that separated it from the first eukaryotic ancestor [16]. The last common ancestor of eukaryotes had sophisticated metabolic capabilities, elaborate endomembranes, a cytoskeletal system based on actinomysin and tubulin, meiosis, and a nucleus with nucleocytoplasmic transport [51].

Unicellular and multicellular eukaryotes share most protein-coding gene families and regulatory control of gene expression by transcription factors (see Chapter 15). Multicellularity has evolved many times (**FIGURE 17.5**). Simple multicellular

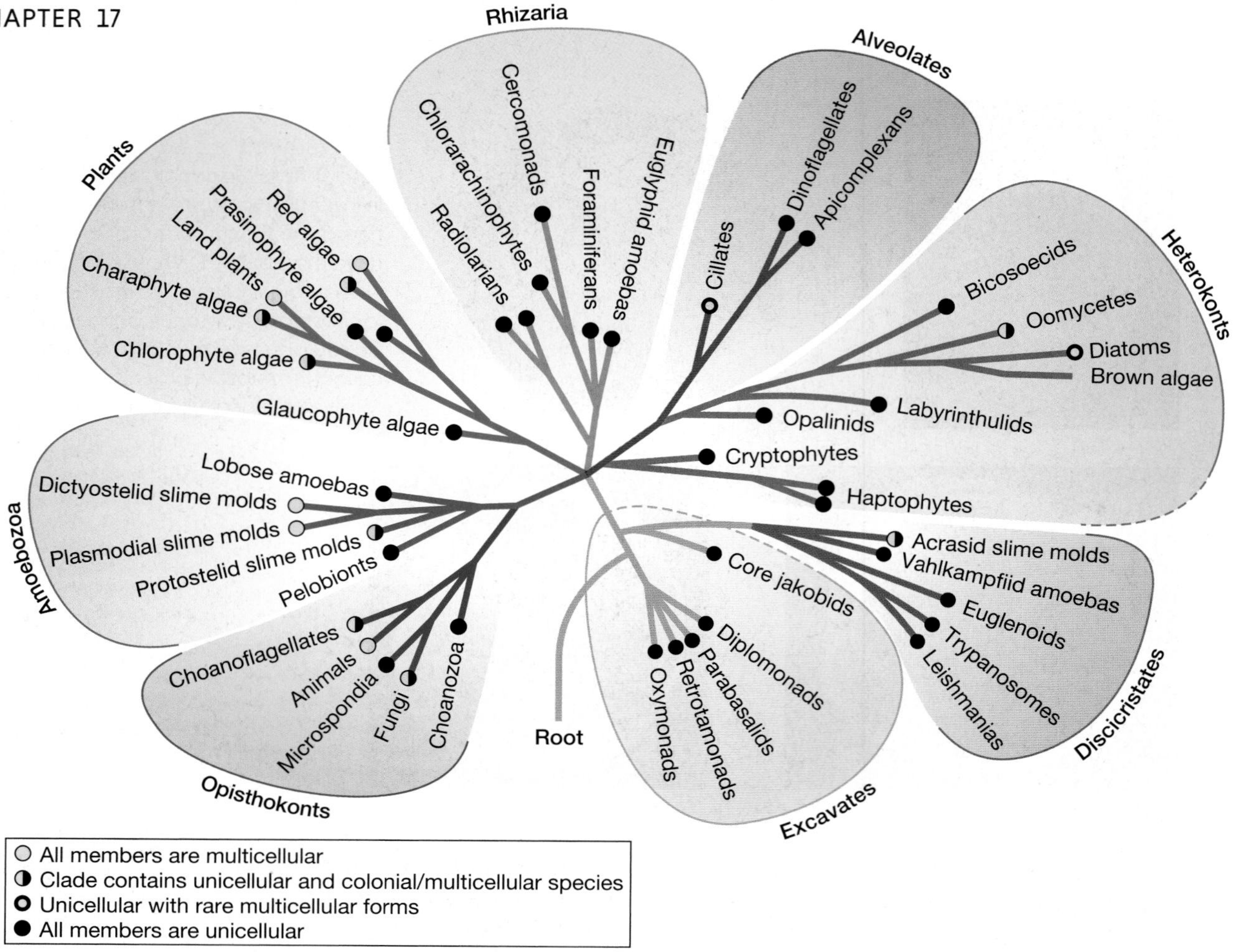

FIGURE 17.5 Multicellularity has evolved many times from unicellular ancestors. Five taxa (yellow circles) are entirely multicellular (red algae, land plants, dictyostelid slime molds, plasmodial slime molds, and animals). Nine taxa (half-yellow circles) include some multicellular or colonial species, and two (open circle) include a few multicellular species. Some currently understood relationships differ from those shown in this phylogeny, which was published in 2003. (From [34], with phylogeny from [6].)

organisms, with only one cell type, may have evolved because large size protected them from being swallowed by unicellular predators. The advantage of more complex multicellularity was almost surely the division of labor between different cell types with different functions [34, 65]. Multicellularity enabled the evolution of large size and elaborate organ systems. In the origin of animals and plants, and perhaps in the other multicellular lineages as well, the first step seems to have been the evolution of cell adhesion, followed by the evolution of new signaling molecules and transcription factors, as well as intercellular bridges that facilitate the movement of nutrients and signaling molecules [49]. Simple multicellularity, based on adhesion of cells formed by cell division, has evolved in laboratory cultures of yeast [75].

The Cambrian Explosion and the Origins of Animal Diversity

Animals are most closely related to the unicellular choanoflagellates (Choanozoa), which have cell adhesion proteins and form colonies by cell division. They resemble certain cells in sponges (**FIGURE 17.6**). Choanoflagellates and animals

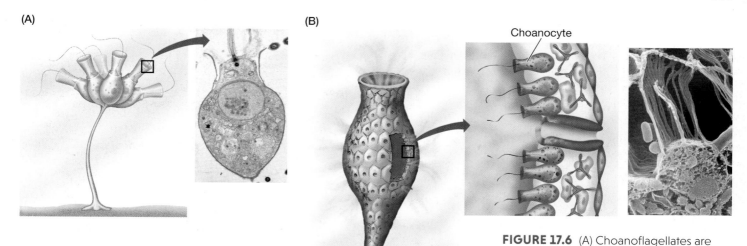

(A)

(B)

Choanocyte

FIGURE 17.6 (A) Choanoflagellates are unicellular eukaryotes that can form simple colonies. They are the closest known relatives of animals, and structurally resemble (B) choanocyte cells of sponges. (B, photo from [23b].)

share many genes that underlie cell and developmental processes [76], suggesting that the genetic toolkit of animals evolved during the Ediacaran period (635–541 Mya) [28, 49]. Among living animals (Metazoa), sponges (phylum Porifera) are thought by many researchers to be the sister group of the other animals, although the relationships among the sponges, the radially symmetrical Cnidaria (jellyfishes, corals), the Ctenophora (comb jellies), and the Bilateria are still uncertain [25]. The Bilateria—bilaterally symmetrical animals with a head, often equipped with mouth appendages, sensory organs, and a brain—include all the other animal phyla. The origin of all these phyla is one of the biggest unsolved problems in the study of evolution.

The final two periods of the Proterozoic are the Cryogenian (720–635 Mya), when Earth experienced lowered temperatures, and the Ediacaran. From about 575 to 541 Mya, there are fossils of a variety of enigmatic animals known as the *Ediacaran fauna*. Most of them were soft-bodied and appear to have been flat creatures that crept or stood on the sea floor (**FIGURE 17.7**). They are thought to have become extinct without leaving any post-Cambrian descendants. Ediacaran animals seem to have lacked features, such as mouthparts or locomotory appendages, that might be used in interacting with other animals, and they left no burrows in sediments. Nor did the bilaterians that are thought to have existed at this time. Based on calibrated DNA sequence divergence, *the phyla of living animals stem from a common ancestor*

(A)

(B)

FIGURE 17.7 Members of the Ediacaran fauna. (A) *Tribrachidium heraldicum*. The triradial form of this animal differs from that of any Phanerozoic animals. (B) The relationship of the wormlike *Dickinsonia costata* to later animals is unknown.

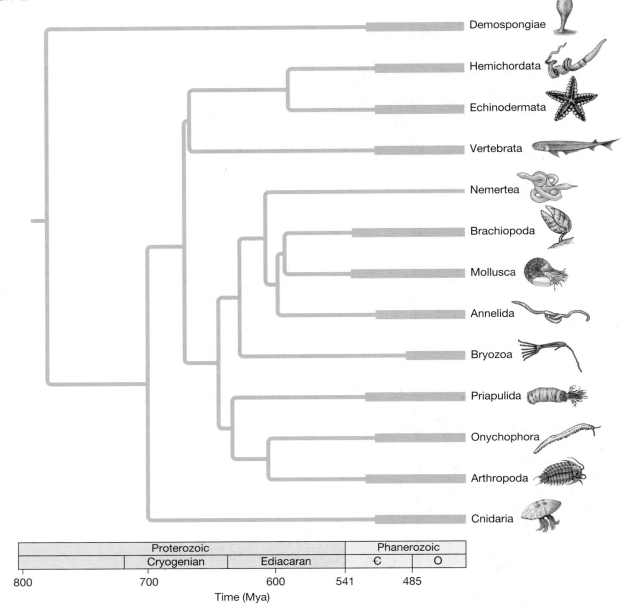

Proterozoic			Phanerozoic	
	Cryogenian	Ediacaran	Є	O

800   700   600   541   485

Time (Mya)

FIGURE 17.8 Phylogeny of some animal phyla and their occurrence in the fossil record. The thick portion of each branch represents time from the earliest known specimen to the present. The time estimates for the branch points when lineages diverged from common ancestors are based on time-calibrated DNA sequence differences. Note that no fossil specimens of any modern phylum have been recorded before the Cambrian, although DNA sequences imply that some lineages diverged long before that. (After [29].)

more than 700 Mya, when sponges and cnidarians formed distinct lineages (**FIGURE 17.8**). Sequence data also show that divergence among many bilaterian lineages occurred well before 541 Mya [29]. But *none of these living phyla*, except for sponges, *has been recorded in the fossil record before the start of the Cambrian period, 541 Mya.*

The Cambrian period begins the Paleozoic era—with a bang. For the first 10 My or so of the Cambrian period, starting about 541 Mya, animal diversity was low. Then, during a period of about 20 My, almost all the modern phyla and classes of skeletonized marine animals, as well as many extinct classes, appeared in the fossil record. This interval marks the first appearance of brachiopods, trilobites and

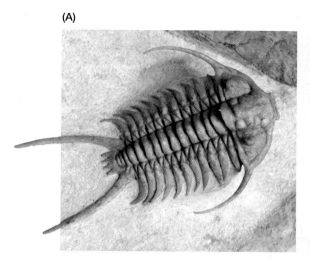 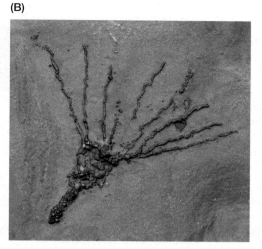

FIGURE 17.9 Two animal groups that first appeared during the Cambrian explosion. Both of these fossils were uncovered in the sandy shales of southern Utah, an area that once was covered by shallow seas. (A) A Cambrian trilobite (*Paraceraurus*), phylum Arthropoda. More than 17,000 species of trilobites have been described from the Paleozoic. The group became extinct at the end of the Permian. (B) An echinoderm (*Gogia spiralis*) from the early Cambrian. Several groups of echinoderms—which, along with chordates, are deuterostome animals—flourish in the modern fauna.

other classes of arthropods, molluscs, and echinoderms, as well as animals that are hard to classify into later phyla (**FIGURE 17.9**). This diversification, surely the most dramatic adaptive radiation in the history of life [92], is called the **Cambrian explosion** because it transpired over such a short time ("only" 20 My). The Cambrian explosion is a conundrum: how can the long prior history of the phyla, revealed by molecular divergence, be reconciled with their absence, and then sudden appearance, in the fossil record?

The bilaterian lineages that existed before the Cambrian probably fed on detritus and plankton (by filter-feeding). Only at the start of the Cambrian did they evolve hard parts and acquire novel ways of living, such as predation and burrowing into sediments. A combination of *genetic and ecological causes may account for this diversification* [29, 49, 60]. Environmental changes, such as an increase in atmospheric oxygen, may have played a role [48]. Mechanisms of gene regulation may have undergone major evolutionary changes at this time, leading to new morphologies. For example, microRNAs, which affect the precision of translation of mRNAs into proteins, are more diverse in morphologically more complex animals [71a]. Some of the resulting morphological changes may have led to novel interactions among different organisms, such as predation, that further enhanced diversity by selecting for protective skeletons and new ways of overcoming such defenses. Other changes, such as those that enabled animals to burrow, provided access to new environments. Some of these activities modified physical and chemical aspects of the environment, providing ecological opportunities for yet other novel ways of life.

Paleozoic Life

Between the start of the Paleozoic era 541 Mya and its end 252 Mya (**FIGURE 17.10**), life on Earth became wonderfully diverse. The era begins with the first evidence of the modern phyla, and ends with seas populated with great predators and with dense forests on the continents, inhabited by some familiar insects and the early ancestors of mammals.

Many animal lineages diversified during the Cambrian, including early chordates such as *Haikouichthys*, which had eyes, gill pouches, a notochord, and segmented musculature, but no jaws or limbs (**FIGURE 17.11A**) [84]. Conodonts, first

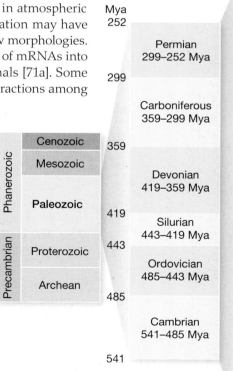

FIGURE 17.10 Time line of the Paleozoic era, illustrating a noteworthy organism from each period except the Silurian.

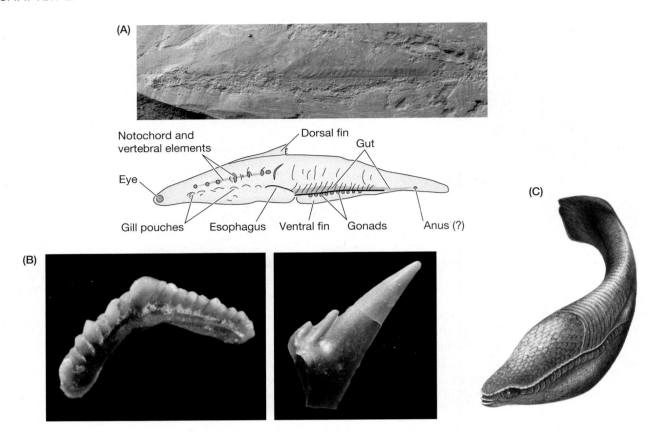

FIGURE 17.11 Cambrian vertebrates. (A) Photo and drawing of one of the earliest known vertebrates, *Haikouichthys*, of the early Cambrian. The drawing calls attention to features that are characteristic of vertebrates. (B) Bony, toothlike structures of Cambrian conodonts. Conodonts were slender, finless chordates believed to be related to agnathans (jawless vertebrates such as lampreys). (C) Reconstruction of a jawless, limbless ostracoderm, *Arandaspis*, as it may have appeared in life. Note the heavy armor on the front part of the body. Ostracoderm armor has been found in late Cambrian rocks. (A courtesy of D.-G. Shu, from [84].)

appearing in the late Cambrian, are the earliest fossils with cellular bone, a distinctive feature of vertebrates (**FIGURE 17.11B**). The earliest definitive vertebrates, also from the late Cambrian, are ostracoderms, jawless fishlike vertebrates that had bony armor and lacked paired fins (**FIGURE 17.11C**) [41].

Especially at its end, the Cambrian was marked by considerable extinction. For example, many of the more than 90 Cambrian families of trilobites became extinct. Afterward, many of the *animal phyla diversified greatly in the Ordovician (485–443 Mya), giving rise to many new classes and orders that included new ways of life.* The major large predators were sea stars and nautiloids (shelled cephalopods; that is, molluscs related to squids). The first reefs were built by two groups of corals, with contributions from sponges, bryozoans, and cyanobacteria. The Ordovician ended with a mass extinction, perhaps caused by a drop in temperature and a drop in sea level, that in proportional terms may have been the second largest of all time.

Among the groups that survived this extinction event were the nautiloids, which gave rise to the ammonoids, shell-bearing cephalopods that are among the most diverse groups of extinct animals (**FIGURE 17.12**). During the Silurian (443–419 Mya), most vertebrates were armored agnathans (jawless vertebrates; **FIGURE 17.13A**). *The first known gnathostomes*, vertebrates with jaws and two pairs of fins, *appeared at this time* (**FIGURE 17.13B,C**). (How paired appendages first evolved is

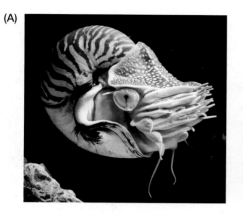

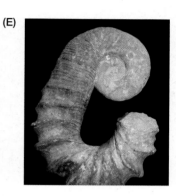

FIGURE 17.12 Nautiloids and ammonoids. Shells housed the squidlike bodies of these cephalopod molluscs. (A) The chambered nautilus (*Nautilus pompilius*), one of the few living species of nautiloid. (B) An orthoconic nautiloid, believed to belong to the same group as the modern *Nautilus*. These animals had noncoiled shells. (C) A Jurassic ammonoid, *Craspedites*, showing the intricate sutures that evolved in many later ammonoids. (D) *Kosmoceras*, a Jurassic ammonoid. (E) *Australoceras*, a Cretaceous ammonoid with a very different form of shell.

FIGURE 17.13 Extinct Paleozoic classes of fishlike vertebrates. (A) A jawless ostracoderm, class Heterostraci (*Pteraspis*, Devonian). (B) A gnathostome (jawed vertebrate), class Acanthodii (*Climatius*, Devonian). Gnathostomes evolved from jawless ancestors. (C) A placoderm (*Bothriolepis*, Devonian).

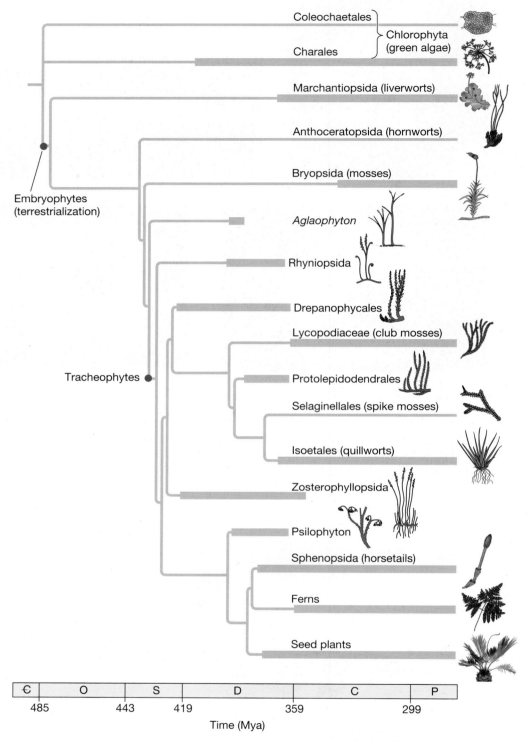

FIGURE 17.14 Phylogeny and Paleozoic fossil record of major groups of terrestrial plants and their closest relatives among the green algae (Chlorophyta). The broad bars show the known temporal distribution of each group in the Paleozoic fossil record. The green algae, liverworts, hornworts, mosses, club mosses, spike mosses, quillworts, horsetails, ferns, and seed plants have living representatives. (After [45].)

one of the great unknowns of vertebrate evolution.) It is also during the Silurian that the bony fishes (Osteichthyes) arose. During the Devonian (419–359 Mya), two subclasses of bony fishes flourished: the ray-finned fishes that would later diversify into the largest group of modern fishes (the teleosts), and the lobe-finned fishes (Sarcopterygii), which included lungfishes and our own ancestors. Immense coral reefs developed during the Silurian and Devonian, but these were among the victims of a major extinction in the late Devonian.

The colonization of land

Terrestrial plants, including mosses, liverworts, and vascular plants, are a monophyletic group that evolved from green algae (Chlorophyta) (**FIGURE 17.14**) [43]. Living on land required the evolution of an external surface and spores that are resistant to loss of water, structural support, vascular tissue to transport water within the plant body, and internalized sexual organs, protected from desiccation. The first known terrestrial organisms are mid-Ordovician spores and spore-bearing structures (sporangia) of very small plants, which were apparently related to today's liverworts [96]. By the mid-Silurian there were small vascular plants, less than 10 cm tall, that lacked true roots and had sporangia at the ends of short, leafless, dichotomously branching stalks (**FIGURE 17.15A**). A staggering amount of adaptive evolution ensued: by the end of the Devonian, about 75 My later, terrestrial plants had evolved deep root systems, wood, leaves, and complex, diverse reproductive structures. These plants included ferns, club mosses, horsetails, and seed plants: all the major groups of plants except the flowering plants. Many were large trees (**FIGURE 17.15B**). Over the course of the Devonian, *the amount of terrestrial biomass increased enormously, and it had huge effects*: it increased atmospheric oxygen, created organic soil, increased the weathering and erosion of rocks, and consumed carbon dioxide, resulting in lowered temperature [3]. Life continued to alter the planet.

The earliest terrestrial arthropods are known from the Silurian. They fall into two major groups, which both arose in the ocean. The chelicerates included spiders, mites, scorpions, and several other groups that still exist. The earliest mandibulates included detritus-feeding millipedes from the late Silurian, followed in the Devonian by predatory centipedes and primitive wingless insects, which evolved from crustaceans. Later, the importance of insects in terrestrial ecosystems became immense: as herbivores, insects have profoundly affected plant evolution; as predators, they affect the evolution of other insects; and as prey, they support the majority of the terrestrial animals that do not feed on plants.

The first terrestrial vertebrates evolved from lobe-finned fishes late in the Devonian. The Sarcopterygii, or lobe-finned fishes, appeared in the early Devonian, about 408 Mya. They include coelacanths and lungfishes, a few of which are still alive, and the osteolepiforms, which had distinctive tooth structure and skull bones (**FIGURE 17.16A**). Osteolepiforms, such as *Eusthenopteron*, had a tail fin and fleshy paired fins, with a central axis of several large bones to which lateral bones and slender, jointed rays (radials) articulated. They could not flex their head relative to the body, and their braincase had a joint between the anterior and posterior sections, as it does in living sarcopterygians. The first definitive tetrapods (four-legged vertebrates), such as *Ichthyostega* from the very late Devonian, had the same tail fin and distinctive teeth and skull, but the gill cover bones at the rear of the skull had been lost, and the head could now be moved on a more flexible neck. Most importantly, they had larger pectoral and pelvic girdles and fully developed tetrapod limbs that bore more than five digits (unlike almost all later tetrapod vertebrates) [20].

Clearly, *ichthyostegids show a mosaic of sarcopterygian and tetrapod features, and are intermediates in the evolution of a major new clade of vertebrates*. Until recently, only a few fossils provided evidence of intermediate steps in the transition from

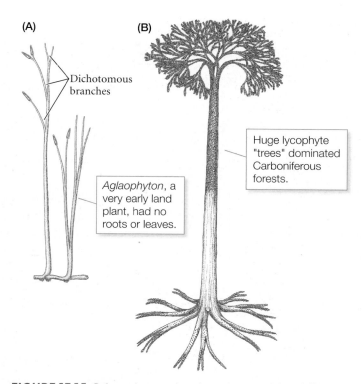

(A) (B)

Dichotomous branches

Aglaophyton, a very early land plant, had no roots or leaves.

Huge lycophyte "trees" dominated Carboniferous forests.

FIGURE 17.15 Paleozoic vascular plants, portrayed at different scales. (A) *Aglaophyton*, from the Devonian, was probably less than 15 cm tall. (B) *Lepidodendron*, a Carboniferous lycophyte tree, was as tall as 30 m. (A from [47]; B, from [89].)

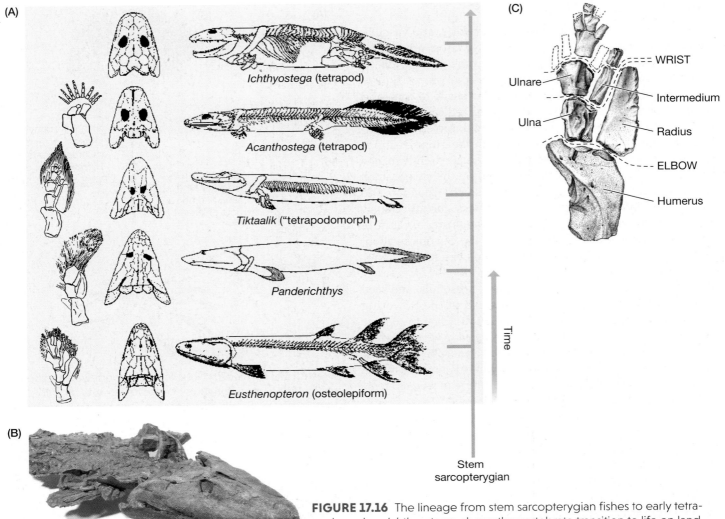

(A)

Ichthyostega (tetrapod)

Acanthostega (tetrapod)

Tiktaalik ("tetrapodomorph")

Panderichthys

Eusthenopteron (osteolepiform)

Time

Stem
sarcopterygian

(C)

WRIST

Ulnare

Intermedium

Ulna

Radius

ELBOW

Humerus

(B)

FIGURE 17.16 The lineage from stem sarcopterygian fishes to early tetrapods, such as *Ichthyostega*, shows the vertebrate transition to life on land. The recently discovered intermediate *Tiktaalik* and the tetrapods have a flatter skull than the fish *Eusthenopteron*, but its structure (as seen from above, at left) is very similar, except that the gill cover bones at the rear have been lost. Among the drawings of forelimbs at far left, note the intermediate structure of the forelimb of *Tiktaalik*. (B) An articulated skeleton of *Tiktaalik*. (C) Drawing of the pectoral fin, or forelimb, of *Tiktaalik*, showing positions of joints and the homologues of the limb bones of tetrapods. Tetrapod digits probably evolved from the many small bones (radials) at the end of the fin. (A after [1, 85]; C after [85].)

fin to limb. Hoping to fill in more of this evolutionary sequence, Neil Shubin and colleagues explored promising Devonian deposits in northern Canada and found just what they sought: a rich fossil deposit of a new "tetrapodomorph" which they named *Tiktaalik roseae* [23a, 85]. Like ichthyostegids, *Tiktaalik* had a flat, mobile head and elongate snout and lacked gill cover bones; it also had overlapping ribs, which would have provided the support that the body of a partly terrestrial animal requires (**FIGURE 17.16B**). Most important, the pectoral (shoulder) girdle and fins of *Tiktaalik* are intermediate between those of the sarcopterygian and tetrapod conditions. The forelimb and wrist bones are clearly homologous to those of early tetrapods and reveal a critical feature: the limb could be flexed at the elbow and wrist (**FIGURE 17.16C**). All the anatomical details of girdle, limbs, and ribs show that *Tiktaalik* could hold its body off the ground—it could do push-ups. Its skull also has several intermediate features, including a less mobile braincase joint [24]. Its mode of breathing was intermediate between that of lungfishes and that of terrestrial tetrapods. These features are more pronounced in recently discovered fossils of *Ventastega*, a very early tetrapod that is intermediate between *Tiktaalik* and *Ichthyostega*. Its limbs and girdles resemble those of *Ichthyostega*, and its skull is like that of *Tiktaalik*, but it is solid, without the braincase joint [2]. *Tiktaalik*, *Ventastega*, and their relatives are transitional forms that make the distinction between fishes and the earliest tetrapods difficult to draw.

Paleozoic life on land

Imagine that you have traveled back through time to the Carboniferous, about 325 Mya, and go for a walk. Needless to say, there are no paths; you have to make your way through the ferns and horsetails, probably in one of the widespread swamps, where you may see gigantic dragonflies or millipedes (**FIGURE 17.17**). You are a few million years too late to see a *Tiktaalik*, but you see its relatives, some up to 1–2 m long, crawling about and occasionally catching a large insect or crustacean. You are pleasantly surprised not to be harassed by mosquitoes or other biting insects, and then you realize that you don't see any butterflies, or ants, or bees; in fact, there aren't any flowers. And it is strangely quiet: no birdsong—not even any cricket song, because there aren't any crickets. No birds are in the sky because there are no birds to fly [17].

But some changes soon followed. During the Carboniferous (359–299 Mya), land masses became aggregated into the supercontinent Gondwana in the Southern Hemisphere and into several smaller continents in the Northern Hemisphere. Widespread tropical climates favored the development of extensive swamp forests dominated by horsetails, ferns, and lycophyte trees, which were preserved as the coal beds that we mine today. *The seed plants diversified in the late Paleozoic.* Unlike earlier plants that depended on water for their swimming sperm to fertilize ovules, some seed plants had wind-dispersed pollen. The evolution of the seed provided the embryo with protection against desiccation as well as a store of nutrients that enabled the young plant to grow rapidly and overcome adverse conditions. Bear in mind that none of these plants had flowers.

The first winged insects evolved during the Carboniferous, and they rapidly diversified into many orders, including primitive dragonflies, orthopteroids (roaches, grasshoppers, and relatives), and hemipteroids (leafhoppers and their relatives). Some Carboniferous insects and other arthropods were gigantic (see Figure 17.17). In the Permian (299–252 Mya), the first insect groups with complete metamorphosis (distinct larval and pupal stages) evolved, including beetles, hymenopterans (wasps and their relatives), primitive flies (Diptera), and the ancestors of the Lepidoptera (moths and butterflies). The DNA-based phylogeny of living insects suggests that insects evolved wings in the Devonian, before the earliest winged fossils. But the sequence of fossil appearances corresponds to the phylogeny, in which groups without complete metamorphosis are the basal branches and the orders with complete metamorphosis form a derived clade (**FIGURE 17.18**).

FIGURE 17.17 Giant arthropods of the Carboniferous included *Arthropleura*, a millipede up to 7 feet (2.3 m) long, and the griffenfly *Meganeura*, related to dragonflies, with a wingspan of 2 feet (25 cm). The human is for scale only; no human has ever seen these arthropods alive.

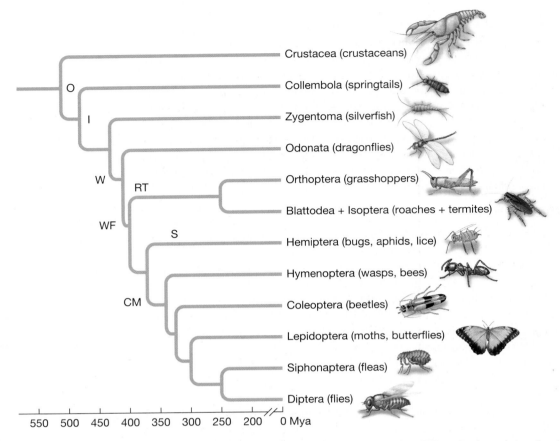

FIGURE 17.18 A phylogeny of some of the orders of living insects, showing estimated dates of branching events. The origin of insects from crustaceans is marked O at the basal branch point. The true insects, as defined by the structure of the mouthparts, diverged from the springtails and related forms at the branch marked I. Basal lineages of insects, such as silverfish, lack wings. W marks the evolution of wings in the ancestor of all the other orders, and WF the evolution of wing folding: the ability to fold the wings over the back instead of holding them permanently outstretched, as in dragonflies. RT marks the origin of a clade that includes the grasshoppers, roaches, and many other forms. The termites (Isoptera) evolved from roaches. The Hemiptera, or true bugs and relatives, have mouthparts modified for sucking (S) plant sap or animal body fluids. CM marks the evolution of a profoundly important feature: complete metamorphosis. The wasps, beetles, moths, and true flies, with distinct larval and pupal stages, include huge numbers of species. (After [66].)

Tetrapod lineages ("amphibians") *diversified in the Carboniferous.* Some of them (anthracosaurs) are intermediate between amphibians and reptiles, and have been classified as both by different researchers. *Anthracosaurs gave rise to the first known amniotes*[3], the captorhinomorphs. By the late Permian these primitive amniotes gave rise to the *synapsids,* which *included the ancestors of mammals and increasingly evolved mammal-like features* (see Chapter 20). The first amniotes also gave rise to the diapsids, a major reptilian lineage whose descendants, as we will see, dominated the Mesozoic landscape.

The end-Permian mass extinction

During the Permian, the continents approached one another and *formed a single world continent,* **Pangaea** (**FIGURE 17.19A**). Collisions between land masses built the Appalachian, Ural, and some other mountain ranges, sea level dropped to its lowest point in history, and climates were greatly altered by the arrangement of land and sea. The Permian ended, 252 Mya, with a catastrophe: the

---

[3] Amniotes are those vertebrates—reptiles, birds, and mammals—with a major adaptation for life on land: the amniotic egg, with its tough shell, protective membranes (chorion and amnion), and a membranous sac for storing waste products.

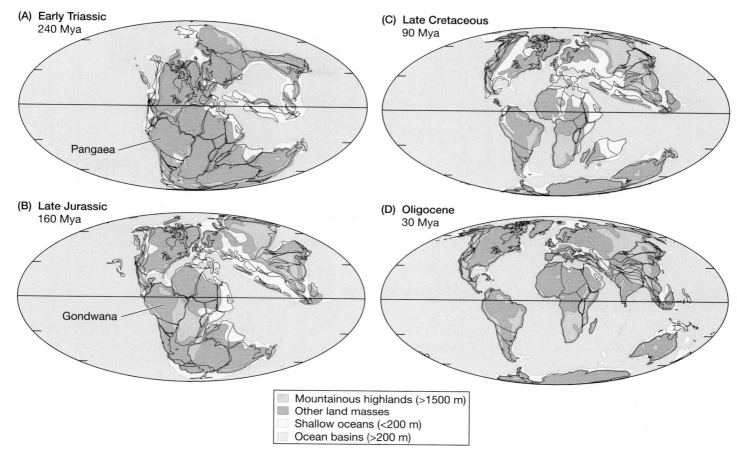

(A) Early Triassic
240 Mya

Pangaea

(B) Late Jurassic
160 Mya

Gondwana

(C) Late Cretaceous
90 Mya

(D) Oligocene
30 Mya

Mountainous highlands (>1500 m)
Other land masses
Shallow oceans (<200 m)
Ocean basins (>200 m)

FIGURE 17.19 Distribution of land masses in geological time. The outlines of the modern-day continents are shown in maps C and D.; other black lines delineate important plate boundaries. (A) In the early Triassic, most land was aggregated into a single mass (Pangaea). (B) Eurasia and North America were separated by the late Jurassic. (C) Gondwana had become fragmented into most of the major southern land masses by the late Cretaceous. (D) By the Oligocene, the land masses were close to their present positions. (Maps © 2004 by C. R. Scotese/ PALEOMAP Project.)

end-Permian mass extinction, one of the most significant events in the history of life. It is estimated that in this, *the most massive extinction event in the history of Earth*, at least 56 percent of the genera and more than 80 percent of all species of skeleton-bearing marine invertebrates became extinct within less than 200,000 years [7, 83]. Groups such as ammonoids, stalked echinoderms, brachiopods, and bryozoans declined greatly, and major taxa such as trilobites and several groups of corals disappeared entirely. Some orders of insects and many families of amphibians and mammal-like reptiles became extinct, and the composition of plant communities changed greatly [63]. The extinction was probably triggered by vast volcanic eruptions in Siberia that covered 7 million km^2 (2.7 million mi^2) with layers of basalt as much as 6500 m (4 mi) deep [14, 46]. These eruptions are thought to have released poisonous gases such as hydrogen sulfide and vast quantities of carbon dioxide, which in turn caused global warming, aridity, and increased acidity of ocean water (which interferes with the formation of calcium carbonate shells and skeletons). The temperature change may have caused turnover in the water column and the reduced oxygen level that is evident in the geological record from this time. On land, there were massive wildfires, deforestation, and soil erosion [27, 50, 83].

FIGURE 17.20 Time line of the Mesozoic era, with illustrations of some noteworthy species. Triassic: *Ginkgo* and phytosaur. Jurassic: cycad and *Apatosaurus*. Cretaceous: magnolia and *Tyrannosaurus*.

Mesozoic Life

Now we come to history's most romantic era, the one that most grips our imagination. Divided into the Triassic (252–201 Mya), Jurassic (201–145 Mya), and Cretaceous (145–66 Mya) periods, the Mesozoic era is often called the "Age of Reptiles," so named for some of the most extraordinary creatures of all time (**FIGURE 17.20**).

During the Mesozoic, Pangaea began to break up, beginning with the formation of the Tethyan Seaway between Asia and Africa, and then the full separation of a northern land mass, called **Laurasia**, from a southern continent known as **Gondwana**. Laurasia began to separate into several fragments during the Jurassic (**FIGURE 17.19B**), but northeastern North America, Greenland, and western Europe remained connected until well into the Cretaceous. The southern continent, Gondwana, consisted of Africa, South America, India, Australia, New Zealand, and Antarctica. These land masses slowly separated in the late Jurassic and the Cretaceous, but even then the South Atlantic formed only a narrow seaway between Africa and South America (**FIGURE 17.19C**). Throughout the Mesozoic, sea level rose, and many continental regions were covered by shallow seas. Although the polar regions were cool, most of Earth enjoyed warm climates: Antarctica had forests and dinosaurs. Global temperatures reached an all-time high in the mid-Cretaceous, after which substantial cooling occurred.

MARINE LIFE   Extinctions continued during the earliest Triassic, but diversity slowly recovered. Many of the marine groups that had been decimated during the end-Permian extinction again diversified. Ammonoids, for example, increased from 2 to more than 100 genera by the middle Triassic (see Figure 17.12). Planktonic foraminiferans (shelled protists) and modern corals evolved, and bony fishes continued to radiate. Another *mass extinction occurred at the end of the Triassic*, associated with a massive release of carbon into the atmosphere and global warming [79]. Marine biodiversity decreased by about half, and groups such as ammonoids and bivalves were devastated, but then recovered and experienced yet another adaptive radiation. The teleosts, today's dominant group of bony fishes, evolved and began to diversify. *During the Mesozoic* and continuing into the early Cenozoic, *predation seems to have escalated* [35, 93]. During this so-called Mesozoic marine revolution, crabs and bony fishes evolved mechanisms for crushing mollusc shells, and molluscs evolved protective mechanisms such as thick shells and spines (**FIGURE 17.21**).

During the Jurassic and Cretaceous, modern groups of gastropods (snails and relatives), bivalves, and bryozoans rose to dominance; gigantic sessile bivalves (rudists) formed reefs; and the seas harbored several groups of large marine reptiles.

FIGURE 17.21 Features of marine predators and prey that escalated during and after the Mesozoic marine revolution. (A) The huge claws of modern lobsters represent a trait found in several crustacean groups. Such claws enable some lobsters and crabs to crush and rip mollusc shells. (B) Spines on both bivalves and gastropods (such as this *Murex*) prevent some fishes from swallowing these prey and may reduce the effectiveness of crushing predators. (C) Thick shells and narrow apertures, as in the gastropod *Cypraea mauritiana*, deter predators.

TERRESTRIAL PLANTS AND ARTHROPODS   For most of the Mesozoic, the flora was dominated by gymnosperms (seed plants that lack flowers). The major groups were the cycads (**FIGURE 17.22A**) and the conifers and their relatives—including *Ginkgo*, a Triassic genus that has left a single surviving species as a "living fossil" (**FIGURE 17.22B,C**). *A major event in life's history is the rise of the angiosperms*, the flowering plants, which evolved from a gymnosperm ancestor in the late Jurassic and became fairly diverse by the early Cretaceous (**FIGURE 17.22D**) [31]. A well-calibrated DNA phylogeny shows that the major modern groups of angiosperms were diversifying rapidly in the early Cretaceous; for example, monocot lineages such as orchids, palms, and grasses had diverged by about 130 Mya [58]. By the mid-Cretaceous, about 108 Mya, the world's forests were

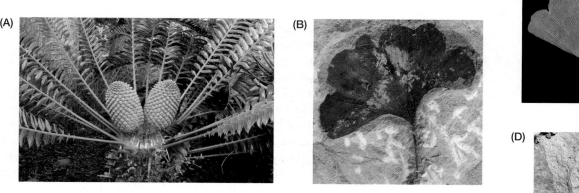

FIGURE 17.22 Seed plants. (A) A living cycad (*Encephalartos* sp.). Gymnosperms were abundant and highly diverse during the Mesozoic, but only about 130 species survive today. (B) A fossilized *Ginkgo* leaf from the Paleocene (66–56 Mya). (C) A leaf of the sole surviving ginkgo species, *Ginkgo biloba*. (D) *Protomimosoidea*, a Paleocene/Eocene fossil member of the legume family, an angiosperm group that includes mimosas and acacias. (D courtesy of W. L. Crepet.)

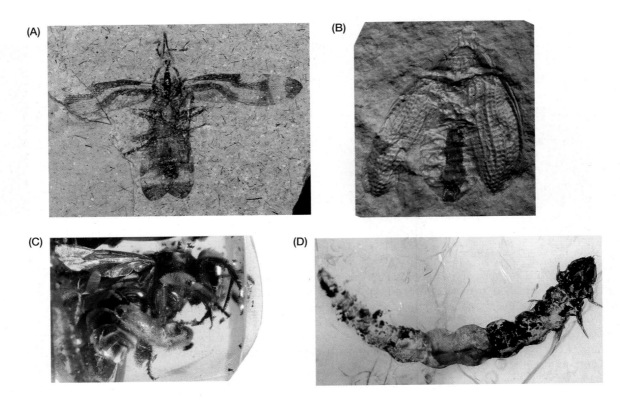

FIGURE 17.23 Some fossil insects. (A) A Jurassic relative of roaches (*Rhipidoblattina*) was a predator, unlike modern roaches. (B) An early Cretaceous beetle from one of the morphologically most primitive beetle families (Cupedidae), which still has a few "living fossil" species. (C) One of the earliest known fossil bees (*Protobombus*, Eocene) was a member of the social bee family, Apidae. (D) Among living moths, the family Micropterigidae has the most ancestral features, such as biting mandibles. This Cretaceous micropterigid larva (100 Mya) is among the earliest lepidopteran fossils. The bee and the moth larva are preserved in amber (fossilized plant resin). (From [33], courtesy of D. Grimaldi.)

dominated by angiosperms, which formed the environment in which diverse groups of plants (e.g., ferns) and animals (e.g., ants, beetles, amphibians, mammals) radiated [95]. The diversity of terrestrial eukaryotes became higher than the diversity of marine species at that time, perhaps because some angiosperms evolved certain characteristics (e.g., leaf vein density) that increased the productivity of terrestrial ecosystems [94].

The anatomically most "advanced" groups of insects made their appearance in the Mesozoic (**FIGURE 17.23**). By the late Cretaceous, most families of living insects, including ants and social bees, had evolved. Throughout the Cretaceous and thereafter, insects and angiosperms affected each other's evolution and may have augmented each other's diversity (see Chapter 13). As different groups of pollinating insects evolved, adaptive modification of flowers to suit different pollinators gave rise to the great floral diversity of modern plants. *Largely because of the spectacular increase of angiosperms and insects, terrestrial diversity is greater today than ever before.*

VERTEBRATES   Amniote vertebrates—the reptiles, birds, and mammals—became very diverse in the Mesozoic. The major groups are distinguished by different openings in the temporal region of the skull (at least in the stem members of each lineage; **FIGURE 17.24**). One such group included marine reptiles that

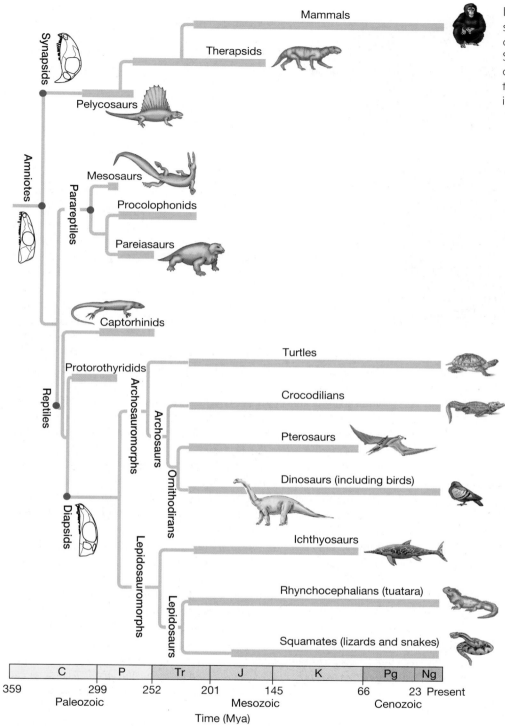

FIGURE 17.24 Phylogenetic relationships and temporal duration (thick bars) of major groups of amniote vertebrates. Some authors define "reptiles" as one of the two major lineages of amniotes, the other being the synapsids, which includes mammals. (After [53].)

flourished from the late Triassic to the end of the Cretaceous, among them the dolphin-like ichthyosaurs, which gave birth to live young (**FIGURE 17.25A**).

The *diapsids*, with two temporal openings, became one of the most diverse groups of amniotes. One major diapsid lineage, the lepidosaurs, includes the lizards, which became differentiated into modern families in the late Cretaceous. Among several lineages of lizards in which legs became reduced or lost, one evolved into the snakes. Snakes became ecologically very diverse during the Cretaceous, and again during the Cenozoic.

(A) An ichthyosaur

FIGURE 17.25 Some Mesozoic reptiles. (A) A marine ichthyosaur (Greek, "fish lizard"), convergent in form with sharks and porpoises. (B) *Lagosuchus*, a Triassic thecodont archosaur (Greek, "ruling lizard"), showing the body form of the stem group from which dinosaurs evolved. (C) A pterosaur (Greek, "wing lizard").

(B) *Lagosuchus talampayensis*

(C) *Pterodactylus*

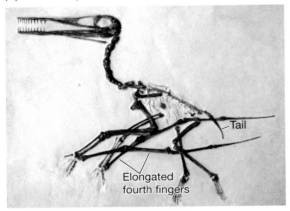

The other major diapsid group, the archosaurs, includes the most spectacular and diverse of the Mesozoic amniotes. Most of the late Permian and Triassic archosaurs were fairly generalized predators 1 m or so in length (**FIGURE 17.25B**). From this generalized body plan, numerous specialized forms evolved. Among the most highly modified archosaurs are the pterosaurs, one of the *three major vertebrate groups that evolved powered flight*. The wing consisted of a membrane extending to the body from the rear edge of a greatly elongated fourth finger (**FIGURE 17.25C**). The pterosaurs diversified greatly: one, with a wingspan of 11 m (36 ft), was the largest flying vertebrate known, while others were as small as sparrows.

Dinosaurs evolved from archosaurs related to the one pictured in Figure 17.25B. Dinosaurs are not simply any old large, extinct reptiles, but members of the orders Saurischia and Ornithischia, which differ in the form of the pelvis. Both orders included bipedal forms and quadrupeds that were derived from bipedal ancestors. Both orders arose in the Triassic and became diverse in the Jurassic.

More than 39 families, about 550 genera, and well over 1000 species of dinosaurs are recognized (**FIGURE 17.26**). The Ornithischia—herbivores with specialized, sometimes very numerous, teeth—included the well-known stegosaurs, with dorsal plates that probably served for thermoregulation, and the ceratopsians (horned dinosaurs), of which *Triceratops* is the most widely known. Among the Saurischia, the sauropods, herbivores with small heads and long necks, include the largest animals that have ever lived on land, such as *Apatosaurus* (= *Brontosaurus*); *Brachiosaurus*, which weighed more than 80,000 kg; and *Argentinosaurus*, which reached about 40 m in length. Saurischians also included carnivorous, bipedal theropods, such as *Velociraptor*, the renowned *Tyrannosaurus rex* (late Cretaceous), which stood 15 m high and weighed about 7000 kg, and many smaller theropods. *All dinosaurs became extinct at the end of the Cretaceous except for a single group of theropods. They radiated extensively in the Cenozoic and today include about 10,000 species that we know as the birds.* Aside from birds, the only living archosaurs are the 22 species of crocodilians.

The late Paleozoic also included the *synapsids* (see Figure 17.24), with a single temporal opening. They gave rise to the therapsids, sometimes called "mammal-like reptiles," which flourished until the middle Jurassic, and *gave rise in the late Triassic and early Jurassic to forms that are almost fully mammalian in their features* (see Chapter 20). *Many clades of mammals*, including more than 300 known genera, *arose and became extinct* in the Jurassic and especially the Cretaceous (**FIGURE 17.27**). Although most were small in size, many were ecologically and morphologically specialized, and convergent with living mammals that evolved similar traits independently (**FIGURE 17.28**) [56]. Only three of the many Mesozoic mammal

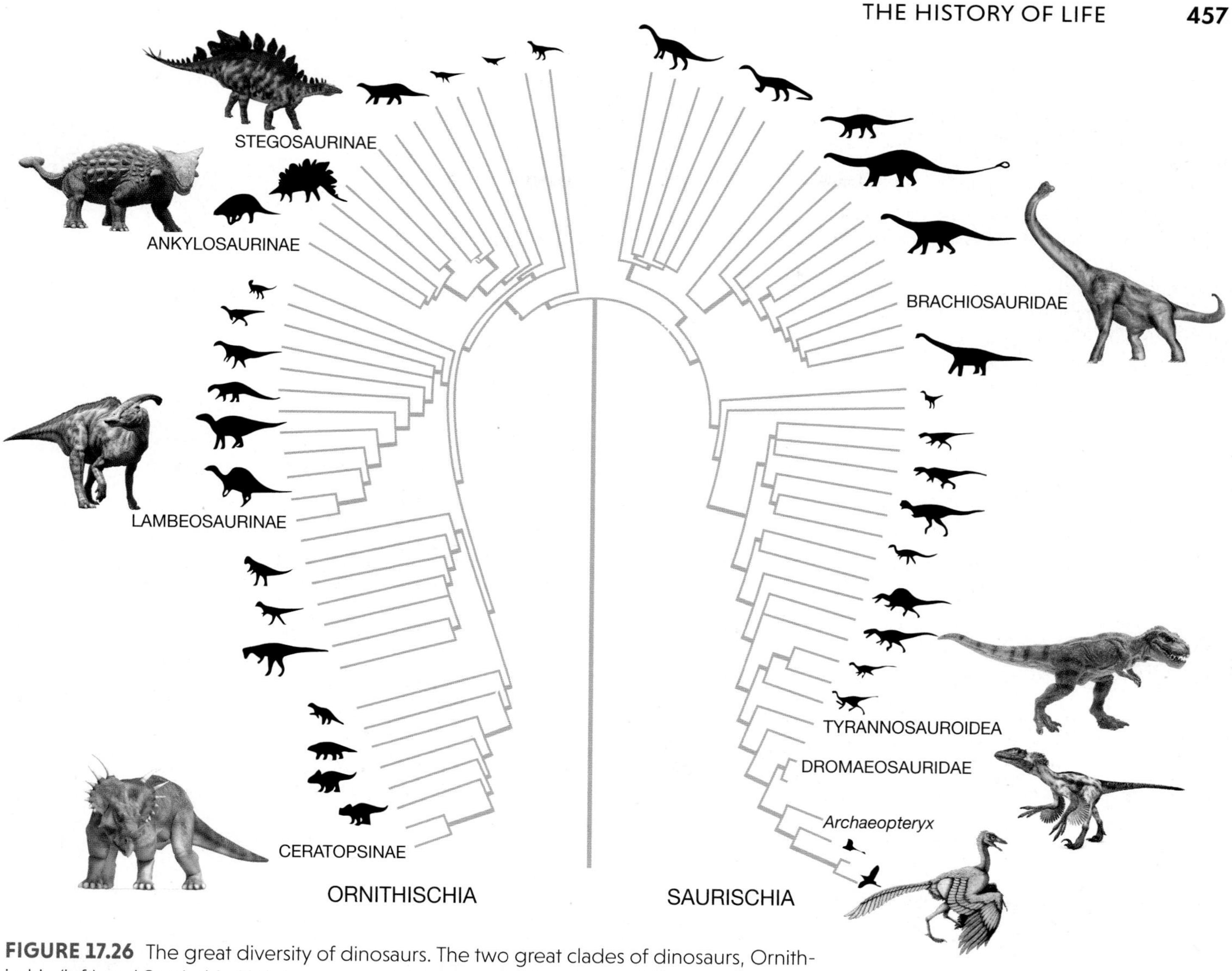

FIGURE 17.26 The great diversity of dinosaurs. The two great clades of dinosaurs, Ornithischia (left) and Saurischia (right), are curved downward to fit the page. The names of most lineages have been omitted, for simplicity. The exceptions are for the few pictured dinosaurs. These are the ornithischian Lambiosaurinae (here, *Parasaurolophus*), Stegosaurinae, Ankylosaurinae, and Ceratopsinae, which includes *Styracosaurus* (shown) and *Triceratops* (not shown). Among the Saurischians are the sauropods, represented here by Brachiosaurinae, and the theropods, illustrated by Tyrannosauroidea (*Tyrannosaurus*), Dromaeosauridae (*Deinonychus*), and *Archaeopteryx*, a close relative of the ancestors of birds. All ornithischian lineages are extinct. (From [82].)

lineages have living representatives: the monotremes (the egg-laying platypus and echidnas), the marsupials, and the placental (eutherian) mammals.

The end of the Cretaceous is marked by the best-known mass extinction, caused by the great environmental disruption that resulted from the impact of an asteroid or some other extraterrestrial body. The site of this impact, the Chicxulub crater, has been discovered off the coast of Mexico's Yucatán Peninsula. There was surely a staggering explosion on impact, probably followed by a worldwide cloud of dust and vapor that blocked sunlight and created a long-lasting winter. This event was formerly called the K/T extinction, using the abbreviations for Cretaceous and Tertiary, but is now called the K/Pg extinction, with Pg referring to Paleogene. The K/Pg event extinguished the last nonavian dinosaurs, many mammalian groups, and

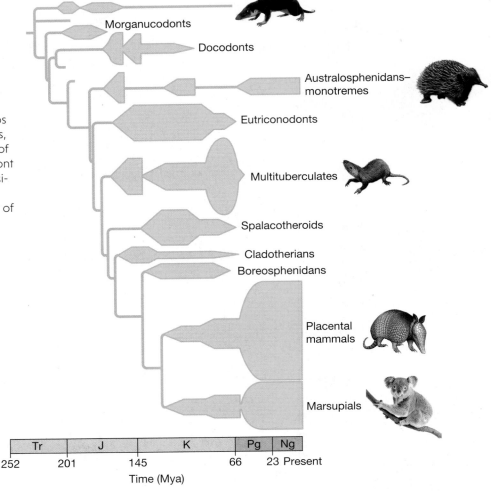

FIGURE 17.27 Phylogenetic relationships and temporal distribution of major groups of mammals. The morganucodonts, arising in the Jurassic, are important intermediates between mammal-like reptiles and true mammals (see Chapter 20). The only groups that survive to the present are monotremes, placentals, and marsupials. The coloration of the reconstructed figures of morganucodont and multituberculate are based on supposition, not data. The changing width of each branch represents changes in the diversity of known species in that group. (After [56].)

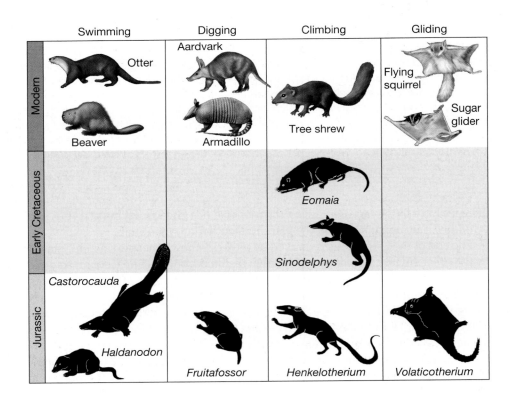

FIGURE 17.28 The lower (black) figures are reconstructions of the likely form of diverse Mesozoic mammals that are inferred, from their anatomy, to have had different ecological features. The upper figures portray living mammals that are ecologically similar to the Mesozoic forms below, a result of convergent evolution. Swimming, digging, climbing, and gliding forms are shown. (After [56].)

about 15 percent of the families and 47 percent of the genera of marine animals [36]. Ammonoids, rudists, most marine reptiles, and many families of invertebrates and planktonic protists became extinct.

The Cenozoic Era

The Cenozoic era, which started 66 Mya, marks the start of modern times (**FIGURE 17.29**). Both in the sea and on land, the flora and fauna have a more familiar cast than in the Mesozoic. To be sure, many groups of animals that originated and flourished in the Cenozoic have become extinct, but even these were closely related enough to living forms that they seem less foreign to us than the plesiosaurs, sauropods, and pterosaurs that they replaced.

By the beginning of the Cenozoic, North America had moved westward, becoming separated from Europe in the east, but forming the broad Bering Land Bridge between Alaska and Siberia (**FIGURE 17.19D**). The Bering Land Bridge remained above sea level throughout most of the Cenozoic (see Figure 17.33A). Gondwana broke up into the separate island continents of South America, Africa, India, and far to the south, Antarctica plus Australia (which separated in the Eocene). About 18–14 Mya, during the Miocene, Africa made contact with southwestern Asia, India collided with Asia (forming the Himalayas), and Australia moved northward, approaching southeastern Asia. During the Pliocene, the Isthmus of Panama arose, and fully connected North and South America for the first time about 2.8 Mya [54, 68].

This reconfiguration of continents and oceans contributed to major climate changes. In the late Eocene and Oligocene there was global cooling and drying; extensive savannahs (sparsely forested grasslands) formed for the first time, and Antarctica acquired glaciers. Sea level fluctuated, dropping drastically in the late Oligocene (about 25 Mya). In the Miocene, cacti and other plant groups that were adapted to arid conditions diversified in several parts of the world [5]. During the Pliocene, temperatures increased but then dropped again, and the Pleistocene epoch, which started about 2.6 Mya, was marked by a series of about 11 glacial-interglacial cycles. The most recent such "ice age" ended only about 12,000 years ago.

The modern world takes shape

Over the course of more than 10 My, marine animal diversity recovered from the end-Cretaceous extinction event, although the speed and amount of recovery differed among animal groups and geographic regions [37]. Some major groups, such as ammonoids, were gone forever, but groups such as gastropods (snails, whelks) flourished, and marine diversity eventually reached new heights [15]. Animals such as crinoids (sea lilies) that stand on the ocean floor were reduced, but the diversity and ecological importance of deep burrowers and especially of predators increased. The Mesozoic marine revolution continued, with more effective predatory crustaceans becoming more prominent [93]. Marine mammals and a great

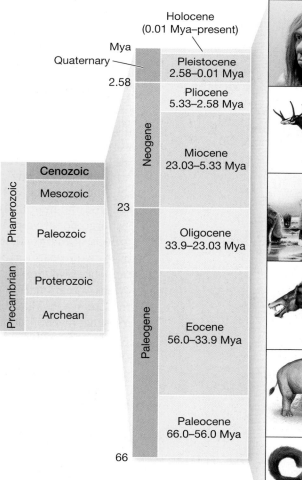

FIGURE 17.29 Time line of the Cenozoic era, with illustrations of some noteworthy organisms.

diversity of fishes took the place of the great Mesozoic marine reptiles. Late in the Cenozoic, modern coral reefs became prevalent in the tropics. Today they are the marine equivalent of tropical rainforests because of their extraordinarily rich diversity of fishes, sponges, and other animals—and they are just as threatened by human activities.

On land, many of the modern families of angiosperms and insects had become differentiated by the late Cretaceous, and many more evolved in the Paleocene or Eocene. Many fossil insects of the late Eocene and Oligocene belong to genera that still survive. The savannahs that developed in the Oligocene because of the more arid climate were populated by grasses (Poaceae), which underwent a major adaptive radiation at this time, and herbaceous plants, many groups of which evolved in the Paleogene from woody ancestors. Among the most important of these groups is the family Asteraceae, which includes sunflowers, daisies, ragweeds, and many others. It is one of the two largest plant families today.

The most dramatic biotic change between the Cretaceous and the Paleogene is the utter absence of the great dinosaurs and other archosaurs that had ruled the world, and in their stead, the rapid proliferation of even more diverse birds and mammals. Calibrated phylogenies of living birds indicate that most of the orders (e.g., pigeons, pelicans, owls) originated in the Paleocene [42, 73], and many living orders and families of birds are recorded from the Eocene (56–33.9 Mya) and Oligocene (33.9–23 Mya). The songbirds (Passeriformes), which account for half of the living species, first displayed their great diversity in the Miocene (23–5.3 Mya). Another great adaptive radiation was the snakes, which began an exponential increase in diversity in the Oligocene. Snakes today feed on a great variety of animal prey, from worms and termites to bird eggs and wild pigs, and they include marine, burrowing, and arboreal forms. Some can even glide between trees.

The adaptive radiation of mammals

Because you may be more familiar with mammals than with other animals or plants (and because you are a mammal), we describe their Cenozoic ups and downs in a little more detail. Although the marsupial and placental mammals originated in the Cretaceous, most of the fossils that can be assigned to modern orders occur after the K/Pg boundary (66 Mya). It has often been suggested that the extinction of the large nonavian dinosaurs at the end of the Cretaceous relieved the mammals from competition and predation, and allowed them to undergo adaptive radiation—but this overlooks the great diversity of extinct groups of mammals that coexisted with dinosaurs (see Figure 17.28). Moreover, two research groups that used multiple fossils to calibrate the rate of DNA sequence evolution concluded that the stem lineages leading to many of the living orders of mammals originated in a burst of diversification at least 80 Mya, during the Cretaceous (**FIGURE 17.30**) [10, 64]. They found no evidence that the rate of mammal diversification increased after the dinosaurs' demise.

The marsupial families that include kangaroos, wombats, and other living Australian marsupials evolved in the Eocene and Oligocene. Marsupials probably arose in Asia: they are known as fossils from all the continents, including Antarctica. Today they are restricted to Australia and South America (except for the North American opossum, which evolved from South American ancestors). In South America, marsupials experienced a great adaptive radiation; some resembled kangaroo rats, others saber-toothed cats (**FIGURE 17.31A,B**). Most South American marsupials became extinct by the end of the Pliocene.

In addition to marsupials, many groups of placental mammals evolved in South America during its long isolation from other continents. These mammals included an ancient placental group, the Xenarthra (or Edentata), which includes the giant

FIGURE 17.30 A phylogeny of living groups of mammals, based on DNA sequence data. The timing of branch points is based on sequence divergence, calibrated by many different fossils. The data indicate that most orders diverged from one another during the Cretaceous. Some very short branches imply that some groups of orders diverged within a relatively short time, as is characteristic of many adaptive radiations. (After [64].)

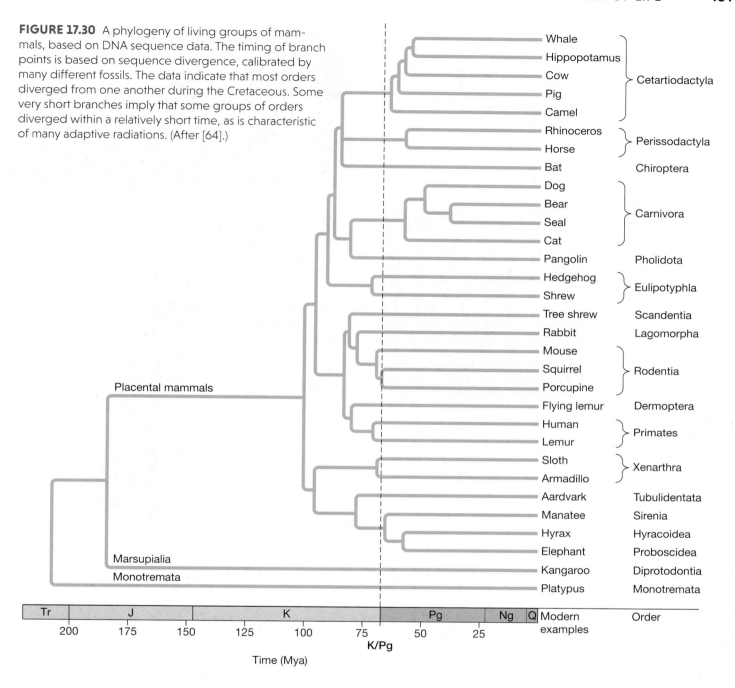

ground sloths and armadillo-like glyptodonts, some as large as a car, that survived until the late Pleistocene (**FIGURE 17.31C,D**). A few species of armadillos, anteaters, and sloths survive in tropical America. At least six orders of hoofed mammals that resembled sheep, rhinoceroses, camels, elephants, horses, and rodents evolved in South America, but declined and became extinct after South America became connected to North America in the late Pliocene. The extinction of many South American mammals might have been caused by the incursion of North American mammal groups, such as cats, bears, weasels, and camels, that moved into South America at this time.

One clade of mammals includes such very different-looking creatures as the rodentlike hyraxes, the aquatic manatees, and the elephants. The elephants diversified greatly, into a range of fascinating forms (**FIGURE 17.32**). Woolly mammoths

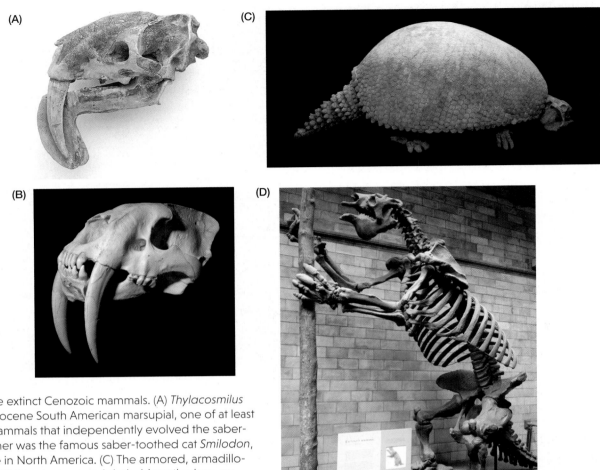

FIGURE 17.31 Some extinct Cenozoic mammals. (A) *Thylacosmilus* was a Miocene to Pliocene South American marsupial, one of at least seven lineages of mammals that independently evolved the saber-tooth form. (B) Another was the famous saber-toothed cat *Smilodon*, from the Pleistocene in North America. (C) The armored, armadillo-like *Glyptodon* and (D) this giant ground sloth, *Megatherium*, are Pleistocene representatives of the Xenarthra.

survived through the most recent glaciation, until about 10,000 years ago, and two genera (the African and Indian elephants) persist today.

In the Eocene, a group of archaic hoofed mammals gave rise to a great radiation of carnivores and hoofed mammals (ungulates). The Carnivora proliferated on land and gave rise to the marine seals. Among the ungulates, the Perissodactyla, or odd-toed ungulates, were very diverse from the Eocene to the Miocene, then dwindled to the few extant species of rhinoceroses, horses, and tapirs. The artiodactyls (order Cetartiodactyla) are first known in the Eocene as rabbit-sized animals, but otherwise bore little similarity to the pigs, camels, and ruminants that appeared soon afterward. In the Miocene, the ruminants began a sustained radiation, mostly in the Old World, that is correlated with the increasing prevalence of grasslands. Among the families that proliferated are the deer, the giraffes and relatives, and the Bovidae, the diverse family of antelopes, sheep, goats, and cattle. During the Eocene, an artiodactyl lineage related to today's hippopotamuses became aquatic and evolved into the cetaceans: the dolphins and whales (see Figure 20.3).

Rodents (Rodentia) are recorded first from the late Paleocene. They became the most diverse order of mammals, in part because of an extraordinary proliferation of rats and mice within the last 10 My. They are related to one of the oldest and, in many ways, structurally most primitive placental orders: the Primates. The earliest fossils assigned to this order are so similar to early placental mammals that it is

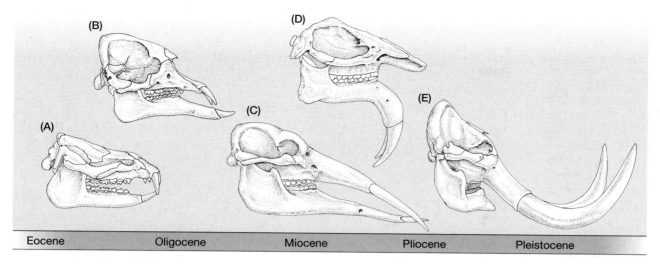

| Eocene | Oligocene | Miocene | Pliocene | Pleistocene |

FIGURE 17.32 Proboscidea, the order of elephants, has only two living genera but was once very diverse. A few of the extinct forms are (A) an early, generalized proboscidean, *Moeritherium* (late Eocene–early Oligocene); (B) *Phiomia* (early Oligocene); (C) *Gomphotherium* (Miocene); (D) *Deinotherium* (Miocene); and (E) *Mammuthus*, the woolly mammoth (Pleistocene). (After [77].)

arbitrary whether they are called primates or not. The first monkeys are known from the Oligocene, the first apes from the Miocene, the earliest hominins from the late Miocene (about 7 Mya), and the first *Homo sapiens* from about 200,000 years ago. In Chapter 21 we shall return to the story of human evolution.

Pleistocene events

Because of its recency and drama, the last Cenozoic epoch before ours, *the Pleistocene, is critically important for understanding today's organisms*. Most living species evolved in the Pleistocene, or shortly before.

By the beginning of the Pleistocene, the continents were situated as they are now. North America was connected in the northwest to eastern Asia by the Bering Land Bridge, in the region where Alaska and Siberia almost meet today (**FIGURE 17.33A**). North and South America were connected by the Isthmus of Panama.

Global temperatures began to drop during the Pliocene, about 3 Mya, and then, in the Pleistocene, underwent violent fluctuations at intervals of about 100,000 years. When temperatures cooled, continental glaciers as thick as 2 km formed at high latitudes, and then receded during the warmer intervals. At least four major glacial advances, and many minor ones, occurred. The most recent glacial episode, termed the Wisconsin in North America and the Würm in Europe, reached its maximum about 20,000 years ago (see Figure 17.33A), and the ice melted back between 15,000 and 8000 years ago. During glacial episodes, sea level dropped as much as 100 m below its present level. This drop exposed parts of the continental shelves, extending many continental margins beyond their present boundaries and connecting many islands to nearby land masses (**FIGURE 17.33B**). Temperatures in equatorial regions were apparently about as high as they are today, so the latitudinal temperature gradient was much steeper than at present. *The global climate during glacial episodes was generally drier.* Thus mesic and wet forests became restricted to relatively small favorable areas. Grasslands expanded, contributing to the diversification of grazing mammals in Africa. During interglacial episodes, the climate became warmer and generally wetter.

FIGURE 17.33 Pleistocene glaciers lowered sea level by 100 m or more, so that many terrestrial regions that are now separated by oceanic barriers were connected. These maps show the configuration of land in two parts of the world about 15,000 years ago. (A) Eastern Asia and North America were joined by the Bering Land Bridge. Note the extent of the glacier in North America. (B) Indonesia and other islands were connected to either southeastern Asia or Australia. (After [13].)

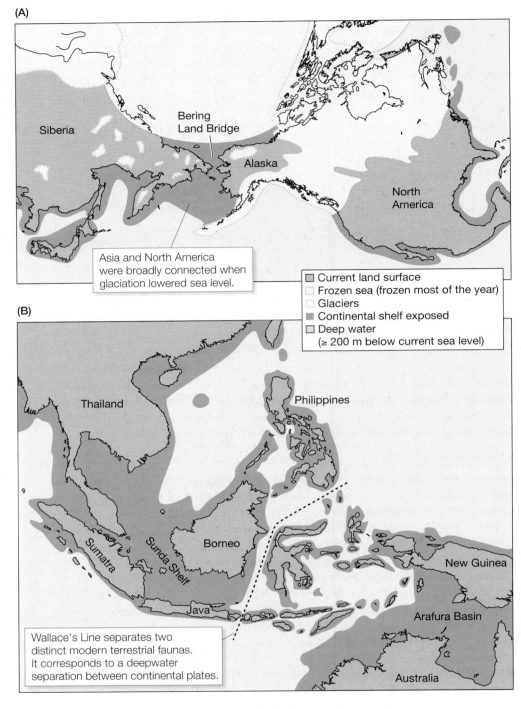

(A)

Siberia

Bering Land Bridge

Alaska

North America

Asia and North America were broadly connected when glaciation lowered sea level.

■ Current land surface
□ Frozen sea (frozen most of the year)
┈ Glaciers
■ Continental shelf exposed
□ Deep water
   (≥ 200 m below current sea level)

(B)

Thailand

Philippines

Sumatra

Sunda Shelf

Borneo

New Guinea

Java

Arafura Basin

Australia

Wallace's Line separates two distinct modern terrestrial faunas. It corresponds to a deepwater separation between continental plates.

These events profoundly affected the distributions of organisms (see Chapter 18). When sea level was lower, many terrestrial species moved freely between land masses that are now isolated; for example, the ice-free Bering Land Bridge was a conduit from Asia to North America for species such as woolly mammoths, bison, and humans. The distributions of many species shifted toward lower latitudes during glacial episodes and toward higher latitudes during interglacial episodes, when tropical species extended far beyond their present limits. These repeated shifts in geographic distributions resulted from normal processes of dispersal and establishment in new favorable areas, coupled with extinction of populations in areas that became climatically unfavorable to a species. Many species were extirpated over broad areas; for instance, beetle species that occurred in England during the Pleistocene are now restricted to such far-flung areas as northern Africa and eastern Siberia [21]. Many species that had been broadly and rather uniformly distributed became isolated in separated areas (refuges, or **refugia**) where favorable conditions persisted during glacial episodes. Some such isolated populations diverged genetically and phenotypically, in some instances becoming different species. In some cases, populations have remained in their glacial refugia to this day, isolated from the major range of their species (see Figure 18.8). However, many species have rapidly spread over broad areas from one or a few local refugia and have achieved their present distributions only in the last 8000 years or fewer. Studies of fossil pollen show that since the glaciations, the geographic distributions of plant species have changed incessantly, and that the species composition of ecological communities has changed kaleidoscopically, mostly because of fluctuations in climate [40].

Aside from changes in species' geographic distributions, the most conspicuous effects of the changes in climate were extinctions. At the end of the Pliocene, *many shallow-water marine invertebrate species became extinct, especially tropical species*, which may have been poorly equipped to withstand even modest cooling [39, 88]. No major taxa of marine animals became entirely extinct, but on land the story was different. Except in Africa, *a very high proportion of large-bodied mammals and birds became extinct* in the late Pleistocene and Holocene. These animals included mammoths, saber-toothed cats, giant bison, giant beavers, giant wolves, ground sloths, and all the endemic South American ungulates. Archaeological evidence, mathematical population models, and the timing of extinctions relative to human population movements and climate change indicate that both human hunting and climate change were major causes of this **megafaunal extinction** [4, 61, 72].

The most recent glaciers had hardly retreated when major new disruptions began. The advent of human agriculture about 11,000 years ago began yet another reshaping of the terrestrial environment. For the last several thousand years, deserts have expanded under the impact of overgrazing, forests have succumbed to fire and cutting, and climates have changed as vegetation has been modified or destroyed. At present, under the impact of an exponentially growing human population and its modern technology, species-rich tropical forests face almost complete annihilation, temperate zone forests and prairies have been eliminated in much of the world, marine communities suffer pollution and appalling overexploitation, and global warming caused by combustion of fossil fuels is changing climates and habitats so rapidly that many species are unlikely to adapt [44, 74]. An analysis of the numbers of threatened species of terrestrial vertebrates suggests that biological diversity could be reduced by 75 percent (a proportion comparable to the K/Pg mass extinction) within the next 900, and perhaps as few as 240, years (see Box 22.B) [8]. Even if these estimates are twice as pessimistic as they should be, one of the greatest ecological disasters, and one of the greatest extinctions of all time, appears to be under way, and can be mitigated only if humans act decisively and quickly.

Go to the
─────────── **Evolution Companion Website** ───────────
EVOLUTION4E.SINAUER.COM
for data analysis and simulation exercises, quizzes, and more.

SUMMARY

- Evidence from living organisms indicates that all living things are descended from a single common ancestor. Some progress has been made in understanding the origin of life, but a great deal remains unknown.

- The first fossil evidence of life dates from about 3.5 Gya, about 1 Gy after the formation of Earth. The earliest life forms of which we have evidence were prokaryotes.

- Eukaryotes evolved about 1.8 Gya. Their mitochondria and chloroplasts evolved from endosymbiotic bacteria.

- Although stem lineages of some modern phyla evolved long before the Cambrian period, the fossil record displays an explosive diversification of the animal phyla near the beginning of the Cambrian, about 541 Mya. The causes of this rapid diversification are debated, but may include a combination of genetic and ecological events. Jawless, limbless vertebrates evolved by the late Cambrian.

- Terrestrial plant and arthropod fossils are found first in the Silurian, and insects in the Devonian. Vertebrates (fishes) with jaws and limbs (fins) evolved in the Silurian, and tetrapods evolved in the late Devonian from lobe-finned fishes.

- The most devastating mass extinction of all time occurred at the end of the Permian (about 252 Mya). It profoundly altered the taxonomic composition of Earth's biotas.

- Seed plants and amniotes became diverse and ecologically dominant during the Mesozoic era (252–66 Mya). Early mammaliaforms evolved in the Triassic, and archosaurs, including especially the dinosaurs, dominated Jurassic and Cretaceous landscapes. Flying dinosaurs, the antecedents of birds, evolved in the Jurassic, and gave rise to some lineages of modern birds in the late Cretaceous. Flowering plants and plant-associated insects diversified greatly from the middle of the Cretaceous onward. A mass extinction (the K/Pg or K/T extinction) at the end of the Mesozoic included the extinction of the last nonavian dinosaurs.

- The climate became drier during the Cenozoic era, favoring the development of grasslands and the evolution of herbaceous plants and grassland-adapted animals.

- Most orders of placental mammals originated in the late Cretaceous, but underwent adaptive radiation in the early Paleogene. Many groups of mammals were once more diverse than they are now, and some are extinct. A few groups, such as rodents and artiodactyls, maintained high diversity.

- A series of glacial and interglacial episodes occurred during the Pleistocene (the last 2.6 My), during which some extinctions occurred and the distributions of species were greatly altered.

- Humans have caused species extinctions since the spread of agriculture or earlier. Human population growth and technology have had an accelerating impact on biological diversity, and have initiated another major extinction.

TERMS AND CONCEPTS

Cambrian explosion	Laurasia	Pangaea	strata
endosymbiont	mass extinction	radiometric dating	
Gondwana	megafaunal extinction	refugia	

SUGGESTIONS FOR FURTHER READING

S. M. Stanley, *Earth and Life through Time*, second edition (W. H. Freeman, New York, 1993), is a comprehensive introduction to historical geology and the fossil record. In the fourth edition of *Life of the Past* (Prentice-Hall, Upper Saddle River, NJ, 1999), W. I. Ausich and N. G. Lane provide a well-illustrated introduction to the theme of this chapter. A good overview of the origin of life is by A. Lazcano in *Evolution Since Darwin: The First 150 Years*, M. A. Bell et al. (eds.) (Sinauer Associates, Sunderland, MA, 2010), pp. 353–375. The Cambrian explosion is the subject of a comprehensive article by D. E. Erwin et al., "The Cambrian conundrum: Early divergence and later ecological success in the early history of animals" (*Science* 334: 1091–1097, 2011). J. Maynard Smith and E. Szathmáry's *The Major Transitions in Evolution* (W. H. Freeman, San Francisco, 1995) is an interpretation by leading evolutionary theoreticians of major events, ranging from the origin of life to the origins of societies and languages. The history of ecological diversification of animals is the subject of "Paleoecologic megatrends in marine Metazoa" by A. M. Bush and R. K. Bambach (*Annu. Rev. Earth Planet. Sci.* 39: 241–269, 2011). Paleontologist and developmental biologist Neil Shubin traces the history of evolution of the human body in *Your Inner Fish: A Journey into the 3.5-Billion-Year History of the Human Body* (Allen Lane/Pantheon, New York, 2008).

Terrestrial Ecosystems through Time: Evolutionary Paleoecology of Terrestrial Plants and Animals (edited by A. K. Behrensmeyer et al., University of Chicago Press, 1992) presents detailed summaries of changes in terrestrial environments and communities in the past. E. C. Pielou's *After the Ice Age: The Return of Life to North America* (University of Chicago Press, 1991) describes the effects of Pleistocene climate change on today's ecology and distribution of species.

Useful books on the evolution of major taxonomic groups include J. W. Valentine, *On the Origin of Phyla* (University of Chicago Press, 2004); P. Kenrick and P. R. Crane, *The Origin and Early Diversification of Land Plants* (Smithsonian Institution Press, Washington, D.C., 1997); R. L. Carroll, *Vertebrate Paleontology and Evolution* (W. H. Freeman, New York, 1988); M. J. Benton, *Vertebrate Palaeontology* (Blackwell, Malden, MA, 2008); M. J. Benton (ed.), *The Phylogeny and Classification of the Tetrapods* (Clarendon, Oxford, 1988); D. B. Weishampel, P. Dodson, and H. Osmolska, *The Dinosauria* (University of California Press, Berkeley, 1990); and D. R. Prothero, *After the Dinosaurs: The Age of Mammals* (Indiana University Press, Bloomington, 2006).

PROBLEMS AND DISCUSSION TOPICS

1. Why, in the evolution of eukaryotes, might it have been advantageous for separate organisms to become united into a single organism? Can you describe analogous, more recently evolved, examples of intimate symbioses that function as a single integrated organism?

2. Early in the origin of life, as it is presently conceived, there was no distinction between genotype and phenotype. What characterizes this distinction, and at what stage of organization may it be said to have come into being?

3. If we employ the biological species concept (see Chapter 9), when did species first exist? What were organisms before then, if not species? What might the consequences of the emergence of species have been for processes of adaptation and diversification?

4. How would you determine whether the morphological diversity of animals has increased, decreased, or remained the same since the Cambrian? What might bias your analysis?

5. Compare terrestrial communities in the Devonian and the Cretaceous periods. Discuss the differences between them in the diversity of plants and animals, and develop some hypotheses as to why those differences existed.

6. Read some papers (find them using any of several literature-search engines that your instructor can suggest) that make different estimates of the timing of either the origin of bilaterian animal phyla or the orders of mammals. How different are the estimates based on molecular clock evidence versus paleontological evidence? What might account for these differences, and how might they be resolved?

7. Animals that are readily classified into extant phyla, such as Mollusca and Arthropoda, appeared in the Cambrian without transitional forms that show how their distinctive body plans evolved. This "explosion" in fossil diversity had to come from somewhere. What are some of the best hypotheses explaining why animal fossils are not found before the Cambrian, despite molecular evidence suggesting divergence in the much more distant past than that?

8. What is the evidence that the megafaunal extinction in the Pleistocene was partly caused by humans?

9. Many species expanded or changed their geographic range after the last (Wisconsin) glacier retreated, about 12,000 years ago. What would the consequences have been for the evolution of those species? What were the effects of range changes on the species composition of ecological communities? (See also Chapters 9, 18, and 19.)

10. What are some ways in which the evolution of new life forms changed environments, from a very local scale to a planetary scale?

The Geography of Evolution

Some of the most avidly read books of the nineteenth century were the tales that explorers recounted of their travels in exotic lands. They were the *Discover* and *National Geographic* equivalents of their day. Some of them were written by naturalists, such as Alexander von Humboldt, Alfred Russel Wallace, and Charles Darwin, who climbed snow-clad mountains, traveled unexplored rivers, and suffered tropical diseases. They told stories not only of hardships and of encounters with indigenous peoples, but also of amazing animals and plants, utterly unlike any in Europe or North America. Some of these organisms—African elephants, Australian kangaroos, South American orchids—were already in zoos, botanical gardens, and museums, but these were only a few of the many thousands of species that naturalists retrieved from around the world. Every visitor to a zoo or garden might marvel at the giraffes and be intrigued by giant cacti or Bornean pitcher plants, but the naturalists went further: they asked why these creatures were found only in these remote regions. Why should apes be in Africa but not South America, and sloths in tropical America only? If European plants were found growing near American seaports, having been accidentally transported across the Atlantic, why had they not already occupied America? If those species could prosper in what proved to be a suitable new region, why weren't they already there?

Based both on their own travels and on specimens brought to Europe by explorers, naturalists started to describe the faunal and floral differences among regions of the world in the eighteenth century, and initiated the study of **biogeography**, the geographic distributions of organisms. By placing this information

Hoofed mammals in the order Cetartiodactyla are among the many groups of animals that do not extend east of Wallace's line, which separates a largely Asian fauna from a characteristic Australian fauna. This bearded pig (*Sus barbatus*), found in Borneo, is one of the easternmost hoofed mammals.

in an evolutionary context, Wallace and Darwin expanded, and largely created, a scientific framework for understanding geographic distributions. In some instances, the geographic distribution of a taxon may best be explained by historical circumstances; in other cases, ecological factors operating at the present time may provide the best explanation. Hence the field of biogeography may be roughly divided into **historical biogeography** and **ecological biogeography** [25, 29].

Biogeographic Evidence for Evolution

The geographic distributions of organisms provided both Darwin and Wallace with inspiration and with evidence that evolution had occurred. To us, today, the reasons for certain facts of biogeography seem so obvious that they hardly bear mentioning. If someone asks us why there are no elephants in the Hawaiian Islands, we will naturally answer that elephants couldn't get there. This answer assumes that elephants originated somewhere else: perhaps on a continent. But in a pre-evolutionary world view, the view of special divine creation that Darwin and Wallace were combating, such an answer would not do: the Creator could have placed each species anywhere, or in many places at the same time.

Darwin devoted two chapters of *On the Origin of Species* to showing that many biogeographic facts that make little sense under the hypothesis of special creation make a great deal of sense if a species (1) has a definite site or region of origin, (2) achieves a broader distribution by dispersal, and (3) becomes modified and gives rise to descendant species in the various regions to which it disperses. Darwin emphasized the following points:

First, he said, *"neither the similarity nor the dissimilarity of the inhabitants of various regions can be wholly accounted for by climatal and other physical conditions."* Similar climates and habitats, such as deserts and rainforests, occur in both the Old and the New World, yet the organisms inhabiting them are unrelated. For example, members of diverse plant families have adapted to deserts by convergent evolution; the cacti (family Cactaceae) are almost entirely restricted to the New World, but the cactuslike plants in Old World deserts are members of other families (**FIGURE 18.1**). All the monkeys in the New World belong to one anatomically distinguishable group (Platyrrhini), and all Old World monkeys to another (Catarrhini), even though they have similar habitats and diets.

Darwin's second point is that *"barriers of any kind, or obstacles to free migration, are related in a close and important manner to the differences between the productions [organisms] of various regions."* Darwin noted, for instance, that marine species on the eastern and western coasts of South America are very different.

FIGURE 18.1 Convergent growth form in desert plants. These plants, all leafless succulents with photosynthetic stems, belong to three distantly related families. (A) A cactus, *Stenocereus* (Cactaceae), in Oaxaca, Mexico. (B) A carrion flower of the genus *Stapelia* (Apocynaceae). These fly-pollinated succulents can be found from southern Africa to eastern India. (C) A member of the Euphorbiaceae (*Euphorbia candelabrum*) in Ethiopia, Africa. (A, C by D. J. Futuyma.)

(A)

(B)

(C)

Darwin's "third great fact" is that *inhabitants of the same continent or the same sea are related, although the species themselves differ from place to place.* He cited as an example the aquatic rodents of South America (the coypu and capybara), which are structurally similar to, and related to, South American rodents of the mountains and grasslands, not to the aquatic rodents (beaver, muskrat) of the Northern Hemisphere.

"We see in these facts," said Darwin, "some deep organic bond, throughout space and time, over the same areas of land and water, independently of physical conditions. ... The bond is simply inheritance [i.e., common ancestry], that cause which alone, as far as we positively know, produces organisms quite like each other."

For Darwin, it was important to show that a species had not been created in different places, but had a single region of origin, and had spread from there. He drew particularly compelling evidence from the inhabitants of islands. First, remote oceanic islands generally have precisely those kinds of organisms that are capable of long-distance dispersal and lack organisms that do not. For example, the only native mammals on many islands are bats. (Island species with poor dispersal ability, such as the dodo and other flightless birds, are closely related to strong flyers, and descended from them.) Second, many continental species of plants and animals have flourished on oceanic islands to which humans have transported them. Thus, said Darwin, "he who admits the doctrine of the creation of each separate species, will have to admit that a sufficient number of the best adapted plants and animals were not created for oceanic islands." Third, most of the species on islands are clearly related to species on the nearest mainland, implying that that was their source. This is the case, as Darwin said, for almost all the birds and plants of the Galápagos Islands. Island species often bear marks of their continental ancestry. For example, as Darwin noted, hooks on seeds are an adaptation for dispersal by mammals, yet on oceanic islands that lack mammals, many endemic plants nevertheless have hooked seeds. Fourth, the proportion of species that are restricted to an island is particularly high when the opportunity for dispersal to the island is low.

Wallace made a special study of species distributions, especially on islands, and is sometimes called the father of biogeography. He came to many of the same conclusions as Darwin—points that hold true today, after more than a century and a half of research. Our greater knowledge of the fossil record and of geological events such as continental drift and sea level changes has added to our understanding, but has not negated any of Darwin and Wallace's major conclusions.

Major Patterns of Distribution

The geographic distribution of almost every species is limited to some extent, and many higher taxa are likewise restricted (**endemic**) to a particular geographic region. For example, the salamander genus *Plethodon* is limited to North America, and *Plethodon caddoensis* occupies only the Caddo Mountains of western Arkansas. Some higher taxa are narrowly endemic (e.g., the kiwi family, Apterygidae, which is restricted to New Zealand), whereas others, such as the pigeon family (Columbidae), are almost cosmopolitan (found worldwide).

Wallace and other early biogeographers recognized that many higher taxa have roughly similar distributions, and that the taxonomic composition of the biota is more uniform within certain regions than between them. For example, Wallace discovered a sharp break in the taxonomic composition of animal species among the islands that lie between southeastern Asia and Australia: as far east as Borneo, most vertebrates belong to Asian families and genera, whereas the fauna to the east has Australian affinities. This faunal break has been called **Wallace's line** ever since. Based on these observations, Wallace designated several **biogeographic realms**—major regions that have characteristic animal and plant taxa—for terrestrial and freshwater organisms

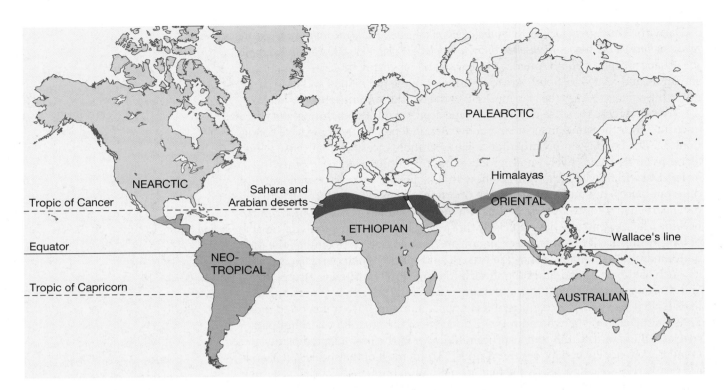

FIGURE 18.2 Biogeographic realms. The biogeographic realms recognized by Wallace are the Palearctic, Ethiopian, Oriental, Australian, Nearctic, and Neotropical. Note the position of Wallace's line.

that are still widely recognized today (**FIGURE 18.2**). These realms are the *Palearctic* (Eurasia and northern Africa), the *Nearctic* (North America), the *Neotropical* (Central and South America), the *Ethiopian* (sub-Saharan Africa), the *Oriental* (India and Southeast Asia), and the *Australian* (Australia, New Guinea, New Zealand, and nearby islands). These realms are more the result of Earth's history than of its current climate or land mass distribution. For example, Wallace's line separates islands that, despite their close proximity and similar climate, differ greatly in their fauna. The islands to the west were connected to the Asian mainland during periods of low sea level; those to the east were not.

Each biogeographic realm is inhabited by many higher taxa that are much more diverse in that realm than elsewhere, or are even restricted to that realm. For example, the endemic (or nearly endemic) taxa of the Neotropical realm include the Xenarthra (anteaters and allies), platyrrhine primates (such as spider monkeys and marmosets), most hummingbirds, a large clade of suboscine birds such as flycatchers and antbirds, many families of catfishes, and plant families such as the bromeliads (**FIGURE 18.3**).

The borders between biogeographic realms cannot be sharply drawn because some taxa infiltrate neighboring realms to varying degrees. In the Nearctic realm (North America), for instance, some species are related to, and have been derived from, Neotropical stocks: examples include an armadillo, an opossum, and Spanish moss (*Tillandsia usneoides*), a bromeliad that festoons southern trees.

Some taxa have **disjunct distributions**; that is, their distributions have gaps (**FIGURE 18.4**). Disjunctly distributed higher taxa typically have different representatives in each area they occupy. For example, many taxa are represented on two or more southern continents, including lungfishes, marsupials, cichlid fishes, and *Araucaria* pines. Another common disjunct pattern is illustrated by alligators (*Alligator*), skunk cabbages (*Symplocarpus*), and tulip trees (*Liriodendron*), which are among the many genera that are found both in eastern North America and in temperate eastern Asia, but nowhere in between [43]. Understanding how taxa became disjunctly distributed has long been a preoccupation of biogeographers.

FIGURE 18.3 Examples of taxa endemic to the Neotropical biogeographic realm. (A) A giant anteater (order Xenarthra). (B) The chestnut-crowned antpitta (Formicariidae) represents a huge evolutionary radiation of suboscine birds in the Neotropics. (C) This armored catfish belongs to the Callichthyidae, one of many families of freshwater catfishes restricted to South America. (D) Most bromeliads (Bromeliaceae) live on branches of trees in wet forests in tropical America. (Bromeliaceae are almost Neotropical endemics, but one species has dispersed to western Africa.)

FIGURE 18.4 Examples of disjunct distributions. (A) Among the many genera of plants found in both eastern North America (on left) and eastern Asia (on right) are lady-slipper orchids (*Cypripedium*). (B) Boas (family Boidae) are distributed in tropical America and in southern Pacific islands. On the left is the South American emerald tree boa (*Corallus caninus*); on the right is a Pacific boa, *Candoia aspera*, from New Guinea. (A, B by D. J. Futuyma.)

FIGURE 18.5 History of range expansion of the European starling (*Sturnus vulgaris*) following its introduction into New York City in 1896 (After [3].)

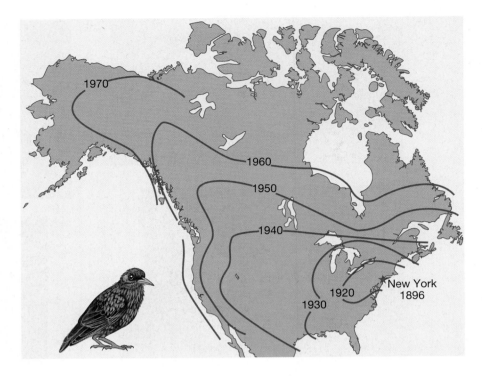

Historical factors affecting geographic distributions

The geographic distribution of a taxon is affected by both current and historical factors. The limits to its distribution may be set by geological barriers that it has not crossed or by current ecological conditions to which it is not adapted. In this section we focus on the historical processes that have led to the current distribution of a taxon: extinction, dispersal, and vicariance.

The distribution of a species may have been reduced by the extinction of some populations, and that of a higher taxon by the extinction of some constituent species. For example, the horse family (Equidae) originated and became diverse in North America, but it later became extinct there; only the African zebras and the Asian wild asses and horses have survived. (Horses returned to North America with European colonists.) Likewise, extinction is the cause of the disjunction between related taxa in eastern Asia and eastern North America. During the Paleogene, many plants and animals spread throughout the northern regions of North America and Eurasia. Their spread was facilitated by a warm, moist climate and by land connections from North America to both Europe and Siberia. Many of these taxa became extinct in western North America in the Neogene as a result of mountain uplift and a cooler, drier climate, and were later extinguished in Europe by Pleistocene glaciations [40, 43].

Species expand their ranges by **dispersal** (movement of individuals). Some species of plants and animals can expand their ranges very rapidly. Within the last 200 years, many species of plants accidentally brought from Europe by humans have expanded across most of North America from New York and New England, and some birds, such as the European starling (*Sturnus vulgaris*) and the house sparrow (*Passer domesticus*), have done the same within a century (**FIGURE 18.5**). Other species have crossed major barriers on their own. The cattle egret (*Bubulcus ibis*) was found only in tropical and subtropical parts of the Old World until about 140 years ago, when it arrived in South America, apparently unassisted by humans (**FIGURE 18.6**). It has since spread throughout the warmer parts of the New World.

Vicariance refers to the separation of populations of a widespread species by barriers arising from changes in geology, climate, or

FIGURE 18.6 A cattle egret (*Bubulcus ibis*) accompanying a longhorn cow. This heron feeds on insects stirred up by grazing ungulates both in the Old World and in the New World, to which it dispersed about 140 years ago.

A. nuttingi

A. millsae

FIGURE 18.7 The snapping shrimps *Alpheus nuttingi*, found on the Atlantic side of the Isthmus of Panama, and *A. millsae*, found on the Pacific side, are sister species that evolved from a common ancestor that became divided into two populations as the isthmus formed. Their geographic distributions illustrate vicariance. (Courtesy of Arthur Anker.)

habitat. The separated populations diverge, and they often become different subspecies, species, or higher taxa. For example, in many fishes, shrimps, and other marine animal groups, the closest relative of a species on the Pacific side of the Isthmus of Panama is a species on the Caribbean side of the isthmus. Each pair of species has descended from a broadly distributed ancestral species that was sundered by the rise of the isthmus during the Pliocene, about 3 Mya (**FIGURE 18.7**) [22, 23, 30]. Vicariance sometimes accounts for the presence of related taxa in disjunct areas.

In many cases, dispersal, vicariance, and extinction together explain distributions. For example, during the Pleistocene glaciations, species shifted their ranges by dispersing into new regions. Some northern, cold-adapted species became distributed far to the south. When the climate became warmer, these species recolonized northern regions, and southern populations became extinct, except for populations of some such species that survived on cold mountaintops (**FIGURE 18.8**).

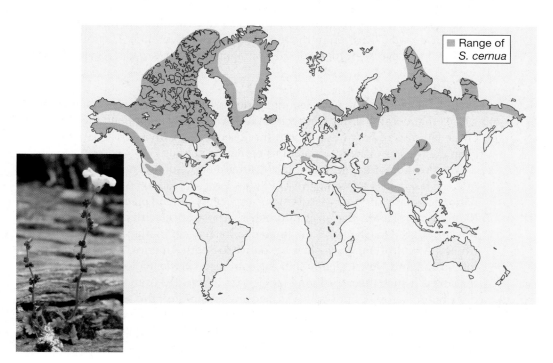

Range of
S. cernua

FIGURE 18.8 The disjunct distribution of a saxifrage (*Saxifraga cernua*) in northern and mountainous regions of the Northern Hemisphere. Relict populations have persisted at high elevations following the species' retreat from the southern regions that it occupied during glacial periods. (After [2]; photo courtesy of Egil Michaelsen and the Norwegian Botanical Association.)

(Populations or species that have been left behind in this way may be called *relicts*). In this case, dispersal expanded the range, and extinction of populations in intervening habitat caused the vicariant disjunction of populations.

Historical Explanations of Geographic Distributions

Biogeographers have used a variety of guidelines for inferring the histories of distributions. Some of these guidelines are well founded. For example, the distribution of a taxon cannot be explained by an event that occurred before the taxon originated: a genus that originated in the Miocene, for example, cannot have achieved its distribution by continental movements that occurred in the Cretaceous. Some other guidelines are more debatable. Some authors in the past assumed that a taxon originated in the region where it is presently most diverse. But this need not be so; as we have seen, wild species of the horse family today are found only in Africa and Asia, even though the fossil record shows that the family originated in North America. Changes in environment can radically alter a taxon's distribution. Species of beetles that were present in England during the Pleistocene became extinct there (probably due to glacial changes in climate) and are restricted today to various remote parts of Asia and Africa [5].

Several sources of evidence cast light on the historical causes of geographic distributions. The fossil record can show that a taxon proliferated in one area before appearing in another, and geological data may describe the appearance or disappearance of barriers [24]. For example, fossil armadillos are limited to South America throughout the early Cenozoic and are found in North American deposits only from the Pliocene and Pleistocene, after the Isthmus of Panama was formed. We may infer that armadillos dispersed into North America from South America. Paleontological data must be interpreted cautiously, because a taxon may be much older, and have inhabited a region longer, than a sparse fossil record shows.

Phylogenetic methods are the foundation of most modern studies of historical biogeography. Inferring ancestral distributions from a phylogeny is much like inferring ancestral character states (see Figure 2.16), although biologists are continuing to develop phylogenetic methods for determining the roles of dispersal, vicariance, and extinction in the history of distributions [35]. The following sections include several examples of biogeographic inferences made from phylogenies.

Vicariance

Changes in climate and in the configuration of land and sea have separated populations that became different species. We have already mentioned the emergence of the Isthmus of Panama, which fully closed about 3 Mya and separated Caribbean and western Pacific populations of marine species [30].

The breakup of Pangaea first into Laurasia and Gondwana, and later into the modern land masses (see Figure 17.19), has long seemed to be a wonderful explanation for many disjunct distributions, especially among pieces of Gondwana in the Southern Hemisphere. Cichlid fishes, whose spectacular adaptive radiation was introduced in Chapter 9, are limited to fresh water in Madagascar, Africa, and tropical America. *Araucaria* pines are native to South America, Australia, and Norfolk Island and New Caledonia in the southern Pacific. Biologists hypothesized that in each such case, the living species are descended from ancestors that were distributed across Gondwana and became isolated on the several land masses they now occupy, after Gondwana started splitting apart. But this scenario can apply only to clades that are older than the split between the land masses they now

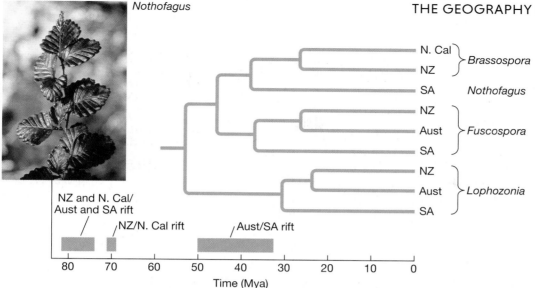

FIGURE 18.9 A simplified phylogeny of the four major lineages (subgenera) of southern beeches (*Nothofagus*), showing branching dates estimated by DNA sequence difference. In the subgenera *Fuscospora* and *Lophozonia*, closely related species are found in New Zealand (NZ) and Australia (Aust), even though these land masses separated long before the rift between Australia and South America (SA). The brown bars show estimated times at which Gondwanan land masses separated, antedating the divergence of *Nothofagus* lineages. Consequently, vicariance by continental drift does not explain the disjunct distribution of these plants. N. Cal, New Caledonia. (After [4]; photo by D. J. Futuyma.)

inhabit. Most of the fragmentation of Gondwana happened in the Mesozoic (see Figure 17.19C), before the K/Pg mass extinction 66 Mya [39].

In many cases, disjunct distributions must be attributed to dispersal [7, 8], and it appears likely that dispersal explains more disjunct distributions than vicariance. Explaining the African/South American disjunction of the cichlid fishes would require that the family arose more than 110 Mya, but both fossil and DNA evidence indicates that they are at most 65 My old [14]. They must have dispersed, somehow, across the Atlantic Ocean when it was considerably narrower than it is now. (These freshwater fishes do not tolerate salt water, so they present a conundrum.) The southern beeches (*Nothofagus*), distributed in southern South America, Australia, New Zealand, and on the island of New Caledonia have a classic Gondwanan distribution, but the fossil-calibrated sequence divergence of several genes indicates that two subgenera that are in both New Zealand and Australia evolved more than 30 My after these land masses separated (**FIGURE 18.9**) [4]. However, some clades are indeed older than the continents on which they occur. These include many groups of insects, and the marsupials—opossums in tropical America and kangaroos, koalas, and many others in Australia—which seem to have spread through Antarctica between South America and Australia before these separated, about 50 Mya [39]. The cypresses (Cupressaceae) originated in the Triassic, when Pangaea was intact; one of the two major clades is distributed mostly in Eurasia and North America (formerly parts of Laurasia), and the other is distributed mostly in South America and other parts of Gondwana (**FIGURE 18.10**) [26].

Dispersal

The normal dispersal processes that occur every generation account for the gradual spread of species via more or less suitable habitat into new areas, where they may differentiate into distinct species (allopatric or parapatric speciation;

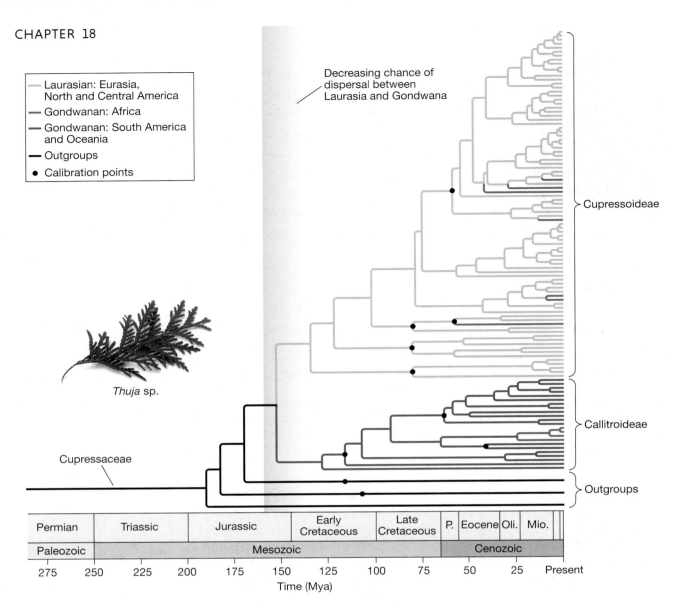

Thuja sp.

FIGURE 18.10 A phylogeny of the cypresses (family Cupressaceae), with a distribution that is ascribed partly to vicariance. Most of the species in the two major subfamilies are located in the continents derived from Laurasia (Eurasia and North and Central America) and from Gondwana (South America and the South Pacific islands [Oceania]). Calibrated by multiple fossils, this DNA-based phylogeny implies that the family originated in the Triassic and divided into the two major subfamilies about 153 Mya, when Laurasia and Gondwana were separating. The orange bar marks the time when floristic interchange between Laurasia and Gondwana was becoming unlikely. The blue branches show that dispersal has also occurred, into Africa from Laurasian and South American regions. (After [26].)

see Chapter 9). For example, as a land corridor gradually formed between North America and South America in the late Miocene and Pliocene, North American groups such as deer, horses, camelids, and cats spread deep into South America, and a few South American mammal families—armadillos, opossums, and porcupines, among others—colonized North America. Some groups of plants, such as the Malpighiaceae, gradually dispersed from tropical America through North America and Europe into Africa, during warm Paleogene intervals (**FIGURE 18.11A**) [6].

(A)

(B)

FIGURE 18.11 Some disjunct distributions have resulted from dispersal. (A) The family Malpighiaceae, represented here by the South American *Malpighia glabra* (left) and the African *Acridocarpus natalitius* (right), moved by progressive dispersal from tropical America through North America and Europe to Africa. (B) The giant kapok tree (*Ceiba pentandra*) emerges above the canopy of rainforests in both tropical America and western Africa.

Organisms sometimes disperse long distances over unsuitable habitat, resulting in rare, even unlikely, colonization. Often a colonizing species gives rise to diverse descendant species, sometimes forming an adaptive radiation. The ancestor of the Galápagos finches (see Figure 2.2) in the Galápagos Islands was a member of a group of South American species known as grassquits. Many insects and nonflying animals such as lizards, as well as seeds of diverse plants, are carried to oceanic islands by masses of vegetation that have been swept to sea by floods; some of these species are transported as water-resistant eggs. A group of South American rodents that includes porcupines and chinchillas stems from a species that arrived by transoceanic dispersal from Africa in the Miocene, as did the ancestor of South American primates. At least 110 genera of plants that occur on both sides of the tropical Atlantic are too young to have occupied pre-rifting Gondwana, and have dispersed across the ocean, most by floating [36]. Even some individual species have dispersed from South America to Africa, such as the kapok tree (*Ceiba pentandra*), a giant of tropical rainforests (**FIGURE 18.11B**) [10].

Biogeographers have studied species in the Hawaiian Islands extensively, because the islands are so remote and have an interesting geological history. The islands have formed as a tectonic plate has moved northwestward, like a conveyor belt, over a "hot spot," causing the sequential formation of volcanic peaks. This process has been going on for tens of millions of years, and a string of submerged volcanoes that once projected above the ocean surface lies northwest of the present islands. Of the current islands, Kauai, at the northwestern end of the archipelago, is about 5.1 My old; the southeasternmost island, the "Big Island" of Hawaii, is the youngest and is less than 500,000 years old (**FIGURE 18.12A**).

Given the geological history of the archipelago, the simplest phylogeny expected of a group of Hawaiian species would be a "comb," in which the most basal lineages occupy Kauai and the youngest lineages occupy Hawaii. This pattern would

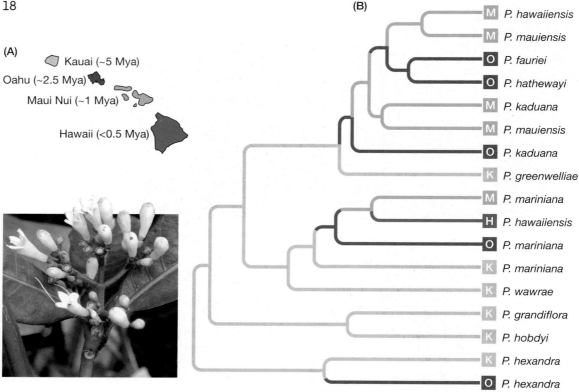

FIGURE 18.12 Dispersal accounts for the distribution of species in the Hawaiian Islands. (A) The present Hawaiian archipelago, showing the approximate dates of each island's formation. (B) A phylogeny of Hawaiian species of *Psychotria* trees shows successive shifts from older islands to younger islands, as expected if younger islands were colonized after they formed. (After [34].)

occur if species successively dispersed to new islands as they were formed, did not disperse from younger to older islands, and did not suffer extinction. A phylogenetic analysis of trees in the genus *Psychotria* revealed just this pattern (**FIGURE 18.12B**) [34]. A similar pattern has been found for other Hawaiian taxa, such as the cricket genus *Laupala* [27].

Phylogeography

Phylogeography is the description and analysis of the processes that govern the geographic distribution of lineages of genes, especially within species and among closely related species [1, 16, 21]. These processes include the dispersal of the organisms that carry the genes, so phylogeography provides insight into the past movements of species—including humans (see Chapter 21)—and the history by which dispersal and vicariance have determined their present distributions. Phylogeographic studies find the phylogenetic relationships among populations, and can be used to infer their history of spread. For example, the tree *Symphonia globulifera* was shown to have dispersed from West Africa to South America more than 15 Mya, after which it crossed the Andes and spread through western South America to Central America (**FIGURE 18.13**).

Phylogeographic studies have traced the expansion of species from Pleistocene refugia after glacial periods, and have shown how Pleistocene events such as sea level changes have shaped some geographic patterns of genetic variation. For example, the genetic differences among populations of many species of freshwater fishes and other species in the coastal plain of the southeastern United States show that they were separated into western and eastern populations in the past (**FIGURE 18.14**), probably by high sea level during interglacial episodes.

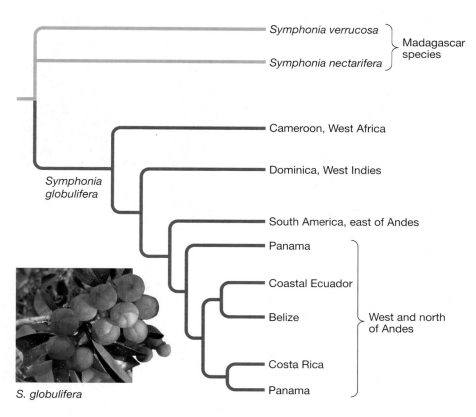

S. globulifera

FIGURE 18.13 Relationships among populations of the widespread tree *Symphonia globulifera*, a member of an Old World genus with species in Madagascar. *S. globulifera* spread from West Africa to the West Indies and eastern South America, and from there to the trans-Andean region, including coastal Ecuador and Central America. (After [9], courtesy of C. W. Dick.)

Geographic Range Limits: Ecology and Evolution

The geographic distribution of a species results not only from the history of its ancestors, but also from current factors, a major subject of ecological biogeography. Several difficulties can retard or prevent a species from expanding its range. It must disperse to the new region. Individuals in sexually reproducing species must find mates, which may be difficult if there are few individuals and the initial population density is very low. They must be able to survive physical conditions, find suitable resources (food, habitat), and contend with other species, such as competitors, predators, and parasites.

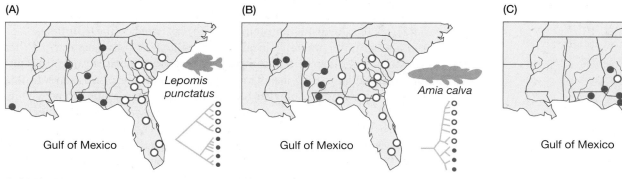

FIGURE 18.14 Patterns of genetic divergence among populations may show that many species have a similar history of subdivision (vicariance). Gene trees of these species show sharp division between eastern and western populations of both freshwater and terrestrial species in the coastal plain of the southeastern United States. These populations are thought to have been isolated and to have diverged in two refugia during the Pleistocene. (A) Spotted sunfish (*Lepomis punctatus*). (B) Bowfin (*Amia calva*). (C) Sister species of mints in the genus *Dicerandra*. (After [41].)

FIGURE 18.15 Self-fertilizing plants ("selfers") have larger geographic ranges than outcrossers. The histogram shows the relative range sizes of selfers compared with outcrossers in 20 clades of flowering plants. Each column that extends above the horizontal axis shows a clade in which selfers have larger ranges than outcrossers. In all but one clade, selfers have larger ranges. (Clades are arranged in descending order of relative range size of the selfers.) *Inset:* The flowers of outcrossing and selfing individuals of a single species of water hyacinth (*Eichhornia paniculata*). (After [15]; photo courtesy of S. C. H. Barrett.)

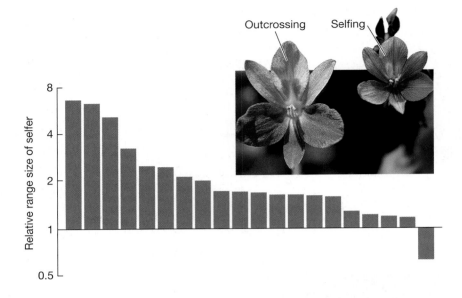

DISPERSAL LIMITATIONS   The border of a species' geographic range is sometimes set by utterly unfavorable conditions, as when the distribution of a terrestrial organism stops at the ocean's edge. This may be a nonequilibrial (i.e., temporary) border, for the species might thrive beyond the barrier if individuals eventually manage to cross it. The plants with trans-Atlantic distributions show that this can happen. In some cases, obtaining mates is a problem. Plant species that reproduce by self-fertilization have consistently broader latitudinal ranges than congeneric outcrossing species (**FIGURE 18.15**), probably because they do not need another plant's pollen [15]. This suggests that the outcrossing species might still be slowly spreading, but are retarded in their progress.

ECOLOGICAL NICHES   A species can persist only if the organism can tolerate each of several environmental conditions, such as the range of temperatures, the amount of available water, and the availability of suitable food items. That is, both abiotic and biotic aspects of the environment can affect the species' distribution. G. Evelyn Hutchinson [18], a leading ecologist, defined the **fundamental ecological niche** of a population as the set of all those environmental conditions in which a species can have positive population growth. A particular locality falls within the species' fundamental niche if all the relevant environmental factors fall within the organism's tolerance limits, but will fall outside the niche if any one variable, such as lowest winter temperature, falls outside these limits.

Even in a potentially habitable locality, competitors or predators may further restrict a species' distribution. The **competitive exclusion principle** holds that species that are too similar in their use of food or other limiting resources cannot coexist indefinitely. Accordingly, ecologists have described many examples in which one species occupies a broader range of elevation or habitat where a related species is absent than where it is present [3]. Competition can affect whether or not two species that have formed by allopatric speciation can spread into each other's range (see Chapter 9). Species of tropical American woodcreepers and ovenbirds that have originated by allopatric speciation become sympatric much faster if they differ morphologically and forage in different ways for different prey (**FIGURE 18.16**) [32].

The distributions of many species are correlated with climate variables (especially aspects of temperature and rainfall). Some species have shifted their geographic or elevational ranges in recent decades, apparently in response to

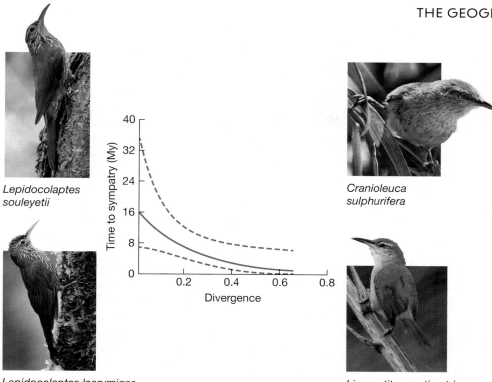

Lepidocolaptes souleyetii

Lepidocolaptes lacrymiger

Cranioleuca sulphurifera

Limnoctites rectirostris

FIGURE 18.16 Since speciation occurred, pairs of Neotropical ovenbird species have taken much longer to become sympatric if their ecological niches are very similar than if they have diverged more appreciably in their niches. This suggests that competition tends to prevent ecologically similar species from becoming sympatric. The difference in their niches (i.e., how they feed) is inferred from the difference in morphological features such as the bill. For example, two morphologically similar *Lepidocolaptes* woodcreepers (left) both feed by probing for insects that hide on tree trunks; the two birds at right (a spinetail, *Cranioleuca*, and a reedhaunter, *Limnoctistes*) both inhabit reedy marshes, but use their differently shaped bills to glean insects from different plant structures. (Graph after [32].)

human-caused climate change; for instance, both the northern (or higher) and southern (or lower) range limits of several butterflies have shifted to higher latitudes (or elevations) [31]. The northernmost latitude reached by amphibians and nonavian reptiles introduced into North America by humans matches the species' northernmost limits in their native ranges, suggesting that climate tolerance determines which species can successfully become established (**FIGURE 18.17**) [45].

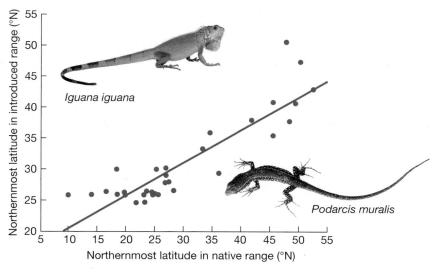

Iguana iguana

Podarcis muralis

FIGURE 18.17 The correlation between the northernmost limit of the range of 35 species of amphibians and reptiles in their native region and in North America, where they are introduced. Most have not spread farther north than they occur in their native range. For example, the tropical American green iguana (*Iguana iguana*, left) has not spread north of Florida, whereas the European wall lizard (*Podarcis muralis*, right) is in New York. (After [45].)

Related species often exhibit **phylogenetic niche conservatism**: similar ecological requirements that they have inherited from their common ancestor. For example, Robert Ricklefs and colleagues have found that congeneric plants in eastern Asia and eastern North America have similar latitudinal and climate distributions, as do genera shared between North America and Europe [33, 38]. Similarly, many lineages of herbivorous insects have remained associated with the same genus or family of food plant; some of these associations have remained unchanged for more than 40 My [47].

Niche conservatism contributes to our understanding of the geographic distributions of many clades [45]. For instance, oaks (*Quercus*) and dogwoods (*Cornus*) are among the many plant taxa that occur in temperate regions of eastern North America, Asia, and Europe but have not adapted to warm tropical environments [11]. Niche conservatism underlies the observation that many species shifted their geographic ranges during the Pleistocene, rather than adapting in situ to changes in climate (see Chapter 17). These observations raise important questions about the ability of species to adapt to new environmental conditions.

What, however, accounts for niche conservatism? Why do species not evolve broader tolerances, and steadily expand their geographic range or the range of habitats or resources they use? We discussed these questions in Chapters 8 and 11 and saw that they have been only partly answered.

Geographic Patterns of Diversity

The field of community ecology is concerned with explaining the species diversity, species composition, and trophic structure of assemblages of coexisting species (often called communities). Both ecological and historical biogeography bear on these topics, since the geographic ranges of species determine whether or not they might coexist.

A long-standing topic in community ecology is a pattern called the *latitudinal diversity gradient*: the numbers of species (and of higher taxa such as genera and families) decline with increasing latitude, both on land and in the ocean (**FIGURE 18.18**). Most taxa of terrestrial animals and plants are far more diverse in tropical regions, especially in lowlands with abundant rainfall, than in extratropical regions.

Three major hypotheses have been proposed to account for this pattern [12, 28]. First, ecological factors might enable more tropical species to coexist in a stable community (**FIGURE 18.19A**). These factors might include high productivity because of abundant solar energy, or fine partitioning of food resources among many species. Alternatively, the pattern might be explained by evolutionary dynamics over many millions of years [37, 44]. One of the leaders of the evolutionary synthesis, the botanist G. Ledyard Stebbins, took this perspective when he suggested that tropical areas might be a "cradle," in which new species arise at a high rate, or a "museum," in which ancient lineages persist [42]. Related to the "cradle" idea, the "diversification rate hypothesis" holds that the rate of increase in diversity has been greater in the tropics for a long time because of a higher speciation rate, a lower extinction rate, or both (**FIGURE 18.19B**). For example, David Jablonski and colleagues determined that new genera of marine bivalves have arisen mostly in tropical areas throughout the last 11 My and have spread from there toward higher latitudes while persisting in tropical regions as well [19].

The "museum" idea is expressed today by the "time and area hypothesis" (**FIGURE 18.19C**), which holds that most lineages have been accumulating species for a

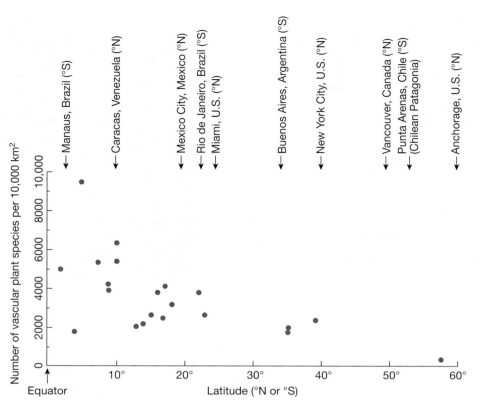

FIGURE 18.18 An example of the latitudinal diversity gradient. The number of species of vascular plants in various regions of North and South America drops more than tenfold between the Equator and high-latitude regions. The cities provide a latitudinal frame of reference; they do not correspond to the data points. (After [17].)

longer time in tropical than in extratropical environments. During the Cretaceous and the first 60 My of the Cenozoic, Earth was warmer than it is today, and much more of the globe had a tropical climate than now. For that reason, most lineages originated in tropical climates, and the relatively few lineages that have evolved adaptations to the stressful temperatures and seasonal fluctuations in food supply that are typical of the temperate zone are younger lineages that have not had time to become as diverse. Thus this hypothesis is based on phylogenetic niche conservatism [44].

Although productivity seems to have an important effect, the time and area hypothesis has been supported by many recent studies. For example, the phylogeny of tree frogs (Hylidae) indicates that all the major lineages of tree frogs and their common ancestors were distributed in tropical America (**FIGURE 18.20A**). The temperate zone has been invaded by only three lineages. Moreover, the number of species in each region is positively correlated with the time since tree frog

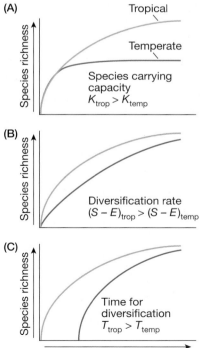

FIGURE 18.19 Three models of species accumulation that have been proposed to account for the latitudinal diversity gradient. (A) Some ecological hypotheses propose that tropical locations can support a higher equilibrium number of species ("carrying capacity," K) than temperate localities. (B) The diversification rate (difference between the speciation rate, S, and the extinction rate, E) might be higher in the tropics. Species numbers have not necessarily reached an equilibrium carrying capacity. (C) Lineages diversify at the same rate, but started to diversify more recently in the temperate zone than in the tropics, perhaps because they originated in tropical environments and only recently adapted to the temperate zone. (After [28].)

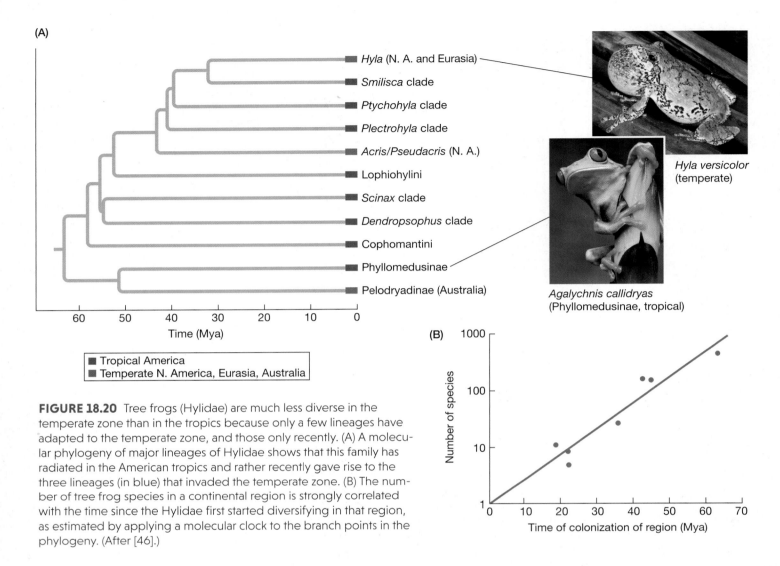

(A)

Hyla (N. A. and Eurasia)

Smilisca clade

Ptychohyla clade

Plectrohyla clade

Acris/Pseudacris (N. A.)

Lophiohylini

Scinax clade

Dendropsophus clade

Cophomantini

Phyllomedusinae

Pelodryadinae (Australia)

Time (Mya)

Hyla versicolor
(temperate)

Agalychnis callidryas
(Phyllomedusinae, tropical)

■ Tropical America
■ Temperate N. America, Eurasia, Australia

(B)

Number of species

Time of colonization of region (Mya)

FIGURE 18.20 Tree frogs (Hylidae) are much less diverse in the temperate zone than in the tropics because only a few lineages have adapted to the temperate zone, and those only recently. (A) A molecular phylogeny of major lineages of Hylidae shows that this family has radiated in the American tropics and rather recently gave rise to the three lineages (in blue) that invaded the temperate zone. (B) The number of tree frog species in a continental region is strongly correlated with the time since the Hylidae first started diversifying in that region, as estimated by applying a molecular clock to the branch points in the phylogeny. (After [46].)

clades first inhabited the region (**FIGURE 18.20B**). It appears that tropical regions have simply had more time to accumulate species [46]. In agreement with this hypothesis, Paul Fine and Richard Ree determined that the number of tree species in tropical, temperate, and boreal ecosystems on each continent is correlated with an index that integrates the area that each of these ecosystems has occupied since the Miocene, Oligocene, or even as far back as the Eocene [13]. A similar model accounts for much of the regional variation in the species richness of vertebrates across the world [20]. Tropical environments and vegetation have occupied larger areas for a longer time than other environments, so that is where most genera and species arose.

Go to the
Evolution Companion Website
EVOLUTION4E.SINAUER.COM
for data analysis and simulation exercises, quizzes, and more.

SUMMARY

- Biogeography, the study of organisms' geographic distributions, has both historical and ecological components. Certain distributions are the consequence of long-term evolutionary history; others are the result of current ecological factors.

- The geographic distributions of organisms provided Darwin and Wallace with some of their strongest evidence for the reality of evolution.

- The historical processes that affect the distribution of a taxon are extinction, dispersal, and vicariance (fragmentation of a continuous distribution by the emergence of a barrier). These processes may be affected or accompanied by environmental change, adaptation, and speciation.

- Histories of dispersal or vicariance can often be inferred from phylogenetic data.

- Disjunct distributions are attributable in some instances to vicariance, but dispersal seems to be the more common cause.

- Genetic patterns of geographic variation within species can provide information on historical changes in a species' distribution.

- The local distribution of species is affected by ecological factors, including both abiotic aspects of the environment and biotic features such as competitors and predators. Why species do not enlarge their ranges indefinitely, by incrementally adapting to conditions farther and farther away, is a major question in evolutionary biology.

- Geographic patterns in the number and diversity of species may stem partly from current ecological factors, but long-term evolutionary history also may explain them.

TERMS AND CONCEPTS

biogeographic realm

biogeography

competitive exclusion principle

disjunct distribution

dispersal

ecological biogeography

endemic

fundamental ecological niche

historical biogeography

phylogenetic niche conservatism

phylogeography

vicariance

Wallace's line

SUGGESTIONS FOR FURTHER READING

M. V. Lomolino, B. R. Riddle, R. J. Whittaker, and J. H. Brown, *Biogeography*, 5th edition (Sinauer Associates, Sunderland, MA, 2017), is the leading textbook of biogeography.

A. de Queiros has written about the role of dispersal in the evolution of distributions in a book for a general audience, *The Monkey's Voyage: How Improbable Journeys Shaped the History of Life* (Basic Books, New York, 2014).

Phylogeography is treated in depth by J. C. Avise in *Phylogeography* (Harvard University Press, Cambridge, MA, 2000), and human phylogeography is included in E. E. Harris's *Ancestors in Our Genome: The New Science of Human Evolution* (Oxford University Press, Oxford, UK, 2015).

PROBLEMS AND DISCUSSION TOPICS

1. Until recently, the plant family Dipterocarpaceae was thought to be restricted to tropical Asia, where many species are ecologically dominant trees. However, a new species of tree in this family was discovered in the rainforest of Colombia, in northern South America. What hypotheses could account for this tree's presence in South America, and how could you test those hypotheses?

2. Except for birds and bats, there are almost no native land vertebrates in New Zealand. There is one native frog species, a few lizard species, and several species of flightless birds. There are no snakes, freshwater fishes, or terrestrial mammals. What might explain this situation? Is this biota more likely derived by dispersal from another region or by vicariance? What is the significance of the missing elements, such as the freshwater fishes?

3. The ratites are a very old clade of flightless birds that include the ostriches in Africa, rheas in South America, emu and cassowaries in Australia, and kiwis and recently extinguished moas in New Zealand. South American tinamous, which are capable of flight, are closely related to the ratites. The "Gondwanan distribution" of these birds has often been attributed to vicariance, but some researchers have questioned this phylogeny and distributional history. Read several phylogenetic studies of the ratites and discuss how best to explain their distribution: A. Cooper et al., 2001, *Nature* 409: 704–707; O. Haddrath and A. J. Baker, 2001, *Proc. Royal Soc. Lond.* B 268: 939–945; S. J. Hackett et al., 2008, *Science* 320: 1763–1768; and A. J. Baker et al., 2014, *Mol. Biol. Evol* 31: 1686–1696.

4. In Chapters 4 and 6 you saw that many characteristics of most species have the genetic variation that is required for those characteristics to evolve, and that many examples of rapid adaptation to human-altered environments have been documented. Discuss whether or not this observation is inconsistent with the fact that many organisms display phylogenetic niche conservatism.

5. In some cases, it can be shown that species are physiologically incapable of surviving temperatures that prevail beyond the borders of their range. Do such observations prove that cold regions have low species diversity because of their harsh physical conditions?

6. By far the most effective way of saving endangered species is to preserve large areas that include their habitat. For social, political, and economic reasons, the number and distribution of areas that can be allocated as preserves are highly limited. It might be easier to save more species if areas of endemism were correlated among different taxa, such as plants, birds, and mammals. Are they correlated? (See, for example, N. Myers et al., 2000, *Nature* 403: 853–858; J. R. Prendergast et al., 1993, *Nature* 365: 335–337; and A. P. Dobson et al., 1993, *Science* 275: 550–553.)

7. Would you expect large numbers of species in a region to have had similar histories of geographic distribution? Why or why not? How could you use phylogeographic analyses, such as illustrated in Figure 18.14, to address this question?

8. In what ways have human activities influenced the biogeographic distribution of animals? How have humans caused animal ranges to expand or contract?

The Evolution of Biological Diversity

19

The exploration of space must count as one of humanity's greatest, most astonishing achievements. Spacecraft have landed on the moon and Mars and have passed close by Jupiter and even Pluto. Yet even as we plumb the secrets of our solar system, we remain surprisingly ignorant of what the entomologist Howard Evans [15] called "life on a little known planet"—our own.

No one can tell you how many species are on Earth—even to the nearest million! Some biologists have estimated that about 1.5 million species of eukaryotes have been discovered and named, but even this is a rough estimate. Experts on insects, mites, nematodes, fungi, and many other groups of organisms know that a far greater number of species have yet to be described and named. The best recent estimate of the number of existing species is about 5 million—with a margin of error of 3 million [12a]. This doesn't include prokaryotes—archaea and bacteria. We know from DNA samples that there are thousands of distinct and unnamed prokaryote genomes in any sample of soil or seawater, or for that matter, on and within a human body. And even among animals, the number of existing species is probably less than 1 percent of all the species that have ever lived on Earth. The diversity of life is truly overwhelming.

What we do know about this diversity is that it is very unevenly distributed among major groups (higher taxa). There are about 220 living species in the pine family, but the orchid family has about 18,000 species and the sunflower family about 23,000. Among the orders of insects, there are 500 known species

A diverse collection of weevils, scarabs, long-horned and metallic wood-borers, and other beetles. Beetles are by far the largest order of insects, with more than 350,000 described species, and untold numbers yet to be described or discovered. What accounts for their amazing diversity?

FIGURE 19.1 Contrasts in species richness. (A) The single species of Ginkgoaceae (*Ginkgo biloba*) and (B) one of the more than 18,000 species of Orchidaceae (*Ophrys apifera*). (C) The webspinners, order Embioptera, are far less diverse than (D) the beetles, order Coleoptera (here *Trachelophorus giraffa*). (E) The order Tubulidentata has a single living member, the African aardvark (*Orycteropus afer*). (F) The order Rodentia includes more than 2280 species, among them this greater Egyptian jerboa, *Jaculus orientalis*.

of webspinners (Embioptera), but 350,000 beetles (Coleoptera), with possibly a million others awaiting description [27]. The orders of mammals range from the single species of aardvark to more than 2280 species of rodents (**FIGURE 19.1**).

These contrasts raise a host of questions. Has the number of species on Earth increased steadily since the origin of life? Do we happen to live when diversity is at its highest point ever, or has diversity fluctuated? Have some groups dwindled even as others have increased, and if so, why? Why are there so many more kinds of beetles than webspinners—or almost anything else? Do groups with more species produce new species at a higher rate, or are they more resistant to extinction? Is there any limit to the possible number of species, and has that limit been reached?

Biodiversity can be studied from the complementary perspectives of ecology and evolutionary history. Ecologists focus primarily on factors that operate over short time scales to influence diversity within local habitats or regions. But factors that operate on longer time scales also affect diversity. On a scale of millions of years, extinction, adaptation, speciation, climate change, and geological change create the potential for entirely new assemblages of species. Understanding factors that have altered biodiversity in the past may help us predict how diversity will be affected by current and future environmental changes, such as the global climate change that is now under way as a result of human use of fossil fuels.

Estimating and Modeling Changes in Biological Diversity

In most evolutionary studies, "diversity" refers to the number of taxa, such as genera or species. (The latter is often called **species richness**.) Over long time scales or large areas, diversity is often estimated by compiling records, such as the publications or museum specimens that have been accumulated by many investigators, into faunal or floral lists of species. Changes in diversity over time are analyzed in two major ways: by paleontology and by phylogenetic analysis of living species.

Both approaches begin with a simple model of change in diversity over time. The number of taxa (N) changes over time by speciation and extinction. These events are analogous to the births and deaths of individual organisms in a population, so models of population growth have been adapted to describe changes in taxonomic diversity. Suppose there are N species alive at a given time. We use S to represent the speciation rate, that is, the probability that one of the species "gives birth" to a second species in a short time period that is dt long. (For these purposes, dt is often 1 year.) E represents the extinction rate. Then on average, the number of new species that appear by speciation during that short time period equals the product of the speciation rate, the number of species that can speciate, and the length of the interval: $S\,N\,dt$. Following the same logic, the number of species that become extinct is $E\,N\,dt$. The change in the number of species during that interval is the number of new species minus the number of extinctions. Putting this together and rearranging the terms, we find that the rate of change in the number of species per unit of time is

$$\frac{dN}{dt} = (S - E)N = DN \qquad 19.1$$

Here D is the net diversification rate, which is the speciation rate minus the extinction rate: $D = S - E$. The number of species will on average increase if the speciation rate is greater than the extinction rate, that is, if $D > 1$. If D is negative, the number of species will decline. This model can also be used to describe changes in the number of higher taxonomic categories such as genera or families. In that case, S represents the rate of origination of new taxa rather than the rate of speciation for individual species. Once again, D is the diversification rate: the average rate per taxon of an increase or decrease in diversity.

If the diversification rate D remains constant, then the number of species will grow or shrink exponentially (**FIGURE 19.2**). But just as competition for resources can act as a density-dependent brake on growth of a population, the diversification rate may decrease as the result of **diversity-dependent factors**, factors such as competition for food or space that become more intense as the diversity (number) of competing taxa increases. The diversity may then attain an equilibrium at K species. Of course, this model is a great oversimplification of reality because changes in the environment and in organisms themselves are likely to change rates of origination and extinction, and consequently the rate of diversification, over time. Nevertheless, it provides a framework for thinking about differences in diversity.

In Chapter 18 we noted that two world regions might differ in species diversity because of differences in the time since diversification began, in the rate

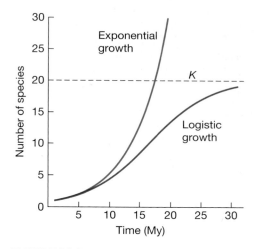

FIGURE 19.2 Two models for the change in species diversity through time. In both, we follow the number of species in a clade that starts with just a single species. In this example, the diversification rate is $D = 0.2$/million years, which means there is a 20 percent chance that one species will have two descendant species after 1 My. With exponential growth, the diversification rate stays constant and the number of species in the clade grows exponentially. With logistic growth, the diversification rate decreases as the number of species increases. In this example, the equilibrium is $K = 20$ species in the clade.

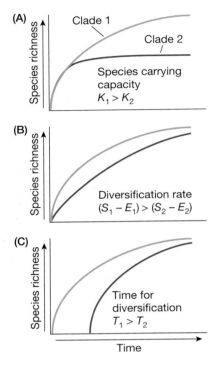

FIGURE 19.3 Just as for contrasts in species richness between different geographic regions (see Figure 18.19), two clades can differ in species richness because of differences in (A) carrying capacity, the equilibrium number that can stably coexist; (B) their rate of diversification (speciation rate minus extinction rate); or (C) their age, meaning the time they have had to diversify.

of species diversification (D), or in the maximum number of species (K) the regions can support at equilibrium (see Figure 18.19). Exactly the same possible explanations could account for why some clades have more species than others (**FIGURE 19.3**).

Studying diversity in the fossil record

Most paleontological studies of diversity employ counts of higher taxa, such as families and genera, because they generally provide a more complete fossil record than individual species do. Although paleobiologists have used several expressions for rates of origination, extinction, and diversification, the most useful are the numbers per taxon per unit of time.

Because the fossil record is a very incomplete sample of past life, paleobiologists have developed correction factors to estimate accurately the number of species alive at different points in the past [20, 57, 67]. For example, rare species are more likely to be included in large samples, which include more individual organisms, than in small samples. If we want to compare the species diversity in two samples that differ in size, we must correct for this problem, perhaps by picking the same number of specimens at random from all the samples.

In addition, the geological or stratigraphic *stages* into which each geological period is divided vary in duration, and more recent geological times are represented by greater volumes and areas of fossil-bearing rock. Therefore it may be necessary to adjust the count of taxa by the amount of time and rock volume represented. Because fossils constitute a small sample of the organisms that actually lived at the time they were formed, a taxon is often recorded from several separated time horizons, but not from those in between. This means the fossil record of these species is incomplete. In turn, that suggests that the actual origination of a taxon may have occurred before its earliest fossil record, and its extinction after its latest record. It follows that if many taxa actually became extinct in the same time interval, the last recorded occurrences of some are likely to be earlier, so that their *apparent* times of extinction will be spread out over time. Conversely, if many taxa actually originated at the same time, some of them may appear to have originated at later times.

Since our count of living species is much more complete than our count of past species, taxa that are still alive today appear to have longer durations and lower extinction rates than they would if they had been recorded only as fossils. That is, we can list a living taxon as present throughout the last 10 My, let's say, even if its only fossil occurrence was 10 Mya. Because the more recently a taxon arose, the more likely it is to still be extant, diversity will seem to increase as we approach the present, even if it didn't actually increase. This artifact, or bias, is called the **pull of the Recent**. (The Recent epoch, more commonly referred to now as the Holocene [see Table 17.1] began 12,000 years ago.) The bias can be reduced by counting only fossil occurrences of each living taxon and not listing it for time intervals between its last fossil occurrence and the Holocene.

Because of unusually favorable preservation conditions at certain times or other chance events, a taxon may be recorded from only a single geological stage, even though it lived longer than that. Such "singletons" make up a higher proportion of taxa as the completeness of sampling decreases and therefore bias the sample; moreover, they can create a spurious correlation between rates of origination and rates of extinction because they appear to originate and become extinct in the same time interval. Diversity may be more accurately estimated by ignoring such singletons and counting only those taxa that cross the border from one stage to another.

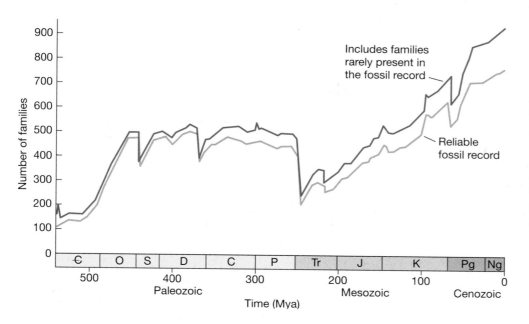

FIGURE 19.4 Taxonomic diversity of skeletonized marine animals during the Phanerozoic. The number of families entered for each geological stage includes all those whose known temporal extent includes that stage. The blue curve includes families that are rarely preserved; the green curve represents only families that have a more reliable fossil record. There are approximately 1900 marine animal families alive today, including those rarely or never preserved as fossils. (After [63].)

This discussion illustrates a fundamentally important aspect of every scientific discipline, including evolutionary biology: scientists discuss the ways in which their data could possibly be misinterpreted and lead to false conclusions, and they devise ways of avoiding that.

Diversity through the Phanerozoic

The most complete fossil record has been left by skeletonized marine animals (those with hard parts such as shells or skeletons). Jack Sepkoski accomplished the heroic task of compiling data from the paleontological literature on the stratigraphic ranges of more than 4000 skeletonized marine families and 30,000 genera throughout the 541 My since the beginning of the Cambrian period [63, 64]. Using this database, he plotted the diversity of families throughout the Phanerozoic, creating one of the most famous graphs in the literature of paleobiology (**FIGURE 19.4**). The graph shows a rapid increase in the Cambrian and Ordovician, a plateau throughout the rest of the Paleozoic, and a steady, almost fourfold increase throughout the Mesozoic and Cenozoic. This pattern is interrupted by decreases in diversity caused by mass extinction events (see Chapter 17). Similar studies of the terrestrial fossil record show that the end-Permian extinction was followed, after a delay of more than 15 My, by a great diversification of dinosaurs, crocodilians, and synapsid proto-mammals [8]. In the Cretaceous, flowering plants proliferated and largely replaced gymnosperms, and insects exploded in diversity. Life on land became more diverse than in the sea [74].

Since Sepkoski first summarized the history of marine diversity, other paleobiologists have applied various corrections for sampling errors and the biases that we have noted (**FIGURE 19.5**) [2, 21]. They have found a decline in diversity in the Devonian instead of a Paleozoic plateau, an increase in the Permian before the end-Permian extinction, and a less steep, but still pronounced, increase through the Mesozoic and Cenozoic to the present time [10].

Much of the increase in the number of taxa reflects the evolution of morphologically and ecologically new forms of life. This aspect of diversity has increased from the early Paleozoic to the present; among marine animals, for example, the

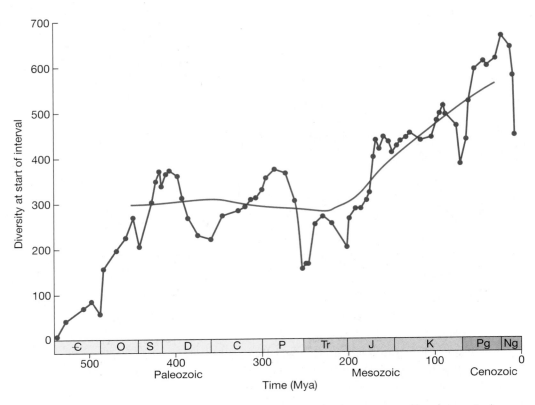

FIGURE 19.5 Numbers of skeletonized marine animal genera over time, corrected for biases such as temporal differences in rock volume and the pull of the Recent. The smooth curve, a running average from the late Ordovician to the mid-Cenozoic, suggests that, aside from mass extinctions and subsequent recoveries, diversity showed a stronger increasing trend in the Mesozoic and Cenozoic than in the Paleozoic. (After [21].)

variety of different modes of life and associated adaptations is far greater now than in the Cambrian (**FIGURE 19.6**). The evolution of ecologically novel life forms, such as dinosaurs, snakes, birds, and bats, likewise accounts for much of the increasing diversity of tetrapods on land [8].

Rates of origination and extinction

The increase in diversity during the Mesozoic and Cenozoic tells us that on average, the rate of origination of marine animal taxa has been greater than the rate of extinction. However, both rates have fluctuated throughout Phanerozoic history (**FIGURE 19.7**). Some lineages have become extinct during every geological time interval—so-called normal or **background extinction**. But the fossil record reveals several dramatic crashes in diversity, **mass extinctions**, when a great many or even most species became extinct. What caused these global catastrophes is one of the most fascinating questions in paleontology. Five mass extinctions are generally recognized (see Chapter 17): at the end of the Ordovician, in the late Devonian, at the Permian/Triassic (P/Tr) boundary (the end-Permian extinction), at the end of the Triassic, and at the Cretaceous/Paleogene (K/Pg) boundary (the K/Pg extinction).

David Raup and Jack Sepkoski discovered that the background extinction rate has declined during the Phanerozoic [57], a conclusion supported by subsequent studies (see Figure 19.7A). The rate of origination of new genera and families (see Figure 19.7B–E) also declined after the Paleozoic, although it increased at certain times, including after the "big five" mass extinctions, when diversity recovered, usually within 10–15 My [1].

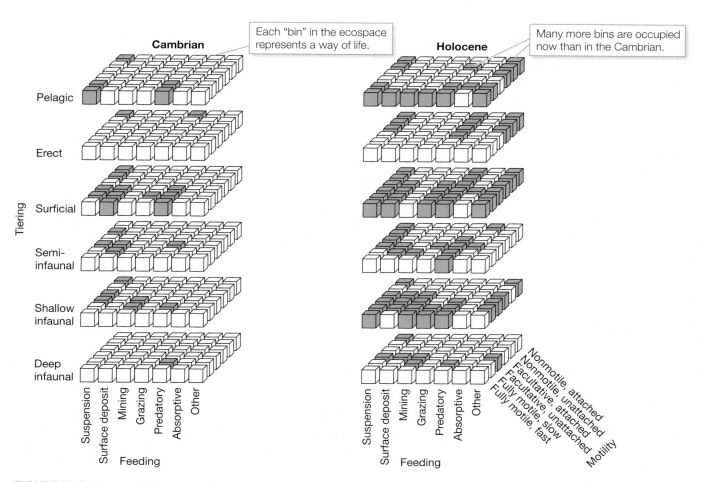

FIGURE 19.6 The use of "ecospace" by marine animals in the Cambrian compared with the present. Each layer represents the vertical space used by animals, from open ocean (pelagic) to deep in the sediment (deep infaunal). In each layer, the "bins" from left to right represent different modes of feeding, and those from front to rear represent different habits of movement (motility). Far more bins, or ways of life, are filled (indicated by green) by animals now than in the Cambrian. (After [9].)

What might account for these declines? Our first thought might be that species become better adapted by natural selection over time, and so should become less vulnerable to changing environments. But natural selection, having no foresight, cannot prepare species for novel changes in the environment. If the environmental changes that threaten extinction are numerous in kind, we should not expect much carryover of "extinction resistance" from one change to the next. So extinction rates should vary randomly over time if changes in the environment occur at random.

Two other hypotheses have been suggested for the decline in extinction rates. The average number of species per genus and per family seems to be greater in the Cenozoic than in earlier eras. This would lower the probability of extinction of higher taxa because a genus or family does not become extinct until all its constituent species are extinct [1, 19]. Another explanation is that some clades are more volatile than others: they have a higher turnover rate, evolving new families and losing others before the entire clade becomes extinct. The extinction of such taxa leaves the less volatile taxa, those that have longer life spans and lower extinction rates.

FIGURE 19.7 Rates of extinction and origination of lineages have declined over geological time, as shown by two studies. (A) Extinction rates of marine animal families, expressed as the number of families per My. The solid regression line fits the blue points, which represent fewer than 8 extinctions per My, and which are interpreted as background extinction. The red points mark the five major mass extinction events, at the end of the Ordovician (O), Devonian (D), Permian (P), Triassic (Tr), and Cretaceous (K). The extinction rates are given as absolute numbers of extinctions, not per capita rates, and thus do not control for differences in diversity at different times. (B–E) Rates of origination of genera in two phyla of animals in the sea, Mollusca (B) and Echinodermata (C), and of families in two terrestrial groups, insects (D) and nonflowering vascular plants (E). The origination rates are expressed as the proportion of lineages in a geological stage that originated (are first seen) in that stage. (A after [58]; B–E after [12b].)

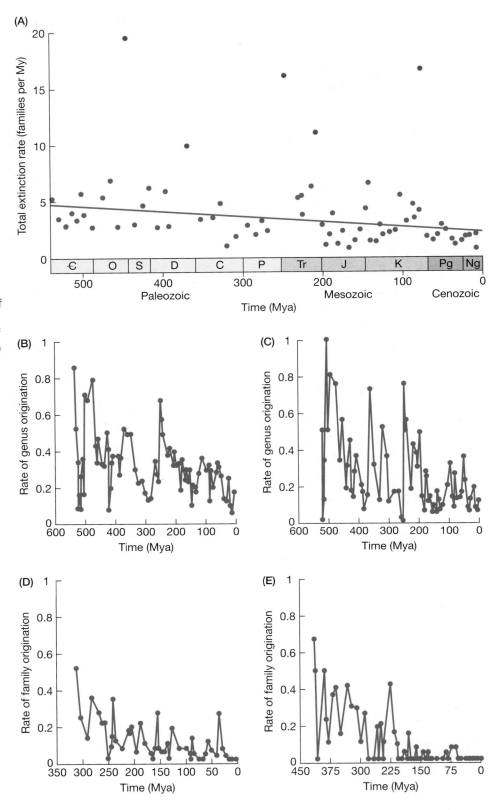

Rates of extinction (E) and origination (S) are correlated (**FIGURE 19.8**); trilobites, for example, were more volatile than gastropods, turning over more rapidly.

Why are extinction and origination rates correlated? Steven Stanley suggested that certain features of organisms influence both rates [70]. First, ecologically specialized species of mammals and other groups show higher extinction rates than generalized species [43, 68] because they are more vulnerable to changes in their environment (see Chapter 11) [35]. They may also be more likely to speciate because they require specific environments and so may have patchier distributions. Second, species with broad geographic ranges tend to have a lower risk of extinction because they are not extinguished by local environmental changes [24]. They also have lower rates of speciation [34], probably because they have a high capacity for dispersal and perhaps broader environmental tolerances.

Major groups characteristically differ in how long species persist before they become extinct. For example, the average duration of a species of bivalve (clams and relatives) is 23 My, whereas it is only 10 My for a gastropod (snails and relatives) and 7 My for a sea urchin [43, 69]. Leigh Van Valen wondered if within any such group, older species might be better adapted than young ones (because they had had more times to adapt) and less prone to extinction [73]. The result would be a declining extinction rate [41]. This can be determined by plotting the fraction of taxa (e.g., the fraction of genera in a family) that survive for different lengths of time. This approach is different from asking whether or not extinction rates have changed over the course of geological time (e.g., whether they were lower in the Jurassic than in the Devonian). If new taxa have the same probability of extinction as older ones, then the proportion of component taxa surviving to increasingly greater ages should decline exponentially. Plotted logarithmically, the survivorship curve would become a straight line. If taxa become increasingly resistant to causes of extinction as they age, the logarithmic plot should be upwardly concave, with a long tail (**FIGURE 19.9A**).

When Van Valen plotted taxon survivorship in this way, he found a surprising result: rather straight curves, implying that the probability of extinction is roughly constant (**FIGURE 19.9B**). There was no evidence that these animals became more resistant to extinction over time. Instead, this is what

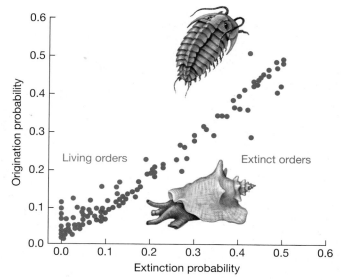

FIGURE 19.8 Groups of marine organisms vary in volatility. The rate (probability) of origination of new families within an order, per time interval, is correlated with the rate (probability) of extinction of families. More volatile orders, with higher rates, have higher turnover, and so are more likely to decrease greatly or become extinct, as the trilobites (upper drawing) did. Most living orders, such as orders of gastropods (lower drawing), have low rates of origination and extinction compared with extinct orders (After [25].)

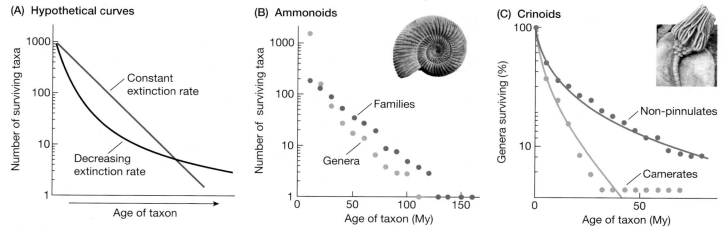

FIGURE 19.9 Taxonomic survivorship curves. Each curve or series of points represents the number of taxa that persisted in the fossil record for a given duration, irrespective of when they originated during geological time. (A) Hypothetical survivorship curves. In a semilogarithmic plot, the curve is linear if the probability of extinction is constant. It is concave if the probability of extinction declines as a taxon ages, as it might if adaptation lowered the long-term probability of extinction. (B) Taxonomic survivorship curves for families and genera of ammonoids. The plot for families suggests an extinction rate that is constant with age, whereas the plot for genera suggests that older survivors might have a lower rate of extinction. (C) Taxonomic survivorship curves for two groups of Paleozoic crinoids (sea lilies), sessile echinoderms that filter plankton with branched arms. Older genera in these two groups, which differ in their filter structure, have lower extinction rates. (B after [73]; C after [6].)

we would expect if organisms are continually assaulted by new environmental changes, each carrying a risk of extinction. One possibility, Van Valen suggested, is that the environment of a taxon is continually deteriorating because of the evolution of other organisms. He proposed the **Red Queen hypothesis**, which states that, like the Red Queen in Lewis Carroll's *Through the Looking-Glass*, each species has to run (i.e., evolve) as fast as possible just to stay in the same place (i.e., survive), because its competitors, predators, and parasites also continue to evolve (see Chapter 13). There is always a roughly constant chance that a species will fail to do so. But some studies have found patterns that match what Van Valen had first expected (**FIGURE 19.9C**). In many clades, older genera (those that originated many stages earlier) have higher survival rates than younger genera. This is what we might expect if similar changes in the physical environment frequently recur; if so, lineages that survived such episodes earlier would have characteristics that enabled survival through later, similar episodes [17].

Mass extinctions

The history of extinction is dominated by the "big five" mass extinctions at the end of the Ordovician, Devonian, Permian, Triassic, and Cretaceous periods (see Figure 19.7B) [4]. The end-Permian extinction was the most drastic, eliminating about 56 percent of genera and more than 80 percent of species of skeleton-bearing marine invertebrates (see Chapter 17). On land, major changes in plant assemblages occurred, several orders of insects became extinct, and the dominant tetrapods were replaced by new groups that included the ancestors of mammals and dinosaurs. This extinction probably resulted, at least partly, from massive volcanic eruptions in the region of Siberia. Less severe, but much more famous, was the K/Pg extinction at the end of the Cretaceous, which marked the demise of many marine and terrestrial plants and animals, including the dinosaurs (except for birds). The mass extinction events, especially the end-Permian and K/Pg extinctions, had an enormous effect on the subsequent history of life because, to a great extent, they wiped the slate clean.

Mass extinctions were selective—some taxa were more likely than others to survive. Survival of gastropods through the end-Permian extinction was greater for species with wide geographic and ecological distributions and for genera consisting of many species [13]. Extinction appears to have been random with respect to other characteristics, such as mode of feeding. Patterns of survival through the K/Pg extinction differed from those during "normal" times [31]. During times of background extinction, survivorship of late-Cretaceous bivalves and gastropods was greater for taxa with planktotrophic larvae (those that feed while being dispersed by currents) and for genera consisting of numerous species, especially if those genera had broad geographic ranges. In contrast, taxa with both planktotrophic and nonplanktotrophic larvae had the same extinction rates during the K/Pg extinction, and the survival of genera, although enhanced by broad distribution, was not influenced by their species richness. Thus the characteristics that were correlated with survival seem to have differed from those during "normal" times.

During mass extinction events, taxa with otherwise superb adaptive qualities succumbed because they happened not to have some critical feature that might have saved them from extinction under those circumstances. Evolutionary trends initiated in "normal" times were cut off at an early stage. For example, the ability to drill through bivalve shells and feed on the animals inside evolved in a Triassic gastropod lineage, but that lineage became extinct in the late-Triassic mass extinction [23]. The same feature evolved again 120 My later, in a different lineage that gave rise to diverse oyster drills. A new adaptation that might have led to a major adaptive radiation in the Triassic was strangled in its cradle, so to speak.

Both abiotic and biotic environmental conditions were probably very different after mass extinctions than before. Perhaps for this reason, many taxa continued to dwindle long after the main extinction events [32], while others, often members of previously subdominant groups, diversified. For example, the rate of origination of genera of bivalves increased after the K/Pg extinction and has remained high ever since. New genera have arisen mostly in tropical latitudes, so ongoing recovery from the mass extinction has affected the geographic pattern of diversity that exists today (see Chapter 18) [37].

Stephen Jay Gould suggested that there are "tiers" of evolutionary change, each of which must be understood in order to comprehend the full history of evolution [26]. The first tier is microevolutionary change *within populations and species*. The second tier is "species selection," the *differential proliferation and extinction of species* during "normal" geological times, which affects the relative diversities of lineages with different characteristics. The third tier is the *shaping of the biota by mass extinctions*, which can extinguish diverse taxa and reset the stage for new evolutionary radiations, initiating evolutionary histories that are largely decoupled from earlier ones.

Richard Bambach and colleagues found some support for Gould's idea when they classified Phanerozoic marine animal genera by three functional criteria: whether they were motile or nonmotile, whether they were "buffered" against physiological stress (with well-developed gills and circulatory system, such as crustaceans) or not (such as echinoderms), and whether or not they were predatory [5]. With respect to all three kinds of functional groupings, the proportions of taxa with alternative characteristics remained stable over intervals as long as 200 My, even though the total diversity and the taxonomic composition of the marine fauna changed greatly (**FIGURE 19.10**). However, shifts from one stable configuration to another occurred at the ends of the Ordovician, Permian, and Cretaceous, suggesting that the extinction of some taxa permitted the emergence of new community structures.

No truly massive extinction has occurred for 66 My; even the great climate oscillations of the Pleistocene, though they altered geographic distributions and ecological assemblages, had a relatively small impact on the diversity of life. But it is depressingly safe to say that a major extinction—perhaps the next mass extinction—has begun (see Box 22B). The course of biodiversity has been altered for the foreseeable future by human domination of Earth, and altered for the worse. Without massive, dedicated action, humanity will suffer profoundly, and much of the glorious variety of the living world will be extinguished.

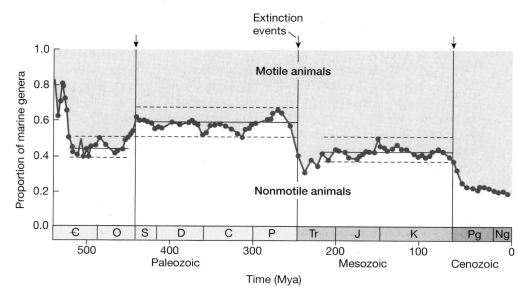

FIGURE 19.10 Changes in the proportions of genera of motile versus nonmotile marine animals during the Phanerozoic. The proportions were roughly stable (dashed horizontal lines) between mass extinctions, but shifted rapidly to a new stable state after mass extinction events at the end of the Ordovician, Permian, and Cretaceous (solid vertical lines). Similar changes (not shown here) occurred in the proportions of predators versus nonpredators and of animals thought to be physiologically buffered versus unbuffered, based on anatomical criteria. (After [5].)

FIGURE 19.11 Adaptive radiation of Hawaiian honeycreepers (family Fringillidae). The species vary greatly in diet, reflected in their bill shapes. Some feed on insects (short, thin bill), some on seeds and fruit (thick bill), and some on nectar (slender, curved bill), matching diverse unrelated birds that fill these ecological niches on continents. This group is descended from an Asian ancestor related to the common rosefinch (*Carpodacus erythrinus*), the bird in the center. The Hawaiian honeycreepers diversified as the several islands in the archipelago were sequentially formed.

Phylogenetic Studies of Diversity

Zoologists and botanists who study living organisms have long recognized certain conditions that appear to have fostered high diversity, which is often manifested as adaptive radiations. For example, many clades have radiated in species richness and ecological disparity where they found **ecological opportunity**, that is, many open ecological niches [71]. Taxa on oceanic islands provide many examples. The Hawaiian honeycreepers, derived from an ancestor in the diverse finch family Fringillidae (rosefinches, goldfinches, and others), are a spectacular example. About 60 species (of which 18 survive) are almost the only songbirds native to the Hawaiian archipelago, where they faced almost no competition and diversified greatly in diet and in bill morphology (**FIGURE 19.11**) [38]. The diversification of many clades accelerated when they expanded into new geographic regions [46].

Research on both extinct and living organisms has pointed to certain **key adaptations** that have enhanced species diversity [3, 51]. These are features that enable a lineage to interact with the environment in a new way and to use new resources. During insect evolution, for example, rates of origination and extinction of families were first accelerated when wings evolved, and later by the evolution of complete metamorphosis: the distinct larval and pupal stages that characterize the immensely diverse beetles (Coleoptera), wasps (Hymenoptera), true flies (Diptera), and moths (Lepidoptera) (see Figure 17.18) [50]. Among the sea urchins (Echinoidea), three orders increased greatly in diversity beginning in the early Mesozoic (**FIGURE 19.12**). The order Echinacea evolved stronger jaws that enabled its members to use a greater variety of foods. The heart urchins (Atelostomata) and sand dollars (Gnathostomata) became specialized for burrowing in sand, where they feed on fine particles of organic sediment. The key adaptations allowing this major shift of habitat and diet include a flattened form and a variety of highly modified tube feet that can capture fine particles and transfer them to the mouth.

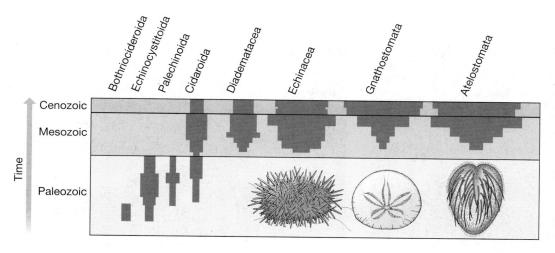

FIGURE 19.12 Changes in the diversity of several groups of echinoid echinoderms over time. The width of the symmetrical profile of each group represents the number of families in that group. The diversity of sea urchins (order Echinacea), sand dollars (Gnathostomata), and heart urchins (Atelostomata) greatly increased during the Mesozoic and Cenozoic, probably because of the key adaptations described in the text. (After [3].)

The immense diversity of flowering plants (angiosperms), compared with that of other vascular plants, has been ascribed to features such as animal-mediated pollination, closed carpels that protect the developing seeds, and more efficient water-conducting vasculature of angiosperm leaves [40]. But the angiosperms reveal a problem: from among the several distinctive features of a diverse clade, any of which might have enhanced speciation or reduced the likelihood of extinction, how can we identify the characteristic that was the key to the group's high diversity? Demonstrating the cause of a single event is always difficult. Stronger evidence is provided if the rate of diversification is consistently associated (correlated) with a particular character that has evolved independently in several different clades. Such tests have been applied mostly to living organisms. The diversity of several clades that independently evolved a similar novel character can be compared with the diversity of their sister groups that retain the ancestral character state. Since sister taxa are equally old, the difference between them in number of species cannot be ascribed to age. If the convergently evolved character is consistently associated with high diversity, we have support for the hypothesis that it has caused a higher rate of origination or has lowered the extinction rate, by allowing a greater number of species to coexist (as in Figure 19.3A and B).

Charles Mitter and colleagues applied this method, called replicated sister-group comparison, to herbivorous insects and plants [16, 45]. The habit of feeding on the vegetative tissues of green plants has evolved at least 50 times in insects, usually from predatory or detritus-feeding ancestors. Phylogenetic studies identified the nonherbivorous sister groups of 13 herbivorous clades. In 11 of these cases, the herbivorous lineage has more species than its sister group (**FIGURE 19.13**). This

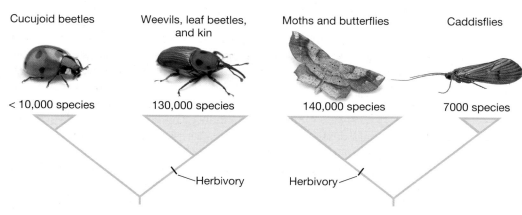

Cucujoid beetles   Weevils, leaf beetles, and kin   Moths and butterflies   Caddisflies

< 10,000 species   130,000 species   140,000 species   7000 species

Herbivory   Herbivory

FIGURE 19.13 Two replicated sister-group comparisons of herbivorous clades of insects with their sister clades that feed on animals, fungi, or detritus. Herbivorous clades are consistently more diverse, demonstrating higher rates of diversification. (Data from [45].)

(A)

(B)

FIGURE 19.14 Two kinds of plant defenses against herbivores that have increased species richness. (A) Extrafloral nectaries (nectar-producing glands) on the leaf petioles in some leguminous trees (here *Acacia*) attract bodyguards such as ants. (B) Milkweeds (*Asclepias*) are among the many groups of plants that produce latex, a sticky liquid that deters many herbivores (but not this milkweed leaf beetle, *Labidomera clivicollis*).

significant correlation supports the hypothesis that the ability to eat plants has promoted diversification. More elaborate subsequent analyses by other authors supported Mitter and colleagues' conclusion that herbivory has increased insect diversity, in part because species diverge in diet, becoming specialized on different plant species [22, 28, 77]. The diversity of many clades of plants, conversely, has been enhanced by effective defenses against herbivores. Groups that evolved rubbery latex (as in milkweeds) or resin (as in pines), both of which deter attack by herbivorous insects, are more diverse than their sister groups [16], and diversification rates have been twice as high in the many plant families with extrafloral nectaries than in families that lack this feature [76]. Extrafloral nectaries are sugar-secreting organs that attract mutualistic arthropods, such as ants, that protect the plant against herbivores (**FIGURE 19.14**).

In some cases, higher diversity seems to result simply from factors that increase the probability or rate of speciation. For example, many groups of birds and other terrestrial organisms are very diverse in mountainous regions, where allopatric speciation rates are elevated due to isolation of populations on different slopes [18]. The nectar spurs on the petals of columbines are associated with greater diversification, probably because they enabled different species to use morphologically different insects and birds as pollinators, which would contribute to reproductive isolation (**FIGURE 19.15**). Probably for the same reason, plant clades with bilaterally symmetrical flowers have had higher speciation rates than those with radially symmetrical flowers (see Figure 16.4) [56]. Of course, different diversity-enhancing factors may act in combination. The phenomenal diversification of cichlid fishes in the large lakes of eastern Africa (see p. 213) is mirrored by smaller radiations in many other African lakes. Comparing the cichlids in many lakes, Catherine Wagner and colleagues found that species numbers were enhanced both by the availability of habitat (measured by lake depth) and by characteristics associated with sexual selection, such as sexual dimorphism in coloration (**FIGURE 19.16**) [75].

In these examples, there is a correlation between a trait and the rate at which the number of species increases or decreases. A consistent difference of this kind is termed **species selection**, and results in certain characteristics becoming more prevalent than others, among all species taken together.

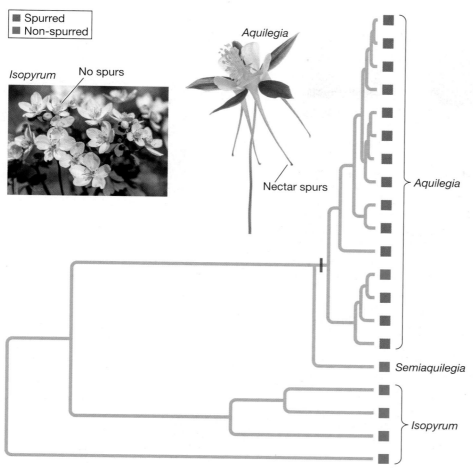

Isopyrum

No spurs

Aquilegia

Nectar spurs

Aquilegia

Semiaquilegia

Isopyrum

FIGURE 19.15 Enhanced diversification attributable to a key adaptation. The evolution of nectar spurs in the ancestor of columbines (*Aquilegia*), shown by the red crossbar, was followed by the origin of numerous species within a short time, as shown by the shortness of the branches between speciation events. The sister group (*Isopyrum*) that lacks spurs did not diversify as abundantly or as quickly. Columbines that differ in spur length have different pollinators, which serve as different resources, but also reduce gene exchange among diverging columbine populations and contribute to reproductive isolation. (After [59]; data from [30].)

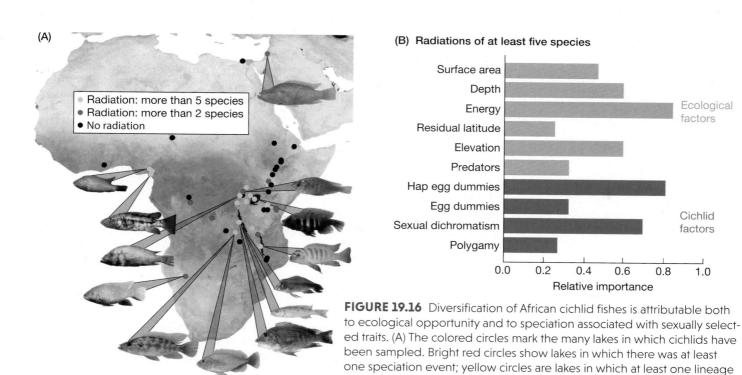

(A)

Radiation: more than 5 species
Radiation: more than 2 species
No radiation

(B) **Radiations of at least five species**

Surface area
Depth
Energy
Residual latitude
Elevation
Predators
Hap egg dummies
Egg dummies
Sexual dichromatism
Polygamy

Ecological factors

Cichlid factors

0.0   0.2   0.4   0.6   0.8   1.0
Relative importance

FIGURE 19.16 Diversification of African cichlid fishes is attributable both to ecological opportunity and to speciation associated with sexually selected traits. (A) The colored circles mark the many lakes in which cichlids have been sampled. Bright red circles show lakes in which there was at least one speciation event; yellow circles are lakes in which at least one lineage has five or more species. (B) Factors that have been important in fostering speciation within lakes. The most significant factors that are correlated with speciation are lake depth, amount of incident solar energy, sexual color difference, and egg dummies in haplochromine cichlids. Egg dummies are colored spots on the males' fins that attract females. (From [75].)

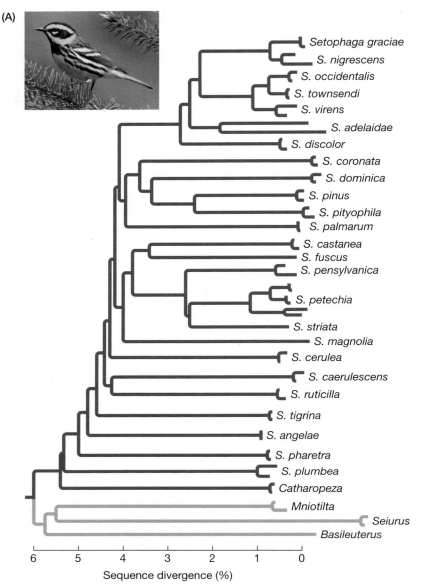

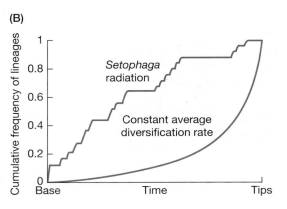

FIGURE 19.17 A phylogeny showing a decline in the diversification rate over time. (A) Phylogeny of a clade of American wood-warblers, *Setophaga*, and the closely related genus *Catharopeza*, based on mtDNA sequences. The green branches represent outgroup genera in the wood-warbler family. The branch lengths are proportional to degree of sequence divergence, which is assumed to indicate time since speciation. (B) A lineage-through-time (LTT) plot of the *Setophaga* clade, on an arithmetic axis. Note that the increase slows down and starts to level off. A theoretical curve in which the number of lineages grows exponentially (with a constant diversification rate) is shown for comparison. The LTT plot accurately portrays the history of diversity if extinction has been constant (or zero). (After [39].)

The shapes of phylogenies

We can also learn about patterns of diversification through time by studying the phylogenies of living species. Using a molecular clock, the ages of nodes on the phylogeny can be estimated. Changes in the rates of splitting on the tree (corresponding to speciation events) can then be followed through time, forming a **lineage-through-time (LTT) plot** [49, 52]. This approach is complicated by the fact that extinction events cannot be seen on the phylogeny, and for that reason the approach is somewhat controversial.

For example, **FIGURE 19.17A** is a time-calibrated phylogeny of wood-warblers in the genus *Setophaga*. In this case, most of the branch points are close to the base of the tree. This is reflected in the shape of the LTT plot, which increases steeply at first but then increases at a lower rate (i.e., it tends to level off; **FIGURE 19.17B**). This decline in the diversification rate (*D*) contrasts with the exponentially increasing plot we would expect if the rate of diversification had been constant (see Figure 19.2). A declining rate of diversification could be caused by a declining rate of speciation (*S*), an increasing rate of extinction (E), or both. Several methods have been proposed to estimate S and E, but this is a very difficult problem because the phylogeny of living species does not reflect the past existence of species that were not ancestors of living species [47]. For example, the fossil record shows that cetaceans (whales and dolphins) were much more diverse in the Miocene than they are today, and have suffered more extinction that can be inferred from the phylogeny of living species [53].

Many clades show a declining rate of lineage accumulation; relatively few show a pattern of increasing diversification [44]. Mathematical analysis and computer simulations suggest that this pattern is likely to be caused by a decreasing rate of speciation, not by an increasing extinction rate [55]. This pattern strongly suggests that diversification has been diversity-dependent, but we will see that this is a matter of some controversy. And it is likely that most new species become extinct soon after, or even during, the lengthy process of speciation, so the roles of speciation and extinction may be hard to distinguish, even conceptually [61].

Does Species Diversity Reach Equilibrium?

A huge ecological literature is concerned with whether or not the number of coexisting species (of some group such as plants or mammals) tends toward an equilibrium. This question is complex and not entirely resolved, but ecologists agree that some factors tend to limit species diversity. The space that plants compete for and the energy fluxes that organisms depend on are finite, so they can be divided among a limited number of species populations that are still large enough to persist. At a local level, the number of species is sometimes directly correlated with the number in a larger region—a pool of species, of which only a sample is found at any one place. This pattern suggests that the number of coexisting species is limited only by the number available to colonize a local site [11]. If the species richness in local assemblages shows little variation despite access to more diverse species pools, some limiting factor, such as competition for resources, is likely to place an upper bound on the number of coexisting species. Both patterns have been found in different situations. Phenomena such as competitive exclusion of species from each other's ranges suggest that interactions among species can limit local species diversity.

The ecological factors that determine the number of locally coexisting species may differ from those that determine the number of species in a clade or taxonomic group. Researchers differ as to whether or not the diversity of most clades has approached limits set by competition or other diversity-dependent factors (see [29] vs. [54]). Paleontologists have found some evidence that the per taxon rate of increase in the number of species (or higher taxa) is diversity-dependent: it decreases as the number grows. For example, Michael Foote calculated the rates of origination (S), extinction (E), and diversification (D) of marine genera from one stratigraphic stage to the next, then correlated these short-term changes with the number of genera present (N) at the beginning of the stage (**FIGURE 19.18**) [21]. Both the diversification rate ($D = S - E$) and the origination rate (S) declined as

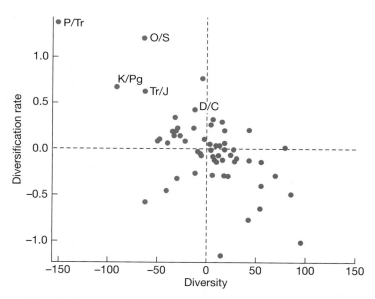

FIGURE 19.18 The per lineage rate of diversification of skeletonized marine invertebrate genera during the Phanerozoic (i.e., since the start of the Cambrian) is diversity-dependent. Each point plots the rate of change during a stratigraphic stage against the diversity of taxa at the start of that stage. The higher the diversity, the lower the rate of diversification. Further analysis showed that this pattern is attributable to the reduced rate at which new genera arise. The pattern suggests that higher diversity imposes stronger competition and prevents new genera from evolving. For statistical reasons, the points are shown on scales that are centered at zero. Points representing mass extinction events are labeled (O/S, end-Ordovician; D/C, late Devonian; P/Tr, end-Permian; Tr/J, end-Triassic; K/Pg, Cretaceous/Paleogene boundary). (After [21].)

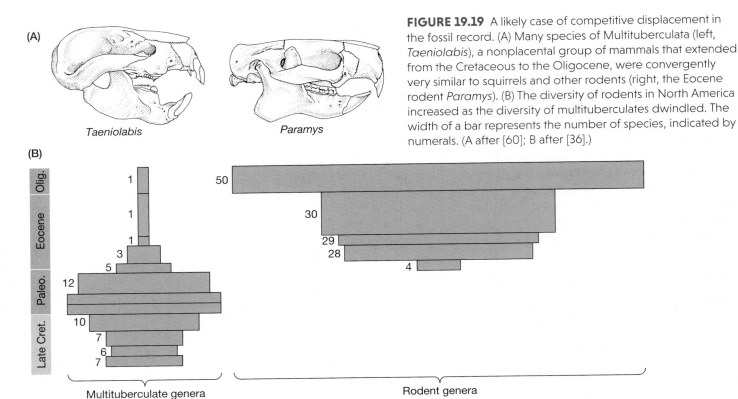

(A)

Taeniolabis   *Paramys*

(B)

FIGURE 19.19 A likely case of competitive displacement in the fossil record. (A) Many species of Multituberculata (left, *Taeniolabis*), a nonplacental group of mammals that extended from the Cretaceous to the Oligocene, were convergently very similar to squirrels and other rodents (right, the Eocene rodent *Paramys*). (B) The diversity of rodents in North America increased as the diversity of multituberculates dwindled. The width of a bar represents the number of species, indicated by numerals. (A after [60]; B after [36].)

Multituberculate genera   Rodent genera

diversity increased. In a similar analysis, John Alroy found evidence that extinction rates (*E*) were higher if diversity at the start of an interval was higher [1]. Moreover, a high extinction rate in one interval was correlated with a high origination rate in the following interval. These and other such analyses imply that the diversity of taxa tends to be stabilized and approach an equilibrium.

Likewise, the tendency of LTT plots from phylogenies to level off with time, as seen in the wood-warbler data (see Figure 19.17B), is generally interpreted to mean that the earliest species in a new clade rapidly adapt to different resources or environments (i.e., adaptive radiation), and that fewer subsequently formed species can persist because fewer vacant ecological opportunities remain available. Nevertheless, both the fossil record and some phylogenetic studies suggest that diversity is still increasing, even though the rate slows down over time [48]. One reason is that throughout the history of life, as we have seen, evolutionary innovations have enabled clades to break through into new ecological modes of resource use. And diversity promotes diversity. For example, an entire family of fishes (the pearl fishes, Carapidae) lives inside sea squirts and sea cucumbers. Many plant lineages in the American tropics have adapted to hummingbird pollination by evolving long, tubular flowers—but these are a resource for the flowerpiercers (*Diglossa*) that "rob nectar" by biting through the base of the flower (see Figure 11.17). On a global scale, moreover, diversification has undoubtedly been augmented by the separation of Pangaea into separate land masses and by the greater temperature gradient between low and high latitudes that developed during the Cenozoic [72].

Competition is generally thought to be the chief brake on increasing diversity. The fossil record provides many instances in which the reduction or extinction of one group of organisms has been followed or accompanied by the proliferation of an ecologically similar group. For example, the diversity of rodents in North America increased as that of the ecologically similar multituberculate mammals declined (**FIGURE 19.19**) [36].

Two major hypotheses that involve competition can account for these patterns [7, 65]. One possibility is that the later clade *caused* the extinction of the earlier clade by competition, a process called competitive displacement (**FIGURE 19.20A**). By contrast, an incumbent taxon may have *prevented* an ecologically similar taxon from diversifying because it already occupied resources. Extinction of the incumbent taxon may then have vacated ecological niche space, *permitting* the second taxon to radiate (**FIGURE 19.20B**). This process has been called incumbent replacement [62].

Jack Sepkoski and colleagues developed a mathematical model in which two clades increase in diversity, but in which the increase in each clade is inhibited by both its own diversity and that of the other clade [66]. They applied the model to data on the number of genera of two groups of bryozoans ("moss animals"), the cyclostomes and the cheilostomes. These sessile colonial animals spread over rocks or other surfaces by budding. When colonies of these two groups meet, cheilostomes generally overgrow cyclostomes (**FIGURE 19.21A**). Especially since the end-Cretaceous extinction, the diversity of cheilostomes has increased, whereas cyclostomes have not recovered (**FIGURE 19.21B**). When Sepkoski and colleagues simulated the end-Cretaceous drop in the diversity of both clades, their model rendered a profile of subsequent diversity change that closely matches the data (**FIGURE 19.21C**). This result does not prove that competition determined the history of bryozoan diversity, but it is consistent with that hypothesis.

In general, a pattern of replacement is consistent with competitive displacement if the earlier and later taxa lived in the same place at the same time, if they used the same resources, if the earlier taxon was not decimated by a mass extinction event, and if the diversity and abundance of the later taxon increased as the earlier taxon

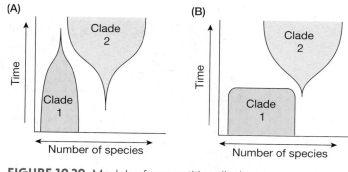

FIGURE 19.20 Models of competitive displacement and incumbent replacement. In each diagram, the width of a "spindle" represents the number of species. (A) Competitive displacement, in which the increasing diversity of clade 2 causes a decline in clade 1 by direct competitive exclusion. Compare with Figure 19.19B. (B) Incumbent replacement, in which the extinction of clade 1 enables clade 2 to diversify later. (After [65].)

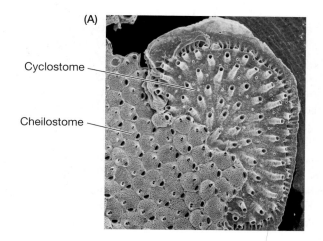

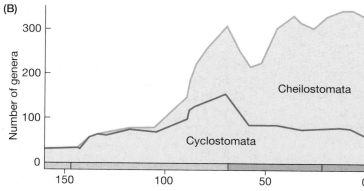

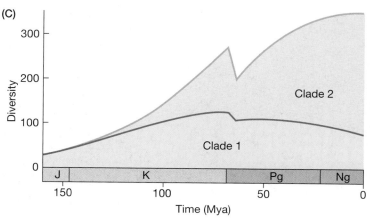

FIGURE 19.21 Competitive displacement among bryozoans, sessile colonial animals that spread over hard surfaces. (A) Colonies of cheilostome bryozoans (left) can overgrow colonies of cyclostome bryozoans (right), causing their death. (B) Cheilostomes appeared in the late Jurassic and soon increased greatly in diversity, whereas the increase of cyclostomes was reversed in the Cenozoic. (C) The changes in diversity in a model of competition among species in two clades, assuming that clade 2 species are competitively superior and that the diversity of both clades was reduced by an external perturbation 60 Mya. (A from [42], courtesy of Frank K. McKinney; B, C after [66].)

(A)

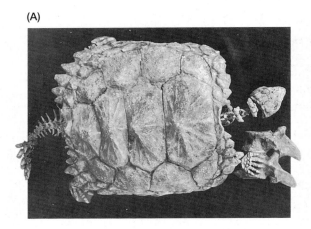

(B)

(C)

FIGURE 19.22 Pleurodiran and cryptodiran turtles replaced incumbent amphichelydian turtles, which became extinct. (A) Amphichelydians, represented here by the reconstructed skeleton of an early turtle (*Proganochelys quenstedti*, Triassic), could not retract their head for protection. (B) Snakeneck turtles such as *Chelodina longicollis* are pleurodiran turtles, which flex the neck sideways beneath the edge of the carapace. (C) Cryptodiran turtles, represented here by an eastern box turtle (*Terrapene carolina*), fully retract the head into the shell by flexing the neck vertically. (A courtesy of E. Gaffney, American Museum of Natural History.)

declined [40]. By these criteria, the rodents seem to have displaced the multituberculates (see Figure 19.19B). Vascular plants, which certainly compete for space and light, showed this pattern during the Cretaceous, when flowering plants increased in diversity and abundance at the expense of nonflowering plants.

Incumbent replacement has probably been more common than competitive displacement [7, 8, 33]. The best evidence of incumbency and release is supplied by repeated replacements. For instance, amphichelydians, the "stem group" of turtles, could not retract their head and neck into their shell (**FIGURE 19.22**). Two groups of modern turtles, which protect themselves by bending the neck within the shell or under its edge, replaced the amphichelydians in different parts of the world four or five times, especially after the K/Pg extinction event. The modern groups evidently could not radiate until the amphichelydians had become extinct. That this replacement occurred in parallel in different places and times makes it a likely example of release from competition [62].

What can we conclude? Is diversity of species constrained by ecological limits? Probably both sides in the debate are partly right. Most clades, viewed individually, seem to have approached a diversity limit. But new clades, with new ways of living, have arisen throughout life's history, seeming to push the diversity ceiling higher and higher—which, to advocates of boundless diversity, looks like no ceiling at all [29].

Go to the
Evolution Companion Website
EVOLUTION4E.SINAUER.COM
for data analysis and simulation exercises, quizzes, and more.

SUMMARY

- The per taxon rate of diversification equals the rate of origination (or speciation) minus the rate of extinction. Analyses of diversity in the fossil record require procedures to correct for biases caused by the incompleteness of the record. Some inferences about rates of diversification and speciation can also be made from time-calibrated phylogenies of living species.

- The diversity of skeletonized marine animals has increased during the Phanerozoic, but at varying rates. Diversity appears to have increased in the Cambrian to an approximate equilibrium that lasted for most of the Paleozoic; then, after the mass extinction at the end of the Permian, it has increased, with interruptions and at varying rates, ever since.

- The background rate of extinction (in between mass extinctions) has declined during the Phanerozoic, perhaps because higher taxa that were particularly susceptible to extinction became extinct early.

- Five mass extinctions (at or near the ends of the Ordovician, Devonian, Permian, Triassic, and Cretaceous) are recognized. These periods of high extinction rates have been followed by intervals of rapid origination of new taxa. Their diversification was probably released by the extinction of taxa that had occupied similar ecological space. Newly diversifying groups have sometimes replaced other taxa by direct competitive displacement, but more often they have replaced incumbent taxa after those taxa became extinct.

- The increase in diversity over time appears to have been caused mostly by adaptation to vacant or underused adaptive zones ("ecological space"), and by the evolution of key adaptations. Diversity has also been affected by biological interactions, whereby new species are often used as resources by other species.

- Both paleontological and phylogenetic evidence shows that the increase in diversity in most clades has been diversity-dependent. Such observations imply that diversity tends toward an equilibrium, but diversity seems nevertheless to increase, partly because new and specialized ways of living continue to evolve.

TERMS AND CONCEPTS

background extinction

diversity-dependent factor

ecological opportunity

key adaptation

lineage-through-time (LTT) plot

mass extinction

pull of the Recent

Red Queen hypothesis

species richness

species selection

SUGGESTIONS FOR FURTHER READING

Many of the topics in this chapter are treated clearly in *Principles of Paleontology* (third edition) by M. Foote and A. I. Miller (W. H. Freeman, New York, 2007). See also D. Jablonski et al. (eds.), *Evolutionary Paleobiology* (University of Chicago Press, 1996).

The end-Permian mass extinction is the subject of a popular book by D. H. Erwin, *Extinction: How Life on Earth Nearly Ended 250 Million Years Ago* (Princeton University Press, Princeton, NJ, 2006). The consequences of mass extinctions are reviewed by R. K. Bambach, in "Phanerozoic biodiversity mass extinctions" (*Annu. Rev. Earth Planet. Sci.* 34: 127–155, 2006), and D. Jablonski, in "Mass extinctions and macroevolution" (*Paleobiology* 31 [Supp.]: 192–210, 2005).

Adaptive radiation is treated by R. E. Glor's "Phylogenetic insights on adaptive radiation" (Annu. Rev. Ecol. Evol. System. 41: 251–270, 2010), and at greater length by D. Schluter in The Ecology of Adaptive Radiation (Oxford University Press, Oxford, UK, 2010).

Current phylogenetic methods for studying diversification are reviewed by R. A. Pyron and F. T. Burbrink in "Phylogenetic estimates of speciation and extinction rates for testing ecological and evolutionary hypotheses" (*Trends Ecol. Evol.* 28: 729–736, 2013). Contrasting views on whether or not diversity is constrained by ecological limits are presented by L. J. Harmon and S. Harrison in "Species diversity is dynamic and unbounded at local and continental scales" (*Am. Nat.* 185:

584–593, 2015), and by D. L. Rabosky and A. H. Hurlbert in "Species richness at continental scales is dominated by ecological limits" (*Am. Nat.* 185: 572–583, 2015). See also D. L. Rabosky, "Diversity-dependence, ecological speciation, and the role of competition in macroevolution" (*Annu. Rev. Ecol. Evol. Syst.* 44: 481–502, 2013), and J. J. Wiens, "The causes of species richness patterns across space, time, and clades and the role of ecological limits" (*Quart. Rev. Biol.* 86: 75–96, 2011).

PROBLEMS AND DISCUSSION TOPICS

1. Distinguish between the rate of speciation in a higher taxon and its rate of diversification. What are the possible relationships between the present number of species in a taxon, its rate of speciation, and its rate of diversification?

2. Many pairs of sister taxa differ markedly in their numbers of extant species. In this chapter we saw huge disparities—for example, between lepidopterans and their sister group, caddisflies. What factors (both general and specific) might account for differences among taxa in their numbers of extant species? Suggest methods for determining which factor might actually account for an observed difference.

3. Ehrlich and Raven (1964, *Evolution* 18: 586–608) suggested that coevolution with plants was a major cause of the great diversity of herbivorous insects, and Mitter et al. (1988, *American Naturalist* 132: 107–128) presented evidence that the evolution of herbivory was associated with increased rates of insect diversification. However, the increase in the number of insect families in the fossil record was not accelerated by the explosive diversification of flowering plants (Labandeira and Sepkoski, 1993, Science 261: 310–315). Suggest some hypotheses to account for this apparent conflict and some ways to test the hypotheses.

4. A factor that might contribute to increasing species numbers over time is the evolution of increased specialization in resource use, whereby more species coexist by more finely partitioning resources. Discuss ways in which, using either fossil or extant organisms, one might test the hypothesis that a clade is composed of increasingly specialized species over the course of evolutionary time.

5. In several phyla of marine invertebrates, lineages classified as new orders appear first in the fossil record in shallow-water environments and are recorded from deep-water environments only later in their history (Jablonski and Bottjer 1990, in R. M. Ross and W. D. Allmon [eds.], *Causes of Evolution: A Paleontological Perspective* [Chicago: University of Chicago Press], pp. 27–75). What might explain this observation? (Note: No one has offered a definitive explanation so far, so use your imagination.)

6. The analysis by McPeek and Brown (2007, *American Naturalist* 169: E97–E106) suggests that clades with few living species may be very young. Is this necessarily the case? Are there alternative hypotheses? Can you find evidence for any of these hypotheses? What would constitute evidence?

7. The method of replicated sister-group comparison of species richness has been used to implicate certain adaptive characteristics as contributors to higher species richness. Is there any way, conversely, to test hypotheses on what factors may have contributed to the decline or extinction of groups? For commentary and examples, see Vamosi and Vamosi (2005, *Evolutionary Ecology Research* 7[4]: 567–579) or Wiegmann et al. (1993, *American Naturalist* 142 [5]: 737–754).

8. Scientific debate continues about the history and interpretation of the diversity patterns of many taxa. Analyze such a debate, and decide whether either side has settled the issue. If not, what further research would be needed to do so? An example is whether or not the enormous diversity of leaf beetles (Chrysomelidae) is due to a long history of co-diversification with their host plants. See Farrell 1998 (*Science* 281: 555–559), Farrell and Sequeira 2004 (*Evolution* 58: 1894–2001), and Gómez-Zurita et al. 2007 (*PLoS ONE* 2[4], e360).

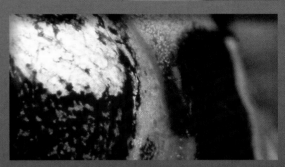

20

Macroevolution: Evolution above the Species Level

In earlier chapters, we have seen that bighorn sheep have evolved smaller horns because of selection imposed by human hunting, and that environmental changes caused selection for larger bills in a Galápagos finch. We followed Darwin in noting that species of finches differ slightly in bill size, and inferring that evolution has proceeded by successive slight changes. We learned that although some evolutionary changes may entail mutations with rather large effects, many or most phenotypic characteristics vary quantitatively, based on small effects of alleles at many loci—alleles that arise by spontaneous mutation, without regard for their possible adaptive utility. We also learned that such evolutionary changes may differ among populations of a species, and that some such changes result in reproductive barriers that mark the emergence of different species, which may become the ancestors of different clades of descendant species. We saw that some evolutionary changes are based on genome changes, such as the evolution of new genes by duplication and divergence of ancestral genes. We glimpsed some of the current research on how phenotypic changes may result from mutations that alter coding sequences or gene regulation. And we marveled, in surveying the history of life, that in the fullness of time, not only slightly different species evolve, but also forms that come to differ profoundly from their ancestors: tetrapods such as *Tiktaalik* evolved from lobe-finned fishes, and winged insects from wingless ancestors that themselves had evolved from crustaceans.

The abundant evidence on these points supports the fundamental tenets of the evolutionary synthesis that emerged in the 1930s and 1940s (see Chapter 1),

The eye of a South American horned frog (*Ceratophrys ornata*). Despite their complexity, eyes have evolved independently in many groups of animals, often based on changes in some of the same genes. The black and white patterning of the horned frog's iris makes the circular black pupil harder for predators to detect.

especially (1) that adaptive evolution results from natural selection on random (i.e., not adaptively directed) mutations; (2) that the selected mutations are mostly those with small effects; (3) that these kinds of genetic variations arise and persist in large populations, so that adaptive evolution need not await new mutations but instead can be very rapid; and (4) that large evolutionary changes, transpiring over long periods of time, have occurred gradually, by the accumulation of small changes.

But although these points are well established, we can pose many more questions, especially about evolution over long periods of time. Is the rate of evolution, based on the supply of genetic variation and the strength of natural selection, fast enough to account for the emergence of major new kinds of organisms, such as birds and whales? What are the steps by which such new forms (higher taxa) have evolved? The size of a beak or horn is genetically variable and can evolve readily, but where did beaks and horns come from in the first place? That is, how do we account for novel characteristics? Has the evolution of higher taxa been entirely a history of gradual change, or might there have been discontinuities—big changes without intermediates? And is the history of evolution a history of random changes, triggered by random environmental events, or is there some predictability? Have there been any grand trends in the history of life?

These questions pertain to **macroevolution**, which is often defined as "evolution above the species level," whereas **microevolution** refers mostly to processes that occur within species. Before the evolutionary synthesis, some authors proposed that these levels of evolution involved different processes. In contrast, the paleontologist George Gaylord Simpson [99, 100], who focused on rates and directions of evolution perceived in the fossil record, and the zoologist Bernhard Rensch [90], who inferred patterns of evolution from comparative morphology and embryology, argued convincingly that macroevolution is based on microevolutionary processes, and differs only in scale. Although their arguments have largely been accepted, this remains a somewhat controversial question.

The Origin of Major New Forms of Life

Paleontologists have documented intermediate steps in the origin of many major forms of life, or higher taxa (see Chapters 16 and 17). Chapter 16 described the origin of birds from theropod dinosaurs. Recall that many feathered theropods have been found (see Figure 16.15), with features such as long, clawed fingers, elongate tail, and leg structure that closely resemble those of other theropods. Feathers almost certainly first provided insulation and helped maintain body temperature; the modifications that enabled flight, as in *Archaeopteryx*, came later. A key feature in the evolution of birds was smaller size, which may have enabled the evolution of other features, such as more paedomorphic skulls with relatively large eyes and brains [4, 58]. A genomic study found that almost all the genes known to affect feather development are also present in crocodilians, and were therefore almost certainly present in all dinosaurs [63]. Since these genes do not produce feathers in crocodiles, feather development must have required some changes in the network of interactions among these genes. Thus, some of the features of modern birds, such as feathers and hollow limb bones, evolved in theropods long before *Archaeopteryx*, and other characters, such as the keeled breastbone, loss of teeth, and loss of claws on the hands that typify modern birds, evolved later.

Because creationists often claim that there is no evidence of the evolution of new "kinds" of organisms, every student of biology should be aware of the strong and growing evidence. For that reason, and because such cases provide details of macroevolution, and simply because it is a wonderfully interesting story, we describe one more example.

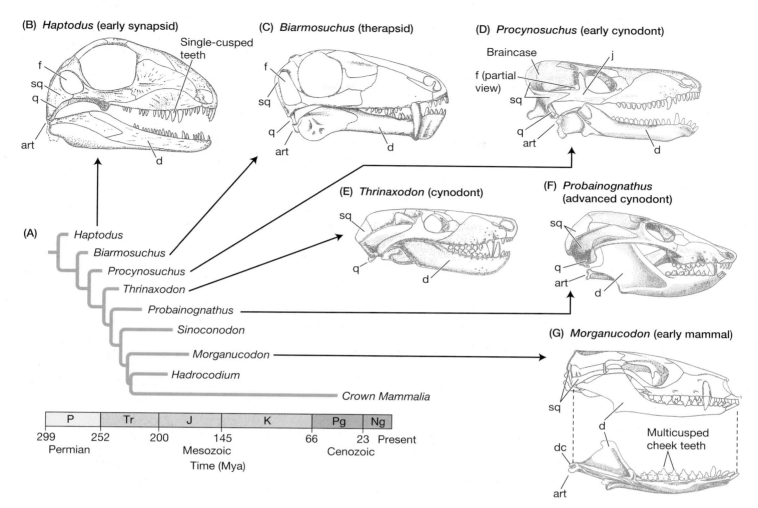

FIGURE 20.1 (A) A phylogeny of a few of the many mammalian and related lineages known from the fossil record. (B–G) Skulls from some early synapsids and early mammals. (B) An early synapsid, *Haptodus*. Note the temporal fenestra (f), multiple bones in the lower jaw, single-cusped teeth, and articular/quadrate (art/q) jaw joint. (C) An early therapsid, *Biarmosuchus*. Note the enlarged temporal fenestra. (D) An early cynodont, *Procynosuchus*. The side of the braincase is now vertically oriented, separated by a large temporal fenestra from a lateral arch formed by the jugal (j) and squamosal (sq). Note the enlarged dentary (d). (E) A later cynodont, *Thrinaxodon*. Note multiple cusps on the rear teeth, the large upper and lower canine teeth, and the greatly enlarged dentary with a vertical extension to which powerful jaw muscles were attached. (F) An advanced cynodont, *Probainognathus*. The cheek teeth had multiple cusps, and two bones of the lower jaw articulated with the skull. (G) *Morganucodon*, often considered to be a mammal. Note the multicusped cheek teeth (including inner cusps) and double articulation of the lower jaw, including articulation of a dentary condyle (dc) with the squamosal (sq). Abbreviations: art, articular; d, dentary; dc, dentary condyle; f, fenestra; q, quadrate; sq, squamosal. (A after [5]; B–G after [28], based on [10] and various sources.)

The origin of mammals

The origin of mammals from earlier amniotes (see Figure 17.24) is one of the most fully documented examples of the evolution of a major taxon [51, 64, 98]. Some features of living mammals, such as hair and mammary glands, do not usually become fossilized, but the evolution of the skeleton has been well documented, including changes in the skull and jaw that we describe here.

Soon after the first amniotes originated, during the Carboniferous, they gave rise to the Synapsida, which developed into diverse mammal-like lineages (**FIGURE 20.1A**). From among these arose the crown Mammalia, containing all the lineages that descended from the common ancestor of the living mammals: the Monotremata (egg-laying echidna and platypus), the Metatheria (marsupials), and the Eutheria (placental mammals). The crown Mammalia also includes many extinct groups. Some of the critical characteristics of the Mammalia are these:

- The lower jaw in reptiles consists of the articular and several other bones, but in mammals it is only a single bone (the dentary).
- The primary (and in all except the earliest mammals, the exclusive) jaw articulation is between the dentary and the squamosal skull bone, rather than between the articular and the quadrate bone, as in other tetrapods.
- Early amniotes have a single sound-transmitting bone in the middle ear, the stirrup (or stapes). Mammals have three bones: not only the stirrup, but also the hammer (malleus) and anvil (incus).
- Mammals' teeth are differentiated into incisors, canines, and multicusped (multipointed) cheek teeth (premolars and molars), whereas most other tetrapods have uniform, single-cusped teeth.

Other features that distinguish most mammals from other amniotes include an enlarged braincase, a large space (temporal fenestra) behind the eye socket, and a secondary palate that separates the breathing passage from the mouth cavity.

The early synapsids had a temporal fenestra that provided space for jaw muscles to expand into when contracted (**FIGURE 20.1B**). The temporal fenestra became progressively enlarged in later synapsids (**FIGURE 20.1C–E**). Permian synapsids in the order Therapsida (see Figure 20.1C) had large canine teeth, and the center of the palate was recessed, suggesting that the breathing passage was partially separated from the mouth cavity. The hind legs were held rather vertically, more like a mammal than a reptile.

Among the Therapsida, the cynodonts, which lived from the late Permian to the late Triassic, represent several steps in the approach toward mammals. The rear of the skull was compressed, giving it a doglike appearance (see Figure 20.1D,E); the dentary was enlarged relative to the other bones of the lower jaw; the cheek teeth had a row of several cusps rather than only one; and a bony shelf formed a secondary palate that was incomplete in some cynodonts and complete in others. The quadrate was smaller and looser than in previous forms and occupied a socket in the squamosal (**FIGURE 20.2A**).

In the advanced cynodonts of the middle and late Triassic (**FIGURE 20.1F**), the cheek teeth had not only a linear row of cusps, but also a cusp on the inner side of the tooth. This seemingly trivial, but actually profoundly important, innovation begins a history of complex cheek teeth of mammals, which are modified in different lineages for chewing different kinds of food. In some late Triassic and Jurassic cynodonts, the lower jaw had not only the old articular/quadrate articulation with the skull, but also an articulation between the dentary and the squamosal, marking a critical transition between the ancestral condition and the mammalian state (**FIGURE 20.2B**). All these features—molars, a strong lower jaw composed mostly of a single bone, an enlarged fenestra to accommodate large jaw muscles, and a secondary palate that enabled the animal to breathe while consuming and chewing large prey—imply increasingly active, efficient predators, probably with a heightened metabolism.

Morganucodon, of the late Triassic and very early Jurassic (**FIGURE 20.1G**), is considered an early mammal. *Morganucodon* had two jaw-skull joints: a weak articular/quadrate hinge and a fully developed mammalian articulation between the dentary and the squamosal. The articular and quadrate bones were sunk into the ear region, similar to the condition in modern mammals (**FIGURE 20.2C,D**). *Hadrocodium*, from the early Jurassic, was very similar to *Morganucodon*, but the lower jaw consisted entirely of the dentary, and the articular and quadrate bones were fully separated from the jaw joint and fully lodged in the middle ear, where they are now called the hammer and anvil, and transmit sound, together with the stirrup. *Morganucodon*,

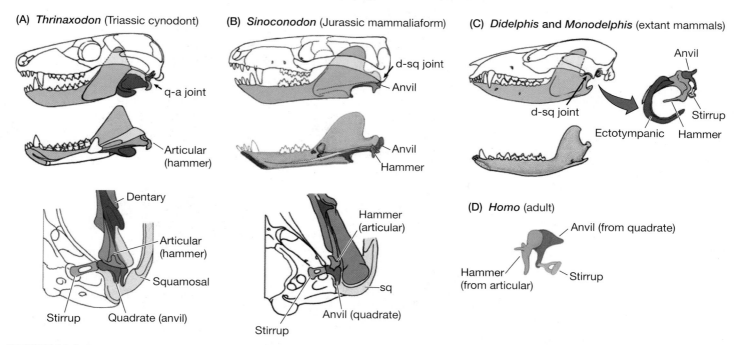

(A) *Thrinaxodon* (Triassic cynodont)

q-a joint

Articular (hammer)

Dentary

Articular (hammer)

Squamosal

Stirrup   Quadrate (anvil)

(B) *Sinoconodon* (Jurassic mammaliaform)

d-sq joint

Anvil

Anvil

Hammer

Hammer (articular)

sq

Anvil (quadrate)

Stirrup

(C) *Didelphis* and *Monodelphis* (extant mammals)

Anvil

d-sq joint

Stirrup

Ectotympanic   Hammer

(D) *Homo* (adult)

Anvil (from quadrate)

Hammer (from articular)   Stirrup

FIGURE 20.2 Evolution of the middle-ear bones, the anvil (or incus) and hammer (or malleus), from the jaw-joint bones (quadrate, articular) of cynodonts. The third middle-ear bone, the stirrup (stapes), was present in ancestral tetrapods. (A) A Triassic cynodont, *Thrinaxodon*. Three views, from top: skull, lower jaw, and a slice through the left jaw joint, viewed from below. Note the joint formed by the quadrate (q) and articular (a). The lower jaw includes the dentary, as well as the angular (ectotympanic) and surangular bones, which became reduced and then lost in advanced mammals. (B) The same views of the mammal-like Jurassic *Sinoconodon*, a relative of the late Triassic *Morganucodon*. The dentary makes up most of the lower jaw, the other bones having been reduced or lost. There is a double joint, between the quadrate and articular, and also between the large dentary (d) and the squamosal (sq) bone of the skull. (C) The skull and lower jaw of living marsupials, here the opossums *Didelphis* and *Mondelphis*. The dentary and squamosal form the single jaw joint (d-sq). The quadrate and articular are now called the anvil and hammer, two of the middle-ear bones (shown in close-up). (D) The middle-ear bones in humans. The ectotympanic, or angular, persisted in marsupials but has been lost in placental mammals. (From [64].)

Hadrocodium, and later mammals also show sequential steps in enlargement of the brain, especially in the olfactory bulb and the neocortex [92].

This description touches on only a few highlights of a complex history. For example, several of the changes in the lower jaw and middle ear occurred independently in different cynodont lineages [64]. Nonetheless, *the emergence of the class Mammalia illustrates some themes that are common in the evolution of higher taxa*. Most mammalian characters (e.g., posture, tooth differentiation, skull changes associated with jaw musculature, secondary palate, brain size, reduction of the elements that became the small bones of the middle ear) evolved gradually. Evolution was mosaic, with different characters evolving at different rates. No new bones evolved; in fact, many bones have been lost in the transition to modern mammals [97], and all the bones that persist are modified from those of the stem amniotes (and in turn, from those of early tetrapods and even lobe-finned fishes). Some repeated elements, such as teeth, became individualized: each molar in the mouth of a human has a distinct identity. Some major changes in the form of structures are associated with changes in their function. The most striking example is the articular and quadrate bones, which serve for jaw articulation in all other tetrapods, but became the sound-transmitting middle-ear bones of mammals [64]. Because the evolution of mammals from synapsids, over the course of more than 130 My, has been gradual, there is no cutoff point for recognizing mammals: the definition of "Mammalia" in a temporal context is arbitrary.

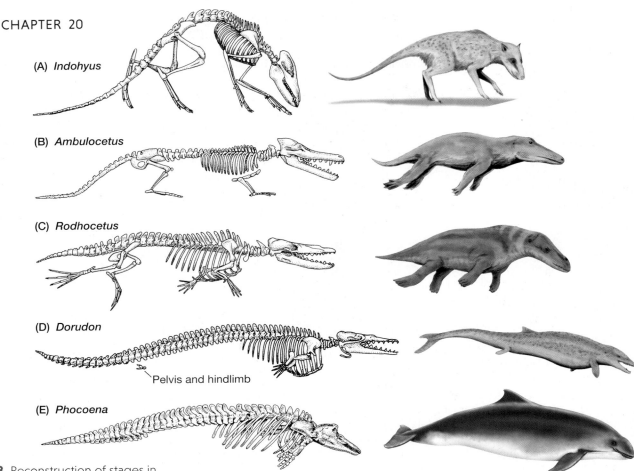

(A) *Indohyus*

(B) *Ambulocetus*

(C) *Rodhocetus*

(D) *Dorudon*

Pelvis and hindlimb

(E) *Phocoena*

FIGURE 20.3 Reconstruction of stages in the evolution of cetaceans from terrestrial artiodactyl ancestors. (A) Eocene raoellids, perhaps the sister group of Cetacea, were terrestrial but show some evidence of semiaquatic life. (B) The amphibious *Ambulocetus*. (C) The middle Eocene *Rodhocetus* had the distinctive ankle bones of artiodactyls, but had numerous cetacean characters. (D) *Dorudon*, of the middle to late Eocene, had most of the features of modern cetaceans, although its nonfunctional pelvis and hindlimb were larger. (E) A modern toothed whale, the harbor porpoise (*Phocoena phocoena*). The nostrils, forming a blowhole, are far back on the top of the head, accounting for the peculiar shape of the skull. (A skeleton after [107]; B–D skeletons after [18, 32]; E, skeleton drawing by Nancy Haver.)

Gradualism and Saltation

Darwin proposed that evolution proceeds gradually, by small steps. His ardent supporter Thomas Henry Huxley, however, cautioned that Darwin's theory of evolution would be just as valid even if evolution proceeded by leaps (sometimes called **saltations**). Some later biologists proposed just this. The geneticist Richard Goldschmidt argued in *The Material Basis of Evolution* [34] that species and higher taxa arise not from the genetic variation that resides within species, but instead "in single evolutionary steps as completely new genetic systems." He postulated that major changes of the chromosomal material, or "systemic mutations," would give rise to highly altered creatures. Most would have little chance of survival, but some few would be "hopeful monsters" adapted to new ways of life. Goldschmidt's genetic system hypothesis has been completely repudiated, but the possibility of evolution by more modest jumps remains one of the most enduring controversies in evolutionary theory. Quite different species are often connected by intermediate forms, so that it becomes arbitrary whether the complex is classified as two genera (or subfamilies, or families) or as one (see Chapter 2). Nonetheless, there exist many conspicuous gaps, especially among higher taxa such as orders and classes. No living species bridge the gap between cetaceans (dolphins and whales) and other mammals, for example.

The most obvious explanation of phenotypic gaps among living species is extinction of intermediate forms that once existed—as the cetaceans themselves illustrate (**FIGURE 20.3**). DNA sequences imply that, among living animals, whales are most closely related to hippopotamuses. The earliest known fossilized members of the cetacean lineage (see Figure 20.3A) were terrestrial, but not particularly similar to living hippopotamuses—implying that the common ancestor of two quite different forms need not have appeared precisely intermediate between them, because the two phyletic lines may have undergone quite different modifications. Extinct cetaceans do show several of the steps leading to modern cetacean morphology, such as reduction of the pelvis and hindlimbs, the shift in the nostrils to the top of the head, and the greater uniformity of the teeth. Toothless baleen whales, such as the blue whale, evolved from the toothed whales (such as today's dolphins and sperm whale), but intermediate extinct forms had both baleen and reduced teeth [17].

Some organisms have puzzling features that seem to call for a saltational origin, because it is hard to see how an intermediate step from their ancestor could have been advantageous. Turtles are a striking example. Their carapace (upper shell), largely formed from the vertebrae and modified ribs, encloses the entire pectoral girdle, including the scapula (shoulder blade). In all other tetrapods, the pectoral girdle lies outside (above) the rib cage. (Check your own shoulder blades.) A combination of paleontological and developmental studies has begun to show that this difference could have evolved gradually [57, 60, 80]. In most amniotes, such as birds, the developing ribs grow laterally and then downward; above them, a muscle plate does the same (**FIGURE 20.4A**). The forelimbs (including the pectoral girdle) grow outward from this dorsal muscle plate, in response to inductive signals from a lateral ridge (Wolffian ridge). In turtles, however, the developing ribs grow laterally and then stop, instead of growing downward, partly because of signals from another external ridge (carapacial ridge), which is a novel feature in modern turtles (**FIGURE 20.4B**). As a result, the ribs lie above the muscle plate and the developing pectoral girdle. The evolution of this developmental transition may have been easy if a recently discovered fossil turtle is a reliable guide. The late Triassic *Odontochelys semitestacea*, one of the oldest members of the turtle lineage yet found, had a lower shell (plastron), but instead of a carapace, it had only standard-issue ribs. But the anterior ribs were deflected backward, so that the pectoral girdle and forelimbs lay in front of the rib cage, instead of above or below it (**FIGURE 20.4C**). If the alteration of rib development occurred at this stage in turtle evolution, and the ribs became directed forward later in evolution, they would lie above the pectoral girdle.

Goldschmidt could point to many mutations that cause large, discontinuous changes that he envisioned might be the basis of saltational evolution. For example, mutation of the *Ultrabithorax* (*Ubx*) gene in *Drosophila* transforms a fly with halteres into a fly with two pairs of wings (see Figure 15.8). It may be tempting to think that a *Ubx* mutation in the ancestor of the Diptera caused the evolutionary transformation of the second pair of insect wings into halteres. But the *Ubx* mutation does not restore "real" hindwings; it transforms the third thoracic segment into a replicated second segment, including a replicated set of forewings. The ancestors of flies did not have identical second and third segments, and the hindwings differ from forewings in all four-winged insects. Mutations that reduce the function of this master control gene interfere with a complex developmental pathway, and development is routed into a "default" pathway that produces the features of the second thoracic segment (including wings). The whole system can be shut down in a single step by turning a master switch, but that does not mean the system came into existence by a single step. And—a critical point—this mutation, like many other such "large-effect" mutations, drastically reduces survival because it so profoundly disturbs normal development.

Certainly, mutations that have fairly large (but not huge) effects can contribute to evolution. For example, variation within and among species in characters such as bristle number in *Drosophila* is often caused by a mixture of quantitative trait loci with both small and large effects [84]. Alleles with large effects contribute importantly to Müllerian mimetic phenotypes in butterflies such as *Heliconius* (see Figures 6.23 and 13.10). Genetic analysis of the color patterns in *Heliconius* suggests that the evolution of one mimetic pattern from another was probably initiated by a mutation large enough to provide substantial resemblance to a different model species, followed by selection of alleles with smaller effects that "fine-tuned" the phenotype [3].

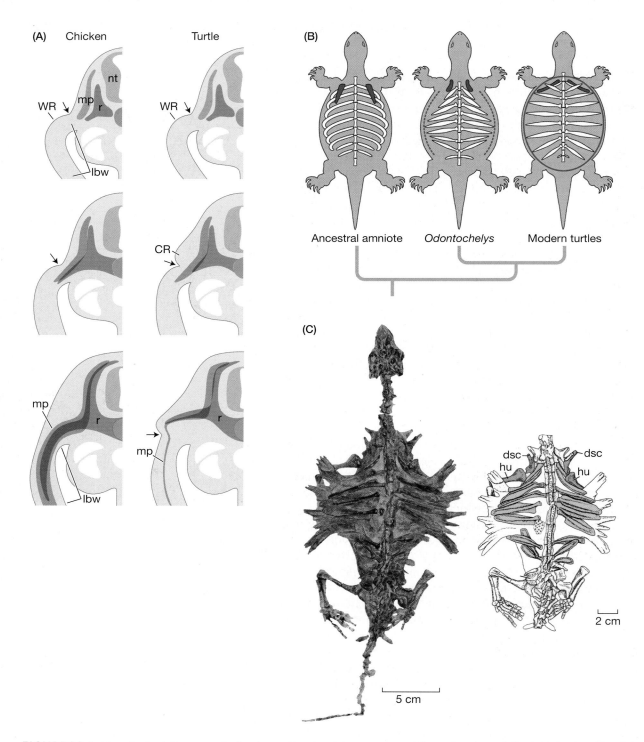

FIGURE 20.4 How turtles' ribs came to lie above the shoulder blade (scapula) instead of below. (A) Cross sections through three embryonic stages of a typical amniote (chicken, left) and a modern turtle (right). Only the left half of the body is shown. Part of the neural tube (nt) provides orientation. Arrows show a groove, with the Wolffian ridge (WR) below, at the upper part of the developing body wall (lbw). Cells destined to become ribs (r) and a muscle plate (mp) invade the body wall, due to signals from the Wolffian ridge. The muscle plate will give rise to the pectoral girdle. In the chicken, rib cells grow down into the body wall, beneath the muscle plate, and form ribs underneath the muscle plate and developing pectoral girdle. But the turtle (right) develops a novel carapacial ridge (CR), which emits signals that arrest the extension of ribs into the body wall (see lower right diagram), so that the ribs develop above the girdle. (B) A phylogenetic hypothesis of how the arrangement of ribs and pectoral girdle evolved from ancestral amniotes to modern turtles, via a stage represented by the extinct turtle *Odontochelys*, in which the anterior ribs were deflected toward the rear. After the changes described in part (A) evolved, the ribs became directed forward and over the girdle. The scapula and carapacial ridge are shown in red. (C) Dorsal view of the Triassic turtle *Odontochelys*, which lacked a carapace. The line drawing distinguishes bones. The scapula (dsc) and humerus (hu), highlighted in red, are clearly visible, lying in front of the ribs (highlighted in blue), which are directed toward the rear. (A after [57]; B from [80]; C from [60].)

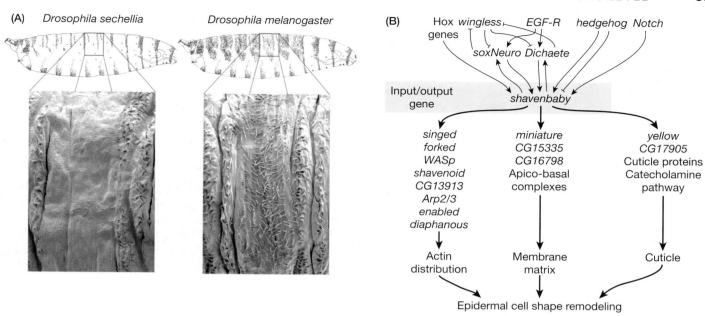

FIGURE 20.5 How a morphological difference between species may be caused by repeated evolution in a single gene. (A) Larval *Drosophila sechellia* (left) lack the dorsal trichomes that are present in *D. melanogaster* (right) and other relatives. (B) The absence of trichomes is caused by several mutations in the *cis*-regulatory control region of the *shavenbaby* gene, which determines trichome development in different sectors of each body segment. This diagram shows the complex gene regula-tory network, in which *shavenbaby* is a key player. Transcription of *shavenbaby* in different parts of the cuticle is regulated by a complex of developmental patterning genes (Hox genes, *wingless*, etc.), and *shavenbaby* in turn regulates the expression of downstream genes that determine actin distribution, membrane matrix, and cuticle, which together transform an epidermal cell into a trichome. (From [105].)

Differences among species in the activity of regulatory genes, such as Hox genes, or in alternative splicing of key regulatory genes, are sometimes associated with substantial morphological differences [38]. It is likely that such genes accumulated their divergent effects gradually, perhaps by successively recruiting different downstream genes and pathways. We have seen, for example, that a new spot pattern in the wing of *Drosophila biarmipes* resulted from novel activation of the *yellow* gene by the Distal-less transcription factor, based on an evolutionary change in the sequence of a *cis*-regulatory element (see Figure 15.17). Without detailed evidence, we cannot tell if a big, discontinuous difference in phenotype was caused simply by a single change in a key regulatory gene, or by multiple changes in the interactions between the *trans*-regulator and *cis*-regulatory changes in many downstream genes.

Some large evolutionary changes do have simple genetic foundations. In some cases, the gene merely extends or truncates a developmental trajectory without causing harmful side effects. For example, the few genetic changes that determine the heterochronic difference between metamorphosing tiger salamanders (*Ambystoma tigrinum*) and their paedomorphic relative the axolotl (*A. mexicanum*; see Figure 15.3) do not engender an entirely new complex morphology, but merely truncate a complex, integrated pathway of development that presumably evolved originally by many small steps. In other cases, as David Stern and Virginie Orgogozo proposed [104, 105], there has been stepwise, gradual change by the accumulation of successive mutations—but these mutations have occurred mostly in a single "hotspot" gene that controls a key point in a developmental pathway. The important feature of such a gene is that it has few pleiotropic effects on other characters, so mutations are less likely to have deleterious side effects that would prevent them from increasing by natural selection. For example, larvae of *Drosophila sechellia* lack the dorsal trichomes (hairlike extensions of cell cuticle) possessed by its relatives, such as *D. melanogaster* (**FIGURE 20.5A**). The absence of trichomes (a derived trait) is caused by mutations in three different *cis*-regulatory regions of a single gene, *shavenbaby*, that regulates expression of three downstream batteries of genes that are necessary for trichome development. The *shavenbaby* gene itself is regulated by an array of upstream genes that determine where it is expressed (**FIGURE 20.5B**). Mutation of any single downstream gene would not suffice to alter trichome development, and mutation of the upstream genes, all of which have multiple functions, might alter other organs as well.

The Evolution of Novelty

How do major changes in characters evolve, and how do new features originate? These questions have two distinct meanings. First, we can ask what role natural selection plays in the evolution of such changes. For instance, we may well ask whether each step, from the slightest initial alteration of a feature to the full complexity of form displayed by later descendants, could have been guided by selection. Second, we can ask what the genetic and developmental bases of such changes are (see Chapter 15). Both questions bear on the problem of how complex characters could have evolved if their proper function depends on the mutually adjusted form of their many components.

Incipient and novel features: Permissive conditions and natural selection

Many features are modifications of ancestral structures that have been shaped by natural selection for new functions. This principle, already recognized by Darwin, is one of the most important in macroevolution [70], and every group of organisms presents numerous examples. A bee's sting is a modified ovipositor, or egg-laying organ. The wings of auks and several other aquatic birds are used in the same way in both air and water; in penguins, the wings have become entirely modified for underwater flight (see Figure 3.13). Many proteins have been co-opted or modified for new functions, such as a heat-shock protein and other proteins that, with little or no modification, form the crystallin lens in vertebrate eyes (see Chapter 14). In some cases, a feature may be an initially nonadaptive by-product of other adaptive features and has been recruited or modified to serve an adaptive function. For instance, by excreting nitrogenous wastes as crystalline uric acid, insects lose less water than if they excreted ammonia or urea. Excreting uric acid is surely an adaptation, but the white color of uric acid is not. However, pierine butterflies such as the cabbage white butterfly (*Pieris rapae*) sequester uric acid in their wing scales, imparting to the wings a white color that plays a role in thermoregulation and probably in sexual interactions.

By their behavior, animal species often affect or even determine the sources of natural selection on morphological and physiological traits [29, 70, 83]. Aquatic mammals would not have started to evolve adaptations for swimming unless their ancestors had selected wet habitats to live in; insects are selected to adapt to a plant's toxic chemicals only if some fraction of the population chooses to eat that species of plant [26, 89]. Behaviorally flexible species are frequently seen doing things they are not specifically adapted for; some species of gulls, for example, will feed on swarms of flying ants or termites, even though they normally eat aquatic animals. Changes in behavior may often be the first step in the evolution of a new ecological niche, to which other features become adapted [70].

Some aspects of organisms' form and function permit or facilitate the evolution of new characteristics. For example, decoupling the multiple functions of an ancestral feature relieves functional constraints, so the feature may be free to evolve in new ways. The loss of lungs in the largest family of salamanders (Plethodontidae) may have relieved a functional constraint on the evolution of the tongue [113]. In other salamanders, the bones that support the tongue are also used for moving air into and out of the lungs. In plethodontids, these bones, no longer used for ventilating the lungs, have been modified into a set of long elements that can be greatly extended from a folded configuration. This modification enables plethodontids to catch prey by projecting the tongue, in some species to extraordinary lengths at extraordinary speed (**FIGURE 20.6**).

Marc Kirschner and John Gerhart draw on cell and developmental biology in developing their hypothesis of "facilitated variation" [53]. They suggest that the "core processes" of protein activity and cell and organ development have properties of robustness and adaptability that cause some variation to arise in ways that facilitate evolution. For example, developing muscles, nerves, and blood vessels in a limb respond to signals from developing bone and dermis, and so grow into their proper positions. Thus genetic changes in the limb skeleton result in altered but functional limbs, without the need for independent genetic changes in musculature and vasculature.

Discoveries in evolutionary developmental biology, such as recruitment of genes and signaling pathways for new functions, are helping biologists understand the origin of novelties. For example, an entire developmental pathway may be triggered heterotopically in a different part of the body. A Mexican plant, *Lacandonia schismatica*, has perfectly formed stamens in the center of the flower, surrounded by pistils—the reverse of the usual arrangement [56]. A fascinating case is the anteriormost digit of a bird's hand, which is morphologically equivalent to digit 1 in the hand of related dinosaurs, and expresses the genes characteristic of a first digit (or "thumb"). However, it develops in digit position 2 and is phylogenetically homologous to the dinosaur's second digit. During the evolution of birds from nonavian dinosaurs, the thumb was lost and digits 2, 3, and 4 underwent a shift in developmental identity, taking on the features of digits 1, 2, and 3 (**FIGURE 20.7**) [110, 114]. In these cases, entire genetic-developmental pathways are deployed in new locations on the body (heterotopy), and produce developmentally coherent, functional phenotypes.

FIGURE 20.6 A lungless bolitoglossine salamander (*Hydromantes supramontis*) captures prey with its extraordinarily long tongue. The rapid tongue extension is accomplished with a modified hyobranchial apparatus, which in other families of salamanders plays an important role in ventilating the lungs. (From [16], courtesy of S. Deban.)

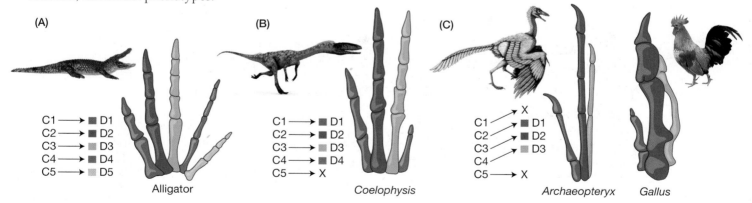

FIGURE 20.7 A "frameshift" in development of the hand in birds is thought to have transformed the identity of the digits. The ancestral state in archosaurs is illustrated by the alligator (A), and the state in theropod dinosaurs by *Coelophysis* (B). The state in *Archaeopteryx* and modern birds such as chickens (*Gallus*) is shown in (C). The developing hand of the embryo has five groups of cells that form cartilaginous digital condensations (C1 through C5). In (A), these differentiate into the digits D1 through D5, which are distinct in form and therefore in identity (signaled by different colors). (B) In theropods such as *Coelophysis*, C5 failed to develop, and only digits D1 through D4 were formed. (C) In birds, only condensations C2, C3, and C4 develop, but the digits have the form and identity of D1, D2, and D3 (which develop from condensations C1, C2, and C3 in the alligator and theropod). The hypothesis is that the gene networks that specify the form of digits D1, D2, and D3 are activated in different C condensations in birds, relative to their ancestors. (After [110].)

Complex characteristics

A common argument against Darwinian evolution is based on so-called irreducible complexity: the proposition that a complex organismal feature cannot function effectively except by the coordinated action of all its components, so that the feature must have required all of its components from the beginning. Since they could not have all arisen in a single mutational step, the feature (it is claimed) could not have evolved.

Needless to say, the first person to recognize this potential problem was Darwin himself, in *On the Origin of Species*: "That the eye, with all its inimitable contrivances for adjusting the focus to different distances, for admitting different amounts of light, and for the correction of spherical and chromatic aberration, could have been formed by natural selection seems, I freely confess, absurd in the highest possible degree." But he then proceeded to supply examples of animals' eyes as evidence that "if numerous gradations from a perfect and complex eye to one very imperfect and simple, each grade being useful to its possessor, can be shown to exist; if further, the eye does vary ever so slightly, and the variations be inherited, which is certainly the case; and if any variation or modification in the organ be ever useful to an animal under changing conditions of life, then the difficulty of believing that a perfect and complex eye could be formed by natural selection, though insuperable by our imagination, can hardly be considered real."

Darwin's claim has been fully supported by later research [81, 82, 85]. The eyes of various animals range from small groups of merely light-sensitive cells (in some flatworms, annelid worms, and others), to cuplike or "pinhole camera" eyes (in cnidarians, molluscs, and others), to the "closed" eyes, capable of registering precise images, that have evolved independently in cnidarians, snails, bivalves, polychaete worms, arthropods, and vertebrates (**FIGURE 20.8**). The evolution of eyes is apparently not so improbable! Each of the many grades of photoreceptors, from the simplest to the most complex, serves an adaptive function. Simple epidermal

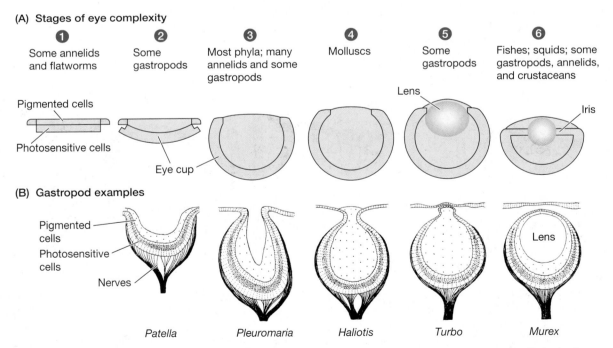

(A) Stages of eye complexity

① Some annelids and flatworms
② Some gastropods
③ Most phyla; many annelids and some gastropods
④ Molluscs
⑤ Some gastropods
⑥ Fishes; squids; some gastropods, annelids, and crustaceans

Pigmented cells
Photosensitive cells
Eye cup
Lens
Iris

(B) Gastropod examples

Pigmented cells
Photosensitive cells
Nerves
Lens

Patella   Pleuromaria   Haliotis   Turbo   Murex

FIGURE 20.8 Intermediate stages in the evolution of complex eyes. (A) Schematic diagrams of stages of eye complexity in various animals, from a simple photosensitive epithelium, through the deepening of the eye cup (providing progressively more information on the direction of the light source), through gradual evolution toward a "pinhole camera" eye, eventually including a refractive lens and a pigmented iris for sharper focusing. (B) Most of these stages can be found among various gastropod species (snails and relatives), as shown in these drawings. (A after [85]; B after [94].)

photoreceptors and cups are most common in slowly moving or burrowing animals; highly elaborate structures are typical of more mobile animals. The molecular basis of vision has also evolved by comprehensible steps. All the major molecular components, such as phototransduction cascades, the screening pigments needed for directional photoreception, the cell membrane elaborations that enable low-resolution spatial vision, and lens crystallin proteins, were present in the common ancestors of animals and even fungi, and most of these were independently recruited for vision in several animal lineages [82]. Neither at the morphological nor the molecular level is the notion of "irreducible complexity" a barrier to evolution.

The antievolutionary argument also fails to recognize that a component of a functional complex that was initially merely superior can become indispensable because other characters evolve to become functionally integrated with it. Although the eyes of many annelid worms and other animals do not have a lens, those animals do quite well without the visual acuity that a lens can provide. But a lens is indispensable for eagles, since their way of hunting prey has been made possible only by such acuity, and evolved after lenses did. Eagles and monkeys have *acquired* dependence on the elements of a complex eye. Such dependence is often lost: many burrowing and cave-dwelling vertebrates have degenerate eyes (see Figures 3.8 and 6.20).

Homology and the emergence of novel characters

"Novel" characteristics, in a broad sense, include both new modifications of ancestral structures, such as the elongated incisors that are the tusks of an elephant, and what may be considered truly new structures. For example, sesamoids are bones that develop in connective tissue. Such bones are the origin of novel skeletal elements, such as the extra "finger" of giant pandas and moles (**FIGURE 20.9**) and the patella (kneecap) in mammals, which is lacking in reptiles [79]. Günter Wagner distinguishes character states from character identity: the tusks of elephants are one of many states that the character "incisor" displays among mammals, but "incisor" and "patella" are

FIGURE 20.9 False fingers ("thumbs") have evolved from sesamoid bones in the giant panda (A) and in moles (C). The unmodified hands of these animals' close relatives, a bear (B) and a shrew (D), are shown for comparison. Developmental constraints probably prevented the evolution of a sixth true digit. These structures exemplify what has been called "tinkering" by natural selection: evolution of adaptations from whatever variable characters a lineage happens to already have. (From [15, 77]; shrew and mole hand photos courtesy of Marcelo Sánchez-Villagra.)

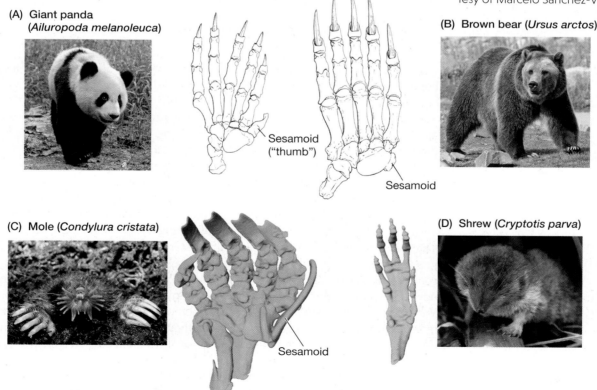

(A) Giant panda
(*Ailuropoda melanoleuca*)

Sesamoid
("thumb")

Sesamoid

(B) Brown bear (*Ursus arctos*)

(C) Mole (*Condylura cristata*)

Sesamoid

(D) Shrew (*Cryptotis parva*)

(A)

(B)

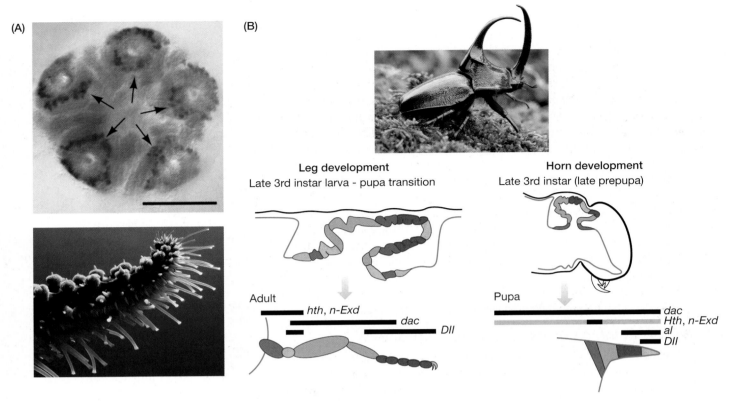

Leg development
Late 3rd instar larva - pupa transition

Horn development
Late 3rd instar (late prepupa)

Adult

hth, n-Exd

dac

Dll

Pupa

dac
Hth, n-Exd
al
Dll

FIGURE 20.10 The genetic toolkit that is shared by many animal phyla includes regulatory gene circuits that underlie the development of diverse features. A "limb program" that includes the *Distal-less* (*Dll*) gene contributes to the development of diverse outgrowths from the body wall. (A) Arrows point to developing tube feet in a larval sea urchin. Each dark spot is the expression of *Dll* in a tube foot. The lower photograph shows the developed tube feet of a mature echinoderm, here a sea star. (B) *Dll* is one of several genes that regulate the development of both legs and horns in scarabaeid beetles. The upper diagrams show that beneath the body wall of the pupa (black line), the adult body wall (blue line) develops, and forms outgrowths (in multiple colors) that become a leg or a horn. The lower diagrams show the corresponding parts of the fully developed leg in the adult and horn in the pupa. The horizontal black bars indicate where various genes are expressed. Some of the genes that are expressed in leg development are also expressed in the horn. (A upper photo courtesy of G. Boekhoff-Falk; B drawings from [78].)

characters with distinct identities, distinguishable from other mammalian teeth or bones [112]. Variation in some features may be hard to classify as one or the other—but this is true of many distinctions in biology, such as cell types or species.

This distinction bears on one of the fundamental concepts of evolution: homology. The forelimbs of various tetrapods, including the one-toed legs of horses, the flippers of whales, and the wings of birds and bats, have long been recognized as homologous, despite their many differences in form and function (see Figure 2.9). Their structure is shared among tetrapods and is attributed to common ancestry. But the same fundamental structure is seen in hindlimbs (where the femur corresponds to the humerus, the tibia to the radius, and so on). The similarity between these serially repeated structures (which has long been called serial homology) suggests that a similar genetic-developmental program is expressed in different parts of the body, just as it is expressed in different species that have inherited this program from their common ancestor. Wagner referred to this similarity, based on a common genetic-developmental program, as **biological homology** [79, 111]. This is a broader concept than the more commonly used concept of homology in systematics, where characters are defined as homologous if and only if they have been inherited from common ancestors. In fact, a character might be biologically homologous in two species even if their common ancestor did not express the feature. Recall (from Chapters 2 and 16) that characters that have been "lost" in a lineage sometimes have been regained (e.g., the aquatic larval stage in salamanders). These cases may represent the reexpression of a genetic-developmental program that had been reduced or unexpressed in the species' ancestors. The same phenomenon is seen in "deep homology," a term Neil Shubin and colleagues use to describe genetic regulatory pathways that may be widely inherited and independently expressed in different evolving lineages, and that may contribute to morphologically disparate features [96]. For example, homologues of the *Pax6* gene initiate the development of eyes in arthropods, vertebrates, and other animals, and a genetic pathway governed by the *Distal-less* gene is the basis for many structures (such as legs and horns) that originate in diverse phyla as evaginations from the body wall (**FIGURE 20.10**).

Novel characteristics are generally thought to be based on new regulatory interactions among previously unconnected genes [10, 55], which Wagner terms character identity networks [112]. Different states of a biologically homologous character, then, share the same fundamental network, together with some genes that may be included in the network in some species but not others. A striking example is the evolution of sex combs in the genus *Drosophila* [55]. These are groups of highly modified bristles on the tarsus (foot) of male flies that are used to hold or contact females. Species vary greatly in the number, shape, and spatial arrangement of the bristles and in how they are used (**FIGURE 20.11**). The development of cells into sex combs depends on the male-specific splice form of the transcription factor doublesex (dsx), which is expressed in parts of the developing tarsus due to information provided by many signal pathway genes. Within these regions, *dsx* and a Hox gene called *Sex combs reduced* (*Scr*) are expressed and form a positive feedback loop: *Scr* activates *dsx* expression, and *dsx* affects the expression level of *Scr*. *Scr* is required to specify the exact position of the sex comb and the number of teeth (modified bristles), and *dsx* specifies the male-specific morphology of the bristles. The expression of these genes differs among species, and corresponds to the size and morphology of the sex combs. These key genes regulate many downstream genes that determine exactly where and how the modified bristles are formed. Thus, a new structure is based on a novel genetic network of regulatory interactions among several key genes, their control by upstream genes, and their interactions with many downstream genes.

From Microevolution to Macroevolution

Rates of evolution

The rate of evolutionary change (i.e., change per unit of time) varies greatly among characters, among evolving lineages, and within the same lineage over time. In general, the longer the time interval over which rates of evolution are measured, the lower the rates are, because when the interval is long, many short periods of rapid and reversing evolution are averaged into a much slower net rate of evolution (**FIGURE 20.12A**). Many characteristics have been seen to evolve very rapidly in living populations, in which

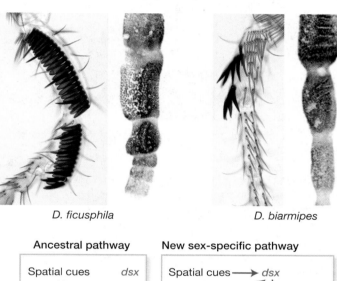

D. ficusphila   *D. biarmipes*

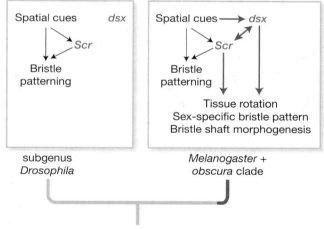

FIGURE 20.11 A model of evolutionary change in gene interactions that underlie the differentiation of specific bristles into sex combs in the *melanogaster + obscura* clade of *Drosophila*. The photos are from two species that are related to *D. melanogaster*. The left-hand image in each pair of photos shows transverse sex combs on the tarsal segments; the right-hand image in each pair, without sex combs, shows developing segments in which the distribution of the Scr protein, detected by antibody staining, is indicated by the dark areas. In the diagram of phylogenetic history, ancestral regulatory interactions are indicated in black, and new interactions in red. Ancestrally, drosophilid flies lack sex combs; *Scr* is expressed in specific parts of the developing tarsus, and organizes the development of the same bristle pattern in both sexes. In the *melanogaster + obscura* clade, interaction of *Scr* with *dsx* results in male-specific differentiation, and both *dsx* and *Scr* regulate downstream genes involved in the processes listed. (After [55]; photos courtesy of Artyom Kopp.)

(A)

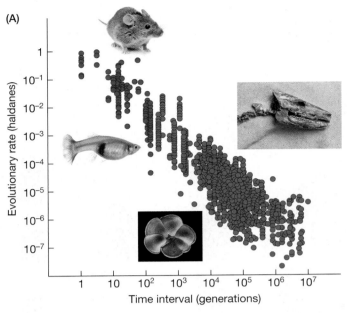

(B)

FIGURE 20.12 Rates of phenotypic evolution vary greatly, especially depending on the time interval over which they are measured. (A) This graph shows two important patterns. The first is that the rates measured on living species (red dots) are hundreds to millions of times faster than those seen in the fossil record (blue dots). Rates are measured in haldanes (on a log scale). A haldane is a rate of evolution equal to the change in the mean of a trait, in units of standard deviations of change per generation. (See the Appendix for a discussion of standard deviations.) The second pattern is that the evolutionary rate is inversely correlated with the time interval over which the rate is measured (shown on the x-axis, again on a log scale). The inset images show some of the species that appear in these data: for living species, house mouse and guppy; for extinct species, *Hyracotherium* (an ancestral horse) and *Globorotalia* (a foraminiferan). (B) A similar analysis of changes in body mass in lineages of terrestrial vertebrates. Each point is the ratio of mean body mass of a later (descendant) to earlier (ancestral) sample. Field studies represent changes within individual populations; fossil time series are ancestor-descendant pairs of fossils; phylogenetic divergence is the difference between two living taxa, divided by time since their common ancestor. Remarkably, changes in body mass accumulate and become steadily greater only over time intervals greater than about 1 million years. Before then, size may fluctuate rapidly, but with little net change. The authors of the study suggest that cumulative change may occur only after a lineage adapts to a substatntially different ecological niche. (A after [31]; B after [108].)

rates are measured by comparing trait means over just a few generations (see Chapter 6). It has been calculated that at these rates, evolution of body mass from a 20-g mouse to a 2-million-g elephant would take fewer than 100,000 generations [22, 31]. But these high rates are not sustained for very long, and the average rates of change calculated for fossil lineages are much lower. These rates are measured by comparing means in the same lineage sampled at two points in geological time that may be separated by many thousands or even millions of years. Extinct populations were likely evolving as fast as living populations, but their rates are averaged over longer periods, giving the appearance of slower evolution. Josef Uyeda and colleagues compiled data on body mass of terrestrial vertebrates, and found that the net change in body size within a lineage is the same for both very short and much longer time intervals, up to 1 million years (**FIGURE 20.12B**) [108].

But even these low rates can produce big changes, because the time spans are so long. The body mass of the largest species in the horse family (Equidae) increased by a factor of ten during the last 25 My (**FIGURE 20.13**). Data from the fossil record show that the body mass of terrestrial mammals has been able to increase 100-fold in about 1.6 million generations, and 5000-fold in 10 million generations [22].

The long-term rates of morphological evolution, as measured either in fossil lineages or among living species, are usually so low that they almost always could be explained simply by mutation and random genetic drift, without a push from natural selection [65]. So the question that requires an answer is not "Can rates of mutation and natural selection explain the rate of long-term evolution?" The problem, rather, is to explain why evolution is often so slow. One major reason (as explained in the caption to Figure 20.12) is that the direction of selection may fluctuate, so that a character mean changes rapidly in the short term but averages little change over a longer time. But other explanations may also be important.

In the extreme, biologists are challenged to understand the existence of "living fossils"—organisms such as the ginkgo (see Figure 17.22), the horseshoe crab (*Limulus*), and the coelacanth that have changed so little over many millions of years that they closely resemble their Mesozoic or even Paleozoic relatives (**FIGURE 20.14**). The synapomorphies (shared derived characters) of large clades also represent conservatism: almost all mammals, no matter how long or short their necks, have seven neck vertebrae (see Figure 3.21), and almost no tetrapods have had more than five digits per limb. (The earliest tetrapods had more, but soon settled on five.) The hypotheses proposed to explain phylogenetic conservatism include stabilizing selection and internal constraints.

One important reason why the optimal condition of a character may remain unchanged is **phylogenetic niche conservatism**: long-continued dependence of related species on much the same resources and environmental conditions [43, 115]. This is manifested by **habitat tracking**: a shift in the geographic distribution of species along with changes in the geographic distribution of habitat to which the species are adapted. For example, diverse molluscs experienced the

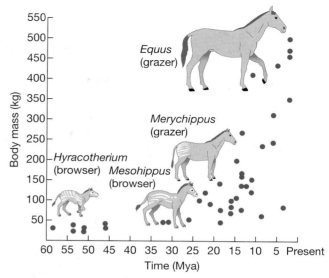

FIGURE 20.13 Estimated body masses of 40 species in the horse family, Equidae, plotted against geological time. Although some lineages decreased in size, more lineages increased, so the average body size in the family increased over time. (After [66].)

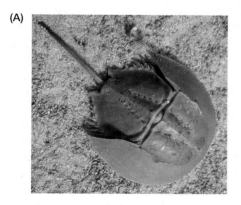

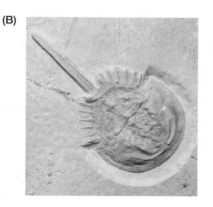

FIGURE 20.14 Two "living fossils" and their extinct relatives. (A) The horseshoe crab *Limulus polyphemus*, found today along the Atlantic coast of North America, closely resembles fossils (B) as far back as the Triassic. (C) Coelacanths are lobe-finned fishes that originated in the Devonian and were thought to have become extinct in the Cretaceous, until the living species (*Latimeria chalumnae*) was discovered in 1938. (D) An extinct coelacanth from the Jurassic period.

same temperature regime throughout the last 3 My, despite the profound climate changes during the Pleistocene [95]. Many groups of herbivorous insects have phylogenetically conservative diets; the larvae of all species of the butterfly tribe Heliconiini feed on plants in the passionflower family (Passifloraceae), and apparently have done so since the tribe originated in the Oligocene. By occupying one niche (e.g., host plant group, climate zone) rather than another, a species subjects itself to some selection pressures and screens off others; it may even be said to "construct" or determine its own niche, and therefore many aspects of its potential evolutionary future [59, 83]. Niches also may remain conservative because other species, often by acting as competitors, may prevent a species from shifting or expanding its niche.

More generally, if there is gene exchange among individuals that inhabit the ancestral niche (e.g., microhabitat) and those that inhabit a novel niche, and if there is a fitness trade-off between character states that improves fitness in the two environments, then selection will generally favor the ancestral character state (i.e., stabilizing selection will prevail) simply because most of the population occupies the ancestral environment (see Chapter 11) [43]. As the degree of adaptation to any one environment increases, adaptation to an alternative environment may become steadily less likely. In some cases, a species may lose the ability to vary in features that would be necessary for a substantial ecological shift, even back to its ancestral niche. For example, in a bird-pollinated clade of morning glories (*Ipomoea*) with red flowers, the genes required to synthesize the blue and purple pigments that typify most other morning glories have become nonfunctional pseudogenes [117]. In general, unused genes acquire disabling mutations and become pseudogenes, as shown by the reduced number of functional olfactory receptors in primates and the degeneration of many genes and phenotypic functions in parasites.

Loss of functional genes is one of several reasons why there may exist genetic constraints on the evolution of some characteristics or on adaptation to novel environments (see Chapters 6, 8, and 15). Some features, such as tolerance to higher temperatures, may have very limited genetic variation [42]. Because of genetic correlations stemming from pleiotropy, genetic variation may exist for certain combinations of characters, but not others, even if each character individually is variable [7, 52]. If genes have consistent patterns of pleiotropy, such character correlations may be very long lasting, and they could be an important determinant, in the long run, of the pattern of genetic variation that is available for evolutionary change [50]. Perhaps for this reason, researchers are increasingly finding that the directions of evolutionary differences among species are correlated with the "genetic lines of least resistance" that have been estimated from genetic or phenotypic correlations in living populations (see Figures 6.21 and 15.20). For example, the evolution of shape characteristics in multiple lineages of the early Paleogene ostracod crustacean *Poseidonamicus* proceeded largely in the direction of those character combinations that showed the greatest variation within living populations (**FIGURE 20.15**) [45]. Similarly, the pattern of divergence in wing shape among various clades of *Drosophila* is broadly similar to the genetic variance-covariance matrix in *D. melanogaster*, suggesting that the pattern of genetic variation has been fairly consistent for more than 50 My [40]. Moreover, laboratory populations selected for a different wing shape rapidly reverted to the normal shape when artificial selection was alleviated, because of deleterious pleiotropic effects [8]. However, experiments have shown that in some cases, pleiotropic correlations can be reduced by selection at other loci that compensate for pleiotropic effects (see Chapter 15) [87]. Just how long genetic constraints may persist is not yet known.

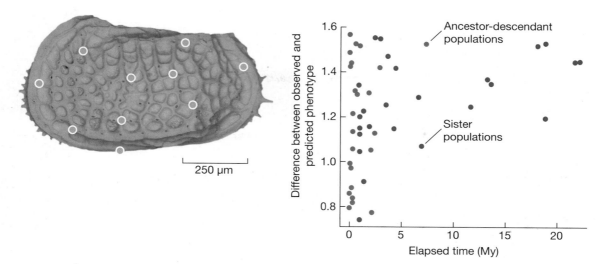

250 μm

FIGURE 20.15 The relationship between evolutionary changes in shape in fossilized lineages of the marine ostracod crustacean *Poseidonamicus* and the elapsed time during which evolution has occurred. The figure shows the ostracod's shell; the colored circles are landmarks used for measurments that were used to characterize the shape. In the graph, blue dots show changes between various ancestral and descendant populations in individual lineages and red dots show differences between sister populations that diverged from a common ancestor. The vertical axis shows the difference between the shape change that actually occurred ("observed" change) and how the shape would have changed if it had evolved strictly along the direction of greatest phenotypic (and presumably genetic) variation within species (i.e., evolution along lines of least resistance, as described in Chapter 6). Especially for pairs of sister populations, changes that deviate more greatly from the presumed line of least genetic resistance occur over longer time spans. (After [45]; photo courtesy of Gene Hunt.)

Gradualism and punctuated equilibria

Paleontologists have held two rather different views of evolutionary rates, which were expressed during a controversy that is not entirely over. Gradual transitions through intermediate states have been described for many morphological transitions, but the fossil record of many other groups is marked by gaps rather than continuous change (**FIGURE 20.16A**). Most paleontologists have followed Darwin in supposing that evolution was actually gradual but that the fossil record is incomplete. In 1972, however, Niles Eldredge and Stephen Jay Gould proposed a more complicated, and much more controversial, explanation, which they called **punctuated equilibria** [20].

Eldredge and Gould said that species in the fossil record often show long periods of little or no detectable phenotypic change, interrupted by rapid shifts from one such "equilibrium" state to

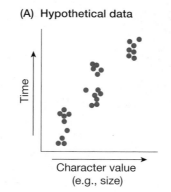

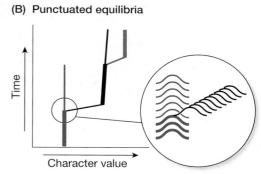

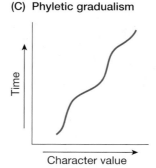

FIGURE 20.16 Two models of evolution, as applied to a hypothetical set of fossils. (A) Hypothetical values for a character in fossils from different time periods. These data might correspond to either of the models shown in panels B and C. (B) The punctuated equilibria model of Eldredge and Gould, in which morphological change occurs in new species. Morphological evolution, although rapid, is still gradual, as shown in the inset by the shift in the frequency distribution of the phenotype. (C) The traditional gradualism model. The character mean changes gradually within a single species, faster at some times than others.

FIGURE 20.17 An example of stasis: specimens of the bivalve *Macrocallista maculata* from a living population and from fossil deposits dated at 1, 2, 4, and 17 Mya. All are from Florida. Scale bars = 1 cm. (Photos courtesy of Steven M. Stanley.)

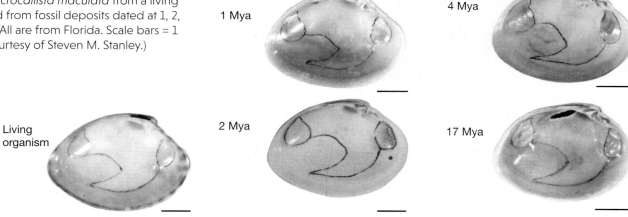

another; that is, **stasis** that is "punctuated" by rapid change (**FIGURE 20.16B**). They contrasted this pattern with what they called **phyletic gradualism**, the traditional notion of slow, incremental change (**FIGURE 20.16C**). Paleontologists agree that stasis is a common pattern (**FIGURE 20.17**). For example, eight living lineages of bivalves with a fossil record all show as much or more variation among geographic populations as they do over the course of 4 My [103]. Most fossil lineages fit a model of either stasis or random fluctuations [44, 47].

Eldredge and Gould proposed that the rapid shifts ("punctuations") represent speciation (especially founder effect speciation), whereby a reproductively isolated new species that originated "offstage" expands its range and replaces the ancestral species. One of the few examples of morphological change associated with speciation is in a Miocene genus of bryozoans, or "moss animals," that persisted with little change for several million years, while new species appeared abruptly, without evident intermediates (**FIGURE 20.18**) [11]. Eldredge and Gould's hypothesis that evolutionary change requires speciation is not widely accepted [46], but it stimulated interest in rates of evolution and posed the question of whether or not speciation might facilitate, or be correlated with, phenotypic evolution.

Speciation and phenotypic evolution

Might speciation enhance evolutionary change? The fossil record can provide evidence only when it is exceptionally complete. Cenozoic planktonic Foraminifera (shelled protists) have an outstanding fossil record, and in these, morphological evolution is almost always accompanied by speciation [106]. Evidence has also been sought in phylogenies of living species, using statistical tests to determine if the amount of morphological or DNA sequence difference among species is attributable mostly to evolution within lineages, or is enhanced by the number of branching (speciation)

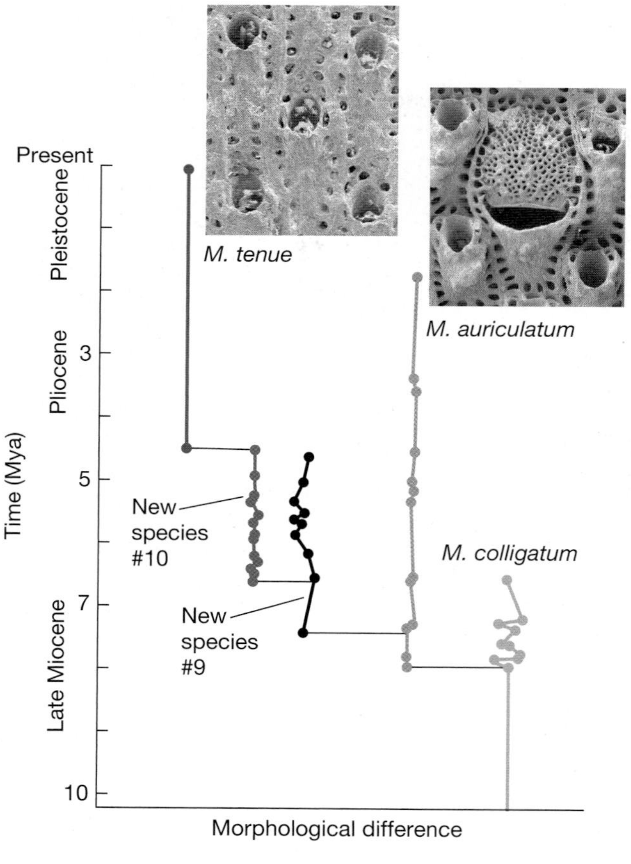

FIGURE 20.18 Punctuated equilibria: the phylogeny and temporal distribution of a lineage of bryozoans (*Metrarabdotos*). The horizontal distance between points represents the amount of morphological difference between samples. The general pattern is one of abrupt shifts to new, rather stable morphologies. Only a part of the full phylogeny of the genus, which has many more species, is shown. (After [11]; photos courtesy of A. Cheetham.)

FIGURE 20.19 Models of phyletic gradualism (A) and punctuated equilibria (B) suggest how phylogenetic data might be used to determine whether speciation is associated with enhanced evolution of molecular or morphological characters. In both models, lineages 1 and 2 differ in the number of speciation events (branch points), but not in the time elapsed since the common ancestor of all the species in the clade. The total number of DNA nucleotide substitutions along all the segments from the common ancestor to a tip (living species) is the path length of the branch lineage. (C) With phyletic gradualism, the path length to any living species is proportional to elapsed time, and is not affected by the number of speciation events. The correlation between the total path length and number of branch points in the phylogeny is expected to be zero, as indicated by the red horizontal line. In the punctuated equilibria model, evolution is accelerated (or occurs only) at speciation, so the path length, shown as the green line, is expected to be correlated with the number of branch points. (C after [86].)

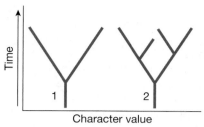

(A) Phyletic gradualism

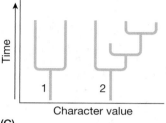

(B) Punctuated equilibria

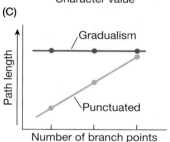

(C)

events [47, 48]. The rate of evolution of a lineage, and therefore the amount of evolutionary change from the root of a phylogenetic tree to any extant species (path length), is expected to increase with the number of speciation events in the punctuated equilibria model, but not in the phyletic gradualism model (**FIGURE 20.19**). Mark Pagel and colleagues found that in many phylogenies of animals, fungi, and plants, the numbers of nucleotide substitutions between taxa were significantly correlated with the number of branch points (speciation events) between them [86], as predicted if speciation accelerates evolution. Molecular evolution in Foraminifera is strongly correlated with the origin of new morphologically recognizable species [23]. Similarly, speciation, rather than gradual evolution within lineages, accounts for more than two-thirds of the variance in body mass among species of mammals [68].

What might cause these patterns? Perhaps the additional evolutionary change, over and above change within lineages, reflects simply the adaptive changes associated with the evolution of reproductive isolation and ecological divergence that generally occurs during speciation. A different possibility is that speciation may enable differences between populations to persist in the long term. Geographic populations of a broadly distributed species commonly inhabit a mosaic of different environments but are connected by gene flow (see Chapter 8). Any one population may adapt to a change in its local environment, but an adaptive change will sweep through the species as a whole only if an environmental change affects all the populations, and this may rarely occur [102, 109]. Moreover, different mutations or genes often provide adaptation in different populations of a species that experience similar selection; thus, the species adapts as a mosaic of different, convergent adaptations (see Chapter 6) [88]. Hence, a species will seldom evolve as a unified whole [61]: change of an entire species may be rather rare.

The interplay between gene flow and spatially variable selection led Futuyma to suggest that speciation may enhance adaptive evolutionary change by stabilizing local adaptations that would otherwise be short-lived [27]. Different populations of a species are adapted to local environments: populations of a plant to wetter versus drier soil, or of an insect to different host plants. But the geographic location of these kinds of environments shifts as the climate changes. Then divergent populations move about (by colonization and extinction), and come into contact sooner or later. Much of the divergence that has occurred between them may then be lost by interbreeding—unless reproductive isolation has evolved (**FIGURE 20.20**). Reproductive isolation captures and stabilizes an adaptive set of genes that can track a geographically moving habitat, or that can disperse from one patch to another of such habitat, without being broken down by interbreeding. A succession of speciation events, each "capturing" further change in a character, may result in a long-term trend. Speciation might act like a piton for a climber who scales an adaptive

FIGURE 20.20 A model of how speciation might facilitate long-term evolutionary change in morphological and other phenotypic characters. (A) An allopatric population diverges in characteristics that adapt it to a different habitat. It does not evolve prezygotic or postzygotic reproductive isolation, so when the barrier breaks down and the populations meet, the divergence may be lost due to interbreeding. (B) If the divergent population evolves reproductive isolation (red bar) while allopatric, it can become sympatric with the other form and retain its distinctive character. (C) The same process might occur again and again, so that each speciation event enables more divergence to evolve and persist indefinitely.

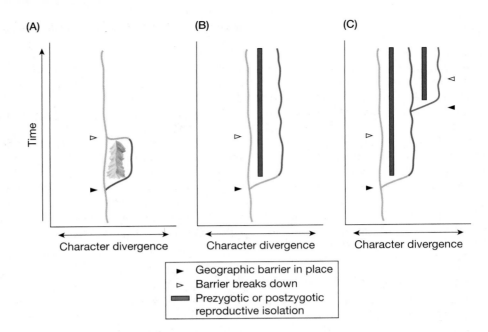

peak, stabilizing new adaptations that otherwise would slip back due to gene flow and have only an ephemeral existence.

Trends, Predictability, and Progress

For decades after the publication of *On the Origin of Species*, many of those who accepted the historical reality of evolution viewed it as a cosmic history of progress. As humanity had been the highest earthly link in the pre-evolutionary Great Chain of Being, just below the angels (see Chapter 1), so humans were seen as the supreme achievement of the evolutionary process (and Western Europeans as the pinnacle of human evolution). Darwin distinguished himself from his contemporaries by denying the necessity of progress or improvement in evolution [24], but almost everyone else viewed progress as an intrinsic, even defining, property of evolution.

In this section, we examine the nature and possible causes of trends in evolution and ask whether the concept of evolutionary progress is meaningful. A trend may be described objectively as a directional shift over time. "Progress" implies improvement or betterment, which requires a criterion for judging improvement, or a value judgment of what "better" might mean.

Trends: Kinds and causes

A **trend** is a persistent, directional change in the average value of a feature, or perhaps its maximal (or minimal) value, in a clade over the course of time. It may describe evolution in a particular clade, in diverse clades, or in all of life. Trends can also be classified as passive or driven [73]. In a **passive trend**, lineages in the clade evolve in both directions with equal probability, but if there is a strong constraint in one direction (e.g., a minimal possible body size), the variation among lineages can expand only in the other direction. Because the variance expands, so do the mean and the maximum. Although the mean increases, some lineages may remain near the ancestral character state (**FIGURE 20.21A**). In a **driven**, or **active**, **trend**, changes in one direction are more likely than changes in the other (i.e., there is a "bias" in direction), so both the maximal and the minimal character values change along with the mean (**FIGURE 20.21B**).

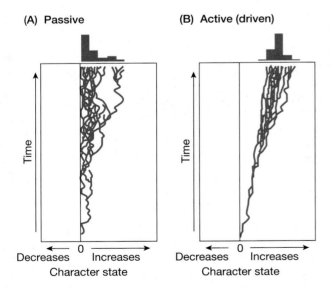

FIGURE 20.21 Computer simulations of the diversification of a clade. The original character state is marked 0. (A) A passive trend. A character shift in either direction is equally likely, but the character state cannot go beyond the boundary at left. The mean increases, but many lineages retain the original character value. (B) A driven trend. The entire distribution of character states is shifted by a bias in the direction of change. (After [73].)

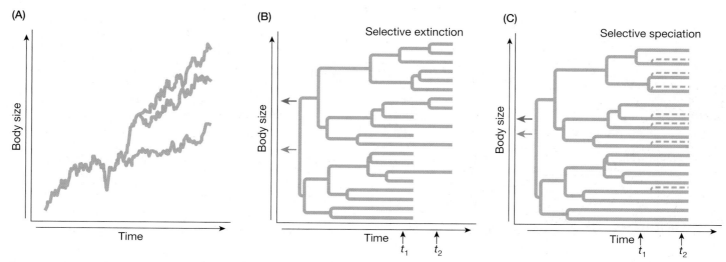

FIGURE 20.22 Three of the possible causes of driven trends. (A) Individual selection within all three species in this clade favors larger size, on average. There are short-term fluctuations in size around the overall trend. (B) Smaller species have a higher extinction rate, so species selection causes a trend toward larger size. (C) Recent speciation events (shown by broken branches) have been more frequent in the upper clades (with larger body size) than in the lower clades, so species selection results in a trend toward larger size. In (B) and (C), blue arrows indicate the average body size before selective extinction or speciation (t_1) and red arrows indicate the average body size afterward (t_2). (After [48].)

Both driven and passive trends could have several causes [48]. Individual selection within lineages could produce both kinds of trends (**FIGURE 20.22A**), or the mean character of species in a clade could change as a result of species selection: a correlation with speciation or extinction rates (**FIGURE 20.22B,C**). A trend could also occur if changes are easier in one direction than the other; for example, genetic or developmental pathways may act as ratchets—mechanisms that make reversal unlikely. Losses of complex features may be irreversible in some cases [9].

Paleontologists noticed long ago that the maximal body size in many animal groups has tended to increase over time, a trend dubbed Cope's rule [41]. The same pattern has been found in phylogenetic analyses of living mammals [2]. An analysis of 1534 diverse species of late Cretaceous and Cenozoic mammals shows a passive trend: mammals were small before the end-Cretaceous mass extinction, and the lower size limit has remained nearly the same ever since (**FIGURE 20.23**) [1]. However, mean and maximal sizes have increased, especially since the end-Cretaceous mass extinction, when the explosive diversification of mammals began. Matched pairs of older and younger fossilized species in the same genera (likely ancestor-descendant pairs) increased in body size more often than they decreased, which suggests that the trend was caused by individual selection within species, rather than by species selection.

Species selection has been identified as the cause of a trend in the mode of larval development in several clades of Cenozoic gastropods (**FIGURE 20.24**). Species that lack a planktotrophic dispersal stage are more susceptible to extinction than are planktotrophic species (species that feed as planktonic larvae). However, the nonplanktotrophic species more than compensate by their higher rate of speciation, probably because their lower rate of dispersal reduces the rate of gene flow among populations [39, 49]. Individual selection, species selection, and irreversibility all affect the proportions of self-compatible (SC) versus self-incompatible (SI) species in the family Solanaceae (tomato, tobacco, nightshade, and others) [33]. Self-fertilization has a long-term disadvantage because inbreeding reduces genetic variation. But individual selection often favors selfing for several reasons; for example, reproduction occurs

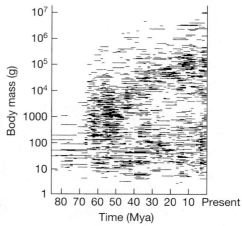

FIGURE 20.23 A passive trend: Cope's rule in late Cretaceous and Cenozoic North American mammals. Each of 1534 species is plotted as a line showing its temporal extension and its body mass (estimated from tooth size). Although small mammals persist throughout the Cenozoic, there is an increasing number of large species over the course of time. Although the trend is called "passive," the evolutionary changes of individual lineages may have been caused by natural selection. (After [1].)

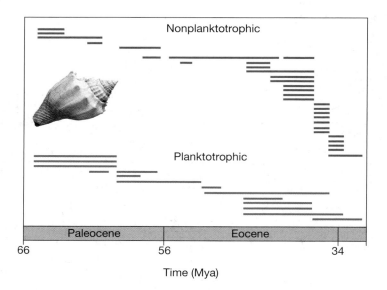

FIGURE 20.24 A trend caused by species selection. Each bar shows the stratigraphic distribution of a fossil species of volutid snail. Although nonplanktotrophic species had shorter durations, they arose by speciation at a higher rate, so the ratio of nonplanktotrophic to planktotrophic species increased over time. (After [39].)

even if pollinators are scarce. The SI system almost never re-evolves in SC lineages. Thus, both a genetic ratchet and individual selection (sometimes) bias evolution toward self-compatibility. However, a phylogenetic analysis showed that the rate of diversification is greater in SI lineages than in SC lineages, apparently because SC lineages have a higher extinction rate (**FIGURE 20.25**). The great diversity of SI species seems to be maintained by species selection.

Are there major trends in the history of life?

Do any trends or directions characterize the entire evolutionary history of life? Although many have been postulated, all have exceptions. Still, one might ask if there is any feature that, on the whole, has evolved with enough consistency of direction that one would be able to tell, from snapshots of life at different times in the past, which was taken earlier and which later [54, 74]. Let's consider two promising possibilities.

EFFICIENCY AND ADAPTEDNESS There are innumerable examples of improvements in the form of features that serve a specific function. The mammal-like reptiles, for example, show trends in feeding and locomotion associated with higher metabolism and activity levels, which culminated in the typical body plan of mammals. There might well be a global trend toward greater efficiency [30]. But efficiency and effectiveness must always be defined relative to the task set by the organism's environment and way of life. We cannot meaningfully compare the level of adaptedness

Solanum melongena (eggplant)

FIGURE 20.25 A phylogeny of 356 species of Solanaceae (tomato family). Blue tips indicate self-incompatible (SI) species, and red tips self-compatible (SC) species that can reproduce by self-fertilization. Primarily blue clades, in which self-incompatibility is ancestral, often give rise to SC species (in *Solanum*, for example), but primarily red SC clades (such as *Nicotiana*) seldom give rise to blue SI species. The predominance of self-incompatibility in this family can be attributed, at least partly, to species selection. (From [33]; phylogeny courtesy of B. Igić.)

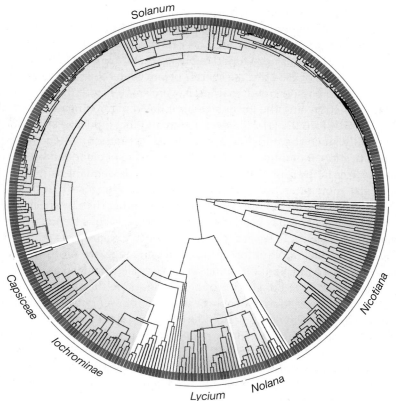

of a shark and a falcon, or even of a bird-hunting falcon and a rodent-hunting hawk, since they are as adapted to different tasks as are flat-head and Phillips-head screwdrivers.

If efficiency of design has increased, does that mean that organisms are more highly adapted than in the past? Darwin imagined that if long-extinct species were revived, they would lose in competition with today's species. If he were right, we might expect the fossil record to document many examples of competitive *displacement* of early by later taxa—but this pattern is less common than *replacement* by later taxa, well after the earlier ones became extinct (see Chapter 19).

We might suppose that species longevity would be a measure of increase in adaptedness, but environments are almost ceaselessly changing, and natural selection does not imbue a species with insurance against future environmental change. We have seen that in many clades, the age of a genus or family does not influence its probability of extinction, implying that a lineage does not become more extinction-resistant over time (see Figure 19.9).

COMPLEXITY   John Maynard Smith and Eörs Szathmáry have proposed a list of major transitions in the history of life, most of them marked by increasing hierarchical organization (see Table 17.2) [69]. That is, entities have emerged that consist of functionally integrated associations of lower-level individuals [75]. The first cells arose from compartments of replicating molecules; the eukaryotic cell evolved from an association of prokaryotic cells; multicellular organisms with different cell types evolved from unicellular ancestors that formed clonal aggregations of undifferentiated cells. The highly integrated colonies of a few kinds of multicellular organisms, among them the social insects, certain social mammals, and humans (**FIGURE 20.26**), have been called the "pinnacles of social evolution" [116]. The difficulty to be overcome in all these transitions was that selection at the level of the component units (e.g., individual cells) could threaten the integrity of the larger unit (e.g., multicellular organism). In general, such conflict has been suppressed by development through a stage (e.g., the unicellular egg) that establishes high relatedness (and thus the power of kin selection) among the component units (e.g., the genetic identity of the cells of a multicellular organism) [69, 76]. Lineages of clonal multicellular organisms (such as animals and brown algae) consistently are more complex—they have more cell types—than nonclonal multicellular lineages (such as cellular slime molds) [25]. Multicellular organisms have many defenses against disruption by rogue cells [37].

The major changes in hierarchical organization represent only a few evolutionary events, in which the great majority of lineages did not participate, so this is not a universal trend. It is difficult to define, measure, or compare complexity among very different organisms. The anatomical complexity of Cambrian animals was arguably as great as that of living forms. Certainly, complexity has increased in some clades; for example, the number of types of appendages has increased in many lineages of crustaceans. However, many characteristics have evolved toward simplification or loss in innumerable clades (**FIGURE 20.27**) [67]. This is true of both morphology and behavior. For example, eusociality has evolved many times in the Hymenoptera but has also been frequently lost [14]. The advantage of eusociality, or probably the advantage of any complex behavior, must depend on the environment, and there is no guarantee that it will always increase.

FIGURE 20.26 Two "pinnacles of social evolution" and their technology. African termites of the genus *Macrotermes* cooperate to build mounds in which a constant temperature is maintained by air conditioning: cool air flows into the base of the mound, passes through vertical tunnels as it warms, and flows out at the top. Author Futuyma, shown for scale, has been transported from New York to Ethiopia by, and in every way depends on, the technological products that social cooperation has made possible. (Photo by D. J. Futuyma's camera.)

(A) *Eusthenopteron* (lobe-finned fish)

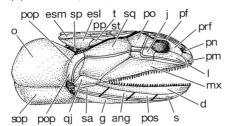

(B) *Milleretta* (early amniote)

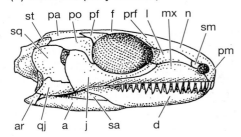

FIGURE 20.27 An example of decreasing complexity during evolution. (A) Skull of an early lobe-finned fish (the Devonian *Eusthenopteron*), the clade from which tetrapods evolved. (B) An early amniote, or reptile (*Milleretta*, from the Permian), similar to the ancestors of mammals. (C) A modern mammal (the domestic dog, *Canis*). The great number of labels indicates that the lobe-finned fish had more skull bones than the early amniotes, which in turn had far more bones than their mammalian descendants. The reduction in the number of bones in the lower jaw (from six labeled in A to one in C) is particularly notable. (After [91].)

(C) *Canis* (modern mammal)

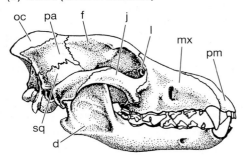

Genome size, meaning the amount of DNA in the cell nucleus, varies 60,000-fold among eukaryotes [21, 36]. Although genome size has decreased in some lineages, such as *Arabidopsis*, increases have been more prevalent because of mechanisms such as polyploidy and the proliferation of transposable elements. However, the number of coding sequences does not appear to be correlated with our traditional (but perhaps erroneous) impressions of phenotypic complexity; we like to think that mammals such as humans, with about 20,000 genes, are more complex than water fleas (*Daphnia*) with 31,000, or rice, with 60,000 (see Figure 14.2). The number of functional genes both increases and decreases in evolution; for example, it is often much lower in parasites and in endosymbiotic bacteria than in free-living relatives [72]. But we do not yet know if there has been a trend in the information content of genomes throughout evolution, because the variety of functions that reside in a genome may be greatly amplified by alternative splicing of genes, multiple binding sites for different transcription factors, and other processes.

The bottom line is that although all organisms taken together show a passive trend toward greater complexity, no characteristic displays a consistent driven trend among all, or even most, branches of the tree of life.

Predictability and contingency in evolution

A course of events (a history) is said to be "predictable" if it proceeds by lawlike principles that determine the sequence of events (e.g., A, B, …, E), each caused by the preceding event. Physics is the epitome of a predictive science. For many people, predictability has the implication that the realized course of events was inevitable from the start. In this view, as the paleontologist Stephen Jay Gould noted [35], the evolutionary history of life was inevitable, including the evolution of humans—or at least a comparably intelligent life form (a "humanoid," perhaps). Among evolutionary biologists, paleontologist Simon Conway Morris [12, 13] is a proponent of this viewpoint (see also [19]).

Historical contingency, in contrast, means that although each event (e.g., E) is caused by a preceding event (D), the outcome of the history would be different (say E', not E) if any of many antecedent events had been different (C' rather than C, or B' rather than B). Perhaps, in principle, we could know the series of events that caused B' and C', rather than B and C, and therefore E' rather than E, to occur, but realistically, we can never know all of the incredibly vast number of possible causal chains. In this view, which Gould championed, the history of life would be very different if it were to start again from any point in the past. The concept of contingency has been familiar in human history at least since the seventeenth-century philosopher Blaise Pascal mused about the effect of Cleopatra's nose: if Mark Antony, one of the triumvirate that ruled Rome from 43 to 33 BC, had not been smitten by Cleopatra's beauty and become her lover, Octavian would not have battled and vanquished him, and there never would have been a Roman Empire. On a slightly less grand scale, most of us can think of "chance" events that changed the course of our own lives.

There is certainly some predictability in evolution, and it is the basis of a great deal of the evolutionary theory presented in this book. Some of the selection equations in population genetics, for example, deterministically predict allele frequency changes in large populations (see Chapter 5). Organisms conform to physical principles, so massive terrestrial vertebrates such as elephants have, predictably, disproportionately thick leg bones. Many features of organisms are more or less successfully predicted by "optimality" theories of life history evolution, behavioral ecology, and functional morphology (see Chapters 10–12). These theories are successful largely because of the high incidence of convergent evolution: the many instances in which similar adaptations to similar environmental selection pressures have evolved independently. Conway Morris depends on convergent evolution to support his argument that humans, or humanoids, were an inevitable outcome of evolution [12, 13]. Along similar lines, many people are convinced that there must exist intelligent humanoids elsewhere in the universe; this conviction is the basis of SETI (Search for Extraterrestrial Intelligence) and similar projects.

Some—probably most—evolutionary biologists reject this position and argue for a strong role of contingency in the history of life. The course of evolution can depend on rare mutations or combinations of interacting mutations [6] and on the sequence of environmental changes. Consequently, convergence between closely related lineages may be close, but convergence between remotely related lineages is usually more superficial [62].

Although examples of convergence abound, so do unique events in the history of life. As far as we know, life originated only once, as did flowers, vertebrates, terrestrial vertebrates, the amnion, the feather, the mammalian diaphragm, and countless other examples. Moreover, extinction, including mass extinction events, has cut short the possible evolutionary future of the vast majority of lineages that have ever lived, and no equivalents of trilobites, ammonoids, dinosaurs, and many other extinct groups have ever replaced them. As Gould emphasized [35], if any species in our long line of ancestors, back to the first vertebrate or even beyond, had become extinct, intelligent hominids would probably never have evolved. Among the billion or more species of organisms in Earth's history, only one evolved human intelligence, and this happened only after at least 3 billion years of cellular life. We have no reason to suppose that any human equivalent would have evolved in our stead. For these reasons, George Gaylord Simpson [101] and Ernst Mayr [71], two of the most influential biologists of the twentieth century, argued that the probability that there exists another intelligent life form in the universe that we have the faintest hope of detecting, much less communicating with, is, for all intents and purposes, zero, and that our own evolutionary history was far from inevitable.

The question of progress

Many people who accept evolution conceive it as a purposive, progressive process, culminating in the emergence of consciousness and intellect. Even some evolutionary biologists have seen in evolution a history of progress toward the emergence of humankind (see [93]).

The word "progress" usually implies movement toward a goal, as well as improvement or betterment. But the processes of evolution, such as mutation and natural selection, cannot imbue evolution with a goal. Moreover, progress in the sense of betterment implies a value judgment, and there is no objective basis for calling human features better than those of other species. A conscious, reflective rattlesnake or knifefish (if such existed) would probably measure evolutionary progress by the elegance of an animal's venom-delivery system or its ability to communicate by electrical signals. But these features are advantageous *in the context* of the environment and lifestyle of these organisms. The great majority of animal lineages—to say nothing of plants and fungi—show no evolutionary trend toward greater "intelligence" (however it might be defined and measured), which must be seen as a special adaptation appropriate to some ways of life, but not others. It is difficult, if not impossible, to specify a universal criterion by which to measure "improvement" that is not laden with our human-centered values.

Many evolutionary biologists have therefore concluded that we cannot objectively find progress in evolutionary history, except in the sense of context-dependent adaptive improvements [93]. The most characteristic feature of the history of evolution, rather, is the unceasing proliferation of new forms of life, of new ways of living, of seemingly boundless, exquisite diversity. The majesty of this history inspired Darwin to end *On the Origin of Species* by reflecting on the "grandeur in this view of life," that "whilst this planet has gone cycling on according to the fixed law of gravity, from so simple a beginning endless forms most beautiful and most wonderful have been, and are being, evolved."

Go to the
——————— **Evolution Companion Website** ———————
EVOLUTION4E.SINAUER.COM
for data analysis and simulation exercises, quizzes, and more.

SUMMARY

- Steps by which some higher taxa have evolved (e.g., birds, mammals) have been well documented in the fossil record. Intermediate forms give evidence of both mosaic evolution of different characters and of changes in the form and function of specific characteristics.

- Major changes in characteristics evolve not by large jumps (saltations), but generally evolve gradually, through intermediate stages. The evolution of some characters does include effects of mutations with moderately large effects. Complex structures such as eyes evolve by rather small, individually advantageous steps. They may acquire functional integration with other features so that they become indispensable.

- Homologous characters may be based on similar or the same networks of regulatory gene interactions. Novel features have arisen, in at least some cases, by the recruitment of integrated genetic and developmental pathways in new contexts or combinations.

- Some fundamental characteristics of developmental processes and organismal integration may enhance evolvability, the capacity of a genome to produce variants that are potentially adaptive.

- The fossil record provides examples of both gradual change and the pattern called punctuated equilibria a rapid shift from one static phenotype to another. The hypothesis that such shifts require speciation is not widely accepted because responses to selection do not depend on speciation.

- The long-term average rate of evolution of most characters is very low because long periods of little change (stasis) are averaged with short periods of rapid evolution, or because the character mean fluctuates without long-term directional change. The highest rates of character evolution in the fossil record are comparable to rates observed in current populations and can readily be explained by known processes such as mutation, genetic drift, and natural selection.

- Stasis and low rates of character evolution can be explained by genetic constraints, stabilizing selection (owing largely to habitat tracking), or gene flow among divergently selected populations that may prevent or reverse the evolution of divergent phenotypes. In some cases, speciation appears to be correlated with higher rates of phenotypic and molecular evolution.

- Long-term trends may result from individual selection, species selection, or constraints that bias the direction of evolution between character states. Driven trends, whereby the entire frequency distribution of a character among species in a clade shifts in a consistent direction over time, are distinguished from passive trends, in which variation among species (and therefore the mean of the clade) expands from an ancestral state that is located near a boundary (such as a minimal body size).

- Probably no feature exhibits a trend common to all clades in the tree of life. Features such as genome size and structural complexity display passive trends, in that the maximum has increased since very early in evolutionary history, but such changes have been inconsistent among lineages. There is no clear evidence of trends in measures of adaptedness, such as the longevity of species or higher taxa, in geological time.

- Certain aspects of evolution are predictable, especially in the short term, and may be manifested by convergent evolution. However, long evolutionary histories are probably contingent: that is, particular evolutionary events would have differed, or would not have occurred, if any of a great many previous events had been different. Unique events such as the emergence of human intelligence may have been highly contingent and improbable.

- If "progress" implies movement toward a goal, then there can be no progress in evolution. If "progress" implies betterment or improvement, improvement can be seen only relative to a species' environment or way of life.

TERMS AND CONCEPTS

biological homology

driven trend (= active trend)

habitat tracking

macroevolution

microevolution

passive trend

phyletic gradualism

phylogenetic niche conservatism

punctuated equilibria

saltation

stasis

SUGGESTIONS FOR FURTHER READING

Tempo and Mode in Evolution, by G. G. Simpson (Columbia University Press, New York, 1944), and *Evolution above the Species Level*, by B. Rensch (Columbia University Press, New York, 1959), are classic works of the evolutionary synthesis, in which the authors reconcile macroevolutionary phenomena with neo-Darwinian theory. *Punctuated Equilibrium*, by Stephen Jay Gould (Harvard University Press, Cambridge, MA, 2007), is the posthumously published central chapter of his magnum opus, *The Structure of Evolutionary Theory* (Harvard University Press, Cambridge, MA, 2002).

The relation of macroevolution to microevolutionary processes is discussed by D. J. Futuyma, "Can modern evolutionary theory explain macroevolution?" (pp. 29–85 in *Macroevolution: Explanation, Interpretation and Evidence*, E. Serrelli and N. Gontier [eds.], Springer International Publishing, Heidelberg, 2015).

The evolution of eyes, and why it is not a mystery, is the subject of a special issue of the journal *Evolution: Education and Outreach* (vol. 1, issue 4, October 2008). Molecular aspects of this topic are summarized by T. H. Oakley and D. I. Speiser, "How complexity originates: The evolution of animal eyes" (*Annu. Rev. Ecol. Evol. System.* 46: 237–260, 2015). N. Shubin recounts the evolutionary history of the human body in *Your Inner Fish: A Journey into the 3.5-Billion-Year History of the Human Body* (Allen Lane/Pantheon, New York, 2008).

Developmental and genetic aspects of the origin of novel characteristics are treated by N. Shubin, C. Tabin, and S. Carroll in "Deep homology and the origins of evolutionary novelty" (*Nature* 457: 818–823, 2009), and by G. P. Wagner in *Homology, Genes, and Evolutionary Innovation* (Princeton University Press, Princeton, NJ, 2014).

Modern paleontological and phylogenetic approaches to some aspects of macroevolution are reviewed by G. Hunt and D. L. Rabosky in "Phenotypic evolution in fossil species: pattern and process" (*Annu. Rev. Earth Planet. Sci.* 42: 421–441, 2014), and by G. Hunt and G. Slater in "Integrating paleontological and phylogenetic approaches to macroevolution" (*Annu. Rev. Ecol. Evol. System.* 47: 189–213, 2016).

PROBLEMS AND DISCUSSION TOPICS

1. Snapdragons (*Antirrhinum*) and their relatives in the traditionally recognized family Scrophulariaceae have bilaterally symmetrical flowers, derived from the radially symmetrical condition of their ancestors. A mutation in the *cycloidea* gene makes snapdragon flowers radially symmetrical. Should we conclude that evolution from radial to bilateral symmetry was caused by change in this one gene? Would that be a saltation? Are there other possible histories of the evolution of bilaterally symmetrical flowers that are not saltational but are compatible with this observation?

2. Suppose you are studying a genus of living organisms that have a fairly short generation time and can be bred in captivity. A certain characteristic does not differ among these species, and looks like a case of evolutionary stasis. How might one test whether the stasis is best explained by genetic constraints, stabilizing selection, or gene flow among divergently selected populations?

3. Would you expect "living fossils," such as horseshoe crabs, to differ from other species in amount

characters, canalization (see Chapter 15), or any other feature that might affect evolvability? Why or why not?

4. Many creationists will allow that microevolution (for example, changing gene frequencies in a population) has occurred, and will even acknowledge that species adapt to different environments. However, they deny that macroevolution (they define it as the evolution of new "kinds") is supported by evidence. Explain how they are mistaken. How might you help a creationist friend better grasp how modern scientists consider micro- versus macroevolution?

5. Rates of evolutionary change measured over short time intervals are often very high. However, rates of evolutionary change measured over long time intervals are generally much lower. For example, see the discussion of body mass in mammals in this chapter. Should we expect this pattern to hold true for most characters? Does this pattern imply that the evolution of major new characters, such as the wings of bats, should occur very rapidly?

UNIT V
Evolution
and
Homo sapiens

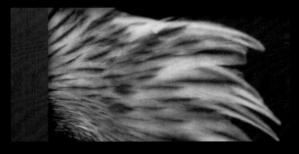

21

The Evolutionary Story of *Homo sapiens*

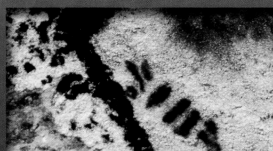

Picture a Neanderthal.

Perhaps you are thinking of a hulking, stooping, club-wielding, hairy cave-man—the defining image of brawn, not brains. That's the cartoon we get from popular culture. But the brain of a real Neanderthal was 11 percent bigger than ours. Neanderthals walked with their heads held high, used sophisticated stone tools, cooked food, used language, and made art. And unless all of your recent ancestors were African, you share some of their genes. Neanderthals and early humans interbred at several different times. Some of those genes helped humans adapt to new diets, diseases, and environmental conditions as they spread throughout the world.

Many of us want to know about our ancestry. We want to know our personal past and that of our group, back to the origin of our species and beyond. Many cultures have origin myths. Darwin began to replace such stories about the origin of humans with testable, scientific hypotheses. In the time since his book *The Descent of Man* was published in 1871 [9], evolutionary studies in paleontology, anatomy, developmental biology, genetics, behavior, and now genomics have revealed more and more about our origins, our relationships to other species, and the evolution of our extraordinary characteristics.

All of evolutionary biology—everything in this book—helps illuminate where we came from and who we are.

After 3,500,000 years of evolution, art first appeared on Earth. These exquisite paintings of animals in a cave in Lascaux, France, are about 17,000 years old. Rock art and carved figures dating back to 38,000 years ago reveal that mental capacities in humans had evolved to a level unprecedented in the history of

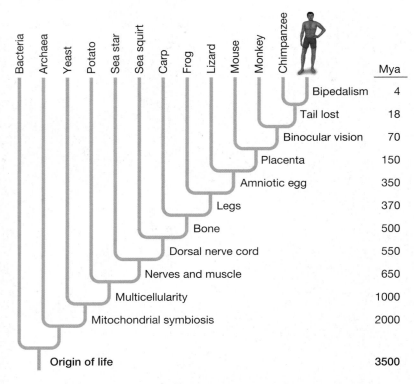

FIGURE 21.1 The path of evolution leading from the origin of life to *Homo sapiens*. Some of the key traits we gained along the way are indicated. Several of the dates for the origins of the traits (shown at right) are very uncertain because fossils showing those transitions are not available.

Trait	Mya
Bipedalism	4
Tail lost	18
Binocular vision	70
Placenta	150
Amniotic egg	350
Legs	370
Bone	500
Dorsal nerve cord	550
Nerves and muscle	650
Multicellularity	1000
Mitochondrial symbiosis	2000
Origin of life	3500

Where Did We Come From?

Humans are descended from the last universal common ancestor (LUCA) of all living organisms on Earth. We share fundamental features with them: inheritance based on nucleic acids (DNA and RNA), the genetic code, proteins composed of L amino acids, and much more. Among the grand events in our earliest ancestry was the origin of the eukaryotes, a symbiosis between an archaean and a bacterium (**FIGURE 21.1**). That momentous event led to the evolution of diverse unicellular forms and several multicellular lineages, which then evolved tissues and organs. One lineage was the progenitor of animals. A descendant then gave rise, some 550 Mya, to the sea stars and other echinoderms on the one hand, and to the chordates on the other hand. Much later, from among diverse vertebrate chordates, arose the ancestor of tetrapods, with legs that evolved from fins. About 150 My after the first land-dwelling tetrapod, some of its descendants stood on the brink of mammalhood. By about 70 Mya, some familiar groups of mammals had appeared, including the first primates.

Our closest living relatives

About 35 Mya, the Old World monkeys and apes (the catarrhine primates) arose. Humans are in a group called the apes, which are most visibly distinguished from other primates by the lack of an external tail (**FIGURE 21.2**). Our closest relatives are the chimpanzee (*Pan troglodytes*) and the bonobo (*Pan paniscus*). The lineage that includes our own species is called

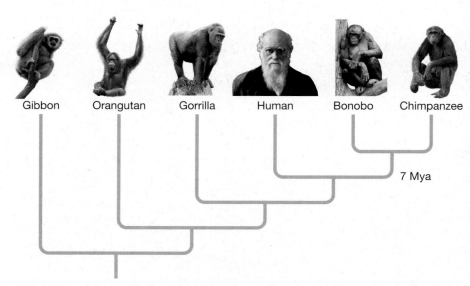

FIGURE 21.2 Phylogeny of some of the living apes, illustrated by the white-handed gibbon (*Hylobates lar*), orangutan (*Pongo pygmaeus*), western gorilla (*Gorilla gorilla*), human (*Homo sapiens*), bonobo (*Pan paniscus*), and chimpanzee (*Pan troglodytes*).

the **hominins**, which diverged from the chimpanzee lineage about 7 Mya[1] [1, 36]. The hominin lineage gave rise to many more species, all of which went extinct save for one: *Homo sapiens*.

We have learned much about the last few million years of human evolution by comparing hominins with our closest living relatives [48]. The other apes have arms that are longer than their legs. All can walk more or less upright for short distances. Their feet are like hands, with opposable first (big) toes, but their thumbs are not as opposable as ours. The African apes are highly social, and males are considerably larger than females, a consequence of sexual selection. Bonobos use sexual interactions, in all possible heterosexual and homosexual combinations, to resolve conflicts and maintain bonds. Chimpanzees have more conflict-ridden societies. They use tools, cracking open nuts on stone anvils and using twigs to fish termites out of their mounds (**FIGURE 21.3**). They hunt cooperatively for monkeys. Some investigators think chimpanzees have cognitive abilities and emotions like those of humans, although less developed.

FIGURE 21.3 Chimpanzees (*Pan troglodytes*) learn how to use tools by imitation. As a female and her infant watch, a male cracks nuts using a rock as a hammer.

They may have a rudimentary "theory of mind"—the ability to infer the intentions and emotions of others [10]. Although chimpanzees have a great variety of vocalizations, they do not use language in nature. Captive apes, however, can learn to use sign language or sets of symbolic objects to express rudimentary language abilities [67], suggesting these abilities were present in our common ancestor.

Humans and chimpanzees differ by less than 2 percent in the DNA sequences of our protein-coding genes[2] [72]. The two species are so closely related that at several loci they share polymorphic alleles that have persisted since our common ancestor because of balancing selection, perhaps related to resistance to pathogens [40]. Both the human and chimpanzee lineages have evolved in many ways since they diverged [14]. Our common ancestor surely shared many features with living chimpanzees, such as an opposable big toe, longer arms than legs, a projecting lower face, large canine teeth, plentiful body hair, and a relatively small brain. Because chimpanzees and bonobos are our closest relatives, they provide critical insights into how humans evolved.

How humans differ from other apes

One of the most conspicuous differences between modern humans and other living apes is that we are fully bipedal. We are well adapted for walking and running [41]. Our pelvis has a different shape, anchoring muscles that stabilize the body. Our legs are relatively longer than in other apes, and are angled inward so that they are directly below our center of gravity. This improves balance and

---

[1] Several factors contribute to uncertainty about the ages of speciation events in apes, including the hominins. The mutation rates used to estimate the dates from molecular data are not known with great accuracy. Hybridization between emerging species may have continued for long periods after the lineages began to diverge. Last, the geological ages of fossils tell us when their lineages were living, but not when they diverged from other lineages. For simplicity, we use 7 Mya as the approximate date for the split between the human and chimpanzee lineages.

[2] This number is based on the genes that humans and chimpanzees share in common. If we take into account the genes that have been deleted or duplicated since the lineages diverged, the difference between the two species increases to about 6 percent [11].

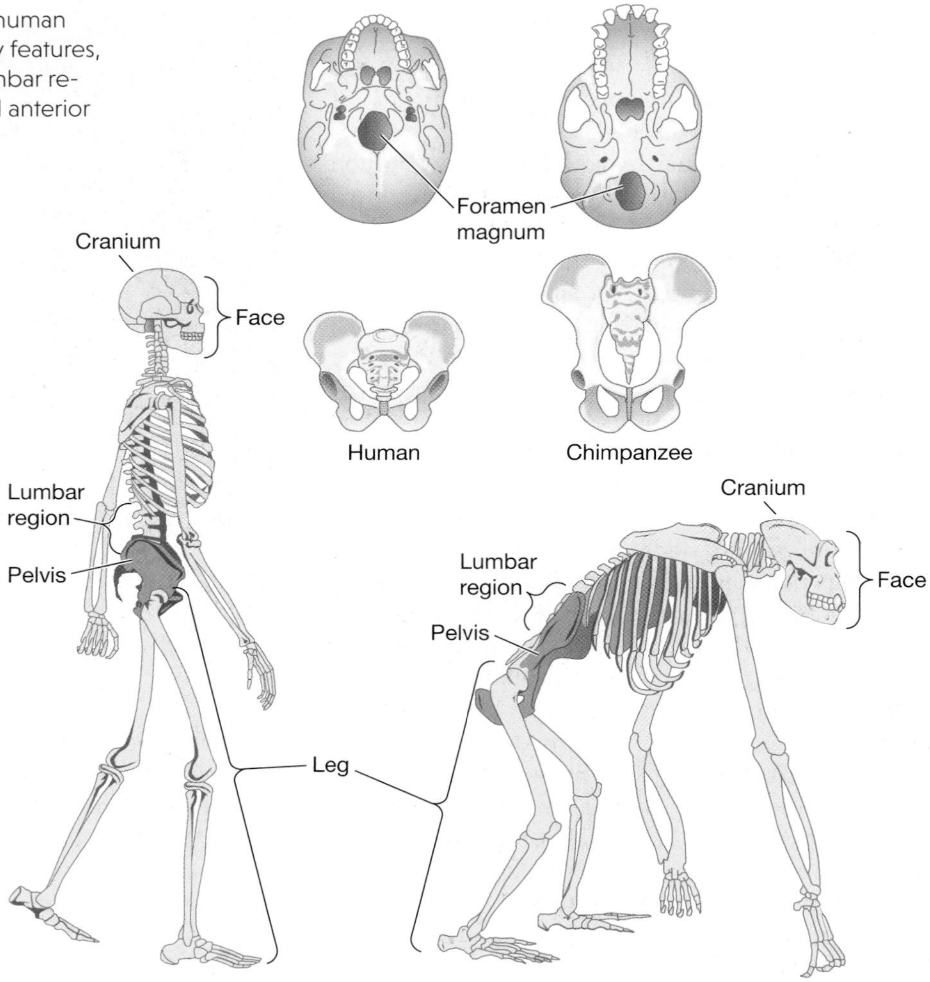

FIGURE 21.4 Comparison of chimpanzee and human skeletons. Human bipedality is reflected in many features, including the very different pelvis, S-shaped lumbar region of the vertebral column, foot structure, and anterior position of the foramen magnum of the skull.

efficiency compared with other apes, which sway from side to side when they occasionally walk. The curvature of our spine, especially in the lumbar region, also improves stability (**FIGURE 21.4**). The foramen magnum, the hole at the base of the skull through which the spinal cord exits from the brain, is shifted forward so that we more easily face straight ahead when standing. Our feet are highly modified for running: the big toe is not opposable, but instead is enlarged and directed forward. Together with the rigid, curved arch of the foot and the toe joints that flex upward, this helps push us forward and upward at the end of each stride.

By the time bipedality evolved, the African climate had become drier, and hominins inhabited open woodland instead of wet forests. Natural selection favored walking rather than climbing in this new environment. An erect posture may also have aided in picking fruit on low trees and running while hunting prey. When hominins became runners, sweating was important for evaporative cooling. This probably selected for reduced body hair in the species that an anthropologist dubbed the "naked ape" [47].

There are many important differences between the hands of humans and other apes. We have shorter fingers with straighter phalanges, and longer, more opposable thumbs. Strong muscles provide our hands with both strength and precision—no other ape has the dexterity to play a guitar. Other distinctive features are our small teeth (especially the canines) and our flat, nonprojecting face.

Most important is our enormous brain. Relative to body mass, the human brain is three times the size of other primate brains, and five times the size of most mammalian brains (**FIGURE 21.5**).

Human babies are larger than those of other primates, but they are unusually helpless. Infancy is followed by a long period of childhood that requires continued parental care. In contrast to humans, female chimpanzees care for their offspring for about 5 years and do not ovulate or have more offspring during that time. But human females can give birth to more children while their older children are still dependent on them. The potential growth rate of human populations is therefore much greater than that of other apes. That may help explain the demise of our competitors, as you will see shortly.

Our ancestry: Hominins through time

At least 7 My of evolution separate the single living species of hominin, *Homo sapiens*, from the other living species of apes. Critical clues to the story of how our physical differences evolved come from the fossil record. There are few fossils of other ape lineages, but fortunately there are many hominin fossils. Some hominin species are known from only a few fragments, such as a jaw. Few are represented by enough specimens to determine whether one or several species were alive at the same time, or what the evolutionary relationships were among fossils from different times. But while there is uncertainty about some of those details, there is broad agreement about the major features of hominin evolution.

In testimony to the predictive power of evolutionary science, Darwin wrote in *The Descent of Man* [9]:

> In each great region of the world the living mammals are closely related to the extinct species of the same region. It is therefore probable that Africa was formerly inhabited by extinct apes closely allied to the gorilla and chimpanzee; and as these two species are now man's nearest allies, it is somewhat more probable that our early progenitors lived on the African continent than elsewhere.

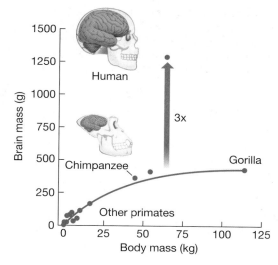

FIGURE 21.5 Plot of average brain mass against average body mass for primates. Relative to body mass, the human brain is three times larger than the average primate brain. (After [41].)

More than 50 years later, the first hominin fossils were found in South Africa. The entire history of paleoanthropology since then has shown that Darwin's prediction was right: the origin and most of the later evolution of hominins, including *Homo sapiens*, played out in Africa.

Fossils show that after diverging from the chimpanzee lineage, hominins proliferated into several species. Most of them were not our direct ancestors, but instead were on closely related lineages that later became extinct. They give important clues to human evolution, however, because those extinct species are more closely related to us than to any living species.

The species of hominins that are generally agreed on are shown in **FIGURE 21.6**. Most anthropologists are confident that *Sahelanthropus*, *Orrorin*, and *Ardipithecus* are members of the hominin clade. If so, the 6- to 7-My-old *Sahelanthropus* marks the minimal age of the split between hominins and the chimpanzee lineage. A key link between hominins and their common ancestor with other apes may be *Ardipithecus ramidus* (**FIGURE 21.7A**), from 4.4-My-old deposits in Ethiopia [80]. It had many apelike features, such as a brain the size of a chimpanzee's and adaptations for climbing such as an opposable big toe. But it also had hominin features, such as small canine teeth (which are enlarged in male apes for fighting) and a pelvis adapted for walking upright.

FIGURE 21.6 The dates and relationships of fossil hominins. Uncertain relations are indicated by dotted lines. Species named in red are discussed in the text. (Courtesy of Ian Tattersall, artwork by Patricia Wynne.)

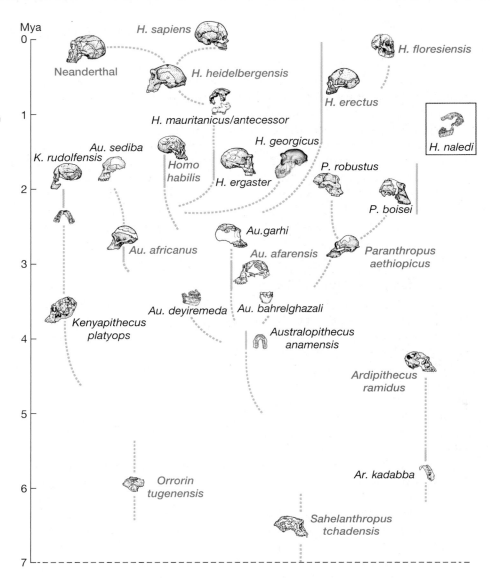

Among the earliest well-fossilized hominins is *Australopithecus afarensis*, dated at about 3.5 Mya (**FIGURE 21.7B**). Its many ancestral features show that it had much in common with the ancestor of humans and chimpanzees, including a lower face that projected far beyond the eyes, large canine teeth, long arms relative to the legs, and a small brain, with a volume of about 400 cc (**FIGURE 21.8**). However, the limb structure shows that *afarensis* not only could climb trees, but also could walk. In fact, fossilized footprints have been found in rock formed from volcanic ash near an *afarensis* site in Tanzania dating to about 3.5 Mya. Bipedalism seems to have been the first distinctively human trait to have evolved.

Following *A. afarensis*, hominin species proliferated, and several coexisted. About 3.3 Mya, one of them mastered the technology of making stone tools that could butcher animals, opening up an important new food source [24]. The tool maker may have been one of three hominin species (the "robust" australopithecines, *Paranthropus*) that became extinct without having contributed to the ancestry of modern humans. A slender species called *Australopithecus africanus*, which is thought to have descended from *A. afarensis*, had a greater cranial capacity (see Figure 21.8).

The earliest fossil from our own genus, *Homo*, dates to about 3 Mya [78]. One early species in the genus was *H. habilis* [82]. It resembled modern humans more than earlier hominins, with a flatter face, shorter tooth row, humanlike hand, and greater cranial capacity. Although its limbs suggest an ability to climb, its legs and feet show that its walk was nearly human. *H. habilis* made stone tools (*habilis* means "handy man"), and animal bones with cut marks have been found with its fossils.

(A)

(B)

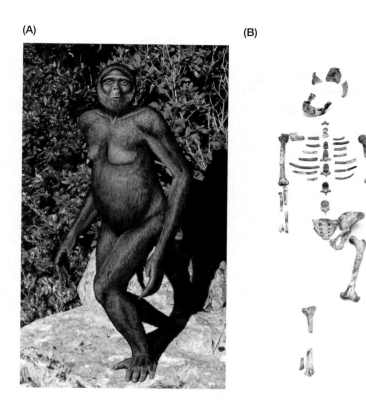

FIGURE 21.7 (A) *Ardipithecus ramidus* as it may have appeared in life. The small braincase, long fingers, and opposable big toe are ancestral features, shared with other African apes, but the bipedal posture is a hominin feature. (B) Skeletal remains of the Pliocene hominin *Australopithecus afarensis*. This famous specimen, nicknamed "Lucy," is unusually complete. This key fossil shows that bipedal locomotion preceded the evolution of a large brain.

Later hominin fossils, from about 1.9 to about 0.2 Mya, are often referred to a species called *Homo erectus*. Most authorities think that *habilis* and then *erectus* were the ancestors of our own species. In many respects, *erectus* had the anatomy and behavior of modern humans. Its skull was rounded, its face projected less than in earlier species, and its teeth were smaller. Importantly, its cranial capacity was larger, about 1000 cc (see Figure 21.8 and 21.11).

Homo erectus made evolutionary history as the first hominin to leave Africa. Almost 2 Mya, it spread into the Middle East. Later, it pushed eastward all the way to China and Java and westward into Europe (**FIGURE 21.9A**). It used stone tools that were more sophisticated than those of *H. habilis*. A million years ago, it made fire in southern Africa, and by 500 Kya fire was widely used across its range. *H. erectus* may have been the ancestor of an extraordinary species called *Homo floresiensis* which lived on a small Indonesian island about 700 Kya. It stood only a meter tall and had a tiny brain, but it used stone tools [19].

Chimpanzee

Australopithecus afarensis

Australopithecus africanus

Homo erectus

FIGURE 21.8 Skulls of a chimpanzee (*Pan troglodytes*) and three hominins. Note the chimpanzee's large canines, low forehead, prominent face, and brow ridge. The skull of *Australopithecus afarensis* shows several similarities with that of the chimpanzee. *A. africanus* had smaller canines and a higher forehead. *Homo erectus* had a more vertical face and rounded forehead. (From [41].)

(A)

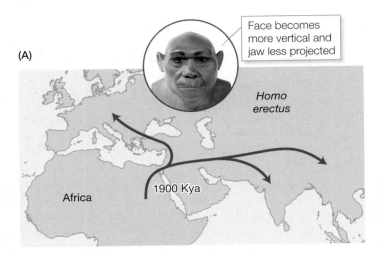

Face becomes more vertical and jaw less projected

Homo erectus

1900 Kya

Africa

(B)

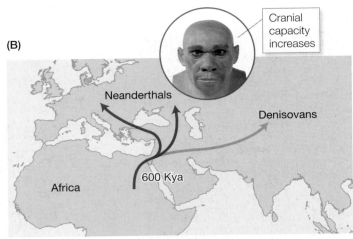

Cranial capacity increases

Neanderthals

Denisovans

600 Kya

Africa

FIGURE 21.9 Before humans did so, other hominins spread out of Africa at least twice. (A) Starting about 1900 Kya, *H. erectus* spread to the Middle East, Europe, and Asia. It became extinct without contributing to human ancestry outside Africa. (B) The ancestor of Neanderthals (*H. heidelbergensis*) left Africa about 600 Kya. It spread into Europe and Asia, where it gave rise to the mysterious Denisovans. Later, both Neanderthals and Denisovans hybridized with humans outside Africa.

Starting about 600 Kya, a second wave of hominins spread out of Africa and across Europe and Asia (**FIGURE 21.9B**). By about 500 Kya, that species (*H. heidelbergensis*) gave rise to the **Neanderthals**, named after the Neander Valley[3] of western Germany, where their fossils were first discovered. Neanderthals had dense bones, a thick skull, and a projecting brow (**FIGURE 21.10**). Their brains were larger than ours (up to 1500 cc), and they had an elaborate culture that included stone tools, art, and burial of the dead [65].

In 2010, a research group led by Svante Pääbo published a remarkable paper. They sequenced a Neanderthal genome using DNA extracted from fossils. They confirmed that humans and Neanderthals are very closely related but genetically distinct [22]. Pääbo's group then sequenced DNA from a 50,000-year-old finger bone found in a cave in Siberia. Astonishingly, its genome is sufficiently distinct that it must have belonged to another group of hominins that diverged from Neanderthals perhaps 400 Ky earlier [64]. Named for the cave where the fossil was found, this group is called Denisovan. The phylogeny of the hominins, on which humans are a leaf, more closely resembles a densely tangled bush than an erect sequoia.

---

[3] The German word *Thal* ("valley") is pronounced "tal." Today it is spelled *Tal,* and the name of the hominin is sometimes spelled "Neandertal" in English.

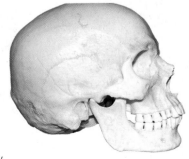

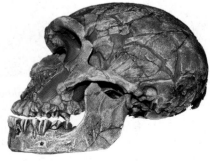

FIGURE 21.10 Skulls of a human from 28 Kya and a Neanderthal from 60 Kya. Neanderthals had an even bigger brain than living humans, but other features of their skull, such as the brow ridge, were more like those of other apes.

Human

Neanderthal

The Arrival of *Homo sapiens*

About 200,000 years ago (200 Kya), the first hominins appeared in Africa that were so similar to us that we recognize them as members of our own species, *Homo sapiens*. The evolutionary path to reach that point was not smooth: different physical features evolved at different rates. Average brain size increased throughout hominin history (**FIGURE 21.11**). Along the lineage leading from *afarensis* to *africanus* to *erectus* and finally to *sapiens*, there were many other changes in the teeth, face, pelvis, hands, and feet. Although some details remain unresolved, one key point has been proven: modern humans evolved from an apelike ancestor.

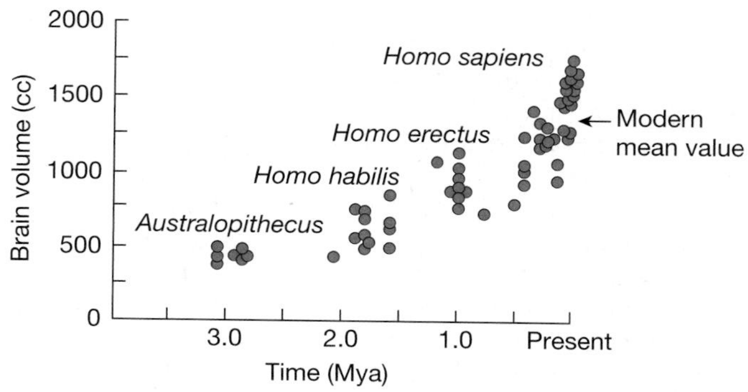

Humans first spread out of Africa roughly 60 Kya (**FIGURE 21.12**) [23]. Remarkably, analyses of DNA from living humans (using methods described in Chapter 7) suggest that only about 2000 individuals dispersed. They were the ancestors of almost all of the 6 billion people now living outside Africa [21]. Humans colonized Europe, East Asia, and Australia between 60 and 40 Kya. They walked from Siberia into Alaska about 20 Kya, when the sea level was low and the Bering Strait was dry. It then took them only about 8 Ky to spread throughout North and South America [63]. In less than 50 Ky, our species colonized the entire planet. Perhaps no other species, except those that travel with humans, has ever gone so far so fast.

The story of these adventures is written in our genes. Gene trees of mtDNA sequences make several key points (**FIGURE 21.13**). The gene trees decisively confirm the fossil evidence that humans are closely related to the extinct Neanderthals. Among living humans, the deepest branches in the tree (representing many nucleotide changes) are from Africans, showing that our species originated there. The data allow us to estimate the age of the most recent common ancestor of mitochondrial DNA in living humans. The woman who carried that mitochondrion is thought to have lived about 125 Kya [5]. (As we discussed in Chapter 7, many other humans were also living then, and they contributed other genes to modern humans.) Figure 21.13 also shows that all mitochondria outside Africa descended from just one branch of the African gene tree, corresponding to the expansion

FIGURE 21.11 The brain volume of hominins has increased through time. This figure assigns specimens to *Homo habilis* and *H. erectus*, but some of those specimens are intermediate and cannot be classified with certainty. (After [34].)

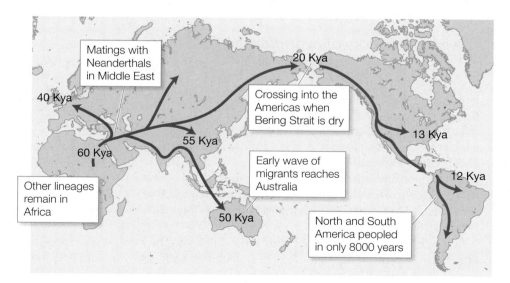

FIGURE 21.12 The colonization of Earth by humans. The arrows show paths of colonization, and the numbers show the times of arrival in years before present. Several of the paths and dates are not known with great certainty.

FIGURE 21.13 Mitochondrial gene trees reveal relations among humans and closely related species. (A) A gene tree based on the mtDNA from living humans, Neanderthal fossils, chimpanzees, and bonobos. The similarity of sequences from Neanderthals and living humans is a dramatic confirmation of the conclusion made earlier from fossils that the two lineages are very closely related. In this tree, the lengths of branches are proportional to the number of changes in the DNA sequences. (B) A gene tree of mtDNA from living humans shows that its deepest branches are found in Africa, corresponding to where our species originated. All non-African lineages are descendants of a single ancestral mtDNA that left Africa about 60 Kya. The shallower branches connecting all non-African lineages indicate that they are more recently diverged than lineages in Africa. (A after [5]; B after [33].)

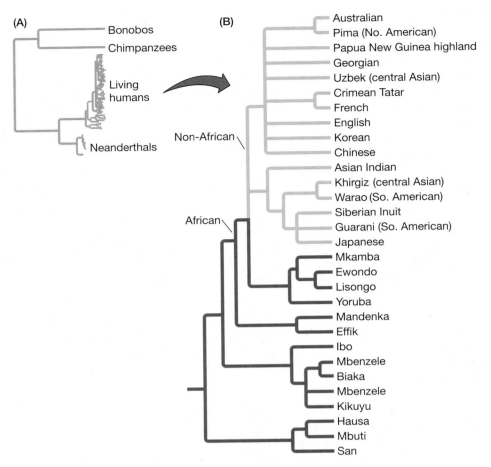

of humans out of Africa. Most of the branches connecting non-Africans are relatively short, consistent with humans' rapid and relatively recent colonization of the planet. Further genetic signatures of the expansion of humans across Earth are seen in nuclear genes: the highest nucleotide diversity and the lowest linkage disequilibrium in humans are found in populations in southern Africa [26], and heterozygosity declines the farther a population is from Africa (see Figure 7.7). That is just the pattern we expect, since genetic bottlenecks resulted as small groups of intrepid ancient explorers colonized new regions.

The human history of hybridization

As humans spread out across the Middle East, Europe, and Asia, they encountered the Neanderthals and Denisovans, whose ancestors had left Africa more than 500 Ky earlier. Although we don't know much about how humans interacted with those other two groups, one thing is clear: they hybridized, and did so more than once [50, 52, 77]. Pulses of hybridization happened at different times and in different places. These liaisons left modern humans a checkered genetic legacy (**FIGURE 21.14**). Since the matings occurred outside Africa, living Africans have little or no DNA from Neanderthals or Denisovans. In contrast, about 2 percent of the DNA in modern Europeans and Asians comes from Neanderthals, as a result of at least two bouts of hybridization. Living Melanesians have DNA inherited from both Neanderthals and Denisovans.

Neanderthals contributed advantageous alleles to the human gene pool that affect skin and immune traits [76, 77]. In Chapter 7 you saw that an allele in the gene *EPAS1* helped Tibetans adapt to life at high elevations (see Figure 7.23). That

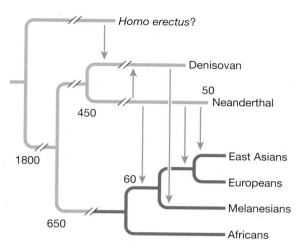

FIGURE 21.14 Comparison of DNA of living humans from different populations with DNA from Neanderthal and Denisovan fossils reveals there was hybridization among these lineages at several times and in several places. Lineages of modern humans are shown in blue. Africans have no DNA from Neanderthals or Denisovans, indicating that the matings with Neanderthals happened outside Africa. East Asians, Europeans, and Melanesians have DNA sequences from Neanderthals that were inherited from a pulse of hybridization in their common ancestor. Melanesians have DNA inherited from Denisovans that is not found in other modern human populations. East Asians and Europeans share DNA from a second pulse of Neanderthal hybridization that occurred in their common ancestor. East Asians also have Neanderthal DNA not found in other populations that is inherited from a third period of hybridization that happened after their ancestors diverged from European populations. Denisovan DNA shows evidence of much older hybridization with another species of hominin, possibly *Homo erectus*. The numbers show approximate dates of branch points in thousands of years ago. The breaks in the phylogeny indicate that the deeper branches are not shown to scale. The phylogeny is simplified, and the timing and number of hybridization events are not yet certain. (After [50, 61, 77].)

mutation originated in the Denisovans, and introgressed into humans by hybridization with them [31]. Intriguingly, some parts of the human genome are significantly more free of Neanderthal DNA than average. One is around the *FOXP2* gene, which is associated with speech and language. These regions of the genome may have been important in adapting to the new human lifestyle, and so resisted introgression of Neanderthal genes.

After coexisting with humans for tens of thousands of years, Neanderthals became extinct about 40 Kya [28]. While the reasons are not known, it seems very likely that growing competition with our ancestors was a major cause of their demise.

The diversity of human populations

As our ancestors colonized the planet, populations began to diverge genetically. The genetic differences persist today despite the greatly increased mobility of people during the last 100 years. Differences in skin color, height, facial bones, body fat, and innumerable other traits evolved. Those traits, which are so obvious and distinctive to us, in fact give a misleading impression of genetic differences among modern human populations. Comparing the genomes of East Asians, the Yoruba of Nigeria, and Europeans shows they are genetically very similar. Recall from Chapter 8 that F_{ST} measures the fraction of genetic variation that results from differences between populations. F_{ST} among these three populations is 0.12 [73]. That is, the genetic differences among them account for only 12 percent of all the genetic variation found in those populations combined. The vast bulk of genetic variation, a full 88 percent, is found within each of those populations. The differences between human populations that are so striking to us are not representative of our genomes as a whole. The genetic similarity among populations today is the result of the small number of generations since our species spread across the planet. Many other species are much more genetically fragmented, even over much smaller geographic ranges (see Chapter 8).

Many of the striking phenotypic differences among human populations are adaptations to the different environments in which we live. Convergent evolution of similar phenotypes in similar environments strengthens the case for adaptation (see Chapter 16). For example, the light skin color of Europeans and East Asians evolved by mutations at different loci as these populations adapted to the limited sunlight at northern latitudes.

The terms "race" and "ethnic group" are often used in common speech. From a biological perspective, however, they are not useful. Those terms suggest that phenotypic variation falls into discrete categories. In most cases, however, there is a continuous range of genetic and phenotypic variation linking different populations. Although Dutch people tend to be taller and lighter skinned than Spaniards, there are smooth clines of body height and skin color that connect those extremes. Populations that are truly distinctive, such as the very small Biaka hunter-gatherers of central Africa, are uncommon. For that reason, in this book we refer to the people living in a given region as a "population." Using that word connects our discussions of humans with those of other species, and it avoids the emotional and political baggage that comes with the terms "race" and "ethnic group."

Brain and Language

The most extraordinary physical characteristic of modern humans is our enormous brain. Relative to our body mass, our brain is three times larger than that of other primates (see Figure 21.5). Our brain gives us cognitive abilities ("intelligence") that far surpass those of any other species. These cognitive abilities enable humans to make and understand language, which in humans is vastly more complex than in any other species.

Two hypotheses have been proposed for the evolution of our unique brain. The first is ecological. This hypothesis suggests that selection favored learning how to function in complex environments, for example while hunting. A second, proposed by anthropologist Robin Dunbar, is the **social brain hypothesis** [15]. Dunbar reasoned that living in complex social groups selected for large brains, particularly enlargement of a region called the neocortex, which is responsible for learning, memory, and cognition. In early human societies, individuals formed alliances with key social partners for help in hunting and in resolving conflicts. In support of this idea, Dunbar showed that primate species with a larger neocortex live in larger social groups. The size of our brain corresponds to that of a species that lived in groups with 100–200 members (**FIGURE 21.15**), a plausible group size for humans early in our evolutionary history.

Homo sapiens has a broader geographic distribution, inhabits a greater variety of environments, and consumes a greater variety of foods than any other species on Earth. This ecological success depends on elaborate social cooperation, on causal reasoning, and on accumulated knowledge. In turn, all of those abilities depend on language. Language originally may have been advantageous because it mediated social interactions, but it also enabled humans to transmit and receive information that was important in many other contexts.

The key elements of language, according to the evolutionary psychologist and linguist Steven Pinker, are metaphorical abstraction and combinatorial structuring [55, 56]. Words are abstract symbols associated with objects, actions, and concepts. Meaning is conveyed by words and how they are arranged. Language transmits information efficiently and at low cost to the transmitter, and it allows information to be pooled among individuals and accumulated across generations. An individual's ability to use language can increase fitness, and this was very likely important in the evolution of our large brains.

Some of the great complexity of human language is made possible by the great variety of sounds made in human speech. Our vocal versatility is possible only because of changes in our vocal tract that evolved after humans diverged from other apes. Our larynx is deeper in the throat, and the tongue curves down into the throat. This produces an L-shaped vocal tract that enables us to produce a remarkable diversity of sounds. An unfortunate side effect of this arrangement is that we

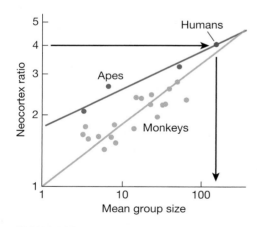

FIGURE 21.15 Living in large social groups may have caused the natural selection that favored the evolution of the very large human brain. Across species of monkeys and apes, there is a strong correlation between the mean size of social groups and the relative size of the neocortex, a part of the brain important to learning, memory, and cognition. The relative size of the human neocortex suggests that our species may have evolved in groups with 100–200 individuals. (After [15].)

can choke to death on food—as you may know if you have ever used the Heimlich maneuver. This fitness cost of our reshaped vocal tract testifies to a strong fitness advantage of spoken language [42].

There is no clear picture of when and how human language evolved. No other primate uses verbal language in the wild, but other species have been taught to use sign language and symbols. The most extraordinary case is a bonobo named Kanzi, studied by Sue Savage-Rumbaugh and colleagues [67]. The researchers did not try to teach Kanzi to associate symbols with objects when he was being reared by his mother, but they did try to teach his mother. Kanzi spontaneously learned the symbols' meaning by observation. He later learned about 200 other symbols, how to associate English words with the symbols, and most important, how to create meaningful combinations of symbols (**FIGURE 21.16**). Chimpanzees can learn simple requests, such as "Give banana," in which the person addressed is the giver and the chimpanzee is the recipient. But Kanzi formed more complex requests in which he was neither the giver nor recipient, but referred instead to

FIGURE 21.16 The bonobo Kanzi and researcher Sue Savage-Rumbaugh having a conversation using a set of plastic symbols. Kanzi can form simple requests that refer to other individuals and can make statements about what he will then do.

other individuals. The exact criteria for what constitutes language are imprecise, so there is little point in debating whether or not bonobos and chimpanzees are capable of true language. But at least rudimentary cognitive abilities for language—though not the physical apparatus for speech—must date back to our common ancestor that lived some 7 Mya.

Tool using and tool making were once claimed to be unique to humans, but they no longer are. Several species of birds, including one of the Galápagos finches, use twigs and spines to extricate insects from crevices. The New Caledonian crow (*Corvus moneduloides*) fashions hooked and barbed tools from twigs and leaves [32], and in experiments it can use three tools in the required sequence to get a reward [71]. Orangutans, chimpanzees, and bonobos can accomplish similar tasks [45]. All the great apes use tools, both in the wild and in captivity (see Figure 21.3). Kanzi and other bonobos learned how to strike stones together in order to make sharp flakes that they used to cut strings in order to obtain food [75, 81]. This is just what our ancestors were doing 3.3 Mya. As hominin brains evolved, technology and culture generally grew in complexity.

Diet and Agriculture: A Revolution in Our World

With our large brain come large costs. Our big head makes childbirth difficult and dangerous—humans are the only species in which helpers assist with birth. Even more important is the vast quantity of energy that our brain uses [41]. It consumes about 20 percent of an adult's basal metabolism, and up to 60 percent of an infant's. Growing the large adult brain requires about 18 years, longer than in any other primate. During that time, neural connections are formed, cognitive abilities increase, and social skills are shaped.

Humans are paradoxical primates. At first look, our life histories do not follow the trade-offs one might expect (see Chapter 11). We reproduce more often and have offspring that are larger at birth than other primates. Yet despite those higher reproductive costs, we have a longer life span. To pay the energetic price of a big

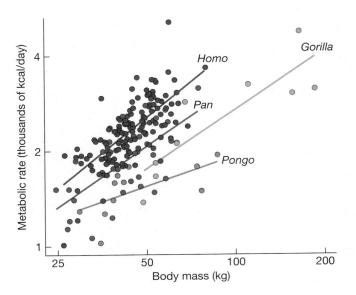

FIGURE 21.17 Humans have a higher metabolic rate relative to fat-free body mass than do chimpanzees (*Pan*), gorillas (*Gorilla*), and orangutans (*Pongo*). (After [58].)

brain and a high reproductive rate, we evolved a higher metabolic rate and larger energy budget than any other primate (**FIGURE 21.17**) [58]. Even the leanest humans have much more body fat than other apes. Humans evolved special cells to store fat not found in any other primate. These were critical to maintain our brain and high reproductive rate when food was scarce [41].

The high metabolic rate of humans was made possible by an evolutionary shift in diet from fruits to meat and tubers. Humans first obtained meat by scavenging carcasses left by lions and other predators [57], and later by active hunting. In turn, hunting selected for the endurance to run long distances, the ability to throw spears, and many other characters [41]. Judging by living hunter-gatherers, such as the Hadza of Tanzania, a hunter with this lifestyle walks more than 15 km per day. He expends almost twice as much energy above the basal metabolic rate as the average American or European today. Humans learned how to extract even more energy from the meat they hunted by slicing, pounding, and cooking it [83, 84].

Humans became such proficient hunters that each time they arrived on a new continent, they extinguished many of the large mammals and birds (the "megafauna"). Mammoths, woolly rhinoceros, giant bison, giant beaver, diverse South American ungulates, and the giant ostrich-like moas of New Zealand were among the many victims [44, 59].

Perhaps the most profound change in human history was the invention of *agriculture*, which first appeared in the Middle East at least 11 Kya. Over the next 5000 years, agriculture was independently invented in China, Mexico, New Guinea, the central Andes, northern Africa, and the Mississippi Valley of North America (**FIGURE 21.18**) [3, 13]. During that time, humans domesticated animals as livestock and plants as crops (**BOX 21A**). Some species, such as rice, were domesticated independently more than once. Agriculture became widespread across Earth only about 300 generations ago. In evolutionary time, that is the blink of an eye, and

FIGURE 21.18 Agriculture was invented independently at several places around the world. Some of the crops that were domesticated are listed. (After [62].)

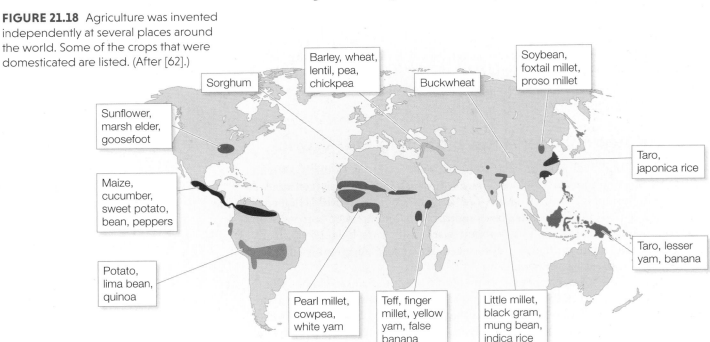

BOX 21A

Domesticated Plants and Animals

Domestication is an evolutionary process, by which plant and animal species used by humans become different from their wild ancestors. The earliest domesticated species is the dog, which had evolved from wolves by 16 Kya in Eurasia—although where and how is controversial [39, 54]. After that, the earliest evidence of domesticated plants and animals dates from about 11 Kya (possibly 13 Kya), in the Middle East, one of several areas in which plant cultivation arose (**TABLE A1**) [62].

Archaeological and genetic approaches provide evidence on the time and place of origin and subsequent spread of domestic forms, on gene exchange with wild relatives, and on the evolution of distinctive traits [17]. It is thought that in most cases the earliest stages of domestication were more accidental than deliberately planned. Wolves may in effect have domesticated themselves by natural selection of less fearful individuals that lurked near humans to eat food scraps. Their descendants, dogs, may represent the commensal pathway to domestication [38]. Most domesticated ungulates, such as goat, sheep, and cattle, were initially hunted as prey, and may have been held captive and bred in order to supplement overhunted natural prey. Camels, horses, and asses (donkeys), which were domesticated later, may have been deliberately bred as beasts of burden. (Honeybees

and silkworm moths probably were also domesticated deliberately.) In most domesticated animals, a key selected trait was behavior: domesticated forms are tame, not fearful.

In most species, the various familiar breeds were developed, largely by deliberate selection, only in the last 300 years, long after the original domestication process. Genomic studies show that some genes were positively selected (e.g., color in pigs). Domestication of many species involved a reduction in effective population size, allowing many slightly deleterious alleles to drift to high frequency. This has resulted in reduced reproductive fitness and increased susceptibility to diseases that are features of most domesticated animals and plants [79].

Many crop plants display similar, convergently evolved "domestication traits" [62]. These include large seeds (which can grow when deeply buried by plowing) and ready germination (in contrast to the obligate seed dormancy in most wild ancestors). Harvesting and replanting cereals, such as wheat and rice, automatically selects for seeds that remain attached to the plant rather than dropping off before harvest. This makes the plant dependent on humans for seed dispersal and germination.

TABLE A1 Where and when some species were domesticated

REGION	DATE (thousands of years ago)	PLANTS	ANIMALS
Eurasia	16		Dog
Western Asia (Middle East)	11–10	Wheat, barley, lentil, chickpea	Goat, sheep, cattle, pig, cat
China	10.5	Rice, millets	
Mexico	10	Corn (maize), squash, peppers	
New Guinea	10	Taro, yam, banana	
Central Andes	10	Squash, potato, quinoa	
Amazonian South America	8	Manioc, peanut	
Sahel (Africa)	7	Sorghum	
Eurasian steppe	6		Horse
Southern Asia	6		Water buffalo, cattle
Andes	6		Llama
North America	5	Squash, sunflower	
Northeast Africa	5		Ass (donkey)
Asia	5		Camels (2 species)
Andes	5		Guinea pig
Southeast Asia	5.5		Chicken
Tibet	4		Yak
Northern Eurasia	2		Reindeer

Sources: [38, 62, 69, 79].

major genetic change can occur in so short a time only if selection is extremely strong. But as you will soon see, the change to agricultural societies caused such a radical transformation of our environment and lifestyle that it left evolutionary skid marks across our genome.

Agriculture has great benefits: it is the foundation for human civilization. But agriculture also has great costs, to both the health of humans and the planet on which we live. The biologist and anthropologist Jared Diamond suggested that agriculture is "the worst mistake in the history of the human race" [12].

Agriculture required changing from a nomadic life to a sedentary one. The cultural and ecological consequences were enormous. Humans established permanent settlements that increased in size and political complexity. Reproductive rates skyrocketed, and the human population has been growing exponentially ever since. Agriculture began the devastation of habitats that today threatens countless species. It introduced many plants and insects to new regions where they have become invasive enemies of native species. The dense and sedentary populations in villages, which grew into towns and then cities, triggered outbreaks of infectious diseases. The most devastating was malaria, which became widespread about 15 Kya and still kills more people than any other disease. More than 50 other diseases, including influenza, tuberculosis, and diphtheria, were acquired from domesticated animals.

Agriculture radically changed people's diet, and not entirely for the better [37]. Populations became dependent on a few foods (such as potatoes, rice, and corn), and they suffered famine when crops failed. Most of the food was, and still is, low in vitamins and important nutrients but high in carbohydrates. Access to almost limitless calories—at least when harvests were good—created conditions for obesity, diabetes, and other diet-related diseases to develop.

Natural Selection, Past and Present

Agriculture profoundly changed the environment in which humans were living. It is no surprise that it caused new types of selection on our ancestors. What is a surprise, however, is how strong and widespread that selection was across the genome.

Very recently, human geneticists have developed an evolutionary time machine. It is now possible to sequence the ancient DNA from skeletons unearthed by archeologists. By comparing them with sequences from people still living in the same place, we can see for the first time directly how gene frequencies changed in time. Mathieson and colleagues compared DNA from 230 individuals that lived between 8500 and 2300 years ago in Europe and western Asia with samples from over two thousand living humans [46]. Twelve genes show large swings in allele frequencies between the past and present, the "smoking gun" of adaptive evolution (**FIGURE 21.19**). Four of the 12 genes are involved in adapting to the new diet that came with an agricultural lifestyle, which arrived in Europe less than 10 Kya.

The strongest signal of selection is on a mutation responsible for lactase persistence. Lactase is the enzyme that digests lactose in mother's milk. In our ancestors and in many populations today, the lactase gene turns off after weaning. But in populations that domesticated livestock and consumed dairy products, there was a strong fitness advantage to keeping that gene turned on so that adults could digest lactose. Mutations with that effect became established independently several times in Europe, Africa, and the Middle East [74]. Analysis of the ancient DNA from Europe showed that the mutation appeared there just 4500 years ago [46]. It rapidly swept nearly to fixation in northern Europe, where today about 98 percent

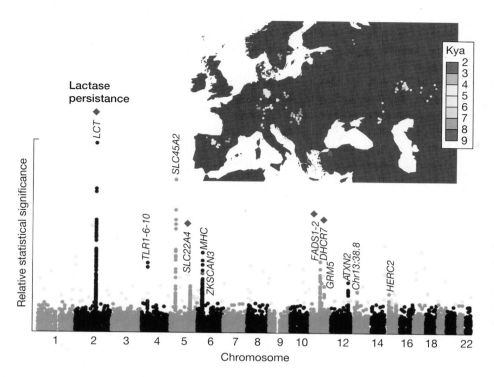

FIGURE 21.19 Twelve genes show strong signals of adaptive evolution in human populations in Europe and western Asia over the last 8500 years. Ancient DNA was collected at points shown on the map (inset). Comparison of single nucleotide polymorphisms in those samples with ones from modern humans living in the same locations pinpointed loci that evolved under positive selection. In the main panel, the genome is displayed from chromosome 1 to 22. The *y*-axis shows the strength of evidence for positive selection among more than 1 million single nucleotide polymorphisms. Four genes involved in adaptation to an agricultural diet are indicated by red diamonds. The strongest signal is for the mutation that causes lactase persistence, which allows adults to digest milk and other dairy products. It appeared in Europe just 4500 years ago. (After [46].)

of people carry the mutation (see Figure 5.11). It must have conferred a very large fitness advantage indeed.

Three of the other genes that show strong signs of selection are also involved in adaptation to the new agricultural diet. One, called *SLC22A4*, carries a mutation that increases absorption of the amino acid ergothioneine. That amino acid was abundant in the European diet before agriculture, but it is present only at low levels in wheat. Ancient farmers with the mutation gained a large fitness advantage. The remaining eight genes that show rapid recent adaptation affect skin and eye color, the immune system, and tooth morphology. Other genes that affect height also show signatures of selection. Body height was selected to increase in some times and some places, and to decrease in others. The reasons for those selective pressures are not yet known.

The study by Mathieson and colleagues adds to the dozens of examples of recent adaptation in humans that have been revealed by the analysis of DNA over the last decade. Other cases discussed earlier in this book include skin color (see Figure 5.12), malaria resistance (see Figure 5.18), *EPAS1* (see Figure 7.23), *BRCA1* (see Figure 7.20), and amylase (see Figure 14.4). Still more are shown in Figure 6.28. We have the fantastic fortune to live in a great age of genetic discovery whose revelations are opening whole new vistas on human evolution.

Our genetic loads

Not all the genetic changes in our recent past have been for the best. Some of the beneficial mutations that spread by positive selection dragged along with them deleterious mutations by genetic hitchhiking (see Chapter 5). Genetic variants tightly linked to the beneficial *SLC22A4* mutation that we discussed in the last section are associated with two digestive disorders, celiac disease and irritable bowel syndrome. Northern Europeans who benefit from more efficient absorbtion of ergothioneine are also at greater risk from those diseases.

FIGURE 21.20 The number of deleterious mutations in the human genome increases with the distance of a population from southern Africa. (A) Samples from seven populations were analyzed from the points shown on the map. Pathways by which ancient humans may have colonized Earth are shown by the arrows. (B) The *y*-axis shows the number of mutations in samples of people from each population that alter protein structure and are estimated to have negative fitness effects. The colors for each population correspond to the points on the map. The populations are arranged from left to right with increasing distance from Africa along the hypothesized migration routes. For each population, the horizontal bar shows the median, the box shows the middle 50 percent of the distribution, and the whiskers show the full range of values. (After [27].)

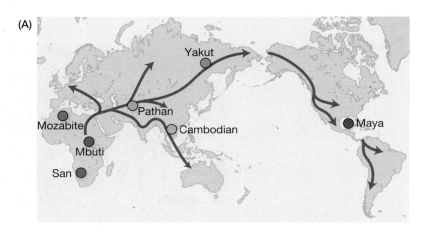

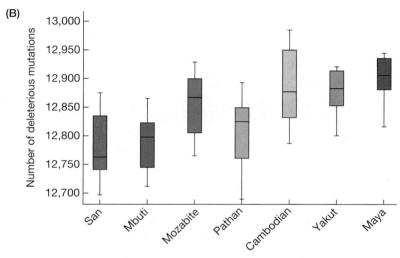

There are now more than 7 billion humans on Earth, but our numbers were very much smaller just a few dozen generations ago. The human population living in Africa 100 Kya had an effective population size of only about 10,000 individuals, so few that we would now consider them an endangered species. Genetic drift was intense. As a result, the large majority of amino acid substitutions in the proteins of the human lineage were fixed by drift, not by positive selection [35]. And many of them were deleterious.

The effects of drift were even more severe for the humans who left Africa. Recall from Chapter 7 that the effective size of a population can be estimated from the level of heterozygosity (genetic variation). DNA sequence variation in human populations outside Africa suggests that the original exodus may have involved only about 2000 people. Each time humans spread to an even more remote part of the planet, a small number of bold colonists set out. This caused a series of genetic bottlenecks that increased in number the farther from Africa they went. Population genetic theory shows that bottlenecks cause heterozygosity to be lost, so we expect variation to decline with distance from Africa. That is exactly the pattern seen in native populations around Earth today (see Figure 7.7). But the smaller a population is, the less effective natural selection becomes (see Chapter 7). Those bouts of intense drift caused by repeated episodes of colonization fixed many deleterious mutations. Again, the pattern seen in modern humans matches what theory predicts (**FIGURE 21.20**). Africans have the fewest deleterious mutations, while Europeans and Asians have more. People whose ancestors managed to spread all the way across Asia, the Bering Strait, and finally into the New World, are burdened with even more deleterious mutations.

Natural selection and evolution in real time

Many people think that humans living in the twenty-first century are largely free of natural selection. Don't modern medicine, hygiene, and diet flatten the fitness landscape? No, not entirely.

The Framingham Heart Study is the longest-running longitudinal study in medical history. It has collected a wealth of physical and demographic data on three generations of Americans. We can ask how fitness changes with the values of the traits the study measured [6]. **FIGURE 21.21A** shows the fitness function for total cholesterol level in the blood, based on data on lifetime reproductive success from 1948 to 2008. The implications are dramatic. Individuals with low total cholesterol (15 mg/l) had an average of just over three children in their lifetime. Those with high cholesterol levels (35 mg/l) had only two children. There is strong directional selection favoring lower cholesterol levels, acting right now, on people living in the United States.

An obvious next question is: Will selection on cholesterol levels cause evolutionary change? The answer is yes. The heritability of total cholesterol is $h^2 = 0.61$ in the population studied by the Framingham Heart Study. Thus we have the two ingredients needed for evolution of a quantitative trait: directional selection and heritable variation (see Chapter 6).

Many other traits are also currently under selection, even in industrialized societies [68]. One is body height. **FIGURE 21.21B** shows the fitness function (again based on lifetime reproductive success) for height in Finland between 1935 and 1967. Stabilizing selection acted on females: women with average height had the largest number of children over their lifetime, while short and tall individuals had fewer. In males, however, directional selection was acting, favoring even taller people in what is already one of the world's tallest populations. Height is highly heritable (h^2 averages about 0.8 across human populations), so we can expect that it too will evolve. How height affects fitness is not clear.

Many people think that modern hygiene and medicine have alleviated or even ended natural selection on humans. This is true for some genes and for some traits in some populations. But in much of the world, many people do not have access to sophisticated health care. And even in affluent and technologically advanced societies, natural selection continues to act on modern *Homo sapiens*, and to shape the evolution of our species.

Evolutionary mismatches

Environments are constantly changing, and adaption always lags behind. While this is true for all species, it is a particularly conspicuous fact in our own species. Some of the biggest challenges to human health are the results of bodies that have not yet adapted to the agricultural revolution that began just a few hundred generations ago.

Obesity is a growing epidemic in many countries. It is strongly correlated with major causes of mortality, including diabetes and heart disease. Americans with a body mass index[4] over 40 have a 2.8 times greater risk of developing diabetes than do those with average weight [49]. Diabetes brings with it greatly increased risks of blindness, kidney failure, neuropathy, and other maladies. In 2015, more than 400 million people worldwide had diabetes. The tragedy of this situation is that most of these cases are preventable. They result largely from low physical activity and bad diet, such as consumption of the high-fructose corn syrup used in processed foods and soft drinks.

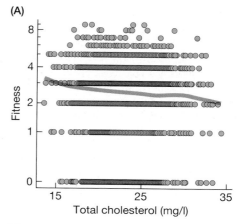

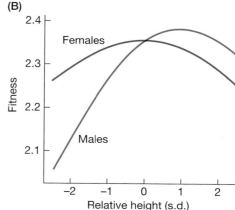

FIGURE 21.21 Selection is still acting on humans in modern industrial societies. (A) The red curve shows the fitness function acting on total cholesterol for a U.S. population between 1948 and 2008. Fitness was measured as lifetime reproductive success; each circle represents an individual. (B) The fitness function acting on body height in Finland between 1935 and 1967. Height is measured in standard deviations (s.d.) from the mean, with shorter individuals to the left and taller ones to the right. (The meaning of a standard deviation is discussed in the Appendix.) (A after [6]; B after [70].)

---

[4] The body mass index (BMI) is one measure of obesity. It is defined as a person's mass divided by the square of height, and is expressed in units of kilograms per square meter (kg/m^2). In many populations, a BMI in the range of 18–25 is considered normal.

FIGURE 21.22 Restoring diverse microbes is sometimes necessary to eliminate harmful bacteria that proliferate in the intestine after antibiotic treatment. Antibiotics reduce the normal diversity of bacteria, often leaving a few resistant harmful species, such as *Clostridium difficile*, to proliferate. Reintroducing the full community of other species suppresses the harmful species. (From [53].)

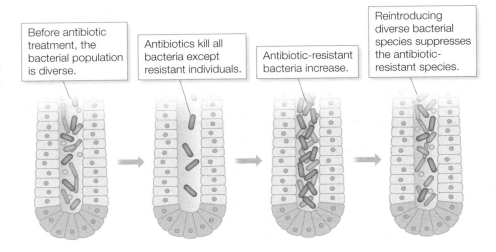

Before antibiotic treatment, the bacterial population is diverse.

Antibiotics kill all bacteria except resistant individuals.

Antibiotic-resistant bacteria increase.

Reintroducing diverse bacterial species suppresses the antibiotic-resistant species.

Why do so many of us eat so much that is so bad? Our appetites, digestion, and metabolism are geared largely to the diet of our preagriculture ancestors. The arrival of agriculture opened up a world of almost unlimited calories. Suddenly, the ability to store energy for lean times was maladaptive. Many diseases soared in frequency with the abrupt changes in lifestyle caused by agriculture, among them heart attacks, stroke, stomach ulcers, certain cancers, and tooth decay [41]. Some genes have evolved that partly compensate. In several populations, mutations have spread at a gene called *TCF7L2* that decreases the risk of developing diabetes by affecting appetite, fat storage, and metabolism [25]. But many, many more generations will go by before we exhibit full adaptation to our current diet—adaptation that will result from natural selection caused by higher death rates. By the time populations have adapted to high-calorie diets, we probably will be eating very different things. Diet is not the only aspect of modern life that is mismatched with our evolved physiology and anatomy [41]. Exercise strengthens not only muscles but also bones. Osteoporosis has increased because of our modern sedentary lifestyle.

Humans did not evolve in a sterile environment. Throughout our evolutionary history, we have been exposed to diverse bacteria and other microbes from birth, and we naturally harbor a microbiome of thousands of species throughout our body. The *hygiene hypothesis* ascribes the great recent increase in the incidence of allergies and other autoimmune maladies to insufficient exposure to diverse microbes. This exposure is necessary, in the first years of life, for proper development of our immune system [18, 51]. Humans who grow up in rural environments and are exposed to farm animals are less likely than city dwellers to develop autoimmune diseases, ranging from allergies to asthma and inflammatory bowel diseases. Even parasitic worms have some beneficial effects, reducing the risk of developing diabetes [2]. Excessive exposure to antibiotics in infancy has many harmful effects: development of autoimmune disorders, lowered immunity to viruses, and increased risk of lifelong obesity [20, 60]. Restoring the normal intestinal microbiome by fecal therapy is sometimes necessary to correct a dangerous imbalance caused by antibiotics (**FIGURE 21.22**). Basic hygiene is unquestionably good for us, but a sterile environment is unquestionably bad.

The Evolution of Culture

Anthropologists have uncovered a revolution in human culture that began more than 50 Kya (**FIGURE 21.23**). Sophisticated stonework became increasingly widespread. Sublime cave paintings show an aesthetic sense that Picasso admired. Abstract artifacts suggest mystic or perhaps religious beliefs. Flutes carved from bone document the invention of music.

What caused this cultural explosion? The enormous change in the capacity for culture was made possible by evolutionary changes in the brain that may have occurred much earlier. But the cultural elements themselves, such as paintings on cave walls, do not evolve by changes in allele frequencies. Just as genetic evolution begins with a single mutation, change in cultural elements—**cultural evolution**—begins with a single variant. These cultural "mutations," or **memes**, spread by different rules than do genetic mutations [4, 7, 66]. Most important, cultural traits can spread quickly by horizontal transmission, that is, between individuals of the same generation. Horizontal transmission does occur rarely with genes (see Chapter 4), but it can be orders of magnitude faster with cultural traits. Politicians and advertisers exploit that fact in efforts to convert the opinion of an entire population in less than one generation. In many ways, the horizontal transmission of cultural traits resembles that of a disease more than the inheritance of a gene. It is no coincidence that we say a video can "go viral."

A second key difference between cultural and genetic inheritance concerns the forces that cause traits to spread. A genetic mutation spreads by natural selection if it improves fitness. Many cultural innovations, however, spread with no help from natural selection, but because they are preferentially copied or learned. This results in **biased transmission**. Individuals can imitate a behavior because of its content (e.g., it is perceived to be advantageous, or is simply easy to remember), because it gives psychological rewards (e.g., consuming alcohol), because of features of the individuals who already exhibit the trait (e.g., copying prestigious or successful persons), or because people prefer to conform to the norm. Consequently, not only useful traits, but also traits that decrease fitness, can spread. Tobacco and alcohol decrease survival and fertility, but both are used by societies across the globe. Smoking and drinking are socially attractive and physiologically addictive. In evolutionary terms, those behaviors spread because they have an advantage in horizontal transmission, not because they increase fitness.

A third way in which cultural and genetic inheritance differ is that at least some "mutations" can be intentional. Over much of human history, many of the cultural variants that spread and became the norm in particular populations were not consciously planned, but others were, such as improvements in tools. In the modern world, cultural mutations are intentional more often than not. Unlike the mutation that causes lactose tolerance in adults, the next version of the Internet will not appear at random.

Cultural evolution also happens by processes that are similar to those in genetic evolution. The frequencies of cultural variants may change by random fluctuation, or "cultural drift" (e.g., the few practitioners of a special craft in a small population may die before having trained apprentices). Cultural differences among groups, such as tribes, can evolve by group selection. Group selection can be more effective in cultural than in genetic evolution because cultural traits are often very homogeneous within groups, since group norms may be forcibly maintained. Some religious cultures, for example, have been maintained not only by the inheritance of parents' beliefs, but also by policing (think of the Inquisition) and by social exclusion of nonconformists.

(A)

(B)

FIGURE 21.23 Sophisticated culture in human societies began more than 50 Kya. The magnificent cave paintings found in France and Spain date from 30 to 10 Kya. This painting is from Lascaux, France. (B) The earliest flutes, made from the bones of bears, birds, and mammoths, are about 40 Ky old. These are from Hohle Fels cave in Germany.

(A)

(B)

(C)

FIGURE 21.24 Wardrobe and technology are two of the cultural differences among current human populations such as (A) the Huli of montane Papua New Guinea, (B) the Inuit of northern Canada, and (C) urban professionals in the United States and western Europe. (A by D. J. Futuyma.)

Evolutionary biology helps us understand human traits by focusing on our diversity. People have a remarkable variety of customs, beliefs, and ways of life (**FIGURE 21.24**). The comparative method (see Chapters 3 and 16) is one evolutionary approach to understand this variation. We hypothesize that a certain cultural trait is advantageous under some conditions, and compare that trait among cultures that experience different conditions. For example, if males monopolize resources required for successful reproduction, we predict that females will sometimes benefit if they bond with a male that already has other mates. That results in a polygynous mating system—one male with multiple females. This hypothesis has been tested and supported by comparing species of birds. The prediction is also supported in human cultures. Polygyny is more common in societies where wealth is based on cattle and which are typically controlled by males [29, 30, 43]. In contrast, monogamy is the norm in the agricultural societies of Eurasia, where land is divided among heirs and not monopolized by a single male [16].

A fascinating but unanswered question is why some cultural traits are widespread across human cultures. Music and religion are universal, but we do not understand how or why they evolved. The comparative method gives us power to understand how differences among populations evolved, but because it relies on variation, it has limited power to unravel the origin of behaviors and instincts that are universal.

Human behavior results from cultural influences that are overlaid on our biological tendencies. While those tendencies result from genetic evolution, they allow an immense range of cultural expressions and individual potentialities. Whatever the biological foundations of our psychology may be, they do not tell us what we must, much less what we should, do with our lives.

Go to the
Evolution Companion Website
EVOLUTION4E.SINAUER.COM
for data analysis and simulation exercises, quizzes, and more.

SUMMARY

- Humans evolved from arboreal, social primate ancestors with binocular vision, grasping hands, and cognitive abilities associated with social life—features that were important foundations for later human evolution.

- Important evolutionary changes in the evolution of the human body include adaptations for bipedality, opposable thumbs, alteration of the vocal tract, a higher reproductive rate, a higher metabolic rate, and a longer childhood. The most important change is our very large brain, with its unparalleled cognitive abilities.

- Hominins are the lineage that includes humans and that diverged from chimpanzees about 7 Mya. Fossil hominins show that humans originated in Africa. Fossils of species in the genus *Homo* date from about 3 Mya. *H. erectus* was the first hominin to leave Africa, and spread through Europe and Asia. About 600 Kya, a second wave left Africa and gave rise to Neanderthals and Denisovans. Finally, *Homo sapiens* spread out of Africa 60 Kya, hybridized and acquired genes from Neanderthals and Denisovans, and spread across the entire Earth by 12 Kya.

- Human populations do not show much divergence across the genome: among Africans, East Asians, and Europeans, only 12 percent of the total genetic variation is caused by differences in allele frequencies among populations. Some of those differences, however, are responsible for variation in skin color, metabolism, and other traits that adapt humans to different environments. Many traits show continuous ranges of variation, and the concept of discrete races does not apply to the human species.

- Two distinctly human traits are our enormous brain and our use of language. A larger brain was likely selected for by ecological factors and by social interactions in groups. Our high metabolic rate supports both the brain's huge energy consumption and our high rate of reproduction. Speech is enabled by both the large brain and a modified vocal tract.

- Other ape species, especially chimpanzee and bonobo, make and use tools in the wild, and can learn and use elements of language in captivity. This suggests that the common ancestor of humans and African apes had rudimentary capacities for language, tool making, and reasoning.

- Culture enabled humans to occupy more different environments, over a broader geographic area, and to use a greater variety of food and other resources than any other species. Agriculture began about 11 Kya. It had profound impacts on our diet, social organization, and population growth; the prevalence of diseases; and the fates of countless other species. The changes caused by agriculture altered the course of human evolution, as shown by several genetic adaptations to diet and changed conditions that came with agricultural societies.

- Natural selection and evolution are ongoing in human populations, even in industrialized societies. Height and cholesterol level are two of the many traits that affect fitness and that are heritable. Many traits are mismatched to our agricultural diet, which has been widespread for only several hundred generations, and to other aspects of modern life.

- Culture is a pronounced human feature that has enabled our species to inhabit and dominate almost all of Earth. Cultural traits change in ways that have some similarities to genetic evolution, but there are also important differences between cultural and genetic evolution. The most important is horizontal transmission: by imitation and learning, a cultural trait can spread across a population (and today, even the entire globe) within a single generation.

TERMS AND CONCEPTS

biased transmission

cultural evolution

meme

Neanderthal

social brain hypothesis

SUGGESTIONS FOR FURTHER READING

An introduction to almost all aspects of human evolution is *Basics in Human Evolution*, edited by M. P. Muehlenbein (Academic Press, London, 2015), with contributions by 48 authors. *The Story of the Human Body: Evolution, Health, and Disease,* by D. E. Lieberman (Vintage Books, NY, 2014), is an outstanding treatment of the topics indicated by its title. Genetic and genomic aspects of human evolution are comprehensively treated in the textbook *Human Evolutionary Genetics* by M. Jobling and colleagues (Garland Science, NY,

2014). Health-related aspects of human evolution are the subject of *Evolutionary Medicine* by S. C. Stearns and R. Medzhitov (Sinauer, Sunderland, MA, 2016). A very readable and personal account of the Neanderthal genome project is *Neanderthal Man: In Search of Lost Genomes* (Basic Books, New York, 2014) by Svante Pääbo, the project's leader. Pääbo's article "The human condition—a molecular approach" (*Cell* 157: 212–226, 2014) reviews human evolution from a genetic perspective. A recent review of the fascinating story of how humans colonized the planet is "Tracing the peopling of the world through genomics," by R. Nielsen, J. M. Akey, M. Jakobsson, J. K. Pritchard, S. Tishkoff, and E. Willerslev (*Nature* 541: 302–310, 2017).

A valuable introduction to cultural evolution and gene-culture coevolution is *Not by Genes Alone: How Culture Transformed Human Evolution* by P. J. Richerson and R. Boyd (University of Chicago Press, Chicago, 2005). A broad variety of articles on cultural evolution is found in *Philosophical Transactions of the Royal Society B*, vol. 336, issue 1567 (April 12, 2011).

PROBLEMS AND DISCUSSION TOPICS

1. What evidence shows that the most recent common ancestor of chimpanzees and humans was much more arboreal than modern humans are? What changes in environmental conditions in Africa might have selected for a less arboreal lifestyle in the human lineage? Why did the same changes not evolve in the chimpanzee lineage?

2. How has the effective population size of humans changed over the last 100,000 years? How have these changes altered the relative contributions that natural selection and genetic drift make to human evolution?

3. The first modern humans evolved in Africa. Give two kinds of evidence that support that conclusion, one based on data from living individuals and one from some other source of data.

4. Discussions of human ancestry sometimes refer to the "Mitochondrial Eve" and "Y-chromosome Adam." Who were these individuals, and why do they have those names? Did they live in the same place and at the same time? Explain.

5. Many people assume that modern medicine has eliminated natural selection in humans. What are three traits that are currently under natural selection in one or more human populations? What form of selection is acting on those traits? What kinds of data show how selection is acting? You might consider examples discussed in other chapters in addition to those described in Chapter 21.

6. Neanderthal fossils were first discovered in the nineteenth century. Study of their morphology suggested that Neanderthals were more closely related to humans than any living species of primate was. Much later, it became possible to sequence DNA from Neanderthal fossils and compare the sequences to those from other primates. Did the results confirm or refute the earlier conclusions based on morphology? Explain.

7. What did Jared Diamond mean when he called agriculture "the worst mistake in the history of the human race"? Provide arguments both against and in favor of this statement.

8. Humans spread out of Africa and across the rest of Earth starting about 60,000 years ago. As they did so, species that are commensals and parasites on humans spread with them. What geographic patterns might you expect to see in the genetic variation of those species?

9. The "aquatic ape hypothesis" is a discredited hypothesis about human evolution. It proposes that several features of the modern human phenotype (including hairlessness, upright posture, and subcutaneous fat) result from descent from a semiaquatic ape that was adapted to life in shallow water. Research the aquatic ape hypothesis and explain why it has been rejected by anthropologists.

22

Evolution and Society

"There are many generalizations in biology, but precious few theories," wrote François Jacob [62], who shared the Nobel Prize for discovering how the transcription of genes is regulated. He continued, "Among these, the theory of evolution is by far the most important." Theodosius Dobzhansky [38], one of the greatest contributors to evolutionary science, went further: "Nothing in biology makes sense except in the light of evolution." But biologists are not alone in proclaiming the profound importance of evolution. Andrew Dickson White, historian, diplomat, and co-founder of Cornell University, wrote in 1896, "Whatever additional factors may be added to natural selection—and Darwin admitted that there might be others—the theory of an evolution process in the formation of the universe and of animated nature is established, and the old theory of direct creation is gone forever. In place of it science has given us conceptions far more noble, and opened the way to an argument from design infinitely more beautiful than any ever developed by theology" [117]. Philosophers are drawn to evolution as to an intellectual feast; Daniel Dennett, who has grappled with the meaning and implications of consciousness, notes that "the Darwinian revolution is both a scientific and a philosophical revolution" [35]. He goes on: "If I were to give an award for the single best idea anyone has ever had, I'd give it to Darwin, ahead of Newton and Einstein and everyone else. In a single stroke, the idea of evolution by natural selection unifies the realm of life, meaning, and purpose with the realm of space and time, cause and effect, mechanism and physical law."

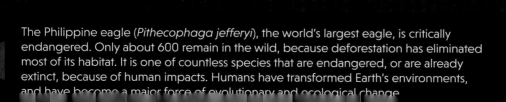

The Philippine eagle (*Pithecophaga jefferyi*), the world's largest eagle, is critically endangered. Only about 600 remain in the wild, because deforestation has eliminated most of its habitat. It is one of countless species that are endangered, or are already extinct, because of human impacts. Humans have transformed Earth's environments, and have become a major force of evolutionary and ecological change.

Throughout this book, we have seen how evolution by natural selection sheds light on almost every area of biology, including the origin and diversification of humans. Any science that explains so much must have practical uses and implications for our lives—the main topic of this last chapter. But many people, especially in the United States, do not accept evolution. So we will come to grips with antievolutionary arguments, and how to refute them (**BOX 22A**). What is at stake is not only society's appreciation of "the single best idea anyone has ever had," but more: society's acceptance of science in general, and of the role of evidence and reason.

BOX 22A

Refuting Antievolutionary Arguments

Because there is no evidence for supernatural creation of organisms, opponents of evolution usually try to demonstrate the falsehood or inadequacy of evolutionary science and to show that biological phenomena must, by default, be the products of intelligent design. Here are some of the most commonly encountered creationist arguments, together with capsule counterarguments. We emphasize that these are not arguments against religion as such.

1. **Evolution is outside the realm of science because it cannot be observed.**

   Evolutionary changes have indeed been observed, as we have noted throughout this book. In any case, most of science depends not on direct observation, but on testing hypotheses against the predictions they make about data.

2. **Evolution cannot be proved.**

   Nothing in science is ever absolutely proved. "Facts" are hypotheses in which we can have very high confidence because of massive evidence in their favor and the absence of contradictory evidence. Abundant evidence from every area of biology and paleontology supports the fact of evolution, and there exists no contradictory evidence.

3. **Evolution is not a scientific hypothesis because it is not testable: no possible observations could refute it.**

   Many conceivable observations could refute or cast serious doubt on evolution, such as finding incontrovertibly mammalian fossils in incontrovertibly Precambrian rocks. In contrast, any puzzling quirk of nature could be attributed to the inscrutable will and infinite power of a supernatural intelligence, so creationism is untestable.

4. **The orderliness of the universe, including the order manifested in organisms' adaptations, is evidence of intelligent design.**

   Order in nature, such as the structure of crystals, arises from natural causes and is not evidence of intelligent design. The order displayed by the correspondence between organisms' structures and their functions is the consequence of natural selection acting on genetic variation. Darwin's realization that the combination of a random process (the origin of genetic variation) and a nonrandom process (natural selection) can account for adaptations provided a natural explanation for the apparent design and purpose in the living world and made a supernatural account unnecessary and obsolete.

5. **Evolution of greater complexity violates the second law of thermodynamics, which holds that entropy (disorder) increases.**

   The second law applies only to closed systems, such as the universe as a whole. Order and complexity can increase in local, open systems as a result of an influx of energy. This is evident in the development of complex individual organisms, in which biochemical reactions are powered by energy derived ultimately from the Sun.

6. **It is almost infinitely improbable that even the simplest life could arise from nonliving matter. The probability of random assembly of a functional nucleotide sequence only 100 bases long is $1/4^{100}$, an exceedingly small number. And scientists have never synthesized life from nonliving matter.**

   It is true that a fully self-replicating system of nucleic acids and replicase enzymes has not yet arisen from simple organic constituents in the laboratory, but the history of scientific progress shows that it would be foolish and arrogant to assert that what science has not accomplished in a few decades cannot be accomplished. (And even if, given our human limitations, we should never succeed in this endeavor, why should that require us to invoke the supernatural?) Some critical steps in the probable origin of life have been demonstrated in the laboratory (see Chapter 17). And there is no reason to think that the first self-replicating or polypeptide-encoding

Refuting Antievolutionary Arguments (continued)

nucleic acids had to have had any particular sequence. If there are many possible sequences with such properties, the probability of their formation rises steeply. Moreover, we do not need to know anything about the origin of life in order to understand and document the evolution of different life forms from their common ancestor.

7. Mutations are harmful and do not give rise to complex new adaptive characteristics.

Most mutations are indeed harmful and are purged from populations by natural selection. Some, however, are beneficial, as shown in many experiments (see Chapters 5 and 6). Complex adaptations are usually based not on single mutations, but on combinations of mutations that jointly or successively increase in frequency as a result of natural selection.

8. Natural selection merely eliminates unfit mutants, rather than creating new characters.

"New" characters, in most cases, are modifications of pre-existing characters, which are altered in size, shape, developmental timing, or organization (see Chapters 2 and 20). This is true at the molecular level as well (see Chapter 14). Natural selection "creates" such modifications by increasing the frequencies of alleles at several or many loci so that combinations of alleles, initially improbable because of their rarity, become probable (see Chapter 6). Observations and experiments on both laboratory and natural populations have demonstrated the efficacy of natural selection.

9. Chance could not produce complex structures.

This is true, but natural selection is a deterministic, not a random, process. The random processes of evolution—mutation and genetic drift—do not result in the evolution of complexity, as far as we know. When natural selection is relaxed, complex structures, such as the eyes of cave-dwelling animals, slowly degenerate, due in part to selection for antagonistic pleiotropic effects.

10. Complex adaptations such as wings, eyes, and biochemical pathways could not have evolved gradually because the first stages would not have been adaptive. The full complexity of such an adaptation is necessary, and it could not arise in a single step by evolution.

This was one of the first objections that greeted *On the Origin of Species*, and it has been christened "irreducible complexity" by advocates of intelligent design. Our

answer has two parts. First, many such complex features, such as hemoglobins and eyes, do show various stages of increasing complexity and functional advantage among different organisms (see Chapters 2, 14, and 20). Second, many structures have been modified for a new function after being elaborated to serve a different function (see Chapters 2 and 20).

11. If an altered structure, such as the long neck of the giraffe, is advantageous, why don't all species have that structure?

This naïve question ignores the fact that different species and populations have different ecological niches and environments, for which different features are adaptive. This principle holds for all features, including "intelligence."

12. If gradual evolution had occurred, there would be no phenotypic gaps among species, and classification would be impossible.

Many disparate organisms are connected by intermediate species, and in such cases, classification into higher taxa is indeed rather arbitrary (see Chapter 2). In other cases, gaps exist because of the extinction of intermediate forms (see Chapters 17 and 20). Moreover, although much of evolution is gradual, some advantageous mutations with large, discrete effects on the phenotype have probably played a role (see Chapter 20). Whether or not evolution has been entirely gradual is an empirical question, not a theoretical necessity.

13. The fossil record does not contain any transitional forms representing the origin of major new forms of life.

This very common claim is flatly false, for there are many such intermediate forms (see Chapters 2, 17, and 20).

14. Vestigial structures are not vestigial, but functional.

According to creationist thought, an intelligent Creator must have had a purpose, or design, for each element of His creation. Thus all features of organisms must be functional. For this reason, creationists view adaptations as support for their position. However, nonfunctional, imperfect, and even maladaptive structures are expected if evolution is true, especially if a change in an organism's environment or way of life has rendered them superfluous or harmful. As noted earlier, organisms display many features, at both the

(continued)

morphological and molecular levels, that are very unlikely to have any function.

15. **The classic examples of evolution are false.**

Some creationists have charged that some of the best-known studies of evolution are flawed and that evolutionary biologists have dishonestly perpetuated these supposed falsehoods. For example, H. B. D. Kettlewell, who performed the classic study of industrial melanism in the peppered moth, was accused of having obtained spurious evidence for natural selection by predatory birds because he pinned moths to unnatural resting sites (tree trunks). Later research tested and strongly validated Kettlewell's conclusions—an example of the classic tradition of the scientific method. But suppose that Kettlewell's study had been flawed. First, it does not follow that textbook authors and other current biologists have deliberately perpetuated falsehood; they simply might have relied on earlier sources, since no textbook author can check every study in depth. Second, whether or not Kettlewell's work was flawed is irrelevant to the validity of the basic claims involved. Both natural selection and rapid evolutionary changes have been demonstrated in so many species that these principles would stand firmly even if the peppered moth story were completely false.

16. **Disagreements among evolutionary biologists show that Darwin was wrong.**

Disagreements among scientists exist in every field of inquiry and are, in fact, the fuel of scientific progress. They stimulate research and are thus a sign of vitality. There are plenty of unresolved, debated questions about evolution, but they do not at all undermine the strength of the evidence for the historical fact of evolution—that is, descent, with modification, from common ancestors. On this point, there is no disagreement among evolutionary biologists.

17. **There are no fossil intermediates between humans and other apes; australopithecines were merely apes. And there exists an unbridgeable gap between humans and all other animals in cognitive abilities.**

This is a claim about one specific detail in evolutionary history, but it is the issue about which creationists care most. This claim is simply false. See Chapter 21 for evidence of stages in morphological evolution revealed by fossil hominids; DNA sequence similarities among modern humans, Neanderthals, and African apes; and evidence that although the cognitive abilities of humans are indeed developed to a far greater degree than those of other species, many of our mental faculties seem to be present in more rudimentary form in other primates and mammals.

18. **As a matter of fairness, alternative theories, such as supernatural creation and intelligent design, should be taught, so that students can make their own decisions.**

This train of thought, if followed to its logical conclusion, would have teachers presenting hundreds of different creation myths, in fairness to the peoples who hold them, and it would compel teachers to entertain supernatural explanations of everything in earth science, astronomy, chemistry, and physics, because anything explained by these sciences, too, could be argued to have a supernatural cause. It would imply teaching students that to do a proper job of investigating an airplane crash, federal agencies should consider the possibility of mechanical failure, a terrorist bomb, a missile impact—and supernatural intervention [4]. Science teachers should be expected to teach the content of current science—which means the hypotheses that have been strongly supported and the ideas that are subjects of ongoing research. That is, they should teach what scientists do. Several scientists have searched the scientific literature for research reports on intelligent design and "creation science" and have found no such reports. Nor is there any evidence that "creation scientists" have carried out scientific research that a biased community of scientists has refused to publish. That means that the subject should not be taught in a science course.

Creationism and Science

Creationism

More than 50 percent of people in the United States deny or doubt evolution, and most of those believe that the human species was created directly by God (Harris Poll, December 2013).[1] People who hold this belief are often referred to as **creationists**. In contrast, a great majority of people in Europe do not question the reality of evolution (even in countries such as Italy that have an officially established religion), and they are often astonished that antiscientific attitudes on evolution flourish in the technologically and scientifically most prominent country in the world. There is wide opposition to evolution in many Muslim countries, but evolution (although not human evolution) is widely taught [6, 55]. Among 34 Western countries (and Japan), Turkey ranks lowest in public acceptance of evolution, and the United States second lowest (**FIGURE 22.1**). Creationist pressure has greatly weakened science education in the United States, for even teachers who accept evolution often compromise their teaching, or minimize their coverage, in order to avoid controversy [11]. Some high-school biology textbooks and teachers convey the impression that the evidence for evolution (and even for the great age of Earth) is doubtful.[2]

Most disbelievers in evolution reject the idea because they think it conflicts with their religious beliefs. For Christian and Jewish fundamentalists, evolution conflicts with their literal interpretation of the Bible, especially the first chapters of Genesis, which portray God's creation of the heavens, Earth, plants, animals, and humans in six days. However, many Western religions understand these biblical descriptions to contain symbolic truths, not literal or scientific ones. Many deeply religious people accept evolution, viewing it as the natural mechanism by which God has enabled creation to proceed. Some scientists, including some researchers in evolutionary biology and some of the most impassioned opponents of creationism, subscribe to this view (see [76, 77]). Some religious leaders have made clear that they accept evolution. (See an array of such statements in the book "Voices for Evolution," available for free download at www.ncse.com, the website of the National Center for Science Education.) For example, Pope John Paul II affirmed the validity of evolution in 1996, although he reserved a divine origin for the human soul. (The text of his letter was reprinted in the *Quarterly Review of Biology* 72: 381–396 [December 1997].) The pope's position was close to the argument generally known as **theistic evolution**, which holds that God established natural laws (such as natural selection) and then let the universe run on its own, without further supernatural intervention.

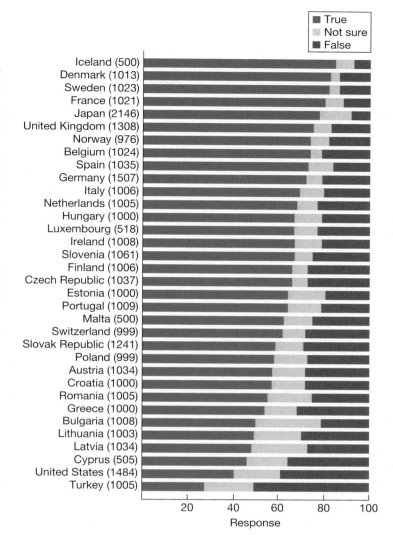

FIGURE 22.1 Public acceptance of evolution in the United States, Japan, and 32 European countries in 2005, according to opinion polls. Blue versus red segments indicate the percentage of people who said that evolution is true versus false. The numbers for each country are sample sizes. (After [75].)

---

[1] See https://ncse.com/news/2013/12/evolution-new-harris-poll-0015255.

[2] See, for example, https://en.wikipedia.org/wiki/Creation_and_evolution_in_public_education_in_the_United_States#Recent_developments_in_state_education_programs.

The beliefs of creationists vary considerably [85]. The most extreme creationists interpret every statement in the Bible literally. They include "young Earth" creationists who believe in **special creation** (the doctrine that each species, living or extinct, was created independently by God, essentially in its present form) and in a young universe and Earth (less than 10,000 years old), a deluge that drowned Earth, and an ark in which Noah preserved a pair of every living species. They must therefore deny not only evolution, but also most of geology and physics (including radiometric and astronomical evidence of the great age of the universe). Some other creationists allow that mutation and natural selection can occur, and even that very similar species can arise from a common ancestor. However, they deny that higher taxa have evolved from common ancestors, and they assert that the human species was specially created by God.

Most of the efforts of activist creationists are devoted to combating the teaching of evolution in schools, or at least insisting on "equal time" for their views. In the United States, however, the constitutional prohibition of state sponsorship of religion has been interpreted by the courts to mean that any explicitly religious version of the origin of life's diversity cannot be promulgated in public schools. The activists have therefore adopted several forms of camouflage. One is **intelligent design (ID)** theory. ID proponents generally do not publicly invoke special creation by God, but they argue that many biological phenomena are too complicated ("irreducibly complex") to have arisen by natural processes and can therefore be explained only by an intelligent designer. (See [89] for an analysis of ID.) In an important court case in 2005, the judge ruled that intelligent design is not science, but rather thinly disguised religious doctrine.[3] Since then, the creationist strategy in the United States has been to invoke "critical analysis" or "academic freedom," urging that students be encouraged to examine the evidence for both sides of scientific issues—usually limited to evolution, global warming, and perhaps one or two other socially controversial topics. Of course, every thoughtful person should develop the habit of critical thinking, but these efforts mainly serve to sow doubt where, from a scientific perspective, there is little or none.

The nature of science

Science is a *process* of acquiring an understanding of natural phenomena. This process consists largely of posing hypotheses and testing them with observational or experimental evidence. *The most important feature of scientific hypotheses is that they are testable*, at least in principle. Sometimes we can test a hypothesis by direct observation, but more often we infer objects (e.g., atoms) and processes (e.g., DNA replication) by comparing the outcome of observations or experiments with predictions made from competing hypotheses. In order to make such inferences, we must assume that the processes obey **natural laws**: statements that certain patterns of events will occur if certain conditions hold. In contrast, supernatural events or agents are supposed to suspend or violate natural laws.

Scientists can test (and have falsified) many specific creationist claims, such as the occurrence of a worldwide flood or the claim that Earth and all organisms are less than 10,000 years old. (This claim about the age of Earth was refuted before *On the Origin of Species* was published.) But scientists cannot test the hypothesis that an omnipotent God exists, or that He created anything, because we do not know what consistent patterns these hypotheses might predict.

Despite loose talk about "proving" hypotheses, most scientists agree that the hypothesis that currently best explains the data is *provisionally* accepted, with the understanding that it may be altered, expanded, or rejected if subsequent evidence warrants doing so, or if a better hypothesis, not yet imagined, is devised.

---

[3] See https://en.wikipedia.org/wiki/Kitzmiller_v._Dover_Area_School_District.

For instance, Mendel's laws of assortment and independent segregation, which initiated modern genetics, were modified when phenomena such as linkage and meiotic drive were discovered, but Mendel's underlying principle of inheritance based on "particles" (genes) holds true today.

This process reflects one of the most important and valuable features of science: even if individual scientists may be committed to a hypothesis, scientists as a group are not irrevocably committed to it; they must, and do, change their minds if the evidence so warrants. Indeed, much of science consists of seeking chinks in the armor of established ideas. Thus science, as a social process, is tentative; it questions belief and authority; it continually tests its views against evidence. Scientific claims, in fact, are the outcome of a process of natural selection, for ideas (and scientists) compete with one another, so that the body of ideas in a scientific field grows in explanatory content and power [60]. Science differs in this way from creationism, which does not use evidence to test its claims, does not allow evidence to shake its a priori commitment to certain beliefs, and does not grow in its capacity to explain the natural world.

The ideal of democracy doesn't extend to ideas—some are simply wrong, and as a purely practical matter, it is imperative that we recognize them as such [91]. In everyday life, we assume and depend on natural, not supernatural, explanations. Unlike the Puritans of Salem, Massachusetts, who in 1692 condemned people for witchcraft, we no longer seriously entertain the notion that someone can be victimized by a witch's spell or possessed by devils, and we would be outraged if a criminal successfully avoided conviction because he claimed "the Devil made me do it." We depend on scientific explanations, and we know that science has proven its ability—because it works.

Is evolution a fact or a theory? Both. Recall from Chapter 1 that a theory, as the word is used in science, doesn't mean an unsupported speculation or hypothesis (the popular use of the word). A theory is instead a big, well-supported idea that encompasses other ideas and hypotheses and weaves them into a coherent fabric. The word "fact" applies to hypotheses that have become so well supported by evidence that we feel safe in acting as if they were true. To use a courtroom analogy, they have been "proven" beyond reasonable doubt. Not beyond any conceivable doubt, but reasonable doubt. By this criterion, evolution is a scientific fact. That is, the descent of all species, with modification, from common ancestors is a hypothesis that in the last 150 years or so has been supported by so much evidence, and has so successfully resisted all challenges, that it has become a fact. This history of evolutionary change—and the diversity of life—is explained by evolutionary theory, the body of statements (about mutation, selection, genetic drift, developmental constraints, and so forth) that together account for the various changes that organisms have undergone.

It is chiefly the fact of evolution that creationists deny. Everyone who has studied biology should be able to counter creationist arguments and present evidence for the fact of evolution.

The Evidence for Evolution

The evidence for evolution has been presented throughout the preceding chapters of this book (see especially Box 2B, p. 44–45). In this section we simply review the sources of evidence and refer back to earlier chapters for detailed examples.

The fossil record

Even though the fossil record is known to be very incomplete, paleontologists have found many examples of transitional stages in the origin of higher taxa, such as tetrapods (see Chapter 17), birds (see Chapter 16), and mammals (see Chapter 20). Critically important intermediates are still being found, such as ancestors of modern turtles (see Chapter 20). The fossil record, moreover, documents two important aspects of character evolution: mosaic evolution and gradual change of individual features (both illustrated by hominin evolution; see Chapter 21).

Many discoveries in the fossil record fit predictions based on phylogenetic or other evidence. For example, the age of groups estimated from the fossil record often matches phylogenetically predicted sequences (see Chapters 17 and 19). Prokaryotes precede eukaryotes in the fossil record, wingless insects precede winged insects, fishes precede tetrapods, ferns and gymnosperms precede flowering plants.

Phylogenetic and comparative studies

Even if we had no fossil record at all, many other kinds of information would provide incontrovertible evidence for evolution. Common ancestry of, for example, birds and crocodiles, is implied by both anatomical characteristics and DNA sequences. Molecular phylogenetic trees support many relationships that have long been implied by entirely independent morphological data (see Chapters 2 and 17).

We are confident today that all known living things stem from a single ancestor because of the many features that are universally shared (see Chapter 17), such as the genetic code, the mechanisms of transcription and translation, and proteins composed only of "left-handed" (L isomer) amino acids. Many genes are shared among all organisms, including the three major domains (Bacteria, Archaea, and Eucarya), and these genes have been successfully used to infer the deepest branches in the tree of life. Systematists have demonstrated the common origin, or homology, of characteristics that may differ greatly among taxa (see Chapters 2, 15, and 20). Hox genes and other developmental mechanisms are shared among animal phyla that diverged from common ancestors more than a half-billion years ago (see Chapters 15 and 20).

Genes and genomes

Molecular biology and genomics show the extraordinary commonality of all living things. Common ancestry is the only scientific rationale for learning about human biology by studying yeast, flies, rats, or monkeys (**FIGURE 22.2**).

Molecular studies show that the genomes of most organisms have similar elements, such as a great abundance of noncoding pseudogenes and satellite DNA and a plethora of "selfish" transposable elements that generally provide no advantage to the organism. These and other features are readily understandable under evolutionary theory, but lack any evidence of intelligent design [7]. Some DNA polymorphisms are shared among species, so that, for example, some major histocompatibility sequences of humans are more similar and more closely related to chimpanzee sequences than to other human sequences (**FIGURE 22.3**). What more striking evidence of common ancestry could there be?

Biogeography

We noted in Chapter 18 that the geographic distributions of organisms provided Darwin with abundant evidence of evolution, and they have continued to do so. For example, the distributions of many taxa correspond to geological events such as the formation and dissolution of connections between land masses. We saw that the phylogenies of Hawaiian species match the sequence by which the islands came into existence. We saw, as did Darwin, that an isolated region such as an island commonly lacks whole groups of organisms, and that human-introduced species often come to dominate.

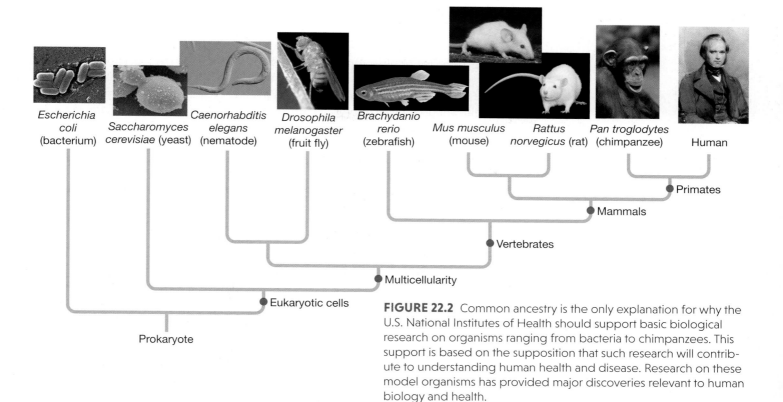

Escherichia coli (bacterium)
Saccharomyces cerevisiae (yeast)
Caenorhabditis elegans (nematode)
Drosophila melanogaster (fruit fly)
Brachydanio rerio (zebrafish)
Mus musculus (mouse)
Rattus norvegicus (rat)
Pan troglodytes (chimpanzee)
Human

Primates
Mammals
Vertebrates
Multicellularity
Eukaryotic cells
Prokaryote

FIGURE 22.2 Common ancestry is the only explanation for why the U.S. National Institutes of Health should support basic biological research on organisms ranging from bacteria to chimpanzees. This support is based on the supposition that such research will contribute to understanding human health and disease. Research on these model organisms has provided major discoveries relevant to human biology and health.

Failures of the argument from design

Since God cannot be known directly, theologians such as Thomas Aquinas have long attempted to infer His characteristics from His works. Theologians have argued, for instance, that order in the universe, such as the predictable movement of celestial bodies, implies that God must be orderly and rational, and that He creates according to a plan. From the observation that organisms have characteristics that serve their survival, it could similarly be inferred that God is a rational, intelligent designer who, furthermore, is beneficent, having equipped living things for all their needs. Such a beneficent God would not create an imperfect

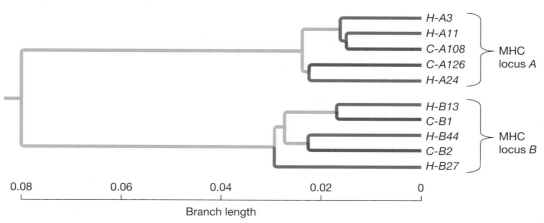

H-A3
H-A11
C-A108
C-A126
H-A24
} MHC locus *A*

H-B13
C-B1
H-B44
C-B2
H-B27
} MHC locus *B*

0.08   0.06   0.04   0.02   0
Branch length

FIGURE 22.3 Due to incomplete lineage sorting (see Chapter 16), a polymorphism may be inherited by two or more species from their common ancestor, and certain haplotypes in each species may be most closely related to haplotypes in the other species. This gene tree shows the relationships among six alleles in humans and four alleles in chimpanzees at the major histocompatibility (MHC) loci *A* and *B*. Both species have loci *A* and *B*, which form monophyletic clusters, indicating that the two loci arose by gene duplication before speciation gave rise to the human and chimpanzee lineages. At both loci, each chimpanzee allele is more closely related (and has a more similar nucleotide sequence) to a human allele than to other chimpanzee alleles. Thus polymorphism at each locus in the common ancestor has been carried over into both descendant species. (After [78].)

world; so, as the philosopher Leibniz said, this must be "the best of all possible worlds." (His phrase was mercilessly ridiculed by Voltaire in his satire *Candide*.) The adaptive design of organisms, in fact, has long been cited as evidence of an intelligent designer. This was the thrust of William Paley's famous example: as the design evident in a watch implies a watchmaker, so the design evident in organisms implies a designer of life [88]. This **argument from design** has been renewed in the "intelligent design" version of creationism, and it is apparently the most frequently cited reason people give for believing in God [90].

Of course, Darwin made this particular theological argument passé by providing a natural mechanism of design: natural selection. Moreover, Darwin and subsequent evolutionary biologists have described innumerable examples of biological phenomena that are hard to reconcile with beneficent intelligent design. Just as Voltaire showed (in *Candide*) that cruelties and disasters make a mockery of the idea that this is "the best of all possible worlds," biology has shown that organisms have imperfections and anomalies that can be explained only by the contingencies of history, and characteristics that make sense only if natural selection has produced them. If "good design" were evidence of a kindly, omnipotent designer, would "inferior design" be evidence of an unkind, incompetent, or handicapped designer?

Only evolutionary history can explain vestigial organs—the rudiments of once-functional features, such as the human appendix, the reduced wings under the fused wing covers of some flightless beetles, and the nonfunctional stamens or pistils of plants that have evolved separate-sexed flowers from an ancestral hermaphroditic condition. Only history can explain why the genome is full of "fossil" genes: pseudogenes that have lost their function.

Because characteristics evolve from pre-existing features, often undergoing changes in function, many features are poorly engineered, as anyone who has suffered from lower back pain or wisdom teeth can testify. Once the pentadactyl limb became developmentally canalized, tetrapods could not evolve more than five digits even if they would be useful: the extra "thumb" of the giant panda and of moles is not a true digit at all, and it lacks the flexibility of true fingers because it is not jointed (see Figure 20.9). And it is a pity that humans, unlike salamanders, cannot regenerate lost limbs or digits.

If a designer were to equip species with a way to survive environmental change, it might make sense to devise a Lamarckian mechanism, whereby genetic changes would occur in response to need. Instead, adaptation is based on a combination of a random process (mutation) that cannot be trusted to produce the needed variation (and often does not) and a process that is the very epitome of waste and seeming cruelty (natural selection, which requires that great numbers of organisms fail to survive or reproduce). It would be hard to imagine a crueler instance of natural selection than sickle-cell disease, whereby part of the human population is protected against malaria at the expense of countless other people, who are condemned to die because they are homozygous for a gene that happens to be worse for the malarial parasite than for heterozygous carriers (see Chapter 5). And, of course, this process often does not preserve species in the face of change: more than 99 percent of all species that have ever lived are extinct. Were those species the products of an incompetent designer? Or one that couldn't foresee that species would have to adapt to changing circumstances?

Many species become extinct because of competition, predation, and parasitism. Some of these interactions are so appalling that Darwin was moved to write, "What a book a devil's chaplain might write on the clumsy, wasteful, blundering, low, and horribly cruel works of Nature!" Darwin knew of maggots that work their way into the brains of sheep, and of wasp larvae that, having consumed the internal organs of a living caterpillar, burst out like the monster in the movie *Alien*. The life histories of parasites, whether parasitic wasp or human immunodeficiency virus (HIV), ill fit our concept of an intelligent, kindly designer, but they are easily explained by natural selection (see Chapter 13).

No one has yet demonstrated a characteristic of any species that serves only to benefit a different species, or only to enhance the so-called balance of nature—for, as Darwin saw, "such could not have been produced through natural selection." Because natural selection consists only of differential reproductive success, it results in "selfish" genes and genotypes, some of which have results that are inexplicable by intelligent design (see Chapter 12). We have seen that genomes are brimming with sequences such as transposable elements that increase their own numbers without benefiting the organism. We have seen maternally transmitted cytoplasmic genes that cause male sterility in many plants, and nuclear genes that have evolved to override them and restore male fertility (see Chapter 12). Such conflicts among genes in a genome are widespread. Are they predicted by intelligent design theory? Likewise, no theory of design can predict or explain features that we ascribe to sexual selection, such as males that remove the sperm of other males from the female's reproductive tract (see Chapter 10). Nor can we rationalize why a beneficent designer would shape the many other selfish behaviors that natural selection explains, such as cannibalism, siblicide, and infanticide.

Evolution, and its mechanisms, observed

Anyone can observe erosion, and geologists can measure the movement of continental plates, which travel up to 10 cm per year. No geologist doubts that these mechanisms, even if they accomplish only slight changes on the scale of human generations, have shaped the Grand Canyon and have separated South America from Africa over the course of millions of years. Likewise, biologists do not expect to see anything like the origin of mammals played out on a human time scale, but they have documented the processes that will yield such grand changes, given enough time.

We know that genetic variation in all kinds of phenotypic characters originates by mutation (see Chapters 4 and 6). This variation has been used by humans for millennia to develop strains of domesticated plants and animals that differ in morphology more than whole families of natural organisms do. In experimental studies of laboratory populations of microorganisms, we have seen new advantageous mutations arise and enable rapid adaptation to temperature changes, toxins, or other environmental stresses. Evolutionary biologists have documented hundreds of examples of natural selection acting on genetic and phenotypic variation, and of rapid adaptation to new environmental factors (see Chapters 3, 5, and 6)—including resistance to pesticides and antibiotics (**FIGURE 22.4**). Within the past century, some populations (e.g., the apple maggot; see Chapter 9) have almost become different species.

In summary, the major causes of evolution have been extensively documented. The two

FIGURE 22.4 Cartoonist Garry Trudeau affirms the importance of evolutionary science.

major processes of long-term evolution, anagenesis (changes in characters within lineages) and cladogenesis (origin of two or more lineages from a common ancestor), are abundantly supported by evidence from every possible source, ranging from molecular biology to paleontology. Over the past century we have certainly learned of evolutionary processes that were formerly unknown: we now know, for example, that some DNA sequences are mobile and can cause mutations in other genes. But no scientific observations have ever cast serious doubt on the reality of the basic mechanisms of evolution, such as natural selection, or on the reality of the basic historical patterns of evolution, such as transformation of characters and the origin of all known forms of life from common ancestors. Contrast this mountain of evidence with the evidence for supernatural creation or intelligent design: *there is no such evidence.*

The Uses and Implications of Evolutionary Science

Like any other science, evolutionary biology has practical applications—in health science, food production, and other areas—that can affect our lives. And to a far greater degree than most other sciences, it has enormous implications for understanding ourselves as individuals, as societies, as a species. No other subject in biology has more profound philosophical implications. We touch here on some of these topics, although only superficially; hundreds of books have been written on these subjects.

Evolution by natural selection: A broad and flexible concept

Darwin already recognized that evolution—that is, descent with modification from common ancestors—doesn't describe only the history of species: languages and other cultural elements also evolve. Since Darwin's time, the basic idea of evolution, as well as the specific concept of evolution by natural selection of variants, has been widely applied [94]. Karl Popper, David Hull, and other philosophers have proposed that knowledge, especially scientific knowledge, changes by a process of selection among competing ideas or schools of thought—an idea that is at the core of evolutionary epistemology [60]. There exist academic fields of evolutionary social science, evolutionary political science, evolutionary economics, evolutionary anthropology, and applications of the ideas of evolution and selection to technological innovation and even to literature and creative thought [79, 116]. In many of these areas, the analogy to biological evolution by natural selection of random variation is inexact and perhaps forced; for example, the variations that may or may not survive are often not random in the sense that describes gene mutations. Still, evolution by natural selection, in a broad sense, can describe a great range of human experiences and activities [35], and in some areas, it has a real payoff. An enormous field of *evolutionary computation*, founded on the principle that complex adaptations can evolve by natural selection among randomly generated variations, uses *genetic algorithms* that mimic natural selection to address a huge range of real-world problems. There even exist evolutionary algorithms that enable robots to adapt to unforeseen changes [1].

Practical applications of evolutionary science

Almost all of the principles and methods of evolutionary biology bear on a range of practical applications [56]. Among these are human use of other organisms' adaptations, food production, management of natural resources, conservation, and human health.

Many of the methods and theoretical models developed to study evolution have been broadly useful. Evolutionary geneticists developed the concept of linkage disequilibrium and the dynamics of association between genes or markers; these models are the basis of mapping genes and mutations in human and other populations, for example in genome-wide association studies (GWAS). Population genetics, likewise, is the basis for matching DNA evidence with individual suspects in forensics, paternity identification, and other applications. Evolutionary biologists developed the models and statistical tests for natural selection in genomes, including both adaptive and deleterious variants. Phylogenetic methods, applied to variant DNA sequences, are used to trace the origin and spread of pathogens such as HIV, and have been used to identify criminals who deliberately infected victims (such as unwanted sexual partners) with HIV [103].

Using organisms' adaptations

A great many things that humans would like to do or make have already been done or made—by natural selection, in one or another of the millions of species, living or extinct. Some of these adaptive characteristics are, or could be, used directly. The antibiotic penicillin is an adaptation of the *Penicillium chrysogenum* fungus for suppressing competing bacteria. The polymerase chain reaction (PCR) that is the foundation of modern molecular biology uses the heat-stable DNA polymerase isolated from the bacterium *Thermus aquaticus*, which lives in hot springs and hydrothermal vents. The CRISPR-Cas9 mechanism, which evolved in bacteria to combat foreign DNA such as bacteriophage, is an enormously important tool for surgically altering genes to meet our purposes. Artemisinin, a drug that suppresses malaria, was discovered in sweet wormwood (*Artemisia annua*); it is one of many thousands of compounds with antibiotic properties that plants have evolved as defenses against herbivores and parasites. The silkworm *Bombyx mori* has been used for silk production for thousands of years, and various spider silks, with diverse tensile features, are now being studied for a range of possible industrial uses.

Other adaptations provide inspiration, or serve as models, for useful inventions, an approach called biomimetics [106]. The inventor of Velcro was inspired by the hooks on the burrs that have evolved for seed dispersal in many plants. Bioengineers have studied the feet of geckos, which can run across ceilings and climb on glass, as models for dry adhesion, and have found in the byssal threads of mussels a chemical basis for underwater adhesion. The naked mole-rat (*Heterocephalus glaber*; see Figure 12.10A) is exceptionally long-lived yet appears to be almost completely free of cancer, perhaps because of its high concentration of hyaluronan, a major component of the extracellular matrix [112]. And a recent, potentially very important, discovery is that a bacterium that normally inhabits the human nose, *Staphylococcus lugdunensis*, produces an antibiotic that kills diverse multidrug-resistant pathogens, including methicillin-resistant *Staphylococcus aureus* (MRSA) [125]. Much of what is known about the countless adaptive characteristics of diverse organisms has been learned by botanists, entomologists, microbiologists, and other such experts on specific taxa, and testifies to the immense value of taxonomy and natural history. Understanding these adaptations has often been enhanced by evolutionary research, such as studies of form and function, of life histories, and of coevolution of plants and herbivores.

Agriculture and natural resources

The development of improved varieties of domesticated plants and animals is evolution by artificial selection (see Chapter 6). Evolutionary genetics and plant and animal breeding have had an intimate, mutually beneficial relationship for more than a century. Theoretical evolutionary methods and experimental studies of *Drosophila* and other model organisms have contributed both to traditional breeding and to modern quantitative trait locus (QTL) analysis, which is used to locate and characterize genes that contribute to traits of interest (see Chapter 6).

Agronomists should heed what evolutionary biologists have long known: that genetic diversity is essential for a population's long-term success. For example, an epidemic of the southern corn leaf blight fungus in 1970 destroyed much of the U.S. corn crop because so much of the corn being grown was a single genetic strain that carried an allele that increased susceptibility (see Chapter 13) [114]. Experiments have shown that genetically diverse plots of rice suffer much less disease than do single-genotype plots [124]. The corporations that dominate much of modern agriculture may profit from propagating a single strain across broad landscapes, but this approach courts disaster.

Many wild plants have characteristics that can improve crop species. Before modern genetic technology, the genes underlying these traits were typically introduced into the crop by hybridization. For example, at least 20 genes for resistance to various diseases have been crossed into commercial tomato stocks from wild species of tomatoes. Today, molecular methods of genetic engineering can move a gene from any organism into a crop plant, once a useful gene is identified. For example, genes for tolerating salt can be transferred from plant species that are naturally adapted to saline conditions. Genes that confer resistance to insects or other crop pests are already in use; widely planted strains of corn and cotton carry genes derived from the soil bacterium *Bacillus thuringiensis* (Bt), which produces Bt toxin that kills the larvae of Lepidoptera such as the corn earworm. Many other genes in wild plants and microbes that confer resistance to pests will surely be isolated and introduced into crops in the future.

Evolutionary biology aids this revolution in agronomics by contributing to gene mapping methods, by identifying likely sources of genes for useful characteristics, and by evaluating possible risks posed by transgenic organisms (often called genetically modified organisms, or GMOs). Food crops obviously must be tested for safety for humans, but there are also potential ecological risks; for example, Bt toxin and other natural insecticides might kill nontarget species, and there is concern that transgenes may spread from crop plants to wild species, which then could become more vigorous weeds. Phylogenetic studies can identify wild species that might hybridize with crop plants, and population genetic methods can estimate the fitness effects of transgenes and the chances of gene flow into natural plant populations [40]. "Darwinian agriculture" may provide other guidelines for crop improvement as well [33, 34]. For example, crop yield may be improved by understanding allocation trade-offs among growth, survival, and reproduction (see Chapter 11), and by selecting characteristics that reduce the competitive ability of individual plants but enhance the productivity of the group. Individual plant fitness is often enhanced when plants grow taller than their neighboring plants and shade them; artificial selection for reduced height in cereal crops may reallocate energy from growth to seed production.

Insects, weeds, and other organisms cause billions of dollars' worth of crop losses. Much of this loss is caused when crop pests evolve resistance to chemical insecticides and herbicides (see Chapter 3). This resistance not only increases the costs of agriculture, but also results in a steady increase in the amount of toxic chemicals sprayed on the landscape (some of which find their way up the food chain, affecting humans and other consumers). In some places, regulations on the use of pesticides follow recommendations made by evolutionary biologists about how to manage pest populations in order to keep them susceptible to pesticides [53, 54]. *Biological control* of pests also benefits from evolutionary analysis. When a new pest species suddenly appears, phylogenetic systematics is the first approach to identifying the pest and determining where in the world it has come from. That is where entomologists will search for natural enemies, scrutinizing in particular those that are related to known enemies of species that are related to the new pest. Likewise, herbivorous insects used to control weeds or invasive plants must be screened to be sure they will not also attack crops or native plants. A good approach is to see whether they have the potential to feed on or adapt to plants that are related to the target plant species [45].

Conservation

There is little doubt that a major extinction event has been initiated by the huge and accelerating impact of human activities on every aspect of the environment (**BOX 22.B**). So far, the main human threat to other species has been elimination of their habitats by land use and climate change. In this context, the most important means of conservation are obvious and require mostly ecological, political, legal, and economic expertise: save natural habitats in preserves, establish and enforce limits on the exploitation of fish populations and other biological resources, reduce pollution and global warming. But evolutionary biologists also

---

BOX 22B

The Current Extinction Crisis

For the first time in the history of life, a single species has precipitated a major extinction. Within the next few centuries, the diversity of life will almost certainly plummet at a pace that may well equal any mass extinction in Earth's history.

The human threat to Earth's biodiversity has accelerated steadily with the advent of ever more powerful technology and the exponential growth of the world's human population, which has surpassed 7 billion. The per capita rate of population growth is greatest in the developing countries, which are chiefly tropical and subtropical, but the per capita impact on the world's environment is greatest in the most highly industrialized countries. An average American, for example, has perhaps 140 times the environmental impact of an average Kenyan, because the United States is so profligate a consumer of resources (harvested throughout the world) and of energy (with impacts ranging from strip mines, fracking, and oil spills to greenhouse gases that cause global warming).

Some species are threatened by hunting or overfishing and others by species that humans have introduced into new regions. But by far the greatest cause of extinction, now and probably over the course of the twenty-first century, is the destruction of habitat [101]. It is largely for this reason that 29 percent of North American freshwater fishes are endangered or already extinct, and that about 10 percent of the world's bird species are considered endangered by the International Council for Bird Preservation.

The numbers of species likely to be lost are highest in tropical forests, which are being destroyed at a phenomenal and accelerating rate in Asia, Africa, and tropical America. As E. O. Wilson said, "in 1989 the surviving rainforests occupied an area about that of the contiguous forty-eight states of the United States, and they were being reduced by an amount equivalent to the size of Florida each year" [120]. Several authors have estimated that 10–25 percent of tropical rainforest species—accounting for as much as 5–10 percent of Earth's species diversity—will become extinct in the next 30 years. To this toll must be added extinctions caused by the destruction of species-rich coral reefs, pollution of other marine habitats, and losses of habitat in areas such as Madagascar and the Cape Province of South Africa, which harbor unusually high numbers of endemic species. Even if extinction rates are much lower than Wilson estimated, they will still equal or exceed those described by paleontologists, such as the huge end-Permian mass extinction [104].

In the long run, an even greater threat to biodiversity may be global warming caused by high and increasing consumption of fossil fuels and production of carbon dioxide and other greenhouse gases. Earth's climate has warmed by a global average of 0.6°C during the last century, and the rate of warming is much faster than most of the climate changes that have occurred in the past. The effects of climate change vary geographically; for example, some regions are becoming much drier.

Some species may adapt by genetic change, but the rate of climate change is so high that the rate of evolution of species' "climate niches"—the range of climate conditions they actually occupy—would have to be more than 10,000

(continued)

BOX 22B

The Current Extinction Crisis (continued)

times faster than niche evolution has typically been in the past, in order to keep pace [96]. There is already evidence that many species will shift their ranges instead of genetically adapting, but such shifts are difficult or impossible for most mountaintop and Arctic species, and for many others that lack the habitat "corridors" along which they might disperse. (Most forest-dwelling species, for example, will not disperse through cities or cornfields.) Computer simulations, based on various scenarios of warming rate and species' capacity for dispersal, suggest that within the next 50 years, between 18 and 35 percent of species will become "committed to extinction"—that is, they will have passed the point of no return [111].

If mass extinctions have happened naturally in the past, why should we be so concerned? Different people have different answers, ranging from utilitarian to aesthetic to spiritual. Some point to the many thousands of species that are used by humans throughout the world today, ranging from familiar foods to fiber, herbal medicines, and spices. Others cite the economic value of ecotourism and the enormous popularity of bird-watching in some countries. Biologists will argue that thousands of species may prove useful (as many already have) as pest control agents or as sources of medicinal compounds or industrially valuable materials. Except in a few well-known groups, such as vertebrates and vascular plants, most species have not even been described, much less been studied for their ecological and possible social value.

The rationale for conserving biodiversity is only partly utilitarian, however. Many people (including the authors) cannot bear to think that future generations will be deprived of tigers, sea turtles, and macaws. They share with millions of others a deep renewal of spirit in the presence of unspoiled nature. Still others feel that it is in some sense cosmically unjust to extinguish, forever, the species with which we share Earth.

Conservation is an exceedingly complicated topic; it requires not only a concern for other species, but compassion and understanding of the very real needs of people whose lives depend on clearing forests and making other uses of the environment. It requires that we understand not only biology, but also global and local economics, politics, and social issues ranging from the status of women to the reactions of the world's peoples and their governments to what may seem like elitist Western ideas. Anyone who undertakes work in conservation must deal with these complexities. But everyone can play a helpful role, however small. We can try to waste less; influence people about the need to reduce population growth (surely the most pressing problem of all); support conservation organizations; patronize environment-conscious businesses; stay aware of current environmental issues; and communicate our concerns to elected officials at every level of government. Few actions of an enlightened citizen of the world can be more important.

make indispensable contributions to conservation efforts. They use phylogenetic information to determine where potential nature reserves should be located to protect the greatest variety of biologically different species; they use evolutionary biogeography to identify regions with many endemic species (e.g., Madagascar); they use genetic methods and theory to prevent inbreeding load in rare species and to distinguish genetically unique populations [3, 44]; and they use genetic markers to identify illegal traffic in endangered species (see [8]).

Climate change is real and is being caused by humans. (See *Climate Change: Evidence and Causes,* written for a general audience by the world's most prestigious scientific assemblies [110].) Evolutionary biologists are increasingly concerned with the risk of species extinction due to climate change. Their studies range from analyses of shifts in species geographic ranges in the past, to genetic variation and other factors that will determine the likelihood of evolutionary adaptation [69, 109, 115]. Some researchers offer a more optimistic outlook than others, at least for certain kinds of species in some parts of the world, but all agree that more research is urgently needed.

Health and medicine

The direct and indirect applications of evolutionary biology are probably more numerous, and more important, in medicine and public health than in any other area [5]. Depending on the topic, evolutionary theory may provide new conceptual approaches to medically relevant research (e.g., the evolution of senescence), principles that medical research and practice should take into account (e.g., natural selection for drug resistance in pathogens), or methods for making inferences and discoveries (e.g., phylogenetic methods for tracing the spread of pathogens). *Some education in evolutionary biology is essential for every medical researcher*, because it bears directly on almost every field of biological study, and *is useful for every clinical practitioner*, because it deepens one's understanding both of the human body and its ills and of the organisms that cause harm [80]. *Evolutionary Medicine*, by S. C. Stearns and R. Medzhitov, is the best introduction to this large subject [108].

As you approach this topic, bear in mind that owing to genetic differences (as well as epigenetic and direct environmental effects), people vary in everything from susceptibility to inherited or pathogenic diseases to their reactions to drugs and other therapies. Moreover, many characteristics show genotype × environment interactions (see Chapter 15); for example, genetic variation in N-acetyl transferases (enzymes that break down some environmental toxins) affects the risk of developing cancer from smoking. Every physician should be aware that a particular therapy may have to be modified, or may be unsuitable, for some patients.

The term "disease" includes many kinds of ills. Below we mention (1) evolutionary legacies that characterize the human species; (2) mismatches between modern environments and those that prevailed during most of human evolution; (3) genetic diseases, caused by mutations—including both the ones we inherit and the ones that arise within our cells (somatic mutations); and (4) interactions with other organisms, including both our symbiotic microbiome and pathogens and parasites.

EVOLUTIONARY LEGACIES   Like almost all other eukaryotes, we have evolved a limited life span and functional breakdown as we age (senescence). The theory of life history evolution explains senescence mostly as the result of antagonistic pleiotropy: genes that enhance fitness in younger age classes but reduce function in older individuals (see Chapter 11). The theory predicts, and evidence shows, that many genes, affecting many functions, have this effect, so there is no single cause of senescence and ultimate death.

Many of our adaptations can go awry due to various stresses. For example, there is evidence that fever is an adaptation to suppress or kill pathogens, but extreme fever is dangerous; the same holds for allergic reactions such as sneezing and inflammation. The sensation of pain is a necessary adaptation for withdrawing from certain dangers, but it can become intolerable. Obesity, type II diabetes, and addictive behaviors represent normal, necessary functions carried to abnormal, harmful extremes. And our lives are subject to inherent genetic conflicts (see Chapter 12). The level of expression of paternal alleles, which enhance fetal extraction of nutrients through the placenta, is opposed by maternal alleles that are selected to prevent the fetus from extracting too many nutrients and lowering the mother's fitness. Imbalance between the expression of these genes can result in abnormalities, including Beckwith-Wiedemann and Prader-Willi syndromes.

MISMATCH WITH MODERN ENVIRONMENTS   Humans today inhabit environments that they have largely constructed. As we described in Chapter 21, the agricultural revolution created conditions to which the human body was not adapted, and it still isn't. Agriculture resulted in dietary changes that caused nutritional disorders and tooth decay; it increased exposure to new infectious diseases, acquired from domesticated animals; and sedentary life in villages created environmental conditions such as standing water, which increased the abundance of mosquitoes and of mosquito-borne diseases such as malaria. The Industrial Revolution, and especially the modern technological life that it made possible, has certainly had some positive effects on health (such as modern science and medicine), but it has given rise to "diseases of civilization," including cardiovascular disease, obesity, and type II diabetes. The best recipe for escaping these diseases is plenty of exercise, eating in moderation, and avoiding nutrient-poor, calorie-rich processed foods, with their starch and refined sugar.

We also noted in Chapter 21 that a diverse microbiome is essential for normal development of the immune system, and that urban environments and an obsession with "germs" reduce exposure to the diverse bacteria and other antigens that humans have experienced throughout evolutionary history [100]. As a result, autoimmune diseases such as inflammatory bowel disease, asthma, and severe allergies have become more prevalent. Similarly, disrupting the gut microbiome by antibiotics, especially in young children when the immune system is developing, greatly increases the risk of developing autoimmune diseases later [12]. Members of the human microbiome have some direct beneficial effects as well. Some directly benefit the human host; for example, the gut bacterium *Bacteroides fragilis* triggers gut cells to produce lymphoid tissue that prevents bacteria from penetrating the gut wall. Some microbial species suppress harmful bacteria, such as *Clostridium difficile*, which often takes over the intestine after excessive antibiotic treatment (see Figure 21.22).

GENETIC DISEASES   Although some inherited diseases have epigenetic causes, the great majority are caused by mutations in DNA sequences and chromosomal aberrations [105]. Most base pair mutations in humans enter the population through sperm, and the average mutation rate (about 10^{-8} per base pair) increases considerably during a man's life. Most nonsynonymous mutations slightly reduce fitness [41]. Many inherited maladies, such as psychiatric illnesses, have a polygenic basis [50], but specific genes and mutations have been identified for many inherited defects. Many of them are rare mutations that represent the opposition of mutation and purifying selection, but some—most famously those that enhance resistance to malaria—are alleles that have—or had, in the past—countervailing advantageous effects (see Chapter 5). In traditional societies, some antagonistically pleiotropic mutations that increased fitness at an early age may have expressed injurious effects late in life, at ages that few people attained in the past. Today, when the average human life span is much greater, the deleterious effects of these mutations are more commonly encountered. Possible examples include the *BRCA1* and *BRCA2* mutations that cause breast cancer and the *APOE4* mutation that may be advantageous for children's cognitive development but is associated with a greater risk of developing Alzheimer's disease later in life [87, 108].

CANCER   Somatic mutations—those that occur in cells that do not give rise to gametes—are the primary causes of cancers [72, 105]. Cancer is an evolutionary process: selection among genetically different cell lineages (clones) favors rapidly increasing lineages. Somatic mutations occur at a fairly high rate; by middle age, skin cells exposed to sun carry thousands of mutations. Cancers result only from mutations in certain "driver" genes that increase cell division, and several such mutations are required. As a tumor grows, mutations continue and mark descendant cells, so phylogenetic analysis can trace the history of a cancer, just as it can trace gene trees in populations and species (see Chapter 16). Such studies have shown that driver mutations may initiate cancers early in life, long before

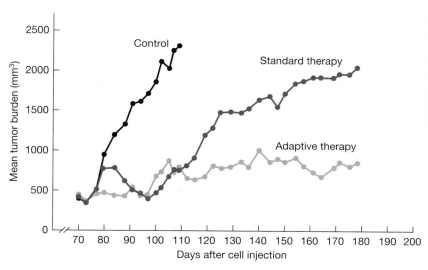

FIGURE 22.5 Tumor development in mice that were given a lower dose of therapeutic chemicals (adaptive therapy) was slower than in mice given the standard high-dose therapy. Adaptive therapy is based on the hypothesis that within a tumor, competition among clones may reduce the growth of aggressive clones if less aggressive clones are allowed to survive. (From [108], after [48].)

they become apparent; furthermore, a cancer often has had multiple origins. Animals have evolved several mechanisms that reduce the incidence of cancer, such as tumor suppressor genes and immune surveillance, but selection among cells favors mutations in a cancer genome that help cells evade these mechanisms, just as it does in pathogenic microbes. Several ideas about possible cancer therapies that are based on evolutionary principles are being investigated. One of these, *adaptive therapy*, aims to slow down the development of a cancer by using low doses of drugs to control the cancer, thus allowing less aggressive clones within the cancer to compete against more aggressive clones (**FIGURE 22.5**).

INFECTIOUS DISEASES   Some of the diverse bacteria, viruses, protists, and helminths that cause infectious diseases have been associated with human ancestors for millions of years; others (emerging diseases) are quite new. Pathogens that cause infectious diseases in humans range from those that are mostly endemic to other animals and only occasionally infect humans (e.g., rabies), through those that are animal-borne but can be transmitted among humans to a greater or lesser extent, to those that are specific to humans, such as the agents that cause smallpox, syphilis, and measles (**FIGURE 22.6**).

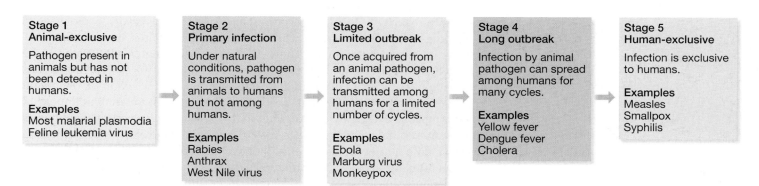

Stage 1
Animal-exclusive

Pathogen present in animals but has not been detected in humans.

Examples
Most malarial plasmodia
Feline leukemia virus

Stage 2
Primary infection

Under natural conditions, pathogen is transmitted from animals to humans but not among humans.

Examples
Rabies
Anthrax
West Nile virus

Stage 3
Limited outbreak

Once acquired from an animal pathogen, infection can be transmitted among humans for a limited number of cycles.

Examples
Ebola
Marburg virus
Monkeypox

Stage 4
Long outbreak

Infection by animal pathogen can spread among humans for many cycles.

Examples
Yellow fever
Dengue fever
Cholera

Stage 5
Human-exclusive

Infection is exclusive to humans.

Examples
Measles
Smallpox
Syphilis

FIGURE 22.6 The ecological relationships among pathogens, humans, and nonhuman hosts can be expressed as five stages of increasing cycling within human populations. Whether or not all pathogens in the higher stages have evolved through the lower stages is not yet known. Some exemplar pathogens are given for each stage. (After [122].)

In Chapter 13 we described the conditions under which parasites (including pathogenic microbes) are expected to evolve lower virulence (trade-off between virulence and transmission, vertical transmission from parent hosts to offspring, single-strain rather than multiple infections). Hosts suffer little mortality in some old parasite-host associations, such as between some primates and their simian immunodeficiency viruses (SIVs), due probably to evolution of both lower virulence and enhanced host resistance [27].

Four perspectives from evolutionary biology bear special mention. First, phylogenetic methods are now a standard tool in tracing the time and region of a pathogen's origin and its subsequent spread. For example, phylogenetic research showed that the 2014–2015 Ebola virus outbreak originated from a natural reservoir in Sierra Leone and was spread by human contact, undergoing rapid genetic diversification in the process (see Chapter 1) [25, 51, 113]. Second, mutations in influenza that change their antigenic phenotype enable such strains to increase rapidly and cause outbreaks. The problem is to predict which of the many genotypes in the virus population is likely to break out, so that vaccines can be constructed in advance. Researchers have made progress toward this goal, using a combination of fitness models and phylogenetic analysis [19, 71].

Third, progress is being made in using evolution to control the spread of infectious diseases, such as dengue fever and malaria, by mosquitoes or other vectors. One mode of selection at the level of the gene is segregation distortion by mechanisms such as meiotic drive (see Chapter 12). The idea is to introduce into a mosquito population a driving genetic element that will increase in frequency by gene-level selection, linked to a gene that will interfere with the ability of the mosquito to transmit the pathogen. The aim is not to eliminate the mosquito vector (which is nearly impossible in most situations) but to transform it so that it no longer carries the pathogen. Several model systems have been developed, such as using *Wolbachia*, a bacterial parasite of many insects, as the driving agent [57]. Some strains of *Wolbachia* spread rapidly through host populations by mechanisms that have the same effect as meiotic drive. The best progress, so far, has been the creation of a genetic construct with CRISPR-Cas9 that has been inserted into the genome of experimental mosquitoes [47]. The construct makes the mosquitos unable to carry malaria and is transmitted to 99.5 percent of their progeny.

Finally, and probably most important, is the evolution of antibiotic resistance, to which we have referred repeatedly in this book. Public health officials have warned that multidrug-resistant pathogens are a "nightmare" that poses a "catastrophic threat" to humans throughout the world [73]. Some bacteria that are common in hospitals have added resistance to carbapenems—drugs of last resort—to their resistance repertoire. (This has occurred in *Klebsiella pneumoniae*, one of several multidrug-resistant bacteria that are common in hospitals, which are environments that select for resistance. These bacteria caused about 90,000 deaths in the United States in 2004 [108].) There is some hope that negative genetic correlations in resistance to different drugs might be discovered; if so, selection for one might cause resistance to the other(s) to be lost [10]. But the most urgent and important priority must be to reduce the enormous, and often pointless, overuse of antibiotics that create natural selection for resistance. Antibiotics, developed for controlling bacteria, do not affect viruses, yet millions of antibiotic prescriptions are written for viral ailments such as the common cold, by doctors who respond to pressure from patients to "do something." Worse, and completely inexcusable, is that as much as half the antibiotic use in the United States is applied to farm animals—and 80 percent of that usage is to promote growth, not to improve the animals' health. There is powerful natural selection for mutations that provide resistance. These mutations can easily spread among species of bacteria by horizontal gene transfer (see Chapter 2) and end up in the worst human pathogens. (In December

2016, the U.S. Food and Drug Administration instituted a policy to phase out the use of antibiotics for promoting growth in livestock.)

INDIVIDUAL HEALTH AND PUBLIC HEALTH   Many characteristics that are advantageous to individual organisms are harmful at the population level, and vice versa (see Chapters 3 and 12). This conflict arises in public health. Farmers who use antibiotics to promote livestock growth may profit economically, but perhaps at great cost to the entire population. If most (say, 90 percent) of the members of a population are vaccinated against an infectious disease such as measles, the pathogen will not spread, and the few unvaccinated members will remain healthy, protected by "herd immunity." But if the proportion of unvaccinated individuals is too great, the pathogen will spread readily, leading to an outbreak. Outbreaks of some childhood diseases are correlated with geographic regions where a large percentage of people refused to let their children be vaccinated [86]. These situations are among the many in which there must be arbitration between what individuals consider to be their rights, and the greater good.

Evolution and Human Behavior

No topic in evolutionary biology is more intriguing or more controversial than the genetic and evolutionary foundations of human behavior. Resistance to hypotheses about human behavioral evolution, or even research on the subject, is widespread for several reasons. Many people are emotionally reluctant to see human abilities as extensions of those of other species, and they justify making a sharp distinction by pointing to the immense difference between the mental capacities of humans and those of any other mammal (see Chapter 21). Almost every aspect of our behavior varies greatly among individuals and among populations because of learning and cultural differences, so the hypothesis that there is any genetic basis for human behaviors is viewed with skepticism, especially by many social scientists. Finally, any intimation of biological determinism summons memories of *Social Darwinism*, a political philosophy that ascribed poverty, illiteracy, and crime to genetic inferiority, rather than the social conditions that exclude much of society from education and economic self-sufficiency [58]. Social Darwinism was developed by the philosopher Herbert Spencer, who coined the term "survival of the fittest" before Darwin wrote *On the Origin of Species*. Spencer was a fierce individualist who believed that competition was the driving force for improvement in nature and society. Darwin, who voiced the optimistic view that the evolution of cooperativeness in humans would lead to greater compassion and inclusiveness, did not espouse Spencer's view, but his idea of natural selection, and therefore his name, became applied to Spencer's social philosophy. Social Darwinism became linked with the idea of eugenics, a movement (largely initiated by one of Darwin's cousins) that advocated encouraging "superior" people to have more children and discouraging or preventing "inferior" people from doing so [63]. There is no basis for the common belief that Hitler used evolution to support his racist, anti-Semitic rhetoric; instead, he rejected the idea of Darwinian evolution and invoked racist ideologies that had a long pre-Darwinian history and were rampant in Europe [98]. It is true, though, that evolution has been used to support racist beliefs that some populations are "higher" or "superior" to others. These abuses were based on misunderstanding or twisting of the data and theory of evolution and genetics. To their credit, some evolutionary biologists and geneticists said so at the time, and in the last few decades they have been prominent in warning of misinterpretations of scientific information. A proper understanding of evolutionary biology, as of any science, is necessary to prevent it from being misused.

Today, both biologists and many social scientists recognize that human behaviors are affected both by our genetic evolutionary heritage and by culture, the product of the extraordinary human ability to think, learn, imagine, and speak.

Variation in cognitive and behavioral traits

Because genetic variation is the basis of evolution of any characteristic, questions immediately arise about the extent to which behavioral traits have a genetic foundation. These questions apply both to variation among individuals and among populations, and to supposedly "universal" human traits—"human nature." Variable traits can be analyzed by methods that partition the variance into genetic and environmental components (see Chapters 6 and 15). Learning and culture would contribute environmental variance. Among the traits studied this way, two are especially controversial: cognitive abilities, or "intelligence," and sexual orientation.

VARIATION IN COGNITIVE ABILITIES Cognitive abilities described as "intelligence" are measured by IQ ("intelligence quotient") scores. Psychologists distinguish various cognitive abilities (e.g., spatial ability, vocabulary), which contribute to a "general" factor (g) that is analogous to an overall "size" factor that captures correlated variation in different measurements on animal bodies (see Appendix regarding principal components analysis). Genetic and environmental components of variation can be difficult to distinguish in humans because family members typically share not only genes but also environments. For this reason, studies of people adopted as children are critically important. The genetic component of variation is estimated by correlations between twins or other siblings reared apart, or by adoptees' correlations with their biological parents. Because adoption agencies often place children in homes that are similar to those of their siblings, many modern studies try to measure such environmental correlations and take them into account. The environmental component of variance (V_E) can then be broken down into the variance due to shared environment and a residual environmental variance due to other, unmeasured influences. Twins have been important in human genetic studies, since monozygotic ("identical") twins should be more similar than dizygotic ("nonidentical") twins if variation has a genetic component. In addition to these correlational studies, researchers are starting to use genome-wide association studies (GWAS) to find genetic markers that are correlated with differences in IQ scores (e.g., [64]).

Several conclusions have arisen from many such studies [15, 82, 93]. First, general cognitive ability (g) has quite high heritability (h^2, or G/P, see Chapter 6). For example, h^2 was estimated as 0.70 in a study of 11-year-olds in Scotland and as 0.67 in 11-year-olds in Minnesota, and the variance ascribed to shared environments was 0.21 and 0.26, respectively, in the two studies. In these studies, IQ has about the same heritability as height. Second, in these and many other studies, h^2 increases with age: the effects of childhood environment are eroded over time. Third, the various cognitive abilities that have been compiled into a general g factor are strongly correlated with each other, with genetic correlations of about 0.60.

As we know from studies of phenotypic plasticity, a genetic basis for a trait does not mean that the trait is fixed or unalterable. Twin studies have suggested that the heritability of human height is 0.8 or more, yet in many industrial nations, mean height has increased considerably within one or two generations as a result of nutritional and other improvements. Similarly, IQ scores are increased greatly in children who have been adopted into homes that provide a richer, more stimulating learning environment [82]. By the same token, high heritability of variation *within* populations does not mean that differences *among* populations have a genetic basis, because the populations may have very different environments. This is important for interpreting the differences in average IQ scores among so-called "racial" or ethnic groups, such as the 15 points (1 standard deviation) that, until recently, separated the average scores of European-Americans and African-Americans. This difference is due to the populations' very different social, economic, and educational environments [74, 81]. That gap has been reduced by one-third in recent years, and studies of adopted children show that black and white children reared in the same environment have similar IQ scores [82]. One study found that the average IQ of German children fathered by white American soldiers during World War II was nearly identical to the IQ of those with black American fathers. All the evidence says that people of different ethnicities do not differ genetically in cognitive ability.

VARIATION IN SEXUAL ORIENTATION   Humans exhibit a bimodal continuum of sexual orientation, from exclusively heterosexual through bisexual to exclusively homosexual. (Sexual orientation has been studied much more in men than women.) At least a small percentage of men in almost all cultures studied, throughout the world and throughout history, have expressed homosexuality, and in some cultures this has been the norm [39]. It seems likely, then, that sexual orientation has been variable since before our species spread throughout the world. Homosexual behavior has been recorded in diverse animals, including more than 100 species of mammals, and is a subject of increasing attention among evolutionary biologists [95, 107]. Sexual orientation is not chosen, nor determined by childhood experience [121]; it has a largely biological basis, and has a heritability of about 0.2–0.4 [61, 68]. A genomic scan found two chromosome regions associated with male sexual orientation, one of which had been tentatively identified in an earlier study [102]. However, several lines of evidence suggest that homosexuality may not be based in DNA sequence, but might result if epigenetic marks that canalize sexual development in one sex became inherited by the opposite sex [97]. Epigenetic marks that influence a fetus's response to testosterone might "feminize" certain pathways in the brains of males, or "masculinize" those of females, and result in same-sex orientation.

To the extent that homosexual orientation has a genetic basis, the rather high frequency of homosexuality is evolutionarily enigmatic, since homosexuals are generally supposed to reproduce less than heterosexuals. (However, there is little evidence that this has been true for most societies throughout human history; even today, social expectations cause many homosexual people to marry and have children.) Assuming that male homosexuals do have a low average reproductive rate, several population genetic models could account for a stable polymorphism in sexual orientation [22]. The models that match the data best ascribe homosexual inclination to at least two loci, including at least one on the X chromosome, with effects such that the reproductive disadvantage of male homosexuality is balanced by the increased fecundity of females with the same X-linked allele. This situation would be an example of a polymorphism maintained [by sexually antagonistic selection (see Chapter 12)]. In northern Italy, gay men reported a significantly higher proportion of homosexual maternal relatives (5 percent) than paternal relatives (2 percent), consistent with inheritance on the X chromosome. Moreover, both the mothers and maternal aunts of gay men had significantly more children than those of heterosexual men (**FIGURE 22.7**), and had greater reproductive health [22, 24]. Genetic factors inclining men toward homosexual orientation may provide a reproductive advantage to women.

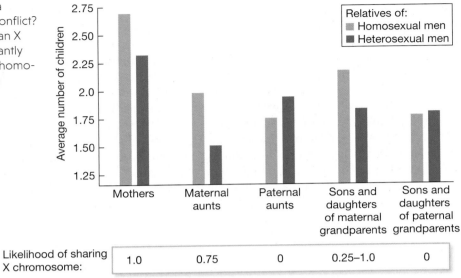

FIGURE 22.7 Does male homosexuality reflect a polymorphism maintained by intralocus sexual conflict? In northern Italy, women who are likely to share an X chromosome with a male relative have a significantly higher number of children if the male relative is homosexual than if he is heterosexual. (Data from [24].)

Likelihood of sharing X chromosome:	1.0	0.75	0	0.25–1.0	0

Human behavior: Evolution and culture

Darwin devoted a book, *The Expression of the Emotions in Man and Animals* [31], to the thesis that rudimentary homologues of human mental abilities and emotions can be seen in other species. Most animals are capable of learning, and we have seen that some nonhuman primates make tools and have a rudimentary ability to use language (see Chapter 21). Social interactions, including reciprocal aid and alliances, are quite elaborate in some primates, leading researchers to describe "baboon metaphysics" [28] and "chimpanzee politics" [36].

It would be contrary to everything we know about evolution to suppose that the hominid lineage has diverged so far from other apes that there should remain no trace of homology in the inherited components of our brain organization and behavior, or that the human brain should be a completely blank slate, without genetically determined predispositions. Evolutionary psychologists have compiled a long list of supposedly universal, and thus perhaps genetically determined, human behavioral tendencies and capacities, such as fear of snakes, body adornment, cooperation, death rituals, division of labor, sex differences in aggression and dominance, marriage, and the capacity for language [17, 92]. But everyone recognizes that the expression of most such behaviors is culturally highly variable. All humans speak a grammatical language, for example, but grammar and vocabulary vary greatly. In almost all species, many traits express phenotypic plasticity (see Chapter 6), the capacity of a genotype to express different phenotypes depending on environmental conditions. Thus, it could be the case that our ancestors evolved a *propensity* to respond to certain conditions in specific ways—aggressively or cooperatively, for example—but that the expressed behavior depends on cultural and other aspects of the environment.

Several schools of evolutionary research on human behavior have developed in the last few decades [66]. *Human sociobiology,* announced by E. O. Wilson in 1975, proposed to interpret a wide range of behaviors (such as conflicts between offspring and parents) as adaptations that had evolved by natural selection, especially kin selection and reciprocity [119]. This approach has been succeeded by several other movements. One is *human behavioral ecology,* which uses adaptive models to predict and explain a variety of behaviors, including cultural norms. For instance, we may ask why, in polygynous societies, a woman would choose to marry a man who already has one or more wives. Adopting a model that was first developed to explain polygyny in birds, Monique Borgerhoff-Mulder proposed, and found evidence, that women decide on the basis of the resources they may expect a man to provide [14]. Human behavioral ecologists do not necessarily assume that a specific behavior is genetically determined; they may assume, instead, that humans have evolved the cognitive abilities to respond adaptively to various environmental circumstances. The response may have been learned, perhaps from a cultural norm, but this norm is itself adaptive [2].

Complementing this approach, Leda Cosmides, John Tooby, and others have developed the active (and controversial) field of *evolutionary psychology* [9, 21, 92, 123]. Evolutionary psychologists propose that over the course of human evolution, especially during the Pleistocene, specific adaptations, conceived as mental "organs" or modules, evolved to solve different classes of problems, especially social challenges such as choosing mates and detecting cheaters. These psychological mechanisms are assumed (as in psychology generally [83]) to underlie universal human capabilities that have evolved by natural selection. Some of these cognitive capacities are highly flexible, enabling people to adjust in novel ways to novel circumstances.

Some research in evolutionary psychology uses psychological methods to test hypotheses about the postulated adaptive modules. For example, Cosmides and Tooby proposed that in social exchanges among unrelated individuals (reciprocity), cheating is an ever-present threat, so that evolved mechanisms for detecting cheaters should have design features that are not activated in nonsocial contexts [29]. They presented college students with problems that had the same logical form but different content, and found that the students solved the problem more often if it described cheating than if it did not. Many evolutionary psychologists use a cross-cultural approach, on the supposition that a postulated behavioral adaptation that is much the same among culturally very different populations may be an evolved trait. For example, David Buss reasoned that because reproduction entails a far greater commitment and investment of resources by women than by men, women should have evolved to seek mates who are likely to provide resources, while men, as a consequence of sexual selection, might be expected to place more value on young, physically attractive mates who are likely to be fertile [20]. This sounds like the epitome of sexism, yet Buss reported that a large majority of 37 diverse cultures conformed to the predictions.[4] For example, women are said to value earning potential more than men do in 97 percent of the cultures, and men prefer women younger than themselves in 100 percent. Similarly, many of the physical features that are considered appealing by the opposite sex are consistent across very different cultures, and some seem to be indicators of reproductive fitness [46]. A match between data and theoretical prediction is used to accrue support for hypotheses in many evolutionary studies, and in science generally. Many researchers do not accept the concept of mental modules that Cosmides and Tooby introduced, but these are not a critically important part of a broader evolutionary approach to psychology. Still, it remains difficult, in many cases, to determine whether human behaviors are best explained by evolutionary (genetic) adaptation or by cultural effects.

We have seen (see Chapter 21) that evolutionary concepts and some methods of evolutionary analysis have been applied to cultural evolution. Cultural evolution and genetic evolution of behavior have been joined in models of *gene-culture coevolution* [26, 43], or *dual-inheritance theory* [16, 99]. At one level, this field studies how culture affects the selective milieu within which genetic evolution occurs; for example, lactase persistence has evolved in several cultures with milk-based diets,

---

[4] An interesting essay that attempts to answer skeptics about this topic is "Yes, but... Answers to Ten Common Criticisms of Evolutionary Psychology," by D. Schmitt (https://evolution-institute.org/wp-content/uploads/2016/03/20160307_evopsych_ebook.pdf).

and genome studies are revealing many genes that have undergone recent selection that is thought to stem from changes in diet, the invention of cooking, sexual selection, and migration into novel climates (see Chapter 21) [67]. This approach also includes mathematical analyses of how cultural and genetic variation might interact.

Surely the most interesting problem addressed by these models is what accounts for the unique level of organization and cooperation in human societies [65]. Complex social relationships must have required that an individual be able to recognize many other individuals and remember their past interactions; they may have required the ability to interpret the state of mind and intentions of others; they were probably facilitated by punishing cheaters; and they may have been fostered by selection for the ability to form coalitions against stronger "bullies" [42, 49]. In this context, natural selection may have led to humans' extraordinary capacity for altruism, which extends beyond the limits that kin selection or reciprocity readily explain [13]. Many authors, beginning with Darwin, have proposed that multi-level selection—selection among groups or tribes, stemming from competition and warfare—played a major role in the evolution of the extraordinary human capacity for cooperation and altruism. Peter Richerson and Robert Boyd propose that such group selection was facilitated by processes of cultural evolution, such as pressure to conform to the group's norms, that reduce behavioral variation within groups and increase the variance among groups [99]. These processes would make group selection more effective than it usually is in purely genetic models (see Chapters 3 and 12).

Natural selection may have favored genotypes with a greater predisposition to adopt certain kinds of cultural traits (such as greater cooperativeness or tendency to conform to group norms), if these cultural traits enhanced survival or reproduction of individuals or of groups. To some extent, then, some behavioral features such as those postulated by evolutionary psychologists might have evolved, but in cultural contexts that themselves could change. But such features are predispositions only, and they allow for an immense range of cultural expressions and individual potential.

Cultural evolution and genetic evolution together may help us understand some of the most distinctly human traits: art, music, and religion [18, 37, 52, 118]. Many hypotheses have been suggested to account for them. Music, for example, has been considered a nonadaptive side effect of other mental features, or as an adaptation shaped by sexual selection or by natural selection to reinforce social bonds within groups [59]. Religious belief may be grounded in innate cognitive mechanisms that seek causes for events, and envision supernatural agents to account for natural dangers and to protect against them. When an idea of this kind becomes prevalent in a culture, it may in turn strengthen social bonds and cultural identity, reinforcing cooperation and enabling the group to compete with other groups [84].

Understanding nature and humanity

"All art," said Oscar Wilde, "is perfectly useless." Many people may disagree with him, but his larger point was that art is a human creation that needs no utilitarian justification, a creation that is justified simply by being an expression—one of the defining characteristics—of humanity.

Much of what is most meaningful to us is "perfectly useless": music, sunsets, walking on a beach, baseball, soccer, movies, gardening, spiritual inspiration—and understanding. Whether the subject is mathematics, the natural world, philosophy, or human nature, attempting to understand is rewarding in itself, aside from whatever practical consequences it may yield.

To know about the extraordinary diversity of organisms, about the complexities of the cell, of development, or of our brains, and about how these marvels came to be, is deeply rewarding to anyone with a sense of curiosity and wonder. To have achieved such knowledge is, like other advances in science and technology, among humanity's great accomplishments. Likewise, to have some understanding, however imperfect, of what we humans are and how we came into existence is richly rewarding. It is fascinating and ennobling to learn of our 3.5-billion-year-old pedigree, of when and how and possibly why our ancestors evolved the characteristics that led to our present condition, of how and when modern humans emerged from Africa and colonized the rest of Earth, of how genetically unified all humans are with one another, and yet how genetically diverse we are. It is both challenging and important to try to understand "human nature"—to understand how our behavior is shaped by our genes and therefore by our evolutionary past, and how it is shaped by culture, social forces, and our unique individual history of learning and experience. Evolution may challenge our view of life and its meaning. Some see in it a dark denial of purpose, while others find that Darwin "re-enchants the world," providing "a way of knowing that is deeply human, saturated with value and feeling, and rigorously honest" [70].

Evolution is the unifying theme of the biological sciences and an important foundation for the "human sciences" of medicine, psychology, and sociology. Psychologists and anthropologists may differ among themselves on the role of evolution in determining "human nature," but most will agree that some knowledge of evolutionary principles is essential for understanding their subject. And although evolutionary biologists and social scientists do not set social policy, they can speak out against abuses of their science. They can point out misunderstandings of evolutionary theory, such as racist interpretations of differences among human populations, or the "naturalistic fallacy" that what is natural is good: the false justification of Social Darwinism, of the belief that homosexuality is wrong because it does not lead to reproduction, and of the ideology that women should be subservient to men. Science can play an important role in the ever-necessary defense of human rights and justice.

Evolution has neither moral nor immoral content, and evolutionary biology provides no philosophical basis for aesthetics or ethics. But evolutionary science, like other knowledge, can serve the cause of human dignity by helping us relieve disease and hunger, and appreciate both the unity and the diversity of humankind. And it can enhance our appreciation of life in all its magnificent diversity.

Go to the
——— **Evolution Companion Website** ———
EVOLUTION4E.SINAUER.COM
for data analysis and simulation exercises, quizzes, and more.

SUMMARY

- Evolution is a fact—a hypothesis that is so thoroughly supported that it is extremely unlikely to be false. The theory of evolution is not a speculation, but rather a complex set of well-supported hypotheses that explain how evolution happens.

- There is a great range of views on whether or not religion and evolution—or religion and science generally—are compatible. Especially in the United States, many reject evolution and instead accept divine creation because they think evolution conflicts with their religious beliefs. The positions taken by creationists on issues such as the age of Earth and of life vary.

- Science is tentative; it accepts hypotheses provisionally and changes them in the face of convincing new evidence. It is concerned only with testable hypotheses; it depends on empirical studies that are subject to peer scrutiny and that can be verified and repeated by others. Creationism has none of the features of science, so it has no claim to be taught in science classes.

- The evidence for evolution comes from all realms of biology and geology, including comparative studies of morphology, development, life histories, and other features, as well as molecular biology, genomics, paleontology, and biogeography. Evolutionary principles can explain features of organisms that would not be expected of a beneficent intelligent designer, such as imperfect adaptation, useless or vestigial features, extinction, selfish DNA, sexually selected characteristics, conflicts among genes within the genome, and infanticide. Furthermore, all the proposed mechanisms of evolution have been thoroughly documented, and evolution has been observed.

- It is important to understand evolution not only because it has broad implications for how we think about nature and humanity, but also because it has many practical ramifications. Evolutionary science contributes to many aspects of medicine and public health, agriculture and natural resource management, pest management, and conservation.

- One of the most difficult and controversial challenges is to join biological and social science in order to understand how the distinctively human cognitive and behavioral characteristics evolved, the extent to which human behaviors have an evolved genetic foundation, how that foundation interacts with cultural and other environmental factors to shape individual behavior, and how genes and culture have coevolved.

TERMS AND CONCEPTS

argument from design
creationist

intelligent design (ID)
natural laws

special creation
theistic evolution

SUGGESTIONS FOR FURTHER READING

Books on creationism, ID, and defending evolution and science:

Why Evolution Is True, by Jerry A. Coyne (Oxford University Press, New York, 2009), and *The Greatest Show on Earth*, by Richard Dawkins (Free Press, New York, 2009), are outstanding, well-written descriptions of the evidence for evolution and its mechanisms by leading evolutionary biologists.

Evolution: What the Fossils Say and Why it Matters, by D. R. Prothero (Columbia University Press, New York, 2007), is a valuable introduction to the subject by an authority on vertebrate paleontology.

Evolution vs. Creationism: An Introduction, second edition (Greenwood Press, Westport, CT, 2009), was written by Eugenie C. Scott, director of the National Center for Science Education, who probably knows more about the evolution-creationism controversy than anyone else.

Defending Evolution in the Classroom: A Guide to the Creation/Evolution Controversy, by B. J. Alters and S. M. Alters (Jones and Bartlett, Sudbury, MA, 2001), is a perceptive analysis of how to present and teach evolution.

Denying Evolution: Creationism, Scientism, and the Nature of Science, by M. Pigliucci (Sinauer Associates, Sunderland, MA, 2002), shows why creationism fails as science, but addresses the limits of science as well.

Science, Evolution, and Creationism (National Academies Press, Washington, D.C., 2008) is a 70-page booklet issued by the most prestigious scientific organization in the United States, the National Academy of Sciences, and the Institute of Medicine of the National Academies. It is available at www.nap.edu.

Books and articles that address the relationship between religion and evolution:

Finding Darwin's God: A Scientist's Search for Common Ground between God and Evolution (HarperCollins, New York, 1999), by Kenneth Miller, a cell biologist at a leading university, argues that one can fully accept evolution and reconcile it with religion. See also his later book, *Only a Theory: Evolution and the Battle for America's Soul* (Viking, New York, 2008).

Faith vs. Fact: Why Science and Religion Are Incompatible, by Jerry Coyne (Viking, New York, 2015), rigorously argued and well written by a prominent evolutionary biologist, is the best exposition of this thesis.

Living with Darwin: Evolution, Design, and the Future of Faith, by Philip Kitcher (Oxford University Press, New York, 2007), is a short reflection by an eminent philosopher on reconciling evolution with our need for meaning in life, whether this is provided by religion or other sources of fulfillment.

Denial of unwelcome science applies not only to evolution, but also to climate change. For the science, see *Climate Change: Evidence and Causes*, by The Royal Society and the U.S. National Academy of Sciences (National Academies Press, Washington, D.C., 2014; http://www.nap.edu/catalog/18730/climate-change-evidence-and-causes).

Several excellent websites provide information about evolution and can serve as valuable teaching aids:

BioInteractive: Evolution (www.hhmi.org/biointeractive/evolution-collection), from the Howard Hughes Medical Institute, is an outstanding collection of short films and interactive features about evolution.

Understanding Evolution (http://evolution.berkeley.edu) is an outstanding site, developed by the University of California Museum of Paleontology to provide content and resources for teachers at all grade levels.

The National Center for Science Education (www.ncse.com) actively supports the teaching of evolution and combats creationism, as well as climate-change denial. This is the most comprehensive website on the conflict, and it provides links to a great range of resources.

The TalkOrigins Archive (www.talkorigins.org) has a wealth of material on many aspects of evolution and the social controversy. It includes a comprehensive list of creationist claims and rebuttals to them, by Mark Isaak, that is also available in book form (M. Isaak, *The Counter-Creationism Handbook*, University of California Press, Berkeley, 2007).

Ken Miller's Evolution Resources (www.millerandlevine.com/km/evol/) includes text, video clips of interviews, and other material, especially on debunking creationist claims and on Miller's position that religion and evolution are compatible.

David Sloan Wilson's site This View of Life (https://evolution-institute.org/this-view-of-life/) is largely devoted to implications of evolution for the social sciences and humanities.

The rap artist Baba Brinkman offers a quite different approach to the topic of evolution and creationism at www.bababrinkman.com.

International Darwin Day (www.darwinday.org) describes annual educational events held around the world on or near the anniversary of Darwin's birth (February 12).

Some references on applications of evolutionary biology are:

The Evolving World: Evolution in Everyday Life, by D. P. Mindell (Harvard University Press, Cambridge, MA, 2006), and *Pragmatic Evolution: Applications of Evolutionary Theory*, edited by A. Poiani (Cambridge University Press, Cambridge, UK, 2012).

On evolution and health, leading textbooks include *Evolutionary Medicine*, by S. C. Stearns and R. Medzhitov (Sinauer, Sunderland, MA, 2016), *Evolution and Medicine*, by R. L. Perlman (Oxford University Press, Oxford, 2013), and *Principles of Evolutionary Medicine*, by P. Gluckman, A. Beadle, and M. Hanson (Oxford University Press, Oxford, 2009). Also see *Missing Microbes: How the Overuse of Antibiotics Is Fueling Our Modern Plagues*, by M. J. Blaser (Henry Holt and Co., New York, 2014).

A huge literature concerns the evolution of human behavior. A very good introduction to the major current approaches is *Sense and Nonsense: Evolutionary Perspectives on Human Behaviour*, by K. N. Laland and G. R. Brown (Oxford University Press, Oxford, 2011). *Evolutionary Psychology: An Introduction*, by L. Workman and W. Reader (Cambridge University Press, Cambridge, 2008), is a comprehensive textbook on this subject. *The Blank Slate:*

The Modern Denial of Human Nature, by Steven Pinker (Penguin Books, New York, 2002), is a well-written exposition and defense of evolutionary psychology, for a general audience. A very readable, convincing treatment of culture and human evolution is *Not by Genes Alone: How Culture Transformed Human Evolution*, by P. J. Richerson and R. Boyd (University of Chicago Press, 2005), leading figures in this field. Simon Reader has compiled a useful annotated bibliography on the evolution of cognition at www.oxfordbibliographies.com/view/document/obo-9780199941728/obo-9780199941728-0028.xml.

Appendix:
A Statistics Primer

Variation is the stuff of evolution. Darwin made the breakthrough discovery that natural selection cannot work without variation among individuals. Understanding evolution therefore requires ways to measure and analyze variation. That need stimulated R. A. Fisher and other evolutionary biologists to lay the foundations of modern statistics. Since then, statistics has grown immensely, both in what it is able to do and the number of areas where it is used.

We use statistics in two basic ways. The first is to describe things. When we say that the mean weight of male African elephants (*Loxodonta africana*) is about 7000 kg, we are conveying information about what a typical elephant is like. A second use of statistics is to test hypotheses. Are female elephants smaller than males? They are: females weigh on average 3600 kg. Statistics tells us that we can be very confident that the difference between male and female weights is real, and not simply because we happened to measure some unusually large males and some unusually small females.

This appendix starts by introducing the key concept of a probability distribution. We then briefly review how statistics are used to describe populations, estimate quantities, and test hypotheses. We end with a brief overview of two major frameworks of statistical analysis, likelihood and Bayesian inference. While statistics is a branch of mathematics, this appendix keeps the math to a minimum. For more details, with many examples drawn from evolutionary biology, we recommend the excellent text by Whitlock and Schluter [2].[1]

---

[1] Statistics can be intimidating because it is such a large and technical field. Luckily for evolutionary biologists, the text by Whitlock and Schluter [2] is clear, friendly, and (best of all) filled with examples from evolutionary biology.

Probability Distributions

Physics tells us that all protons in the universe are identical. There is absolutely no uncertainty about the properties of the next proton you will encounter. But the same is not true of kangaroos.

A **probability distribution** describes how frequent different kinds of things are, or how likely the different outcomes for some future event are. About 90 percent of people are right-handed, and 10 percent are left-handed. This is an example of a **discrete distribution** because it describes the frequencies of distinct (discrete) categories. A discrete distribution can have more than two categories. The frequencies of people with blue, dark brown, brown, green, and hazel eyes are described by a distribution with five categories. From the distribution of eye colors shown in **FIGURE A.1**, we see that the frequency in the United States of individuals with dark brown eyes is 0.25, or one-quarter of the population. The frequencies of all of the outcomes in a distribution must sum to 1.

A different kind of distribution is needed to describe traits such as body height that do not fall into discrete categories. These are described by **continuous distributions**. A familiar example is the *normal distribution*, sometimes called a Gaussian distribution or bell curve (**FIGURE A.2**). We saw an example that was visualized with living students in Figure 6.2. A normal distribution has a single peak, and it falls off symmetrically to the left and right of the peak. The distribution shown in Figure A.2C tells us (for example) that there are many more women who are about 165 cm tall than individuals who are about 180 cm tall. The normal distribution is just one type of continuous distribution. A continuous distribution can be asymmetrical, and it can have more than one peak. In some human populations, the distribution of heights has two peaks because it is a mixture of females, who tend to be shorter, and males, who tend to be taller.

Continuous distributions are interpreted differently than discrete distributions. In Figure A.2C, we see that the value of the probability density for female heights at 164 cm is 0.055. That does not imply, however, that 5.5 percent of females are that height. In fact, *no* females are *exactly* 164.0000 cm tall. What the distribution conveys is the relative probability that a height will be close to a given value. For example, the probability is roughly twice as large that a female's height is close to 165 cm than that it is close to 155 cm. The distribution can be used to find the probability that the height of a randomly chosen female will fall in a given range. For example, the probability that her height will be between 164 cm and 170 cm tall is given by the corresponding area beneath the curve (that is, the integral from 164 cm to 170 cm). For females in the United States, that probability is 0.29 (or 29 percent; see Figure A.2C). The total area under the curve must again sum to 1.

Descriptive Statistics

A distribution often has more information than we really need. We might want to know how large male elephants typically are, but not care how often the weight of a male elephant falls between 7123 kg and 7125 kg. A **descriptive statistic** is a number that summarizes a useful fact about a distribution.

The most common (and most familiar) descriptive statistic is the **mean**, also called the arithmetic mean or average. Dutch men are the tallest in Europe, with a mean height of about 183 cm (6 feet). That statistic is enough to immediately convey the fact that many of us will spend a lot of time looking up at tall people if we visit

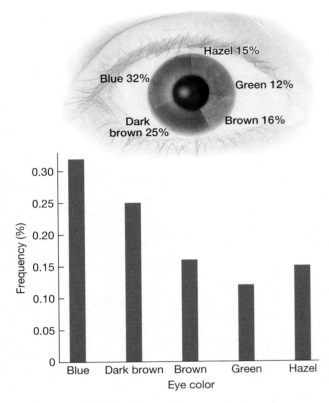

FIGURE A.1 The frequencies of eye colors can be represented by a discrete distribution, shown here for eye color in the United States.

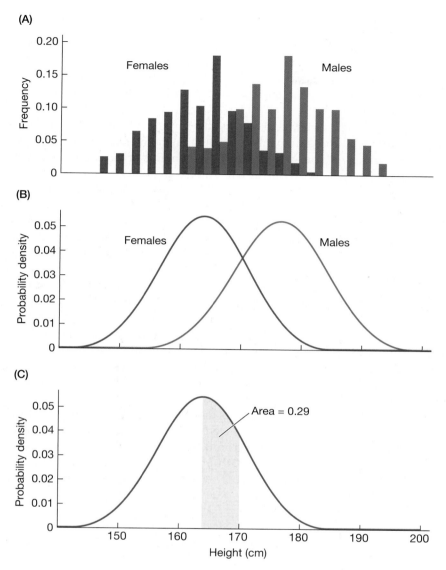

FIGURE A.2 The frequencies of height in humans can be represented by a continuous distribution. (A) Histograms for the heights of a sample of women and men ages 20–29 in the United States. Here the data are represented as a discrete distribution, with the height of each bar showing the frequency of individuals that fell within a range of heights. (B) The distributions of heights of women and men in the U.S. population are well represented by normal distributions. Here the y-axis represents the relative probabilities that an individual has a height very close to a given value on the x-axis. The general equation for a normal distribution with mean $\bar{x}$ and variance σ^2 is

$$f(x) = \frac{1}{\sqrt{2\pi\sigma^2}} \exp\left\{\frac{-(x-\bar{x})^2}{2\sigma^2}\right\}$$

In women, the mean height is 164 cm, and the variance is 54 cm². In men, the mean height is 176 cm, and the variance is 58 cm². (C) The area under the female distribution between 164 cm and 170 cm equals 0.29. This shows that 29 percent of females in this population have heights within that range.

Amsterdam. To calculate the mean of a set of measurements, we simply add them together and then divide that total by the number of measurements. By convention, we often symbolize the mean of a distribution by putting a bar over the symbol used for the measurement. Say that we use x to represent the height of an individual. In a group of Dutch men, if half of them have a height of $x = 180$ cm and the other half have a height of $x = 186$ cm, then their mean height is $\bar{x} = 183$ cm.

The **variance** is a statistic that measures the spread of a distribution around the mean. Evolution depends critically on variation, and so variance plays a central role in evolutionary biology. (In fact, the word "variance" was invented by R. A. Fisher in the first scientific publication in population genetics [1].) A variance is often symbolized by σ^2. Variance is defined as

$$\sigma^2 = \text{Mean value of } (x-\bar{x})^2 \qquad\qquad \text{(A.1)}$$

Continuing with the Dutch men, for the shorter individuals we have $(x-\bar{x})^2 = (180 \text{ cm} - 183 \text{ cm})^2 = 9 \text{ cm}^2$. For the taller individuals, we have $(x-\bar{x})^2 = (180 \text{ cm} - 183 \text{ cm})^2 = 9 \text{ cm}^2$. By taking the mean of 9 cm² and 9 cm², we find that the variance in height for this group of men is $\sigma^2 = 9 \text{ cm}^2$.

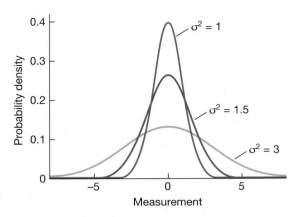

FIGURE A.3 These three distributions have the same mean but different variances. As the variance increases, the spread of the distribution around the mean increases.

Just what does a variance tell us? A variance equal to zero says that there is no spread around the mean: all the measurements are identical. If one distribution has a larger variance than another, then it has more dispersion around the mean (**FIGURE A.3**). Notice the units in the example above: they are the units of measurement *squared*. A variance therefore can never be negative.

Often several factors contribute to the variance of a variable. Whenever we work with measurements, the data inevitably include measurement error, which will be small if the measurements are accurate, but large if not. The variance in our measurements then has contributions from two sources: the variance in the actual variable, and the variance in the errors. Statistics provides methods to estimate and correct for the errors (for example, by remeasuring some of the individuals). In Chapter 6 we discuss another situation in which multiple factors contribute to a variance: the phenotypic variance for a quantitative trait has both genetic and environmental components.

A second useful statistic that describes variation is the **standard deviation**. It is the square root of the variance and is often symbolized as σ. The standard deviation in the example of Dutch men is $\sigma = \sqrt{9 \text{ cm}^2} = 3$ cm. A standard deviation has the same units as the original measurements. For a normal distribution, about two-thirds (68 percent) of the distribution falls within 1 standard deviation of the mean, that is, between $(\bar{x} - \sigma)$ and $(\bar{x} + \sigma)$, and 95 percent falls within 2 standard deviations of the mean, between $(\bar{x} - 2\sigma)$ and $(\bar{x} + 2\sigma)$ (**FIGURE A.4A**).

The standard deviation is useful for measuring the difference between the means of two distributions. If the difference is much less than a standard deviation of one of the distributions, then it can be difficult for us to see by eye that the distributions really are different (**FIGURE A.4B**). A difference of more than 1 standard deviation is typically easy to see. The means of the distributions for male

FIGURE A.4 The standard deviation, σ, is a measure of variation. (A) For normal distributions, about 2/3 (68 percent) of the distribution falls within 1 standard deviation above and below the mean (top), and 95 percent falls within 2 standard deviations of the mean (bottom). (B) The difference between the means of two distributions can be measured in terms of their standard deviations. When the difference is much less than σ, the distributions overlap greatly (top). When the difference in their means is much greater than σ, much of the distributions do not overlap (bottom).

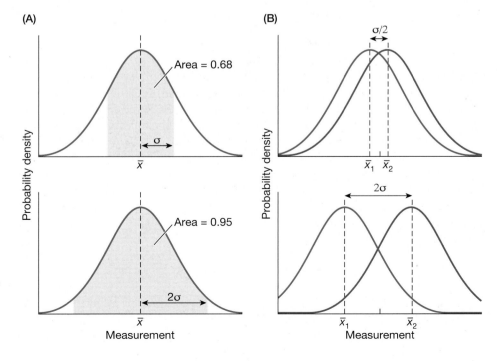

and female height shown in Figure A.2 differ by about 1.5 standard deviations. For some purposes, the standard deviation is a more useful unit than the original units of measurement. If a species evolves to become 1 mm larger, is that a lot of change or not? It is a microscopic change for a species of elephant, but an enormous change for a species of ant. In contrast, if a species has evolved to become 5 standard deviations larger, we know immediately that it is very different than what it used to be, whether it is an elephant or an ant.

We often are interested in the relationship between variation in two or more traits. Humans with long arms tend also to have long legs, simply because those individuals are larger than average overall. **Covariance** is a basic measure of the association between two measurements. The covariance between variables x and y is often written as σ_{xy}, and is defined as

$$\sigma_{xy} = \text{Mean value of } (x - \overline{x})(y - \overline{y}) \tag{A.2}$$

where x and y represent the two measurements, and $\overline{x}$ and $\overline{y}$ are their means. A covariance is positive if the measurements tend to increase and decrease together, and it is negative if one measurement tends to get smaller as the other becomes larger.

Another way to measure the association between two variables is the **correlation**, symbolized as r. A correlation is a covariance that has been rescaled so that it has no units, and ranges from a minimum value of –1 to a maximum value of 1. The correlation between variables x and y is defined as

$$r = \frac{\sigma_{xy}}{\sigma_x \sigma_y} \tag{A.3}$$

where σ_x and σ_y are the standard deviations of x and y. A positive correlation means that individuals that are larger for one trait also tend to be larger for the second, as with arm and leg length. A negative correlation implies that individuals that are larger than average for one trait tend to be smaller than average for the second trait. A correlation of $r = 0$ means there is no simple relation between the two measurements: individuals that are larger than average for the first are equally likely to be either smaller or larger than average for the second. At the other extreme, a correlation of $r = 1$ tells us that the value of one measurement is perfectly associated with the value of the second. Examples of correlations are shown in **FIGURE A.5**.

A **regression** predicts the value of one variable from the value of another. The most common kind of regression fits a line to the points in a plot of the two

FIGURE A.5 Correlation measures how two variables vary together. (A) A correlation of $r = 0$ means there is no simple relation between the two variables. (B) A negative value of r means that larger than average values of one variable tend to be associated with smaller than average values of the other. (C) A positive value of r means that larger than average values of one variable are associated with larger than average values of the other. The largest possible value for a correlation is $r = 1$, which means the two variables are perfectly associated.

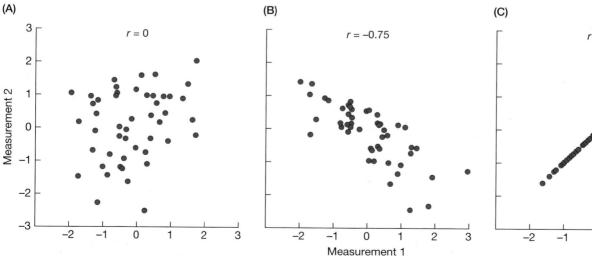

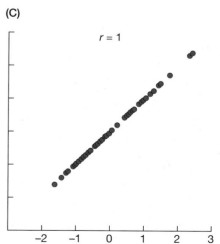

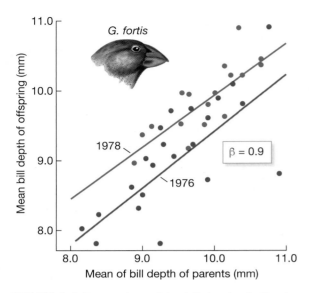

FIGURE A.6 Regressions of the bill depth of offspring on the bill depth of their parents in the Galápagos finch *Geospiza fortis* in 1976 and 1978. Each point represents a family, with the average value of the parents plotted on the *x*-axis and the average of their offspring on the *y*-axis. Although offspring were larger in 1978, the regression coefficients (the slopes of the lines) were nearly the same in both years: β = 0.9. (This value estimates the heritability (h^2) of bill depth—see Chapter 6.) The bill depth of an offspring can be predicted from that of its parents: find the mean of the two parents on the *x*-axis, move upward from there to the regression line, then move left to the *y*-axis. That value is the expected bill depth of the offspring. (After Grant 1986, based on Boag 1983.)

variables (**FIGURE A.6**). We saw in Chapter 6 that regression is used to estimate selection gradients (see Figure 6.11) and heritability (see Figure 6.14). A regression line is fit so that the sum of the squared differences between the line and the values of the variable on the *y*-axis is minimized. The slope of this line, which is called the regression coefficient and is often symbolized as β, tells us how rapidly the value of the second variable typically increases or decreases with the value of the first variable. The regression coefficient of measurement *y* on measurement *x* is defined as

$$\beta = \frac{\sigma_{xy}}{\sigma_x^2} \tag{A.4}$$

The range of possible values that a regression coefficient can take is unbounded: in principle it can range from $-\infty$ to $+\infty$. The units of a regression coefficient are units of the *y* variable divided by units of the *x* variable. A regression predicts the average value of *y* for a given value of *x*, while a correlation conveys how tight the association is between *x* and *y* without making a prediction about one from the other.

Principal components are another tool used to describe and analyze how different measurements vary together. The first principal component, or PC1, is a line fit to the data so that its orientation is along the direction that has the greatest amount of variation. (This line is not equal to the regression line but often is close to it.) The second principal component, or PC2, is fit using two rules: it must be perpendicular to the first principal component, and it must run in the direction that has the greatest amount of remaining variation. **FIGURE A.7** shows an example in which the measurements are arm length and leg length, which are strongly correlated ($r = 0.8$).

Principal components have several uses. One is to simplify data analysis by reducing the number of variables that need to be analyzed. When two variables are strongly correlated, as in Figure A.7, the location of any point can be quite accurately described by one number

FIGURE A.7 Principal components describe how two or more measurements vary together. This example shows the relation between arm length and leg length in a hypothetical population of humans. The first principal component, PC1, runs along the direction that has the greatest amount of variation. In this case, it corresponds to overall size: individuals with small values of PC1 are small overall (lower left), while those with large values of PC1 are large overall (upper right). The second principal component, PC2, is perpendicular to PC1. Here this axis of variation corresponds to the relative sizes of the arm and leg: individuals with small values of PC2 have long arms and short legs (lower right), while those with large values have long legs and short arms (upper left). The lengths of the lines for PC1 and PC2 are proportional to the amount of variation (specifically, the standard deviation) in that direction.

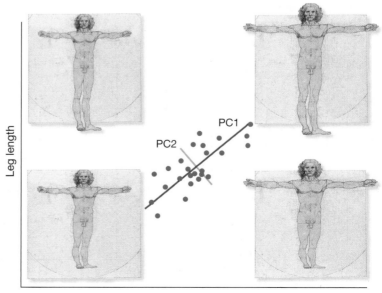

rather than two: the distance between the point and the mean along PC1. The description is not perfect because the points fall a little way off the line. By working with those distances rather than the original pairs of measurements, we reduce the size of the data set by half. In **FIGURE A.8**, an individual is identified who has a much larger than average value for PC1 (and a slightly smaller than average value of PC2). Using just his value for PC1 describes most of the differences between him and the average individual in the population.

In some situations we have more than two measurements on each individual (for example, measures of arm length, leg length, and body mass). In that case, additional principal components are fit in the same way as the first two were. Each new principal component must be perpendicular to those that have already been fit, and it must run in the direction that has the most remaining variation (**FIGURE A.9**). With three variables (as in this example), if we represent each individual in the data set only by his value for PC1, we shrink the number of measurements for each individual from three to one. That decreases the number of variables we need to analyze by two-thirds. If there are hundreds of measurements per individual, the savings are much larger yet.

Estimation

Asian elephants (*Elephas maximus*) are smaller than African elephants. The mean weight of a male Asian elephant is 5000 kg, which is about 2000 kg less than its African kin. How do we know those facts? Of course, nobody has weighed all the elephants, which is what would be needed to know the true mean weight of these two species. Instead, researchers have taken the weights of a number of elephants, and from those data they **estimate** the means for the two species.

When we estimate a mean or other quantity, the group of individuals that are measured is called the **sample**, and the group from which the sample comes is called the **population**. Without measuring all individuals in the population, there is always some uncertainty about the actual mean of the population. Statistics lets us quantify that uncertainty.

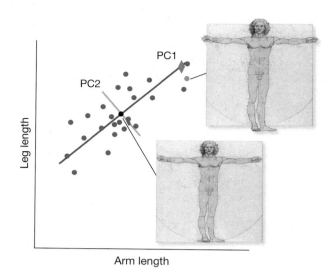

FIGURE A.8 The position of a point along PC1 gives a good approximation of its location. An example is shown for the point highlighted in red, which represents the man shown at upper right. The population's mean is shown by the black point, which corresponds to a man that looks like the one shown at lower right. The position of the red dot along PC1 is shown by the green diamond. By using only the distance along PC1 from the mean to the green diamond, rather than the two original measurements of arm length and leg length, we can reduce by half the number of variables needed to describe the man at upper right.

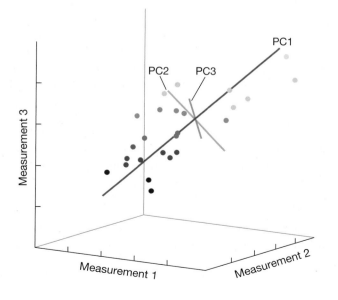

FIGURE A.9 Principal components can be used with three or more variables. To make their location in space more clear, the points are colored according to their distance from the viewer (darker points are closer). The first principal component (PC1) runs in the direction in which there is the most variation. The second principal component (PC2) is perpendicular to PC1, and runs in the direction in which there is the most remaining variation. The third principal component (PC3) is perpendicular to PC1 and PC2. The lengths of the lines for the principal components are again proportional to the amount of variation (the standard deviation) in that direction.

FIGURE A.10 Our confidence in detecting a difference between two populations grows with the size of the samples. The three panels show sample sizes (*n*) of 5, 25, and 250 females and males drawn from the distributions shown in Figure A.2. The horizontal lines show the means of the samples.

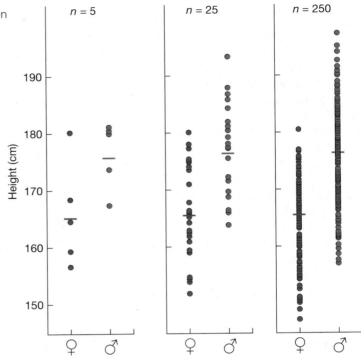

The larger a sample we have, the more confident we can be about the estimates we make about the population. This is seen in **FIGURE A.10**, which shows samples of heights drawn from the distributions for females and males shown in Figure A.2. With only five individuals of each sex, it is not clear whether the difference in their means is because males really are taller than females, or because by chance we happened to measure women who are shorter and men who are taller than average. But with 250 individuals of each sex, it's obvious that the difference is real.

Testing Hypotheses

White-tailed deer (*Odocoileus virginianus*) live north to Alaska and south to Peru. Like many other mammals, deer tend to be larger the farther they are from the equator, a pattern known as Bergmann's rule (see Chapter 8). One hypothesis to explain this pattern is that populations in colder climates have evolved larger body sizes as a way to conserve heat.

To test that hypothesis, we might see if the body size of deer also varies with elevation. The heat conservation hypothesis predicts that deer populations at high elevations should be larger than those at low elevations because temperature tends to decrease with elevation. Imagine that you have a friend in Colorado who reports that the weight of a single adult male deer living at 1600 m above sea level is 75 kg. Another friend living near sea level in Texas reports that a single adult male there weights 60 kg. Are those data strong support for the hypothesis? No, because there is a plausible alternative hypothesis. Perhaps on average there is no difference between deer in Colorado and Texas, but by chance the deer that was weighed in Texas happened to be smaller than the one weighed in Colorado. After requesting that your friends each weigh a total of 20 deer, you find that all but 1 of the deer weighed in Colorado are heavier than the deer weighed in Texas. You would now be very confident that the deer living in Colorado are heavier, which is consistent with the heat conservation hypothesis.

Statistics can quantify how certain we are that a difference between two sets of measurements represents a real difference between the two populations from which the measurements came, rather than an accident of sampling. A calculation shows that the probability is less than 1 in a billion that by chance so many of the deer from Colorado would be heavier than those from Texas if the two populations in fact did have the same mean.

This example illustrates a fundamental principle about how statistics is used to make inferences. Statistics cannot prove a hypothesis, it can only *reject* one. Our samples do not let us calculate the true mean weights of all the deer living in Colorado and Texas. It is therefore impossible to know with certainty that deer in Texas are heavier on average. We can, however, determine the probability that they are different.

To do that, we begin with a **null hypothesis** that we will seek to reject. We then calculate the probability that if the null hypothesis were true, then our data would produce a result as extreme as or more extreme than what our data show. This probability is called the P value. The smaller the P value, the more confident we are that the null hypothesis is false. In evolutionary biology, it is conventional to reject a null hypothesis if the P value is less than 0.05. We say that a conclusion is **statistically significant** if it meets that threshold.

To see how this idea is used, let's return to the deer example. Our null hypothesis is that the weights of deer in Texas and Colorado have the same distribution. If that were true, the probability that what we see in the two samples—all but one of the deer weighed in Colorado are heavier than those in Texas—turns out to be 1.5×10^{-10}. An even more extreme outcome would be that *all* the deer weighed in Colorado are heavier than those in Texas, and under the null hypothesis the probability of that outcome is 1.5×10^{-30}. Adding together the two probabilities gives us the P value, which is 1.5×10^{-10}. In other words, there is a 99.999999985 percent chance that the null hypothesis is wrong. The P value is far smaller than the threshold of 5 percent, so we can conclude that deer in Colorado are highly significantly heavier than those in Texas.

Several approaches are used to test null hypotheses. The most common strategy is to use statistical tests that make assumptions about the distributions in the populations that are being sampled. Tests of this sort that you may have encountered already are the chi-square test and the t-test. The appropriate choice of which test to use depends on the nature of the null hypothesis and the data. (The text by Whitlock and Schluter has the details [2].) Returning to the example of heights shown in Figure A.10, a t-test for the samples of five females and males (left panel) reports the probability that those two distributions have the same mean is $P = 0.06$, which is not statistically significant. In contrast, the probability that the samples of 250 females and 250 males (right panel) have the same mean is $P = 3.7 \times 10^{-50}$, which is highly statistically significant.

A second strategy for testing hypotheses statistically is called **randomization**. Say that we have weights of 14 deer from Texas and 19 deer from Colorado, and the difference in their mean weights is 10 kg. We can use a computer to randomly assign the 33 weights in this data set to two groups, one of size 14 (representing a sample of deer from Texas) and the other of size 19 (representing a sample from Colorado). We then record the difference between the weights in these two groups. The aim of this procedure is to simulate what we might see under the null hypothesis that there in fact is no difference in the distributions of weights in Colorado and Texas. By repeating this randomization thousands of times, we determine how often the difference in means of the two groups of randomized data is as large as, or larger than, what we actually observed (**FIGURE A.11**). If the difference in the randomized data is as big as in the real data less than 5 percent of the time, we conclude that the difference in our sample is statistically significant.

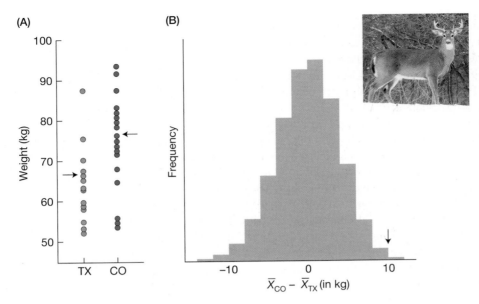

FIGURE A.11 Randomization is a powerful way to test statistical hypotheses. In this example, we ask whether deer in Colorado are heavier than deer in Texas. (A) The weights of 14 deer from Texas and 19 deer from Colorado are shown. The mean weight of the Texas deer is $\bar{x}_{TX} = 67$ kg, and the mean weight of the Colorado deer is $\bar{x}_{CO} = 77$ kg (means indicated by the two arrows). (B) The null hypothesis is that that the distribution of weights is the same in Texas and Colorado. Randomizing the data 10^6 times produces the distribution of the difference between the means $(\bar{x}_{CO} - \bar{x}_{TX})$ under that null hypothesis. The actual difference observed, shown by the arrow, is extremely unlikely. The probability of a difference greater than what is actually seen in the data is given by the area under the histogram to the right of the arrow. That is much less than 5 percent, the standard threshold for statistical significance. We reject the null hypothesis that deer in Texas and Colorado have the same weight on average, and conclude that the population of deer in Colorado is heavier on average than the population in Texas.

A final point is that it is critical to distinguish between *statistical significance* and *biological significance*. Two populations of deer might have very different mean sizes, but with a small sample size we would not be able to prove statistically that they are different. Conversely, with enormous sample sizes it is possible to prove that two populations have different mean sizes, even if the difference is so small that it is irrelevant to the biological question of interest. Deciding how large an effect must be in order to qualify as "biologically significant" is the job of the investigator, and no statistical analysis can determine that. The most useful inferences are made when an effect is statistically significant and also large enough to be biologically interesting.

Likelihood

Likelihood is an important branch of statistics used to estimate properties of a population and to test hypotheses. In statistics, "likelihood" is defined as the probability of observing the data that we have, given assumptions for how the data were generated.

Imagine that we sample ten platypuses from a river and find that they have 4 copies of allele A_1 and 16 copies of allele A_2. We can use likelihood to find the probability of that sample if the actual frequency of allele A_1 in the population is a given value, for example $p_1 = 0.5$. Probability theory tells us that, if the allele frequency in the population is p_1, then the likelihood that in a random sample we would get n_1 copies of A_1 and n_2 copies of A_2 is:

$$L(n_1, n_2 \,|\, p_1) = \frac{(n_1 + n_2)!}{n_1! n_2!} \, p_1^{n_1} (1 - p_1)^{n_2} \tag{A.5}$$

where $x!$ stands for $x \times (x-1) \ldots 3 \times 2 \times 1$. (This equation comes from the binomial distribution, which often appears in statistics.) With that formula, we find that if the actual frequency of A_1 in the population is $p_1 = 0.5$, then the likelihood of our data is $L = 0.0046$.

FIGURE A.12 shows how the likelihood (that is, probability of the observed data) varies as p_1 ranges from 0 to 1. This is called the *likelihood function*. The likelihood function reaches its greatest value, $L = 0.22$, with $p_1 = 0.2$. That is, the data are most likely if that is the true frequency of A_1 in the population we sampled. This is called the **maximum likelihood estimate** of the allele frequency. In this example, the maximum likelihood estimate corresponds to common sense: it equals the frequency of A_1 in our actual sample of genes ($4/20 = 0.2$). In other situations, the maximum likelihood estimate cannot be found from an average or other simple summary statistic.

The complete likelihood function gives us more information than just its maximum. It also conveys the range of values of p_1 that are plausible. The maximum likelihood estimate suggests that the frequency of A_1 is somewhere near 0.2, but it is almost certainly not exactly equal to 0.2. It is often useful to consider the **confidence interval**, which is the range of values in which the real value of p_1 is very likely to lie. A rule commonly used is to determine the range of values of p_1 for which the likelihood L is no more than seven times smaller than the maximum likelihood. We can be 95 percent certain that the true value of p_1 lies within that range. In our example, the maximum likelihood is $L = 0.22$, so we seek the value of p_1 that gives a likelihood that is at least equal to $0.22 / 7 = 0.031$. Figure A.12 shows that range of values is from $p_1 = 0.07$ to $p_1 = 0.41$. We are 95 percent confident that the true value of the allele frequency lies somewhere in that range.

This example illustrates two of the major applications of the likelihood approach: using maximum likelihood to estimate something about the population (such as its mean), and finding the confidence interval for that quantity. Likelihood is used for a broad range of problems in evolutionary biology, such as estimating phylogenies and effective population sizes. The key requirement is that we be able to calculate the probability of the data given assumptions about how they were produced.

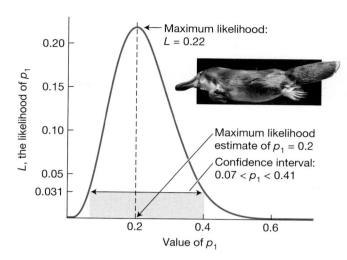

FIGURE A.12 The likelihood function for p_1, the frequency of allele A_1 in the population, given that we have a sample with 4 copies of allele A_1 and 16 copies of allele A_2. The likelihood reaches a maximum value of $L = 0.22$ when $p_1 = 0.2$. The confidence interval is the range of values in which we are 95 percent sure that the true value of the allele frequency lies (the shaded box). It corresponds approximately to values of p_1 that give likelihoods no less than seven times smaller than the maximum likelihood, that is, L greater than $0.22 / 7 = 0.031$. The confidence interval ranges from $p_1 = 0.07$ to $p_1 = 0.41$.

Bayesian Inference

An alternative to likelihood that is increasingly used in many areas of evolutionary biology is **Bayesian inference**. The goal here is to find the probability that the allele frequency (or other variable) in the population is equal to any given value. There are two main motivations for using the Bayesian approach. The first is to make use of information that we already have. Likelihood has no way of combining prior information with new data, but Bayesian inference does. With little or no new data, Bayesian estimates rely heavily on the prior information. But as more and more new data are gathered, they are given more and more weight. With enough new data, the prior information has a negligible effect on our estimate.

Say, for example, that after sampling ten alleles from platypuses living in the first river, we move to a second river nearby. We think that platypuses migrate back and forth between the rivers, so we expect allele frequencies in the two populations to be similar. We can therefore use our first sample to form a **prior**

probability distribution, which says how likely we think different values are for the allele frequency at locus A in the second population. In this example, we'll use the likelihood function shown in Figure A.12 for that purpose. The choice of the prior distribution is made by the investigator, and in some cases it can be based simply on an intuition about what values are most plausible.

We now sample alleles from the second river. We calculate the likelihood function for this new sample in the same way as we did earlier for the first river. We then calculate the **posterior probability distribution**, which is simply the product of the new likelihood function and the prior probability distribution. This posterior distribution is the main goal of a Bayesian analysis. It tells us the relative probability that the actual allele frequency in the population is any given value. This is the key difference with likelihood estimation (discussed in the previous section), which gives us the likelihood that we would obtain our sample if the actual allele frequency were a given value.

As with the likelihood function, the posterior probability distribution does not require us to rely on a single estimate for the allele frequency. It says how probable any given value is, and we can evaluate that information however we like. If we want a single estimate of the allele frequency, a good value to use is the one corresponding to the maximum of the posterior distribution. (This is the Bayesian analog of the maximum likelihood estimate.)

To make the platypus example more specific, imagine that (unknown to us) the actual frequency in the second population is $p_1 = 0.4$. After sampling just four alleles, we find one copy of A_1 and three copies of A_2. Because the sample size is so small, this does not give us much new information. The likelihood function, again calculated using Equation A.5, is quite flat (**FIGURE A.13A**). The posterior probability distribution is found by multiplying this likelihood function and the prior distribution. Because the likelihood function is so flat, this posterior distribution is very similar to the prior distribution. The peak in the posterior distribution corresponds to $p_1 = 0.21$, which we can use as the estimate for the frequency of the A_1 allele in the second river.

Now say that we capture more platypuses, and find a total of 8 copies of A_1 and 12 copies of A_2 from the second river. The likelihood function for the data is more strongly peaked because the larger sample size gives us more confidence in the actual frequency in the second population (**FIGURE A.13B**). The posterior distribution (again given by the product of the prior distribution and the likelihood function) now estimates that the frequency of A_1 is $p_1 = 0.3$, closer to the true value of $p_1 = 0.4$. Finally, after sampling still more platypuses, we have 37 copies of A_1 and 63 copies of A_2 from the second river. The posterior distribution is now even more strongly peaked, and nearly centered on the true allele frequency of $p_1 = 0.4$ (**FIGURE A.13C**). Our estimate for the frequency of A_1 is now $p_1 = 0.34$. With even more data, the estimate would tend to move even closer to the true value of the allele frequency.

This example shows how the Bayesian approach combines prior information with new data. Another motivation for using Bayesian methods is simply practical. Many problems in evolutionary biology, such as estimating phylogenetic trees, involve extremely complicated likelihood functions. They cannot be analyzed in the relatively straightforward way that we used in the discussion of allele frequencies in the previous section. A strategy based on Bayesian methods can save the situation. The basic idea is illustrated in the following example.

Say that we are interested in using DNA sequences to estimate the age of a node (branching point) in a phylogeny of several species. We use a computer to randomly sample a possible value for that age from a prior distribution that we assume. We then calculate the likelihood of our sequence data using that value (along with assumptions about how rapidly the sequences evolve). We now draw a second random value for the age of the node, and again calculate the likelihood.

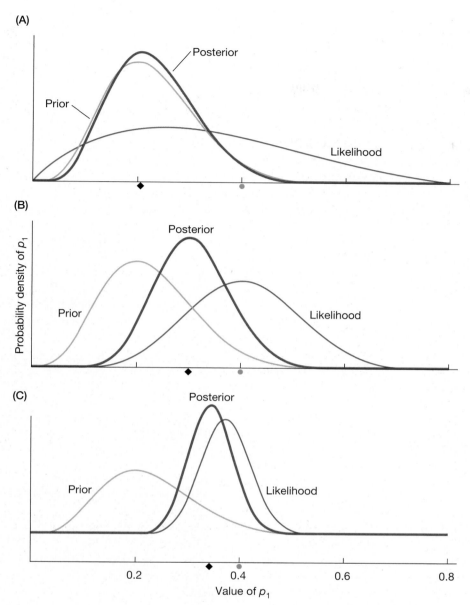

FIGURE A.13 Bayesian estimates for the frequency of allele A_1 in a second population of platypuses. The likelihood function from the first population (Figure A.12) is used for the prior distribution. The actual frequency in the second population is $p_1 = 0.4$ (the red circle). (A) A sample of just four alleles from the second population has one copy of A_1 and three copies of A_2. The resulting likelihood function is quite flat. The posterior distribution (equal to the product of the prior distribution and the likelihood function) is very similar to the prior distribution. The peak in the posterior distribution, is $p_1 = 0.21$ (the black diamond). (B) With a sample of 20 alleles, we have 8 copies of A_1 and 12 copies of A_2. The likelihood function is more strongly peaked because of the larger sample size. The posterior distribution now estimates that the frequency of A_1 is $p_1 = 0.3$. (C) With a sample of 100 alleles, we have 37 copies of A_1 and 63 copies of A_2. The posterior distribution is now even more strongly peaked, and nearly centered on the true allele frequency of $p_1 = 0.4$. Our estimate for the frequency of A_1 is now $p_1 = 0.34$.

We compare the two likelihoods, and then use a probabilistic rule (see https://en.wikipedia.org/wiki/Metropolis_Hastings_algorithm) that tells us whether to keep the first value for the node's age and discard the second, or to do the reverse. We record the age that is retained, then repeat the process. After thousands (or even millions) of repetitions, the distribution of ages that we retained will very closely resemble the posterior probability distribution for the node's age. We can use the distribution of retained values to estimate the node's age (the age that has the greatest probability) and the confidence interval for that estimate.

This method is one of several that collectively are called Markov Chain Monte Carlo, often abbreviated as MCMC. In this example, the aim is to estimate a single quantity (the age of a node). In practice, the method is typically used to do much more ambitious jobs, such as simultaneously estimating the branching pattern of the phylogeny, the ages of all its nodes, and the rates of sequence evolution.

SUMMARY

- A probability distribution describes the frequencies of different events or kinds of things. A distribution can be either discrete or continuous.
- Two of the most commonly used descriptive statistics are the mean, which describes the average individual in a population, and the variance, which describes the amount of dispersion around the mean.
- A correlation measures the degree to which two kinds of measurements vary together.
- A regression predicts the value of one variable from the value of another. The regression coefficient is the slope of a regression line.
- When we have two or more measurements on each individual, principal components are used to simplify analyses by reducing the size of the data set.
- Statistics is used to estimate properties of a population (such as the mean and variance) based on a sample from it.
- Statistics can measure our confidence in a null hypothesis. We reject the null hypothesis if there is less than a 5 percent probability that the data would result from it. If we reject the null hypothesis that the means of two populations are equal, we say that they are significantly different.
- The likelihood is the probability that the data would be produced, given a specific assumption for how the data were produced. Likelihood can be used to estimate properties of a distribution such as its mean and variance, to test hypotheses, and to determine confidence intervals (the range of plausible values for a property of a population, such as its mean).
- Bayesian inference combines prior information about the distribution of a variable with new data. As with likelihood, it is used to make estimates, test hypotheses, and so on. A second use of Bayesian inference is to estimate quantities when the likelihood function is too complex to analyze directly.

TERMS AND CONCEPTS

Bayesian inference
confidence interval
continuous distribution
correlation
covariance
descriptive statistic

discrete distribution
estimate
maximum likelihood estimate
mean
null hypothesis
population

posterior probability distribution
principal component
prior probability distribution
probability distribution

randomization
regression
sample
standard deviation
statistically significant
variance

Glossary

Most of the terms in this glossary appear at several or many places in the text of this book. Many terms that are used broadly in biology or are used in this book only near their definition in the text are not included here.

A

absolute fitness   *See* **relative fitness**.

active trend   *See* **driven trend**.

adaptation   A process of genetic change in a population whereby, as a result of natural selection, the average state of a character becomes improved with reference to a specific function, or whereby a population is thought to have become better suited to some feature of its environment. Also, *an* adaptation: a feature that has become prevalent in a population because of a selective advantage conveyed by that feature in the improvement in some function.

adaptive landscape   A metaphor for the relationship, or mathematical function, between mean fitness of a population and the allele frequencies at one or more loci that affect fitness. Possible populations with allele frequencies that maximize mean fitness are represented as peaks on the metaphorical landscape.

adaptive peak   That allele frequency, or combination of allele frequencies at two or more loci, at which the mean fitness of a population has a (local) maximum. Also, the mean phenotype (for one or more characters) that maximizes mean fitness. An **adaptive valley** is a set of allele frequencies at which mean fitness has a minimum.

adaptive radiation   Evolutionary divergence of members of a single phylogenetic lineage into a variety of different adaptive forms; usually the taxa differ in the use of resources or habitats, and have diverged over a relatively short interval of geological time. The term **evolutionary radiation** describes a pattern of rapid diversification without assuming that the differences are adaptive.

adaptive valley   *See* **adaptive peak**.

adaptive zone   A set of similar **ecological niches** occupied by a group of (usually) related species, often constituting a higher taxon.

additive effect   The magnitude of the effect of an allele on a character, measured as half the phenotypic difference between homozygotes for that allele compared with homozygotes for a different allele.

additive genetic variance   That component of the **genetic variance** in a character that is attributable to additive effects of alleles.

allele   One of several forms of the same gene, presumably differing by mutation of the DNA sequence. Alleles are usually recognized by their phenotypic effects; DNA sequence variants, which may differ at several or many sites, are usually called **haplotypes**.

allele frequency   The proportion of gene copies in a population that are a given allele; i.e., the probability of finding this allele when a gene is taken randomly from the population; also called **gene frequency**.

allometric growth   Growth of a feature during ontogeny at a rate different from that of another feature with which it is compared.

allopatric   Of a population or species, occupying a geographic region different and separated from that of another population or species. *Cf.* **parapatric**, **sympatric**.

allopatric speciation   Speciation by genetic divergence of allopatric populations of an ancestral species; contrasted with **parapatric** and **sympatric speciation**, in which divergence occurs in parapatry or sympatry (q.v.).

allopolyploid   A **polyploid** in which the several chromosome sets are derived from more than one species.

allozyme   One of several forms of an enzyme encoded by different alleles at a locus, that are distinguished by gel electrophoresis.

alternative mating strategies   Different mating behaviors and morphologies that are maintained as a stable polymorphism by negative frequency-dependent selection.

alternative splicing   Splicing of different sets of exons from RNA transcripts to form mature transcripts that are translated into different proteins (thus allowing the same gene to encode different proteins).

altruism   Conferral of a benefit on other individuals at an apparent cost to the donor.

anagenesis   Evolutionary change of a feature within a lineage over an arbitrary period of time.

ancestral character state   An evolutionarily older character state, relative to another (derived) state that has evolved from it in one or more lineages.

aneuploid   Of a cell or organism, possessing too many or too few homologous chromosomes, relative to the normal (**euploid**) set.

anisogamy   The condition of having two types of gametes of different sizes and forms, one large and immobile (the egg), the other small and usually mobile (the sperm).

antagonistic pleiotropy   Contrasting effects of a gene on two different characters, such that the effect of an allele substitution on one character increases fitness, but the effect on the other character decreases fitness.

antagonistic selection   A source of natural selection that opposes another source of selection on a trait.

apomixis   Parthenogenetic reproduction in which an individual develops from one or more mitotically produced cells that have not experienced recombination or syngamy.

apomorphic   Having a **derived** character or state, with reference to another character or state. *See* **synapomorphy**.

aposematic   Coloration or other features that advertise noxious properties; warning coloration.

artificial selection   Selection by humans of a deliberately chosen trait or combination of traits in a (usually captive) population; differing from natural selection in that the criterion for survival and reproduction is the trait chosen, rather than fitness as determined by the entire genotype.

asexual   Pertaining to reproduction that does not entail meiosis and syngamy.

assortative mating   Nonrandom mating on the basis of phenotype; usually refers to positive assortative mating, the propensity to mate with others of like phenotype.

autopolyploid   A **polyploid** in which the several chromosome sets are derived from the same species.

autosome   A chromosome other than a sex chromosome.

B

back mutation   Mutation of an allele back to the allele from which it arose by an earlier mutation.

background extinction   A long-prevailing rate at which taxa become extinct, in contrast to the highly elevated rates that characterize mass extinction.

background selection   Elimination of deleterious mutations in a region of the genome; may explain low levels of neutral sequence variation.

balancing selection   A form of natural selection that maintains polymorphism at a locus within a population.

base pair substitution   As usually used in this book, a base pair that, because of genetic drift or natural selection, has replaced another base pair at a specific DNA site in a population or species.

behavioral ecology   The study of the evolution of behaviors, often in relation to the environment, including other members of the same species.

behavioral isolation   *See* **sexual isolation**.

benthic   Inhabiting the bottom, or substrate, of a body of water. *Cf.* **planktonic**.

biogeographic realm   Major geographic regions of Earth that have characteristic animal and plant taxa.

biogeography   The study of the geographic distribution of organisms.

biological homology   Commonality of different traits, among or within species, based on a shared genetic basis and developmental pathway; the traits are often, but not always, homologous in the usual phylogenetic sense. *See* **homology**.

biological species   A population or group of populations within which genes are actually or potentially exchanged by interbreeding, and which are reproductively isolated from other such groups.

bottleneck   A severe, temporary reduction in population size.

breeder's equation   The equation that predicts that the evolutionary change in the mean of a quantitative trait resulting from selection in one generation is equal to the product of the trait's **heritability** and the **selection differential**.

C

C-value paradox   The lack of correlation between the DNA content of eukaryotic genomes and a given organism's phenotypic complexity (i.e., the genome of a less complex eukaryotic organism, such as a plant, may contain far more DNA than that of a more complex organism, such as a human). The paradox is explained by the amount of noncoding DNA sequences in a genome.

Cambrian explosion   The first appearance in the fossil record of many animal phyla, within a relatively short (<20 million years) interval.

canalization   The evolution of internal factors during development that reduce the effect of perturbing environmental and genetic influences, thereby constraining variation and consistently producing a particular (usually wild-type) phenotype.

candidate gene   A gene postulated to be involved in the evolution of a particular trait based on its mutant phenotype or the function of the protein it encodes.

carrying capacity   The population density that can be sustained by limiting resources.

category   In taxonomy, one of the ranks of classification (e.g., genus, family). *Cf.* **taxon**.

cDNA (complementary DNA)   A DNA copy of an mRNA made using reverse transcriptase isolated from a retrovirus.

cDNA library   A collection of cDNAs, representing the transcriptome (all of the mRNAs expressed) of a tissue or whole organism at a particular life history stage, created by isolating cDNA, cloning it into circular DNA plasmids and propagating it in bacterial cells.

character   A feature, or trait. *Cf.* **character state**.

character displacement   As originally used, a pattern of geographic variation in which a character differs more greatly between sympatric than between allopatric populations of two species; now often used for the evolutionary process of accentuation of differences between sympatric populations of two species as a result of the reproductive or ecological interactions between them.

character state   One of the variant conditions of a character (e.g., yellow versus brown as the state of the character "color of snail shell").

cheat   In **behavioral ecology**, a behavior that allows individuals to receive the fitness benefit of the altruistic behavior of others without paying the fitness cost that the altruistic individuals pay.

chimeric gene   A gene that consists of parts of two or more different ancestral genes.

chronospecies   A segment of an evolving lineage preserved in the fossil record that differs enough from earlier or later members of the lineage to be given a different binomial (name). Not equivalent to biological species.

cis-regulatory element   A noncoding DNA sequence in or near a gene required for proper spatiotemporal expression of that gene, often containing binding sites for transcription factors. *Cf.* **control region**, *trans*-**regulatory element**.

clade   The set of species descended from a particular ancestral species.

cladistic   Pertaining to branching patterns; a cladistic classification classifies organisms on the basis of the historical sequences by which they have diverged from common ancestors.

cladogenesis   Branching of lineages during phylogeny.

cladogram   A branching diagram depicting relationships among taxa; i.e., an estimated history of the relative sequence in which they have evolved from common ancestors.

cline   A gradual change in an allele frequency or in the mean of a character over a geographic transect.

clonal interference   During the increase in frequency of two different beneficial alleles or genotypes, the elimination of one by another that has greater fitness.

clone   A lineage of individuals reproduced asexually, by mitotic division.

coadapted gene pool   A population or set of populations in which prevalent genotypes are composed of alleles at two or more loci that confer high fitness in combination with each other, but not with alleles that are prevalent in other such populations.

coalescence   Derivation of the gene copies in one or more populations from a single ancestral copy, viewed retrospectively (from the present back into the past).

codon   A nucleotide triplet that encodes an amino acid or acts as a "stop" signal in translation.

codon bias   Nonrandom usage of synonymous codons to encode a given amino acid.

coefficient of relationship   The probability that an allele carried by one individual is also carried by a related individual.

coefficient of selection   The proportion by which the average fitness of individuals of one genotype differs from that of a reference genotype.

coevolution   Strictly, the joint evolution of two (or more) ecologically interacting species, each of which evolves in response to selection imposed by the other. Sometimes used loosely to refer to evolution of one species caused by its interaction with another, or simply to a history of joint divergence of ecologically associated species.

commensalism   An ecological relationship between species in which one is benefited but the other is little affected.

common ancestor   A lineage (often designated as a taxon) from which two or more descendant lineages evolved.

common garden   A place in which (usually conspecific) organisms, perhaps from different geographic populations, are reared together, enabling the investigator to ascribe variation among them to genetic rather than environmental differences. Originally applied to plants, but now more generally used to describe any experiment of this design.

comparative genomics   Analysis of similarities and differences between the genomes of different species.

comparative method   A procedure for inferring the adaptive function of a character by correlating its states in various taxa with one or more variables, such as ecological factors hypothesized to affect its evolution.

competition   An interaction between individuals of the same species or different species whereby resources used by one are made unavailable to others.

competitive exclusion   Extinction of a population due to competition with another species.

competitive exclusion principle   The theoretical assertion that one of two ecologically identical species will eventually replace the other by competition.

concerted evolution   Maintenance of a homogeneous nucleotide sequence among the members of a gene family, which evolves over time.

condition-dependent indicator   A characteristic, usually used in behavioral display, that is correlated with, and therefore indicates, the health or physiological vigor ("condition") of an individual.

conflict   In **behavioral ecology**, interactions between individuals that increase the fitness of one individual at a cost to the fitness of the other. **Altruistic** and **spiteful** interactions include conflict.

congeneric   Belonging to the same genus.

conservative characters   Features that evolve slowly and are retained with little or no change for long periods of evolutionary time.

conspecific   Belonging to the same species.

constraints   Properties of organisms or their environment that tend to retard evolution of a feature or to direct its evolution along some paths rather than others.

control regions   Untranscribed regions of the genome to which products of other genes bind, and which determine transcription of specific genes.

convergent evolution (convergence)   Evolution of similar features independently in different evolutionary lineages, usually from different antecedent features or by different developmental pathways.

cooperation   In **behavioral ecology**, interactions between individuals that are either **mutualistic** or **altruistic**, so that the actor enhances the fitness of the recipient.

co-option   The evolution of a function for a gene, tissue, or structure other than the one it was originally adapted for. At the gene level, used interchangeably with **recruitment** and, occasionally, **exaptation**.

Cope's rule   A proposed generalization that individual body size in animals tends to increase during evolution.

copy number variants   Refers to variation among conspecific individuals in the number of duplicates (copies) of a DNA sequence.

correlated selection   Natural selection for specific combinations of traits, such that selection on one trait is correlated with selection on the other.

correlation   A statistical relationship that quantifies the degree to which two variables are associated. For *phenotypic correlation, genetic correlation, environmental correlation* as applied to the relationship between two traits, see Chapter 6.

cost   A reduction in fitness caused by a correlated effect of a feature that provides an increment in fitness (i.e., a benefit).

cost of reproduction   Reduction of an individual's future fitness (survival and/or future reproduction) caused by reproductive activity.

cost of sex   Usually refers to a reduced rate of population growth of a sexual compared to an asexual population, owing to production of males.

creationism   The doctrine that each species (or perhaps higher taxon) was created separately, essentially in its present form, by a supernatural Creator.

crown group   A taxon, distinguished by derived character states, that has descended from an ancestral group (**stem group**) that may bear a different name.

cultural evolution   Changes in the frequency of nongenetic cultural traits within and among populations, based on processes such as nonrandom imitation.

cytoplasmic male sterility (CMS)   The sterilization of male function in an otherwise hermaphroditic individual as the result of factors transmitted with the cytoplasm, typically mitochondrial in origin.

D

deleterious mutation   A mutation that reduces fitness.

deme   A local population; usually, a small, panmictic population.

demographic   Pertaining to processes that change the size of a population (i.e., birth, death, dispersal).

de novo genes   Coding DNA sequences that originate from noncoding DNA.

density-dependent   Affected by population density.

derived character (state)   A character (or character state) that has evolved from an antecedent (ancestral) character or state.

deterministic   Causing a fixed outcome, given initial conditions. *Cf.* **stochastic**.

developmental arrest   A halting of the development of a morphological structure, resulting in a final adult phenotype that lacks the structure or bears an immature form of the structure. This can also refer to developmental arrest at the level of the entire organism, resulting in an adult that resembles the juvenile form of an ancestral or related species (i.e., **paedomorphosis**).

developmental circuit   *See* **developmental pathway**.

developmental constraint   A restriction that prevents the appearance of certain structures or traits due to the inability of an organism's developmental system to produce them.

developmental pathway   A sequence of gene expression through developmental time, involving both gene regulation and the expression of gene products that provide materials for and regulate morphogenesis, resulting in the normal development of a tissue, organ, or other structure. Also called **developmental circuit**.

differential gene expression   Differences in the time, location, and/or quantitative level at which a gene expresses the protein it encodes. Differential gene expression involves differences between species, developmental stages, or physiological states in the specific cells, tissues, structures, or body segments that express a given gene.

dioecious   Of a species, consisting of distinct female and male individuals.

diploid   Of a cell or organism, possessing two chromosome complements. *See also* **haploid**, **polyploid**.

direct benefit (direct fitness benefit)   A fitness increment accrued by an individual by performing an action or receiving the action.

direct development   A life history in which the intermediate larval stage is omitted and development proceeds directly from an embryonic form to an adult-like form. *Cf.* **indirect development**.

direct fitness   *See* **inclusive fitness**.

direct response to selection   The component of evolutionary change in the mean of a trait that results from selection acting directly on that trait. *See also* **indirect response to selection**.

direct selection   Selection that acts directly on a locus. *See also* **indirect selection**.

directional selection   Selection for a value of a character that is higher or lower than its current mean value.

disparity   The magnitude of variation in morphological or other phenotypic characters among species in a clade or taxon.

dispersal   In population biology, movement of individual organisms to different localities; in biogeography, extension of the geographic range of a species by movement of individuals.

disruptive selection   Selection in favor of two or more phenotypes and against those intermediate between them; also called **diversifying selection**.

divergence   The evolution of increasing difference between lineages in one or more characters.

diversification   An evolutionary increase in the number of species in a clade, usually accompanied by divergence in phenotypic characters.

diversifying selection   *See* **disruptive selection**.

diversity-dependent factors   Processes that have a stronger effect on per capita rates of speciation or extinction when the diversity of species is greater.

d_N/d_S ratio   The ratio of the number of nonsynonymous substitutions per nonsynonymous site (d_N) and the number of synonymous substitutions per synonymous site (d_S). Values of this ratio smaller than one are consistent with **purifying selection**, while values greater than one suggest the action of **positive selection**.

Dobzhansky-Muller (DM) incompatibility   Reduction in the fitness of a hybrid because of interaction between certain alleles in one parent population with specific alleles at other loci in the other parent population.

Dollo's law   A biological generalization positing that complex characters, once lost in evolution, are extremely unlikely to reappear and thus the loss of complex characters is virtually always irreversible.

domain   A relatively small protein segment or module (usually 100 amino acids or less) that can fold into a specific three-dimensional structure independently of other domains.

dominance   Of an allele, the extent to which it produces when heterozygous the same phenotype as when homozygous; may be contrasted with a **recessive** allele, one that is phenotypically detectable only when homozygous. Dominance of a species describes the extent to which it is numerically or otherwise predominant in a community.

driven trend   Also called **active trend**. A prolonged shift in the mean of a character among the species in a clade, owing to more frequent changes within species in one direction than the other. In a **passive trend**, changes in both directions would be equally likely, but are constrained by a boundary in one direction.

duplication   The production of another copy of a locus (or other sequence) that is inherited as an addition to the genome.

E

ecological biogeography   *See* **historical biogeography**.

ecological niche   The range of combinations of all relevant environmental variables under which a species or population can persist; often more loosely used to describe the "role" of a species, or the resources it utilizes.

ecological release   The expansion of a population's niche (e.g., range of habitats or resources used) where competition with other species is alleviated.

ecological speciation   Speciation caused by divergent selection, by ecological factors, on characteristics that contribute to reproductive isolation.

ecotype   A genetically determined phenotype of a species that is found as a local variant associated with certain ecological conditions.

effective population size   The effective size of a real population is equal to the number of individuals in an ideal population (i.e., a population in which all individuals reproduce equally) that produces the rate of genetic drift seen in the real population.

electrophoresis   A method of separating genetically different forms of a protein, once an important way to detect variation in the encoding genes.

endemic   Of a taxon, restricted to a specified region or locality.

endosymbiont   An organism that resides within the cells of a host species.

enhancer   A DNA sequence that, when acted on by **transcription factors** controls transcription of an associated gene. *Cf.* **cis-regulatory element**, **control region**, **promoter**.

environment   Usually, the complex of external physical, chemical, and biotic factors that may affect a population, an organism, or the expression of an organism's genes; more generally, anything external to the object of interest (e.g., a gene, an organism, a population) that may influence its function or activity. Thus, other genes within an organism may be part of a gene's environment, or other individuals in a population may be part of an organism's environment.

environmental correlation (r_E)   *See* **genetic correlation**.

environmental sex determination   The condition in which an individual's sex is determined by the environmental conditions it experiences during development, rather than (for example) its genotype. *See also* **genetic sex determination**.

environmental variance   Variation among individuals in a phenotypic trait that is caused by variation in the environment rather than by genetic differences.

epigenetic inheritance   Inherited changes in gene expression or phenotype that are not based on changes in DNA sequence.

epistasis   An effect of the interaction between two or more gene loci on the phenotype or fitness whereby their joint effect differs from the sum of the loci taken separately.

equilibrium   An unchanging condition, as of population size or genetic composition. Also, the value (e.g., of population size, allele frequency) at which this condition occurs. An equilibrium need not be stable. *See* **stability**, **unstable equilibrium**.

ESS   *See* **evolutionarily stable strategy**.

essentialism   The philosophical view that all members of a class of objects (such as a species) share certain invariant, unchanging properties that distinguish them from other classes.

euploid   Of a cell or organism, possessing the normal, balanced, number of chromosomes.

eusociality   Animal societies characterized by overlapping generations, cooperative care of offspring (including those of other individuals), and a division of labor between reproductive and non-reproductive groups of adults.

evolution   In a broad sense, the origin of entities possessing different states of one or more characteristics and changes in the proportions of those entities over time. *Organic evolution*, or *biological evolution*, is a change over time in the proportions of individual organisms differing genetically in one or more traits. Such changes transpire by the origin and subsequent alteration of the frequencies of genotypes from generation to generation within populations, by alteration of the proportions of genetically differentiated populations within a species, or by changes in the numbers of species with different characteristics, thereby altering the frequency of one or more traits within a higher taxon.

evolutionarily stable strategy (ESS)   A phenotype such that, if almost all individuals in a population have that phenotype, no alternative phenotype can invade the population or replace it.

evolutionary constraint   A property of organisms that tends to retard evolution of a feature or to direct its evolution along some paths rather than others.

evolutionary developmental biology (EDB)   The study of evolutionary changes in the developmental bases of phenotypic characteristics.

evolutionary radiation   *See* **adaptive radiation**.

evolutionary reversal   The evolution of a character from a derived state back toward a condition that resembles an earlier state.

evolutionary synthesis   The reconciliation of Darwin's theory with the findings of modern genetics, which gave rise to a theory that emphasized the coaction of random mutation, selection, genetic drift, and gene flow; also called the **modern synthesis**.

evolutionary trade-off   The existence of both a fitness benefit and a fitness cost of a mutation or character state, relative to another.

evolutionary trend   A bias in the direction of repeated changes in a character, within one lineage or among multiple lineages, over an extended period of time.

evolvability   Can refer either to a measure of additive genetic variation that enables response to selection, or to the ability of genetic and developmental processes to generate potentially adaptive variation.

exaptation   The evolution of a function of a gene, tissue, or structure other than the one it was originally adapted for; can also refer to the adaptive use of a previously nonadaptive trait.

exon   That part of a gene that is translated into a polypeptide (protein). *Cf.* **intron**.

exon shuffling   The formation of new genes by assembly of exons from two or more preexisting genes. The classical model of exon shuffling generates new combinations of exons mediated via recombination of intervening introns; however, exon shuffling can also come about by retrotransposition of exons into pre-existing genes.

exponential growth   Nonlinear increase (or decrease) of a property (e.g., body size, population size) over time, described by an exponential equation.

F

fecundity   The quantity of gametes (usually eggs) produced by an individual.

female choice   Differential response of females to phenotypic variation in male traits, that may result in sexual selection on males; female choice does not require or imply active cognition.

Fisher's runaway   A process postulated by R. A. Fisher in which a mating preference and a sexual display become very rapidly exaggerated as the result of a genetic correlation that develops between them. *See also* **indirect selection**.

fitness   The success of an entity in leaving descendants to the next generation. Most often refers to the average contribution of an allele, genotype, or phenotype; can also refer to the contribution of a specific entity (e.g., an individual). *See also* **relative fitness**.

fitness component   One of several events in the life cycle of many organisms that contributes to the determination of fitness, such as survival to maturity, mating success, and fecundity.

fitness function   The function that relates the phenotypic value for a trait to the average fitness of individuals with that trait value.

fixation   Attainment of a frequency of 1 (i.e., 100 percent) by an allele in a population, which thereby becomes **monomorphic** for the allele.

founder effect   The principle that the founders of a new population carry only a fraction of the total genetic variation in the source population.

founder event   A **population bottleneck** that results when a new population is founded by a small number of individuals.

founder-flush speciation   A hypothesis for speciation, in which genetic change is enhanced in populations that grow rapidly ("flush") after being founded by a few individuals.

frameshift mutation   An insertion or deletion of base pairs in a translated DNA sequence that alters the reading frame, resulting in multiple downstream changes in the potential gene product.

frequency   In this book, usually used to mean *proportion* (e.g., the frequency of an allele is the proportion of gene copies having that allelic state).

frequency-dependent selection   A mode of natural selection in which the fitness of each genotype varies as a function of its frequency in the population.

functional constraint   Limitation on the variation expressed in a phenotype (perhaps a protein) because many variants have impaired function and reduce fitness.

fundamental theorem of natural selection   A mathematical result derived by R. A. Fisher stating that under certain conditions, the mean fitness of a population will increase in each generation by an amount equal to the additive genetic variance for relative fitness.

G

game theory   In **behavioral ecology**, a theoretical framework for analyzing the evolution of social interactions. *See* **evolutionarily stable strategy (ESS)**.

gametic selection   Natural selection among alleles based on their effects in gametes.

gene   The functional unit of heredity.

gene conversion   A process involving the unidirectional transfer of DNA information from one gene to another. In a typical conversion event, a gene or part of a gene acquires the same sequence as the other allele at that locus (intralocus or intra-allelic conversion), or the same sequences as a different, usually paralogous, locus (interlocus conversion). One consequence of gene conversion may be the homogenization of sequences among members of a gene family.

gene copy   Refers to a representative of a particular gene in an individual or cell (e.g., one copy in a haploid cell, two copies in a diploid).

gene duplication   The process whereby new genes arise as copies of preexisting gene sequences. The result can be a **gene family**.

gene family   Two or more loci with similar nucleotide sequences that have been derived from a common ancestral sequence.

gene flow   The incorporation of genes into the gene pool of one population from one or more other populations.

gene frequency   *See* **allele frequency**.

gene pool   The totality of the genes of a given sexual population.

gene swamping   The loss of a locally advantageous allele cause by the influx of other alleles from other populations.

gene trafficking   The movement of a locus between locations in the genome that results when a gene is duplicated and the original copy of the gene is deleted or becomes a pseudogene.

gene tree   A diagram representing the history by which gene copies have been derived from ancestral gene copies in previous generations.

genetic assimilation   A process whereby a phenotype whose development is triggered by an environmental stimulus evolves to be constitutively expressed (i.e., no longer requires the stimulus).

genetic conflict   Antagonistic fitness relationships between alleles, either at the same locus (intralocus conflict) or at different loci (interlocus conflict).

genetic constraint   A restriction that prevents a lineage from evolving along a particular evolutionary trajectory because genetic variation enabling that trajectory is not available.

genetic correlation   Correlated differences among genotypes in two or more phenotypic characters, due to **pleiotropy** or **linkage disequilibrium**. Genetic correlation, together with character correlation caused by different environmental conditions (**environmental correlation**), accounts for the correlation that may be observed between phenotypic characters within a population (**phenotypic correlation**).

genetic covariance   The component of the phenotypic covariance between two quantitative traits that results from genetic causes; genetic covariances result from pleiotropy and linkage disequilibrium. *See also* **genetic correlation**.

genetic distance   Any of several measures of the degree of genetic difference between populations, based on differences in allele frequencies.

genetic drift   Random changes in the frequencies of two or more alleles or genotypes within a population.

genetic line of least resistance   The combination of quantitative traits for which additive genetic variance is maximized and so will show the maximal response to directional selection.

genetic load   Any reduction of the mean fitness of a population resulting from the existence of genotypes with a fitness lower than that of the most fit genotype.

genetic marker   A readily detected genetic variant (such as a visible mutation or a polymorphic nucleotide) that is used to trace variation and inheritance of a closely linked region that may include a gene of interest.

genetic sex determination   The condition in which an individual's sex is determined by its genotype, rather than (for example) environmental conditions. *See also* **environmental sex determination**.

genetic toolkit   The set of genes and proteins, often conserved across distantly related organisms, and the developmental pathways that they comprise, by which multicellular organisms are constructed during development.

genetic variance   Variation in a trait within a population, as measured by the variance that is due to genetic differences among individuals.

genic selection   A form of selection in which the single gene is the unit of selection, such that the outcome is determined by fitness values assigned to different alleles. *See* **individual selection**, **kin selection**, **natural selection**.

genome   The entire complement of DNA sequences in a cell or organism. A distinction may be made between the nuclear genome and organelle genomes, such as those of mitochondria and plastids.

genotype × environment interaction   Phenotypic variation arising from the difference in the effect of the environment on the expression of different genotypes.

genotype frequency   The proportion of individuals in a population that carry a specific genotype at one or more loci.

genotype   The set of genes possessed by an individual organism; often, its genetic composition at a specific locus or set of loci singled out for discussion.

geographic variation   Variation in a characteristic or allele frequency among spatially distributed populations of a species.

Gondwana   The southern of the two large continents that existed in the early Mesozoic.

good genes   A term used for a mechanism that may cause the evolution of mating preferences for extreme mating displays. When these displays are correlated with genetic variation for lifetime fitness, natural selection acting on those "good genes" can cause the mating preferences to become exaggerated by **indirect selection**.

grade   A group of species that have evolved the same state in one or more characters and typically constitute a **paraphyletic** group relative to other species that have evolved further in the same direction.

gradualism   The proposition that large differences in phenotypic characters have evolved through many slightly different intermediate states. *See* **phyletic gradualism**.

green beard effect   The evolution of altruistic behavior through the evolution of a trait that simultaneously allows individuals to recognize and to help others with the same trait (e.g., a "green beard").

group selection   The differential rate of origination or extinction of whole populations (or species, if the term is used broadly) on the basis of differences among them in one or more characteristics. May also refer to differences among populations in their contribution of genes to the combined gene pool. *See also* **interdemic selection**, **species selection**.

H

habitat selection   The capacity of an organism (usually an animal) to choose a habitat in which to perform its activities. Habitat selection is not a form of natural selection.

habitat tracking   The tendency for the geographic range of a species to shift in accordance with changes in the location of its ecological requirements, rather than adapting to environmental changes in its former range.

Haldane's rule   The generalization that when only one sex manifests sterility or inviability in hybrids between species, it is the heterogametic sex (with two different sex chromosomes) that does so.

Hamilton's rule   The theoretical principle that an altruistic trait can increase if the benefit to recipients, multiplied by their relationship to the altruist, exceeds the fitness cost to the altruist.

haplodiploid sex determination   A form of sex determination (found in several groups of arthropods) in which unfertilized eggs develop as males and fertilized eggs as females.

haploid   Of a cell or organism, possessing a single chromosome complement, hence a single gene copy at each locus.

haplotype   A DNA sequence that differs from homologous sequences at one or more base pair sites.

Hardy-Weinberg   Pertaining to the genotype frequencies expected at a locus under ideal equilibrium conditions in a randomly mating population.

heritability   The proportion of the **variance** in a trait among individuals that is attributable to differences in genotype. Heritability in the narrow sense is the ratio of **additive genetic variance** to phenotypic variance.

hermaphroditic   Performance of both female and male sexual functions by a single individual.

heterochrony   An evolutionary change in phenotype caused by an alteration of timing of developmental events.

heterokaryotype   A genome or individual that is heterozygous for a chromosomal rearrangement such as an inversion. *Cf.* **homokaryotype**.

heterotopy   Expression of a gene or character in a different location on the body of a descendant than in its ancestor.

heterozygosity   In a population, the proportion of loci at which a randomly chosen individual is heterozygous, on average. Applied to a single locus, it refers to the proportion of heterozygotes in a population. In both senses, Hardy-Weinberg equilibrium is often assumed.

heterozygote   An individual organism that possesses different alleles at a locus.

heterozygous advantage   The manifestation of higher fitness by heterozygotes than by homozygotes at a specific locus.

higher taxon   A taxon above the species level, such as a named genus or phylum.

historical biogeography   The study of historical changes in the geographic distribution of organisms, including those that affect their present distribution; **ecological biogeography** addresses current factors that affect present distributions.

historical contingency   Of a dynamic system (such as a locus evolving under selection), the situation in which the course and outcome of change depend in part on initial conditions.

hitchhiking   Change in the frequency of an allele due to linkage with a selected allele at another locus.

homeobox genes   A large family of eukaryotic genes that contain a DNA sequence known as the homeobox. The homeobox sequence encodes a protein domain about 60 amino acids in length that binds DNA. Most homeobox genes are transcriptional regulators. *Cf.* **domain**; **Hox genes**.

homeostasis   Maintenance of an equilibrium state by some self-regulating capacity of an individual.

homeotic mutation   A mutation that causes a transformation of one structure into another of the organism's structures.

homeotic selector genes   Genes whose expression is required for the development of an entire organ, segment, or compartment of an organism.

homokaryotype   A genome or individual that is homozygous for a chromosomal rearrangement such as an inversion. *Cf.* **heterokaryotype**.

homologous chromosome   *See* **homology**.

homology   Possession by two or more species of a character state derived, with or without modification, from their common ancestor. **Homologous chromosomes** are those members of a chromosome complement that bear the same genes.

homonymous   Pertaining to biological structures that occur repeatedly within one segment of the organism, such as teeth or bristles.

homoplasy   Possession by two or more species of a similar or identical character state that has not been derived by both species from their common ancestor; embraces **convergence**, **parallel evolution**, and **evolutionary reversal**.

homozygote   An individual organism that has the same allele at all of its copies of a genetic locus.

horizontal transmission   Movement of genes or symbionts (such as parasites) between individual organisms other than by transmission from parents to their offspring (which is **vertical transmission**). Horizontal transmission of genes is also called **lateral gene transfer**.

Hox genes   A subfamily of **homeobox genes**, conserved in all metazoan animals, that controls anterior-posterior segment identity by regulating the transcription of many genes during development.

hybrid   An individual formed by mating between unlike forms, usually genetically differentiated populations or species.

hybrid zone   A region in which genetically distinct populations come into contact and produce at least some offspring of mixed ancestry.

hybridization   Production of offspring by interbreeding between members of genetically distinct populations.

hypermorphosis   An evolutionary increase in the duration of ontogenetic development, resulting in features that are exaggerated compared to those of the ancestor.

hypothesis   An informed conjecture or proposition of what might be true.

hypothetico-deductive method   A scientific method in which a hypothesis is tested by deducing expected data or observations from it, if it were true, and comparing the deduced predictions with real data.

I

identical by descent   Of two or more gene copies, being derived from a single gene copy in a specified common ancestor of the organisms that carry the copies.

inbreeding   Mating between relatives that occurs more frequently than if mates were chosen at random from a population.

inbreeding coefficient   The probability that a random pair of gene copies, inherited by offspring from two parents, is identical by descent.

inbreeding depression   Reduction, in inbred individuals, of the mean value of a character (usually one correlated with fitness) relative to offspring of unrelated individuals.

inbreeding load   The decline in a population's mean fitness that results from the fixation of deleterious mutations by drift. *See also* **inbreeding depression**.

inclusive fitness   The fitness of a gene or genotype as measured by its effect on the survival or reproduction of both the organism bearing it (**direct fitness**) and the genes, identical by descent, borne by the organism's relatives (**indirect fitness**).

incomplete lineage sorting   Persistence of a genetic polymorphism through a speciation event, so that fixation occurs only in the descendant species, or in their descendants after subsequent speciation.

indirect development   A life history that includes a larval stage between embryo and adult stages. *Cf.* **direct development**.

indirect fitness   *See* **inclusive fitness**.

indirect response to selection   The component of evolutionary change in the mean of a trait resulting from selection that acts on other traits that are genetically correlated with it. *See also* **direct response to selection**.

indirect selection   The evolution of an allele (or trait) caused by selection that acts on another locus (or trait) with which it is genetically correlated.

individual selection   A form of natural selection consisting of nonrandom differences in **fitness** among different phenotypes (or genotypes) within a population. *See also* **genic selection**, **natural selection**.

individualization   The evolution of distinct form and identity of each of several structures that were not differentiated from one another in an ancestor.

infanticide   The killing of immature individuals by mature conspecific individuals.

ingroup   *See* **outgroup**.

inheritance of acquired characteristics   The formerly widespread belief that modifications of an individual during its lifetime, due to its behavior or its environment, could be transmitted to its descendants.

intelligent design (ID)   A strain in creationism that claims that the complexity of organisms is too great to have evolved by natural processes and therefore must have been designed by an intelligent being.

inter-, intra-   Prefixes meaning, respectively, "between" and "within." For example, "interspecific" differences are differences between species and "intraspecific" differences are differences among individuals within a species.

interaction   Strictly, the dependence of an outcome on a combination of causal factors, such that the outcome is not predictable from the average effects of the factors taken separately. More loosely, an interplay between entities that affects one or more of them (as in interactions between species). *See also* **genotype × environment interaction**.

interdemic selection   **Group selection** of populations within a species.

intragenic recombination   Recombination within a gene.

intrinsic rate of natural increase   The potential per capita rate of increase of a population with a stable age distribution whose growth is not depressed by the negative effects of density.

introgression   Movement of genes from one species or population into another by hybridization and backcrossing; carries the implication that some genes in a genome undergo such movement, but others do not.

intron   A part of a gene that is not translated into a polypeptide. *Cf.* **exon**.

inversion   A 180° reversal of the orientation of a part of a chromosome, relative to some standard chromosome.

isolating barrier, isolating mechanism   A genetically determined difference between populations that restricts or prevents gene flow between them. The term does not include spatial segregation by extrinsic geographic or topographic barriers.

isolation by distance   A model of population structure in which the likelihood of mating decreases with the geographic distance between individuals, so that local mating causes geographic variation in allele frequencies.

iteroparous   Pertaining to a life history in which individuals reproduce more than once. *Cf.* **semelparous**.

K

karyotype   The chromosome complement of an individual.

key adaptation   An adaptation that provides the basis for using a new, substantially different habitat or resource.

kin selection   A form of selection whereby alleles differ in their rate of propagation by influencing the impact of their bearers on the reproductive success of individuals (kin) who carry the same alleles by common descent.

L

Lamarckism   The theory that evolution is caused by inheritance of character changes acquired during the life of an individual due to its behavior or to environmental influences.

lateral gene transfer   *See* **horizontal transmission**.

Laurasia   The northern of the two large continents that existed in the early Mesozoic.

lek   An aggregation of males who engage in competitive mating displays; females mate at the lek but gain no direct benefits from their mates.

lethal allele   An allele (usually recessive) that causes virtually complete mortality, usually early in development.

levels of selection   The several kinds of reproducing biological entities (e.g., genes, organisms, species) that can vary in fitness, resulting in potential selection among them.

life history   Usually refers to the set of traits that affect changes in numbers of individuals over generations, including age-specific values of survival, female reproduction, and male reproduction.

lineage   A series of ancestral and descendant populations through time; usually refers to a single evolving species, but may include several species descended from a common ancestor.

lineage sorting   The process by which each of several descendant species, carrying several gene lineages inherited from a common ancestral species, acquires a single gene lineage; hence, the derivation of a monophyletic gene tree, in each species, from the paraphyletic gene tree inherited from their common ancestor.

lineage-through-time plot   A graph of the apparent change in number of lineages in a clade, often based on a time-calibrated phylogeny.

linkage   Occurrence of two loci on the same chromosome: the loci are functionally linked only if they are so close together that they do not segregate independently in meiosis.

linkage disequilibrium   The association of two alleles at two or more loci more frequently (or less frequently) than predicted by their individual frequencies.

linkage equilibrium   The association of two alleles at two or more loci at the frequency predicted by their individual frequencies.

local adaptation   Of an allele, trait, or population, the state of being differentially adapted to conditions that prevail in a spatially restricted area.

locus (plural: loci)   A site on a chromosome occupied by a specific gene; more loosely, the gene itself, in all its allelic states.

logistic equation   An equation describing the idealized growth of a population subject to a density-dependent limiting factor. As density increases, the rate of growth gradually declines until population growth stops.

M

macroevolution   A vague term, usually meaning the evolution of substantial phenotypic changes, usually great enough to place the changed lineage and its descendants in a distinct genus or higher taxon. *Cf.* **microevolution**.

major transition   One of several events in the history of life on Earth in which a qualitatively higher level of organization emerged.

male combat   Direct physical contest between males that may result in sexual selection.

male-male competition   Sexual selection that results from males' competing in various ways directly with one another, as distinct from sexual selection by **female choice**. *See also* **male combat**, **sperm (pollen) competition**.

mass extinction   A highly elevated rate of extinction of species, extending over an interval that is relatively short on a geological time scale (although still very long on a human time scale).

maternal effect   A nongenetic effect of a mother on the phenotype of her offspring, stemming from factors such as cytoplasmic inheritance, transmission of symbionts from mother to offspring, or nutritional conditions.

maximum likelihood (ML)   A framework for statistical inference used for tasks such as the estimation of the parameters of a model or the properties of a population (such as its mean and variance, in simple cases) from data.

maximum parsimony   *See* **parsimony**.

McDonald-Kreitman (MK) test   A test for selection at a locus by comparing DNA sequence variation within species with the variation among species.

mean fitness   The arithmetic average fitness of all individuals in a population, usually relative to some standard.

mean   Usually the arithmetic mean or average; the sum of n values, divided by n. The mean value of x, symbolized as $\bar{x}$, equals $(x_1 + x_2 + \ldots + x_n)/n$.

meiotic drive   A form of **segregation distortion** that occurs during meiosis and causes an allele to have greater than 50% probability of being transmitted to a gamete.

metapopulation   A set of local populations, among which there may be gene flow and patterns of extinction and recolonization.

microevolution   A vague term, usually referring to slight, short-term evolutionary changes within species. *Cf.* **macroevolution**.

microsatellite   A short, highly repeated, untranslated DNA sequence.

migration   Used in theoretical population genetics as a synonym for gene flow among populations; in other contexts, refers to directed large-scale movements of organisms that do not necessarily result in gene flow.

migration rate (m)   The fraction of individuals (or gene copies) in a population that immigrated into the population from elsewhere within the current generation.

migration variance (σm^2)   The square root of the mean of the squared distances between the birth places of mothers and their offspring.

mimicry   Similarity of certain characters in two or more species due to convergent evolution when there is an advantage conferred by the resemblance. Common types include *Batesian* mimicry, in which a palatable *mimic* experiences lower predation because of its resemblance to an unpalatable *model*; and *Müllerian* mimicry, in which two or more unpalatable species enjoy reduced predation due to their similarity.

modern synthesis   *See* **evolutionary synthesis**.

modularity   The ability of individual parts of an organism, such as segments or organs, to develop or evolve independently from one another; the ability of developmental regulatory genes and pathways to be regulated independently in different tissues and developmental stages.

molecular clock   The concept of a steady rate of change in DNA sequences over time, providing a basis for dating the time of divergence of lineages if the rate of change can be estimated.

monomorphic   Having one form; refers to a population in which virtually all individuals have the same genotype at a locus. *Cf.* **polymorphism**.

monophyletic   Refers to a taxon, or a branch of a phylogenetic tree or gene tree, that includes all the species (or genes) that descended from a common ancestor. *Cf.* **paraphyletic**, **polyphyletic**.

mosaic evolution   Evolution of different characters within a lineage or clade at different rates, hence more or less independently of one another.

Muller's ratchet   The process postulated by H. J. Muller in which a population's mean fitness declines when genotypes with the fewest deleterious mutations are lost by drift and cannot be recovered without recombination.

multigene family   Also called "gene family," a set of distinct loci in a genome that originated from a single locus in an ancestor by duplication and sequence divergence.

multiple-niche polymorphism   Stable variation at a locus owing to superior fitness of different genotypes under different conditions of a varying environment.

multiple stable equilibria   *See* **stability**.

mutation   An error in the replication of a nucleotide sequence, or any other alteration of the genome that is not manifested as reciprocal recombination.

mutation load   The decrease in a population's mean fitness that results from deleterious mutations, relative to a hypothetical mutation-free population.

mutational variance   The increment in the genetic variance of a phenotypic character caused by new mutations in each generation.

mutualism   A symbiotic relation in which each of two species benefits by their interaction.

N

natural laws   Consistent natural phenomena, described by statements that certain effects will always occur if specific conditions hold.

natural selection   The differential survival and/or reproduction of classes of entities that differ in one or more characteristics. To constitute natural selection, the difference in survival and/or reproduction cannot be due to chance, and it must have the potential consequence of altering the proportions of the different entities. Thus natural selection is also definable as a deterministic difference in the contribution of different classes of entities to subsequent generations. Usually the differences are inherited. The entities may be alleles, genotypes or subsets of genotypes, populations, or, in the broadest sense, species. *See also* **genic selection**, **individual selection**, **kin selection**, **group selection**.

naturalistic fallacy   A frequently used name for the belief that what is "natural" is morally right or good.

Nₑ (the effective population size)   *See* **effective population size**.

neo-Darwinism   Originally, the theory of natural selection of inherited variations, that denied that acquired characteristics might be inherited; often used more broadly to mean the modern theory that natural selection, acting on randomly generated particulate genetic variation, is the major, but not the sole, cause of evolution.

neofunctionalization   Divergence of duplicate genes whereby one acquires a new function. *Cf.* **subfunctionalization**.

neoteny   Heterochronic evolution whereby development of some or all somatic features is retarded relative to sexual maturation, resulting in sexually mature individuals with juvenile features. *See also* **paedomorphosis**, **progenesis**.

neutral alleles   Alleles that do not differ measurably in their effect on fitness.

neutral theory of molecular evolution   The hypothesis that most alleles that are polymorphic within populations and that become fixed do not significantly alter fitness and evolve by genetic drift.

nonadaptive evolution   Evolution by substitution of neutral alleles.

nonsynonymous substitution   A base pair substitution in DNA that results in an amino acid substitution in the protein product; also called **replacement substitution**. *Cf.* **synonymous substitution**.

norm of reaction   The set of phenotypic expressions of a genotype under different environmental conditions. *See also* **phenotypic plasticity**.

normal distribution   A bell-shaped frequency distribution of a variable; the expected distribution if many factors with independent, small effects determine the value of a variable; the basis for many statistical formulations.

nucleotide substitution   The complete replacement of one nucleotide base pair by another within a lineage over evolutionary time.

O

ontogeny   The development of an individual organism, from fertilized zygote until death.

operational sex ratio   The relative numbers of males and females available to mate at any given time.

operon   A segment of DNA containing multiple genes whose transcription is under the control of a single promoter.

optimality theory   Models of adaptive evolution that assume that characters have evolved to nearly their optimum, within limits set by specified constraints.

optimum phenotype   The phenotype that maximizes fitness.

organism   Usually used in this book to refer to an individual member of a species.

orthologous   Refers to corresponding (homologous) members of a gene family in two or more species. *Cf.* **paralogous**.

outcrossing   Mating with another genetic individual. *Cf.* **selfing**.

outgroup   A taxon that diverged from a group of other taxa (the **ingroup**) before they diverged from one another.

overdominance   The expression by two alleles in heterozygous condition of a phenotypic value for some character that lies outside the range of the two corresponding homozygotes.

overlapping gene   A gene whose coding region overlaps with another gene, which is often transcribed in an alternate reading frame.

P

paedomorphosis   Possession in the adult stage of features typical of the juvenile stage of the organism's ancestor.

Pangaea   The single large "world continent" formed by coalescence of land masses in the late Paleozoic.

panmixia   Random mating among members of a population.

parallel evolution (parallelism)   The evolution of similar or identical features independently in related lineages, thought usually to be based on similar modifications of the same developmental pathways.

paralogous   Refers to the evolutionary relationship between two different members of a gene family, within a species or in a comparison of different species. *Cf.* **orthologous**.

parapatric   Of two species or populations, having contiguous but non-overlapping geographic distributions. *Cf.* **allopatric**, **sympatric**.

parapatric speciation   *See* **allopatric speciation**.

paraphyletic   Refers to a taxon, phylogenetic tree, or gene tree whose members are all derived from a single ancestor, but which does not include all the descendants of that ancestor. *Cf.* **monophyletic**.

parent-offspring conflict   A condition in which a character state that enhances fitness of offspring reduces the fitness of a parent (or vice versa).

parental investment   Parental activities or processes that enhance the survival of existing offspring but whose **costs** reduce the parent's subsequent reproductive success.

parsimony   Economy in the use of means to an end (*Webster's New Collegiate Dictionary*); the principle of accounting for observations by that hypothesis requiring the fewest or simplest assumptions that lack evidence; in systematics, the principle of invoking the minimal number of evolutionary changes to infer phylogenetic relationships.

parthenogenesis   Virgin birth; development from an egg to which there has been no paternal contribution of genes.

passive trend   *See* **driven trend**.

PCR (polymerase chain reaction)   A laboratory technique by which the number of copies of a DNA sequence is increased by replication in vitro.

peak shift   Change in allele frequencies within a population from one to another local maximum of mean fitness by passage through states of lower mean fitness.

peramorphosis   Evolution of a more extreme character state by prolongation of development in the descendant, compared to the ancestor.

perceptual bias (sensory bias)   A difference in the ability of an organism to perceive different stimuli (e.g., low vs. high frequency sounds).

peripatric   Of a population, peripheral to most of the other populations of a species.

peripatric speciation   Speciation by evolution of reproductive isolation in peripatric populations as a consequence of a combination of genetic drift and natural selection.

phenetic   Pertaining to phenotypic similarity, as in a phenetic classification.

phenotype   The morphological, physiological, biochemical, behavioral, and other properties of an organism manifested throughout its life; or any subset of such properties, especially those affected by a particular allele or other portion of the **genotype**.

phenotypic correlation   *See* **genetic correlation**.

phenotypic integration   Correlation between the state of two or more functionally related characteristics, so that they are advantageously matched in most individuals.

phenotypic plasticity   The capacity of an organism to develop any of several phenotypic states, depending on the environment; usually this capacity is assumed to be adaptive.

phenotypic variance   The **variance** (q.v.) in a trait within a population; it may include both **genetic variance** and **environmental variance**.

phyletic gradualism   A term for gradual evolutionary change in features over a long period of time.

phylogenetic niche conservatism   Slow evolution of the ecological requirements of a group of organisms, resulting in long-continued dependence of related species on similar resources and environmental conditions.

phylogenetic species concept (PSC)   Species conceived as groups of populations that are distinguishable from other such groups.

phylogenetic tree   A diagram representing the evolutionary relationships among named groups of organisms, i.e., their history of descent from common ancestors.

phylogeny   The history of descent of a group of taxa such as species from their common ancestors, including the order of branching and sometimes the absolute times of divergence.

phylogeography   Description and analysis of the history and processes that govern the geographic distribution of genes within species and among closely related species, analysis that may shed light on the history of the populations.

physical constraint   A restriction that prevents a lineage from evolving a trait due to the properties of biological materials.

planktonic   Living in open water. *Cf.* **benthic**.

pleiotropy   A phenotypic effect of a gene on more than one character.

ploidy   The number of chromosome complements in an organism.

point mutation   A mutation that maps to a specific gene locus; in a molecular context, usually a change of a single base pair.

polygenic character   A character whose variation is based wholly or in part on allelic variation at more than a few loci.

polymerase chain reaction   *See* **PCR**.

polymorphic equilibrium   Of allele frequencies, a stable equilibrium at which more than one allele is maintained by selection.

polymorphism   The existence within a population of two or more genotypes, the rarest of which exceeds some arbitrarily low frequency (say, 1 percent); more rarely, the existence of phenotypic variation within a population, whether or not genetically based. *Cf.* **monomorphic**.

polyphenism   The capacity of a species or genotype to develop two or more forms, with the specific form depending on specific environmental conditions or cues, such as temperature or day length. A polyphenism is distinct from a **polymorphism** in that the former is the property of a single genotype, whereas the latter refers to multiple forms encoded by two or more different genotypes.

polyphyletic   Refers to a taxon, phylogenetic tree, or gene tree composed of members derived by evolution from ancestors in more than one ancestral taxon; hence, composed of members that do not share a unique common ancestor. *Cf.* **monophyletic**.

polyploid   Of a cell or organism, possessing more than two chromosome complements.

population   A group of conspecific organisms that occupy a more or less well defined geographic region and exhibit reproductive continuity from generation to generation; ecological and reproductive interactions are more frequent among these individuals than with members of other populations of the same species.

population bottleneck (bottleneck)   A severe, temporary reduction in population size.

positive selection   Selection for an allele that increases fitness. *Cf.* **purifying selection**.

postzygotic   Occurring after union of the nuclei of uniting gametes; usually refers to inviability or sterility that confers reproductive isolation.

preadaptation   Possession of the necessary properties to permit a shift to a new niche, habitat, or function. A structure is preadapted for a new function if it can assume that function without evolutionary modification.

prezygotic   Occurring before union of nuclei of uniting gametes; usually refers to events in the reproductive process that cause reproductive isolation, including those that occur before mating .

primary sexual trait   The gonads and closely associated structures that distinguish males and females. *See* **secondary sexual trait**.

primordium   A group of embryonic or larval cells destined to give rise to a particular adult structure.

processed pseudogene   A **pseudogene** that has arisen via the retrotransposition of mRNA into cDNA.

progenesis   A decrease during evolution of the duration of ontogenetic development, resulting in retention of juvenile features in the sexually mature adult. *See also* **neoteny**, **paedomorphosis**.

promoter   A region of DNA that initiates transcription of a gene, by binding RNA polymerase II and certain proteins (transcription factors).

provinciality   The degree to which the taxonomic composition of a biota is differentiated among major geographic regions.

pseudogene   A nonfunctional member of a gene family that has been derived from a functional gene. *Cf.* **processed pseudogene**.

pull of the Recent   An artifact in estimating changes in diversity in the fossil record, whereby taxa that are still alive have apparently longer durations than they would if they had been counted only from fossil data, and so will inflate the count of taxa, compared to the more remote past.

punctuated anagenesis   *See* **punctuated gradualism**.

punctuated equilibria   A pattern of rapid evolutionary change in the phenotype of a lineage separated by long periods of little change; also, a hypothesis intended to explain such a pattern, whereby phenotypic change transpires rapidly in small populations, in concert with the evolution of reproductive isolation.

punctuated gradualism   Alternating periods of slow and more rapid gradual change in a single lineage. Also called **punctuated anagenesis**.

purifying selection   Elimination of deleterious alleles from a population. *Cf.* **positive selection**.

Q

quantitative genetics   Genetic analysis of continuously varying characters, often employing statistical descriptions and estimators of variation.

quantitative trait   A phenotypic character that varies continuously rather than as discretely different character states.

quantitative trait locus/loci (QTL)   A chromosome region containing at least one gene that contributes to variation in a quantitative trait. QTL mapping is a procedure for determining the map positions of QTL on chromosomes.

quantitative variation   *See* **quantitative trait**.

R

race   A poorly defined term for a set of populations occupying a particular region that differ in one or more characteristics from populations elsewhere; equivalent to subspecies. In some writings, a distinctive phenotype, whether or not allopatric from others.

radiation   *See* **adaptive radiation**.

radiometric dating   Estimating ages of geological materials and events by the decay of radioactive elements.

random genetic drift   *See* **genetic drift**.

random walk   A mathematical model of a series of random fluctuations, used to describe random genetic drift and some other biological processes.

reaction norm   *See* **norm of reaction**. *See also* **genotype × environment**, **phenotypic plasticity**.

realized heritability   The heritability of a trait as calculated retrospectively from the change in a population's mean phenotype, relative to the selection differential that was applied to the character in an artificial selection experiment.

recessive   *See* **dominance**.

reciprocal translocation   A recombinational exchange of parts of two nonhomologous chromosomes.

reciprocity   Cooperation based on reciprocal aid in a succession of encounters between individuals.

recombinational speciation   The origin of a new species by selection among genotypes formed by hybridization between two ancestral species.

recruitment   (1) In evolutionary genetics, the evolution of a new function for a gene other than the function for which that gene was originally adapted. (2) In population biology, refers to the addition of new adult (breeding) individuals to a population via reproduction (i.e., individuals born into the population that reach reproductive age).

recurrent mutation   Repeated origin of mutations of a particular kind within a species.

Red Queen hypothesis   The proposition that taxa become extinct at an approximately constant rate because they fail to evolve as fast as other taxa with which they have antagonistic interactions. "Red Queen" more generally refers to averting extinction by evolving as fast as possible.

refugia   Locations in which species have persisted while becoming extinct elsewhere.

regression   In geology, withdrawal of sea from land, accompanying lowering of sea level; in statistics, a function that best predicts a dependent from an independent variable.

regulatory modularity   The property of gene regulation that allows gene expression or protein function to vary in different cells, tissues, or developmental stages of the same organism, without affecting the entire morphology or life history of the organism.

reinforcement   Evolution of enhanced reproductive isolation between populations due to natural selection for greater isolation.

relatedness   In **behavioral ecology**, the probability that a given individual carries the same allele as a focal individual at a given locus.

relative fitness   The fitness of a genotype relative to (as a proportion of) the fitness of a reference genotype, which is often set at one; the fitness values before such standardization are **absolute fitness** values.

relict   A species that has been "left behind"; for example, the last survivor of an otherwise extinct group. Sometimes, a species or population left in a locality after extinction throughout most of the region.

replacement substitution   *See* **nonsynonymous substitution**.

reporter construct   A DNA segment in which a putative *cis*-regulatory sequence is spliced upstream of a gene whose expression can be easily assayed, such as β-galactosidase or green fluorescent protein.

reproductive assurance   Mechanisms that increase the probability of successful reproduction when potential mates are rare.

reproductive effort   The proportion of energy or materials that an organism allocates to reproduction rather than to growth and maintenance.

reproductive isolation   Reduction of gene exchange between populations by any of several possible factors, usually those arising from biological differences between the populations.

reproductive success   The fitness of a genotype or other biological entity, often measured by the average per capita number of offspring that a newly formed zygote will have, or by similar measures.

response to selection   The change in the mean value of a character over one or more generations due to selection.

restriction enzyme   An enzyme that cuts double-stranded DNA at specific short nucleotide sequences. Genetic variation within a population results in variation in DNA sequence lengths after treatment with a restriction enzyme, or **restriction fragment length polymorphism (RFLP)**.

reticulate evolution   Union of different lineages of a clade by hybridization.

retrotransposition   The insertion into a chromosome of a DNA sequence that originated by the reverse transcription of an RNA precursor. Used by several types of **transposons** to replicate themselves, retrotransposition is also a mechanism for gene **duplication**. *See also* **reverse transcriptase**.

reverse transcriptase   An enzyme in retroviruses that synthesizes DNA copies of RNA molecules.

RFLP   *See* **restriction enzyme**.

rtPCR (reverse transcriptase PCR, real-time PCR)   A PCR reaction using mRNA as a template in which an initial step converts the mRNA to cDNA using reverse transcriptase and a subsequent step uses PCR to amplify the cDNA.

ruby-in-the-rubbish effect   The loss of a beneficial mutation (the "ruby") that occurs in a genotype that has low fitness because of deleterious mutations (the "rubbish") that it carries at other loci.

runaway sexual selection   A model of sexual selection in which a male display character and female preference for the character reinforce one another so that both evolve to be more extreme.

S

saltation   A jump; a discontinuous mutational change in one or more phenotypic traits, usually of considerable magnitude.

sampling error   The amount of inaccuracy (i.e., random variation) in the estimate of some value of a population, caused by measuring only a portion of the population; by extension, the chance variation in the value of repeated samples from the population.

scala naturae   The "scale of nature," or Great Chain of Being: the pre-evolutionary concept that all living things were created in an orderly series of forms, from lower to higher.

scientific theory   A coherent body of statements, based on reasoning and (usually) evidence, that explains some aspect of nature by recourse to natural laws or processes.

secondary contact   Contact and potential interbreeding between formerly allopatric populations, owing to range expansion.

secondary sexual traits   Traits not directly associated with the gonads and genitalia (the **primary sexual traits**) that differ between the sexes.

segregation distortion   Any of several biological processes that alter the rules of Mendelian inheritance such that some alleles when heterozygous have a greater than 50 percent chance of transmission to the offspring. *See* **meiotic drive**.

selection   Nonrandom differential survival or reproduction of classes of phenotypically different entities. *See* **natural selection**, **artificial selection**.

selection coefficient   The difference between the mean relative fitness of individuals of a given genotype and that of a reference genotype.

selection differential   The difference between the mean character value in a population before selection, and in the subset of individuals that survive and reproduce.

selection gradient   The slope of the relationship between phenotype and fitness, for a quantitative character, usually taking correlations with other characters into account.

selection plateau   The mean character value at which a population ceases to respond to continuing directional selection.

selective advantage   The increment in fitness (survival and/or reproduction) provided by an allele or a character state.

selective (or functional) constraint   A restriction that prevents a lineage from evolving a particular trait because that trait is always disadvantageous or interferes with the function of another trait.

selective interference   Reduction in the spread of an advantageous allele that results from selection acting on other loci. *See also* **clonal interference**, **Muller's ratchet**, and **ruby-in-the-rubbish effect**.

selective sweep   The increase in frequency and fixation of a beneficial allele; often used in reference to the associated reduction or elimination of DNA sequence variation in its vicinity on the chromosome.

self-incompatibility   A condition in which gametes produced by a hermaphroditic individual are unable to unite, due to molecular interactions that prevent self-fertilization.

selfing   Self-fertilization; union of female and male gametes produced by the same genetic individual. *Cf.* **outcrossing**.

selfish   In **behavioral ecology**, a behavior that increases the fitness of the actor and decreases the fitness of one or more others.

"selfish DNA"   A DNA sequence that has the capacity for its own replication, or replication via other self-replicating elements, but has no immediate function (or is even deleterious) for the organism in which it resides.

semelparous   Pertaining to a life history in which individuals (especially females) reproduce only once. *Cf.* **iteroparous**.

semispecies   One of several groups of populations that are partially but not entirely isolated from one another by biological factors (**isolating mechanisms**).

sensory bias   *See* **perceptual bias**.

serial homology   A relationship among repeated, often differentiated, structures of a single organism, defined by their similarity of developmental origin; for example, the several legs and other appendages of an arthropod.

sex-linked   Of a gene, being carried by one of the sex chromosomes; it may be expressed phenotypically in both sexes.

sex ratio   Often described as the proportion of males among offspring, either of an individual ("individual sex ratio") or a population ("population sex ratio").

sex role reversal   A mating system in which females actively court males, often associated with male parental care.

sexual dimorphism   The condition in which males and females are phenotypically distinct.

sexual isolation   Reduction of gene exchange between populations because of preferential mating between individuals from the same population; also called **behavioral isolation**.

sexual reproduction   Production of offspring whose genetic constitution is a mixture of those of two potentially genetically different gametes.

sexual selection   Differential reproduction as a result of variation in the ability to obtain mates.

sexually antagonistic selection   Selection that favors an allele or character state in one sex but a different allele or character state in the other sex.

sibling species   Species that are difficult or impossible to distinguish by morphological characters, but may be discerned by differences in ecology, behavior, chromosomes, genetic markers, or other such features.

silent substitution   *See* **synonymous substitution**.

single nucleotide polymorphism (SNP)   Variation in the identity of a nucleotide base pair at a single position in a DNA sequence, within or among populations of a species.

sister taxa   Two species or higher taxa that are derived from an immediate common ancestor, and are therefore one another's closest relatives.

special creation   The idea that each species was individually created by God in much its present form.

speciation   Evolution of reproductive isolation within an ancestral species, resulting in two or more descendant species.

species   In the sense of biological species, the members of a group of populations that interbreed or potentially interbreed with one another under natural conditions. Also, a fundamental taxonomic category to which individual specimens are assigned, which often but not always corresponds to the biological species. *See also* **biological species**, **phylogenetic species concept**.

species hitchhiking   Increase in the proportion of species with a specific trait because it is correlated with another trait that enhances speciation or reduces extinction.

species selection   A form of **group selection** in which species with different characteristics increase (by speciation) or decrease (by extinction) in number at different rates because of a difference in their characteristics.

sperm (pollen) competition   Competition among male gametes for fertilization.

spite   In **behavioral ecology**, a behavior that decreases the fitness of both the actor and the recipient(s).

stability   Often used to mean constancy; more often in this book, the propensity to return to a condition (a stable equilibrium) or to one of several such conditions (**multiple stable equilibria**) after displacement from that condition.

stabilizing selection   Selection that maintains the mean of a character at or near a constant intermediate value in a population.

standard deviation   The square root of the **variance**.

standing genetic variation   Genetic variation that is present in a population before positive or directional selection acts to change allele frequencies. Contrasts with new mutations on which selection may act.

stasis   Absence of substantial evolutionary change in one or more characters for some period of evolutionary time.

stem group   *See* **crown group**.

stochastic   Random. *Cf.* **deterministic**.

strata   Layers of sedimentary rock that were deposited at different times.

subfunctionalization   Divergence of duplicate genes whereby each retains only a subset of the several functions of the ancestral gene. *Cf.* **neofunctionalization**.

subspecies   A named geographic race; a set of populations of a species that share one or more distinctive features and occupy a different geographic area from other subspecies.

substitution   Usually, the complete replacement of one allele for another within a population or species over evolutionary time (*cf.* **fixation**). Sometimes refers to base pair differences in comparisons of homologous DNA sequences.

superspecies A group of **semispecies**.

symbiosis An intimate, usually physical, association between two or more species.

sympatric Of two species or populations, occupying the same geographic locality so that the opportunity to interbreed is presented. *Cf.* **allopatric**, **parapatric**.

sympatric speciation *See* **allopatric speciation**.

synapomorphy A derived character state that is shared by two or more taxa and is postulated to have evolved in their common ancestor.

synonymous substitution Fixation of a base pair change that does not alter the amino acid in the protein product of a gene; also called **silent substitution**. *Cf.* **nonsynonymous substitution**.

T

tandem repeat One of a group of adjacent duplicate copies of a DNA sequence.

taxon (plural: taxa) The named taxonomic unit (e.g., *Homo sapiens*, Hominidae, or Mammalia) to which individuals, or sets of species, are assigned. **Higher taxa** are those above the species level. *Cf.* **category**.

teleology The belief that natural events and objects have purposes and can be explained by their purposes.

tension zone A cline maintained by underdominant selection, even if relative fitnesses are uniform in space.

territory An area or volume of habitat defended by an organism or a group of organisms against other individuals, usually of the same species; territorial behavior is the behavior by which the territory is defended.

theistic evolution The belief that evolution occurs based on natural laws that were established by a deity.

theory *See* **scientific theory**.

threshold trait A characteristic that is expressed as discrete states, although the genetic variation underlying it is polygenic.

time for speciation The amount of time required for reproductive isolation to evolve, once the process starts.

trade-off The existence of both a fitness benefit and a fitness cost of a mutation or character state, relative to another.

transcription factor A protein that interacts with a regulatory DNA sequence and affects the transcription of the associated gene.

transcriptome A specified set of mRNA transcripts, such as those found in a specific cell type, under specific conditions or in the organism as a whole.

transition A mutation that changes a nucleotide to another nucleotide in the same class (purine or pyrimidine). *Cf.* **transversion**.

translocation The transfer of a segment of a chromosome to another, nonhomologous, chromosome; the chromosome formed by the addition of such a segment.

transposable element A DNA sequence, copies of which become inserted into various sites in the genome.

transposition Movement of a copy of a transposable element to a different site in the genome.

***trans*-regulatory element** A soluble molecule, usually a transcription factor protein, that binds to a *cis*-regulatory element of a gene, and is encoded by a gene located elsewhere in the genome.

transversion A mutation that changes a nucleotide to another nucleotide in the opposite class (purine or pyrimidine). *Cf.* **transition**.

trend *See* **evolutionary trend**.

two-fold cost of males The loss of fitness incurred by a sexually-reproducing genotype or population, relative to an asexual genotype, caused by producing male offspring, that themselves do not make descendants.

U

ultraconserved elements Regions of the genome that are highly conserved, sometimes at the level of 100% identity, between distantly related species. Many ultraconserved elements occur in exons that encode proteins, but others occur outside of genes and presumably have a regulatory function.

underdominance Lower fitness of a heterozygote than of both of the homozygotes for the same alleles.

unequal crossing over Recombination between nonhomologous sites on two homologous chromosomes.

uniformitarianism The proposition that natural processes that operated in the past are the same as in the present. (The term has usually implied gradual rather than catastrophic change.)

unstable equilibrium An equilibrium to which a system does not return if disturbed.

V

variability Properly, the ability of a system to vary. Often used to mean "variation."

variance (s^2, s^2, V, σ^2) The average squared deviation of an observation from the arithmetic mean; hence, a measure of variation.

vegetative propagation Production of offspring from somatic tissues, e.g., by buds.

vertical transmission *See* **horizontal transmission**.

vestigial Occurring in a rudimentary condition as a result of evolutionary reduction from a more elaborated, functional character state in an ancestor.

viability Capacity for survival; often refers to the fraction of individuals surviving to a given age, and is contrasted with inviability due to deleterious genes.

vicariance   Separation of a continuously distributed ancestral population or species into separate populations because of the development of a geographic or ecological barrier.

virulence   Usually, the damage inflicted on a host by a pathogen or parasite; sometimes, the capacity of a pathogen or parasite to infect and develop in a host.

W

whole-genome duplication   The origin of a polyploid descendant by either the duplication of the genome of one species (**autopolyploidy**) or by hybridization between two unreduced gametes from different species (**allopolyploidy**).

wild-type   The allele, genotype, or phenotype that is most prevalent (if there is one) in wild populations; with reference to the wild-type allele, other alleles are often termed mutations.

Z

zygote   A single-celled individual formed by the union of gametes. Occasionally used more loosely to refer to an offspring produced by sexual reproduction.

Literature Cited

CHAPTER 1

1. Bowler, P. J. 1989. *Evolution: The History of an Idea*. University of California Press, Berkeley.
2. Bowler, P. J. 1996. *Life's Splendid Drama: Evolutionary Biology and the Reconstruction of Life's Ancestry 1860–1940*. University of Chicago Press, Chicago.
3. Centers for Disease Control (CDC). 2013. Antibiotic resistance threats in the United States, 2013. http://www.cdc.gov/drugresistance/threat-report-2013
4. Dobzhansky, Th. 1937. *Genetics and the Origin of Species*. Columbia University Press, New York.
5. Dobzhansky, Th. 1973. Nothing in biology makes sense except in the light of evolution. *American Biology Teacher* 35: 125–129.
6. Gemmell, N. J., V. J. Metcalf, and F. W. Allendorf. 2004. Mother's curse: The effect of mtDNA on individual fitness and population viability. *Trends Ecol. Evol.* 19: 238–244.
7. Gire, S. K., and 57 others. 2014. Genomic surveillance elucidates Ebola virus origin and transmission during the 2014 outbreak. *Science* 345: 1369–1372.
8. Goldschmidt, R. B. 1940. *The Material Basis of Evolution*. Yale University Press, New Haven, CT.
9. Hahn, B. H., G. M. Shaw, K. M. De Cock, and P. M. Sharp. 2000. AIDS as a zoonosis: Scientific and public health implications. *Science* 287: 607–614.
10. Hey, J. 2011. Regarding the confusion between the population concept and Mayr's "population thinking." *Q. Rev. Biol.* 86: 253–264.
11. Hofstadter, R. 1955. *Social Darwinism in American Thought*. Beacon Press, Boston, MA.
12. Innocenti, P., E. H. Morrow, and D. K. Dowling. 2011. Experimental evidence supports a sex-specific selective sieve in mitochondrial genome evolution. *Science* 332: 845–848.
13. Kimura, M. 1983. *The Neutral Theory of Molecular Evolution*. Cambridge University Press, Cambridge.
14. Laland, K. N., K. Sterelny, J. Odling-Smee, W. Hoppitt, and T. Uller. 2011. Cause and effect in biology revisited: Is Mayr's proximate-ultimate dichotomy still useful? *Science* 334: 1512–1516.
15. Levin, B. R., and R. M. Anderson. 1999. The population biology of anti-infective chemotherapy and the evolution of drug resistance: More questions than answers. In S. C. Stearns (ed.), *Evolution in Health and Disease*, pp. 125–137. Oxford University Press, Oxford.
16. Lovejoy, A. O. 1936. *The Great Chain of Being: A Study of the History of an Idea*. Harvard University Press, Cambridge, MA.
17. Mayr, E. 1942. *Systematics and the Origin of Species*. Columbia University Press, New York.
18. Mayr, E. 1982. *The Growth of Biological Thought: Diversity, Evolution, and Inheritance*. Harvard University Press, Cambridge, MA.
19. Mayr, E., and W. B. Provine (eds.). 1980. *The Evolutionary Synthesis: Perspectives on the Unification of Biology*. Harvard University Press, Cambridge, MA.
20. Palumbi, S. R. 2001. *The Evolution Explosion: How Humans Cause Rapid Evolutionary Change*. W. W. Norton, New York.
21. Paradis, J., and G. C. Williams. 1989. *Evolution and Ethics: T. H. Huxley's Evolution and Ethics with New Essays on Its Victorian and Sociobiological Context*. Princeton University Press, Princeton, NJ.
22. Perron, G. G., R. F. Inglis, P. S. Pennings, and S. Cobey. 2015. Fighting microbial drug resistance: A primer on the role of evolutionary biology in public health. *Evol. Appl.* 8: 211–222.
23. Perros, M. 2015. A sustainable model for antibiotics. *Science* 347: 1062–1064.
24. Rensch, B. 1959. *Evolution above the Species Level*. Columbia University Press, New York.
25. Sharp, P. M., and B. H. Hahn. 2011. Origins of HIV and the AIDS epidemic. *Cold Spring Harbor Perspect. Med.* 1: a006841.
26. Simpson, G. G. 1944. *Tempo and Mode in Evolution*. Columbia University Press, New York.
27. Simpson, G. G. 1953. *The Major Features of Evolution*. Columbia University Press, New York.
28. Smocovitis, V. B. 1996. *Unifying Biology: The Evolutionary Synthesis and Evolutionary Biology*. Princeton University Press, Princeton, NJ.
29. Stebbins, G. L. 1950. *Variation and Evolution in Plants*. Columbia University Press, New York.
30. Székely, T., C. K. Catchpole, A. Devoogd, Z. Marchl, and T. J. Devoogd. 1996. Evolutionary changes in a song control area of the brain (HVC) are associated with evolutionary changes in song repertoire among European warblers (Sylviidae). *Proc. R. Soc. Lond. B* 263: 607–610.

CHAPTER 2

1. Boto, L. 2010. Horizontal gene transfer in evolution: Facts and challenges. *Proc. R. Soc. Lond. B* 277: 819–827.
2. Chippindale, P. T., R. M. Bonett, A. S. Baldwin, and J. J. Wiens. 2004. Phylogenetic evidence for a major reversal of life-history evolution in plethodontid salamanders. *Evolution* 58: 2809–2822.
3. Collin, R., and M. P. Miglietta. 2008. Reversing opinions on Dollo's Law. *Trends Ecol. Evol.* 23: 602–609.
4. Crisp, A., C. Boschetti, M. Perry, A. Tunnacliffe, and G. Micklem. 2015. Expression of multiple horizontally acquired genes is a hallmark of both vertebrate and invertebrate genomes. *Genome Biol.* 16, art. No. 50. doi:10.1186/s13059-015-0607-3.
5. Darwin, C. 1859. *The Origin of Species by Means of Natural Selection, or the Preservation of Favored Races in the Struggle for Life*. Modern Library, New York.

6. Fournier, G. P., C. P. Andam, and J. P. Gogarten. 2015. Ancient horizontal gene transfer and the last common ancestors. *BMC Evol. Biol.* 15, art. No. 70. doi:10.1186/s12862-015-0350-0.

7. Futuyma, D. J. 1995. *Science on Trial: The Case for Evolution.* Sinauer, Sunderland, MA.

8. Gaut, B., L. Yang, S. Takano, and L. E. Eguiarte. 2011. The patterns and causes of variation in plant nucleotide substitution rates. *Annu. Rev. Ecol. Evol. Syst.* 42: 245–266.

9. Gillooly, J. F., A. P. Allen, G. B. West, and J. H. Brown. 2005. The rate of DNA evolution: Effects of body size and temperature on the molecular clock. *Proc. Natl. Acad. Sci. USA* 102: 140–145.

10. Grant, P. R., and B. R. Grant. 2008. *How and Why Species Multiply: The Radiation of Darwin's Finches.* Princeton University Press, Princeton, NJ.

11. Gross, B. L., and L. H. Rieseberg. 2005. The ecological genetics of homoploid hybrid speciation. *J. Hered.* 96: 241–252.

12. Halder, G., P. Callaerts, and W. J. Gehring. 1995. Induction of ectopic eyes by targeted expression of the eyeless gene in *Drosophila. Science* 267: 1788–1792.

13. Hartwell, L. H., L. Hood, M. L. Goldberg, A. E. Reynolds, L. M. Silver, and R. C. Veres. 2000. *Genetics: From Genes to Genomes.* McGraw-Hill Higher Education, Boston, MA.

14. Hillis, D. M. 2010. Phylogenetic progress and applications of the tree of life. In M. A. Bell, D. J. Futuyma, W. F. Eanes, and J. S. Levinton (eds.), *Evolution since Darwin: The First 150 Years,* pp. 421–449. Sinauer, Sunderland, MA.

15. Hoge, M. A. 1915. Another gene in the fourth chromosome of *Drosophila. Am. Nat.* 49: 47–49.

16. Jonsson, K. A., and 10 others. 2012. Ecological and evolutionary determinants for the adaptive radiation of the Madagascan vangas. *Proc. Natl. Acad. Sci. USA* 109: 6620–6625.

17. Juhas, M. 2015. Horizontal gene transfer in human pathogens. *Crit. Rev. Microbiol.* 41: 101–108.

18. Kachroo, A. H., J. M. Laurent, C. M. Yellman, A. G. Meyer, C. O. Wilke, and E. M. Marcotte. 2015. Systematic humanization of yeast genes reveals conserved functions and genetic modularity. *Science* 348: 921–925.

19. Kerney, R. R., D. C. Blackburn, H. Müller, and J. Hanken. 2012. Do larval traits re-evolve? Evidence from the embryogenesis of a direct-developing salamander, *Plethodon cinereus. Evolution* 66: 252–262.

20. Lamichhaney, S., and 13 others. 2015. Evolution of Darwin's finches and their beaks revealed by genome sequencing. *Nature* 518: 371–375.

21. Langley, C. H., and W. M. Fitch. 1974. An examination of the constancy of the rate of molecular evolution. *J. Mol. Evol.* 3: 161–177.

22. Li, W.-H. 1997. *Molecular Evolution.* Sinauer, Sunderland, MA.

23. Li, W.-H., D. L. Ellsworth, J. Khrushkal, B.-J. Chang, and D. H. Emmet. 1996. Rates of nucleotide substitution in primates and rodents and the generation-time effect hypothesis. *Mol. Phyl. Evol.* 5: 182–187.

24. Lodato, M. A., and 13 others. 2015. Somatic mutation in single human neurons tracks developmental and transcriptional history. *Science* 350: 94–98.

25. Michael, G. B., and 7 others. 2015. Emerging issues in antimicrobial resistance of bacteria from food-producing animals. *Future Microbiol.* 10: 427–443.

26. Milá, B., D. J. Girman, M. Kimura, and T. B. Smith. 2000. Genetic evidence for the effect of a postglacial population expansion on the phylogeography of a North American songbird. *Proc. R. Soc. Lond. B* 267: 1033–1040.

27. Moran, N. A., and T. Jarvik. 2010. Lateral transfer of genes from fungi underlies carotenoid production in aphids. *Science* 328: 624–627.

28. Naxerova, K., and R. K. Jain. 2015. Using tumour phylogenetics to identify the roots of metastasis in humans. *Nat. Rev. Clin. Oncol.* 12: 258–272.

29. Porter, M. L., and K. A. Crandall. 2003. Lost along the way: The significance of evolution in reverse. *Trends Ecol. Evol.* 18: 541–547.

30. Reddy, S., and 5 others. 2012. Diversification and the adaptive radiation of the vangas of Madagascar. *Proc. R. Soc. Lond. B* 279: 2062–2071.

31. Schluter, D. 2000. *The Ecology of Adaptive Radiation.* Oxford University Press, Oxford.

32. Ujvari, B., and 10 others. 2015. Widespread convergence in toxin resistance by predictable molecular evolution. *Proc. Natl. Acad. Sci. USA* 112: 11911–11916.

33. Walls, G. L. 1942. *The Vertebrate Eye and Its Adaptive Radiation.* Cranbrook Institute of Science, Bloomfield Hills, MI.

34. Wells, M. J. 1966. Cephalopod sense organs. In K. M. Wilbur and C. M. Yonge (eds.), *Physiology of Mollusca,* vol. 2, pp. 523–545. Academic Press, New York.

35. Young, J. Z. 1971. *The Anatomy of the Nervous System of* Octopus vulgaris. Oxford University Press, London.

CHAPTER 3

1. Antonovics, J., A. D. Bradshaw, and R. G. Turner. 1971. Heavy metal tolerance in plants. *Adv. Ecol. Res.* 7: 1–85.

2. Bradshaw, A. D. 1991. Genostasis and the limits to evolution. The Croonian Lecture, 1991. *Philos. Trans. R. Soc., B* 333: 289–305.

3. Bradshaw, W. E., and C. M. Holzapfel. 2001. Genetic shift in photoperiodic response correlated with global warming. *Proc. Natl. Acad. Sci. USA* 98: 14509–14511.

4. Burt, A., and R. Trivers. 2006. *Genes in Conflict: The Biology of Selfish Genetic Elements.* Harvard University Press, Cambridge, MA.

5. Carroll, S. B., and C. Boyd. 1992. Host race radiation in the soapberry bug: Natural history with the history. *Evolution* 46: 1052–1069.

6. Carroll, S. P., H. Dingle, and S. P. Klassen. 1997. Genetic differentiation of fitness-associated traits among rapidly evolving populations of the soapberry bug. *Evolution* 51: 1182–1188.

7. Castellanos, M. C., P. Wilson, and J. D. Thomson. 2004. "Anti-bee" and "pro-bird" changes during the evolution of hummingbird pollination in *Penstemon* flowers. *J. Evol. Biol.* 17: 876–885.

8. Coltman, D. W., P. O'Donoghue, J. T. Jorgenson, J. T. Hogg, C. Strobeck, and M. Festa-Blanchet. 2003. Undesirable evolutionary consequences of trophy hunting. *Nature* 426: 655–658.

9. Dawkins, R. 1976. *The Selfish Gene.* Oxford University Press, Oxford.

10. Dennett, D. C. 1995. *Darwin's Dangerous Idea: Evolution and the Meanings of Life.* Simon & Schuster, New York.

11. Eldakar, O. T., and D. S. Wilson. 2011. Eight criticisms not to make of group selection. *Evolution* 65: 1523–1526.

12. Endler, J. A. 1986. *Natural Selection in the Wild.* Princeton University Press, Princeton, NJ.

13. Felsenstein, J. 1985. Phylogenies and the comparative method. *Am. Nat.* 125: 1–15.

14. Futuyma, D. J. 2010. Evolutionary constraint and ecological consequences. *Evolution* 64: 1865–1884.

15. Galis, F. 1999. Why do almost all mammals have seven cervical vertebrae? Developmental constraints, Hox genes, and cancer. *J. Exp. Zool.* 285: 19–26.

16. Ghiselin, M. T. 1969. *The Triumph of the Darwinian Method.* University of California Press, Berkeley.

17. Gill, F. B. 1995. *Ornithology,* 2nd ed. W. H. Freeman, New York.

18. Gould, S. J. 1982. The meaning of punctuated equilibrium and its role in validating a hierarchical approach to macroevolution. In R. Milkman (ed.), *Perspectives on Evolution,* pp. 83–104. Sinauer, Sunderland, MA.

19. Gould, S. J. 2002. *The Structure of Evolutionary Theory.* Belknap Press of Harvard University Press, Cambridge, MA.

20. Gould, S. J., and E. S. Vrba. 1982. Exaptation: A missing term in the science of form. *Paleobiology* 8: 4–15.

21. Grant, P. R. 1986. *Ecology and Evolution of Darwin's Finches.* Princeton University Press, Princeton, NJ.

22. Grant, P. R., and B. R. Grant. 2006. Evolution of character displacement in Darwin's finches. *Science* 313: 224–226.

23. Grant, P. R., and B. R. Grant. 2008. *How and Why Species Multiply: The Radiation of Darwin's Finches*. Princeton University Press, Princeton, NJ.

24. Haldane, J. B. S. 1932. *The Causes of Evolution*. Longmans, Green, New York.

25. Harvey, P. H., and M. D. Pagel. 1991. *The Comparative Method in Evolutionary Biology*. Oxford University Press, Oxford.

26. Hendry, A. P., and M. T. Kinnison. 1999. Perspective: The pace of modern life: Measuring rates of contemporary microevolution. *Evolution* 53: 1637–1653.

27. Hurst, G. D. D., and J. H. Werren. 2001. The role of selfish genetic elements in eukaryotic evolution. *Nat. Rev. Genet.* 2: 597–606.

28. Jablonski, D. 2008. Species selection: Theory and data. *Annu. Rev. Ecol. Evol. Syst.* 39: 501–524.

29. Keller, E. F., and E. A. Lloyd (eds.). 1992. *Keywords in Evolutionary Biology*. Harvard University Press, Cambridge, MA.

30. Kuparinen, A., and J. Merilä. 2007. Detecting and managing fisheries-induced evolution. *Trends Ecol. Evol.* 22: 652–659.

31. Lewontin, R. C. 2000. *The Triple Helix: Gene, Organism, and Environment*. Harvard University Press, Cambridge, MA.

32. Macnair, M. R. 1981. Tolerance of higher plants to toxic materials. In J. A. Bishop and L. M. Cook (eds.), *Genetic Consequences of Man Made Change*, pp. 177–207. Academic Press, New York.

33. Mayr, E. 1988. Cause and effect in biology. In E. Mayr (ed.), *Toward a New Philosophy of Biology*, pp. 24–37. Harvard University Press, Cambridge, MA.

34. Metcalf, R. L., and W. H. Luckmann (eds.). 1994. *Introduction to Insect Pest Management*, 3rd ed. Wiley, New York.

35. Normark, B. B., O. P. Judson, and N. A. Moran. 2003. Genomic signatures of ancient asexual lineages. *Biol. J. Linn. Soc.* 79: 69–84.

36. Odling-Smee, F. J., K. N. Laland, and M. W. Feldman. 2003. *Niche Construction: The Neglected Process in Evolution*. Princeton University Press, Princeton, NJ.

37. Okasha, S. 2006. *Evolution and the Levels of Selection*. Oxford University Press, Oxford.

38. Olsen, E. M., and 6 others. 2004. Maturation trends indicative of rapid evolution preceded the collapse of northern cod. *Nature* 428: 932–935.

39. Paley, W. 1928. *Natural Theology*. (Reprinted 2009, Cambridge University Press, Cambridge).

40. Palumbi, S. R. 2001. *The Evolution Explosion: How Humans Cause Rapid Evolutionary Change*. W. W. Norton, New York

41. Porter, K. R. 1972. *Herpetology*. W. B. Saunders, Philadelphia, PA.

42. Rabosky, D. L., and A. R. McCune. 2010. Reinventing species selection with molecular phylogenies. *Trends Ecol. Evol.* 25: 68–74.

43. Reeve, H. K., and P. W. Sherman. 1993. Adaptation and the goals of evolutionary research. *Q. Rev. Biol.* 68: 1–32.

44. Richerson, P. J., and R. Boyd. 2005. *Not By Genes Alone: How Culture Transformed Human Evolution*. University of Chicago Press, Chicago.

45. Ruse, M. 1979. *The Darwinian Revolution*. University of Chicago Press, Chicago.

46. Schluter, D., and P. R. Grant. 1984. Determinants of morphological patterns in communities of Darwin's finches. *Am. Nat.* 123: 175–196.

47. Smith, T. B., and L. Bernatchez. 2008. Evolutionary change in human-altered environments. *Mol. Ecol.* 17: 1–8.

48. Sober, E. 1984. *The Nature of Selection: Evolutionary Theory in Philosophical Focus*. MIT Press, Cambridge, MA.

49. Stearns, S. C. 1986. Natural selection and fitness, adaptation and constraint. In D. M. Raup and D. Jablonski (eds.), *Patterns and Processes in the History of Life*, pp. 23–44. Springer-Verlag, Berlin.

50. Varela-Lasheras, I., A. J. Baker, S. D. van der Mije, J. A. J. Metz, J. van Alphen, and F. Galis. 2011. Breaking evolutionary and pleiotropic constraints in mammals: On sloths, manatees, and homeotic mutations. *EvoDevo* 2, article no. 11. doi:10.1186/2041-9139-2-11.

51. Williams, G. C. 1966. *Adaptation and Natural Selection*. Princeton University Press, Princeton, NJ.

52. Williams, G. C. 1992. *Gaia*, nature worship and biocentric fallacies. *Q. Rev. Biol.* 67: 479–486.

CHAPTER 4

1a. Allison, A. C. 1956. The sickle-cell and haemoglobin-C genes in some African populations. *Ann. Hum. Genet.* 21: 67–89.

1b. Bank, C., R. T. Hietpas, A. Wong. D. N. Bolon, and J. D. Jenson. 2014. A Bayesian approach to assess the complete distribution of fitness effects of new mutations: Uncovering the potential for adaptive walks in challenging environments. *Genetics* 196: 841–852.

2. Bonduriansky, R., and T. Day. 2009. Nongenetic inheritance and its evolutionary implications. *Annu. Rev. Ecol. Evol. Syst.* 40: 103–125.

3. Chandler, V. L. 2007. Paramutation: From maize to mice. *Cell* 128: 641–645.

4. Crosland, M. W. J., and R. H. Crozier. 1986. *Myrmecia pilosula*, an ant with only one pair of chromosomes. *Science* 231: 1278.

5. Darmon, E., and D. R. F. Leach. 2014. Bacterial genome instability. *Microbiol. Mol. Biol. Rev.* 78: 1–39.

6. Dobzhansky, Th. G. 1970. *Genetics of the Evolutionary Process*. Columbia University Press, New York.

7. Eyre-Walker, A., and P. D. Keightley. 2007. The distribution of fitness effects of new mutations. *Nat. Rev. Genet.* 8: 610–618.

8. Eyre-Walker, A., M. Woolfit, and T. Phelps. 2006. The distribution of fitness effects of new deleterious amino acid mutations in humans. *Genetics* 173: 891–900.

9. Feuk, L., and 7 others. 2005. Discovery of human inversion polymorphisms by comparative analysis of human and chimpanzee DNA sequence assemblies. *PLoS Genet.* 1: 489–498.

10. Ford, E. B. 1971. *Ecological Genetics*. Chapman & Hall, London.

11. Keightley, P. D. 2012. Rates and fitness consequences of new mutations in humans. *Genetics* 190: 295–304.

12. Khandelwal, S. 1990. Chromosome evolution in the genus *Ophioglossum* L. *Bot. J. Linn. Soc.* 102: 205–217.

13. Kondrashov, F. A., and A. S. Kondrashov. 2010. Measurements of spontaneous rates of mutations in the recent past and the near future. *Philos. Trans. R. Soc., B* 365: 1169–1176.

14. Lederberg, J., and E. M. Lederberg. 1952. Replica plating and indirect selection of bacterial mutants. *J. Bacteriol.* 63: 399–406.

15. Lee, H., E. Popodi, H. X. Tang, and P. L. Foster. 2012. Rate and molecular spectrum of spontaneous mutations in the bacterium Escherichia coli as determined by whole-genome sequencing. *Proc. Natl. Acad. Sci. USA* 109: E2774–E2783.

16. Lynch, M. 2010. Evolution of the mutation rate. *Trends Genet.* 26: 345–352.

17. Lynch, M., and 6 others. 2016. Genetic drift, selection and the evolution of the mutation rate. *Nat. Rev. Genet.* 17: 704–714.

18. MacDougall-Shackleton, E. A., and S. A. MacDougall-Shackleton. 2001. Cultural and genetic evolution in mountain white-crowned sparrows: Song dialects are associated with population structure. *Evolution* 55: 2568–2575.

19. Ochman, H., J. G. Lawrence, and E. A. Groisman. 2000. Lateral gene transfer and the nature of bacterial innovation. *Nature* 405: 299–304.

20. Richardson, A. O., and J. D. Palmer. 2007. Horizontal gene transfer in plants. *J. Exp. Bot.* 58: 1–9.

21. Sebat, J., and 20 others. 2004. Large-scale copy number polymorphism in the human genome. *Science* 305: 525–528.

22. Sniegowski, P. D., P. J. Gerrish, T. Johnson, and A. Shaver. 2000. The evolution of mutation rates: Separating causes from consequences. *Bioessays* 22: 1057–1066.

23. Sturtevant, A. H. 1923. Inheritance of direction of coiling in *Limnaea*. *Science* 58: 269–270.

24. Swart, E. C., and 28 others. 2013. The *Oxytricha trifallax* macronuclear genome: A complex eukaryotic genome with 16,000 tiny chromosomes. *PLoS Biol.* 11: e1001473.

25a. The International HapMap Consortium. 2005. A haplotype map of the human genome. *Nature* 437: 1299–1320.

25b. Wang, I. J., and H. B. Shaffer. 2008. Rapid color evolution in an aposematic species: A phylogenetic analysis of color variation in the strikingly polymorphic strawberry poison-dart frog. *Evolution* 62: 2742.

26. Ye, X. D., and 6 others. 2000. Engineering the provitamin A (beta-carotene) biosynthetic pathway into (carotenoid-free) rice endosperm. *Science* 287: 303–305.

CHAPTER 5

1. Agrawal, A. F., and M. C. Whitlock. 2012. Mutation load: The fitness of individuals in populations where deleterious alleles are abundant. *Annu. Rev. Ecol. Evol. Syst.* 43: 115–135.

2. Ayala, D., R. F. Guerrero, and M. Kirkpatrick. 2013. Reproductive isolation and local adaptation quantified for a chromosome inversion in a malaria mosquito. *Evolution* 67: 946–958.

3a. Barb, J. G., and 6 others. 2014. Chromosomal evolution and patterns of introgression in *Helianthus*. *Genetics* 197: 969–979.

3b. Barrett, R. D. H., S. M. Rogers, and D. Schluter. 2008. Natural selection on a major armor gene in threespine stickleback. *Science* 322: 255–257.

4. Beja-Pereira, A., and 10 others. 2003. Gene-culture coevolution between cattle milk protein genes and human lactase genes. *Nat. Genet.* 35: 311–313.

5. Bell, M. A., W. E. Aguirre, and N. J. Buck. 2004. Twelve years of contemporary armor evolution in a threespine stickleback population. *Evolution* 58: 814–824.

6. Bersaglieri, T., and 8 others. 2004. Genetic signatures of strong recent positive selection at the lactase gene. *Am. J. Hum. Genet.* 74: 1111–1120.

7. Bodmer, W. F., and L. L. Cavalli-Sforza. 1976. *Genetics, Evolution, and Man*. W. H. Freeman, San Francisco.

8. Burt, A. 1995. Perspective: The evolution of fitness. *Evolution* 49: 1–8.

9. Charlesworth, B. 2015. Causes of natural variation in fitness: Evidence from studies of *Drosophila* populations. *Proc. Natl. Acad. Sci. USA* 112: 1662–1669.

10. Christiansen, F. B. 1984. The definition and measurement of fitness. In B. Shorrocks (ed.), *Evolutionary Ecology*, pp. 65–79. Blackwell Scientific, Oxford.

11. Cook, L. M. 2003. The rise and fall of the *Carbonaria* form of the peppered moth. *Q. Rev. Biol.* 78: 399–417.

12. Cook, L. M., B. S. Grant, I. J. Saccheri, and J, Mallet. Selective bird predation on the peppered moth: The last experiment of Michael Majerus. *Biol. Lett.* (*London, U.K.*) 8: 609–612.

13. Coop, G., and 9 others. 2009. The role of geography in human adaptation. *PLoS Genet.* 5.

14. Elgueroa, E., and 14 others. 2015. Malaria continues to select for sickle cell trait in Central Africa. *Proc. Natl. Acad. Sci. USA* 112: 7051–7054.

15. Eyre-Walker, A., and P. D. Keightley. 2007. The distribution of fitness effects of new mutations. *Nat. Rev. Genet.* 8: 610–618.

16. Gigord, L. D. B., M. R. Macnair, and A. Smithson. 2001. Negative frequency-dependent selection maintains a dramatic flower color polymorphism in the rewardless orchid *Dactylorhiza sambucina* (L.) Soò. *Proc. Natl. Acad. Sci. USA* 98: 6253–6255.

17. Haldane, J. B. S. 1924. A mathematical theory of natural and artificial selection. I. *Trans. Cambridge Philos. Soc.* 23: 19–41.

18. Hartl, D. L., and A. G. Clark. 2007. *Principles of Population Genetics*. Sinauer, Sunderland, MA.

19. Henn, B. M., L. R. Botigué, C. D. Bustamante, A. G. Clark, and S. Gravel. 2015. Estimating the mutation load in human genomes. *Nat. Rev. Genet.* 16: 333–343.

20. Hof, A. E. van't, and 8 others. 2016. The industrial melanism mutation in British peppered moths is a transposable element. *Nature* 534: 102–105.

21. Ingram, C. J. E., C. A. Mulcare, Y. Itan, M. G. Thomas, and D. M. Swallow. 2009. Lactose digestion and the evolutionary genetics of lactase persistence. *Hum. Genet.* 124: 579–591.

22. Jaquiery, J., and 9 others. 2012. Genome scans reveal candidate regions involved in the adaptation to host plant in the pea aphid complex. *Mol. Ecol.* 21: 5251–5264.

23. Johnston, S. E., and 6 others. 2013. Life history trade-offs at a single locus maintain sexually selected genetic variation. *Nature* 502: 93–95.

24. Keightley, P. D. 2012. Rates and fitness consequences of new mutations in humans. *Genetics* 190: 295–304.

25. Kirkpatrick, M. 2010. How and why chromosome inversions evolve. *PLoS Biol.* 8: e1000501.

26. Le Rouzic, A., and 6 others. 2011. Strong and consistent natural selection associated with armour reduction in sticklebacks. *Mol. Ecol.* 20: 2483–2493.

27. Lesecque, Y., P. D. Keightley, and A. Eyre-Walker. 2012. A resolution of the mutation load paradox in humans. *Genetics* 191: 1321–1330.

28. Lynch, M. 2010. Rate, molecular spectrum, and consequences of human mutation. *Proc. Natl. Acad. Sci. USA* 107: 961–968.

29. Macklon, N. S., J. P. M. Geraedts, and B. Fauser. 2002. Conception to ongoing pregnancy: The "black box" of early pregnancy loss. *Hum. Reprod. Update* 8: 333–343.

30. Majerus, M. E. N. 1998. *Melanism: Evolution in Action*. Oxford University Press, Oxford.

31. Mallet, J., and N. H. Barton. 1989. Strong natural selection in a warning color hybrid zone. *Evolution* 43: 421–431.

32. Mathieson, I., and 37 others. 2015. Genome-wide patterns of selection in 230 ancient Eurasians. *Nature* 528: 499–503.

33. Nielsen, R. 2005. Molecular signatures of natural selection. *Annu. Rev. Genet.* 39: 197–218.

34. Palumbi, S. R. 2001. *The Evolution Explosion: How Humans Cause Rapid Evolutionary Change*. Norton, New York.

35. Piel, F. B., and 7 others. 2010. Global distribution of the sickle cell gene and geographical confirmation of the malaria hypothesis. *Nat. Commun.* 1.

36. Ross, M., and J. Silverman. 1995. Genetic studies of a behavioral mutant, glucose aversion, in the German cockroach (Dictyoptera: Blattellidae). *J. Insect Behav.* 8: 825–834.

37. Sharakhov, I. V., and 11 others. 2002. Inversions and gene order shuffling in *Anopheles gambiae* and *A. funestus*. *Science* 298: 182–185.

38. Studer, A., Q. Zhao, J. Ross-Ibarra, and J. Doebley. 2011. Identification of a functional transposon insertion in the maize domestication gene tb1. *Nat. Genet.* 43: 1160–1163.

39. Sturm, R. A., and D. L. Duffy. 2012. Human pigmentation genes under environmental selection. *Genome Biol.* 13: 248.

40. Tutt, J. W. 1891. *Melanism and Melanochroism in British Lepidoptera*. Swan Sonnenschein, London.

41. Tutt, J. W. 1896. *British Moths*. Routledge, London.

42. Wada-Katsumata, A., J. Silverman, and C. Schal. 2013. Changes in taste neurons support the emergence of an adaptive behavior in cockroaches. *Science* 340: 972–975.

43. Wang, R. L., A. Stec, J. Hey, L. Lukens, and J. Doebley. 1999. The limits of selection during maize domestication. *Nature* 398: 236–239.

44. West, S. A., A. S. Griffin, A. Gardner, and S. P. Diggle, 2006. Social evolution theory for microorganisms. *Nat. Rev. Microbiol.* 4: 597–607.

45. Wright, S. 1949. Adaptation and selection. In E. May (ed.) *Genetics, Paleontology, and Evolution*, pp. 365–389. Princeton University Press, Princeton, NJ.

46. Zuidhof, M. J., B. L. Schneider, V. L. Carney, D. R. Corver, and F. E. Robinson. 2014. Growth, efficiency, and yield of commercial broilers from 1957, 1978, and 2005. *Poult. Sci.* 93: 2970–2982.

CHAPTER 6

1a. Allen, C. E., P. Beldade, B. J. Zwaan, and P. M. Brakefield. 2008. Differences in the selection response of serially repeated color pattern characters: Standing variation, development, and evolution. *BMC Evol. Biol.* 8: 94.

1b. Barrett, R. D. H., and D. Schluter. 2008. Adaptation from standing genetic variation. *Trends Ecol. Evolut.* 23: 38–44.

2. Beldade, P., K. Koops, and P. M. Brakefield. 2002a. Developmental conswtraints versus flexibility in morphological evolution. *Nature* 416: 844–847.

3. Beldade, P., K. Koops, and P. M. Brakefield. 2002b. Modularity, individuality, and evo-devo in butterfly wings. *Proc. Natl. Acad. Sci. USA* 99: 14262–14267.

4. Bell, G., and A. Gonzalez. 2009. Evolutionary rescue can prevent extinction following environmental change. *Ecol. Lett.* 12: 942–948.

5. Bell, G., and A. Gonzalez. 2011. Adaptation and evolutionary rescue in metapopulations experiencing environmental deterioration. *Science* 332: 1327–1330.

6. Benkman, C. W. 2003. Divergent selection drives the adaptive radiation of crossbills. *Evolution* 57: 1176–1181.

7. Boag, P. T. 1983. The heritability of external morphology in Darwin's ground finches (*Geospiza*) on Isla Daphne Major, Galápagos. *Evolution* 37: 877–894.

8. Boag, P. T., and P. R. Grant. 1984. The classical case of character release: Darwin's finches (*Geospiza*) on Isla Daphne Major, Galápagos. *Biol. J. Linn. Soc.* 22: 243–287.

9. Brodie, E. D. 1989. Genetic correlations between morphology and antipredator behaviour in natural populations of the garter snake *Thamnophis ordinoides*. *Nature* 342: 542–543.

10. Brodie, E. D. 1992. Correlational selection for color pattern and antipredator behavior in the garter snake *Thamnophis ordinoides*. *Evolution* 46: 1284–1298.

11. Carlson, S. M., C. J. Cunningham, and P. A. H. Westley. 2014. Evolutionary rescue in a changing world. *Trends Ecol. Evolut.* 29: 521–530.

12. Cavalli-Sforza, L. L., and W. F. Bodmer. 1971. *The Genetics of Human Populations.* W. H. Freeman, San Francisco.

13. Chevin, L. M., R. Lande, and G. M. Mace. 2010. Adaptation, plasticity, and extinction in a changing environment: Towards a predictive theory. *PLoS Biol.* 8.

14. Dudley, J. W., and R. J. Lambert. 2004. 100 generations of selection for oil and protein content in corn. *Plant Breed. Rev.* 24: 79–110.

15. Etterson, J. R. 2004. Evolutionary potential of *Chamaecrista fasciculata* in relation to climate change. 1. Clinal patterns of selection along an environmental gradient in the great plains. *Evolution* 58: 1446–1458.

16. Etterson, J. R., and R. G. Shaw. 2001. Constraint to adaptive evolution in response to global warming. *Science* 294: 151–154.

17. Florez, J. C., J. Hirschhorn, and D. Altshuler. 2003. The inherited basis of diabetes mellitus: Implications for the genetic analysis of complex traits. *Annu. Rev. Genomics Hum. Genet.* 4: 257–291.

18. Frary, A., and 9 others. 2000. fw2.2: A quantitative trait locus key to the evolution of tomato fruit size. *Science* 289: 85–88.

19. Gejman, P. V., A. R. Sanders, and K. S. Kendler. 2011. Genetics of schizophrenia: New findings and challenges. *Annu. Rev. Genomics Hum. Genet.* 12: 121–144.

20. Ghalambor, C. K., J. K. McKay, S. P. Carroll, and D. N. Reznick. 2007. Adaptive versus non-adaptive phenotypic plasticity and the potential for contemporary adaptation in new environments. *Funct. Ecol.* 21: 394–407.

21. Gomulkiewicz, R., and R. D. Holt. 1995. When does evolution by natural selection prevent extinction? *Evolution* 49: 201–207.

22. Hansen, T. F., and D. Houle. 2008. Measuring and comparing evolvability and constraint in multivariate characters. *J. Evol. Biol.* 21: 1201–1219.

23. Hill, W. G., and M. Kirkpatrick. 2010. What animal breeding has taught us about evolution. *Annu. Rev. Ecol. Evol. Syst.* 41: 1–19.

24. Houde, A. E. 1987. Mate choice based upon naturally occurring color pattern variation in a guppy population. *Evolution* 41: 1–10.

25. Houle, D. 1992. Comparing evolvability and variability of quantitative traits. *Genetics* 130: 194–204.

26. Hughes, K. A., A. E. Houde, A. C. Price, and F. H. Rodd. 2013. Mating advantage for rare males in wild guppy populations. *Nature* 503: 108–110.

27. Johnson, T., and N. Barton. 2005. Theoretical models of selection and mutation on quantitative traits. *Philos. Trans. R. Soc., B* 360: 1411–1425.

28. Jones, F. C., and 33 others. 2012. The genomic basis of adaptive evolution in threespine sticklebacks. *Nature* 484: 55–61.

29. Joron, M., and 22 others. 2011. Chromosomal rearrangements maintain a polymorphic supergene controlling butterfly mimicry. *Nature* 477: 203–206.

30. Kellermann, V., B. van Heerwaarden, C. M. Sgrò, and A. A. Hoffmann. 2009. Fundamental evolutionary limits in ecological traits drive *Drosophila* species distributions. *Science* 325: 1244–1246.

31. Kellermann, V. M., B. van Heerwaarden, A. A. Hoffmann, and C. M. Sgrò. 2006. Very low additive genetic variance and evolutionary potential in multiple populations of two rainforest *Drosophila* species. *Evolution* 60: 1104–1108.

32. Kingsolver, J. G., and S. E. Diamond. 2011. Phenotypic selection in natural populations: What limits directional selection? *Am. Nat.* 177: 346–357.

33. Kirkpatrick, M. 2009. Patterns of quantitative genetic variation in multiple dimensions. *Genetica* 136: 271–284.

34. Lande, R. 1979. Quantitative genetic analysis of multivariate evolution, applied to brain: Body size allometry. *Evolution* 33: 402–416.

35. Lynch, M., and B. Walsh. 1998. *Genetics and Analysis of Quantitative Traits.* Sinauer Associates, Sunderland, MA.

36. Mundy, N. I. 2005. A window on the genetics of evolution: MC1R and plumage colouration in birds. *Proc. R. Soc. Lond. B* 272: 1633–1640.

37. Musunuru, K., and S. Kathiresan. 2010. Genetics of coronary artery disease. *Annu. Rev. Genomics Hum. Genet.* 11: 91–108.

38. Nienhuis, A. W., A. C. Nathwani, and A. M. Davidoff. 2016. Gene therapy for hemophilia. *Hum. Gene Ther.* 27: 305–308.

39. Pfennig, D. W., A. Mabry, and D. Orange. 1991. Environmental causes of correlations between age and size of metamorphosis in *Scaphiopus multiplicatus*. *Ecology* 72: 2240–2248.

40. Price, T. D., P. R. Grant, H. L. Gibbs, and P. T. Boag. 1984. Recurrent patterns of natural selection in a population of Darwin's finches. *Nature* 309: 787–789.

41. Quintero, I., and J. J. Wiens. 2013. Rates of projected climate change dramatically exceed past rates of climatic niche evolution among vertebrate species. *Ecol. Lett.* 16: 1095–1103.

42. Schluter, D. 1988. Estimating the form of natural selection on a quantitative trait. *Evolution* 42: 849–861.

43. Schluter, D. 1996. Adaptive radiation along genetic lines of least resistance. *Evolution* 50: 1766–1774.

44. Scoville, A. G., and M. E. Pfrender. 2010. Phenotypic plasticity facilitates recurrent rapid adaptation to introduced predators. *Proc. Natl. Acad. Sci. USA* 107: 4260–4263.

45. Shaw, R. G., and J. R. Etterson. 2012. Rapid climate change and the rate of adaptation: insight from experimental quantitative genetics. *New Phytol.* 195: 752–765.

46. Sheldon, B. C., L. E. B. Kruuk, and J. Merila. 2003. Natural selection and inheritance of breeding time and clutch size in the collared flycatcher. *Evolution* 57: 406–420.

47. Smith, T. B. 1993. Disruptive selection and the genetic basis of bill size polymorphism in the African finch *Pyrenestes*. *Nature* 363: 618–620.

48. Stern, D. L., and V. Orgogozo. 2008. The loci of evolution: How predictable is genetic evolution? *Evolution* 62: 2155–2177.

49. Storz, J. F., G. R. Scott, and Z. A. Cheviron. 2010. Phenotypic plasticity and genetic adaptation to high-altitude hypoxia in vertebrates. *J. Exp. Biol.* 213: 4125–4136.

50. Streisfeld, M. A., and M. D. Rausher. 2011. Population genetics, pleiotropy, and the preferential fixation of mutations during adaptive evolution. *Evolution* 65: 629–642.

51. Tishkoff, S. 2015. Strength in small numbers. *Science* 349: 1282–1283.

52. Weber, K. E. 1990. Increased selection response in larger populations. 1. Selection for wing-tip height in *Drosophila melanogaster* at three population sizes. *Genetics* 125: 579–584.

53. Wood, A. R. T. Esko J. Yang, and 461 others. 2014. Defining the role of common variation in the genomic and biological architecture of adult human height. *Nat. Genet.* 46: 1173–1186.

54. Wray, G. A. 2013. Genomics and the evolution of phenotypic traits. *Annu. Rev. Ecol. Evol. Syst.* 44: 51–72.

55. Yoshizawa, M., Y. Yamamoto, K. E. O'Quin, and W. R. Jefferey. 2012. Evolution of an adaptive behavior and its sensory receptors promotes eye regression in blind cavefish. *BMC Biol.* 10: 1–16.

56. Young, K. V., E. D. Brodie, Jr., and E. D. Brodie, III. 2004. How the horned lizard got its horns. *Science* 304: 65–65.

CHAPTER 7

1. Agrawal, A. F., and M. C. Whitlock. 2012. Mutation load: The fitness of individuals in populations where deleterious alleles are abundant. *Annu. Rev. Ecol. Evol. Syst.* 43: 115–135.

2. Balakrishnan, C. N., and S. V. Edwards. 2009. Nucleotide variation, linkage disequilibrium and founder-facilitated speciation in wild populations of the zebra finch (*Taeniopygia guttata*). *Genetics* 181: 645–660.

3. Barrick, J. E., and 7 others. 2009. Genome evolution and adaptation in a long-term experiment with *Escherichia coli*. *Nature* 461: 1243–1247.

4. Bonnell, M. L., and R. K. Selander. 1974. Elephant seals: Genetic variation and near extinction. *Science* 184: 908–909.

5. Buri, P. 1956. Gene frequency in small populations of mutant *Drosophila*. *Evolution* 10: 367–402.

6. Bustamante, C. D., and 13 others. 2005. Natural selection on protein-coding genes in the human genome. *Nature* 437: 1153–1157.

7. Charlesworth, B. 2009. Effective population size and patterns of molecular evolution and variation. *Nat. Rev. Genet.* 10: 195–205.

8. Charlesworth, J., and A. Eyre-Walker. 2006. The rate of adaptive evolution in enteric bacteria. *Mol. Biol. Evol.* 23: 1348–1356.

9. Ellegren, H., and N. Galtier. 2016. Determinants of genetic diversity. *Nat. Rev. Genet.* 17: 422–433.

10. Frankham, R., J. D. Ballou, and D. A. Briscoe. 2002. *Introduction to Conservation Genetics*. Cambridge University Press, Cambridge, UK.

11. Graur, D. 2016. *Molecular and Genome Evolution*. Sinauer, Sunderland, MA.

12. Hershberg, R., and D. A. Petrov. 2008. Selection on codon bias. *Annu. Rev. Genet.* 42: 287–299.

13. Hoelzel, A. R., R. C. Fleischer, C. Campagna, B. J. Le Boeuf, and G. Alvord. 2002. Impact of a population bottleneck on symmetry and genetic diversity in the northern elephant seal. *J. Evol. Biol.* 15: 567–575.

14. Hoffmann, A. A., and L. H. Rieseberg. 2008. Revisiting the impact of inversions in evolution: From population genetic markers to drivers of adaptive shifts and speciation? *Annu. Rev. Ecol. Evol. Syst.* 39: 21–42.

15. Karmin, M. L., and 99 others. 2015. A recent bottleneck of Y chromosome diversity coincides with a global change in culture. *Genome Res.* 25: 459–466.

16. Keightley, P. D., and A. Eyre-Walker. 2012. Estimating the rate of adaptive molecular evolution when the evolutionary divergence between species is small. *J. Mol. Evol.* 74: 61–68.

17. Kimura, M. 1983. *The Neutral Theory of Molecular Evolution*. Cambridge University Press, Cambridge.

18. Kreitman, M. 1983. Nucleotide polymorphism at the alcohol dehydrogenase locus of Drosophila melanogaster. *Nature* 304: 412–417.

19. Li, H., and R. Durbin. 2011. Inference of human population history from individual whole-genome sequences. *Nature* 475: 493–496.

20. Li, J. Z., and 10 others. 2008. Worldwide human relationships inferred from genome-wide patterns of variation. *Science* 319: 1100–1104.

21. Liu, H., F. Prugnolle, A. Manica, and F. Balloux. 2006. A geographically explicit genetic model of worldwide human settlement history. *Am. J. Hum. Genet.* 79: 230–237.

22. Lou, D., R. and 6 others. 2014. Rapid evolution of BRCA1 and BRCA2 in humans and other primates. *BMC Evol. Biol.* 14: 155.

23. Mackay, T. F. and 51 others. 2012. The *Drosophila melanogaster* genetic reference panel. *Nature* 482: 173–178.

24. Madsen, T., R. Shine, M. Olsson, and H. Wittzell. 1999. Conservation biology: Restoration of an inbred adder population. *Nature* 402: 34–35.

25. McDonald, J. H., and M. Kreitman. 1991. Adaptive protein evolution at the ADH locus in *Drosophila*. *Nature* 351: 652–654.

26. McKusick, V. A. 2000. Ellis-van Creveld syndrome and the Amish. *Nat. Genet.* 24: 203–204.

27. Morris, R. M., and 6 others. 2002. SAR11 clade dominates ocean surface bacterioplankton communities. *Nature* 420: 806–810.

28. National Cancer Institute. 2015. BRCA1 and BRCA2: Cancer risk and genetic testing. http://www.cancer.gov/about-cancer/causes-prevention/genetics/brca-fact-sheet

29. Nei, M., Y. Suzuki, and M. Nozawa. 2010. The neutral theory of molecular evolution in the genomic era. *Annu. Rev. Genomics Hum. Genet.* 11: 265–289.

30. Nielsen, R. 2005. Molecular signatures of natural selection. *Annu. Rev. Genet.* 39: 197–218.

31. Nielsen, R., and M. Slatkin. 2013. *An Introduction to Population Genetics: Theory and Applications*. Macmillan Education, London.

32. Pool, J. E., I. Hellmann, J. D. Jensen, and R. Nielsen. 2010. Population genetic inference from genomic sequence variation. *Genome Res.* 20: 291–300.

33. Poznik, G. D., and 10 others. 2013. Sequencing Y chromosomes resolves discrepancy in time to common ancestor of males versus females. *Science* 341: 562–565.

34. Rambaut, A., O. G. Pybus, M. I. Nelson, C. Viboud, J. K. Taubenberger, and E. C. Holmes. 2008. The genomic and epidemiological dynamics of human influenza A virus. *Nature* 453: 615–619.

35. Selander, R. K., and D. W. Kaufman. 1975. Genetic structure of populations of brown snail (*Helix aspersa*). 1. Micro-geographic variation. *Evolution* 29: 385–401.

36. Vitti, J. J., S. R. Grossman, and P. C. Sabeti. 2013. Detecting natural selection in genomic data. *Annu. Rev. Genet.* 47: 97–120.

37. Yi, X., and 68 others. 2010. Sequencing of 50 human exomes reveals adaptation to high altitude. *Science* 329: 75–78.

CHAPTER 8

1a. Barrett, R. D. H., S. M. Rogers, and D. Schluter. 2008. Natural selection on a major armor gene in threespine stickleback. *Science* 322: 255–257.

1b. Barton, N. H., and K. S. Gale. 1993. Genetic analysis of hybrid zones. In R. G. Harrison (ed.), *Hybrid Zones and the Evolutionary Process*, pp. 13–45. Oxford University Press, Oxford.

2. Barton, N. H., and G. M. Hewitt. 1989. Adaptation, speciation and hybrid zones. *Nature* 341: 497–503.

3. Blair, W. F. 1960. *The Rusty Lizard: A Population Study*. University of Texas Press, Austin.

4. Bridle, J. R., S. Gavaz, and W. J. Kennington. 2009. Testing limits to adaptation along altitudinal gradients in rainforest Drosophila. *Proc. R. Soc. Lond. B* 276: 1507–1515.

5. Charlesworth, B., and Charlesworth, D. 2010. *Elements of Evolutionary Genetics*. Roberts and Company, Greenwood Village, CO.

6. Chen, I. C., J. K. Hill, R. Ohlemüller, D. B. Roy, and C. D. Thomas. 2011. Rapid range shifts of species associated with high levels of climate warming. *Science* 333: 1024–1026.

7. Clausen, J., D. D. Keck, and W. Hiesey. 1948. Experimental studies on the nature of species. III: Environmental responses of climatic races of *Achillea. Carnegie Inst. Wash. Publ.* 581.

8. Daday, H. 1954. Gene frequencies in wild populations of *Trifolium repens*. I. Distribution by latitude. *Heredity* 8: 61–78.

9. Ebert, D., C. Haag, M. Kirkpatrick, M. Riek, J. W. Hottinger, and V. I. Pajunen. 2002. A selective advantage to immigrant genes in a *Daphnia* metapopulation. *Science* 295: 485–488.

10. Friedenberg, N. A. 2003. Experimental evolution of dispersal in spatiotemporally variable microcosms. *Ecol. Lett.* 6: 953–959.

11. Haag, C. R., M. Saastamoinen, J. H. Marden, and I. Hanski. 2005. A candidate locus for variation in dispersal rate in a butterfly metapopulation. *Proc. R. Soc. Lond. B* 272: 2449–2456.

12. Hoekstra, H. E., K. E. Drumm, and M. W. Nachman. 2004. Ecological genetics of adaptive color polymorphism in pocket mice: geographic variation in selected and neutral genes. *Evolution* 58: 1329–1341.

13. Hohenlohe, P. A., S. Bassham, P. D. Etter, N. Stiffler, E. A. Johnson, and W. A. Cresko. 2010. Population genomics of parallel adaptation in threespine stickleback using sequenced RAD tags. *PLoS Genet.* 6: e1000862.

14. Jablonski, D., and G. Hunt. 2006. Larval ecology, geographic range, and species survivorship in Cretaceous mollusks: Organismic versus species-level explanations. *Am. Nat.* 168: 556–564.

15. Kellermann, V., B. van Heerwaarden, C. M. Sgrò, and A. A. Hoffmann. 2009. Fundamental evolutionary limits in ecological traits drive *Drosophila* species distributions. *Science* 325: 1244–1246.

16. Kellermann, V. M., B. van Heerwaarden, A. A. Hoffmann, and C. M. Sgrò. 2006. Very low additive genetic variance and evolutionary potential in multiple populations of two rainforest *Drosophila* species. *Evolution* 60: 1104–1108.

17. Kirkpatrick, M., and N. H. Barton. 1997. Evolution of a species' range. *Am. Nat.* 150: 1–23.

18. Lambers, J. H. R. 2015. Extinction risks from climate change. *Science* 348: 501–502.

19. Lenormand, T. 2002. Gene flow and the limits to natural selection. *Trends Ecol. Evolut.* 17: 183–189.

20. Lenormand, T., D. Bourguet, T. Guillemaud, and M. Raymond. 1999. Tracking the evolution of insecticide resistance in the mosquito *Culex pipiens. Nature* 400: 861–864.

21. McNeilly, T. 1968. Evolution in closely adjacent plant populations. 3. *Agrostis tenuis* on a small copper mine. *Heredity* 23: 99–108.

22. McPeek, M. A., and R. D. Holt. 1992. The evolution of dispersal in spatially and temporally varying environments. *Am. Nat.* 140: 1010–1027.

23. Munshi-South, J., Y. Zak, and E. Pehek. 2013. Conservation genetics of extremely isolated urban populations of the northern dusky salamander (*Desmognathus fuscus*) in New York City. *Peerj* 1: e64.

24. Nachman, M. W., H. E. Hoekstra, and S. L. D'Agostino. 2003. The genetic basis of adaptive melanism in pocket mice. *Proc. Natl. Acad. Sci. USA* 100: 5268–5273.

25. Olsen, K. M., N. J. Kooyers, and L. L. Small. 2014. Adaptive gains through repeated gene loss: parallel evolution of cyanogenesis polymorphisms in the genus *Trifolium* (Fabaceae). *Philos. Trans. R. Soc., B* 369: 20130347.

26. Parmesan, C. 2006. Ecological and evolutionary responses to recent climate change. *Annu. Rev. Ecol. Evol. Syst.* 37: 637–669.

27. Phillips, B. L., G. P. Brown, J. K. Webb, and R. Shine. 2006. Invasion and the evolution of speed in toads. *Nature* 439: 803.

28. Ramachandran, S., O. Deshpande, C. C. Roseman, N. A. Rosenberg, M. W. Feldman, and L. L. Cavalli-Sforza. 2005. Support from the relationship of genetic and geographic distance in human populations for a serial founder effect originating in Africa. *Proc. Natl. Acad. Sci. USA* 102: 15942–15947.

29. Raymond, M., and M. Marquine. 1994. Evolution of insecticide resistance in *Culex pipiens* populations: The Corsican paradox. *J. Evol. Biol.* 7: 315–337.

30. Saccheri, I., M. Kuussaari, M. Kankare, P. Vikman, W. Fortelius, and I. Hanski. 1998. Inbreeding and extinction in a butterfly metapopulation. *Nature* 392: 491–494.

31. Sand, H., G. Cederlund, and K. Danell. 1995. Geographical and latitudinal variation in growth patterns and adult body size of Swedish moose (*Alces alces*). *Oecologia* 102: 433–442.

32. Sexton, J. P., P. J. McIntyre, A. L. Angert, and K. J. Rice. 2009. Evolution and ecology of species range limits. *Annu. Rev. Ecol. Evol. Syst.* 40: 415–436.

33. Spitze, K. 1993. Population structure in *Daphnia obtusa*—quantitative genetic and allozymic variation. *Genetics* 135: 367–374.

34. Tallmon, D. A., G. Luikart, and R. S. Waples. 2004. The alluring simplicity and complex reality of genetic rescue. *Trends Ecol. Evolut.* 19: 489–496.

35. The International HapMap Consortium. 2005. A haplotype map of the human genome. *Nature* 437: 1299–1320.

36. Travis, J., and D. J. Futuyma. 1993. Global change: Lessons from and for evolutionary biology. In P. M. Kareiva, J. G. Kingsolver, and R. B. Huey (eds.), *Biotic Interactions and Global Change*, pp. 251–263. Sinauer, Sunderland MA.

37. Wang, X. H., and L. Kang. 2014. Molecular mechanisms of phase change in locusts. *Annu. Rev. Entomol.* 59: 225–244.

38. Whitlock, M. C. 2008. Evolutionary inference from Q_{ST}. *Mol. Ecol.* 17: 1885–1896.

39. Zera, A. J., and L. G. Harshman. 2001. The physiology of life history trade-offs in animals. *Annu. Rev. Ecol. Syst.* 32: 95–126.

CHAPTER 9

1. Abbott, R., and 39 others. 2013. Hybridization and speciation. *J. Evol. Biol.* 26: 229–246.

2. Barton, N. H., and B. Charlesworth. 1984. Genetic revolutions, founder effects, and speciation. *Annu. Rev. Ecol. Syst.* 15: 133–164.

3. Bates, M., and L. W. Hackett. 1939. The distinguishing characteristics of the populations of *Anopheles maculipennis* found in southern Europe. *Proc. 7th Int. Congr. Ent.* 3: 1555–1569.

4. Behm, J. E., A. R. Ives, and J. W. Boughman. 2010. Breakdown in postmating isolation and the collapse of a species pair through hybridization. *Am. Nat.* 175: 11–26.

5. Berlocher, S. H., and J. L. Feder. 2002. Sympatric speciation in phytophagous insects: Moving beyond controversy? *Annu. Rev. Entomol.* 47: 773–815.

6. Bikard, D., and 6 others. 2009. Divergent evolution of duplicate genes leads to genetic incompatibilities within *A. thaliana. Science* 323: 623–626.

7. Bolnick, D. I., and B. M. Fitzpatrick. 2007. Sympatric speciation: Models and empirical evidence. *Annu. Rev. Ecol. Evol. Syst.* 38: 459–487.

8. Bradshaw, H. D., S. M. Wilbert, K. G. Otto, and D. W. Schemske. 1998. Quantitative trait loci affecting differences in floral morphology between two species of monkeyflower. *Genetics* 149: 367–382.

9. Burton, R. S., R. J. Pereir, and F. S. Barreto. 2013. Cytonuclear genomic interactions and hybrid breakdown. *Annu. Rev. Ecol. Evol. Syst.* 44: 281–302.

10. Cooley, J. R., C. Simon, D. C. Marshall, K. Slon, and C. Ehrhardt. 2001. Allochronic speciation, secondary contact, and reproductive character displacement in periodical cicadas (Hemiptera: *Magicicada*

spp.): Genetic, morphological, and behavioural evidence. *Mol. Ecol.* 10: 661–671.

11. Coolon, J. D., C. J. McManus, K. R. Stevenson, B. R. Graveley, and P. J. Wittkopp. 2014. Tempo and mode of regulatory evolution in *Drosophila. Genome Res.* 24: 797–808.

12. Coyne, J. A. 1974. The evolutionary origin of hybrid inviability. *Evolution* 28: 505–506.

13. Coyne, J. A. 1984. Genetic basis of male sterility in hybrids between two closely related species of *Drosophila. Proc. Natl. Acad. Sci. USA* 81: 4444–4447.

14. Coyne, J. A., and H. A. Orr. 1997. "Patterns of speciation in *Drosophila*" revisited. *Evolution* 51: 295–303.

15. Coyne, J. A., and H. A. Orr. 2004. *Speciation.* Sinauer, Sunderland, MA

16. Coyne, J. A., and T. D. Price. 2000. Little evidence for sympatric speciation in island birds. *Evolution* 54: 2166–2171.

17. Cracraft, J. 1989. Speciation and its ontology: The empirical consequences of alternative species concepts for understanding patterns and processes of differentiation. In D. Otte and J. A. Endler (eds.), *Speciation and Its Consequences*, pp. 29–59. Sinauer, Sunderland, MA.

18. Crespi, B., and P. Nosil. 2013. Conflictual speciation: Species formation via genomic conflict. *Trends Ecol. Evol.* 28: 48–57.

19. Cruikshank, T. E., and M. W. Hahn. 2014. Reanalysis suggests that genomic islands of speciation are due to reduced diversity, not reduced gene flow. *Mol. Ecol.* 23: 3133–3157.

20. Dobzhansky, Th. 1937. *Genetics and the Origin of Species.* Columbia University Press, New York.

21. Dodd, D. M. B. 1989. Reproductive isolation as a consequence of adaptive divergence in *Drosophila pseudoobscura. Evolution* 43: 1308–1311.

22. Eberhard, W. G. 1996. *Female Control: Sexual Selection by Cryptic Female Choice.* Princeton University Press, Princeton, NJ.

23. Feder, J. L., and 9 others. 2003. Allopatric genetic origins for sympatric host-plant shifts and race formation in *Rhagoletis. Proc. Natl. Acad. Sci. USA* 100: 10314–10319.

24. Felsenstein. J. 1981. Skepticism towards Santa Rosalia, or why are there so few kinds of animals? *Evolution* 35: 124–138.

25. Fjeldså, J., R. C. K. Bowie, and C. Rahbek. 2012. The role of mountain ranges in the diversification of birds. *Annu. Rev. Ecol. Evol. Syst.* 43: 249–265.

26. Fontaine, M. C., and 18 others. 2015. Extensive introgression in a malaria vector species complex revealed by phylogenomics. *Science* 347: 42.

27. Fowler, N. L., and D. A. Levin. 1984. Ecological constraints on the establishment of a novel polyploid in competition with its diploid progenitor. *Am. Nat.* 124: 703–711.

28. Friesen, V. L., and 6 others. 2007. Sympatric speciation by allochrony in a seabird. *Proc. Natl. Acad. Sci. USA* 104: 18589–18594.

29. Fry, J. D. 2003. Multilocus models of sympatric speciation: Bush vs. Rice vs. Felsenstein. *Evolution* 57: 1735–1746.

30. Futuyma, D. J. 1991. A new species of *Ophraella* Wilcox (Coleoptera: Chrysomelidae) from the southeastern United States. *J. NY Entomol. Soc.* 99: 643–653.

31. Galiana, A., A. Moya, and F. J. Ayala. 1993. Founder-flush speciation in *Drosophila pseudoobscura*: A large-scale experiment. *Evolution* 47: 432–444.

32. Galindo, B. E., V. D. Vacquier, and W. J. Swanson. 2003. Positive selection in the egg receptor for abalone sperm lysin. *Proc. Natl. Acad. Sci. USA* 100: 4639–4643.

33. Grace, J. L., and K. L. Shaw. 2011. Coevolution of male mating signal and female preference during early lineage divergence of the Hawaiian cricket, *Laupala cerasina. Evolution* 65: 2184–2196.

34. Hardwick, K. M., J. M. Robertson, and E. B. Rosenblum. 2013. Asymmetrical mate preference in recently adapted White Sands and black lava populations of *Sceloporus undulatus. Curr. Zool.* 59: 20–30.

35. *Heliconius* Genome Consortium. 2012. Butterfly genome reveals promiscuous exchange of mimicry adaptations among species. *Nature* 487: 94–98.

36. Hoffmann, A. A., and L. H. Rieseberg. 2008. Revisiting the impact of inversions in evolution: From population genetic markers to drivers of adaptive shifts and speciation? *Annu. Rev. Ecol. Evol. Syst.* 39: 21–42.

37. Hopkins, R. 2013. Reinforcement in plants. *New Phytol.* 197: 1095–1103.

38. Husband, B. C. 2000. Constraints on polyploid evolution: A test of the minority cytotype exclusion principle. *Proc. R. Soc. Lond. B* 267: 217–223.

39. Kampen, H. 2005. Integration of *Anopheles beklemishevi* (Diptera: Culicidae) in a PCR assay diagnostic for Palaearctic *Anopheles maculipennis* sibling species. *Parasitol. Res.* 97: 113–117.

40. Kaneshiro, K.Y. 1980. Sexual isolation, speciation and the direction of evolution. *Evolution* 34: 437–444.

41. Katakura, H., and T. Hosogai. 1994. Performance of hybrid ladybird beetles (*Epilachna*) on the host plants of parental species. *Entomol. Exp. Appl.* 71: 81–85.

42. Kirkpatrick, M., and V. Ravigné. 2002. Speciation by natural and sexual selection: Models and experiments. *Am. Nat.* 159: S22–S35.

43. Kisel, Y., and T. G. Barraclough. 2010. Speciation has a spatial scale that depends on gene flow. *Am. Nat.* 175: 316–334.

44. Knowlton, N., L. A. Weigt, L. A. Solórzano, D. K. Mills, and E. Bermingham. 1993. Divergence in proteins, mitochondrial DNA, and reproductive compatibility across the Isthmus of Panama. *Science* 260: 1629–1632.

45. Kocher, T. D. 2004. Adaptive evolution and explosive speciation: The cichlid fish model. *Nat. Rev. Genet.* 5: 288–298.

46. Kocher, T. D., J. A. Conroy, K. R. McKaye, and J. R. Stauffer. 1993. Similar morphologies of cichlid fish in Lakes Tanganyika and Malawi are due to convergence. *Mol. Phylogenet. Evol.* 2: 158–165.

47. Kozak, K. H., and J. J. Wiens. 2006. Does niche conservatism promote speciation? A case study in North American salamanders. *Evolution* 60: 2604–2621.

48. Kronforst, M. R., L. G. Young, D. D. Kapan, C. McNeely, R. J. O'Neill, and L. E. Gilbert. 2006. Linkage of butterfly mate preference and wing color preference cue at the genomic location of *wingless. Proc. Natl. Acad. Sci. USA* 103: 6575–6580.

49. Lessios, H. A. 2008. The great American schism: Divergence of marine organisms after the rise of the Central American Isthmus. *Annu. Rev. Ecol. Evol. Syst.* 39: 63–91.

50. Levin, D. A. 1983. Polyploidy and novelty in flowering plants. *Am. Nat.* 122: 1–25.

51. Mallet, J. 2005. Hybridization as an invasion of the genome. *Trends Ecol. Evol.* 20: 229–237.

52. Martin, M. D., and T. C. Mendelson. 2014. Changes in sexual signals are greater than changes in ecological traits in a dichromatic group of fishes. *Evolution* 68: 3618–3628.

53. Masly, J. P. 2012. 170 Years of "lock-and-key": Genital morphology and reproductive isolation. *Intl. J. Evol. Biol.* 247352. doi:10.1155/2012/247352.

54. Matute, D. 2013. The role of founder effects on the evolution of reproductive isolation. *J. Evol. Biol.* 26: 2299–2311.

55. Mayr, E. 1942. *Systematics and the Origin of Species.* Columbia University Press, New York.

56. Mayr, E. 1954. Change of genetic environment and evolution. In J. Huxley, A. C. Hardy, and E. B. Ford (eds.), *Evolution as a Process*, pp. 157–180. Allen and Unwin, London.

57. Mayr, E. 1963. *Animal Species and Evolution.* Harvard University Press, Cambridge, MA.

58. Mayr, E. 1982. Processes of speciation in animals. In C. Barigozzi (ed.), *Mechanisms of Speciation*, pp. 1–19. Alan R. Liss, New York.

59. Mendelson, T. C. 2003. Sexual isolation evolves faster than hybrid inviability in a diverse and sexually dimorphic genus of fish (Percidae: *Etheostoma*). *Evolution* 57: 317–327.

60. Merrill, R. M., and 6 others. 2012. Disruptive ecological selection on a mating cue. *Proc. R. Soc. Lond. B* 279: 4907–4913.

61. Michel, A. P., S. Sim, T. H. Q. Powell, M. S. Taylor, P. Nosil, and J. L. Feder. 2010. Widespread genomic divergence during sympatric speciation. *Proc. Natl. Acad. Sci. USA* 107: 9724–9729.

62. Muller, H. J. 1942. Isolating mechanisms, evolution and temperature. *Biol. Symp.* 6: 71–125.

63. Navarro, A., and N. H. Barton. 2003. Accumulating postzygotic isolation genes in parapatry: A new twist on chromosomal speciation. *Evolution* 57: 447–459.

64. Nosil, P. 2012. *Ecological Speciation*. Oxford University Press, Oxford.

65. Nosil, P., and D. Schluter. 2011. The genes underlying the process of speciation. *Trends Ecol. Evol.* 26: 160–167.

66. Ochman, H., E. Lerat, and V. Daubin. 2005. Examining bacterial species uner the specter of gene transfer and exchange. *Proc. Natl. Acad. Sci. USA* 102: 6595–6599.

67. Otto, S. P., and J. Whitton. 2000. Polyploid incidence and evolution. *Annu. Rev. Genet.* 34: 401–437.

68. Ownbey, M. 1950. Natural hybridization and amphiploidy in the genus *Tragopogon*. *Am. J. Bot.* 37: 489–499.

69. Palumbi, S. R. 1998. Species formation and the evolution of gamete recognition loci. In D. J. Howard and S. H. Berlocher (eds.), *Endless Forms: Species and Speciation*, pp. 271–278. Oxford University Press, New York.

70. Papadopulos, A. S. T., and 9 others. 2014. Evaluation of genetic isolation within an island flora reveals unusually widespread local adaptation and supports sympatric speciation. *Phil. Trans. R. Soc., B* 369: 20130342.

71. Phadnis, N., and H. A. Orr. 2009. A single gene causes both male sterility and segregation distortion in *Drosophila* hybrids. *Science* 323: 376–379.

72. Pires, J. C., and 9 others. 2004. Molecular cytogenetic analysis of recently evolved *Tragopogon* (Asteraceae) allopolyploids reveal a karyotype that is additive of the diploid progenitors. *Am. J. Bot.* 91: 1022–1035.

73. Polly, P. D., and 9 others. 2013. Phenotypic variation across chromosomal hybrid zones of the common shrew (*Sorex araneus*) indicates reduced gene flow. *PLoS ONE* 8: e67455.

74. Polyakov, A. V., T. A. White, R. M. Jones, P. M. Borodin, and J. B. Searle. 2011. Natural hybridization between extremely divergent chromosomal races of the common shrew (*Sorex araneus*, Soricidae, Soricomorpha): Hybrid zone in Siberia. *J. Evol. Biol.* 24: 1393–1402.

75. Presgraves, D. C. 2010. The molecular evolutionary basis of species formation. *Nat. Rev. Genet.* 11: 175–180.

76. Price, T. D. 2008. *Speciation in Birds*. Roberts and Co., Greenwood Village, CO.

77. Proudfoot, G. A., F. R. Gehlbach, and R. L. Honeycutt. 2007. Mitochondrial DNA variation and phylogeography of the Eastern and Western Screech-Owls. *Condor* 109: 617–627.

78. Ramsey, J. 2011. Polyploidy and ecological adaptation in wild yarrow. *Proc. Natl. Acad. Sci. USA* 108: 7096–7101.

79. Ramsey, J., and D. W. Schemske. 1998. Pathways, mechanisms, and rates of polyploid formation in flowering plants. *Annu. Rev. Ecol. Syst.* 29: 467–502.

80. Ramsey, J., H. D. Bradshaw, and D. W. Schemske. 2003. Components of reproductive isolation between the monkeyflowers *Mimulus lewisii* and *M. cardinalis* (Scrophulariaceae). *Evolution* 57: 1520–1534.

81. Rieseberg, L. H. 2006. Hybrid speciation in wild sunflowers. *Ann. Mo. Bot. Gard.* 93: 34–48.

82. Rieseberg, L. H., and 7 others. 2003. Majory ecological transitions in wild sunflowers facilitated by hybridization. *Science* 301: 1211–1216.

83. Rieseberg, L. H., J. Whitton, and K. Gardner. 1999. Hybrid zones and the genetic architecture of a barrier to gene flow between two sunflower species. *Genetics* 152: 713–727.

84. Rodríguez, D. J. 1996. A model for the establishment of polyploidy in plants. *Am. Nat.* 147: 33–46.

85. Roelofs, W., and 8 others. 1987. Sex pheromone production and perception in European corn borer moths is determined by both autosomal and sex-linked genes. *Proc. Natl. Acad. Sci. USA* 84: 7585–7589.

86. Rosenblum, E. B., and L. J. Harmon. 2011. "Same same but different": Replicated ecological speciation at White Sands. *Evolution* 65: 946–960.

87. Rosenblum, E. B., B. A. J. Sarver, J. W. Brown, and 6 others. 2012. Goldilocks meets Santa Rosalia: An ephemeral speciation model explains patterns of diversification across time scales. *Evol. Biol.* 39: 255–261.

88. Ryan, M. J., and A. S. Rand. 1993. Species recognition and sexual selection as a unitary problem in animal communication. *Evolution* 47: 647–657.

89. Salzburger, W., B. Van Boexlaer, and A. S. Cohen. 2014. Ecology and evolution of the African Great Lakes and their faunas. *Annu. Rev. Ecol. Evol. Syst.* 45: 519–545.

90. Savolainen, V., and 9 others. 2006. Sympatric speciation in palms on an oceanic island. *Nature* 441: 210–213.

91. Schemske, D. W. 2010. Adaptation and the origin of species. *Am. Nat.* 176: S4–S25.

92. Schemske, D. W., and H. D. Bradshaw, Jr. 1999. Pollinator preference and the evolution of floral traits in monkeyflowers (*Mimulus*). *Proc. Natl. Acad. Sci. USA* 96: 11910–11915.

93. Schluter, D. 2000. *The Ecology of Adaptive Radiation*. Oxford University Press, Oxford.

94. Seddon, N., and 9 others. 2013. Sexual selection accelerates signal evolution during speciation in birds. *Proc. Royal Soc. Lond. B* 280: 20131065. doi:10.1098/rspb.2013.1065.

95. Seehausen, O. 2006. African cichlid fish: A model system in adaptive radiation research. *Proc. R. Lond. Soc. B* 273: 1987–1998.

96. Seehausen, O., J. J. M. van Alphen, and F. Witte. 1997. Cichlid fish diversity threatened by eutrophication that curbs sexual selection. *Science* 277: 1808–1811.

97. Slatkin, M. 1996. In defense of founder-effect speciation. *Am. Nat.* 147: 493–505.

98. Stebbins, R. C. 1954. *Amphibians and Reptiles of Western North America*. McGraw-Hill, New York.

99. Swanson, W. J., and V. D. Vacquier. 2002. Reproductive protein evolution. *Annu. Rev. Ecol. Syst.* 33: 161–179.

100. Szymura, J. M. 1993. Analysis of hybrid zones with *Bombina*. In R. G. Harrison (ed.), *Hybrid Zones and the Evolutionary Process*, pp. 261–289. Oxford University Press, New York.

101. Templeton, A. R. 2008. The reality and importance of founder speciation in evolution. *BioEssays* 30: 470–479.

102. Tilley, S. G., P. A. Verrell, and S. J. Arnold. 1990. Correspondence between sexual isolation and allozyme differentiation: A test in the salamander *Desmognathus ochrophaeus*. *Proc. Natl. Acad. Sci. USA* 87: 2715–2719.

103. Tobias, J. A., and 6 others. 2010. Song divergence by sensory drive in Amazonian birds. *Evolution* 64: 2820–2839.

104. Turelli, M., and L. C. Moyle. 2007. Asymmetric postmating isolation: Darwin's corollary to Haldane's rule. *Genetics* 176: 1059–1088.

105. Wagner, C. E., L. J. Harmon, and O. Seehausen. 2014. Cichlid species-area relationships are shaped by adaptive radiations that scale with area. *Ecol. Lett.* 17: 583–592.

106. Weir, J. T., and T. D. Price. 2011. Limits to speciation inferred from times to secondary sympatry and ages of hybridizing species along a latitudinal gradient. *Am. Nat.* 177: 462–469.

107. Wood, T. E., N. Takebayashi, M. S. Barker, I. Mayrose, P. B. Greenspoon, and L. H. Rieseberg. 2009. The frequency of polyploid

speciation in vascular plants. *Proc. Natl. Acad. Sci. USA* 106: 13875–13879.

108. Yukilevich, R. 2012. Asymmetrical patterns of speciation uniquely support reinforcement in *Drosophila*. *Evolution* 66: 1430–1446.

CHAPTER 10

1. Andersson, M. 1982. Female choice selects for extreme tail length in a widowbird. *Nature* 299: 818–820.

2. Andrade, M. C. B. 1996. Sexual selection for male sacrifice in the Australian redback spider. *Science* 271: 70–72.

3. Arnold, S. J., and D. Duvall. 1994. Animal mating systems: A synthesis based on selection theory. *Am. Nat.* 143: 317–348.

4. Bachtrog, D., and 14 others. 2014. Sex determination: Why so many ways of doing it? *PLoS Biol.* 12.

5. Barber, I., S. A. Arnott, V. A. Braithwaite, J. Andrew, and F. A. Huntingford. 2001. Indirect fitness consequences of mate choice in sticklebacks: Offspring of brighter males grow slowly but resist parasitic infections. *Proc. R. Soc. Lond. B* 268: 71–76.

6. Basolo, A. L. 1995. Phylogenetic evidence for the role of a preexisting bias in sexual selection. *Proc. R. Soc. Lond. B* 259: 307–311.

7. Bateman, A. J. 1948. Intra-sexual selection in *Drosophila*. *Heredity* 2: 349–368.

8. Bell, G. 1985. On the function of flowers. *Proc. R. Soc. Lond. B* 224: 223–265.

9. Berglund, A., A. Bisazza, and A. Pilastro. 1996. Armaments and ornaments: An evolutionary explanation of traits of dual utility. *Biol. J. Linn. Soc.* 58: 385–399.

10. Castellano, D., M. Coronado-Zamora, J. L. Campos, A. Barbadilla, and A. Eyre-Walker. 2016. Adaptive evolution is substantially impeded by Hill-Robertson interference in *Drosophila*. *Mol. Biol. Evol.* 33: 442–455.

11. Charlesworth, D., X. Vekemans, V. Castric, and S. Glemin. 2005. Plant self-incompatibility systems: A molecular evolutionary perspective. *New Phytol.* 168: 61–69.

12. Clutton-Brock, T. H. 1982. *Red deer: Behavior and ecology of two sexes.* University of Chicago Press, Chicago.

13. Cortez, D., and 7 others. 2014. Origins and functional evolution of Y chromosomes across mammals. *Nature* 508: 488–493.

14. Crow, J. F., and M. Kimura. 1965. Evolution in sexual and asexual populations. *Am. Nat.* 99: 439–450.

15. Eberhard, W. G. 1985. *Sexual Selection and Animal Genitalia.* Harvard University Press, Cambridge, MA.

16. Emlen, S. T., and L. W. Oring. 1977. Ecology, sexual selection, and evolution of mating systems. *Science* 197: 215–223.

17. Felsenstein, J. 1974. Evolutionary advantage of recombination. *Genetics* 78: 737–756.

18. Fontaneto, D., and 6 others. 2007. Independently evolving species in asexual bdelloid rotifers. *PLoS Biol.* 5: 914–921.

19. Gwynne, D. T. 2008. Sexual conflict over nuptial gifts in insects. *Annu. Rev. Entomol.* 53: 83–101.

20. Hamilton, W. D. 1967. Extraordinary sex ratios. *Science* 156: 477–488.

21. Harcourt, A. H., P. H. Harvey, S. G. Larson, and R. V. Short. 1981. Testis weight, body weight and breeding systems in primates. *Nature* 293: 55–57.

22. Herre, E. A. 1985. Sex-ratio adjustment in fig wasps. *Science* 228: 896–898.

23. Janzen, F. J., and G. L. Paukstis. 1991. Environmental sex determination in reptiles: Ecology, evolution, and experimental design. *Q. Rev. Biol.* 66: 149–179.

24. Jennions, M. D., and M. Petrie. 1997. Variation in mate choice and mating preferences: A review of causes and consequences. *Biol. Rev. Camb. Philos. Soc.* 72: 283–327.

25. Kamran-Disfani, A., and A. F. Agrawal. 2014. Selfing, adaptation and background selection in finite populations. *J. Evol. Biol.* 27: 1360–1371.

26. Kirkpatrick, M., and N. H. Barton. 1997. The strength of indirect selection on female mating preferences. *Proc. Natl. Acad. Sci. USA* 94: 1282–1286.

27. Kirkpatrick, M., and M. J. Ryan. 1991. The evolution of mating preferences and the paradox of the lek. *Nature* 350: 33–38.

28. Komdeur, J., S. Daan, J. Tinbergen, et al. 1997. Extreme adaptive modification in sex ratio of the Seychelles warbler's eggs. *Nature* 385: 522–525.

29. Lively, C. M., and J. Jokela. 2002. Temporal and spatial distributions of parasites and sex in a freshwater snail. *Evol. Ecol. Res.* 4: 219–226.

30. McDonald, M. J., D. P. Rice, and M. M. Desai. 2016. Sex speeds adaptation by altering the dynamics of molecular evolution. *Nature* 531: 233–236.

31. McLean, C. Y., and 12 others. 2011. Human-specific loss of regulatory DNA and the evolution of human-specific traits. *Nature* 471: 216–219.

32. National Library of Medicine Genetics Home Reference. 2015. Y chromosome. ghr.nlm.nih.gov

33. Oliveira, R. F., M. Taborsky, and H. J. Brockmann. 2008. *Alternative Reproductive Tactics.* Cambridge University Press, New York.

34. Otto, S. P. 2009. The evolutionary enigma of sex. *Am. Nat.* 174: S1–S14.

35. Packer, C., and A. E. Pusey. 1983. Adaptations of female lions to infanticide by incoming males. *Am. Nat.* 121: 716–728.

36. Page, R. A., M. J. Ryan, and X. E. Bernal. 2013. Be loved, be prey, be eaten. In K. Yasukawa, ed. *Animal Behavior. Vol. 3: Case Studies: Integration and Application of Animal Behavior*, pp. 123–154. Praeger, New York.

37. Parker, G. A. 1970. Sperm competition and its evolutionary consequences in insects. *Biol. Rev. Camb. Philos. Soc.* 45: 525–567.

38. Peck, J. R. 1994. A ruby in the rubbish—beneficial mutations, deleterious mutations, and the evolution of sex. *Genetics* 137: 597–606.

39. Pietsch, T. W. 2005. Dimorphism, parasitism, and sex revisited: modes of reproduction among deep-sea ceratioid anglerfishes (Teleostei: Lophiiformes). *Ichthyol. Res.* 52: 207–236.

40. Prokop, Z. M., Ł. Michalczyk, S. M. Drobniak, M. Herdegen, and J. Radwan. 2012. Meta-analysis suggests choosy females get sexy sons more than "good genes." *Evolution* 66: 2665–2673.

41. Redfield, R. J. 2001. Do bacteria have sex? *Nat. Rev. Genet.* 2: 634–639.

42. Richards, C. M. 2000. Inbreeding depression and genetic rescue in a plant metapopulation. *Am. Nat.* 155: 383–394.

43. Rodd, F. H., K. A. Hughes, G. F. Grether, and C. T. Baril. 2001. A possible non-sexual origin of mate preference: ERE male guppies mimicking fruit? *Proc. R. Soc. Lond. B* 269: 475–481.

44. Ryan, M. J., and M. E. Cummings. 2013. Perceptual biases and mate choice. *Annu. Rev. Ecol. Evol. Syst.* 44: 437–459.

45. Shuster, S. M. 2007. The evolution of crustacean mating systems. In E. J. Duffy and M. Thiel (eds.), *Evolutionary Ecology of Social and Sexual Systems: Crustaceans as Model Organisms* (pp. 29–47). Oxford University Press, New York.

46. Shuster, S. M., and M. J. Wade. 1991. Equal mating success among male reproductive strategies in a marine isopod. *Nature* 350: 608–610.

47. Stern, C. 1973. *Principles of Human Genetics.* W. H. Freeman, San Francisco.

48. Vergara, D., J. Jokela, and C. M. Lively. 2014. Infection dynamics in coexisting sexual and asexual host populations: Support for the Red Queen hypothesis. *Am. Nat.* 184: S22–S30.

49. Walsh, N. E., and D. Charlesworth. 1992. Evolutionary interpretations of differences in pollen tube growth rates. *Q. Rev. Biol.* 67: 19–37.

50. Westneat, D. F., and I. R. K. Stewart. 2003. Extra-pair paternity in birds: Causes, correlates, and conflict. *Annu. Rev. Ecol., Evol. Syst.* 34: 365–396.

51. Zuk, M., J. T. Rotenberry, and R. M. Tinghitella. 2006. Silent night: Adaptive disappearance of a sexual signal in a parasitized population of field crickets. *Biol. Lett. (London, U.K.)* 2: 521–524.

CHAPTER 11

1. Alsop, D. J., and S. A. West. 2003. Constant relative age and size at sex change for sequentially hermaphroditic fish. *J. Evol. Biol.* 921–929.

2. Angilleta, M. J., R. S. Wilson, C. A. Banav, and R. S. James. 2003. Trade-offs and the evolution of thermal reaction norms. *Trends Ecol. Evol.* 18: 234–240.

3. Augspurger, C. K. 1984. Demography and life history variation of *Puya dasylirioides*, a long-lived rosette in tropical subalpine bogs. *Oikos* 45: 341–352.

4. Auld, J. R., A. A. Agrawal, and R. A. Relyea. 2010. Reevaluating the costs and limits of adaptive phenotypic plasticity. *Proc. R. Soc. Lond. B* 277: 503–511.

5. Ayala, F. J. 1968. Genotype, environment, and population numbers. *Science* 162: 1453–1459.

6. Bassar, R. D., A. Lopez-Sepulcre, D. N. Reznick, and J. Travis. 2013. Experimental evidence for density-dependent regulation and selection on Trinidadian guppy life-histories. *Am. Nat.* 181: 25–38.

7. Bell, G., and V. Koufopanou. 1986. The cost of reproduction. *Oxford Surv. Evol. Biol.* 3: 83–131.

8. Bernays, E. A. 2001. Neural limitations in phytophagous insects: Implications for diet breadth and evolution of host affiliation. *Annu. Rev. Entomol.* 46: 703–727.

9. Bernays, E. A., and D. J. Funk. 1999. Specialists make faster decisions than generalists: Experiments with aphids. *Proc. Soc. Lond. B* 266: 151–156.

10. Bernays, E. A., and M. Graham. 1988. On the evolution of host specificity in phytophagous arthropods. *Ecology* 69: 886–892.

11. Charlesworth, B. 1994. *Evolution in Age-Structured Populations.* Cambridge University Press, Cambridge.

12. Charnov, E., and U. Skuladottir. 2000. Dimensionless invariants for the optimal size (age) of sex change. *Evol. Ecol. Res.* 2: 1067–1071.

13. Cooper, V. S., and R. E. Lenski. 2000. The population genetics of ecological specialization in evolving *Escherichia coli* populations. *Nature* 407: 736–739.

14. Cox, R. M., and R. Calsbeek. 2010. Severe costs of reproduction persist in *Anolis* lizards despite the evolution of a single-egg clutch. *Evolution* 64: 1321–1330.

15. Croft, D. P., L. J. N. Brent, D. W. Franks, and M. A. Cant. 2015. The evolution of prolonged life after reproduction. *Trends Ecol. Evol.* 30: 407–416.

16. Deutsch, C. A., and 6 others. 2008. Impacts of climate warming on terrestrial ectotherms across latitude. *Proc. Natl. Acad. Sci. USA* 105: 6668–6672.

17. Forister, M. L., L. A. Dyer, M. S Singer, J. O. Stireman III, and J. T. Lill. 2012. Revisiting the evolution of ecological specialization, with emphasis on insect-plant interactions. *Ecology* 93: 981–991.

18. Foster, S. A., and C. H. Janson. 1985. The relationship between seed size and establishment conditions in tropical woody plants. *Ecology* 66: 773–780.

19. Foster, E. A., and 6 others. 2012. Adaptive prolonged postreproductive life span in killer whales. *Science* 337: 1313.

20. Franco, M., and J. Silvertown. 1996. Life history variation in plants: An exploration of the fast-slow continuum hypothesis. *Phil. Trans. R. Soc.,B* 351: 1341–1348.

21. Fry, J. D. 1996. The evolution of host specialization: Are trade-offs overrated? *Am. Nat.* 148: S84–S107.

22. Futuyma, D. J., and G. Moreno. 1988. The evolution of ecological specialization. *Annu. Rev. Ecol. Syst.* 19: 207–233.

23. Futuyma, D. J., and T. E. Philippi. 1987. Genetic variation and covariation in responses to host plants by *Alsophila pometaria* (Lepidoptera: Geometridae). *Evolution* 41: 269–279.

24. Ghalambor, C. K., R. B. Huey, P. R. Martin, J. J. Tewksbury, and G. Wang. 2006. Are mountain passes higher in the tropics? Janzen's hypothesis revisited. *Integr. Comp. Biol.* 46: 5–17.

25. Hall, A. R., and N. Colegrave. 2007. Decay of unused characters by selection and drift. *J. Evol. Biol.* 21: 610–617.

26. Holt, R. D., and M. S. Gaines. 1992. Analysis of adaptation in heterogeneous landscapes: Implications for the evolution of fundamental niches. *Evol. Ecol.* 6: 433–447.

27. Hughes, K.A., J. A. Alipaz, J. M. Drnevich, and R. M. Reynolds. 2002. A test of evolutionary theories of aging. *Proc. Natl. Acad. Sci. USA* 99: 14286–14291.

28. Janzen, D. H. 1967. Why mountain passes are higher in the tropics. *Am. Nat.* 101: 233–249.

29. Jones, O. R., and 13 others. 2014. Diversity of ageing across the tree of life. *Nature* 505: 169–173.

30. Kassen, R. 2002. The experimental evolution of specialists, generalists, and the maintenance of diversity. *J. Evol. Biol.* 15: 173–190.

31. Kawecki, T. J., N. H. Barton, and J. D. Fry. 1997. Mutational collapse of fitness in marginal habitats and the evolution of ecological specialization. *J. Evol. Biol.* 10: 407–429.

32. Keller, L. F., J. M. Reid, and P. Arcese. 2008. Testing evolutionary models of senescence in a natural population: Age and inbreeding effects on fitness components in song sparrows. *Proc. R. Soc. Lond. B* 275: 597–604.

33. Lack, D. 1954. *The Natural Regulation of Animal Numbers.* Oxford University Press, Oxford.

34. Lee, C. E., J. L. Remfert, and G. W. Gelembiuk. 2003. Evolution of physiological tolerance and performance during freshwater invasions. *Integr. Comp. Biol.* 43: 439–449.

35. López-Otín, C., M. A. Blasco, L. Partridge, M. Serrano, and G. Kroemer. 2013. The hallmarks of aging. *Cell* 153: 1194–1217.

36. Medawar, P. B. 1952. *An Unsolved Problem of Biology.* H. K. Lewis, London.

37. Messina, F. J., and J. D. Fry. 2003. Environment-dependent reversal of a life history trade-off in the seed beetle *Callosobruchus maculatus*. *J. Evol. Biol.* 16: 501–509.

38. Metcalf, J. C., K. E. Rose, and M. Rees. 2003. Evolutionary demography of monocarpic perennials. *Trends Ecol. Evol.* 18: 471–480.

39. Moles, A. T., and 6 others. 2005. Factors that shape seed mass evolution. *Proc. Natl. Acad. Sci. USA* 102: 10540–10544.

40. Murren, C. J., and 15 others. 2015. Constraints on the evolution of phenotypic plasticity: Limits and costs of phenotype and plasticity. *Heredity* 115: 293–301.

41. Nielsen, J., and 10 others. 2016. Eye lens radiocarbon reveals centuries of longevity in the Greenland shark (*Somniosus microcephalus*). *Science* 353: 702–704.

42. Overgaard, J., T. N. Kristensen, K. A. Mitchell, and A. A. Hoffmann. 2011. Thermal tolerance in widespread and tropical *Drosophila* species: Does phenotypic plasticity increase with latitude? *Am. Nat.* 178: S80–S96.

43. Partridge, L. 2001. Evolutionary theories of ageing applied to long-lived organisms. *Exper. Gerontol.* 36: 641–650.

44. Partridge, L., N. Prowse, and P. Pignatelli. 1999. Another set of responses and correlated responses to selection on age at reproduction in *Drosophila melanogaster*. *Proc. R. Soc. Lond. B* 266: 255–261.

45. Petralia, R. S., M. P. Mattson, and P. J.Yao. 2014. Aging and longevity in the simplest animals and the quest for immortality. *Ageing Res. Rev.*16: 66–82.

46. Poorter, L., P. A. Zuidema, M. Peña-Claros, and R. G. A. Boot. 2005. A monocarpic tree species in a polycarpic world: How can *Tachygali vasquezii* maintain itself so successfully in a tropical rain forest community? *J. Ecol.* 93: 268–278.

47. Promislow, D. E. L., and P. H. Harvey. 1991. Mortality rates and the evolution of mammalian life histories. *Acta Oecologica* 220: 417–437.

48. Reznick, D., and J. Travis. 2002. Adaptation. In C. W. Fox, D. A. Roff, and D. J. Fairbairn (eds.), *Evolutionary Ecology: Concepts and Case Studies*, pp. 44–57. Oxford University Press, New York.

49. Roff, D. A. 2002. *Life History Evolution*. Sinauer, Sunderland, MA.

50. Roughgarden, J. 1971. Density-dependent natural selection. *Ecology* 52: 453–468.

51. Schaffer, W. M. 1974. Selection for optimal life histories: Effects of age structure. *Ecology* 55: 291–303.

52. Schondube, J. E., and C. Martinez del Rio. 2003. The flower-piercers' hook: An experimental test of an evolutionary trade-off. *Proc. R. Soc. Lond. B* 270: 195–198.

53. Shine, R., and E. L. Charnov. 1992. Patterns of survival, growth, and maturation in snakes and lizards. *Am. Nat.* 139: 1257–1269.

54. Singer, M. S., and J. O. Stireman. 2005. The tri-trophic niche concept and adaptive radiation of phytophagous insects. *Ecol. Lett.* 8: 1247–1255.

55. Stearns, S. C. 1992. *The Evolution of Life Histories*. Oxford University Press, Oxford.

56. Sunday, J. M., A. E. Bates, and N. K. Dulvy. 2011. Global analysis of thermal tolerance and latitude in ectotherms. *Proc. R. Soc. Lond. B* 278: 1823–1830.

57. Tinbergen, J. M., and S. Daan. 1990. Family planning in the great tit (*Parus major*): Optimal clutch size as integration of parent and offspring fitness. *Behaviour* 114: 161–190.

58. Tucić, N., D. Cvetković, and D. Milanović. 1988. The genetic variation and covariation among fitness components in *Drosophila melanogaster* females and males. *Heredity* 60: 55–60.

59. van Noordwijk, A. J., and G. deJong. 1986. Acquisition and allocation of resources: Their influence on variation in life history tactics. *Am. Nat.* 128: 137–142.

60. Warner, R. R. 1984. Mating behavior and hermaphroditism in coral reef fishes. *Am. Sci.* 72: 128–136.

61. Whitlock, M. J. 1996. The Red Queen beats the jack-of-all-trades: The limitations on the evolution of phenotypic plasticity and niche breadth. *Am. Nat.* 148: S65–S77.

62. Williams, G. C. 1957. Pleiotropy, natural selection, and the evolution of senescence. *Evolution* 11: 398–411.

63. Wilson, A. M., and K. Thompson. 1989. A comparative study of reproductive allocation in 40 British grasses. *Functional Ecology* 3: 297–302.

CHAPTER 12

1. Abbot, P., and 136 others. 2011. Inclusive fitness theory and eusociality. *Nature* 471: E1–E4.

2. Alcock, J. 2013. *Animal Behavior*. Sinauer, Sunderland, MA.

3. Arnqvist, G., and L. Rowe. 2005. *Sexual Conflict*. Princeton University Press, Princeton, NJ.

4. Axelrod, R., and W. D. Hamilton. 1981. The evolution of cooperation. *Science* 211: 1390–1396.

5. Birch, J., and S. Okasha. 2015. Kin selection and its critics. *BioScience* 65: 22–32.

6. Bolzer, A., and 10 others. 2005. Three-dimensional maps of all chromosomes in human male fibroblast nuclei and prometaphase rosettes. *PLoS Biol.* 3(5): e157. doi: 10.1371/journal.pbio.0030157

7. Bourke, A. F. G. 2011. *Principles of Social Evolution*. Oxford University Press, Oxford.

8. Bourke, A. F. G. 2014. Hamilton's rule and the causes of social evolution. *Phil. Tran. R. Soc., B* 369: 20130362.

9. Bourke, A. F. G., and N. R. Franks. 1995. *Social Evolution in Ants*. Princeton University Press, Princeton, NJ.

10. Brand, C. L., A. M. Larracuente, and D. C. Presgraves. 2015. Origin, evolution, and population genetics of the selfish Segregation Distorter gene duplication in European and African populations of *Drosophila melanogaster*. *Evolution* 69: 1271–1283.

11. Bull, J. J. and A. S. Lauring. 2014. Theory and empiricism in virulence evolution. *PLoS Pathogens* 10.

12. Cavalier-Smith, T. 2013. Symbiogenesis: Mechanisms, evolutionary consequences, and systematic implications. *Annu. Rev. Ecol. Evol. Syst.* 44: 145–172.

13. Chapman, T., L. F. Liddle, J. M. Kalb, M. F. Wolfner, and L. Partridge. 1995. Cost of mating in Drosophila melanogaster females is mediated by male accessory gland products. *Nature* 373: 241–244.

14. Charlesworth, D. and V. Laporte. 1998. The male-sterility polymorphism of *Silene vulgaris*: Analysis of genetic data from two populations and comparison with *Thymus vulgaris*. *Genetics* 150: 1267–1282.

15. Clutton-Brock, T. H. 1991. *The Evolution of Parental Care*. Princeton University Press, Princeton, NJ.

16. Clutton-Brock, T. 2009. Cooperation between non-kin in animal societies. *Nature* 462: 51–57.

17. Clutton-Brock, T., and C. Godfray. 1991. Parental investment. In J. R. Krebs and N. B. Davies (eds.), *Behavioural Ecology: An Evolutionary Approach*, 3rd ed., pp. 234–262. Blackwell Scientific, Oxford.

18. Crozier, R. H., and P. Pamilo. 1996. *Evolution of Social Insect Colonies*. Oxford University Press, Oxford.

19. Dao, D. N., R. H. Kessin, and H. L. Ennis. 2000. Developmental cheating and the evolutionary biology of *Dictyostelium* and *Myxococcus*. *Microbiology* 146: 1505–1512.

20. Dawkins, R. 1989. *The Selfish Gene*, rev. ed. Oxford University Press, Oxford.

21. Duval, E. H. 2007. Adaptive advantages of cooperative courtship for subordinate male Lance-tailed Manakins. *Am. Nat.* 169: 423–432.

22. El-Mouden, C., S. A. West, and A. Gardner. 2010. The enforcement of cooperation by policing. *Evolution* 64: 2139–2152.

23. Emlen, S. T. 1995. An evolutionary theory of the family. *Proc. Natl. Acad. Sci. USA* 92: 8092–8099.

24. Finseth, F. R., Y. Z. Dong, A. Saunders, and L. Fishman. 2015. Duplication and adaptive evolution of a key centromeric protein in *Mimulus*, a genus with female meiotic drive. *Mol. Biol. Evol.* 32: 2694–2706.

25. Fisher, H. S., and H. E. Hoekstra. 2010. Competition drives cooperation among closely related sperm of deer mice. *Nature* 463: 801–803.

26. Fishman, L., and J. K. Kelly. 2015. Centromere-associated meiotic drive and female fitness variation in *Mimulus*. *Evolution* 69: 1208–1218.

27. Frank, S. A. 1998. *Foundations of Social Evolution*. Princeton University Press, Princeton, NJ.

28. Frank, S. A. 2003. Perspective: Repression of competition and the evolution of cooperation. *Evolution* 57: 693–705.

29. Frank, S. A., and B. J. Crespi. 2011. Pathology from evolutionary conflict, with a theory of X chromosome versus autosome conflict over sexually antagonistic traits. *Proc. Natl. Acad. Sci. USA* 108: 10886–10893.

30. Gantz, V. M., and 6 others. 2015. Highly efficient Cas9-mediated gene drive for population modification of the malaria vector mosquito *Anopheles stephensi*. *Proc. Natl. Acad. Sci. USA* 112: E6736–E6743.

31. Gardner, A., and S. A. West. 2010. Greenbeards. *Evolution* 64: 25–38.

32. Gemmell, N. J., V. J. Metcalf, and F. W. Allendorf. 2004. Mother's curse: The effect of mtDNA on individual fitness and population viability. *Trends Ecol. Evol.* 19: 238–244.

33. Godfray, H. C. J. 1999. Parent-offspring conflict. In L. Keller (ed.), *Levels of Selection in Evolution*, pp. 100–120. Princeton University Press, Princeton, NJ.

34. Griffin, A. S., S. A. West, and A. Buckling. 2004. Cooperation and competition in pathogenic bacteria. *Nature* 430: 1024–1027.

35. Haig, D. 1993. Genetic conflicts in human pregnancy. *Q. Rev. Biol.* 68: 495–532.

36. Hamilton, W. D. 1964. The genetical evolution of social behavior, I and II. *J. Theor. Biol.* 7: 1–52.

37. Hamilton, W. D. 1971. Geometry for the selfish herd. *J. Theor. Biol.* 31: 295–311.

38. Hausfater, G., and S. Hrdy (eds.). 1984. *Infanticide: Comparative and Evolutionary Perspectives.* Aldine Publishing Co., New York.

39. Henikoff, S., K. Ahmad, and H. S. Malik. 2001. The centromere paradox: Stable inheritance with rapidly evolving DNA. *Science* 293: 1098–1102.

40. Hughes, W. O. H., B. P. Oldroyd, M. Beekman, and F. L. W. Ratnieks. 2008. Ancestral monogamy shows kin selection is key to the evolution of eusociality. *Science* 320: 1213–1216.

41. Inglis, R. F., A. Gardner, P. Cornelis, and A. Buckling. 2009. Spite and virulence in the bacterium *Pseudomonas aeruginosa*. *Proc. Natl. Acad. Sci. USA* 106: 5703–5707.

42. Innocenti, P., E. H. Morrow, and D. K. Dowling. 2011. Experimental evidence supports a sex-specific selective sieve in mitochondrial genome evolution. *Science* 332: 845–848.

43. Kapheim, K. M., P. Nonacs, A. R. Smith, R. K. Wayne, and W. T. Wcislo. 2014. Kinship, parental manipulation and evolutionary origins of eusociality. *Proc. R. Soc. Lond. B* 282: 20142886.

44. Keller, L. (ed.). 1993. *Queen Number and Sociality in Insects.* Oxford University Press, Oxford.

45. Kerr, P. J. 2012. Myxomatosis in Australia and Europe: A model for emerging infectious diseases. *Antiviral Res.* 93: 387–415.

46. Krakauer, A. H. 2005. Kin selection and cooperative courtship in wild turkeys. *Nature* 434: 69–72.

47a. Lehmann, L., and L. Keller. 2006. The evolution of cooperation and altruism—a general framework and a classification of models. *J. Evol. Biol.* 19: 1365–1376.

47b. Lindholm, A. K., and 21 others. 2016. The ecology and evolutionary dynamics of meiotic drive. *Trends Ecol. Evol.* 31: 315–326.

48. Lukas, D., and E. Huchard. 2014. The evolution of infanticide by males in mammalian societies. *Science* 346: 841–844.

49. Lynch, M. 2007. *The Origins of Genome Architecture.* Sinauer, Sunderland, MA.

50. Manno, T. G., F. S. Dobson, J. L. Hoogland, and D. W. Foltz. 2007. Social group fission and gene dynamics among black-tailed prairie dogs. *J. Mammal.* 88: 448–456.

51. Maynard Smith, J., and E. Szathmáry. 1995. *The Major Transitions in Evolution.* W. H. Freeman, San Francisco.

52. McCracken, G. F., and M. K. Gustin. 1991. Nursing behavior in Mexican free-tailed bat maternity colonies. *Ethology* 89: 305–321.

53. McDonald, D. B., and W. K. Potts. 1994. Cooperative display and relatedness among males in a lek-mating bird. *Science* 266: 1030–1032.

54. Michod, R. E. 1999. *Darwinian Dynamics: Evolutionary Transitions in Fitness and Individuality.* Princeton University Press, Princeton, NJ.

55. Mock, D. W. 2004. *More Than Kin and Less Than Kind: The Evolution of Family Conflict.* Harvard University Press, Cambridge, MA.

56. Moran, N. A. 2007. Symbiosis as an adaptive process and source of phenotypic complexity. *Proc. Natl. Acad. Sci. USA* 104: 8627–8633.

57. Nowak, M. A. 2006. Five rules for the evolution of cooperation. *Science* 314: 1560–1563.

58. Presgraves, D. C. 2010. The molecular evolutionary basis of species formation. *Nat. Rev. Genet.* 11: 175–180.

59. Presgraves, D. C., E. Severance, and G. S. Wilkinson. 1997. Sex chromosome meiotic drive in stalk-eyed flies. *Genetics* 147: 1169–1180.

60. Queller, D. C. 1994. Genetic relatedness in viscous populations. *Evol. Ecol.* 8: 70–73.

61. Queller, D. C., and J. E. Strassmann. 1998. Kin selection and social insects. *BioScience* 48: 165–175.

62. Queller, D. C., E. Ponte, S. Bozzaro, and J. E. Strassmann. 2003. Single-gene greenbeard effects in the social amoeba *Dictyostelium discoideum*. Science 299: 105–106.

63. Ratnieks, F. L. W., and T. Wenseleers. 2007. Altruism in insect societies and beyond: Voluntary or enforced? *Trends Ecol. Evol.* 23: 45–53.

64. Ratnieks, F. L. W., K. R. Foster, and T. Wenseleers. 2006. Conflict resolution in insect societies. *Annu. Rev. Entomol.* 51: 581–608.

65. Rice, W. R. 1996. Sexually antagonistic male adaptation triggered by experimental arrest of female evolution. *Nature* 381: 232–234.

66. Rice, W. R. 2013. Nothing make sense except in light of genomic conflict. *Annu. Rev. Ecol. Evol. Syst.* 44: 217–237.

67. Riley, M. A., and J. E. Wertz. 2002. Bacteriocins: Evolution, ecology, and application. *Annu. Rev. Microbiol.* 56: 117–137.

68. Sachs, J. L., U. G. Mueller, T. P. Wilcox, and J. J. Bull. 2004. The evolution of cooperation. *Q. Rev. Biol.* 79: 135–160.

69. Silk, J. B., and 8 others. 2009. The benefits of social capital: Close social bonds among female baboons enhance offspring survival. *Proc. R. Soc. Lond. B* 276: 3099–3104.

70. Strassmann, J. E., and D. C. Queller. 2007. Insect societies as divided organisms: The complexities of purpose and cross-purpose. *Proc. Natl. Acad. Sci. USA* 104 (Suppl. 1): 8619–8626.

71. Strassmann, J. E., and D. C. Queller. 2011. Evolution of cooperation and control of cheating in a fsocial microbe. *Proc. Natl. Acad. Sci. USA* 108: 10855–10962.

72. Stutt, A. D., and M. T. Siva-Jothy. 2001. Traumatic insemination and sexual conflict in the bed bug *Cimex lectularius*. *Proc. Natl. Acad. Sci. USA* 98: 5683–5687.

73. Taborsky, M., J. C. Frommen, and C. Riehl. 2016. Correlated pay-offs are key to cooperation. *Phil. Trans. R. Soc., B* 371: 20150084.

74. Trivers, R. L. 1971. The evolution of reciprocal altruism. *Q. Rev. Biol.* 46: 35–57.

75. Trivers, R. L. 1974. Parent-offspring conflict. *Am. Zool.* 11: 249–264.

76. Wade, M. J. 1977. An experimental study of group selection. *Evolution* 31: 134–153.

77. Wade, M. J., and D. E. McCauley. 1980. Group selection: The phenotypic and genotypic differentiation of small populations. *Evolution* 34: 799–812.

78. Wenseleers, T., and F. L. W. Ratnieks. 2006. Comparative analysis of worker reproduction and policing in eusocial Hymenoptera supports relatedness theory. *Am. Nat.* 168: E163–E179.

79. Werren, J. H., and R. Stouthammer. 2003. PSR (paternal sex ratio) chromosomes: The ultimate selfish genetic elements. *Genetica* 117: 85–101.

80. West, S. A., A. S. Griffin, and A. Gardner. 2007a. Social semantics: Altruism, cooperation, mutualism, strong reciprocity and group selection. *J. Evol. Biol.* 20: 415–432.

81. West, S. A., A. S. Griffin, and A. Gardner. 2007b. Evolutionary explanations for cooperation. *Curr. Biol.* 17: R661–R672.

82. West, S. A., and A. Gardner. 2010. Altruism, spite, and greenbeards. *Science* 327: 1341–1344.

83. Wilkinson, G. S., G. C. Carter, K. M. Bohn, and D. M. Adams. 2016. Non-kin cooperation in bats. *Phil. Trans. R. Soc., B* 371: 20150095.

84. Wilson, D. S. 1975. A theory of group selection. *Proc. Natl. Acad. Sci. USA* 72: 143–146.

85. Wilson, E. O. 1971. *The Insect Societies.* Harvard University Press, Cambridge, MA.

86. Yagi, N., and E. Hasegawa. 2012. A halictid bee with sympatric solitary and eusocial nests offers evidence for Hamilton's rule. *Nat. Commun.* 3: 939. doi:10.1038/ncomms1939.

CHAPTER 13

1. Abrams, P. A. 2000. The evolution of predator-prey interactions: Theory and evidence. *Annu. Rev. Ecol. Syst.* 31: 79–108.

2. Abrams, P. A., and H. Matsuda. 1994. The evolution of traits that determine ability in competitive contests. *Evol. Ecol.* 8: 667–686.

3. Agrawal, A. A. 2005. Natural selection on common milkweed (*Asclepias syriaca*) by a community of specialized insect herbivores. *Evol. Ecol. Res.* 7: 651–667.

4. Agrawal, A., and C. M. Lively. 2002. Infection genetics: Gene-for-gene versus matching-alleles models and all points in between. *Evol. Ecol. Res.* 4: 79–90.

5. Berenbaum, M. R., A. R. Zangerl, and J. K. Nitao. 1986. Constraints on chemical coevolution: Wild parsnips and the parsnip webworm. *Evolution* 40: 1215–1228.

6. Blossey, B., and R. Notzold. 1995. Evolution of increased competitive ability in invasive nonindigenous plants—a hypothesis. *J. Ecol.* 83: 887–889.

7. Bolnick, D. I., and 6 others. 2003. The ecology of individuals: Incidence and implications of individual specialization. *Am. Nat.* 161: 1–28.

8. Brodie, E. D., Jr., B. J. Ridenhour, and E. D. Brodie III. 2002. The evolutionary response of predators to dangerous prey: Hotspots and coldspots in the geographic mosaic of coevolution between garter snakes and newts. *Evolution* 56: 2067–2082.

9. Bull, J. J. 1994. Perspective: Virulence. *Evolution* 48: 1423–1437.

10. Bull, J. J., and W. R. Rice. 1991. Distinguishing mechanisms for the evolution of cooperation. *J. Theor. Biol.* 149: 63–74.

11. Bull, J. J., I. J. Molineaux, and W. R. Rice. 1991. Selection of benevolence in a host-parasite system. *Evolution* 45: 875–882.

12. Catenazzi, A. 2015. State of the world's amphibians. *Annu. Rev. Environ. Resour.* 40: 91–119.

13. Cavender-Bares, J., K. H. Kozak, P. V. A. Fine, and S. W. Kembel. 2009. The merging of community ecology and phylogenetic biology. *Ecol. Lett.* 12: 693–715.

14. Connell, J. H. 1971. On the role of natural enemies in preventing competitive exclusion in some marine animals and in rain forest trees. In P. J. den Boer and G. Gradwell (eds.), *Dynamics in Populations*, pp. 298–312. Centre for Agricultural Publishing and Documentation, Wageningen, the Netherlands.

15. Costello, E. K., K. Stagaman, L. Dethlefsen, B. J. M. Bohannon, and D. A. Relman. 2012. The application of ecological theory toward an understanding of the human microbiome. *Science* 336: 1255–1262.

16. Davies, N. B., and M. de L. Brooke. 1998. Cuckoos versus hosts: Experimental evidence for coevolution. In S. I. Rothstein and S. K. Robinson (eds.), *Parasitic Birds and Their Hosts: Studies in Coevolution*, pp. 59–79. Oxford University Press, NY.

17. Dayan, T., D. Simberloff, E. Tchernov, and Y. Yom-Tov. 1989. Inter- and intraspecific character displacement in mustelids. *Ecology* 70: 1526–1539.

18. Decaestecker, E., S. Gaba, J. A. M. Raeymaekers, R. Stoks, L. Van Kerckhoven, D. Ebert, and L. De Meester. 2007. Host-parasite "Red Queen" dynamics archived in pond sediment. *Nature* 450: 870–873.

19. Dickman, C. R. 1996. Impact of exotic generalist predators on the native fauna of Australia. *Wildlife Biol.* 2: 185–195.

20. Dybdal, M. F., and A. Storfer. 2003. Parasite local adaptation: Red Queen versus suicide king. *Trends Ecol. Evol.* 18: 523–530.

21. Ebert, D. 1994. Virulence and local adaptation of a horizontally transmitted parasite. *Science* 265: 1084–1086.

22. Edger, P. P., and 23 others. 2015. The butterfly plant arms-race escalated by gene and genome duplications. *Proc. Natl. Acad. Sci. USA* 112: 8362–8366.

23. Ehrlich, P. R., and P. H. Raven. 1964. Butterflies and plants: A study in coevolution. *Evolution* 18: 586–608.

24. Ewald, P. W. 1994. *Evolution of Infectious Disease*. Oxford University Press, Oxford.

25. Fisher, R. A. 1930. *The Genetical Theory of Natural Selection*. Clarendon Press, Oxford.

26. Frank, S. A. 1996. Models of parasite virulence. *Q. Rev. Biol.* 71: 37–78.

27. Futuyma, D. J., and A. A. Agrawal. 2009. Macroevolution and the biological diversity of plants and herbivores. *Proc. Natl. Acad. Sci. USA* 106: 18054–18061.

28. Futuyma, D. J., and M. Slatkin (eds.). 1983. *Coevolution*. Sinauer, Sunderland, MA.

29. Futuyma, D. J., M. C. Keese, and D. J. Funk. 1995. Genetic constraints on macroevolution: The evolution of host affiliation in the leaf beetle genus *Ophraella*. *Evolution* 49: 797–809.

30. Grant, P. R., 1986. *Ecology and Evolution of Darwin's Finches*. Princeton University Press, Princeton, NJ.

31. Hanifin, C. T., E. D. Brodie, Jr., and E. D. Brodie III. 2008. Phenotypic mismatches reveal escape from arms-race coevolution. *PLoS Biol.* 6(3): e60, pp. 0471–0482.

32. Heath, K. D., and P. Tiffin. 2009. Stabilizing mechanisms in legume-rhizobium mutualism. *Evolution* 63: 652–662.

33. Herre, E. A., N. Knowlton, U. G. Mueller, and S. A. Rehner. 1999. The evolution of mutualism: Exploring the paths between conflict and cooperation. *Trends Ecol. Evol.* 14: 49–53.

34. Hooper, L. V., D. R. Littman, and A. J. Macpherson. 2012. Interactions between the microbiota and the immune system. *Science* 336: 1268–1273.

35. Hougen-Eitzman, D., and M. D. Rausher. 1994. Interactions between herbivorous insects and plant-insect coevolution. *Am. Nat.* 143: 677–697.

36. Hueffer, K., J. S. L. Parker, W. S. Weichert, R. E. Geisel, J.-Y. Sgro, and C. R. Parrish. 2003. The natural host range shift and subsequent evolution of canine parvovirus resulted from virus-specific binding to the canine transferrin receptor. *J. Virol.* 77: 1718–1726.

37. Janzen, D. H. 1970. Herbivores and the number of tree species in tropical forests. *Am. Nat.* 104: 501–528.

38. Kapan, D. D. 2001. Three-butterfly system provides a field test of Müllerian mimicry. *Nature* 409: 338–340.

39. Keller, B., C. Feuillet, and M. Messmer. 2000. Genetics of disease resistance: Basic concepts and application in resistance breeding. In A. Slusarenko, R. S. S. Fraser, and L. C. van Loon (eds.), *Mechanisms of Resistance to Plant Diseases*, pp. 101–160. Kluwer Academic Publishers, the Netherlands.

40. Kiers, E. T., R. A. Rousseau, S. A. West, and R. F. Dennison. 2003. Host sanctions and the legume-rhizobium mutualism. *Nature* 425: 78–81.

41. Kursar, T. A., and 9 others. 2009. The evolution of antiherbivore defenses and their contribution to species coexistence in the tropical tree genus *Inga*. *Proc. Natl. Acad. Sci. USA* 106: 18073–18078.

42. Liu, X. B., M. X. Liang, R. S. Etienne, Y. F. Wang, C. Staehelin, and S. X. Yu. 2012. Experimental evidence for a phylogenetic Janzen-Connell effect in a subtropical forest. *Ecol. Lett.* 15: 111–118.

43. Losos, J. B. 2009. *Lizards in an Evolutionary Tree: Ecology and Adaptive Radiation of Anoles*. University of California Press, Berkeley.

44. Losos, J. B. 2010. Adaptive radiation, ecological opportunity, and evolutionary determinism. *Am. Nat.* 175: 623–639.

45. Luijckx, P., H. Fienberg, D. Duneau, and D. Ebert. 2013. A matching-allele model explains host resistance to parasites. *Curr. Biol.* 23: 1085–1088.

46. Mack, R. H., D. Simberloff, W. M. Lonsdale, H. Evans, M. Clout, and F. A. Bazzaz. 2000. Biotic invasions: Causes, epidemiology, global consequences, and control. *Ecol. Appl.* 10: 689–710.

47. Mahler, D. L., T. Ingram, L. J. Revell, and J. B. Losos. 2013. Exceptional convergence on the macroevolutionary landscape in island lizard radiations. *Science* 341: 292–295.

48. Mallet, J. 2010. Shift happens! Shifting balance and the evolution of diversity in warning color and mimicry. *Ecol. Entomol.* 35: 90–104.

49. Mallet, J., and M. Joron. 1999. Evolution of diversity in warning color and mimicry: polymorphisms, shifting balance, and speciation. *Annu. Rev. Ecol. Syst.* 30: 201–233.

50. Martin, W., and 9 others. 2002. Evolutionary analysis of *Arabidopsis*, cyanobacterial, and chloroplast genomes reveals plastid phylogeny and thousands of cyanobacterial genes in the nucleus. *Proc. Natl. Acad. Sci.* 99: 12246–12251.

51. Mittelbach, G. G., and D. W. Schemske. 2015. Ecological and evolutionary perspectives on community assembly. *Trends Ecol. Evol.* 30: 241–247.

52. Moran, N. A. 2007. Symbiosis as an adaptive process and source of phenotypic complexity. *Proc. Natl. Acad. Sci. USA* 104: 8627–8633.

53. Moran, N. A., and P. Baumann. 1994. Phylogenetics of cytoplasmically inherited microorganisms of arthropods. *Trends Ecol. Evol.* 9: 15–20.

54. Muchhala, N., and J. D. Thomson. 2009. Going to great lengths: Selection for long corolla tubes in an extremely specialized bat-flower mutualism. *Proc. R. Soc.Lond. B* 276: 2147–2152.

55. Nicholson, J. K., and 6 others. 2012. Host-gut microbiota metabolic interactions. *Science* 336: 1262–1267.

56. Nuismer, S. L., B. J. Ridenhour, and B. P. Oswald. 2007. Antagonistic coevolution mediated by phenotypic differences between quantitative traits. *Evolution* 61: 1823–1834.

57. Page, R. D. M. 2002. Introduction. In R. D. M. Page (ed.), *Tangled Trees: Phylogeny, Cospeciation, and Coevolution*, pp. 1–21. University of Chicago Press, Chicago.

58. Pauw, A., J. Stofberg, and R. J. Waterman. 2009. Flies and flowers in Darwin's race. *Evolution* 63: 268–279.

59. Pellmyr, O., and J. Leebens-Mack. 1999. Forty million years of mutualism: Evidence for Eocene origin of the yucca-yucca moth association. *Proc. Natl. Acad. Sci. USA* 96: 9178–9183.

60. Pfennig, D. W., and K. S. Pfennig. 2010. Character displacement and the origins of diversity. *Am. Nat.* 176: S26–S44.

61. Porco, T. C., J. O. Lloyd-Smith, K. L. Gross, and A. P. Galvani. 2005. The effect of treatment on pathogen virulence. *J. Theor. Biol.* 233: 91–102.

62. Read, A. F., and 8 others. 2015. Imperfect vaccination can enhance the transmission of highly virulent pathogens. *PLoS Biol.* 13 (7), no. e1002198.

63. Rezende, E. L., P. Jordano, and J. Bascomte. 2007. Effects of phenotypic complementarity and phylogeny on the nested structure of mutualistic networks. *Oikos* 116: 1919–1929.

64. Robinson, B. W., D. S. Wilson, and A. S. Margosian. 2000. A pluralistic analysis of character release in pumpkinseed sunfish (*Lepomis gibbosus*). *Ecology* 81: 2799–2812.

65. Rothstein, S. I., and S. K. Robinson (eds.). 1998. *Parasitic Birds and Their Hosts: Studies in Coevolution*. Oxford University Press, NY.

66. Schluter, D. 2000. *The Ecology of Adaptive Radiation*. Oxford University Press, Oxford.

67. Siepelski, A. M., and C. W. Benkman. 2007. Convergent patterns in the selection mosaic for two North American bird-dispersed pines. *Ecol. Monogr.* 77: 203–220.

68. Slatkin, M. 1980. Ecological character displacement. *Ecology* 61: 163–177.

69. Stuart, Y. E., T. S. Campbell, P. A. Hohenlohe, R. G. Reynolds, L. J. Revell, and J. B. Losos. 2014. Rapid evolution of a native species following Invasion by a congener. *Science* 346: 463–466.

70. Taper, M. L., and T. J. Case. 1992. Coevolution among competitors. *Oxford Surv. Evol. Biol.* 8: 63–109.

71. Terborgh, J. 2012. Enemies maintain hyperdiverse tropical forests. *Am. Nat.* 179: 303–314.

72. Thompson, J. N. 1994. *The Coevolutionary Process*. University of Chicago Press, Chicago.

73. Thompson, J. N. 2009. The coevolving web of life. *Am. Nat.* 173: 125–140.

74. Thompson, J. N. 2013. *Relentless Evolution*. University of Chicago Press, Chicago and London.

75. Timmermans, M. J. T. N., and 12 others. 2014. Comparative genetics of the mimicry switch in *Papilio dardanus. Proc. R. Soc. Lond. B* 281: 20140645.

76. Toju, H., and T. Sota. 2006. Imbalance of predator and prey armament: Geographic clines in phenotypic interface and natural selection. *Am. Nat.* 167: 105–117.

77. Uesugi, A., and A. Kessler. 2013. Herbivore exclusion drives the evolution of plant competitiveness via increased allelopathy. *New Phytol.* 198: 916–924.

78. Ullstrup, A. J. 1972. The impacts of the southern corn leaf blight epidemic of 1970–1971. *Annu. Rev. Phytopathol.* 19: 37–50.

79. Van Valen, L. 1973. A new evolutionary law. *Evol. Theory* 1: 1–30.

80. Webb, C. O. 2000. Exploring the phylogenetic structure of ecological communities: An example for rain forest trees. *Am. Nat.* 156: 145–155.

81. Weese, D. J., K. D. Heath, B. T. M. Dentlinger, and J. A. Lau. 2015. Long-term nitrogen addition causes the evolution of less-cooperative mutualists. *Evolution* 69: 631–642.

82. Winkler, I. S., and C. Mitter. 2008. The phylogenetic dimension of insect-plant interactions: A review of recent evidence. In K. J. Tilmon (ed.), *Specialization, Speciation, and Radiation: The Evolutionary Biology of Herbivorous Insects*, pp. 240–263. University of California Press, Berkeley.

83. Woolhouse, M. E. J., D. T. Haydon, and R. Antia. 2005. Emerging pathogens: The epidemiology and evolution of species jumps. *Trends Ecol. Evol.* 20: 238–244.

84. Yuan, Y., and 7 others. 2013. Enhanced allelopathy and competitive ability of invasive plant *Solidago canadensis* in its introduced range. *J. Plant Ecol.* 6: 253–263.

CHAPTER 14

1. Adams, K. L., and J. D. Palmer. 2003. Evolution of mitochondrial gene content: Gene loss and transfer to the nucleus. *Mol. Phylogenet. Evol.* 29: 380–395.

2. Arnegard, M. E., D. J. Zwickl, Y. Lu, and H. H. Zakon. 2010. Old gene duplication facilitates origin and diversification of an innovative communication system—twice. *Proc. Natl. Acad. Sci. USA* 107: 22172–22177.

3. Batut, B., C. Knibbe, G. Marais, and V. Daubin. 2014. Reductive genome evolution at both ends of the bacterial population size spectrum. *Nat. Rev. Microbiol.* 12: 841–850.

4. Batzer, M. A., and P. L. Deininger. 2002. Alu repeats and human genomic diversity. *Nat. Rev. Genet.* 3: 370–379.

5. Belancio, V. P., D. J. Hedges, and P. Deininger. 2008. Mammalian non-LTR retrotransposons: For better or worse, in sickness and in health. *Genome Res.* 18: 343–358.

6. Bennett, G. M., and N. A. Moran. 2013. Small, smaller, smallest: The origins and evolution of ancient dual symbioses in a phloem-feeding insect. *Genome Biol. Evol.* 5: 1675–1688.

7. Bhutkar, A., S. W. Schaeffer, S. M. Russo, M. Xu, T. F. Smith, and W. M. Gelbart. 2008. Chromosomal rearrangement inferred from comparisons of 12 *Drosophila* genomes. *Genetics* 179: 1657–1680.

8. Bird, D. M., J. T. Jones, C. H. Opperman, T. Kikuchi, and E. G. J. Danchin. 2015. Signatures of adaptation to plant parasitism in nematode genomes. *Parasitology* 142: S71–S84.

9. Bosco, G., P. Campbell, J. T. Leiva-Neto, and T. A. Marko. 2007. Analysis of *Drosophila* species genome size and satellite DNA content reveals significant differences among strains as well as between species. *Genetics* 177: 1277–1290.

10. Brenchley, R., and 28 others. 2012. Analysis of the breadwheat genome using whole-genome shotgun sequencing. *Nature* 491: 705–710.

11. Charlesworth, J., and A. Eyre-Walker. 2006. The rate of adaptive evolution in enteric bacteria. *Mol. Biol. Evol.* 23: 1348–1356.

12. Chen, S. D., B. H. Krinsky, and M.Y. Long. 2013. New genes as drivers of phenotypic evolution. *Nat. Rev. Genet.* 14: 645–660.

13. Deininger, P. L., and M. A. Batzer. 2002. Mammalian retroelements. *Genome Res.* 12: 1455–1465.

14. de Jong, W. W., W. Hendriks, J. W. M. Mulders, H. Bloemendal. 1989. Evolution of eye lens crystallins: The stress connection. *Trends Biochem. Sci.* 14: 365–368.

15a. Demuth, J. P., T. D. Bie, J. E. Stajich, N. Cristianini, and M. W. Hahn. 2006. The evolution of mammalian gene families. *PLoS ONE* 1: e85. doi:10.1371/journal.pone.0000085

15b. Dobzhansky, Th., and C. Epling. 1944. Contribution to the genetics, taxonomy, and ecology of *Drosophila pseudoobscura* and its relatives. *Carnegie Institution Publication* 554: 111–183.

16. Drosophila 12 Genomes Consortium. 2007. Evolution of genes and genomes on the *Drosophila* phylogeny. *Nature* 450: 203–218.

17. Eyre-Walker, A., and P. D. Keightley. 2007. The distribution of fitness effects of new mutations. *Nat. Rev. Genet.* 8: 610–618.

18. Friz, C. T. 1968. Biochemical composition of free-living amoebae *Chaos chaos*, *Chaos dubia*, and *Chaos proteus*. *Comp. Biochem. Physiol.* 26: 81–90.

19. Galvani, A. P., and J. Novembre. 2005. The evolutionary history of the *CCR5-Δ32* HIV-resistance mutation. *Microb. Infect.* 7: 302–309.

20. Ganley, A. R. D., and T. Kobayashi. 2007. Highly efficient concerted evolution in the ribosomal DNA repeats: Total rDNA repeat variation revealed by whole-genome shotgun sequence data. *Genome Res.* 17: 184–191.

21. Giovannoni, S. J., and 13 others. 2005. Genome streamlining in a cosmopolitan oceanic bacterium. *Science* 309: 1242–1245.

22. Gregory, T. R. 2001. Coincidence, coevolution, or causation? DNA content, cell size, and the C-value enigma. *Biol. Rev.* 76: 65–101.

23. Gregory, T. R. 2005. Genome size evolution in animals. In T. R. Gregory (ed.), *The Evolution of the Genome* (pp. 3–87). Elsevier, Burlington, MA.

24. Guerzoni, D., and A. McLysaght. 2015. New genes from non-coding sequence: the role of de novo protein-coding genes in eukaryotic evolutionary innovation. *Phil. Trans. R. Soc. Lond. B* 370: 20140332.

25. Halligan, D. L., F. Oliver, A. Eyre-Walker, B. Harr, and P. D. Keightley. 2010. Evidence for pervasive adaptive protein evolution in wild mice. *PLoS Genet.* 6: e1000825.

26. Hershberg, R., and D. A. Petrov. 2010. Evidence that mutation is universally biased towards AT in bacteria. *PLoS Genet.* 6: e1001115.

27. Hill, R. W., G. A. Wyse, and M. Anderson. 2016. *Animal Physiology*. Sinauer, Sunderland, MA.

28. Hillier, L. W., A. Coulson, J. I. Murray, Z. Bao, J. E. Sulston, and R. H. Waterston. 2005. Genomics in *C. elegans*: So many genes, such a little worm. *Genome Res.* 15: 1651–1660.

29. Hof, A. E. van't, and 8 others. 2016. The industrial melanism mutation in British peppered moths is a transposable element. *Nature* 534: 102–105.

30. Huvet, M., and M. Stumpf. 2014. Overlapping genes: A window on gene evolvability. *BMC Genomics* 15: 721.

31. Innan, H., and F. Kondrashov. 2010. The evolution of gene duplications: Classifying and distinguishing between models. *Nat. Rev. Genet.* 11: 97–108.

32. Jones, F. C., and 28 others. 2012. The genomic basis of adaptive evolution in threespine sticklebacks. *Nature* 484: 55–61.

33. Kasahara, M. 2007. The 2R hypothesis: An update. *Curr. Opin. Immunol.* 19: 547–552.

34. Kasimova, M. A., D. Granata, and V. Carnevale. 2016. Voltage-gated sodium channels: Evolutionary history and distinctive sequence features. *Curr. Top. Membr.* 78: 261–286.

35. Kawahara, R., and M. Nishida. 2007. Extensive lineage-specific gene duplication and evolution of the *spiggin* multi-gene family in stickleback. *BMC Evol. Biol.* 7: 209.

36. Kirkpatrick, M. 2010. How and why chromosome inversions evolve. *PLoS Biol.* 8: e1000501.

37. Kondrashov, F. A. 2012. Gene duplication as a mechanism of genomic adaptation to a changing environment. *Proc. R. Soc. Lond., Ser. B: Biol. Sci.* 279: 5048–5057.

38. Lamichhaney, S., and 20 others. 2015. Structural genomic changes underlie alternative reproductive strategies in the ruff (*Philomachus pugnax*). *Nat. Genet.* 48: 84–88. Epub.

39. Lamichhaney, S., and 20 others. 2016. Structural genomic changes underlie alternative reproductive strategies in the ruff (*Philomachus pugnax*). *Nat. Genet.* 48: 84–88.

40. Li, X. C., M. A. Schuler, and M. R. Berenbaum. 2007. Molecular mechanisms of metabolic resistance to synthetic and natural xenobiotics. *Annu. Rev. Entomol.* 52: 231–253.

41. Lindblad-Toh, K., and 78 others. 2011. A high-resolution map of human evolutionary constraint using 29 mammals. *Nature* 478: 476–482.

42. Long, M., E. Betran, K. Thornton, and W. Wang. 2003. The origin of new genes: Glimpses from the young and old. *Nat. Rev. Genet.* 4: 865–875.

43. Lynch, M. 2006. Streamlining and simplification of microbial genome architecture. *Annu. Rev. Microbiol.* 60: 327–349.

44. Lynch, M. 2007. *The Origins of Genome Architecture*. Sinauer, Sunderland, MA.

45. Lynch, M. 2010. Rate, molecular spectrum, and consequences of human mutation. *Proc. Natl. Acad. Sci. USA* 107: 961–968.

46. Lysak, M. A., M. A. Koch, J. M. Beaulieu, A. Meister, and I. J. Leitch. 2009. The dynamic ups and downs of genome size evolution in Brassicaceae. *Mol. Biol. Evol.* 26: 85–98.

47. McBride, C. S., and 7 others. 2014. Evolution of mosquito preference for humans linked to an odorant receptor. *Nature* 515: 222–227.

48. McCutcheon, J. P., and N. A. Moran. 2012. Extreme genome reduction in symbiotic bacteria. *Nat. Rev. Microbiol.* 10: 13–26.

49. Mudge, J. M., and 9 others. 2011. The origins, evolution, and functional potential of alternative splicing in vertebrates. *Mol. Biol. Evol.* 28: 2949–2959.

50. Nachman, M. W., and J. B. Searle. 1995. Why is the house mouse karyotype so variable? *Trends Ecol. Evolut.* 10: 397–402.

51. Otto, S. P., and J. Whitton. 2000. Polyploid incidence and evolution. *Annu. Rev. Genet.* 34: 401–437.

52. Pääbo, S. 2014. The human condition—A molecular approach. *Cell* 157: 216–226.

53. Palazzo, A. F., and T. R. Gregory. 2014. The case for junk DNA. *PLoS Genet.* 10.

54. Pardo-Manuel de Villena, F., and C. Sapienza. 2001. Female meiosis drives karyotypic evolution in mammals. *Genetics* 159: 1179–1189.

55. Perry, G. H., and 12 others. 2007. Diet and the evolution of human amylase gene copy number variation. *Nat. Genet.* 39: 1256–1260.

56. Piatigorsky, J. 2007. *Gene Sharing and Evolution*. Harvard University Press, Cambridge, MA.

57. Plotkin, J. B., and G. Kudla. 2011. Synonymous but not the same: The causes and consequences of codon bias. *Nat. Rev. Genet.* 12: 32–42.

58. Price, M. N., A. P. Arkin, and E. J. Alm. 2006. The life-cycle of operons. *PLoS Genet.* 2: 859–873.

59. Reneker, J., and 6 others. 2012. Long identical multispecies elements in plant and animal genomes. *Proc. Natl. Acad. Sci. USA* 109: E1183–E1191.

60. Sadava, D. E., D. M. Hillis, H. C. Heller, and S. D. Hacker. 2017. *Life: The Science of Biology*, 11th ed. Sinauer, Sunderland MA.

61. Schlotterer, C. 2015. Genes from scratch—the evolutionary fate of de novo genes. *Trends Genet.* 31: 215–219.

62. Storz, J. F., J. C. Opazo, and F. G. Hoffmann. 2013. Gene duplication, genome duplication, and the functional diversification of vertebrate globins. *Mol. Phylogen. Evol.* 66: 469–478.

63. Thompson, A., H. H. Zakon, and M. Kirkpatrick. 2016. Compensatory drift and the evolutionary dynamics of dosage-sensitive duplicate genes. *Genetics* 202: 765–774.

64. True, J. R., and S. B. Carroll. 2002. Gene co-option in physiological and morphological evolution. *Annu. Rev. Cell. Dev. Biol.* 18: 53–80.

65. Wikipedia. 2017. List of sequenced plant genomes. https://en.wikipedia.org/wiki/List_of_sequenced_plant_genomes#Gymnosperm

66. Yoshida, K., and J. Kitano. 2012. The contribution of female meiotic drive to the evolution of neo-sex chromosomes. *Evolution* 66: 3198–3208.

67. Zhang, J. Z., Y. P. Zhang, and H. F. Rosenberg. 2002. Adaptive evolution of a duplicated pancreatic ribonuclease gene in a leaf-eating monkey. *Nat. Genet.* 30: 411–415.

68. Zhang, P. G., S. Z. Huang, A. L. Pin, and K. L. Adams. 2010. Extensive divergence in alternative splicing patterns after gene and genome duplication during the evolutionary history of *Arabidopsis*. *Mol. Biol. Evol.* 27: 1686–1697.

69. Zipursky, S. L., W. M. Wojtowicz, and D. Hattori. 2006. Got diversity? Wiring the fly brain with Dscam. *Trends Biochem. Sci.* 31: 581–588.

CHAPTER 15

1. Abzhanov, A., M. Protas, B. R. Grant, P. R. Grant, and C. J. Tabin. 2004. Bmp4 and morphological variation of beaks in Darwin's finches. *Science* 305: 1462–1465.

2. Alberch, P., and E. A. Gale. 1985. A developmental analysis of an evolutionary trend: Digital reduction in amphibians. *Evolution* 39: 8–23.

3. Armbruster, W. S., C. Pélabon, G. H. Bolstad, and T. F. Hansen. 2014. Integrated phenotypes: Understanding trait covariation in plants and animals. *Phil. Trans. R. Soc., B* 369: 20130245. doi:10.1098/rstb.2013.0245.

4. Arnoult, L., and 6 others. 2013. Emergence and diversification of fly pigmentation through evolution of a gene regulatory module. *Science* 339: 1423–1426.

5. Berg, R. L. 1960. The ecological significance of correlation pleiades. *Evolution* 14: 171–180.

6. Berven, K. A., D. E. Gill, and S. J. Smith-Gill. 1979. Countergradient selection in the green frog, *Rana clamitans*. *Evolution* 33: 609–623.

7. Bowman, R. I. 1961. *Morphological Differentiation and Adaptation in the Galápagos Finches.* University of California Press, Berkeley.

8. Britten, R. J., and E. H. Davidson. 1971. Repetitive and nonrepetitive DNA and a speculation on the origin of evolutionary novelty. *Quart. Rev. Biol.* 46: 111–133.

9. Carroll, S. B., J. K. Grenier, and S. D. Weatherbee. 2005. *From DNA to Diversity: Molecular Genetics and the Evolution of Animal Design,* 2nd ed. Blackwell Science, Malden, MA.

10. Chan, Y. F., and 15 others. 2010. Adaptive evolution of pelvic reduction in sticklebacks by recurrent deletion of a *Pitx1* enhancer. *Science* 327: 302–305.

11. Cheatle Jarvela, A. M., and V. F. Hinman. 2015. Evolution of transcription factor function as a mechanism for changing metazoan developmental gene regulatory networks. *EvoDevo* 63. http://www.evodevojournal.com/content/g/1/3

12. Cook, C. D. K. 1968. Phenotypic plasticity with particular reference to three amphibious plant species. In V. Heywood (ed.), *Modern Methods in Plant Taxonomy*, pp. 97–111. Academic Press, London.

13. Deng, W. and 7 others. 2012. Controlling long-range genomic interactions at a native locus by targeted tethering of a looping factor. *Cell* 149: 1233–1244.

14. Dessain, S., C. T. Gross, M. A. Kuziora and W. McGinnis. 1992. Antp-type homeodomains have distinct DNA-binding specificities that correlate with their different regulatory functions in embryos. EMBO J. 11: 991–1002.

15. Draghi, J., and G. P. Wagner. 2008. Evolution of evolvability in a developmental model. *Evolution* 62: 301–315.

16. Flatt, T. 2005. The evolutionary genetics of canalization. *Quart. Rev. Biol.* 80: 287–316.

17. Galant, R., and S. B. Carroll. 2002. Evolution of a transcriptional repression domain in an insect Hox protein. *Nature* 415: 910–913.

18. Ghalambor, C. K., J. K. McKay, S. P. Carroll, and D. N. Reznick. 2007. Adaptive versus non-adaptive phenotypic plasticity and the potential for contemporary adaptation in new environments. *Funct. Ecol.* 21: 394–407.

19. Gilbert, S. F., and M. J. F. Barresi. 2016. *Developmental Biology*, 11th ed. Sinauer, Sunderland, MA.

20. Gompel, N., B. Prud'homme, P. J. Wittkopp, V. A. Kassner, and S. B. Carroll. 2005. Chance caught on the wing: Cis-regulatory evolution and the origin of pigment patterns in Drosophila. *Nature* 433: 481–487.

21. Gotoh, H., and 7 others. 2014. Developmental link between sex and nutrition; doublesex regulates sex-specific mandible growth via juvenile hormone signaling in stag beetles. *PLoS Genet.* 10(1): e1004098.

22. Gould, S. J. 1977. *Ontogeny and Phylogeny*. Harvard University Press, Cambridge, MA.

23. Grishkevich, V., and I. Yanai. 2013. The genomic determinants of genotype × environment interactions in gene expression. *Trends Genet.* 29: 479–487.

24. Gupta, A. P., and R. C. Lewontin. 1982. A study of reaction norms in natural populations of *Drosophila pseudoobscura*. *Evolution* 36: 934–948.

25. Hansen, T. F., and D. Houle. 2008. Measuring and comparing evolvability and constraint in multivariate characters. *J. Evol. Biol.* 21: 1201–1219.

26. Hansen, T. F., J. M. Álvarez-Castro, A. J. R. Carter, J. Hermisson, and G. P. Wagner. 2006. Evolution of genetic architecture under directional selection. *Evolution* 60: 1523–1536.

27. He, C. Y., and H. Saedler. 2005. Heterotopic expression of MPF2 is the key to the evolution of the Chinese lantern of *Physalis*, a morphological novelty in Solanaceae. *Proc. Natl. Acad. Sci. USA* 102: 5779–5784.

28. Hermisson, J., and G. P. Wagner. 2004. The population genetic theory of hidden variation and genetic robustness. *Genetics* 168: 2271–2284.

29. Jacob, F., D. Perrin, C. Sánchez, and J. Monod. 1960. L'operon: Groupe de génes à coordonnée par un opérateur. *Comptes Rendus Acad. Sci.* 250: 1727–1729.

30. Kaufman, T. C., M. A. Seeger and G. Olsen. 1990. Molecular and genetic organization of the Antennapedia gene complex of *Drosophila melanogaster*. *Adv. Genet.* 27: 309–362.

31. King, M.-C., and A. C. Wilson. 1975. Evolution at two levels in humans and chimpanzees. *Science* 188: 107–116.

32. Kondo, S., and T. Miura. 2010. Reaction-diffusion model as a framework for understanding biological pattern formation. *Science* 329: 1616–1620.

33. Koshikawa, S., and 6 others. 2015. Gain of cis-regulatory activities underlies novel domains of wingless gene expression in Drosophila. *Proc. Natl. Acad. Sci. USA* 112: 7524–7529.

34. Lande, R. 2009. Adaptation to an extraordinary environment by evolution of phenotypic plasticity and genetic assimilation. *J. Evol. Biol.* 22: 1435–1446.

35. Ledon-Rettig, C. C., D. W. Pfennig, and E. J. Crespi. 2010. Diet and hormonal manipulation reveal cryptic genetic variation: Implications for the evolution of novel feeding strategies. *Proc. Biol. Sci.* 277: 3569–3578.

36. Liao, B.-Y., M.-P. Weng, and J. Zhang. 2010. Contrasting genetic paths to morphological and physiological evolution. *Proc. Natl. Acad. Sci. USA* 107: 7353–7358.

37. Löhr, U., and L. Pick. 2005. Cofactor-interaction motifs and the cooption of a homeotic Hox protein into the segmentation pathway of *Drosophila melanogaster*. *Curr. Biol.* 15: 643–649.

38. Lonfat, N., T. Montavon, F. Darbellay, S. Gitto, and D. Duboule. 2014. Convergent evolution of complex regulatory landscapes and pleiotropy at *Hox* loci. *Science* 346: 1004–1006.

39. Lynch, V. J., and G. P. Wagner. 2008. Resurrecting the role of transcription factor change in developmental evolution. *Evolution* 62: 2131–2154.

40. Lynch, V. J., G. May, and G. P. Wagner. 2011. Regulatory evolution through divergence of a phosphoswitch in the transcription factor CEBPB. *Nature* 480: 383–386.

41. Mallarino, R., P. R. Grant, B. R. Grant, A. Herrel, W. P. Kuo, and A. Abzhanov. 2011. Two developmental modules establish 3-D beak shape variation in Darwin's finches. *Proc. Natl. Acad. Sci. USA* 108: 4057–4062.

42. Martin, A., and 8 others. 2016. CRISPR/Cas9 mutagenesis reveals versatile roles of Hox genes in crustaceasn limb specification and evolution. *Curr. Biol.* 26: 14–26.

43. Maynard Smith, J., and 8 others. 1985. Developmental constraints and evolution. *Q. Rev. Biol.* 60: 265–287.

44. McGinnis, W., and R. Krumlauf. 1992. Homeobox genes and axial patterning. *Cell* 68: 283–302.

45. McKinney, M. L., and K. J. McNamara. 1991. *Heterochrony: The Evolution of Ontogeny*. Plenum, New York.

46. McLean, C.Y., and 12 others. 2011. Human-specific loss of regulatory DNA and the evolution of human-specific traits. *Nature* 471: 216–219.

47. Meinhardt, M. 1998. *The Algorhythmic Beauty of Sea Shells*. Springer, Berlin.

48. Moczek, A. P., and 7 others. 2011. The role of developmental plasticity in evolutionary innovation. *Proc. R. Soc,. B* 278: 2705–2713.

49. Monteiro, A., and 6 others. 2007. The combined effect of two mutations that alter serially homologous color pattern elements on the fore and hindwings of a butterfly. *BMC Genet.* 8, art. 22. doi:10.1186/1471-2156-8-22.

50. Müller, G. B., and G. P. Wagner. 1996. Homology, *Hox* genes, and developmental integration. *Am. Zool.* 36: 4–13.

51. Olson, E. C., and R. L. Miller. 1958. *Morphological Integration*. University of Chicago Press, Chicago.

52. Pavlicev, M., and G. P. Wagner. 2012. A model of developmental evolution: selection, pleiotropy and compensation. *Trends Ecol. Evol.* 27: 316–322.

53. Pavlicev, M., J. M. Cheverud, and G. P. Wagner. 2011. Evolution of adaptive phenotypic patterns by direct selection for evolvability. *Proc. R. Soc., B* 278: 1903–1912.

54. Pavlicev, M., E. A. Norgard, C. C. Roseman, J. B. Wolf, and J. M. Cheverud. 2008. Genetic variation in pleiotropy: Differential epistasis as a source of variation in the allometric relationship between long bone lengths and body weight. *Evolution* 62: 199–213.

55. Pélabon, C., W. S. Armbruster, and T. F. Hansen. 2011. Experimental evidence for the Berg hypothesis: Vegetative traits are more sensitive than pollination traits to environmental variation. *Funct. Ecol.* 25: 247–257.

56. Pfennig, D. W., and P. J. Murphy. 2000. Character displacement in polyphenic tadpoles. *Evolution* 54: 1738–1749.

57. Raff, R. A. 1996. *The Shape of Life: Genes, Development, and the Evolution of Animal Form*. University of Chicago Press, Chicago.

58. Refki, P. N., D. Armisén, A. J. Crumière, S. Viala, and A. Khila. 2014. Emergence of tissue sensitivity to Hox protein levels underlies the evolution of an adaptive morphological trait. *Devel. Biol.* 392: 441–453.

59. Reno, P. L., M. A. McCollum, M. J. Cohn, R. S. Meindl, M. Hamrick, and C. O. Lovejoy. 2008. Patterns of correlation and covariation of anthropoid distal forelimb segments correspond to Hoxd expression territories. *J. Exp. Zool. (Mol. Dev. Evol.)* 310B: 240–258.

60. Riedl, R. 1978. *Order in Living Organisms: A Systems Analysis of Evolution*. Wiley, New York.

61. Rockman, M. V. 2011. The QTN program and the alleles that matter for evolution: All that's gold does not glitter. *Evolution* 66: 1–17.

62. Romer, A. S. 1966. *Vertebrate Paleontology*. University of Chicago Press, Chicago.

63. Ronshaugen, M., N. McGinnis, and W. McGinnis. 2002. Hox protein mutation and macroevolution of the insect body plan. *Nature* 415: 914–917.

64. Rubenstein, M., and F. S. J. de Souza. 2013. Evolution of transcriptional enhancers and animal diversity. *Phil. Trans. R. Soc,. B* 368: 20130017. doi:10.1098/rstb20130017.

65. Sanderson, A. R., R. M. Kirby, C. R. Johnson, and L. Yang. 2006. Advanced reaction-diffusion models for texture synthesis. *J. Graphics Tools* 11: 47.

66. Scharloo, W. 1991. Canalization: Genetic and developmental aspects. *Annu. Rev. Ecol. Syst.* 22: 65–93.

67. Schlichting, C. D., and M. Pigliucci. 1998. *Phenotypic Evolution: A Reaction Norm Perspective*. Sinauer, Sunderland, MA.

68. Schlichting, C. D., and M. A. Wund. 2014. Phenotypic plasticity and epigenetic marking: An assessment of evidence for genetic accommodation. *Evolution* 68: 656–672.

69. Schmalhausen, I. I. 1986. *Factors of Evolution*, reprint. University of Chicago Press, Chicago. (Russian publication, 1946; English translation, 1949.)

70. Scoville, A. G., and M. E. Pfrender. 2010. Phenotypic plasticity facilitates recurrent rapid adaptation to introduced predators. *Proc. Natl. Acad. Sci. USA* 107: 4260–4263.

71. Shapiro, M. D., and 7 others. 2004. Genetic and developmental basis of evolutionary pelvic reduction in threespine sticklebacks. *Nature* 428: 717–723.

72. Siegal, M. L., and A. Bergman. 2002. Waddington's canalization revisited: Developmental stability and evolution. *Proc. Natl. Acad. Sci. USA* 99: 10528–10532.

73. Stern, D. L., and V. Orgogozo. 2008. The loci of evolution: How predictable is genetic evolution? *Evolution* 62: 2155–2177.

74. Stümpke, H. 1957. *Bau und Leben der Rhinogradentia*. (English translation: *The Snouters*). Doubleday, New York. (Reprinted 1981, University of Chicago Press.)

75. Sundaram, V., and 7 others. 2014. Widespread contribution of transposable elements to the innovation of gene regulatory networks. *Genome Res.* 24: 1963–1976.

76. Suzuki, Y., and H. F. Nijhout. 2006. Evolution of a polyphenism by genetic assimilation. *Science* 311: 650–652.

77. ten Broek, C. M. A., A. J. Bakker, I. Varela-Lasheras, M. Buhiani, S. Van Dongen, and F. Galis. 2012. Evo-devo of the human vertebral column: On homeotic transformations, pathologies and prenatal selection. *Evol. Biol.* 39: 456–471.

78. True, J. R., and S. B. Carroll. 2002. Gene co-option in physiological and morphological evolution. *Annu. Rev. Cell. Dev. Biol.* 18: 53–80.

79. Vaughan, T. A. 1986. *Mammalogy*, 3rd ed. Saunders College Publishing, Philadelphia.

80. von Baer, K. E. 1828. *Entwicklungsgeschichte der Thiere: Beobachtung und Reflexion*. Bornträger, Konigsberg, Germany.

81. Waddington, C. H. 1953. Genetic assimilation of an acquired character. *Evolution* 7: 118–126.

82. Wagner, G. P. 1996. Homologues, natural kinds, and the evolution of modularity. *Am. Zool.* 36: 36–43.

83. Wagner, G. P., and L. Altenberg. 1996. Perspective: Complex adaptations and the evolution of evolvability. *Evolution* 50: 967–976.

84. Wagner, G. P., and V. J. Lynch. 2008. The generegulatory logic of transcription factor evolution. *Trends Ecol. Evol.* 23: 377–385.

85. Wagner, G. P., C. H. Chiu, and M. Laubichler. 2000. Developmental evolution as a mechanistic science: The inference from developmental mechanisms to evolutionary processes. *Amer. Zool.* 40: 819–831.

86. West-Eberhard, M. J. 2003. *Developmental Plasticity and Evolution*. Oxford University Press, New York.

87. Young, N. M., G. P. Wagner, and B. Hallgrimsson. 2010. Development and the evolvability of human limbs. *Proc. Natl. Acad. Sci. USA* 107: 3400–3405.

CHAPTER 16

1. Barrett, R. D. H., and D. Schluter. 2008. Adaptation from standing genetic variation. *Trends Ecol. Evol.* 23: 38–44.

2. Barrett, R. D. H., S. M. Rogers, and D. Schluter. 2008. Natural selection on a major armor gene in threespine stickleback. *Science* 322: 255–257.

3. Brusatte, S. L., G. T. Lloyd, S. C. Wang, and M. A. Norell. 2014. Gradual assembly of avian body plan culminated in rapid rates of evolution across the dinosaur-bird transition. *Curr. Biol.* 24: 2386–2392.

4. Carstens, B. C., and L. L. Knowles. 2007. Estimating species phylogeny from gene-tree probabilities despite incomplete lineage sorting: An example from *Melanoplus* grasshoppers. *Syst. Biol.* 56: 400–411.

5. Chatterjee, S. 1985. *Postosuchus*, a new thecodontian reptile from the Triassic of Texas and the origin of tyrannosaurs. *Phil. Trans. R. Soc,. B* 309: 395–460.

6. Colosimo, P. F., and 9 others. 2005. Widespread parallel evolution in sticklebacks by repeated fixation of *Ectodysplasin* alleles. *Science* 307: 1928–1933.

7. Crawford, N. G., B. C. Faircloth, J. E. McCormack, R. T. Brumfield, K. Winker, and T. G. Glenn. 2012. More than 1000 ultraconserved elements provide evidence that turtles are the sister group of archosaurs. *Biol. Lett.* 8: 783–786.

8. Cunningham, C. W., H. Zhu, and D. M. Hillis. 1998. Best-fit maximum-likelihood mdels for phylogenetic inference: Empirical tests with known phylogenies. *Evolution* 52: 978–987.

9. Currie, T. E., S. J. Greenhill, R. D. Gray, T. Hasegawa, and R. Mace. 2010. Rise and fall of political complexity in island South-East Asia and the Pacific. *Nature* 467: 801–804.

10. dos Reis, M., P. C. J. Donoghue, and Z. Yang. 2013. Neither phylogenomic nor palaeontological data support a Palaeocene origin of placental mammals. *Biol. Lett.* 10: 20131003. (Cited in Springer and Gatesy 2015.)

11. Fitch, W. M. 1976. Molecular evolutionary clocks. In F. J. Ayala (ed.), *Molecular Evolution*, pp. 160–178. Sinauer, Sunderland, MA.

12. Fontaine, M. C., and 18 others. 2015. Extensive introgression in a malaria vector species complex revealed by phylogenomics. *Science* 347: 1258524.

13. Gaut, B., L. Yang. S. Takuno, and L. E. Eguiarte. 2011. The patterns and causes of variation in plant nucleotide substitution rates. *Annu. Rev. Ecol. Evol. Syst.* 42: 245–266.

14. Gire, S. K., and 57 others. 2014. Genomic surveillance elucidates Ebola virus origin and transmission during the 2014 outbreak. *Science* 345: 1369–1372.

15. Graur, D. 2016. *Molecular and Genome Evolution*. Sinauer, Sunderland, MA.

16. Harms, M. J., and J. W. Thornton. 2010. Analyzing protein structure and function using ancestral gene reconstruction. *Curr. Opin. Struct. Biol.* 20: 360–366.

17. Harvey, P. H., and A. H. Harcourt. 1984. Sperm competition, testes size, and breeding system in primates. In R. L. Smith (ed.), *Sperm Competition and the Evolution of Animal Breeding Systems*, pp. 589–600. Academic Press, New York.

18. Hillis, D. M., J. J. Bull, M. E. White, M. R. Badgett, and I. J. Molineaux. 1992. Experimental phylogenetics: Generation of a known phylogeny. *Science* 255: 589–592.

19. Jarvis, E. D., and 104 others. 2014. Whole-genome analyses resolve early branches in the tree of life of modern birds. *Science* 346: 1320–1331.

20. Jones, F. C., and 28 others. 2012. The genomic basis of adaptive evolution in threespine sticklebacks. *Nature* 484: 55–61.

21. Kumar, S. 2005. Molecular clocks: Four decades of evolution. *Nat. Rev. Genet.* 6: 654–662.

22. Li, Q., and 9 others. 2012. Reconstruction of *Microraptor* and the evolution of iridescent plumage. *Science* 335: 1215–1219.

23. Mace, R., and F. M. Jordan. 2011. Macro-evolutionary studies of cultural diversity: A review of empirical studies of cultural transmission and cultural adaptation. *Phil. Trans. R. Soc,. B* 366: 402–411.

24. Meyer, A., and R. Zardoya. 2003. Recent advances in the (molecular) phylogeny of vertebrates. *Annu. Rev. Ecol. Evol. Syst.* 34: 311–338.

25. Mindell, D. F., and C. E. Thacker. 1996. Rates of molecular evolution: Phylogenetic issues and applications. *Annu. Rev. Ecol. Syst.* 27: 279–303.

26. Pipoly, I., V. Bókony, M. Kirkpatrick, P. F. Donald, T. Székely, and A. Liker. 2015. The genetic sex-determination system predicts adult sex ratios in tetrapods. *Nature* 527: 91–94.

27. Prum, R. O. 2002. Why ornithologists should care about the theropod origin of birds. *The Auk* 119: 1–17.

28. Prum, R. O., and 6 others. 2015. A comprehensive phylogeny of birds (Aves) using targeted next-generation DNA sequencing. *Nature* 526: 569–573.

29. Przeworski, M., G. Coop, and J. D. Wall. 2005. The signature of positive selection on standing genetic variation. *Evolution* 59: 2312–2323.

30. Scally, A., and 70 others. 2012. Insights into hominid evolution from the gorilla genome sequence. *Nature* 483: 169–175.

31. Sites, J. W., Jr., T. W. Reeder, and J. J. Wiens. 2011. Phylogenetic insights on evolutioanry novelties in lizards andd snakes: sex, birth, bodies, niches, and venom. *Annu. Rev. Ecol. Evol. Syst.* 42: 227–244.

32. Springer, M. S., and J. Gatesy. 2015. The gene tree delusion. *Mol. Phyl. Evol.* 94: 1–33.

33. Steiper, M. E., and N. M. Young. 2006. Primate molecular divergence dates. *Mol. Phyl. Evol.* 41: 384–394.

34. Van Rheede, T., T. Bastiaans, D. S. Boone, S. B, Hedges, W. W. deJong, and O. Madsen. 2006. The platypus is in its place: Nuclear genes and indels confirm the sister-group relation of monotremes and theirans. *Mol. Biol. Evol.* 23: 587–597.

35. Wagner, G. P., and J. A. Gauthier. 1999. 1, 2, 3 = 2, 3, 4: A solution to the problem of the homology of the digits in the avialn hand. *Proc. Natl. Acad. Sci. USA* 96: 5111–5116.

36. Wagner, C. E., and 7 others. 2013. Genome-wide RAD sequence data provide unprecedented resolution of species boundaries and relationships in the Lake Victoria cichlid adaptive radiation. *Mol. Ecol.* 22: 787–798.

37. Yokoyama, S. 2008. Evolution of dim-light and color vision pigments. *Annu. Rev. Genomics Hum. Genet.* 9: 259–282.

38. Yokoyama, S. 2012. Synthesis of experimental molecular biology and evolutionary biology: An example from the world of vision. *BioScience* 62: 939–948.

39. Yokoyama, S., T. Tada, H. Zhang, and L. Britt. 2008. Elucidation of phenotypic adaptations: molecular analyses of dim-light vision proteins in vertebrates. *Proc. Natl. Acad. Sci. USA* 105: 13480–13485.

CHAPTER 17

1. Ahlberg, P. E., and J. A. Clack. 2006. A firm step from water to land. *Nature* 440: 747–749.

2. Ahlberg, P. E., J. A. Clack, E. Lukševičs, H. Blom, and I. Zupins. 2008. *Ventastega curonica* and the origin of tetrapod morphology. *Nature* 453: 1199–1204.

3. Algeo, T. J., and S. F. Scheckler. 2010. Land plant evolution and weathering rate changes in the Devonian. *J. Earth Sci.* 21 (Special Issue): 75–78.

4. Alroy, J. 2001. A multispecies overkill simulation of the end-Pleistocene megafaunal mass extinction. *Science* 292: 1893–1896.

5. Arakaki, M., and 8 others. 2011. Contemporaneous and recent radiations of the world's major succulent plant lineages. *Proc. Natl. Acad. Sci. USA* 108: 8379–8384.

6. Baldauf, S. L. 2003. The deep roots of eukaryotes. *Science* 300: 1703–1706.

7. Bambach, R. K. 2006. Phanerozoic biodiversity mass extinctions. *Annu. Rev. Earth Planet. Sci.* 34: 127–155.

8. Barnosky, A. D., and 11 others. 2011. Has the Earth's sixth mass extinction already arrived? *Nature* 471: 51–57.

9. Benton, M. J., M. A. Wills, and R. Hitchin. 2000. Quality of the fossil record through time. *Nature* 403: 534–537.

10. Bininda-Emonds, O. R. P., and 9 others. 2007. The delayed rise of present-day mammals. Nature 446: 507–513.

11. Bourke, A. F. G. 2011. *Principles of Social Evolution*. Oxford University Press, Oxford.

12. Brasier, M. D., J. Antcliffe, M. Saunders, and D. Wacey. 2015. Changing the picture of Earth's earliest fossils (3.5–1.9 Ga) with new approaches and new discoveries. *Proc. Natl. Acad. Sci. USA* 112: 4859–4864.

13. Brown, J. H., and M. V. Lomolino. 1998. *Biogeography*, 2nd ed. Sinauer, Sunderland, MA.

14. Burgess, S. D., S. Bowring, and S.-Z. Shen. 2014. High-precision timeline for Earth's most severe extinction. *Proc. Natl. Acad. Sci. USA* 111: 3316–3321.

15. Bush, A. M., and R. K. Bambach. 2011. Paleoecologic megatrends in marine Metazoa. *Annu. Rev. Earth Planet. Sci.* 39: 241–269.

16. Butterfield, N. J. 2015. Early evolution of the Eukaryota. *Palaeontology* 58 (part 1): 5–17.

17. Carroll, L. 1871. *Through the Looking-Glass, and What Alice Found There*. MacMillan, London.

18. Carroll, R. L. 1988. *Vertebrate Paleontology and Evolution*. W. H. Freeman, New York.

19. Cavalier-Smith, T. 2006. Cell evolution and Earth history: Stasis and revolution. *Phil. Trans. R. Soc, B* 361: 969–1006.

20. Clack, J. A. 2002. *Gaining Ground: The Origin and Evolution of Tetrapods*. Indiana University Press, Bloomington.

21. Coope, G. R. 1979. Late Cenozoic fossil Coleoptera: Evolution, biogeography, and ecology. *Annu. Rev. Ecol. Syst.* 10: 249–267.

22. Crick, F. H. C. 1968. The origin of the genetic code. *J. Mol. Biol.* 38: 367–379.

23a. Daeschler, E. B., N. H. Shubin, and F. A. Jenkins, Jr. 2006. A Devonian tetrapod-like fish and the evolution of the tetrapod body plan. *Nature* 440: 757–763.

23b. DeVos, L., K. Rützler, N. Boury-Esnault, C. Donadey, and J. Vacelet. 1991. *Atlas of Sponge Morphology*. Smithsonian Institution Press, Washington and London.

24. Downs, J. P., E. B. Daeschler, F. A. Jenkins, and N. H. Shubin. 2008. The cranial endoskeleton of *Tiktaalik roseae*. *Nature* 455: 925–929.

25. Dunn, C. W., G. Giribet, G. D. Edgecombe, and A. Hejnol. 2014. Animal phylogeny and its evolutionary implications. *Annu. Rev. Ecol. Evol. Syst.* 45: 371–395.

26. Embley, T. M., and T. A. Williams. 2015. Steps on the road to eukaryotes. *Nature* 521: 169–170.

27. Erwin, D. H. 2006. *Extinction: How Life on Earth Nearly Ended 250 Million Years Ago*. Princeton University Press, Princeton, NJ.

28. Erwin, D. H. 2009. Early evolution of the bilaterian developmental toolkit. *Phil. Trans. R. Soc,. B* 364: 2253–2261.

29. Erwin, D. H., M. Laflamme, S. M. Tweedt, E. A. Sperling, D. Pisani, and K. J. Peterson. 2011. The Cambrian conundrum: Early divergence and later ecological success in the early history of animals. *Science* 334: 1091–1097.

30. Fournier, G. P., J. Huang, and J. P. Gogarten. 2009. Horizontal gene transfer from extinct and extant lineages: Biological innovation and the coral of life. *Phil. Trans. R. Soc., B* 364: 2229–2239.

31. Friis, E. M., K. R. Pedersen, and P. R. Crane. 2010. Diversity in obscurity: Fossil flowers and the early history of angiosperms. *Phil. Trans. R. Soc., B* 365: 369–382.

32. Gollihar, J., M. Levy, and A. D. Ellington. 2014. Many paths to the origin of life. *Science* 343: 259–260.

33. Grimaldi, D. A., and M. S. Engel. 2005. *Evolution of the Insects*. Cambridge University Press, New York.

34. Grosberg, R. K., and R. R. Strathmann. 2007. The evolution of multicellularity: A minor major transition? *Annu. Rev. Ecol. Evol. Syst.* 38: 621–654.

35. Huntley, J. W., and M. Kowalewski. 2007. Strong coupling of predation intensity and diversity in the Phanerozoic fossil record. *Proc. Natl. Acad. Sci. USA* 104: 15006–15010.

36. Jablonski, D. 1995. Extinctions in the fossil record. In J. H. Lawton and R. M. May (eds.), *Extinction Rates*, pp. 25–44. Oxford University Press, Oxford.

37. Jablonski, D. 2008. Extinction and the spatial dynamics of biodiversity. *Proc. Natl. Acad. Sci. USA* 105 (Suppl. 1): 11528–11535.

38. Jablonski, D., S. J. Gould, and D. M. Raup. 1986. The nature of the fossil record: A biological perspective. In D. M. Raup and D. Jablonski (eds.), *Patterns and Processes in the History of Life*, pp. 7–22. Springer-Verlag, Berlin.

39. Jackson, J. B. C. 1995. Constancy and change of life in the sea. In J. H. Lawton and R. M. May (eds.), *Extinction Rates*, pp. 45–54. Oxford University Press, Oxford.

40. Jackson, S. T., and J. L. Blois. 2015. Community ecology in a changing environment: Perspectives from the Quaternary. *Proc. Natl. Acad. Sci. USA* 112: 4915–4921.

41. Janvier, P. 2015. Facts and fancies about early fossil chordates and vertebrates. *Nature* 520: 483–489.

42. Jarvis, E. D., and 104 others. 2014. Whole-genome analyses resolve early branches in the tree of life of modern birds. *Science* 346: 1320–1331.

43. Judd, W. S., C. S. Campbell, E. A. Kellogg, P. F. Stevens, and M. J. Donoghue. 2016. *Plant Systematics: A Phylogenetic Approach*, 4th ed. Sinauer, Sunderland, MA.

44. Kareiva, P. M., J. G. Kingsolver, and R. B. Huey (eds.). 1993. *Biotic Interactions and Global Change*. Sinauer, Sunderland, MA.

45. Kenrick, P., and P. R. Crane. 1997. The origin and early evolution of plants on land. *Nature* 389: 33–39.

46. Kerr, R. A. 2013. Mega-eruptions drove the mother of mass extinctions. *Science* 342: 1424.

47. Kidston, R., and W. H. Lang. 1921. On Old Red Sandstone plants showing structure from the Rhynie chert bed, Aberdeenshire, Part IV. Restorations of the vascular cryptogams, and discussion of their bearing on the general morphology of Pteridophyta and the origin of the organization of land plants. *Trans. R. Soc. Edinburgh* 32: 477–487.

48. Knoll, A. H. 2003. *Life on a Young Planet*. Princeton University Press, Princeton, NJ.

49. Knoll, A. H. 2011. The multiple origins of multicellularity. *Annu. Rev. Earth Planet. Sci.* 39: 217–239.

50. Knoll, A. H., R. K. Bambach, J. L. Payne, S. Pruss, and W. Fischer. 2007. Paleophysiology and end-Permian mass extinction. *Earth and Planetary Sci. Lett.* 256: 295–313.

51. Koumandou, V. L., B. Wickstead, M. L. Ginger, M. van der Giezen, J. B. Dacks, and M. C. Field. 2013. Molecular paleontology and complexity in the last eukaryotic common ancestor. *Crit. Rev. Biochem. Mol. Biol.* 48: 373–396.

52. Lazcano, A. 2010. The origin and early evolution of life: Did it all start in Darwin's warm little pond? In M. A. Bell, D. J. Futuyma, W. F. Eanes, and J. S. Levinton (eds.), *Evolution since Darwin: The First 150 Years*, pp. 353–375. Sinauer, Sunderland, MA.

53. Lee, M. S. Y., T. W. Reeder, J. B. Slowinski, and R. Lawson. 2004. Resolving reptile relationships: Molecular and morphological markers. In J. Cracraft and M. J. Donoghue (eds.), *Assembling the Tree of Life*, pp. 451–467. Oxford University Press, New York.

54. Lessios, H. A. 2008. The great American schism: Divergence of marine organisms after the rise of the Central American Isthmus. *Annu. Rev. Ecol. Evol. Syst.* 39: 63–91.

55. Lilley, D. M. J., and J. Sutherland. 2011. The chemical origins of life and its early evolution: An introduction. *Phil. Trans. R. Soc., B* 366: 2853–2856.

56. Luo, Z.-X. 2007. Transformation and diversification in early mammal evolution. *Nature* 450: 1011–1019.

57. Lyons, T. W., T. Reinhard, and N. J. Planacsky. 2014. The rise of oxygen in Earth's early ocean and atmosphere. *Nature* 506: 307–315.

58. Magallón, S., S. Gómez-Acevedo, L. L. Sánchez-Reyes, and T. Hernández-Hernández. 2015. A metacalibrated time-tree documents the early rise of flowering plant phylogenetic diversity. *New Phytol.* 207: 437–453.

59. Margulis, L. 1993. *Symbiosis in Cell Evolution*, 2nd ed. W. H. Freeman, San Francisco.

60. Marshall, C. R. 2006. Explaining the Cambrian "explosion" of animals. *Annu. Rev. Earth Planet. Sci.* 34: 355–384.

61. Martin, P. S., and R. G. Klein (eds.). 1984. *Quaternary Extinctions: A Prehistoric Revolution*. University of Arizona Press, Tucson.

62. Maynard Smith, J., and E. Szathmáry. 1995. *The Major Transitions in Evolution*. W. H. Freeman, San Francisco.

63. McElwain, J. C., and S. W. Punyasena. 2007. Mass extinction events and the plant fossil record. *Trends Ecol. Evol.* 22: 548–557.

64. Meredith, R. W., and 22 others. 2011. Impacts of the Cretaceous terrestrial revolution and KPg extinction on mammal diversification. *Science* 334: 521–524.

65. Michod, R. E. 2007. Evolution of individuality during the transition from unicellular to multicellular life. *Proc. Natl. Acad. Sci. USA* 104: 8613–8618.

66. Misof, B., and 100 others. 2014. Phylogenomics resolves the timing and pattern of insect evolution. *Science* 346: 763–767.

67. Moran, N. A. 2007. Symbiosis as an adaptive process and source of phenotypic complexity. *Proc. Natl. Acad. Sci. USA* 104: 8627–8633.

68. O'Dea, A. O., and 34 others. 2016. Formation of the Isthmus of Panama. *Sci. Adv.* 2, article e1600883.

69. Paegel, B. M., and G. F. Joyce. 2008. Darwinian evolution on a chip. *PLoS Biol.* 6(4): e85.

70. Parfrey, L. W., D. J. G. Lahr, A. H. Knoll, and L. A. Katz. 2011. Estimating the timing of early eukaryotic diversification with multigene molecular clocks. *Proc. Natl. Acad. Sci. USA* 108: 13624–13629.

71a. Peterson, K. J., M. R. Dietrich, and M. A. McPeek. 2009. MicroRNAs and metazoan macroevolution: Insights into canalization, complexity, and the Cambrian explosion. *BioEssays* 31: 736–747.

71b. Porter, S. M., and L. A. Riedman. 2016. Systematics of organic-walled microfossils from the ca. 780–740 Ma Chuar Group, Grand Canyon, Arizona. *J. Paleont.* 90: 815–853.

71c. Porter, S. M., R. Meisterfeld, and A. H. Knoll. 2003. Vase-shaped microfossils from the Neoproterozoic Chuar Group, Grand Canyon: A classification guided by modern testate amoebae. *J. Paleont.* 77: 409–429.

72. Prescott, G. W., D. R. Williams, A. Balmford, R. E. Green, and A. Manica. 2012. Quantitative global analysis of the role of climate and people in explaining late Quaternary megafaunal extinctions. *Proc. Natl. Acad. Sci. USA* 109: 4527–4531.

73. Prum, R. O., and 6 others. 2015. A comprehensive phylogeny of birds (Aves) using targeted next-generation DNA sequencing. *Nature* 526: 569–573.

74. Quintero, I., and J. J. Wiens. 2013. Rates of projected climate change dramatically exceed past rates of climatic niche evolution. *Ecol. Lett.* 16: 1095–1103.

75. Ratcliff, W. C., R. F. Denison, M. Borrello, and M. Travisano. 2012. Experimental evolution of multicellularity. *Proc. Natl. Acad. Sci. USA* 109: 1595–1600.

76. Richter, D. J., and N. King. 2013. The genomic and cellular foundations of animal origins. *Annu. Rev. Genet.* 47: 509–537.

77. Romer, A. S. 1966. *Vertebrate Paleontology*. University of Chicago Press, Chicago.

78. Romer, A. S., and T. S. Parsons. 1986. *The Vertebrate Body*. Saunders College Publishing, Philadelphia.

79. Ruhl, M., N. R. Bonis, G.-J. Reichart, J. A. S. Damsté, and W. M. Kürschner. 2011. Atmospheric carbon injection linked to End-Triassic mass extinction. *Science* 333: 430–434.

80. Sala, O. E., and 18 others. 2000. Global biodiversity scenarios for the year 2100. *Science* 287: 1770–1774.

81. Sepkoski, J. 1997. Biodiversity: Past, present, and future. *J. Paleont.* 71: 533–539.

82. Sereno, P. C. 1999. The evolution of dinosaurs. *Science* 284: 2137–2147.

83. Shen, A.-Z., and 21 others. 2011. Calibrating the End-Permian mass extinction. *Science* 334: 1367–1372.

84. Shu, D.-G., and 10 others. 2003. Head and backbone of the Early Cambrian vertebrate *Haikouichthys*. *Nature* 421: 526–529.

85. Shubin, N. H., E. B. Daeschler, and F. A. Jenkins, Jr. 2006. The pectoral fin of *Tiktaalik roseae* and the origin of the tetrapod limb. *Nature* 440: 764–771.

86. Spang, A., and 9 others. 2015. Complex archaea that bridge the gap between prokaryotes and eukaryotes. *Nature* 521: 173–179.

87. Spiegelman, S. 1970. Extracellular evolution of replicating molecules. In F. O. Schmitt (ed.), *The Neuro Sciences: A Second Study Program*, pp. 927–945. Rockefeller University Press, New York.

88. Stanley, S. M., and L. A. Campbell. 1981. Neogene mass extinction of western Atlantic mollusks. *Nature* 293: 457–459.

89. Stewart, W. N. 1983. *Paleobotany and the Evolution of Plants*. Cambridge University Press, Cambridge.

90. Theobald, D. L. 2010. A formal test of the theory of universal common ancestry. *Nature* 465: 219–222.

91. Thomas, C. D., and 18 others. 2004. Extinction risk from climate change. *Nature* 427: 145–148.

92. Valentine, J. W. 2004. *On the Origin of Phyla*. University of Chicago Press, Chicago.

93. Vermeij, G. J. 1987. *Evolution and Escalation: An Ecological History of Life*. Princeton University Press, Princeton, NJ.

94. Vermeij, G. J., and R. K. Grosberg. 2010. The great divergence: When did diversity on land exceed that in the sea? *Integr. Comp. Biol.* 50: 675–682.

95. Wang, H., and 9 others. 2009. Rosid radiation and the rapid rise of angiosperm-dominated forests. *Proc. Natl. Acad. Sci. USA* 106: 3853–3858.

96. Wellman, C. H., P. L. Osterloff, and U. Mohiuddin. 2003. Fragments of the earliest land plants. *Nature* 425: 282–290.

97. Williams, T. A., P. G. Foster, C. J. Cox, and T. M. Embley. 2013. An archaeal origin of eukaryotes supports only two primary domains of life. *Nature* 504: 231–236.

98. Wilson, E. O. 1992. *The Diversity of Life*. Harvard University Press, Cambridge, MA.

99. Woese, C. R. 2000. Interpreting the universal phylogenetic tree. *Proc. Natl. Acad. Sci. USA* 97: 8392–8396.

100. Zimmer, C. 2009. On the origin of life on Earth. *Science* 323: 198–199.

CHAPTER 18

1. Avise, J. C. 2000. *Phylogeography*. Harvard University Press, Cambridge, MA.

2. Brown, J. H., and A. C. Gibson. 1983. *Biogeography*. Mosby, St. Louis.

3. Brown, J. H., and M. V. Lomolino. 1998. *Biogeography*, 2nd ed. Sinauer, Sunderland, MA.

4. Cook, L. G., and M. D. Crisp. 2005. Not so ancient: The extant crown group of *Nothofagus* represents a post-Gondwanan radiation. *Proc. R. Soc. Lond. B* 272: 2535–2544.

5. Coope, G. R. 1979. Late Cenozoic fossil Coleoptera: Evolution, biogeography, and ecology. *Annu. Rev. Ecol. Syst.* 19: 247–267.

6. Davis, C. C., P. W. Fritsch, C. D. Bell, and S. Mathews. 2004. High-latitude Tertiary migrations of an exclusively tropical clade: Evidence from Malpighiaceae. *Int. J. Plant Sci.* 165 (4 Suppl.): S107–S121.

7. de Queiroz, A. 2005. The resurrection of oceanic dispersal in historical biogeography. *Trends Ecol. Evol.* 9: 68–73.

8. de Queiroz, A., 2014. *The Monkey's Voyage: How Improbable Journeys Shaped the History of Life*. Basic Books, New York.

9. Dick, C. W., K. Abdul-Salim, and E. Bermingham. 2003. Molecular systematic analysis reveals cryptic Tertiary diverssification of a widespread tropical rainforest tree. *Am. Nat.* 162: 691–703.

10. Dick, C. W., E. Bermingham, M. R.Lemes, and R. Gribel. 2007. Extreme long-distance dispersal of the lowland tropical rainforest tree *Ceiba pentandra* L. (Malvaceae) in Africa and the Neotropics. *Mol. Ecol.* 16: 3039–3049.

11. Donoghue, M. J., and S. A. Smith. 2004. Patterns in the assembly of temperate forests around the Northern Hemisphere. *Phil. Trans. R. Soc. Lond., B* 359: 1633–1644.

12. Fine, P.V. A. 2015. Ecological and evolutionary drivers of geographic variation in species diversity. *Annu. Rev. Ecol. Evol. Syst.* 46: 369–392.

13. Fine, P.V. A., and R. H. Ree. 2006. Evidence for a time-integrated species-area effect on the latitudinal gradient in species diversity. *Am. Nat.* 168: 796–804.

14. Friedman, M., and 7 others. 2013. Molecular and fossil evidence place the origin of cichlid fishes long after Gondwanan rifting. *Proc. R. Soc. B* 280: 20131733.

15. Grossenbacher, D., R. B. Runquist, E. E. Goldberg, and Y. Brandvain. 2015. Geographic range size is predicted by plant mating system. *Ecol. Lett.* 18: 706–713.

16. Hickerson, M. J., and 8 others. 2010. Phylogeography's past, present, and future: 10 years after Avise, 2000. *Mol. Phyl. Evol.* 54: 291–301.

17. Huston, M. 1994. *Biological Diversity: The Coexistence of Species on Changing Landscapes*. Cambridge University Press, New York.

18. Hutchinson, G. E. 1957. Concluding remarks. *Cold Spring Harbor Symp. Quant. Biol.* 22: 415–427.

19. Jablonski, D., K. Roy, and J. W. Valentine. 2006. Out of the tropics: Evolutionary dynamics of the latitudinal diversity gradient. *Science* 314: 102–106.

20. Jetz, W., and P.V. A. Fine. 2012. Global gradients in vertebrate diversity predicted by historical area-productivity dynamics and contemporary environment. *PLoS Biol.* 10(3): e1001292.

21. Knowles, L. L. 2009. Statistical phylogeography. *Annu. Rev. Ecol. Evol. Syst.* 40: 593–612.

22. Knowlton, N., L. A. Weigt, L. A. Solórzano, D. K. Mills, and E. Bermingham. 1993. Divergence in proteins, mitochondrial DNA, and reproductive compatibility across the Isthmus of Panama. *Science* 260: 1629–1632.

23. Lessios, H. A. 2008. The great American schism: Divergence of marine organisms after the rise of the Central American Isthmus. *Annu. Rev. Ecol. Evol. Syst.* 39: 63–91.

24. Lieberman, B. S. 2003. Paleobiogeography: The relevance of fossils to biogeography. *Annu. Rev. Ecol. Evol. Syst.* 34: 51–69.

25. Lomolino, M.V., B. R. Riddle, and R. J. Whittaker. 2017. *Biogeography*, 5th ed. Sinauer, Sunderland, MA.

26. Mao, K., and 7 others. 2012. Distribution of living Cupressaceae reflects the breakup of Pangaea. *Proc. Nat. Acad. Sci. USA* 109: 7793–7798.

27. Mendelson, T. C., and K. L. Shaw. 2005. Rapid speciation in an arthropod. *Nature* 433: 375.

28. Mittelbach, G. G., and 21 others. 2007. Evolution and the latitudinal diversity gradient: Speciation, extinction and biogeography. *Ecol. Lett.* 10: 315–331.

29. Myers, A. A., and P. S. Giller (eds.). 1988. *Analytical Biogeography*. Chapman & Hall, London.

30. O'Dea, A. and 35 others. 2016. Formation of the Isthmus of Panama. *Science Advances* 2(8), article e1600883.

31. Parmesan, C., S. Gaines, L. Gonzalez, D. M. Kaufman, J. Kingsolver, A. T. Peterson, and R. Sagarin. 2005. Empirical perspectives on species borders: From traditional biogeography to global change. *Oikos* 108: 58–75.

32. Pigot, A. L., and J. A. Tobias. 2013. Species interactions constrain geographic range expansion over evolutionary time. *Ecol. Lett.* 16: 330–338.

33. Qian, H., and R. E. Ricklefs. 2004. Geographical distribution and ecological conservation of disjunct genera of vascular plants in eastern Asia and eastern North America. *J. Ecol.* 92: 253–265.

34. Ree, R. H., and S. A. Smith. 2008. Maximum likelihood inference of geographic range evolution by dispersal, local extinction, and cladogenesis. *Syst. Biol.* 57: 4–14.

35. Ree, R. H., B. R. Moore, C. O. Webb, and M. J. Donoghue. 2005. A likelihood framework for inferring the evolution of geographic range on phylogenetic trees. *Evolution* 59: 2299–2311.

36. Renner, S. 2004. Plant dispersal across the tropical Atlantic by wind and sea currents. *Int. J. Plant Sci.* 165 (4 Suppl.): S23–S33.

37. Ricklefs, R. E. 2004. A comprehensive framework for global patterns in biodiversity. *Ecol. Lett.* 7: 1–15.

38. Ricklefs, R. E., and R. E. Latham. 1992. Intercontinental correlation of geographic ranges suggests stasis in ecological traits of relict genera of temperate perennial herbs. *Am. Nat.* 139: 1305–1321.

39. Sanmartín, I., and F. Ronquist. 2004. Southern hemisphere biogeography inferred by event-based models: Plant versus animal patterns. *Syst. Biol.* 53: 216–243.

40. Sanmartín, I., H. Enghoff, and F. Ronquist. 2001. Patterns of animal dispersal, vicariance and diversification in the Holarctic. *Biol. J. Linn. Soc.* 73: 345–390.

41. Soltis, D. E., A. B. Morris, J. S. McLachlan, P. M. Manos, and P. S. Soltis. 2006. Comparative phylogeography of unglaciated eastern North America. *Mol. Ecol.* 15: 4261–4293.

42. Stebbins, G. L. 1974. *Flowering Plants: Evolution above the Species Level*. Belknap Press of Harvard University Press, Cambridge, MA.

43. Wen, J. 1999. Evolution of eastern Asian and eastern North American disjunct distributions of flowering plants. *Annu. Rev. Ecol. Syst.* 30: 421–455.

44. Wiens, J. J., and M. J. Donoghue. 2004. Historical biogeography, ecology and species richness. *Trends Ecol. Evol.* 19: 639–644.

45. Wiens, J. J., and C. H. Graham. 2005. Niche conservatism: Integrating evolution, ecology, and conservation biology. *Annu. Rev. Ecol. Evol. Syst.* 36: 519–539.

46. Wiens, J. J., C. H. Graham, D. S. Moen, S. A. Smith, and T. W. Reeder. 2006. Evolutionary and ecological causes of the latitudinal diversity gradient in hylid frogs: Treefrog trees unearth the roots of high tropical diversity. *Am. Nat.* 168: 579–596.

47. Winkler, I. S., and C. Mitter. 2008. The phylogenetic dimension of insect-plant interactions: A review of recent evidence. In K. J. Tilmon (ed.), *Specialization, Speciation, and Radiation: The Evolutionary Biology of Herbivorous Insects*, pp. 240–263. University of California Press, Berkeley.

CHAPTER 19

1. Alroy, J. 2008. Dynamics of origination and extinction in the marine fossil record. *Proc. Natl. Acad. Sci. USA* 105 (Suppl. 1): 11536–11542.

2. Alroy, J., and 34 others. 2008. Phanerozoic trends in the global diversity of marine invertebrates. *Science* 321: 97–100.

3. Bambach, R. K. 1985. Classes and adaptive variety: The ecology of diversification in marine faunas through the Phanerozoic. In J. W. Valentine (ed.), *Phanerozoic Diversity Patterns: Profiles in Macroevolution*, pp. 191–253. Princeton University Press, Princeton, NJ.

4. Bambach, R. K. 2006. Phanerozoic biodiversity mass extinctions. *Annu. Rev. Earth Planet. Sci.* 34: 127–155.

5. Bambach, R. K., A. H. Knoll, and J. J. Sepkoski, Jr. 2002. Anatomical and ecological constraints on Phanerozoic animal diversity in the marine realm. *Proc. Natl. Acad. Sci. USA* 99: 6854–6859.

6. Baumiller, T. 1993. Survivorship analysis of Paleozoic Crinoidea: Effect of filter morphology on evolutionary rates. *Paleobiology* 19: 304–321.

7. Benton, M. J. 1996. On the nonprevalence of competitive replacement in the evolution of tetrapods. In D. Jablonski, D. H. Erwin, and J. Lipps (eds.), *Evolutionary Paleobiology*, pp. 185–210. University of Chicago Press, Chicago.

8. Benton, M. J. 2010. The origins of modern biodiversity on land. *Phil. Trans. R. Soc., B* 365: 3667–3679.

9. Bush, A. M., and R. K. Bambach. 2011. Paleoecologic megatrends in marine metazoan. *Annu. Rev. Earth Planet. Sci.* 39: 241–269.

10. Bush, A. M., and R. K. Bambach. 2015. Sustained Mesozoic-Cenozoic diversification of marine Metazoa: A consistent signal from the fossil record. *Geology* 43: 979–982.

11. Cornell, H. V. 1993. Unsaturated patterns in species assemblages: The role of regional processes in setting local species richness. In R. E. Ricklefs and D. Schluter (eds.), *Species Diversity in Ecological Communities: Historical and Geographical Perspectives*, pp. 243–252. University of Chicago Press, Chicago.

12a. Costello, M. J., R. M. May, and N. E. Stork. 2013. Can we name Earth's species before they go extinct? *Science* 339: 413–416.

12b. Eble, G. J. 1999. Originations: Land and sea compared. *Geobios* 32: 223–234.

13. Erwin, D. E. 1993. *The Great Paleozoic Crisis: Life and Death in the Early Permian*. Columbia University Press, New York.

14. Erwin, D. H. 2006. *Extinction: How Life on Earth Nearly Ended 250 Million Years Ago*. Princeton University Press, Princeton, NJ.

15. Evans, H. E. 1983. *Life on a Little Known Planet: A Biologist's View of Insects and Their World*. Lyons Press, Guilford, CT.

16. Farrell, B., D. Dussourd, and C. Mitter. 1991. Escalation of plant defenses: Do latex and resin canals spur plant diversification? *Am. Nat.* 138: 881–900.

17. Finnegan, S., J. L. Payne, and S. C. Wang. 2008. The Red Queen revisited: Re-evaluating the age specificity of Phanerozoic marine genus extinctions. *Paleobiology* 34: 318–341.

18. Fjeldså, J., R. C. K. Bowie, and C. Rahbek. 2012. The role of mountain ranges in the diversification of birds. *Annnu. Rev. Ecol. Evol. Syst.* 43: 249–265.

19. Flessa, K., and D. Jablonski. 1985. Declining Phanerozoic extinction rates: Effect of taxonomic structure? *Nature* 313: 216–218.

20. Foote, M. 2000. Origination and extinction components of diversity: General problems. In D. H. Erwin and S. L. Wing (eds.), *Deep Time: Paleobiology's Perspective*, pp. 74–102. *Paleobiology* 26 (4), supplement.

21. Foote, M. 2010. The geological history of biodiversity. In M. A. Bell, D. J. Futuyma, W. F. Eanes, and J. S. Levinton (eds.), *Evolution since Darwin: The First 150 Years*, pp. 479–510. Sinauer, Sunderland, MA.

22. Fordyce, J. A. 2010. Host shifts and evolutionary radiations of butterflies. *Proc. R. Soc. B* 277: 3735–3743.

23. Fürsich, F. T., and D. Jablonski. 1984. Lake Triassic naticid drillholes: Carnivorous gastropods gain a major adaptation but fail to radiate. *Science* 224: 78–80.

24. Gaston, K. J., and T. M. Blackburn. 2000. *Pattern and Process in Macroecology*. Blackwell Science, Oxford.

25. Gilinsky, N. L. 1994. Volatility and the Phanerozoic decline of background extinction. *Paleobiology* 20: 424–444.

26. Gould, S. J. 1985. The paradox of the first tier: An agenda for paleobiology. *Paleobiology* 11: 2–12.

27. Grimaldi, D., and M. S. Engel. 2005. *Evolution of the Insects*. Cambridge University Press, Cambridge.

28. Hardy, N. B., and S. P. Otto. 2014. Specialization and generalization in the diversification of phytophagous insects: Tests of the musical chairs and oscillation hypotheses. *Proc. R. Soc. B* 281, article 20132960. doi:10.1098/rspb.2013.2960.

29. Harmon, L. J., and S. Harrison. 2015. Species diversity is dynamic and unbounded at local and continental scales. *Am. Nat.* 185: 584–593.

30. Hodges, S. A., and M. L. Arnold. 1994. Floral and ecological isolation between *Aquilegia formosa* and *Aquilegia pubescens*. *Proc. Nat. Acad. Sci. USA* 91: 2493–2496.

31. Jablonski, D. 1995. Extinctions in the fossil record. In J. H. Lawton and R. M. May (eds.), *Extinction Rates*, pp. 25–44. Oxford University Press, Oxford.

32. Jablonski, D. 2002. Survival without recovery after mass extinctions. *Proc. Natl. Acad. Sci. USA* 99: 8139–8144.

33. Jablonski, D. 2008. Biotic interactions and macroevolution: Extensions and mismatches across scales and levels. *Evolution* 62: 715–739.

34. Jablonski, D., and K. Roy. 2003. Geographical range and speciation in fossil and living molluscs. *Proc. R. Soc. Lond. B* 270: 401–406.

35. Jackson, J. B. C. 1974. Biogeographic consequences of eurytopy and stenotopy among marine bivalves and their biogeographic significance. *Am. Nat.* 104: 541–560.

36. Krause, D. W. 1986. Competitive exclusion and taxonomic displacement in the fossil record: The case of rodents and multituberculates in North America. In K. M. Flanagan and J. A. Lillegraven (eds.), *Vertebrates, Phylogeny, and Philosophy*, pp. 95–117. Contributions to Geology, Special Paper 3, University of Wyoming, Laramie.

37. Krug, A. Z., D. Jablonski, and J. W. Valentine. 2009. Signature of the end—Cretaceous mass extinction in the modern biota. *Science* 323: 767–771.

38. Lerner, H. R. L., M. Meyer, H. F. James, M. Hofreiter, and R. C. Fleischer. 2011. Multilocus resolution of phylogeny and timescale in the extant adaptive radiation of Hawaiian honeycreepers. *Curr. Biol.* 21: 1838–1844.

39. Lovette, I. J., and E. Bermingham. 1999. Explosive speciation in the New World *Dendroica* warblers. *Proc. R. Soc. Lond. B* 266: 1629–1636.

40. Lupia, R., S. Lidgard, and P. R. Crane. 1999. Comparing palynological abundance and diversity: implications for biotic replacement during the Cretaceous angiosperm radiation. *Paleobiology* 25: 305–340.

41. McCune, A. R. 1982. On the fallacy of constant extinction rates. *Evolution* 36: 610–614.

42. McKinney, F. K., 1992. Competitive interactions between related clades: Evolutionary implications of overgrowth interactions between encrusting cyclostome and cheilostome bryozoans. *Mar. Biol.* 114: 645–652.

43. McKinney, M. L. 1997. Extinction vulnerability and selectivity: Combining ecological and paleontological views. *Annu. Rev. Ecol. Syst.* 28: 495–516.

44. McPeek, M. A. 2008. The ecological dynamics of clade diversification and community assembly. *Am. Nat.* 172: E270–E284.

45. Mitter, C., B. D. Farrell, and B. Wiegmann. 1988. The phylogenetic study of adaptive zones: Has phytophagy promoted insect diversification? *Am. Nat.* 132: 107–128.

46. Moore, B. R., and M. J. Donoghue. 2007. Correlates of diversification in the plant clade Dipsacales: Geographic movement and evolutionary innovations. *Am. Nat.* 170: S28–S55.

47. Morlon, H., T. L. Parsons, and J. B. Plotkin. 2011. Reconciling molecular phylogenies with the fossil record. *Proc. Nat. Acad. Sci. USA* 108: 16327–16332.

48. Morlon, H., M. D. Potts, and J. B. Plotkin. 2010. Inferring the dynamics of diversification: A coalescent approach. *PLoS Biol.* 8, article e1000493.

49. Nee, S. 2006. Birth-death models in macroevolution. *Annu. Rev. Ecol. Evol. Syst.* 37: 1–17.

50. Nicholson, D. B., A. J. Ross, and P. J. Mayhew. 2014. Fossil evidence for key innovations in the evolution of insect diversity. *Proc. R. Soc. B* 281: 20141823.

51. Niklas, K. J., B. H. Tiffney, and A. H. Knoll. 1983. Patterns in vascular land plant diversification. *Nature* 303: 614–616.

52. Pyron, R. A., and F. T. Burbrink. 2013. Phylogenetic estimates of speciation and extinction rates for testing ecological and evolutionary hypotheses. *Trends Ecol. Evol.* 28: 729–736.

53. Quental, T. B., and C. R. Marshall. 2010. Diversity dynamics: Molecular phylogenies need the fossil record. *Trends Ecol. Evol.* 25: 434–441.

54. Rabosky, D. L., and A. H. Hurlbert. 2015. Species richness at continental scales is dominated by ecological limits. *Am. Nat.* 185: 572–583.

55. Rabosky, D. L., and I. J. Lovette. 2008. Explosive evolutionary radiations: Decreasing speciation or increasing extinction through time? *Evolution* 62: 1866–1875.

56. Rabosky, D. L., and A. R. McCune. 2010. Reinventing species selection with molecular phylogenies. *Trends Ecol. Evol.* 25: 68–74.

57. Raup, D. M. 1972. Taxonomic diversity during the Phanerozoic. *Science* 177: 1065–1071.

58. Raup, D. M., and J. J. Sepkoski Jr. 1982. Mass extinctions in the marine fossil record. *Science* 215: 1501–1503.

59. Ree, R. H. 2005. Detecting the historical signature of key innovations using stochastic models of character evolution and cladogenesis. *Evolution* 59: 257–265.

60. Romer, A. S. 1966. *Vertebrate Paleontology*. University of Chicago Press, Chicago.

61. Rosenblum, E. B., and 8 others. 2012. Goldilocks meets Santa Rosalia: An ephemeral speciation model explains patterns of diversification across time scales. *Evol. Biol.* 39: 255–261.

62. Rosenzweig, M. L., and R. D. McCord. 1991. Incumbent replacement: Evidence for long-term evolutionary progress. *Paleobiology* 17: 202–213.

63. Sepkoski, J. J., Jr. 1984. A kinetic model of Phanerozoic taxonomic diversity. III. Post-Paleozoic families and mass extinctions. *Paleobiology* 10: 246–267.

64. Sepkoski, J. J., Jr. 1993. Ten years in the library: New data confirm paleontological patterns. *Paleobiology* 19: 43–51.

65. Sepkoski, J. J. Jr. 1996. Competition in macroevolution: The double wedge revisited. In D. Jablonski, D. H. Erwin, and J. Lipps (eds.), *Evolutionary Paleobiology*, pp. 211–255. University of Chicago Press, Chicago.

66. Sepkoski, J. J. Jr., F. K. McKinney, and S. Lidgard. 2000. Competitive displacement among post-Paleozoic cyclostome and cheilostome bryozoans. *Paleobiology* 26: 7–18.

67. Signor, P. W., III. 1985. Real and apparent trends in species richness through time. In J. W. Valentine (ed.), *Phanerozoic Diversity Patterns: Profiles in Macroevolution*, pp. 129–150. Princeton University Press, Princeton, NJ.

68. Smits, P. D. 2015. Expected time-invariant effects of biological traits on mammal species duration. *Proc. Nat. Acad. Sci. USA* 112: 13015–13020.

69. Stanley, S. M. 1979. *Macroevolution: Pattern and Process*. W. H. Freeman, San Francisco.

70. Stanley, S. M. 1990. The general correlation between rate of speciation and rate of extinction: Fortuitous causal linkages. In R. M. Ross and W. D. Allmon (eds.), *Causes of Evolution: A Paleontological Perspective*, pp. 103–127. University of Chicago Press, Chicago.

71. Stroud, J. T., and J. B. Losos. 2016. Ecological opportunity and adaptive radiation. *Annu. Rev. Ecol. Evol. Syst.* 47: 507–532.

72. Valentine, J. W., T. C. Foin, and D. Peart. 1978. A provincial model of Phanerozoic marine diversity. *Paleobiology* 4: 55–66.

73. Van Valen, L. 1973. A new evolutionary law. *Evol. Theory* 1: 1–30.

74. Vermeij, G. J., and R. K. Grosberg. 2010. The great divergence: When did diversity on land exceed that in the sea? *Integr. Comp. Biol.* 50: 675–682.

75. Wagner, C. E., L. J. Harmon, and O. Seehausen. 2012. Ecological opportunity and sexual selection together predict adaptive radiation. *Nature* 487: 366–369.

76. Weber, M. G., and A. A. Agrawal. 2014. Defense mutualisms enhance plant diversification. *Proc. Nat. Acad. Sci. USA* 111: 16442–16447.

77. Wiens, J. J., R. T. Lapoint, and N. K. Whiteman. 2015. Herbivory increases diversification across insect clades. *Nat. Comun.* 6: 8370. doi:10.1038/ncomms9370.

CHAPTER 20

1. Alroy, J. 1998. Cope's rule and the dynamics of body mass evolution in North American fossil mammals. *Science* 280: 731–734.

2. Baker, J., A. Meade, M. Pagel, and C. Venditti. 2015. Adaptive evolution toward larger size in mammals. *Proc. Nat. Acad. Sci. USA* 112: 5093–5098.

3. Baxter, S. W., S. E. Johnston, and C. D. Jiggins. 2009. Butterfly speciation and the distribution of gene effect sizes fixed during adaptation. *Heredity* 102: 57–65.

4. Benton, M. J. 2014. How birds became birds. *Science* 345: 508–509.

5. Bi, S., Y. Wang, J. Guan, X. Sheng, and J. Meng. 2014. Three new Jurassic euharamiyidan species reinforce early divergence of mammals. *Nature* 514: 579–584.

6. Blount, Z. D., C. Z. Borland, and R. E. Lenski. 2008. Historical contingency and the evolution of a key innovation in an experimental population of *Escherichia coli*. *Proc. Natl. Acad. Sci. USA* 105: 7899–7906.

7. Blows, M. W., and A. A. Hoffmann. 2005. A reassessment of genetic limits to evolutionary change. *Ecology* 86: 1371–1384.

8. Bolstad, G. H., and 6 others. 2015. Complex constraints on allometry revealed by artificial selection on the wing of *Drosophila melanogaster*. *Proc. Nat. Acad. Sci. USA* 112: 13284–13289.

9. Bull, J. J., and E. L. Charnov. 1985. On irreversible evolution. *Evolution* 39: 1149–1155.

10. Carroll, R. L. 1988. *Vertebrate Paleontology and Evolution*. W. H. Freeman, New York.

11. Cheetham, A. H. 1987. Tempo of evolution in a Neogene bryozoan: Are trends in single morphological characters misleading? *Paleobiology* 13: 286–296.

12. Conway Morris, S. 2003. *Life's Solution: Inevitable Humans in a Lonely Universe*. Cambridge University Press, Cambridge.

13. Conway Morris, S. (ed.) 2008. *The Deep Structure of Biology: Is Convergence Sufficiently Ubiquitous to Give a Directional Signal?* Templeton Foundation Press, West Conshohocken, PA.

14. Danforth, B. N., L. Conway, and S. Ji. 2003. Phylogeny of eusocial *Lasioglossum* reveals multiple losses of eusociality within a primitively eusocial clade of bees (Hymenoptera: Halictidae). *Syst. Biol.* 52: 23–36.

15. Davis, D. D. 1964. The giant panda: A morphological study of evolutionary mechanisms. *Fieldiana Memoirs* 3: 1–399.

16. Deban, S. M., D. B. Wake, and G. Roth. 1997. Salamander with a ballistic tongue. *Nature* 389: 27–28.

17. Demera, T. A., M. R. McGowen, A. Berta, and J. Gatesy. 2008. Morphological and molecular evidence for a stepwise evolutionary transition from teeth to baleen in mysticete whales. *Syst. Biol.* 57: 15–37.

18. de Muizon, C. 2001. Walking with whales. *Nature* 413: 259–261.

19. Dennett, D. C. 1995. *Darwin's Dangerous Idea: Evolution and the Meanings of Life*. Simon & Schuster, New York.

20. Eldredge, N., and S. J. Gould. 1972. Punctuated equilibria: An alternative to phyletic gradualism. In T. J. M. Schopf (ed.), *Models in Paleobiology*, pp. 82–115. Freeman, Cooper and Co., San Francisco.

21. Elliott, T. A., and T. R. Gregory. 2015. What's in a genome? The C-value enigma and the evolution of eukaryotic genome content. *Phil. Trans. R. Soc., B* 370: 20140331.

22. Evans, A. R., and 19 others. 2012. The maximum rate of mammal evolution. *Proc. Nat. Acad. Sci. USA* 109: 4187–4190.

23. Ezard, T. H. G., G. H. Thomas, and A. Purvis. 2013. Inclusion of a near-complete fossil record reveals speciation-related molecular evolution. *Methods Ecol. Evol.* 4: 745–753.

24. Fisher, D. C. 1986. Progress in organismal design. In D. M. Raup and D. Jablonski (eds.), *Patterns and Processes in the History of Life*, pp. 99–117. Springer-Verlag, Berlin.

25. Fisher, R. M., C. K. Cornwallis, and S. A. West. 2013. Group formation, relatedness, and the evolution of multicellularity. *Curr. Biol.* 23: 1120–1125.

26. Futuyma, D. J. 1983. Evolutionary interactions among herbivorous insects and plants. In D. J. Futuyma and M. Slatkin (eds.), *Coevolution*, pp. 209–231. Sinauer, Sunderland, MA.

27. Futuyma, D. J. 1987. On the role of species in anagenesis. *Am. Nat.* 130: 465–473.

28. Futuyma, D. J. 1995. *Science on Trial: The Case for Evolution*. Sinauer, Sunderland, MA.

29. Futuyma, D. J., and G. Moreno. 1988. The evolution of ecological specialization. *Annu. Rev. Ecol. Syst.* 19: 207–233.

30. Ghiselin, M. T. 1995. Perspective: Darwin, progress, and economic principles. *Evolution* 49: 1029–1037.

31. Gingerich, P. D. 2001. Rates of evolution on the time scale of the evolutionary process. *Genetica* 112–113: 127–144.

32. Gingerich, P. D. 2003. Land-to-sea transition of early whales: Evolution of Eocene Archaeoceti (Cetacea) in relation to skeletal proportions and locomotion of living semiaquatic mammals. *Paleobiology* 29: 429–454.

33. Goldberg, E. E., J. R. Kohn, R. Lande, K. A. Robertson, S. A. Smith, and B. Igić. 2010. Species selection maintains self-incompatibility. *Science* 330: 493–495.

34. Goldschmidt, R. B. 1940. *The Material Basis of Evolution*. Yale University Press, New Haven, CT.

35. Gould, S. J. 1989. *Wonderful Life: The Burgess Shale and the Nature of History*. W. W. Norton, New York.

36. Gregory, T. R. 2005. Genome size evolution in animals. In T. R. Gregory (ed.), *The Evolution of the Genome*, pp. 89–162. Elsevier, San Diego.

37. Grosberg, R. K., and R. R. Strathmann. 2007. The evolution of multicellularity: A minor major transition? *Annu. Rev. Ecol. Evol. Syst.* 38: 621–654.

38. Guerroussov, S. and 6 others. 2015. An alternative splicing event amplifies evolutionary differences between vertebrates. *Science* 349: 868–873.

39. Hansen, T. A. 1980. Influence of larval dispersal and geographic distribution on species longevity in neogastropods. *Paleobiology* 6: 193–207.

40. Hansen, T. F., and D. Houle. 2008. Measuring and comparing evolvability and constraint in multivariate characters. *J. Evol. Biol.* 21: 1201–1219.

41. Heim, N., M. Knope, E. K. Schaal, and J. L. Payne. 2015. Cope's rule in the evolution of marine animals. *Science* 347: 867–870.

42. Hoffmann, A. A., S. L. Chown, and S. Clusella-Trullas. 2013. Upper thermal limits in terrestrial ectotherms: how constrained are they? *Func. Ecol.* 27: 934–949.

43. Holt, R. D. 1996. Demographic constraints in evolution: Towards unifying the evolutionary theories of senescence and niche conservatism. *Evol. Ecol.* 10: 1–11.

44. Hunt, G. 2007a. The relative importance of directional change, random walks, and stasis in the evolution of fossil lineages. *Proc. Natl. Acad. Sci. USA* 104: 18404–18408.

45. Hunt, G. 2007b. Evolutionary divergence in directions of high phenotypic variance in the ostracode genus *Poseidonamicus*. *Evolution* 61: 1560–1576.

46. Hunt, G. 2010. Evolution in fossil lineages: Paleontology and *The Origin of Species*. *Am. Nat.* 176 (Suppl.): S61–S76.

47. Hunt, G., and D. L. Rabosky. 2014. Phenotypic evolution in fossil species: pattern and process. *Annu. Rev. Earth Planet. Sci.* 42: 421–441.

48. Hunt, G., and G. Slater. 2016. Integrating paleontological and phylogenetic approaches to macroevolution. *Annu. Rev. Ecol. Evol. Syst.* 47: 189–213.

49. Jablonski, D., and R. A. Lutz. 1983. Larval ecology of marine benthic invertebrates: Paleobiological implications. *Biol. Rev.* 58: 21–89.

50. Jones, A. G., S. J. Arnold, and R. Bürger. 2007. The mutation matrix and the evolution of evolvability. *Evolution* 61: 727–745.

51. Kemp, T. S. 2005. *The Origin and Evolution of Mammals*. Oxford University Press, Oxford.

52. Kirkpatrick, M. 2010. Rates of adaptation: Why is Darwin's machine so slow? In M. A. Bell, D. J. Futuyma, W. F. Eanes, and J. S. Levinton (eds.), *Evolution since Darwin: The First 150 Years*, pp. 177–195. Sinauer, Sunderland, MA.

53. Kirschner, M., and J. Gerhart. 2005. *The Plausibility of Life*. Yale University Press, New Haven, CT.

54. Knoll, A. H., and R. K. Bambach. 2000. Directionality in the history of life: Diffusion from the left wall or repeated scaling of the right? In D. H. Erwin and S. L. Wing (eds.), *Deep Time: Paleobiology's Perspective*, pp. 1–14. The Paleontological Society, Allen Press, Lawrence, KS.

55. Kopp, A. 2011. *Drosophila* sex combs as a model of evolutionary innovations. *Evol. Devel.* 13: 504–522.

56. Kramer, E. M., and M. A. Jaramillo. 2005. Genetic basis for innovations in floral organ identity. *J. Exp. Zool. (Mol. Dev. Evol.)* 304B: 526–535.

57. Kuratani, S., S. Kuraku, and H. Nagashima. 2011. Evolutionary developmental perspective for the origin of turtles: The folding theory for the shell based on the developmental nature of the carapacial ridge. *Evol. Devel.* 13: 1–14.

58. Lee, M. S. Y., A. Can, D. Naish, and G. J. Dyke. 2014. Sustained miniaturization and anatomical innovation in the dinosaurian ancestors of birds. *Science* 345: 562–566.

59. Lewontin, R. C. 2000. *The Triple Helix: Gene, Organism, and Environment*. Harvard University Press, Cambridge, MA.

60. Li, C., X.-C. Wu, O. Rieppel, L.-T. Wang, and L.-J. Zhao. 2008. An ancestral turtle from the Late Triassic of southwestern China. *Nature* 456: 497–501.

61. Lieberman, B. S., and S. Dudgeon. 1996. An evaluation of stabilizing selection as a mechanism for stasis. *Palaeogeog. Lapaeoclimat. Palaeoecol.* 127: 229–238.

62. Losos, J. B. 2010. Adaptive radiation, ecological opportunity, and evolutionary determinism. *Am. Nat.* 175: 623–639.

63. Lowe, C. B., J. A. Clarke, A. J. Baker, D. Haussler, and S. V. Edwards. 2014. Feather development genes and associated regulatory innovation predate the origin of Dinosauria. *Mol. Biol. Evol.* 32: 23–28.

64. Luo, Z.-X. 2011. Developmental patterns in Mesozoic evolution of mammal ears. *Annu. Rev. Ecol. Evol. Syst.* 42: 355–380.

65. Lynch, M. 1990. The rate of morphological evolution in mammals from the standpoint of the neutral expectation. *Am. Nat.* 136: 727–741.

66. MacFadden, B. J. 1986. Fossil horses from "Eohippus" (*Hyracotherium*) to *Equus*: Scaling, Cope's law, and the evolution of body size. *Paleobiology* 12: 355–369.

67. Marcot, J. D., and D. W. McShea. 2007. Increasing hierarchical complexity throughout the history of life: Phylogenetic tests of trend mechanisms. *Paleobiology* 33: 182–200.

68. Mattila, T. M., and F. Bokma. 2008. Extant mammal body masses suggest punctuated equilibrium. *Proc. R. Soc. Lond. B* 275: 2195–2199.

69. Maynard Smith, J., and E. Szathmáry. 1995. *The Major Transitions in Evolution*. W. H. Freeman, San Francisco.

70. Mayr, E. 1960. The emergence of evolutionary novelties. In S. Tax (ed.), *The Evolution of Life*, pp. 349–380. University of Chicago Press, Chicago.

71. Mayr, E. 1985. The probability of extraterrestrial intelligent life. In E. Regis, Jr. (ed.), *Extraterrestrials: Science and Alien Intelligence*, pp. 25–30. Cambridge University Press, Cambridge.

72. McCutcheon, J. P., and N. A. Moran. 2012. Extreme genome reduction in symbiotic bacteria. *Nat. Rev. Microbiol.* 10: 13–26.

73. McShea, D. W. 1994. Mechanisms of large-scale evolutionary trends. *Evolution* 48: 1747–1763.

74. McShea, D. W. 1998. Possible largest–scale trends in organismal evolution: Eight "live hypotheses." *Annu. Rev. Ecol. Syst.* 29: 293–318.

75. McShea, D. W. 2001. The hierarchical structure of organisms: A scale and documentation of a trend in the maximum. *Paleobiology* 27: 405–423.

76. Michod, R. E. 1999. *Darwinian Dynamics: Evolutionary Transitions in Fitness and Individuality*. Princeton University Press, Princeton, NJ.

77. Mitgutsch, C., M. K. Richardson, R. Jiménez, J. E. Martin, P. Kondrashov, M. A. G. de Bakker, and M. R. Sánchez-Villagra. 2012. Circumventing the polydactyly "constraint": The mole's "thumb." *Biol. Lett.* 8: 74–77.

78. Moczek, A. 2006. Integrating micro- and macroevolution of development through the study of horned beetles. *Heredity* 97: 168–178.

79. Müller, G. B., and G. P. Wagner. 1991. Novelty in evolution: Restructuring the concept. *Annu. Rev. Ecol. Syst.* 23: 229–256.

80. Nagashima, H., and 6 others. 2009. Evolution of the turtle body plan by the folding and creation of new muscle connections. *Science* 325: 193–196.

81. Nilsson, D. E., and S. Pelger. 1994. A pessimistic estimate of the time required for an eye to evolve. *Proc. R. Soc. Lond. B* 256: 59–65.

82. Oakley, T. H., and D. I. Speiser. 2015. How complexity originates: The evolution of animal eyes. *Annu. Rev. Ecol. Evol. Syst.* 46: 237–260.

83. Odling-Smee, F. J., K. N. Laland, and M. W. Feldman. 2003. *Niche Construction: The Neglected Process in Evolution*. Princeton University Press, Princeton, NJ.

84. Orr, H. A., and J. Coyne. 1992. The genetics of adaptation: A reassessment. *Am. Nat.* 140: 725–742.

85. Osorio, D. 1994. Eye evolution: Darwin's shudder stilled. *Trends Ecol. Evol.* 9: 241–242.

86. Pagel, M., C. Venditti, and A. Meade. 2006. Large punctuational contribution of speciation to evolutionary divergence at the molecular level. *Science* 314: 119–121.

87. Pavlicev, M., and G. P. Wagner. 2012. A model of developmental evolution: Selection, pleiotropy and compensation. *Trends Ecol. Evol.* 27: 316–322.

88. Ralph, P. L., and G. Coop. 2015. The role of standing variation in geographic convergent adaptation. *Am. Nat.* 186: S5–S23.

89. Ravigné, V., U. Dieckmann, and I. Olivieri. 2009. Live where you thrive: Joint evolution of habitat choice and local adaptation facilitates specialization and promotes diversity. *Am. Nat.* 174: E141–E169.

90. Rensch, B. 1959. *Evolution above the Species Level*. Columbia University Press, New York.

91. Romer, A. S. 1966. *Vertebrate Paleontology*. University of Chicago Press, Chicago.

92. Rowe, T. B., T. E. Macrini, and Z.-X. Luo. 2011. Fossil evidence on origin of the mammalian brain. *Science* 332: 955–957.

93. Ruse, M. 1996. *Monad to Man: The Concept of Progress in Evolutionary Biology*. Harvard University Press, Cambridge, MA.

94. Salvini-Plawen, L. V., and E. Mayr. 1977. On the evolution of photoreceptors and eyes. *Evol. Biol.* 10: 207–263.

95. Saupe, E. E., and 6 others. 2014. Macroevolutionary consequences of profound climate change on niche evolution in marine molluscs over the past three million years. *Proc. R. Soc. B* 281 (1795), art. 20141995.

96. Shubin, N., C. Tabin, and S. Carroll. 2009. Deep homology and the origins of evolutionary novelty. *Nature* 457: 818–823.

97. Sidor, C. A. 2001. Simplification as a trend in synapsid cranial evolution. *Evolution* 55: 1419–1442.

98. Sidor, C. A., and J. A. Hopson. 1998. Ghost lineages and "mammalness": Assessing the temporal pattern of character acquisition in the Synapsida. *Paleobiology* 24: 254–273.

99. Simpson, G. G. 1944. *Tempo and Mode in Evolution*. Columbia University Press, New York.

100. Simpson, G. G. 1953. *The Major Features of Evolution*. Columbia University Press, New York.

101. Simpson, G. G. 1964. *This View of Life: The World of an Evolutionist*. Harcourt, Brace and World, New York.

102. Stanley, S. M. 1979. *Macroevolution: Pattern and Process*. W. H. Freeman, San Francisco.

103. Stanley, S. M., and X. Yang. 1987. Approximate evolutionary stasis for bivalve morphology over millions of years: A multivariate, multilineage study. *Paleobiology* 13: 113–139.

104. Stern, D. L., and V. Orgogozo. 2008. The loci of evolution: How predictable is genetic evolution? *Evolution* 62: 2155–2177.

105. Stern, D. L., and V. Orgogozo. 2009. Is genetic evolution predictable? *Science* 323: 746–751.

106. Strotz, L. C., and A. P. Allen. 2013. Assessing the role of cladogenesis in macroevolution by integrating fossil and molecular evidence. *Proc. Natl. Acad. Sci. USA* 110: 2104–2107.

107. Thewissen, J. G. M., L. N. Cooper, M. T. Clementz, S. Bajpal, and B. N. Tiwari. 2007. Whales originated from aquatic artiodactyls in the Eocene epoch of India. *Nature* 450: 1190–1194.

108. Uyeda, J. C., T. F. Hansen, S. J. Arnold, and J. Pienaar. 2011. The million-year wait for macroevolutionary bursts. *Proc. Nat. Acad. Sci. USA* 108: 15908–15913.

109. Van Valen, L. 1982. Integration of species: Stasis and biogeography. *Evol. Theory* 6: 99–112.

110. Vargas, A. O., and G. P. Wagner. 2009. Frame-shifts of digit identity in bird evolution and Cyclopamine-treated wings. *Evol. Devel.* 11:163–169.

111. Wagner, G. P. 1989. The biological homology concept. *Annu. Rev. Ecol. Syst.* 20: 51–69.

112. Wagner, G. P. 2015. *Homology, Genes, and Evolutionary Innovation*. Princeton University Press, Princeton, N. J.

113. Wake, D. B. 1982. Functional and developmental constraints and opportunities in the evolution of feeding systems in urodeles. In D. Mossakowski and G. Roth (eds.), *Environmental Adaptation and Evolution*, pp. 51–66. G. Fischer, Stuttgart.

114. Wang, Z., R. L. Young, X. H.-L. Xue, and G. P. Wagner. 2011. Transcriptomic analysis of avian digits reveals conserved and derived digit identities in birds. *Nature* 477: 583–586.

115. Wiens, J. J., and C. H. Graham. 2005. Niche conservatism: Integrating evolution, ecology, and conservation biology. *Annu. Rev. Ecol. Evol. Syst.* 36: 519–539.

116. Wilson, E. O. 1975. *Sociobiology: The New Synthesis*. Harvard University Press, Cambridge, MA.

117. Zufall, R. A., and M. D. Rausher. 2004. Genetic changes associated with floral adaptation restrict future evolutionary potential. *Nature* 428: 847–850.

CHAPTER 21

1. Benton, M. J., and P. C. J. Donoghue. 2007. Paleontological evidence to date the tree of life. *Mol. Biol. Evol.* 24: 26–53.

2. Berbudi, A., J. Ajendra, A. P. F. Wardani, A. Hoerauf, and M. P. Hübner. 2016. Parasitic helminths and their beneficial impact on type 1 and type 2 diabetes. *Diabetes Metab. Res. Rev.* 32: 238–250.

3. Blumler, M. A. 2015. Agriculturalism. In M. P. Muehlenbein (ed.), *Basics in Human Evolution*, pp. 349–365. Academic Press, Amsterdam.

4. Boyd, R., and P. J. Richerson. 1985. *Culture and the Evolutionary Process*. University of Chicago Press, Chicago.

5. Briggs, A. W., and 17 others. 2009. Targeted retrieval and analysis of five Neandertal mtDNA genomes. *Science* 325: 318–321.

6. Byars, S. G., D. Ewbank, D. R. Govindaraju, and S. C. Stearns. 2010. Natural selection in a contemporary human population. *Proc. Natl. Acad. Sci. USA* 107: 1787–1792.

7. Cavalli-Sforza, L. L., and M. W. Feldman. 1981. *Cultural Transmission and Evolution: A Quantitative Approach*. Princeton University Press, Princeton, NJ.

8. Conard, N. J., M. Malina, and S. C. Münzel. 2009. New flutes document the earliest musical tradition in southwestern Germany. *Nature* 460: 737–740.

9. Darwin, C. 1871. *The Descent of Man, and Selection in Relation to Sex*. J. Murray, London.

10. de Waal, F. B. M., and P. F. Ferrari (eds.). 2012. *The Primate Mind: Built to Connect to Other Minds*. Harvard University Press, Cambridge, MA.

11. Demuth, J. P., T. De Bie, J. E. Stajich, N. Cristianini, and M. W. Hahn. 2006. The evolution of mammalian gene families. *PLoS ONE* 1: e85. doi:10.1371/journal.pone.0000085

12. Diamond, J. 1987. The worst mistake in the history of the human race. *Discover* 5: 64–66.

13. Diamond, J. 1997. *Guns, Germs, and Steel*. Norton, New York.

14. Duda, P., and J. Zrzavy. 2013. Evolution of life history and behavior in Hominidae: Towards phylogenetic reconstruction of the chimpanzee-human last common ancestor. *J. Human Evol.* 65: 424–446.

15. Dunbar, R. I. M. 2003. The social brain: mind, language, and society in evolutionary perspective. *Annu. Rev. Anthropol.* 32: 163–181.

16. Fortunato, L., and M. Archetti. 2010. Evolution of monogamous marriage by maximization of inclusive fitness. *J. Evol. Biol.* 23: 149–156.

17. Fuller, D. Q., and 7 others. 2014. Convergent evolution and parallelism in plant domestication revealed by an expanding archaeological record. *Proc. Nat. Acad. Sci. USA* 111: 6147–6152.

18. Gensollen, T., S. S. Iyer, D. L. Kasper, and R. S. Blumberg. 2016. How colonization by microbiota in early life shapes the immune system. *Science* 352: 539–544.

19. Gómez-Robles, A. 2016. The dawn of *Homo floresiensis*. *Nature* 534: 188–189.

20. González-Pérez, G., and 5 others. 2016. Maternal antibiotic treatment impacts development of the neonatal intestinal microbiome and antiviral immunity. *J. Immunol.* 196: 3768–3779.

21. Gravel, S., and 9 others. 2011. Demographic history and rare allele sharing among human populations. *Proc. Nat. Acad. Sci.* 108: 11983–11988.

22. Green, R. E., and 55 others. 2010. A draft sequence of the Neandertal genome. *Science* 328: 710–722.

23. Groucutt, H. S., and 16 others. 2015. Rethinking the Dispersal of *Homo sapiens* out of Africa. *Evolutionary Anthropology* 24: 149–164.

24. Harmand, S., and 120 others. 2015. 3-million-year-old stone tools from Lomekwi 3, West Turkana, Kenya. *Nature* 521: 310–315.

25. Helgason, A., and 32 others. 2007. Refining the impact of TCF7L2 gene variants on type 2 diabetes and adaptive evolution. *Nature Genetics* 39: 218–225.

26. Henn, B. M., and 18 others. 2011. Hunter-gatherer genomic diversity suggests a southern African origin for modern humans. *Proc. Nat. Acad. Sci.* 108: 5154–5162.

27. Henn, B. M., and 12 others. 2016. Distance from sub-Saharan Africa predicts mutational load in diverse human genomes. *Proc. Natl. Acad. Sci. USA* 113: E440–E449.

28. Higham, T., and 48 others. 2014. The timing and spatiotemporal patterning of Neanderthal disappearance. *Nature* 512: 306–309.

29. Holden, C., and R. Mace. 1997. Phylogenetic analysis of the evolution of lactose digestion in adults. *Hum. Biol.* 69: 605–628.

30. Holden, C. J., and R. Mace. 2003. Spread of cattle led to the loss of matrilineal descent in Africa: a coevolutionary analysis. *Proc. R. Soc. Lond. B* 270: 2425–2433.

31. Huerta-Sánchez, E., and 26 others. 2014. Altitude adaptation in Tibetans caused by introgression of Denisovan-like DNA. *Nature* 512: 194–197.

32. Hunt, G. R. and Gray, R. D. (2004). Direct observations of pandanus-tool manufacture and use by a New Caledonian crow (*Corvus moneduloides*). *Animal Cognition* 7 (2): 114–120.

33. Ingman, M., H. Kaessmann, S. Pääbo, and U. Gyllensten. 2000. Mitochondrial genome variation and the origin of modern humans. *Nature* 408: 708–713.

34. Jones, S., R. Martin, and D. Pilbeam (eds.). 1992. *The Cambridge Encyclopedia of Human Evolution*. Cambridge University Press, Cambridge.

35. Keightley, P. D., and A. Eyre-Walker. 2012. Estimating the rate of adaptive molecular evolution when the evolutionary divergence between species is small. *Journal of Molecular Evolution* 74: 61–68.

36. Langengraber, K. E., and 19 others. 2012. Generation times in wild chimpanzees and gorillas suggest earlier divergence times in great ape and human evolution. *Proc. Nat. Acad. Sci.* 109: 15716–15721.

37. Larsen, C. S. 2015. *Bioarchaeology: Interpreting Behavior from the Human Skeleton* Cambridge University Press, Cambridge, UK.

38. Larson, G., and D. Q. Fuller. 2014. The evolution of animal domestication. *Annu. Rev. Ecol. Evol. Syst.* 45: 115–136.

39. Larson, G., and 19 others. 2012. Rethinking dog domestication by integrating genetics, archeology, and biogeography. *Proc. Nat. Acad. Sci.* 109: 8878–8883.

40. Leffler, E. M., and 12 others. 2013. Multiple instances of ancient balancing selection shared between humans and chimpanzees. *Science* 339: 1578–1582.

41. Lieberman, D. E. 2014. *The Story of the Human Body: Evolution, Health, and Disease*. Vintage Books, New York.

42. Lieberman, P. 2007. The evolution of human speech. *Current Anthropol.* 48: 39–66.

43. Mace, R., and F. M. Jordan. 2011. Macro-evolutionary studies of cultural diversity: a review of empirical studies of cultural transmission and cultural adaptation. *Phil. Trans. R. Soc., B* 366: 402–411.

44. Martin, P. S., and R. G. Klein (eds.). 1984. *Quaternary Extinctions: A Prehistoric Revolution*. University of Arizona Press, Tucson.

45. Martin-Ordas, G., L. Schumacher, and J. Call. 2012. Sequential tool use in great apes. *PLoS ONE*, art. E52074. doi:10.1371/journal.pone.0052074.

46. Mathieson, I., and 37 others. 2015. Genome-wide patterns of selection in 230 ancient Eurasians. *Nature* 528: 499–503.

47. Morris, D. 1967. *The Naked Ape: A Zoologist's Study of the Human Animal*. McGraw-Hill, New York.

48. Muehlenbein, M. P. 2015. *Basics in Human Evolution*. Elsevier, Amsterdam.

49. Must, A., J. Spadano, E. H. Coakley, A. E. Field, G. Colditz, and W. H. Dietz. 1999. The disease burden associated with overweight and obesity. *JAMA* 282: 1523–1529.

50. Nielsen, R., J. M. Akey, M. Jakobsson, J. K. Pritchard, S. Tishkoff, and E. Willerslev. 2017. Tracing the peopling of the world through genomics. *Nature* 541: 302–310.

51. Olszak, T., and 10 others, 2012. Microbial exposure during early life has persistent effects on natural killer T cell function. *Science* 336: 489–493.

52. Pääbo, S. 2014. The human condition—a molecular approach. *Cell* 157: 216–226.

53. Pamer, E. G. 2016. Resurrecting the intestinal microbiota to combat antibiotic-resistant pathogens. *Science* 352: 535–538.

54. Perri, A. 2016. A wolf in dog's clothing: Initial dog domestication and Pleistocene wolf variation. *J. Archaeol. Sci.* 68: 1–4.

55. Pinker, S. 1994. *The Language Instinct*. HarperCollins, NY.

56. Pinker, S. 2010. The cognitive niche: coevolution of intelligence, sociality, and language. *Proc. Natl. Acad. Sci. USA* 107: 8993–8999.

57. Pobiner, B. 2016. Meat-eating among the earliest humans. *Amer. Sci.* 104: 110–117.

58. Pontzer, H., and 16 others. 2016. Metabolic acceleration and the evolution of human brain size and life history. *Nature* 533: 390–392.

59. Prescott, G. W., D. R. Williams, A. Balmford, R. E. Green, and A. Manica. 2012. Quantitative global analysis of the role of climate and people in explaining late Quaternary megafaunal extinctions. *Proc. Natl. Acad. Sci. USA* 109: 4527–4531.

60. Principi, N., and S. Esposito. 2016. Antibiotic administration and the development of obesity in children. *Internat. J. Antimicrob. Agents* 47: 171–177.

61. Prüfer, K., and 44 others. 2014. The complete genome sequence of a Neanderthal from the Altai Mountains. *Nature* 505: 43–49.

62. Purugganan, M. D., and D. Q. Fuller. 2009. The nature of selection during plant domestication. *Nature* 457: 843–848.

63. Raghavan, M., and 100 others. 2015. Genomic evidence for the Pleistocene and recent population history of Native Americans. *Science* 349: 841. doi:10.1126/science.aab3884

64. Reich, D., and 27 others. 2010. Genetic history of an archaic hominin group from Denisova Cave in Siberia. *Nature* 468: 1053–1060.

65. Rendu, W., and 13 others. 2014. Evidence supporting an intentional Neandertal burial at La Chapelle-aux-Saints. *Proc. Nat. Acad. Sci.* 111: 81–86.

66. Richerson, P. J., and R. Boyd, 2005. *Not by Genes Alone: How Culture Transformed Human Evolution.* University of Chicago Press, Chicago.

67. Savage-Rumbaugh, S., K. McDonald, R. A. Sevcik, W. D. Hopkins, and E. Rubert. 1986. Spontaneous symbol acquisition and communicative use by pygmy chimpanzees (*Pan paniscus*). *J. Exp. Psychol. Gen.* 115: 211–235.

68. Stearns, S. C., S. G. Byars, D. R. Govindaraju, and D. Ewbank. 2010. Measuring selection in contemporary human populations. *Nat. Rev. Genet.* 11: 611–622.

69. Storey, A. A., and 18 others. 2012. Investigating the global dispersal of chickens in prehistory using ancient mitochondrial DNA signatures. *PLoS ONE*, doi:10.1371/journal.pone.0039171.

70. Stulp, G., L. Barrett, F. C. Tropf, and M. Mills. 2015. Does natural selection favour taller stature among the tallest people on earth? *Proc. R. Soc. B* 282: 1806:20150211.

71. Taylor, A. H., D. Elliffe, G. R. Hunt, and R. D. Gray. 2010. Complex cognition and behavioural innovation in New Caledonian crows. *Proc. R. Soc. B* 277: 2637–2643.

72. The Chimpanzee Sequencing and Analysis Consortium. 2005. Initial sequence of the chimpanzee genome and comparison with the human genome. *Nature* 437: 69–87.

73. The International HapMap Consortium. 2005. A haplotype map of the human genome. *Nature* 437: 1299–1320.

74. Tishkoff, S. A., and 18 others. 2007. Convergent adaptation of human lactase persistence in Africa and Europe. *Nat. Genet.* 39: 31–40.

75. Toth, N., K. D. Schick, E. S. Savage-Rumbaugh, R. A Savcik, and D. M. Rumbaugh. 1993. Pan the tool-maker: investigations into the stone tool-making and tool-using capabilities of a bonobo (*Pan paniscus*). *J. Archaeol. Sci.* 20: 81–91.

76. Vernot, B., and J. M. Akey. 2014. Resurrecting surviving Neandertal lineages from modern human genomes. *Science* 343: 1017–1021.

77. Vernot, B., and 16 others. 2016. Excavating Neandertal and Denisovan DNA from the genomes of Melanesian individuals. *Science* 352: 235–239.

78. Villmoare, B., and 8 others. 2015. Early *Homo* at 2.8 Ma from Ledi-Geraru, Afar, Ethiopia. *Science* 347: 1352–1355.

79. Wang, G.-D., H.-B. Xie, M.-S. Peng, D. Irwin, and Y.-P. Zhang. 2014. Domestication genomics: evidence form animals. *Annu. Rev. Anim. Biosci.* 2: 65–84.

80. White, T. D., C. O. Lovejoy, B. Asfaw, J. P. Carlson, and G. Suwa. 2015. Neither chimpanzee nor human, *Ardipithecus* reveals the surprising ancestry of both. *Proc. Nat. Acad. Sci.* 112: 4877–4884.

81. Whiten, A., 2015. Experimental studies illuminate the cultural transmission of percussive technologies in *Homo* and *Pan*. *Phil. Trans. R. Soc., B* 370, doi:10.1098/rstb.2014.0359.

82. Wood, B., and J. Baker. 2011. Evolution in the genus *Homo*. *Annu. Rev. Ecol. Evol. Syst.* 42: 47–69.

83. Wrangham, R. W. 2009. *Catching Fire: How Cooking Made Us Human.* Basic Books, New York.

84. Zink, K. D., and D. E. Lieberman. 2016. Impact of meat and Lower Palaeolithic food processing techniques on chewing in humans. *Nature* 531: 500–503.

CHAPTER 22

1. Adami, C. 2015. Robots with instincts. *Nature* 521: 426–427.

2. Alexander, R. D. 1979. *Darwinism and Human Affairs.* Pittman, London.

3. Allendorf, F. W., and G. Luikart. 2007. *Conservation and the Genetics of Populations.* Blackwell, Oxford.

4. Alters, B. J., and S. M. Alters. 2001. *Defending Evolution: A Guide to the Creation/Evolution Controversy.* Jones and Bartlett, Sudbury, MA.

5. Antolin, M. F., and 11 others. 2011. Evolution and medicine in undergraduate education: A prescription for all biology students. *Evolution* 66: 1991–2006.

6. Asghar, A., S. Hameed, and N. K. Farahani. 2014. Evolution in biology textbooks: a comparative analysis of 5 Muslim countries. *Religion & Education* 41: 1–15.

7. Avise, J. C. 2010. Footprints of nonsentient design inside the human genome. *Proc. Natl. Acad. Sci. USA* 107 (suppl. 2): 8969–8976.

8. Baker, C. S., G. M. Lento, F. Cipriano, and S. R. Palumbi. 2000. Predicted decline of protected whales based on molecular genetic monitoring of Japanese and Korean markets. *Proc. R. Soc. Lond. B* 267: 1191–1199.

9. Barkow, J. H., L. Cosmides, and J. Tooby (eds.). 1992. *The Adapted Mind: Evolutionary Psychology and the Generation of Culture.* Oxford University Press, New York.

10. Baym, M., L. K. Stone, and R. Kishony. 2016. Multidrug evolutionary strategies to reverse antibiotic resistance. *Science* 351:40. http://dx.doi.org/10.1126/science.aad3292.

11. Berkman, M. B., and E. Plutzer. 2011. Defeating creationism in the courtroom, but not in the classroom. *Science* 331: 404–405.

12. Blaser, M. J. 2016. Antibiotic use and its consequences for the normal microbiome. *Science* 352: 544–545.

13. Boehm, C. 2012. *Moral Origins: The Evolution of Virtue, Altruism, and Shame.* Basic Books, New York.

14. Borgerhoff Mulder, M. 1990. Kipsigis women's preferences for wealthy men: evidence for female choice in mammals? *Behav. Ecol. Sociobiol.* 27: 255–264.

15. Bouchard, T. J. 2009. Genetic influence on human intelligence (Spearman's *g*): How much? *Ann. Hum. Biol.* 36: 527–544.

16. Boyd, R., and P. J. Richerson. 1985. *Culture and the Evolutionary Process.* University of Chicago Press, Chicago.

17. Brown, D. E. 1991. *Human Universals.* McGraw-Hill, New York.

18. Brown, S. 2000. The "Musiclanguage" model of music evolution. In N. L. Wallin, B. Merker, and S. Brown (eds.), *The Origins of Music*, pp. 271–300. MIT Press, Cambridge, MA.

19. Bush, R. M., C. A. Bender, K. Subbarao, N. J. Cox, and W. M. Fitch. 1999. Predicting the evolution of human influenza A. *Science* 286: 1921–1925.

20. Buss, D. M. 1994. *The Evolution of Desire: Strategies of Human Mating.* HarperCollins, New York.

21. Buss, D. M. 1999. Sex differences in human mate preferences: Evolutionary hypotheses tested in 37 cultures. *Behavioral and Brain Sciences* 12: 1–49.

22. Camperio-Ciani, A., U. Battaglia, and G. Zanzotto. 2015. Human homosexuality: a paradigmatic arena for sexually antagonistic selection? *Cold Spring Harbor Perspectives in Biology* 7: a017657. doi:10.1101/cshperspect.aa017657.

23. Camperio-Ciani, A., P. Cernelli, and G. Zanzotto. 2008. Sexually antagonistic selection in human male homosexuality. *PLoS ONE* 3: 1–8.

24. Camperio-Ciani, A., F. Corna, and C. Capiluppi. 2004. Evidence for maternally inherited factors favouring male homosexuality and promoting female fecundity. *Proc. R. Soc. Lond. B* 271: 2217–2221.

25. Carroll, M. W., and 131 others. 2015. Temporal and spatial analysis of the 2014-2015 Ebola virus outbreak in West Africa. *Nature* 524: 97–101.

26. Cavalli-Sforza, L. L., and M. W. Feldman. 1981. *Cultural Transmission and Evolution: A Quantitative Approach*. Princeton University Press, Princeton, NJ.

27. Chahroudi, A., S. E. Bosinger, T. H. Vanderford, M. Paiardini, and G. Silvestri. 2012. Natural SIV hosts: Showing AIDS the door. *Science* 335: 1188–1193.

28. Cheney, D. L., and R. M. Seyfarth. 2007. *Baboon Metaphysics: The Evolution of a Social Mind*. University of Chicago Press, Chicago.

29. Cosmides, L., and J. Tooby. 1992. Cognitive adaptations for social exchange. In J. H. Barkow, L. Cosmides, and J. Tooby (eds.), *The Adapted Mind*, pp. 163–228. Oxford University Press, New York.

30. Coyne, J. A. 2009. *Why Evolution Is True*. Viking, New York.

31. Darwin, C. 1872. *The Expression of the Emotions in Man and Animals*. (Reprinted 1965, University of Chicago Press, Chicago).

32. Dawkins, R. 2009. *The Greatest Show on Earth*. Free Press, New York.

33. Denison, R. F. 2012. *Darwinian Agriculture. How Understanding Evolution Can Improve Agriculture*. Princeton University Press, Princeton, NJ.

34. Denison. R. F., E. T. Kiers, and S. A. West. 2003. Darwinian agriculture: When can humans find solutions beyond the reach of natural selection? *Quart. Rev. Biol.* 78: 145–168.

35. Dennett, D. C. 1995. *Darwin's Dangerous Idea: Evolution and the Meanings of Life*. Simon & Schuster, New York.

36. de Waal, F. 2007. *Chimpanzee Politics: Power and Sex among Apes*. Johns Hopkins University Press, Baltimore, MD.

37. Dissanayake, E. 1995. *Homo Aestheticus: Where Art Comes From and Why*. University of Washington Press, Seattle.

38. Dobzhansky, Th. 1973. Nothing in biology makes sense except in the light of evolution. *Amer. Biol. Teacher* 35: 125–129.

39. Duberman, M. B., M. Vicinus, and G. Chauncey (eds.). 1989. *Hidden from History: Reclaiming the Gay and Lesbian Past*. Penguin, New York.

40. Ellstrand, N. C. 2003. Current knowledge of gene flow in plants: Implications for transgenic flow. *Phil. Trans. R. Soc., B* 358: 1163–1170.

41. Eyre-Walker, A., M. Woolfit, and T. Phelps. 2006. The distribution of fitness effects of new deleterious amino acid mutations in humans. *Genetics* 173: 891–900.

42. Fehr, E., and U. Fischbacher. 2003. The nature of human altruism. *Nature* 425: 785–791.

43. Feldman, M. W., and K. N. Laland,. 1996. Gene-culture coevolutionary theory. *Trends Ecol. Evol.* 11: 453–457.

44. Frankham, R., J. D. Ballou, and D. A. Briscoe. 2002. *Introduction to Conservation Genetics*. Cambridge University Press, Cambridge.

45. Futuyma, D. J. 2000. Some current approaches to the evolution of plant-herbivore interactions. *Plant Species Biol.* 15: 1–9.

46. Gallup, G. G., Jr., and D. A. Frederick. 2010. The science of sex appeal: An evolutionary perspective. *Rev. Gene. Psychol.* 14: 240–250.

47. Gantz, V. M., and 6 others. 2015. Highly efficient Cas9-mediated gene drive for population modification of the malaria vector mosquito *Anopheles stephensi*. *Proc. Nat. Acad. Sci. USA* 112: E6736–E6743.

48. Gatenby, R. A., A. S. Silva, R. J. Gillies, and B. R. Frieden. Adaptive therapy. *Cancer Research* 69: 4894–4903.

49. Gavrilets, S. 2012. On the evolutionary origins of the egalitarian syndrome. *Proc. Natl. Acad. Sci. USA* 109: 14069–14074.

50. Geschwind, D. H., and J. Flint. 2015. Genetics and genomics of psychiatric disease. *Science* 349: 1489–1494.

51. Gire, S. K., and 57 others. 2014. Genomic surveillance elucidates Ebola virus origin and transmission during the 2014 outbreak. *Science* 345: 1369–1372.

52. Gottschall, J., and D. S. Wilson. 2005. *The Literary Animal: Evolution and the Nature of Narrative*. Northwestern University Press, Evanston, Ill.

53. Gould, F. 1998. Sustainability of transgenic insecticidal cultivars: Integrating pest genetics and ecology. *Annu. Rev. Entomol.* 443: 701–726.

54. Gould, F., N. Blair, M. Reid, T. L. Rennie, J. Lopez, and S. Micinski. 2002. *Bacillus thuringiensis*-toxin resistance management: Stable isotope assessment of alternate host use by *Helicoverpa zea*. *Proc. Natl. Acad. Sci. USA* 99: 16581–16586.

55. Hameed, S. 2014. Making sense of Islamic creationism in Europe. *Public Understanding of Science* 1–12. doi:10.1177/0963662514555055.

56. Hendry, A. P., and 15 others. 2011. Evolutionary principles and their practical application. *Evol. Appl.* 4: 159–183.

57. Hoffmann, A. A., P. A. Ross, and G. Rasic. 2015. *Wolbachia* strains for disease control: Ecological and evolutionary considerations. *Evol. Appl.* 8: 751–768.

58. Hofstadter, R. 1955. *Social Darwinism in American Thought.* Beacon Press, Boston, MA.

59. Honing, H., C. ten Cate, I. Peretz, and S. E. Trehub. 2015. Without it no music: cognition, biology and evolution of musicality. *Phil. Trans. R. Soc., B* 370: 20140088.

60. Hull, D. L. 1988. A mechanism and its metaphysics: an evolutionary account of the social and conceptual development of science. *Biol. Philos.* 3: 123–155.

61. Iemmola, F., and A. Camperio-Ciani. 2009. New evidence of genetic factors influencing sexual orientation in men: female fecundity increase in the maternal line. *Arch. Sexual Behav.* 38: 393–399.

62. Jacob. F. 1974. *The Logic of Life: A History of Heredity*. Random House, New York.

63. Kevles, D. J. 1998. *In the Name of Eugenics: Genetics and the Uses of Human Heredity*. Harvard University Press, Cambridge, MA.

64. Kirkpatrick, R. M., and 5 others. 2014. Results of a "GWAS Plus:" General cognitive ability is substantially heritable and massively polygenic. *PLoS ONE* 9: e112390. doi:10.1371/journal.pone.01122390.

65. Kurzban, R., M. N. Burton-Chellew, and S. A. West. 2015. The evolution of altruism in humans. *Annu. Rev. Psychol.* 66: 575–599.

66. Laland, K. N., and G. R. Brown. 2011. *Sense and Nonsense: Evolutionary Perspectives on Human Behaviour*, 2nd ed. Oxford University Press, Oxford.

67. Laland, K. N., J. Odling-Smee, and S. Myles. 2010. How culture shaped the human genome: Bringing genetics and the human sciences together. *Nat. Rev. Genet.* 11: 137–148.

68. Langstrom, N., O. Rahman, E. Carlstrom, and P. Lichtenstein. 2010. Genetic and environmental effects on same-sex sexual behavior: a population study of twins in Sweden. *Arch. Sexual Behav.* 39: 75–80.

69. Laverge, S., and 4 others. 2012. Are species' responses to global change predicted by past niche evolution? *Phil. Trans. R. Soc., B* 368: 20141995.

70. Levine, G. 2006. *Darwin Loves You: Natural Selection and the Re-enchantment of the World.* Princeton University Press, Princeton, N. J.

71. Luksza, M., and M. Lässig. 2014. A predictive model for influenza. *Nature* 507: 57–61.

72. Martincorena, I., and P. J. Campbell. 2015. Somatic mutation in cancer and normal cells. *Science* 349: 1483–1488.

73. McKenna, M. 2013. The last resort. *Nature* 499: 394–396.

74. Mielke, J. H., L. W. Konigsberg, and J. H. Relethford. 2006. *Human Biological Variation*. Oxford University Press, New York.

75. Miller, J. D., E. C. Scott, and S. Okamoto. 2006. Public acceptance of evolution. *Science* 313: 765–766.

76. Miller, K. 1999. *Finding Darwin's God: A Scientist's Search for Common Ground between God and Evolution.* HarperCollins, New York.

77. Miller, K. R. 2008. *Only a Theory: Evolution and the Battle for America's Soul.* Viking, New York.

78. Nei, M., and A. L. Hughes, 1991. Polymorphism and evolution of the major histocompatibility complex loci in mammals. In R. K. Selander, A. G. Clark, and T. S. Whittam (eds.), *Evolution at the Molecular Level,* pp. 222–247. Sinauer, Sunderland, MA.

79. Nelson, R. R. 2007. Universal Darwinism and evolutionary social science. *Biol. & Phil.* 22: 73–94.

80. Nesse, R. M., and S. C. Stearns. 2008. The great opportunity: Evolutionary applications to medicine and public health. *Evol. Applications* 1: 28–48.

81. Nisbett, R. 1995. Race, IQ, and scientism. In S. Fraser (ed.), *The Bell Curve Wars,* pp. 36–57. BasicBooks, New York.

82. Nisbett, R. E., and 6 others. 2012. Intelligence: New findings and theoretical developments. *Amer. Psychol.* 67: 130–159.

83. Norenzayan, A., and S. J. Heine. 2005. Psychological universals: What are they and how can we know? *Psychol. Bull.* 131: 763–784.

84. Norenzayan, A., and 6 others. 2016. The cultural evolution of prosocial religions. Behavioral and Brain Sciences, pp. 1–19. doi:10.1017/S0140525X14001356.

85. Numbers, R. L. 2006. *The Creationists: From Scientific Creationism to Intelligent Design.* Harvard University Press, Cambridge, MA.

86. Omer, S. B., and D. A. Sanders. 2009. Vaccine refusal, mandatory immunization, and the risks of vaccine-preventable diseases. *New England J. Med.* 360: 1981–1988.

87. Oria, R. B., and 8 others. 2005. APOE4 protects the cognitive development in children with heavy diarrhea burdens in northeast Brazil. *Pediatric Res.* 57: 310–316.

88. Paley, W. 1802. *Natural Theology: Or, Evidences of the Existence and Attributes of the Deity, Collected from the Appearances of Nature.* R. Fauldner, London.

89. Pennock, R. T. 2003. Creationism and intelligent design. *Annu. Rev. Genomics Hum. Genet.* 4: 143–163.

90. Pigliucci, M. 2002. *Denying Evolution: Creationism, Scientism, and the Nature of Science.* Sinauer, Sunderland, MA.

91. Pigliucci, M. 2010. *Nonsense on Stilts: How to Tell Science from Bunk.* University of Chicago Press, Chicago.

92. Pinker, S. 2002. *The Blank Slate: The Modern Denial of Human Nature.* Penguin Books, New York.

93. Plomin, R., and I. J. Deary. 2015. Genetics and intelligence differences: Five special findings. *Mol. Psychiatry* 20: 98–108.

94. Plotkin, H. 2010. *Evolutionary Worlds without End.* Oxford University Press, Oxford.

95. Poiani, A. 2010. *Animal Homosexuality: A Biosocial Perspective.* Cambridge University Press, Cambridge.

96. Quintero, I., and J. J. Wiens. 2013. Rates of projected climate change dramatically exceed past rates of climatic niche evolution. *Ecol. Lett.* 16: 1095–1103.

97. Rice, W. R., U. Friberg, and S. Gavrilets. 2013. Homosexuality via canalized sexual development: A testing protocol for a new epigenetic model. *Bioessays* 35: 764–770.

98. Richards, R. J. 2013. *Was Hitler a Darwinian? Disputed Questions in the History of Evolutionary Theory.* University of Chicago Press, Chicago.

99. Richerson, P. J., and R. Boyd, 2005. *Not by Genes Alone: How Culture Transformed Human Evolution.* University of Chicago Press, Chicago.

100. Rook, G. A. W. 2012. Hygiene hypothesis and autoimmune diseases. *Clinical Rev. Allergy Immunol.* 42: 5–15.

101. Sala, O. E., and 18 others. 2000. Global biodiversity scenarios for the year 2100. *Science* 287: 1770–1774.

102. Sanders, A. R., and 12 others. 2015. Genome-wide scan demonstrates significant linkage for male sexual orientation. *Psychol. Med.* 45: 1379–1388.

103. Scaduto, D. I., J. M. Brown, W. C. Haaland, D. J. Zwickl, D. M. Hillis, and M. L. Metzker. 2010. Source identification in two criminal cases using phylogenetic analysis of HIV-1 DNA sequences. *Proc. Nat. Acad. Sci. USA* 107: 21242–21247.

104. Sepkoski, J. 1997. Biodiversity: past, present, and future. *J. Paleont.* 71: 533–539.

105. Shendure, J., and J. M. Akey. 2015. The origins, determinants, and consequences of human mutations. *Science* 349: 1478–1482.

106. Snell-Rood, E. 2016. Interdisciplinarity: Bring biologists into biomimetics. *Nature* 529: 277–278.

107. Sommer, V., and P. L. Vasey (eds.). 2006. *Homosexual Behaviour in Animals: An Evolutionary Perspective.* Cambridge University Press, Cambridge.

108. Stearns, S. C., and R. Medzhitov. 2016. *Evolutionary Medicine.* Sinauer, Sunderland, MA.

109. Sunday, J. M., P. Collosi, S. Dupont, P. L. Munday, J. H. Stillman, and T. B. Reusch. 2014. Evolution in an acidifying ocean. *Trends Ecol. Evol.* 29: 117–125.

110. The Royal Society and the U. S. National Academy of Sciences. 2014. *Climate Change: Evidence and Causes.* National Academy Press, Washington, D.C. http://www.nap.edu/catalog/18730/climate-change-evidence-and-causes

111. Thomas, C. D., and 18 others. 2004. Extinction risk from climate change. *Nature* 427: 145–148.

112. Tian, X, and 9 others. 2013. High-molecular-mass hyaluronan mediates the cancer resistance of the naked mole-rat. *Nature* 499: 346–350.

113. Tong, Y.-G.,., and 55 others. 2015. Genetic diversity and evolutionary dynamics of Ebola virus in Sierra Leone. *Nature* 524: 93–96.

114. Ullstrup, A. J. 1972. The impacts of the southern corn leaf blight epidemic of 1970–1971. *Annu. Rev. Phytopathol.* 19: 37–50.

115. Visser, M. E. 2008. Keeping up with a warming world: assessing the rate of adaptation to climate change. *Proc. Biol. Sci.* 275: 649–659.

116. Wagner, A., and W. Rosen. 2013. Spaces of the possible: universal Darwinism and the wall between technological and biological innovation. *J. Royal Soc. Interface* 11:20131190.

117. White A. D. 1896. *A History of the Warfare of Science with Theology in Christendom.* D. Appleton and Co., New York.

118. Wilson, D. S. 2002. *Darwin's Cathedral: Evolution, Religion, and the Nature of Society.* University of Chicago Press, Chicago.

119. Wilson, E. O. 1975. *Sociobiology: The New Synthesis.* Harvard University Press, Cambridge, MA.

120. Wilson, E. O. 1992. *The Diversity of Life.* Harvard University Press, Cambridge, MA.

121. Wilson, G., and Q. Rahman. 2005. *Born Gay: The Psychobiology of Sexual Orientation.* Peter Owen Publishers, London.

122. Wolfe, N. D., C. P. Dunavan, and J. Diamond. 2007. Origins of major human infectious diseases. *Nature* 447: 279–283.

123. Workman , L., and W. Reader. 2008. *Evolutionary Psychology: An Introduction.* Cambridge University Press, Cambridge.

124. Zhu, Y. Y., and 13 others. 2000. Genetic diversity and disease control in rice. *Nature* 406: 718–722.

125. Zipperer, A., and 16 others. 2016. Human commensals producing novel antibiotic impair pathogen colonization. *Nature* 535: 511–516.

APPENDIX

1. Fisher, R. A. 1918. The correlation between relatives on the supposition of Mendelian inheritance. *Trans. R. Soc. Edinburgh* 52: 399–433.

2. Whitlock, M. C., and D. Schluter. 2009. *The Analysis of Biological Data.* Roberts and Company Publishers, Greenwood Village, New York.

Illustration and Photo Credits

Index

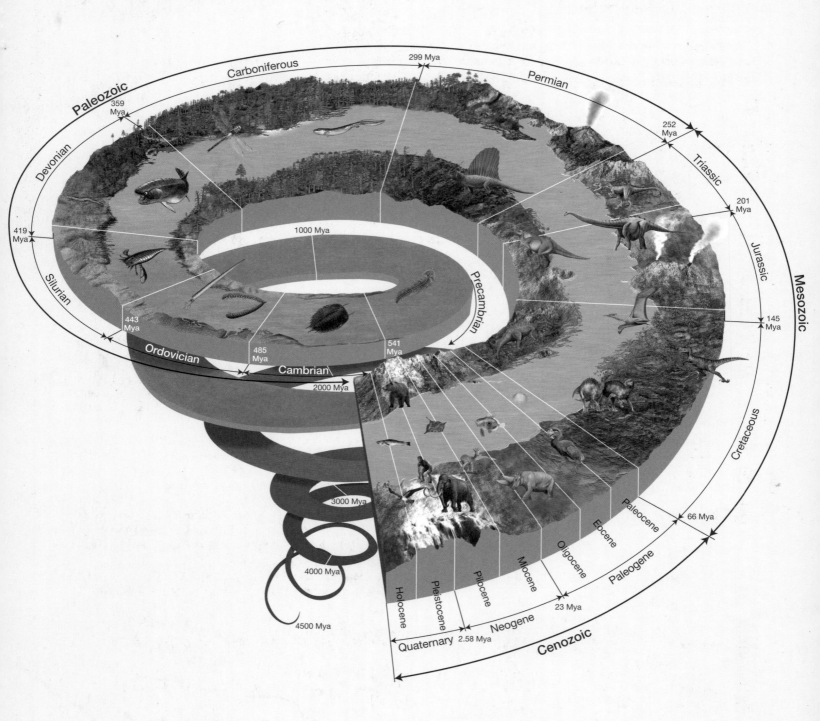